Advanced Series in Physical Chemistry — Vol. 17

CONICAL INTERSECTIONS
Theory, Computation and Experiment

Advanced Series in Physical Chemistry

Editor-in-Charge
Cheuk-Yiu Ng, *Department of Chemistry, University of California at Davis, USA*

Associate Editors
Hai-Lung Dai, *Department of Chemistry, University of Pennsylvania, USA*
James M. Farrar, *Department of Chemistry, University of Rochester, USA*
Kopin Liu, *Institute of Atomic and Molecular Sciences, Taiwan*
David R. Yarkony, *Department of Chemistry, Johns Hopkins University, USA*
James J. Valentini, *Department of Chemistry, Columbia University, USA*

Published

Vol. 6: The Chemical Dynamics and Kinetics of Small Radicals
eds. K. Liu and A. Wagner

Vol. 7: Recent Developments in Theoretical Studies of Proteins
ed. R. Elber

Vol. 8: Charge Sensitivity Approach to Electronic Structure and Chemical Reactivity
R. F. Nolewajski and J. Korchowiec

Vol. 9: Vibration-Rotational Spectroscopy and Molecular Dynamics
ed. D. Papousek

Vol. 10: Photoionization and Photodetachment
ed. C.-Y. Ng

Vol. 11: Chemical Dynamics in Extreme Environments
ed. R. A. Dreššler

Vol. 12: Chemical Applications of Synchrotron Radiation
ed. T.-K. Sham

Vol. 13: Progress in Experimental and Theoretical Studies of Clusters
eds. T. Kondow and F. Mafuné

Vol. 14: Modern Trends in Chemical Reaction Dynamics: Experiment and Theory (Parts I & II)
eds. X. Yang and K. Liu

Vol. 15: Conical Intersections: Electronic Structure, Dynamics and Spectroscopy
eds. W. Domcke, D. R. Yarkony and H. Köppel

Vol. 16: Overviews of Recent Research on Energetic Materials
eds. R. W. Shaw, T. B. Brill and D. L. Thompson

Vol. 17: Conical Intersections: Theory, Computation and Experiment
eds. W. Domcke, D. R. Yarkony and H. Köppel

Advanced Series in Physical Chemistry – Vol. 17

CONICAL INTERSECTIONS
Theory, Computation and Experiment

Editors

Wolfgang Domcke
Technical University of Munich, Germany

David R Yarkony
Johns Hopkins University, USA

Horst Köppel
University of Heidelberg, Germany

NEW JERSEY · LONDON · SINGAPORE · BEIJING · SHANGHAI · HONG KONG · TAIPEI · CHENNAI

Published by

World Scientific Publishing Co. Pte. Ltd.
5 Toh Tuck Link, Singapore 596224
USA office: 27 Warren Street, Suite 401-402, Hackensack, NJ 07601
UK office: 57 Shelton Street, Covent Garden, London WC2H 9HE

British Library Cataloguing-in-Publication Data
A catalogue record for this book is available from the British Library.

Advanced Series in Physical Chemistry — Vol. 17
CONICAL INTERSECTIONS
Theory, Computation and Experiment

Copyright © 2011 by World Scientific Publishing Co. Pte. Ltd.

All rights reserved. This book, or parts thereof, may not be reproduced in any form or by any means, electronic or mechanical, including photocopying, recording or any information storage and retrieval system now known or to be invented, without written permission from the Publisher.

For photocopying of material in this volume, please pay a copying fee through the Copyright Clearance Center, Inc., 222 Rosewood Drive, Danvers, MA 01923, USA. In this case permission to photocopy is not required from the publisher.

ISBN-13 978-981-4313-44-5
ISBN-10 981-4313-44-0

Typeset by Stallion Press
Email: enquiries@stallionpress.com

Printed in Singapore by Mainland Press Pte Ltd.

Foreword

By Ahmed H. Zewail*

In molecular sciences, if we were to reduce into two words the fundamental contributions made over the past century, they would be the *chemical bond* — specifically its *structure* and its *dynamics*. We now know how to visualize static atomic bonding and describe covalent, ionic or hydrogen bond interactions, both theoretically and experimentally. With advanced methods in computations it is also possible to predict the structure of medium-sized molecules in their ground state and, with less certainty, in the excited state.

For dynamics, we have already reached the time scale of molecular vibrations (femtoseconds) and rotations (picoseconds), developed concepts of classical motions of atoms (wave packets) and predicted their persistence (coherence) in different states of molecular interactions. Experimental, theoretical and computational methods have made possible not only the probing of dynamics, but also some control over the outcome of atomic and fragment separations. It is the induced coherence among quantum states that allows for probing and controlling of motion at the atomic scale, without violation of the uncertainty principle!

Bond dynamics of excited states, especially in complex systems, become challenging when considering the myriad of interactions possible among the different vibrations and electrons involved. However, it was possible, because of difference in time scales, to separate electronic and nuclear motions (Born–Oppenheimer approximation) and this approximation has

---

*The author is currently the Linus Pauling Chair Professor and the Director of the Physical Biology Center for Ultrafast Science & Technology at Caltech in Pasadena, California 91125, USA.

become the cardinal concept in molecular spectroscopy. Similarly, the division of the system's modes into "relevant" and "irrelevant," or bath modes, was proven powerful in the description of nonradiative processes. These approximations become invalid when the electronic states become closer in energy and the nuclear motion involves more than one potential-energy surface with possible crossings or avoided crossings.

Early on, theoretical photochemists realized the importance of these nonadiabatic interactions (avoided crossings and conical intersections) in order to account for product formations occurring on different potential energy asymptotes. In the 1980s, the direct experimental visualization of nuclear motions in the adiabatic and nonadiabatic regimes — exemplified by the case of covalent-ionic surfaces crossing in alkali halides — on the femtosecond time scale stimulated the development of numerous theoretical methods for the description of dynamics of systems at far-from-equilibrium geometries, and away from the initially-excited Franck–Condon region (see the contribution by Bonacic-Koutecky and co-workers). Theoretical developments led to the realization that conical intersections are ubiquitous in photochemistry and photobiology (see contributions in Parts I and II), especially among systems excited to high energies with an energetically-dense number of states, and have created a new branch of study that is the subject of this volume.

Nature utilizes conical intersections (CIs) effectively, and in many cases CIs have useful functions. Because they can facilitate an efficient cascade of energy on the femtosecond time scale, photobiological chromophores "utilize" them to dissipate the absorbed excess energy — otherwise bonds would break and give rise to unwanted chemical fragments (see the contributions by Robb, Domcke and collaborators).

In retrospect, the origin of such nonadiabatic behavior can be traced to the geometric-phase property around the intersection, known as Berry phase (see contribution by Althorpe). In molecular systems, the well-known Longuet–Higgins and Herzberg account of degeneracy lifting with change in boundary conditions can provide the needed elucidation of the nature of the wave function (with proper phases) around the intersection in small systems, but new approaches were needed to compute the dynamics "on the fly" and to predict consequences and relevance to experiments in real, complex systems (see the contributions by Yarkony, Martinez and co-workers). Over the past two decades, the combination of experimental and theoretical techniques has uncovered the ubiquity and significance of conical intersections in organic, inorganic and biological systems.

The editors of this volume (Domcke, Yarkony and Köppel) are among the leaders in the theory and computation of conical intersections. The book covers three parts (I, II and III) with contributions in the fundamentals, dynamics and experimental manifestations. The theory and computation span the small and large molecular systems as well as those influenced by the environment. The experimental approaches are highlighted with examples of femtosecond photoelectron spectroscopy (see the contribution by Schuurman and Stolow), femtosecond polarization and vibronic spectroscopy (see the contributions by Jonas, Temps and collaborators), and photodissociation dynamics (see the contribution by Ashfold and coworkers).

In the coming years, efforts should perhaps be directed toward the development (hopefully) of "simple theoretical expressions" that will highlight the key parameters describing energy redistribution, extent of temporal and spatial coherence, and branching of populations. The powerful computational machinery available becomes a tool for reaching this goal. On the experimental side, it is important to know the structure around intersections and it is now feasible to probe such structures in isolated (gas phase) molecules using ultrafast electron diffraction. And, progress is being made to address the question: What is the nature of the electronic distribution on the attosecond time scale? Unlike in femtochemistry, the critical issue in this case is the nature of the initial electronic state prepared, because of the huge energy uncertainty on this time scale. Perhaps theory can become an enlightening guide in this endeavor.

Given the importance of conical intersections in determining the fate of photon-induced reactions, I recommend this volume to all concerned with photochemical and photobiological molecular sciences. Indeed, the book represents an exposé of a unique dimension in the study of the dynamics of the chemical bond.

Preface

The Born–Oppenheimer adiabatic approximation represents one of the cornerstones of molecular physics and chemistry. The concept of adiabatic potential-energy surfaces, defined by the Born–Oppenheimer approximation, is fundamental to our understanding of molecular spectroscopy and chemical reaction dynamics. Many chemical processes can be rationalized in terms of the dynamics of the atomic nuclei on a single Born–Oppenheimer potential-energy surface. Nonadiabatic processes, that is, chemical processes which involve nuclear dynamics on at least two coupled potential-energy surfaces and thus cannot be rationalized within the Born–Oppenheimer approximation, are nevertheless ubiquitous in chemistry, most notably in photochemistry and photobiology. Typical phenomena associated with a violation of the Born–Oppenheimer approximation are the radiationless relaxation of excited electronic states, charge-transfer processes, photoinduced unimolecular decay and isomerization processes of polyatomic molecules.

The last few decades have witnessed a change of paradigms in nonadiabatic chemistry. This paradigm shift is the result of advances in experimental techniques and the concomitant development of new computational tools. First, the remarkable advances achieved in femtosecond laser technology and time-resolved spectroscopy revealed that the radiationless decay of excited electronic states may take place much faster than previously thought. The traditional theory of radiationless decay processes, developed in the 1960s and 1970s, cannot explain electronic decay occurring on a time scale of a few tens of femtoseconds. Second, the development and widespread application of multireference electronic structure methods for the calculation of excited-state potential-energy surfaces and the availability of analytic gradient-based search methods have shown that CONICAL INTERSECTIONS of these multidimensional surfaces, predicted by von Neumann and Wigner in 1929, are the rule rather than the exception in polyatomic molecules. As a result, the concept of conical intersections has become widely known in recent years. That conical intersections may be responsible for ultrafast radiationless processes had been surmised as early as 1937 by Teller. Today, it is increasingly recognized that conical intersections play a key mechanistic role in molecular spectroscopy and chemical reaction dynamics.

This second edited volume on conical intersections in polyatomic molecules complements the first volume (*Conical Intersections: Electronic Structure, Dynamics and Spectroscopy*, Vol. 15 in the Advanced Series in Physical Chemistry, World Scientific) published in 2004. In the past six years, developments in molecular spectroscopy, photochemistry and computational chemistry have considerably extended our insight into the role of conical intersections in nonadiabatic chemistry. For example, significant progress has been achieved with the simulation of ultrafast processes at conical intersections by quantum dynamics calculations and especially also by classical surface-hopping trajectory calculations. These topics are therefore covered by several chapters in this book. This volume further includes four chapters on the experimental detection of ultrafast dynamics at conical intersections, covering the techniques of photofragment translational spectroscopy, time-resolved photoelectron spectroscopy, as well as several variants of femtosecond pump-probe spectroscopy. Various other recent developments, concerning fundamental methodological aspects (more than two-state intersections, spin-orbit coupling, as well as geometric-phase effects) and extensions of established and new fields of application (organic photochemistry, photostability of biomolecules, conical intersections embedded in an environment), are also addressed in the 18 chapters of this book.

The editors hope that this second volume on conical intersections will be of value to a wide readership in the chemical physics and physical chemistry communities.

Wolfgang Domcke, Munich
David R. Yarkony, Baltimore
Horst Köppel, Heidelberg

Contents

Foreword by A. H. Zewail v

Preface ix

Part I. Fundamental Aspects and Electronic Structure 1

1. Conical Intersections in Organic Photochemistry 3
   M. A. Robb

2. Efficient Excited-State Deactivation in Organic Chromophores and Biologically Relevant Molecules: Role of Electron and Proton Transfer Processes 51
   A. L. Sobolewski and W. Domcke

3. Three-State Conical Intersections 83
   S. Matsika

4. Spin-Orbit Vibronic Coupling in Jahn–Teller Systems 117
   L. V. Poluyanov and W. Domcke

5. Symmetry Analysis of Geometric-Phase Effects in Quantum Dynamics 155
   S. C. Althorpe

Part II. Dynamics at Conical Intersections   195

6. Conical Intersections in Electron Photodetachment Spectroscopy: Theory and Applications   197
   M. S. Schuurman and D. R. Yarkony

7. Multistate Vibronic Dynamics and Multiple Conical Intersections   249
   S. Faraji, S. Gómez-Carrasco and H. Köppel

8. Conical Intersections Coupled to an Environment   301
   I. Burghardt, K. H. Hughes, R. Martinazzo, H. Tamura, E. Gindensperger, H. Köppel and L. S. Cederbaum

9. *Ab Initio* Multiple Spawning: First Principles Dynamics Around Conical Intersections   347
   S. Yang and T. J. Martínez

10. Non-Born–Oppenheimer Molecular Dynamics for Conical Intersections, Avoided Crossings, and Weak Interactions   375
    A. W. Jasper and D. G. Truhlar

11. Computational and Methodological Elements for Nonadiabatic Trajectory Dynamics Simulations of Molecules   415
    M. Barbatti, R. Shepard and H. Lischka

12. Nonadiabatic Trajectory Calculations with *Ab Initio* and Semiempirical Methods   463
    E. Fabiano, Z. Lan, Y. Lu and W. Thiel

13. Multistate Nonadiabatic Dynamics "on the Fly" in Complex Systems and Its Control by Laser Fields   497
    R. Mitrić, J. Petersen and V. Bonačić-Koutecký

14. Laser Control of Ultrafast Dynamics at Conical Intersections 569
    Y. Ohtsuki and W. Domcke

Part III. Experimental Detection of Dynamics at Conical Intersections 601

15. Exploring Nuclear Motion Through Conical Intersections in the UV Photodissociation of Azoles, Phenols and Related Systems 603
    T. A. A. Oliver, G. A. King, A. G. Sage and M. N. R. Ashfold

16. Interrogation of Nonadiabatic Molecular Dynamics via Time-Resolved Photoelectron Spectroscopy 633
    M. S. Schuurman and A. Stolow

17. Pump-Probe Spectroscopy of Ultrafast Vibronic Dynamics in Organic Chromophores 669
    N. K. Schwalb, R. Siewertsen, F. Renth and F. Temps

18. Femtosecond Pump-Probe Polarization Spectroscopy of Vibronic Dynamics at Conical Intersections and Funnels 715
    W. K. Peters, E. R. Smith and D. M. Jonas

Index 747

Part I

Fundamental Aspects and Electronic Structure

Chapter 1

Conical Intersections in Organic Photochemistry

Michael A. Robb[*]

1.	Introduction	4
2.	Exploring the Intersection Space: The Extended Conical Intersection Seam	5
	2.1. Theory	5
3.	Extended Seam Benchmarks	14
	3.1. Fulvene	14
	3.2. The $2A_1/1A_1$ conical intersection seam in butadiene	16
4.	Applications of the Extended Seam of a Conical Intersection to Photochemical Mechanisms	18
	4.1. The photoinduced isomerization of 1,3-cyclohexadiene (CHD) to cZc-hexatriene (HT)	18
	4.2. Diarylethylenes	20
	4.3. The keto-enol tautomerism of o-hydroxyphenyl-(1,3,5)-triazine	23
5.	Valence Bond Analysis of Conical Intersections	26
	5.1. Twisted intermolecular charge transfer (T-(ICT)) in aminobenzonitrile (ABN) compounds	27
	5.2. What happens when one does a conical intersection circuit in the branching plane?	32

---

[*]Chemistry Department, Imperial College London, London SW7 2AZ, UK.

6. Exploring the Conical Intersection Seam using Dynamics 40
   6.1. A model cyanine system: The extended seam for cis-trans double bond isomerization 40
   6.2. Benzene 42
   6.3. Biological chromophores: PYP 44
7. Conclusions 46
   Acknowledgments 46
   References 47

1. Introduction

It has now been about 20 years[1] since we published our first paper on a conical intersection in the prototypical organic photochemical problem: the 2 + 2 face-to-face cycloaddition of two ethylenes. In the intervening period, conical intersections have become an essential part of the thought process or paradigm of organic photochemistry (see, for example, the textbooks of Klessinger[2] or Turro[3]). In the previous volume of this series on conical intersections, Migani and Olivucci[4] not only have given an extensive review of the history of the subject and the theory of conical intersections associated with the mechanisms of organic photochemistry, but also discussed many examples that cover the complete range of functional groups that are important in organic photochemistry. The subject is now growing so rapidly that to simply update the Migani article with the photochemical problems that have been studied in the last five years would be almost impossible within the space allowed (but see recent reviews[5-7]). Rather, we limit our discussion to those areas identified by Migani and Olivucci as areas of major growth into the future. These areas include the relationship between the intersection space or the conical intersection seam and photochemical mechanisms, the use of dynamics to investigate organic photochemistry, and the extension to biological chromophores. In addition to these topics, we shall discuss a little about the use of qualitative methods to rationalise computations because this area remains important for the organic chemist.

In writing this section, our target was to produce a more general discussion that seemed to be appropriate for a section focussed on "fundamental aspects". Thus the intention was to be rather broad with most of the examples chosen (mainly from our own work) so as to illustrate conceptual ideas. In the examples, we will choose to discuss the concepts, we will omit the computational details and refer the reader to the original literature. In general, the results are from the CASSCF method where both gradients and hessian can be computed analytically so that we do not need to use "distinguished variables" but rather full geometry optimization is possible. Since most of our discussions are focussed on the general shape of the potential surface near a conical intersection, there is not a requirement for high accuracy.

2. Exploring the Intersection Space: The Extended Conical Intersection Seam

2.1. *Theory*

It is our intention in this section to discuss the mechanistic implications of the extended conical intersection seam.[7–16] We start with a theoretical discussion adapted from the work of Sicila and co-workers.[15]

One often starts a discussion of conical intersections in organic photochemistry with "sand in a funnel" picture for a photochemical mechanism involving excited and ground state branches (with two ground state reaction pathways) and a conical intersection (Fig. 1). We shall use cartoons of this form to illustrate many aspects of nonadiabatic chemistry in this article.

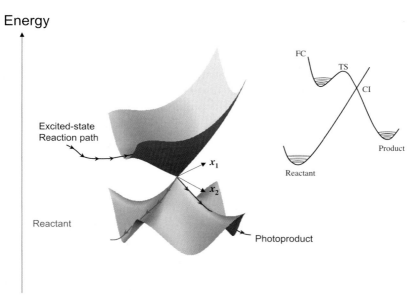

Fig. 1. Cartoon of a "classic" double cone conical intersection, showing the excited-state reaction path and two ground-state reaction paths. (Adapted from Paterson *et al.*[12])

We begin by offering a few comments on how such cartoons should be interpreted. We have plotted the energy in two geometrical coordinates, X_1 and X_2. In general, these two coordinates will be combinations of changes in the bond lengths and bond angles of the molecular species under investigation. We are limited in such cartoons to using two or three combinations of molecular variables. However, we must emphasise that these are just cartoons used to illustrate a mechanistic idea. All the computations we shall discuss are done in the full space of molecular geometries without any constraint.

The coordinates X_1 and X_2 in Fig. 1 correspond to the space of molecular geometrical deformations that lift the degeneracy. These coordinates are precisely defined quantities that can be computed explicitly.[17] Similarly, the apex of the cone corresponds in general to an optimised molecular geometry.[18] (See also the more recent work of Sicila,[13] Martinez[19] and Theil.[20]) The shape or topology in the region of the apex of the double cone will change from one photochemical system to another,[9] and it is the generalities associated with the shape that form part of the mechanistic scenario that we will discuss.

Conical intersections are normally thought of as points on a $(m-2)$-dimensional hyperline. While the degeneracy is lifted by motion in branching space X_1 and X_2 in Fig. 1, motions in the intersection space X_3 shown in Fig. 2 preserve the degeneracy. Thus in Fig. 2, we show the conical intersection hyperline traced out by a coordinate X_3 plotted, this time, in a plane containing the distinguished intersection space coordinate X_3 and one coordinate from the degeneracy-lifting (or branching) space X_1 X_2. We shall call X_3 the "reaction coordinate" because this might be chosen as the path of steepest decent on the potential surface and this would be a gradient-determined choice of this distinguished coordinate. In this figure, the conical intersection line now appears as a seam. In contrast to the "sand in the funnel" model shown in Fig. 1, it is clear that the reaction path could be almost parallel to this seam. Of course this figure can be misunderstood, because each point on the seam line lies in the space of the double-cone X_1 X_2 as suggested by the insert. Thus it is essential to appreciate that decay at the conical intersection is associated with three coordinates, the branching space X_1 X_2 and the reaction path X_3 that may or may not lie in the space X_1 X_2.

In Fig. 3, we have presented another cartoon that attempts to show the potential energy surfaces in the space of the degeneracy lifting coordinates as one traverses a third coordinate X_3. Of course, this is really

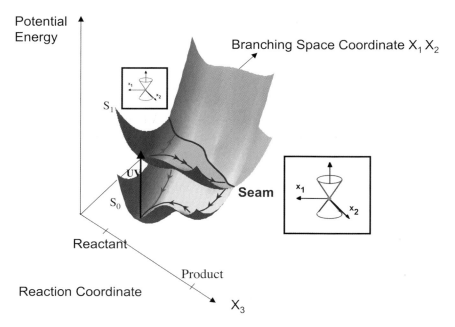

Fig. 2. The conical intersection hyperline traced out by a coordinate X_3 plotted in a space containing the coordinate X_3 and one coordinate from the degeneracy-lifting space X_1 X_2. (Adapted from Paterson et al.[12])

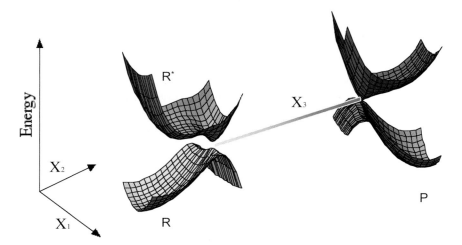

Fig. 3. A cartoon showing the conical intersection hyperline traced out by a degeneracy-preserving coordinate X_3. The system remains degenerate as one traverses the coordinate X_3, but the energy and the shape of the double-cone must change in $X_1 X_2$.

a four-dimensional picture, which is not easy to assimilate. Nevertheless it should be clear that the degeneracy persists along the coordinate X_3. However, the energy will change and so will the "shape" of the double-cone near the apex.

Figures 2 and 3 establish two important mechanistic points:

(1) The important points on a conical intersection hyperline are those where the reaction path meets with the seam (see Fig. 2) and this may not be at the minimum of the seam.
(2) Radiationless decay takes place in the coordinates X_1 X_2 as one passes through the conical intersection diabatically.

In Fig. 2, this second principle appears to be violated since the reaction path appears to pass through the hyperline adiabatically. However, we emphasise — as indicated by the double-cone insert — that as one passes through the hyperline, decay takes place in the coordinates X_1 X_2 and in general their VB structure does not change. We shall use both Figs. 2 and 3 as models in subsequent discussions, but the reader needs to remember the conceptual limitations.

The statement that motion in the intersection space preserves the degeneracy is only true to first order and, in fact, the degeneracy can be lifted at second order by finite steps along coordinates spanning the intersection space. We now develop this idea and show that it can be used to further characterise conical intersections (for further details, the reader is referred to the literature[9-16]).

A second-order description of conical intersections allows one to characterise optimised conical intersection geometries as either minima or saddle points on the crossing hyperline. We will now present[15] a simplified development of the second-order description of conical intersections. However, first we briefly introduce the ideas behind such description in a nonmathematical way with the aid of Fig. 4. In numerical computations, one finds that the degeneracy is, in practice, lifted for a finite displacement along any intersection space coordinate (X_3 in Fig. 2 and Q_i in Fig. 4). Figure 4(a) shows a minimum on the extended seam while Fig. 4(b) shows a saddle point. The curve f_i corresponds to the projection of the seam $U(f_i)$ on the coordinate space consisting of one coordinate from the branching plane, $\bar{Q}_{x_{12}}$, and one from the intersection space, $\bar{Q}_i$. Thus, the crossing seam is curved so that the seam bends towards the branching plane coordinate, with a mixing of branching space and intersection space coordinates. This

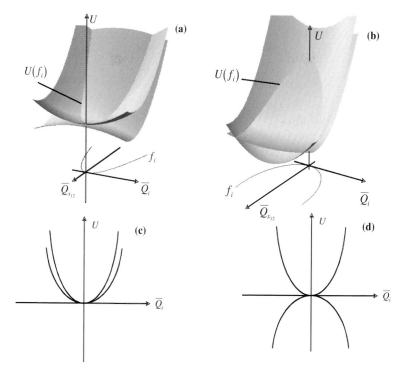

Fig. 4. The locus of the conical intersection seam $U(f_i)$ and the corresponding curvilinear coordinate f_i: (a) minimum, (b) saddle point, (c) cross section of (a) along first-order intersection space coordinate $\bar{Q}_i$, and (d) cross section of (b) along first-order intersection space coordinate $\bar{Q}_i$. (Adapted from Sicilia et al.[15])

curvature is required to describe finite displacements where the degeneracy is preserved.

In Figs. 4(c) and 4(d) we show cuts through Figs. 4(a) and 4(b) in the $(U, \bar{Q}_i)$ plane. It is clear in this figure that the two potential energy curves split apart along any finite displacement lying strictly along the rectilinear first-order intersection modes, $\bar{Q}_i$, that is in the plane containing energy and the intersection coordinate.

Figure 4 shows that a curvilinear coordinate f_i is a convenient way to describe the behaviour of the extended seam. If we define the curvilinear coordinates as f_i, the crossing seam energy can be written as a function of these $(m-2)$ variables $U(f_i)$ rather than the $(m-2)$ rectilinear coordinates. It then becomes clear that the curvature of the seam energy becomes simply the second derivative of the seam energy with respect to such curvilinear coordinates. We will refer to the matrix of second derivatives computed in this way as the intersection space Hessian. The curvilinear coordinates just discussed are the second-order generalization of the intersection-adapted coordinates introduced by Atchity et al.[9]

We now proceed to a more mathematical discussion.[15] We begin with a first order description of a conical intersection. The first-order approximation describes the two intersecting potential energy surfaces in the vicinity of a conical intersection point. In an appropriate region around the intersection point, the electronic two-state potential energy matrix describing the two intersecting *states* A and B can be approximated by a Taylor expansion truncated at the second order[21]:

$$\mathbf{W} = \mathbf{W}^{(0)} + \mathbf{W}^{(1)} + \frac{1}{2}\{\mathbf{W}^{(2)}_a + 2\mathbf{W}^{(2)}_b + \mathbf{W}^{(2)}_c\}. \tag{1}$$

The reference point is assumed to be the conical intersection point. Therefore, the zero-order term $\mathbf{W}^{(0)}$ is a diagonal matrix whose entries are equal; in the following this matrix is set to zero. When the expansion is performed with respect to the m first-order intersection adapted coordinates $\bar{Q}_i$, the potential matrices can be explicitly written as:

$$\mathbf{W}^{(1)} = \left(\frac{\lambda_1}{2}\bar{Q}_{x_1} + \frac{\lambda_2}{2}\bar{Q}_{x_2}\right)\mathbf{1} + \begin{pmatrix} \frac{\delta\kappa}{2}\bar{Q}_{x_1} & \kappa^{AB}\bar{Q}_{x_2} \\ \kappa^{AB}\bar{Q}_{x_2} & -\frac{\delta\kappa}{2}\bar{Q}_{x_1} \end{pmatrix}, \tag{2a}$$

$$\mathbf{W}^{(2)}_a = \left(\sum_{i,j \in BS} \frac{{}^{BS}\omega_{ij}}{2}\bar{Q}_i\bar{Q}_j\right)\mathbf{1}$$

$$+ \begin{pmatrix} \sum_{i,j \in BS} \frac{{}^{BS}\delta\gamma_{ij}}{2}\bar{Q}_i\bar{Q}_j & \sum_{i,j \in BS} {}^{BS}\eta^{AB}_{ij}\bar{Q}_i\bar{Q}_j \\ \sum_{i,j \in BS} {}^{BS}\eta^{AB}_{ij}\bar{Q}_i\bar{Q}_j & -\sum_{i,j \in BS} \frac{{}^{BS}\delta\gamma_{ij}}{2}\bar{Q}_i\bar{Q}_j \end{pmatrix}, \tag{2b}$$

$$\mathbf{W}^{(2)}_b = \left(\sum_{i \in BS, j \in IS} \frac{{}^{BS/IS}\omega_{ij}}{2}\bar{Q}_i\bar{Q}_j\right)\mathbf{1}$$

$$+ \begin{pmatrix} \sum_{i \in BS, j \in IS} \frac{^{BS/IS}\delta\gamma_{ij}}{2} \bar{Q}_i \bar{Q}_j & \sum_{i \in BS, j \in IS} {^{BS/IS}\eta_{ij}^{AB}} \bar{Q}_i \bar{Q}_j \\ \sum_{i \in BS, j \in IS} {^{BS/IS}\eta_{ij}^{AB}} \bar{Q}_i \bar{Q}_j & -\sum_{i \in BS, j \in IS} \frac{^{BS/IS}\delta\gamma_{ij}}{2} \bar{Q}_i \bar{Q}_j \end{pmatrix},$$

(2c)

$$\mathbf{W}_c^{(2)} = \left(\sum_{i,j \in IS} \frac{{}^{IS}\omega_{ij}}{2} \bar{Q}_i \bar{Q}_j \right) \mathbf{1}$$

$$+ \begin{pmatrix} \sum_{i,j \in IS} \frac{{}^{IS}\delta\gamma_{ij}}{2} \bar{Q}_i \bar{Q}_j & \sum_{i,j \in IS} {}^{IS}\eta_{ij}^{AB} \bar{Q}_i \bar{Q}_j \\ \sum_{i,j \in IS} {}^{IS}\eta_{ij}^{AB} \bar{Q}_i \bar{Q}_j & -\sum_{i,j \in IS} \frac{{}^{IS}\delta\gamma_{ij}}{2} \bar{Q}_i \bar{Q}_j \end{pmatrix}.$$

(2d)

In the above equations, the potential constants are defined as:

$$\lambda_i \equiv \nabla_{\bar{Q}_{x_i}} (U_{AA}^0 + U_{BB}^0), \tag{3a}$$

$$\delta\kappa \equiv \nabla_{\bar{Q}_{x_1}} (U_{BB}^0 - U_{AA}^0), \tag{3b}$$

$$\kappa^{AB} \equiv \nabla_{\bar{Q}_{x_2}} H_{AB}^0, \tag{3c}$$

$$\omega_{ij} \equiv \nabla^2_{Q_i Q_j} (U_{AA}^0 + U_{BB}^0), \tag{3d}$$

$$\delta\gamma_{ij} \equiv \nabla^2_{Q_i Q_j} (U_{BB}^0 - U_{AA}^0), \tag{3e}$$

$$\eta_{ij}^{AB} \equiv \nabla^2_{Q_i Q_j} H_{AB}^0. \tag{3f}$$

In Eq. (3), we use the nabla, ∇, to indicate the vector differential operator of first derivatives with respect to nuclear displacements. Therefore, when applied to a scalar, it will give rise to a vector. In the context discussed here, it indicates exclusively the gradient vector. The subscript is introduced to specify which component of the entire vector is considered. Thus, for instance, the element of the gradient calculated with respect to the $\hat{\mathbf{x}}_i$ direction is indicated as ∇_{Q_i}. Note the equivalence $\nabla_{Q_i} \equiv \partial/\partial \bar{Q}_i$. Nabla squared, ∇^2, is an extension of such notation and, therefore, its application to a scalar gives rise to a square symmetrical matrix whose elements are defined as $\nabla^2_{Q_i Q_j} \equiv \partial^2/(\partial \bar{Q}_i \partial \bar{Q}_j)$. The zero superscripts indicate that all the terms are computed at the reference point (a critical point on the seam).

We now introduce the parabolic approximation where the second-order terms within the first-order branching space $^{BS}\delta\gamma_{ij}$ and $^{BS}\eta_{ij}^{AB}$, and the mixing terms between branching plane and first-order intersection space $^{BS/IS}\delta\gamma_{ij}$ and $^{BS/IS}\eta_{ij}^{AB}$, are neglected (the parabolic approximation derives from a three-dimensional case where the approximate intersection coordinate moves along a parabola). In the parabolic approximation, we are left with a simplified electronic Hamiltonian that can be written as:

$$\mathbf{W} = \left(\frac{\lambda_1}{2} \bar{Q}_{x_1} + \frac{\lambda_2}{2} \bar{Q}_{x_2} + \sum_{i,j \in IS} \frac{^{IS}\omega_{ij}}{4} \bar{Q}_i \bar{Q}_j \right) \mathbf{1} + \begin{pmatrix} \frac{\delta\kappa}{2} \bar{Q}_{x_1} & \kappa^{AB} \bar{Q}_{x_2} \\ \kappa^{AB} \bar{Q}_{x_2} & -\frac{\delta\kappa}{2} \bar{Q}_{x_1} \end{pmatrix}$$

$$+ \begin{pmatrix} \sum_{i,j \in IS} \frac{^{IS}\delta\gamma_{ij}}{4} \bar{Q}_i \bar{Q}_j & \sum_{i,j \in IS} \frac{^{IS}\eta_{ij}^{AB}}{2} \bar{Q}_i \bar{Q}_j \\ \sum_{i,j \in IS} \frac{^{IS}\eta_{ij}^{AB}}{2} \bar{Q}_i \bar{Q}_j & -\sum_{i,j \in IS} \frac{^{IS}\delta\gamma_{ij}}{4} \bar{Q}_i \bar{Q}_j \end{pmatrix}. \quad (4)$$

The description of two adiabatic potential energy surfaces around the conical intersection point is obtained from the diagonalisation of the simplified electronic Hamiltonian to give:

$$U_{A,B} = \frac{1}{2} \left\{ \lambda_{x_1} \bar{Q}_{x_1} + \lambda_{x_2} \bar{Q}_{x_2} + \sum_{i,j \in IS} \frac{^{IS}\omega_{ij}}{2} \bar{Q}_i \bar{Q}_j \right\}$$
$$\pm \frac{1}{2} \sqrt{ \left[\delta\kappa \bar{Q}_{x_1} + \sum_{i,j \in IS} \frac{^{IS}\delta\gamma_{ij}}{2} \bar{Q}_i \bar{Q}_j \right]^2 + 4 \left[\kappa^{AB} \bar{Q}_{x_2} + \sum_{i,j \in IS} \frac{^{IS}\eta_{ij}^{AB}}{2} \bar{Q}_i \bar{Q}_j \right]^2 }. \quad (5)$$

Thus the positive energy difference between the two states is:

$$\Delta U \equiv U_B - U_A$$
$$= \sqrt{ \left[\delta\kappa \bar{Q}_{x_1} + \sum_{i,j \in IS} \frac{^{IS}\delta\gamma_{ij}}{2} \bar{Q}_i \bar{Q}_j \right]^2 + 4 \left[\kappa^{AB} \bar{Q}_{x_2} + \sum_{i,j \in IS} \frac{^{IS}\eta_{ij}^{AB}}{2} \bar{Q}_i \bar{Q}_j \right]^2 }. \quad (6)$$

The two coordinates describing the branching space in the parabolic approximation will be a combination of first-order intersection adapted coordinates such that the energy difference Eq. (6) does not vanish. The remaining $(m-2)$ parabolic intersection coordinates will be a set of coordinates where the two intersecting states possess the same energy. This is possible by identifying a set of parameters for which the energy difference Eq. (6) is zero.

Thus the set of curvilinear coordinates f is defined as:

$$f = (f_1, f_2) \oplus (f_3, f_4, \ldots, f_{m-2}). \tag{7}$$

The first bracket includes the two coordinates spanning the parabolic branching space and using the description of the energy difference; these two coordinates may be explicitly defined as:

$$f_1 = \delta\kappa \bar{Q}_{x_1} + \sum_{i,j \in IS} \frac{{}^{IS}\delta\gamma_{ij}}{2} \bar{Q}_i \bar{Q}_j, \tag{8a}$$

$$f_2 = \kappa^{AB} \bar{Q}_{x_2} + \sum_{i,j \in IS} \frac{{}^{IS}\eta_{ij}^{AB}}{2} \bar{Q}_i \bar{Q}_j. \tag{8b}$$

The second subset of coordinates corresponds to the $(m-2)$ parabolic intersection coordinates, which are a combination of the original coordinates where the following conditions are simultaneously fulfilled:

$$\begin{cases} f_1 = 0 \\ f_2 = 0. \end{cases} \tag{9}$$

When moving along one of the first two coordinates (f_1, f_2), the degeneracy is lifted whilst a displacement along the third f_3 guarantees the degeneracy between the two states. The seam hessian can now be written[15] in these parabolic intersection space coordinates as

$$\left.\frac{\partial^2 U_{Seam}}{\partial f^2}\right|_{\mathbf{f}=0} = \begin{bmatrix} \nu_{33} & \nu_{34} & \cdots & \nu_{3,3N-6} \\ \nu_{34} & \nu_{44} & \cdots & \nu_{4,3N-6} \\ \vdots & \vdots & \ddots & \vdots \\ \nu_{3,3N-6} & \nu_{4,3N-6} & \cdots & \nu_{3N-6,3N-6} \end{bmatrix}, \tag{10}$$

where an arbitrary entry ν_{ij} can be rewritten in terms of potential constants as:

$$\nu_{ij} = \frac{1}{2}\left({}^{IS}\omega_{ij} - \frac{\lambda_{x_1}}{\delta\kappa}{}^{IS}\delta\gamma_{ij} - \frac{\lambda_{x_2}}{\kappa^{AB}}{}^{IS}\eta_{ij}^{AB}\right). \tag{11}$$

The curvature of the seam along the i-th coordinate is the i-th eigenvalue of the matrix [Eq. (11)]. All the matrix entries and eigenvalues have units of frequency squared; in atomic units this corresponds to s^{-2}. The eigenvectors obtained by diagonalisation of Eq. (11) are a combination of the $(m-2)$ local axes $\hat{\mathbf{f}}_i$, $i = 3, 4, \ldots, m-2$, tangent to the f_i at a given point.

We now discuss some benchmark applications of the theoretical ideas just presented.

3. Extended Seam Benchmarks

3.1. *Fulvene*

Fulene seems to have become a benchmark molecule for the study of conical intersections.[16,22] In recent work,[16] we have been able to optimise five geometries on an extended conical intersection seam so this provides a nice example to illustrate the theoretical ideas about the extended seam discussed in the preceding subsection.

A schematic two-dimensional schematic potential energy surface is given in Fig. 5. The branching space coordinates are given in the inset of Fig. 5 and correspond to the skeletal deformations of the five-membered ring. So we choose one of these skeletal deformations as $\bar{Q}_{x_1}$ in Fig. 5, while a suitable choice for a reaction coordinate $\bar{Q}_{x_3}$ is the torsional angle $\bar{Q}_\phi$. In Fig. 5 we can see that the extended curve has both a local maxima at a geometry we call CI$_{\text{perp}}$ and a minimum at a twisted geometry. However, the situation is more complicated as one goes to higher dimensions. In fact, CI$_{\text{perp}}$ is a second-order saddle point on the seam with two imagining frequencies corresponding to torsion and pyramidalisation (see the imaginary frequencies illustrated on the left-hand side of the Fig. 5). However, even after a distortion along a pyramidalisation coordinate, the structure CI$_{\text{pym}}$ retains the imaginary frequency corresponding to torsion.

The full set of stationary points on the conical intersection seam, excluding CI$_{\text{perp}}$, is given in Fig. 6 together with the computed imaginary frequencies in the intersection space. The structure CI$_{\text{plan}}$ is a second-order saddle point (like CI$_{\text{perp}}$) and is unstable with respect to both torsion and pyramidalisation. The pyramidalised structure CI$_{\text{pym}}$ is a saddle point

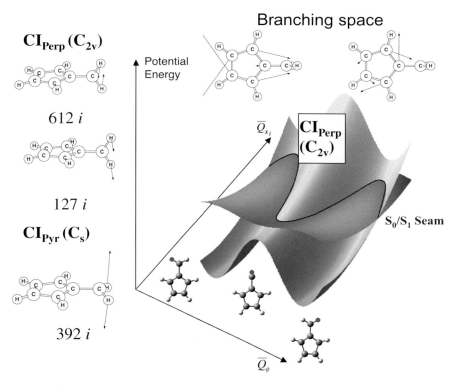

Fig. 5. Potential energy profile for fulvene in the space spanned by Q_{X1} (branching space) and Q_ϕ the X_3 coordinate (torsion). (Adapted from Paterson et al.[11])

with respect to torsion that leads to the partly twisted structure CI_{min}. In contrast, if one follows the torsional coordinate from CI_{plan}, one reaches a structure that is twisted but not pyramidalised that is denoted as CI_{63}.

Of course such a study of the full seam is just an academic exercise. A posteriori results seem obvious. Along the seam the energies of excited and ground state are equal. Thus the electronic effects associated with the π system are in balance. This leads to the conclusion that the stability with respect to torsion is a steric affect. Similarly, the stability with respect to pyramidalisation is the simply expected stereochemistry of an isolated methylene group.

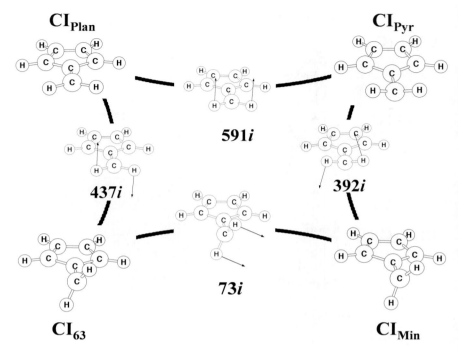

Fig. 6. Schematic representation of the conical intersection hyperline topology (excluding **CI**$_{\text{Perp}}$) in the space of torsion and pyramidalisation. The seam normal modes corresponding to the imaginary frequencies that connect the conical intersection geometries optimised on the S_0/S_1 seam of fulvene are shown as vectors for hydrogen motions only. (Adapted from Sicilia et al.[16])

3.2. *The $2A_1/1A_1$ conical intersection seam in butadiene*

This is another benchmark type problem. While initial photoexcitation takes place at the optically bright B state, the photochemistry of cis-butadiene (Fig. 7) occurs via a conical intersection between the $2A_1$ state and the ground state. Photolysis yields many products, possibly from decay at different points on the seam; the results are summarised in Fig. 7.

Some seven critical points[13,15] on the seam are illustrated in Fig. 8. In our early work,[23] we located CI$_{\text{cis}}$, ^{sp}CI$_{\text{cis/trans}}$ and CI$_{\text{trans}}$. In Fig. 7, we see that ^{sp}CI$_{\text{cis/trans}}$ is confirmed as a saddle point. In fact, it has been possible to use reaction path procedure,[13] constrained to the intersection space, to connect all these structures.

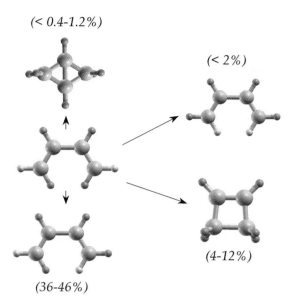

Fig. 7. Photoproduct distribution following irradiation of cis butadiene.

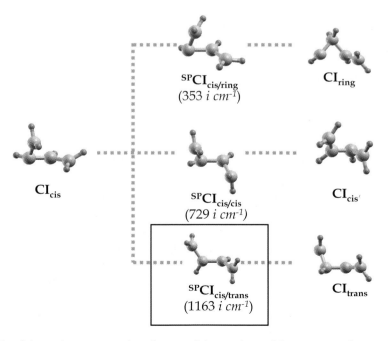

Fig. 8. Schematic representation of some minima and transition states on the extended seam for the S_1/S_0 intersection of butadiene.

4. Applications of the Extended Seam of a Conical Intersection to Photochemical Mechanisms

The extended seam concept is a mechanistic feature of many problems in photochemistry and photophysics, including photochromic systems such as dihydroazulene,[24] the ring opening of cyclohexadiene,[25,26] diarylethylenes,[27] protonated Schiff base rhodopsin models,[28] T-(ICT) compounds,[29,30] cyanine dyes,[31] biological chromophores such as PYP[32] and GFP,[33] excited state proton transfer,[34] the photochemistry of benzene,[14] the DNA bases,[35] and other organic transformations,[36–38] as well as classical problems in photophysics such as the photodissociation of formaldehyde.[39] In this section, we will choose a few examples (mainly from our own work) that illustrate the utility of the extended seam as a mechanistic feature in organic photochemistry.

4.1. *The photoinduced isomerization of 1,3-cyclohexadiene (CHD) to cZc-hexatriene (HT)*[25,26]

The ring opening of CHD is experimentally, as well as theoretically, a prototypical photochemical reaction involving an extended seam of conical intersections. In recent work,[40] we have been able to show (Fig. 9) that the seam is approximately parallel to the reaction path.

We have started with the discussion of this problem because, like fulvene, it is a "benchmark" of organic photochemistry. Nonadiabatic decay during a photochemical reaction was first clarified mechanistically by van der Lugt and Oosterhoff.[41] The central idea uses the concept of an avoided crossing (which provides the photochemical funnel) arising from the ground state and a doubly excited state along a common reaction coordinate (bond breaking, x axis in Fig. 9). However, the reaction path does not intersect with the conical intersection. Rather, as shown in Fig. 9, we have an extended seam lying approximately parallel to the excited state. This seam was computed via a seam MEP[13] (S_1-seam-MEP), which is similar to a conventional MEP but constrained to the intersection space. This motion orthogonal to the reaction path in the direction of the seam must control the ultrafast decay to the ground state.

In Fig. 9, we show the complete minimum energy S_0/S_1-CoIn seam (seam-MEP) for the conrotatory ring-opening reaction of CHD, covering the region from the closed (CHD) to the open ring structure (HT). The conrotatory S_1 IRC-MEP is almost parallel but displaced along a skeletal

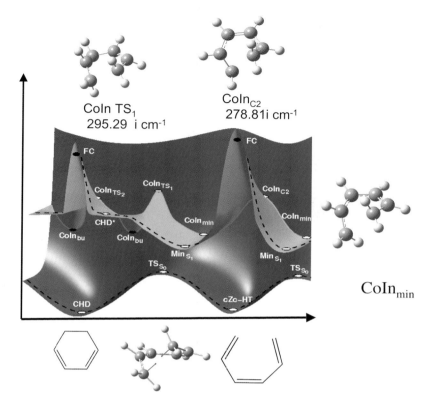

Fig. 9. Schematic representation of the S_0 and S_1 potential energy surfaces for the ring-opening/ring-closure reaction in the CHD/cZc-HT system including the conical intersection seam (seam-MEP). The reaction coordinate (RC) describes the conrotatory CHD-HT conversion along the minimum energy reaction path (IRC-MEP, dashed lines). The orthogonal BS vector is defined at every point along the seam-MEP as a linear combination of the branching space vectors critical points along the MEPs are noted with imaginary frequencies. (Adapted from Nenov et al.[40])

deformation coordinate. The avoided crossing feature (Min_{S_1}/TS_{S_0}) on the MEP is thus displaced from the lowest energy point of the conical intersection seam $CoIn_{min}$. Note that the seam-MEP has local transition state features (such as $CoIn_{TS_1}$ and $CoIn_{C_2}$) and the corresponding imaginary frequencies are also given in Fig. 9. These frequencies were obtained using the second-order representation of the seam and correspond to motion along the intersection space (dominated by bond breaking). The seam

MEP can be mapped out in the intersection space as well, as discussed previously.[13]

In the Woodward–Hoffmann treatment of photochemistry as reformulated by van der Lugt,[41] the excited-state and ground-state reaction paths were assumed to be similar, with the "photochemical funnel" occurring at an avoided crossing. In this classic example, computations show that the ground, state and excited, state reaction paths are indeed very similar. However, CoIn seam is displaced from the excited-state/ground-state MEP along skeletal deformations, i.e. the branching space of the CoIn is orthogonal to the MEP.

4.2. *Diarylethylenes*[27]

These are remarkable photochromic systems where the chemical transformation is single bond breaking (Fig. 10). Because the distribution of π-bonds is different in both isomers, they have distinct absorption spectra. The photophysics and the efficiency of the system are completely controlled by the relationship between the reaction path and the degeneracy-lifting coordinates. From computations,[27] the similarity to the CHD example just discussed is remarkable.

Figure 11 shows the energy profile along a bond-breaking coordinate. We use labels CHD (cyclohexadiene) and HT (hexatriene) to emphasise the role of the central six carbon atoms and the relationship to the previous example. The crosses indicate points optimised on the conical intersection seam. Thus the "seam" does not appear to intersect with the reaction path. In Fig. 12, we show a 3D cartoon of the potential surface. In this figure, we also give the coordinates that lift the degeneracy (X_1 X_2) and the bond breaking reaction coordinate X_3. Like CHD and other examples we will discuss subsequently, the coordinates that lift the degeneracy are just skeletal deformations.

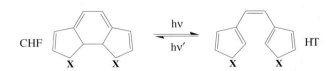

Fig. 10. Diarylethylenes with heterocyclic aryl and bisthienylethylene-based compounds (X = S) exhibit remarkable switching sensitivity (i.e. high quantum yield) and rapid response.

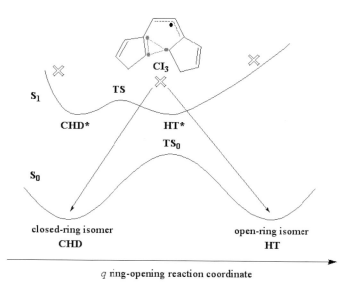

Fig. 11. Diarylethylenes: energy profile along the single bond breaking reaction path.

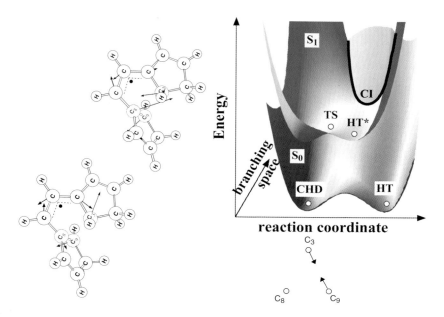

Fig. 12. A cartoon of the diarylethylene potential surfaces that can be distilled from the dynamics computations reported by Boggio–Pasqua et al.[27]

From Fig. 12 it is clear that there is a ground-state thermal reaction path involving a transition state **TS**$_0$. In addition, there is an adiabatic reaction path on the excited state involving a minimum, a transition state **TS**, and another minimum. One can also find three critical points (local minima) on the conical intersection line indicated by crosses in Fig. 11: one near the products, one near the reactants, and also one near the transition state on the adiabatic excited-state reaction path. Thus, apparently disconnected conical intersections lie displaced along the intersection space coordinates (one of which is the transition vector shown in Fig. 12).

The conical intersection seam in Fig. 12 is very similar to CHD in Fig. 9: the reaction path is in the foreground (left to right), and the conical intersection line (**CI**) lies in the background. Thus the reaction coordinate is parallel to the seam and so decay to S$_0$ is controlled by motion orthogonal to the reaction path. The minimum energy point on the conical intersection line (middle cross, **CI**$_3$ in Fig. 11) is the one that is located using gradient-driven optimization. While it appears quite close to the transition state on the excited state, it is actually on the HT side of the barrier.

In summary, we see that like CHD, knowledge of the excited-state reaction path does not yield an understanding of the photochromism of this system. Further, finding the conical intersection points does not yield a complete picture, because they do not lie on the reaction path. Indeed, a large segment of the intersection seam in this region is energetically accessible. However, to demonstrate and understand this, you need to run dynamics calculations.

Experimentally, one observes a fluorescence that is red-shifted, confirming that the position of the minimum is different on the excited state. The cyclization quantum yield (HT* to CHD) is high, which arises from the fact that a trajectory from **HT*** can sample the whole intersection seam, at right angles to the reaction path. On the other hand, for a trajectory starting from **CHD***, the probability of decay to S$_0$ is low, because the main locus of the conical intersection seam appears to be on the HT* side of the transition state. Thus to reach HT from CHD*, you have to pass through the transition state on the excited state reaction path. Thus there is a competition between passing through the transition state to reach the reactive conical intersection on the HT* side of the TS, and decay at a nearby crossing on the CHD* side of the transition state, which does not lead to any reaction. Thus the quantum yield is quite low in this direction.

There remains the question of whether one could design a pulsed laser sequence to "control" this reaction. For the HT isomer, one would want a wavepacket with excess momentum in the branching space direction so that decay would take place quickly. In contrast, for the CHD isomer, one would need a wavepacket with excess momentum in the direction of the reaction coordinate, which would drive the reaction towards the transition state on the excited state and avoid competition with any nearby conical intersection points.

4.3. *The keto-enol tautomerism of o-hydroxyphenyl-(1,3,5)-triazine*

This species provides an efficient photostabilisation system.[34] This is an example where the extended conical intersection seam is one of the contributing factors controlling the efficiency in such species. The enol form [Fig. 13(a)] absorbs light and decays to the keto form [Fig. 13(b)] on the ground state. The ground-state keto form is metastable, and interconverts back to the enol form over a small barrier. Thus we have light absorption followed by no net chemical change and a photostabilising cycle. The low-lying excited states of such species are $\pi - \pi^*$, yet the hydrogen transfer involves the σ electrons. Thus the reaction coordinate X_3, since it involves these σ electrons, must be completely independent from the electronic state changes; the latter clearly involve only the π electrons. This is therefore an example where, *a priori*, the branching space coordinates must be completely different and independent from the reaction path, and one knows from the outset that the surfaces must involve the extended seam topology.

We begin with a VB analysis of ground and excited states at the enol and keto geometries. It is possible to classify the ground state and the two types of $\pi - \pi^*$ excited-state according to the number of π electrons

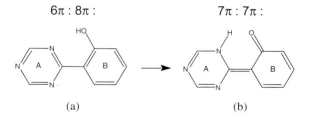

Fig. 13. Enol (a)–keto (b) tautomerism in o-hydroxyphenyl-(1,3,5)-triazine, indicating the number of π electrons in the ground state for each ring A and B. (Adapted from Paterson *et al.*[34])

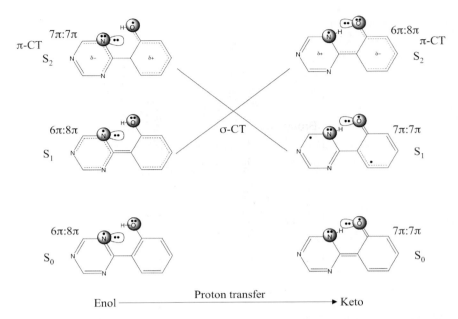

Fig. 14. Correlation diagram for the lowest $\pi-\pi^*$ excited states along a proton transfer coordinate. (Adapted from Paterson et al.[34])

associated with the two rings A and B (as indicated in Fig. 13). In Fig. 14, we show a valence bond correlation diagram for the lowest excited states along a proton transfer coordinate. This correlation diagram was elucidated by analysis of the excited states in recent theoretical calculations.[34] At a given geometry (keto or enol), the locally excited states preserve the number of π electrons in each ring, while the CT states change this population. (Notice that LE and CT, as we use them in this context, are relative to the ground state electronic configuration at a given geometry.) Thus, the state with the configuration $6\pi - 8\pi$ is locally excited at the enol geometry but formally CT at the keto geometry because of the migration of the proton. To avoid ambiguity, we will be classifying the excited states according to the number of π electrons in each ring. Only the ordering of the various states has to be determined from theoretical computations. However, it will be the vertical excitation to the CT state that will be observed experimentally, because of the larger oscillator strength.

If we look at the correlation (Fig. 14) between the enol ground state electronic configuration and the keto ground state configuration, we observe

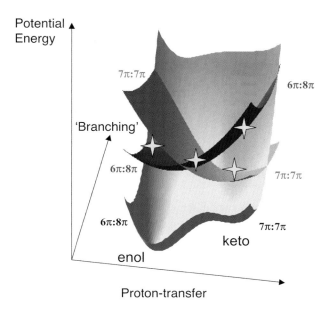

Fig. 15. A cartoon showing the lowest $\pi - \pi^*$ excited states along a proton transfer coordinate and a skeletal deformation coordinate from the branching space. (Adapted from Fig. 4 of Paterson et al.[34])

a change in the number of ring A and B electrons. One might expect an activation barrier due to the change in electronic configuration. In fact, computations[34] suggest that the barrier to back formation of the enol form from the keto form is small (4 kcal mol^{-1}). Thus if the keto form is generated photochemically, the enol form will be rapidly regenerated thermally over a small barrier. It only remains to discuss the photochemical proton transfer to generate a ground state keto form.

The excited state proton transfer can be understood using Fig. 15, where we have labelled the various excited state potential energy surfaces consistent with Fig. 14. In Fig. 15, we show potential energy surfaces in a cartoon involving the proton transfer coordinate and one coordinate from the branching space of the extended conical intersection seam. We have optimized[34] four isolated critical points on the extended seam; three S_1/S_0 conical intersection points in the enol region, in the keto region and the transition state region as well as an S_2/S_1 conical intersection on the keto side as indicated by the four points/stars in Fig. 15. In each case, the branching space coordinates X_1 X_2 involve the skeletal deformations of the two rings and do not include a component along the proton transfer coordinate. Thus in this case, the branching space is rigorously distinct from the reaction coordinate corresponding to proton transfer. Of course, along an adiabatic reaction path from the enol S_1 $6\pi - 8\pi$ minimum to the keto S_1 $7\pi - 7\pi$ minimum, the real crossing will become avoided and generates a transition state.

However, the initial excitation is to the enol S_2 $7\pi - 7\pi$ state. It is clear from Fig. 15 that there is an extended conical intersection seam between the $7\pi - 7\pi$ state/$6\pi - 8\pi$ excited states and the ground state. Thus the system can decay efficiently after photoexcitation at any point along the seam. Since the ground state barrier between the keto and enol form is negligible, the regeneration of the ground state enol form, following photoexcitation, must be exceedingly efficient. Thus the presence of a conical intersection seam along the reaction path, where the branching space coordinates are rigorously orthogonal to the reaction path, can be identified as a desirable design feature for efficient photostabilisers.

In summary, the extended seam of conical intersection, which is parallel to the reaction path, allows for radiationless decay at any point along the proton transfer reaction path, even on the enol side. This topology explains the experimental observation that the proton transfer is in competition with a temperature-dependent deactivation process. For photostability, this paradigm is ideal, since the seam has everywhere a sloped topology (gradients of ground and excited state are approximately parallel) and the ground state enol form is regenerated on an ultrafast timescale. These mechanistic features are independent of the ordering of the locally excited versus charge-transfer configurations. The notion of a seam of intersection explains the high photostability of the o-hydroxyphenyl-triazine class of photostabilisers in particular, but more generally highlights an important photochemical feature that should be considered when designing a photostabiliser.

In recent computations, we have been able to show that this same mechanism operates in other photostabilsers[42] and in a Watson–Crick base pair in DNA.[43] The reader is referred to Chap. 2 for further discussion.

5. Valence Bond Analysis of Conical Intersections

In this section, we would like to consider an example which illustrates how one can understand the occurrence of conical intersections, as well as the directions X_1 X_2 corresponding to the branching space, if one has an understanding of the electronic (VB) structure of the two states involved. We address the following three questions:

(1) What is the connection between the molecular geometry and the electronic (VB) structure?
(2) Can the nature of the adiabatic and nonadiabatic pathways (and the position of the conical intersection) be predicted from VB structures?
(3) What is the VB origin of phase change when one does a circuit of a conical intersection?

5.1. *Twisted intermolecular charge transfer (T-(ICT)) in aminobenzonitrile (ABN) compounds*[29,30]

We will illustrate the first two points above using a T-(ICT) (twisted intermolecular charge transfer) aminobenzonitrile (ABN) compound (Fig. 16) as an example. We have recently completed theoretical work[29] on this

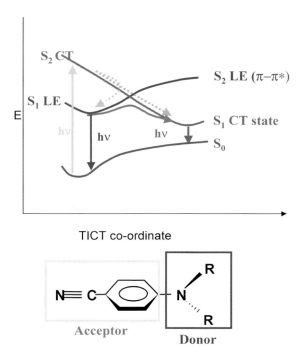

Fig. 16. Schematic representation of the T-(ICT) process and the emission of LE and ICT states.

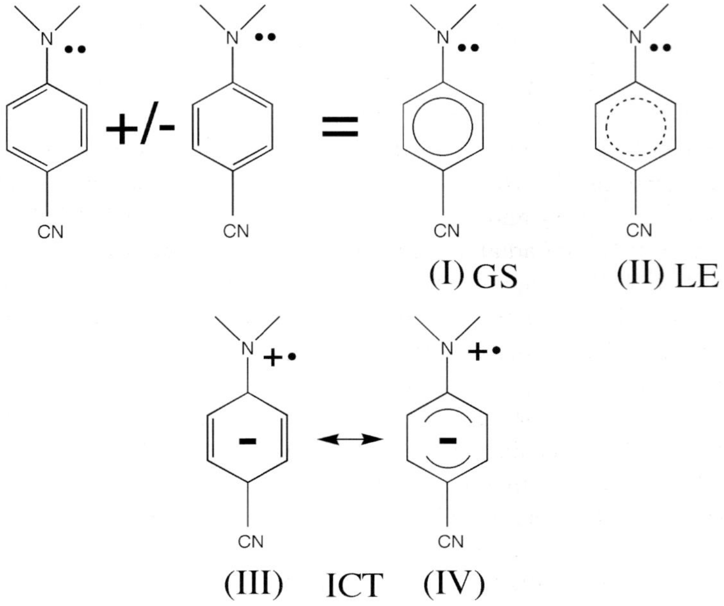

Fig. 17. VB states involved in the ICT LE surface in the T-(ICT) process. (Adapted from Gomez et al.[29])

class of compound and the reader is referred to that work for a complete bibliography.

In ABNs there are two low-lying excited states: a locally excited (LE) state where the excitation is localised on the phenyl ring (the ground state is the sum of the Rumer states while the LE state is the antiaromatic difference state), and the intramolecular charge transfer (ICT) state, where there is a transfer of charge from the amino group to the benzene ring (see Fig. 17). The ICT state is thus similar (electronically) to a benzene radical anion. In spectroscopy, with suitable substitution R and in the appropriate solvent, one can see emission from each state (LE or ICT diabatically, but both on S_1, see Fig. 16). The lowest energy equilibrium geometry of the ICT state is usually assumed to be twisted; hence the acronym T-(ICT). Since one observes dual fluorescence, there must be two S_1 minima, associated with the LE and ICT electronic structures (see Fig. 16). An adiabatic reaction path must therefore connect these two electronic structures on S_1. Thus there is an adiabatic reaction coordinate associated with the electron transfer process. However, the absorption from the ground state to the LE state in the Franck–Condon region will be forbidden. Rather the absorption takes place to S_2, which is the ICT state at the Franck–Condon geometry. Thus there is also a nonadiabatic ICT process associated with the radiationless decay from S_2 (ICT) to S_1 (LE).

The ideas just discussed can be summarised in the potential energy diagram shown in Fig. 16. From the figure it is clear that the (adiabatic) state label S_1 and S_2 and the (diabatic) VB structure label ICT or LE are independent. The adiabatic reaction path involving a transition state (i.e. avoided crossing) appears to be associated with the real crossing. The T-(ICT) coordinate (amino group torsion) is assumed to be the reaction path. The transition state on this reaction path is associated with a state change from LE to ICT. This state change can also be associated with the nonadiabatic process via the real crossing. However, the real crossing and the nature of the branching space and its relationship to the adiabatic reaction path can only be understood by moving to higher dimensions and by consideration of the relationship between the branching space coordinates and the twisting coordinate.

We now give some discussion of the VB states involved in Fig. 16. In the ABN problem, there are four VB structures that are relevant and these are shown in Fig. 17. There are two "dot-dot" (covalent) configurations I and II and two zwitterionic configurations III and IV. Structures I and II are just the Kekulé and anti-Kekulé structures of benzene. The LE structure corresponds to the anti-Kekulé electronic structure, where excitation takes place in the phenyl ring. The zwitterionic structures III and IV are the ICT states. The ICT state has a positive charge on the amino group and an extra electron on the phenyl group and we expect similarities to the benzene radical anion. Thus there will be a quinoid (III) and an anti-quinoid structure (IV). In a theoretical calculation on ABN species, an inspection of the wavefunction will yield the information about which resonance structure dominates.

We are now in a position to discuss the reaction profile outlined in Fig. 16 in the full space of coordinates corresponding to the branching space X_1 X_2 of a conical intersection and the torsional coordinate X_3. This discussion will be focused on four related concepts:

(a) the S_2 to S_1 radiationless decay,
(b) the geometry and electronic structure of the two S_1 minima,
(c) the geometry of the S_1/S_2 conical intersection together with the nature of the X_1 X_2 branching space, and

(d) the nature of reaction path X_3 connecting the LE and ICT regions of S_1. Our objective is to rationalise all the data using the four VB structures in Fig. 17 and to illustrate the overall surface topology according to the models or cartoons given in Figs. 1 and 3.

In Fig. 18(a), we show the geometry of the S_2S_1 ICT/LE crossing (minimum energy crossing point), together with the directions X_1 and X_2. The crossing occurs between the LE and ICT (III quinoid in Fig. 17) VB structures. The most important point about the geometry is that the amino group is not twisted. The directions X_1 and X_2 are mainly the skeletal deformations of the phenyl ring and do not involve torsion. This is completely consistent with the fact that the LE and ICT VB structures differ essentially only in the phenyl ring. Thus we have established that the nonadiabatic decay does not involve the amino group twist, since the directions X_1 and X_2 exclude this coordinate. This in turn follows from the bonding pattern of the two VB states.

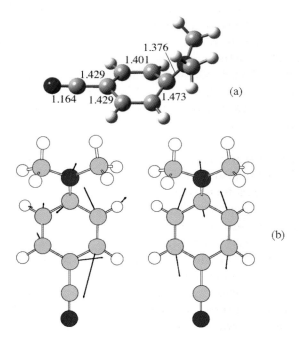

Fig. 18. (a) Geometry of the S_2S_1 ICT/LE crossing MECI, together with (b) the directions X_1 and X_2 for ABN systems. (Adapted from Gomez et al.[29])

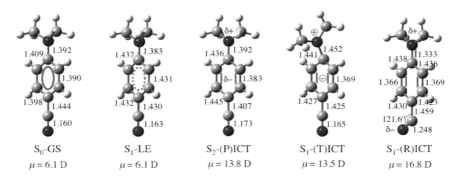

Fig. 19. The geometries of various minima on the ABN S_2 and S_1 states. (Adapted from Gomez et al.[29])

Let us now consider the geometries of the various minima on the S_2 and S_1 states shown in Fig. 19. Notice that we are careful to include both the adiabatic label S_2 S_1 and the diabatic VB state label LE or ICT. One can see that the main difference in the "dot-dot" covalent VB structures associated with S_0-GS and S_1-LE geometries occurs in the C-C bond lengths of the phenyl ring, which are lengthened in S_1-LE because of the anti-Kekulé nature of the VB structure. If we examine the CT structures, we see that we have a planar (P)-ICT structure on S_2 and a twisted (T)-ICT structure on S_1. (There is also a high-energy (R)-ICT structure that has a bent cyano group.) The reaction pathways that connect these structures must include (i) an adiabatic reaction path that connects the S_1-LE and (T)-ICT structures on S_1 along a torsional coordinate and (ii) a nonadiabatic reaction path that connects the S_2 planar (P)-ICT structure with the S_1-LE structure and the S_1 (T)-ICT structure via an extended conical intersection seam, that lies along a torsional coordinate. We now discuss this.

The optimised geometries of the various minima on S_2 and S_1 (Fig. 19), together with the nature of the branching space vectors X_1 X_2 [Fig. 18(b)], suggests that the topology of the potential surface has the form shown in the model surface in Fig. 3. Thus we have a conical intersection seam along X_3 = NR_2 torsion with the branching space X_1 X_2 spanning the phenyl group skeletal deformations shown in Fig. 18(b). We collect together all of the information in Fig. 20, corresponding to the general model given in Fig. 3. At the left-hand side, corresponding to untwisted geometries, one can see the S_1-LE minimum and the planar S_2(P)-ICT state minima. Because the branching space excludes X_3 = NR_2 torsion, the conical intersection seam

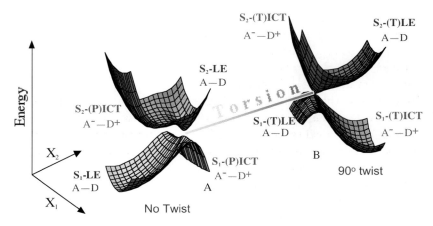

Fig. 20. The geometries from Fig. 19 located in the cone that changes shape along the conical intersection hyperline. (Adapted from Gomez et al.[29])

can persist as an extended seam along this coordinate. The double cone at the twisted geometry is shown on the right-hand side of the figure. Here the double cone shape changes and the twisted $S_1(T)$-ICT state minima develops. Thus we have added the two branching space dimensions X_1 X_2 to Fig. 16 to yield Fig. 20. The origin of the nonadiabatic S_2 to S_1 process is now clear. The initially created state at the Franck–Condon geometry is near the $S_2(P)$-ICT state minimum. This state can decay to either the S_1-LE minimum or the $S_1(T)$-ICT minima along the extended seam. The S_1 adiabatic process can occur following $S_2(P)$-ICT to S_1-LE decay via a path on S_1 involving the $X_3 = NR_2$ torsion.

5.2. What happens when one does a conical intersection circuit in the branching plane?[44–47]

The 1975 paper of Longuet–Higgins[48] states the phase change theorem as:

> "If the wavefunction of a given electronic state changes sign when transported round a loop in nuclear configuration space, then the state must become degenerate with another one at some point within the loop."

This theorem has implications for dynamics[49] and can even provide a method for optimising a geometry.[46] However, it is more interesting when applied using the VB method to understand the chemical nature of the conical intersection. Haas and his co-workers have developed this idea[44] and Vanni et al.[45] have attempted to make these ideas more rigorous. Many years ago, we attempted to rationalise the geometries of conical intersections of hydrocarbons using VB theory.[50,51] It turns out that the phase change rule, when applied to three and four electrons, gives additional insights into the chemical nature of conical intersections. We now give a discussion adapted from the work of Vanni et al.[45]

It turns out that for the case of three orbitals and three electrons or four orbitals and four electrons, where the orbitals are 1s orbitals, one has some simple analytical results that enable one to understand the branching space coordinates and the relationship to valence bond structures. Of course the results are rigorous only in these cases, but they can be applied in a qualitative way to other examples.

Valence bond theory uses a special combination of determinants called Rumer functions. For three orbitals and three electrons, one has three valence bond structures as shown in Fig. 21. The valence bond structures such as A can be defined in terms of determinants as

$$A = \frac{1}{\sqrt{2}}\{|1\ \bar{2}\ 3| - |\bar{1}\ 2\ 3|\}, \quad (12)$$

where we use $||$ to denote the diagonal elements of the determinant. However, it is simpler to formulate arguments directly in the Rumer basis.

In this three-electron example, if the ground state wavefunction is A, then the corresponding excited state wavefunction would be B (since there are only two linearly independent spin functions). There are two complications in practice. Firstly, A and B are not orthogonal if we use Rumer VB functions. The overlap between the Rumer functions A and B is $\langle A|B\rangle = -\frac{1}{2}$. Thus, if we take A to be a ground state wavefunction, the corresponding orthogonal excited state wavefunction B must be Schmidt orthogonalised to give $B'' = \frac{2}{\sqrt{3}}(B + \frac{1}{2}A)$. Secondly, the valence bond structure C is not linearly independent of the other two structures. Thus we have $C = A + B$ and C is linearly dependent on A and B. The excited state partner functions constructed in this way, which correspond to the

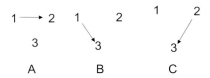

Fig. 21. Rumer VB diagrams for a three orbital three electron system.

VB functions A and B in Fig. 21, are shown in Eq. (13a) below:

$$A'' = \frac{2}{\sqrt{3}}\left(A + \frac{1}{2}B\right),$$
$$B'' = \frac{2}{\sqrt{3}}\left(B + \frac{1}{2}A\right). \qquad (13a)$$

Similarly, the orthogonal partner function (C'') of C can be taken as

$$C'' = \frac{1}{\sqrt{3}}(A - B), \qquad (13b)$$

where the $\sqrt{3}$ in C'' comes from the normalisation requirement since A and B are not orthogonal. (Observe the notation: C'' is the excited state orthogonal state to C itself, while A'' is the orthogonal state of A obtained by Schmidt orthogonalising A to B, and B'' is the orthogonal state of B.)

Now let us define the branching plane, $X_1\ X_2$. We can make any choice of orthogonal states as a starting point, but we shall choose the states $C = A + B$ and $C'' = \frac{1}{\sqrt{3}}(A - B)$ to simplify the algebra.

However, first we must establish an approximate but essential relationship between nuclear configurations and VB structures. The matrix elements between Rumer functions involve exchange integrals K_{ij} ($K_{ij} = [ij|ij] + S_{ij}h_{ij}$ in the case of 2 H atom 1s orbitals). The indices ij are associated with orbitals on nuclear centres i and j. Thus in Fig. 21, the indices 1, 2 and 3 relate to nuclear centres 1, 2 and 3. Thus our discussions are rigorous only for 3 H atoms with a single 1s function on each centre. However, the relationships would be expected to hold approximately for any set of three nuclei with one "active" VB orbital on each centre.

As stated above, for our computation of the branching space, we use the ground $C = A + B$ and excited $C'' = \frac{1}{\sqrt{3}}(A - B)$ orthogonal states. The branching space directions require the computation of the direction of the derivatives of energy difference $\Delta H(H_{CC} - H_{C''C''})$ and off-diagonal matrix element $H_{CC''}$. The AB, BC matrix elements of the Hamiltonian are between the Rumer basis states.[45] A, B and C are collected in Eq. (14).

$$H_{AA} = K_{12} - \frac{1}{2}(K_{13} + K_{23}),$$
$$H_{BB} = K_{13} - \frac{1}{2}(K_{12} + K_{23}),$$
$$H_{CC} = K_{23} - \frac{1}{2}(K_{12} + K_{13}),$$

$$H_{AB} = \frac{1}{2}(K_{12} + K_{13} - 2K_{23}),$$

$$H_{AC} = -\frac{1}{2}(K_{12} + K_{23} - 2K_{13}),$$

$$H_{BC} = \frac{1}{2}(K_{13} + K_{23} - 2K_{12}). \tag{14}$$

All other matrix elements can be derived from these equations. The energy difference is[45]:

$$(X_1):$$

$$\Delta H = H_{C,C} - H_{C''C''} \tag{15}$$

$$= -\frac{5}{3}(K_{12} + K_{13} - 2K_{23}),$$

and the interstate coupling is[45]

$$(X_2):$$

$$H_{C,C''}(q) = \frac{1}{\sqrt{3}}\langle A - B|H|A + B\rangle \tag{16}$$

$$= \frac{\sqrt{3}}{2}(K_{13} - K_{12}).$$

The derivatives can then be expressed qualitatively as follows. The indices ij in the exchange integrals K_{ij} relate to orbitals on centres i and j and $K_{ij} \propto \exp(-bR_{ij})$ where R_{ij} is the distance between centres i and j. Thus the magnitude of the gradient is $dK_{ij}/dR_{ij} \propto \exp(-bR_{ij})$ with direction along a unit vector from centre i to centre j. For this reason we can use the direction of ΔH and $\nabla(\Delta H)$ interchangeably and similarly for H_{AB}.

We now illustrate this idea. For each exchange integral $-K_{ij}$, we draw a vector on atom i heading towards atom j and a vector on atom j heading towards atom i (for K_{ij} two vectors are the opposite of the ones above). We then compute a resultant vector for each atom of the system. The "resultant" will qualitatively describe the gradient difference vectors. For the three-orbital example, we have the result shown in Fig. 22.

For $H_{CC''}(K_{13} - K_{12})$ coordinate we have the following: on atom 1 the resultant vector is the sum between two vectors arising from two terms: $-K_{12}$ and K_{13}, while on atom 2 the resultant vector arises only from $-K_{12}$, and on atom 3 the resultant vector arises only from K_{13}. For the $H_{CC} - H_{C''C''}$ coordinate $(K_{12} + K_{13} - 2K_{23})$ we have the following: on atom

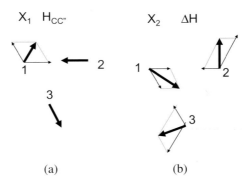

Fig. 22. Derivative coupling and gradient difference associated with ΔH and $H_{CC''}$ coordinates for three orbital three electron systems. Note that we use ΔH and its gradient interchangeably because they are parallel as discussed in the text. (Adapted from Vanni et al.[45])

1 the resultant vector is the sum between two vectors arising from two terms: $-K_{12}$ and $-K_{13}$; on atom 2 the resultant vector is the sum between two vectors arising from two terms: $-K_{12}$ and $2K_{23}$; and on atom 3 the resultant vector is the sum between two vectors arising from two terms: $-K_{13}$ and $2K_{23}$. Thus the $\delta\kappa \equiv \left.\frac{\partial(E_B - E_A)}{\partial Q_{x_1}}\right|_0$ and $\kappa^{AB} \equiv \left.\frac{\partial\langle\Psi_A|\hat{H}|\Psi_B\rangle}{\partial Q_{x_2}}\right|_0$ corresponding to the $H_{CC} - H_{C''C''}$ and $H_{CC''}$ derivatives are shown as the axes in Fig. 22. We can see that the condition $H_{CC} - H_{C''C''} = 0$ is achieved when $K_{12} + K_{13} = 2K_{23}$. Along the $-H_{CC''}$ coordinate at the origin, $K_{13} = K_{12}$. Thus at the apex of the cone, one has

$$K_{12} = K_{13} = K_{23} \tag{17}$$

corresponding to an equilateral triangle, a well-known result.[50]

Now we would like to explore the relationship between molecular structure and VB (i.e. electronic structure) illustrated in Fig. 23. Figure 23 shows the branching space directions (see Fig. 22) as deformations of the three atoms together with three molecular structures where the VB structure (i.e. the electronic structure) is coincident with the nuclear geometry. Notice that they are related by a rotation (ϕ) in the plane of 120° in the space of nuclear coordinates. It is easily demonstrated that the valence bond wavefunctions are obtained by rotation (θ) of ground and excited state of VB structures by 60°. Thus there is a fundamental relationship between electronic structure and nuclear structure in the vicinity of a conical intersection with a special case of three electrons in three hydrogen like 1s orbitals. (The

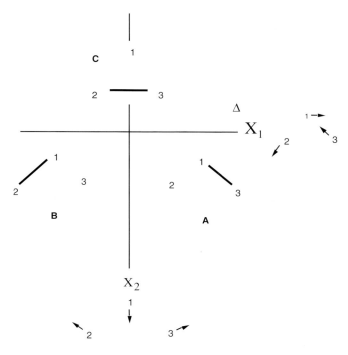

Fig. 23. The ϕ loop in nuclear configuration space showing Valence Bond structures for the three orbital three electron problem and the ΔH and $H_{CC''}$ vectors which define the branching plane. (Adapted from Vanni et al.[45])

same results can be obtained for four orbitals and four electrons.) There is thus a relationship between the polar angle in a closed loop around the apex of the cone (ϕ) (relating the molecular structures) and the mixing angle (θ) (relating the VB structures) of the ground and excited state $\begin{bmatrix} \Psi_A \\ \Psi_B \end{bmatrix}$ wavefunctions under the transformation T.

We now expand on this observation. We define a 2×2 transformation T as

$$\mathbf{T}(\mathbf{q}) = \begin{bmatrix} \cos \theta(\mathbf{q}) & -\sin \theta(\mathbf{q}) \\ \sin \theta(\mathbf{q}) & \cos \theta(\mathbf{q}) \end{bmatrix}. \tag{18}$$

The angle θ depends on the polar angle ϕ only. It is therefore constant along straight lines having their origin at the apex of the double cone. Matrix rotations θ by 60° and 120° [Eq. (19) below] corresponding to the geometry changes ϕ of 120° and 240° involve application of the transformations in Eq. (19) to C and C″

$$\begin{bmatrix} \frac{1}{2} & -\frac{\sqrt{3}}{2} \\ \frac{\sqrt{3}}{2} & \frac{1}{2} \end{bmatrix}, \begin{bmatrix} -\frac{1}{2} & \frac{\sqrt{3}}{2} \\ -\frac{\sqrt{3}}{2} & -\frac{1}{2} \end{bmatrix}, \tag{19}$$

It is easily shown that for a 60° rotation we have

$$\begin{bmatrix} C = A+B \\ C'' = \frac{1}{\sqrt{3}}(A-B) \end{bmatrix} \Longrightarrow \begin{bmatrix} B \\ A'' = \frac{2}{\sqrt{3}}\left(A + \frac{1}{2}B\right) \end{bmatrix}. \tag{20}$$

The whole loop in θ is shown in Fig. 24 (without normalization factors). Thus we have a correspondence between molecular structure and VB structure as we do a circuit around a conical intersection.

It is also instructive to examine a special case. Consider the loop of radius $\rho = a$ passing through $(\phi = 0)\,(a_{x_1}, 0);\ (\phi = \pi)(-a_{x_1}, 0);\ (\phi = 2\pi)$ $(a_{x_1}, 0)$. The corresponding rotation matrices are

$$\begin{bmatrix} 1 & 0 \\ 0 & 1 \end{bmatrix}, \begin{bmatrix} 0 & -1 \\ 1 & 0 \end{bmatrix}, \begin{bmatrix} -1 & 0 \\ 0 & -1 \end{bmatrix}. \tag{21}$$

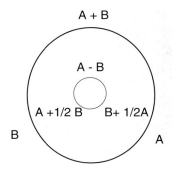

Fig. 24. Loop θ in valence bond structure space for the three orbital three electron problem. The outer loop corresponds to the ground state VB structures while the inner loop corresponds to the excited state VB structures that have been Schmidt orthogonalised but not normalised. (Adapted from Vanni et al.[45])

Thus looking at the first row, states X and Y transform according to Eq. (22) to give

$$X \to \frac{1}{\sqrt{2}}(X-Y) \to -Y \to -\frac{1}{\sqrt{2}}(Y+X) \to -X,$$
$$Y \to \frac{1}{\sqrt{2}}(Y-X) \to -X \to -\frac{1}{\sqrt{2}}(X+Y) \to -Y. \quad (22)$$

There are two important observations: (1) X and Y are interchanged on rotation through 90° and (2) we have a phase change as one rotates a full circle.

Now let us look at an application. The theory we have just developed holds rigorously only for three electrons in three 1s orbitals. However, the principles based upon overlap remained approximately valid. This is illustrated in Fig. 25 where we show a circuit of the prefulvene-like conical intersection in benzene.[52] Our purpose here is to illustrate the circuit of the conical intersection that we have described in the preceding discussion. In Fig. 25 the three valence bond structures A, B and C that lie on the circuit correspond to the couplings of the carbon atoms involved in the "prow" of the prefulvene conical intersection. Notice that the 180° rotation in the geometrical plane exchanges an excited state valence bond structure

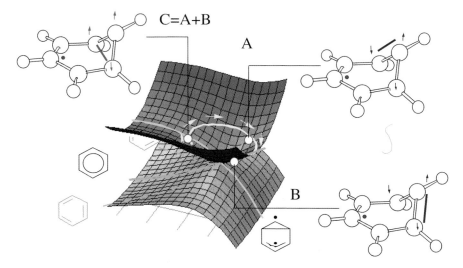

Fig. 25. A circuit of the prefulvene-like conical intersection in benzene.[52] (Adapted from Robb et al.[67])

so that we see the image of structure C on the ground state. Thus for the benzene conical intersection, the circuit about the apex of the cone (Fig. 25) traces out the valence bond structures corresponding to Figs. 21 and 23.

The same type of manipulations can be carried out for four orbitals and four electrons in the same fashion.[45] Unfortunately, for systems larger than four orbitals one cannot do the manipulations analytically. Nevertheless, if one can identify the valence bond structures for ground and excited state at one point on the circuit, then one can still apply the transformation of Eq. (8) to generate remaining valence bond structures that lie on the circuit.

Haas and co-workers have looked at many examples.[44,46,53] In applying such methods qualitatively, one needs to be clear that (a) the circuit of the conical intersection must be in the branching plane and (b) the phase change involves a wavefunction where the VB components involve two linearly independent VB functions corresponding to ground and excited state at one reference point. In the case of six orbitals and six electrons, there are five independent singlet spin functions. In order to apply the phase change method correctly, one needs to choose two linearly independent combinations. As we have shown[45] elsewhere, this is not trivial. Thus the method is probably more useful for three and four electron systems where it can be easily applied qualitatively.

6. Exploring the Conical Intersection Seam using Dynamics

Dynamics methods are becoming essential for the study of nonadiabatic events.[6,54] This subject will be treated by several contributors to this book, so we will limit ourselves to some studies that show how dynamics can sample the extended conical intersection seam and provide mechanistic information that would not be available from reaction path studies.

6.1. *A model cyanine system: The extended seam for cis-trans double bond isomerization*[31,55]

We now turn to the cyanine dye 1,1'-diethyl-4,4'-cyanine (1144-C) shown in Fig. 26(a). Experiments demonstrate[56] that one may control the cis-trans isomerisation by populating vibrational modes orthogonal to an extended seam of intersection. We have[31,55] mapped the potential surface for the model shown in Fig. 26(b).

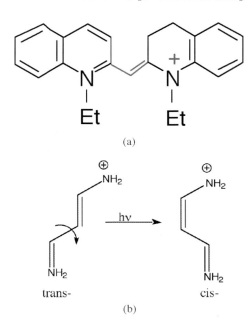

Fig. 26. (a) Cyanine dye 1,1′-diethyl-4,4′-cyanine (1144-C). (b) Three-carbon model of 1144-C.

The reaction coordinate X_3 (Fig. 2) is cis-trans isomerisation. The branching space is symmetric and anti-symmetric skeletal deformation coupled with pyramidal NH_2 distortion. Thus the seam is accessible along the skeletal deformation coordinate before cis-trans isomerisation can occur. Our dynamics results[31] (classical trajectories with a surface hop where the gradient comes from CASSCF) are summarised in Fig. 27. Figure 27 shows the geometries where the system hops. The minimum on the seam occurs at a dihedral angle of around 104°. However, most of the trajectories hop before the molecule has completed the half-rotation. Thus dynamics computations suggest that for the cyanine dye example, the sand flowing through the funnel (decay at the minimum of the conical intersection seam) (Fig. 1) is not applicable and the system samples the seam at all torsion angles (Fig. 2).[a]

---

[a] Since writing this review we have completed quantum dynamics computations [C.S. Allan, B. Lasorne, G.A. Worth and M.A. Robb, A straightforward method of analysis for direct quantum dynamics: application to the photochemistry of a model cyanine, *J. Phys. Chem. A*, **114**(33), 8713–29 (2010)]. The results are qualitatively similar.

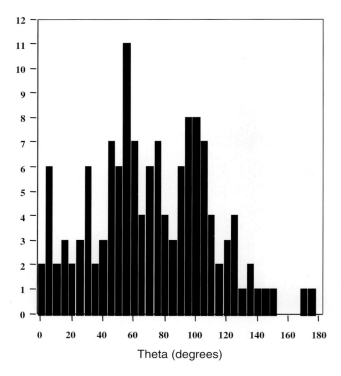

Fig. 27. Surface hop angle along cis-trans isomerisation coordinate for three-carbon model of 1144-C.[31]

6.2. *Benzene*

The photochemistry and photophysics of Benzene is another "benchmark" problem in organic photochemistry.[57,58] Recently,[14,59] dynamics computations have been carried out on this system using variational multi-configuration Gaussian wavepackets.[60] The cartoon shown in Fig. 28 shows the potential surface in the space of the prefulvene distortion that leads to the conical intersection (one of the branching space coordinates) and the ring-breathing mode (an intersection space coordinate). The extended seam develops along this coordinate. Remarkably, the conical intersection is peaked near the prefulvene-like minimum of the conical intersection but sloped further along the seam. This suggests that whether the system decays at the peaked geometry and produces a prefulvene-like product or decays at the sloped part of the seam, regenerating the reactant, might be controlled by the energy in the ring-breathing mode.

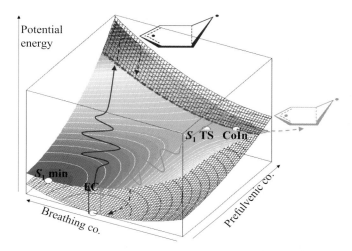

Fig. 28. Controlling the S_1 photochemistry of benzene: targeting the sloped part of the conical intersection leads to regeneration of benzene (photophysics); targeting the opposite end of the seam yields the benzvalene at a peaked conical intersection (photochemistry). (Adapted from Lasorne et al.[59])

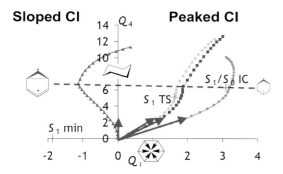

Fig. 29. Four quantum trajectories differing by the value of the initial momentum k_1 along coordinate Q_1 (projection in the (Q_1, Q_4)-subspace). Dashed line is the seam of intersection between diabatic states (prefulvenoid structures of various sizes, as indicated by molecular structures). (Adapted from Lasorne et al.[59])

The results are summarised in Fig. 29.[59] In this figure we show the propagation of the centre of the wavepacket for various initial conditions (arrows) involving the prefunvene (Q_4) and ring-expanding mode (Q_1). It is clear that the point of decay on the seam can indeed be controlled theoretically. Momentum in ring expansion (negative Q_1) leads to decay at the sloped intersection. Alternatively, ring contraction (positive Q_1) leads to decay at prefulvene. Thus, like the cyanine example discussed previously, quantum dynamics[60] may suggest new experiments for controlling photochemistry.

6.3. Biological chromophores: PYP[32,61]

In biological chomophores, nature controls photochemistry via the structure of the protein in which the chromophore is embedded,[32,33,61–63] As the last example, we briefly discuss the cis-trans isomerisation of a double bond in the covalently bound p-coumaric acid chromophore (Fig. 30) in Photoactive Yellow Protein (PYP), an archetypal reversible protein photoreceptor.[32,61]

A combination of *ab initio* (CASSCF) dynamics with surface hopping and classical molecular dynamics (MD) simulation techniques has been used to directly simulate the process of photoisomerisation within the protein.[32,61] We have used CASSCF for the chromophore itself and

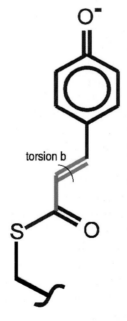

Fig. 30. The p-coumaric acid chromophore in PYP. The chromophore is covalently linked to the side chain of Cys69 through a thioester bond. The p-hydroxyphenyl moiety is deprotonated, but stabilised by hydrogen bonding interactions with the side chains of Tyr42 and Glu46. (Adapted from Groenhof *et al.*[32])

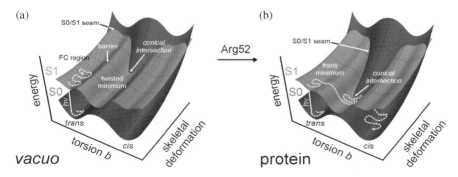

Fig. 31. Potential energy surfaces of the excited and ground states in the *trans*-to-*cis* isomerisation coordinate (torsion b, Fig. 30) and a skeletal deformation of the bonds: in *vacuo* (a) and in the protein (b). (Adapted from Groenhof et al.[32])

molecular mechanics for the remainder of the system. A cartoon of the potential energy surfaces in vacuo and in the protein is shown in Fig. 31.

The extended seam is present for both vacuo and protein along the cis-trans isomerisation coordinate. Again, the coordinates that lift the degeneracy are the skeletal deformations. In vacuo [Fig. 31(a)], there is an excited-state transition state with barrier and minimum where the chromophore is partly twisted, as well as a minimum with a half twist. In the protein [Fig. 31(b)], the S_1 surface is stabilised by the arg52 residue and the conical intersection seam is displaced so that it intersects with the reaction path. Thus the conical intersections in the protein and in the gas phase are significantly different. In the gas phase, substantial additional skeletal deformation motion to reach the seam would be required at the half twist geometry.

In the gas phase dynamics, using the same initial conditions as in the protein simulations, the system never makes it over the first partial twist torsion barrier. In contrast, in the protein, the excited state is specifically stabilised by the charge distribution of the protein (arg52). One observes a decrease of the S_1-S_0 energy gap in the region of the twisted intermediate (from 80 kJ mol^{-1} in vacuo to less than 1 kJ mol^{-1} in the protein), accompanied by a displacement of the crossing seam closer to the global minimum. One also sees a decrease of the energy barrier separating the early planar S_1 minimum and the twisted S_1 minimum. In total, 14 dynamics simulations are discussed in Ref. 32. In the protein, the lifetime of the excited state ranged from 129 to 2293 fs. The ratio of the number of successful isomerisations to the number of excited-state trajectories is ∼0.3, close to the experimental quantum yield of 0.35. Statistically, the number of trajectories is small, but it nevertheless yields a consistent mechanistic picture.

PYP is probably the most dramatic example of a situation where the reaction path is simple (just torsion), and orthogonal to the degeneracy-lifting coordinates (mainly skeletal deformations). In this case, the reactivity is changed when you add the electric field of the protein. Nature has been very careful to position one charged residue in exactly the right place.

7. Conclusions

With developments in theory and computation (see subsequent chapters in this book), quantum dynamics computations[6,14,64] are becoming possible as a method to study nonradiative decay at conical intersections in organic photochemistry. Here a more direct interaction with time-resolved laser methods becomes possible, so that control of photochemistry becomes possible.[65] Recent results show that the extended seam is a common mechanistic feature and this may be useful for the control of photochemistry either by lasers or by chemical substitution. The other major challenge is applications to biological systems.[62,66] Again the extended seam appears to play a role. However, developments in theory are needed to allow for quantum effects in nuclear motion in combination with force field methods and to allow for multi-scale effects.

Acknowledgments

Some of the discussion of the mathematical character of the extended seam (Sec. 2.1) has been adapted from the PhD Thesis of Fabrizio Scilia (University of London, 2008). Our work on photochemistry was started in collaboration with Massimo Olivucci and the late Fernando Bernardi. Much of the work discussed in this article has involved many other senior collaborators (Luis Blancafort, Mike Bearpark and Graham Worth), recent postdocs (Martial Boggio-Pasqua, Benjamin Lasorne, Martin Paterson) who now hold tenured academic positions and recent PhD students (Fabrizio Scilia, Stefan Vanni). Articles such as this, which collect many ideas and thoughts, are only possible through the dedicated hard work of co-workers over many years that helped develop them.

References

1. F. Bernardi, S. De, M. Olivucci and M.A. Robb, *J. Am. Chem. Soc.* **112**(5), 1737 (1990).
2. M. Klessinger and J. Michl, *Excited States and Photochemistry of Organic Molecules* (1995).
3. N.J. Turro, V. Ramamurthy and J. Scaiano, *Modern Molecular Photochemistry of Organic Molecules* (2010).
4. A. Migani and M. Olivucci, in *Conical Intersections*, edited by W. Domke, D.R. Yarkony and H. Koppel p. 271 (2004).
5. M.J. Paterson, M.J. Bearpark, M.A. Robb, L. Blancafort and G.A. Worth, *Phys. Chem. Chem. Phys.* **7**(10), 2100 (2005).
6. G.A. Worth, M.A. Robb and B. Lasorne, *Mol. Phys.* **106**(16–18), 2077 (2008).
7. M.J. Bearpark and M.A. Robb, in *Reviews of Reactive Intermediate Chemistry*, edited by M.S. Platz and R.A. Maitland John Wiley & Sons, Inc, pp. 379 (2007); G.A. Worth, M.J. Bearpark and M.A. Robb, in *Computational Photochemistry* edited by M. Olivucci (Elsevier), pp. 171 (2005).
8. L. Blancafort, B. Lasorne, G.A. Worth and M.A. Robb, in *The Jahn-Teller-Effect Fundamentals and Implications for Physics and Chemistry* edited by H. Köppel, D.R. Yarkony and H. Barentzen (Springer, 2009), pp. 169; D. Yarkony, *J. Chem. Phys.* **123**(13), 134106 (2005); D.R. Yarkony, *J. Chem. Phys.* **123**(20), 204101 (2005).
9. G. Atchity, S. Xantheas and K. Ruedenberg, *J. Chem. Phys.* **95**(3), 1862 (1991).
10. M. Desouter-Lecomte, C. Galloy, J. Lorquet and M. Pires, *J. Chem. Phys.* **71**(9), 3661 (1979); D.R. Yarkony, *Rev. Mod. Phys.* **68**(4), 985 (1996).
11. M.J. Paterson, M.J. Bearpark, M.A. Robb and L. Blancafort, *J. Chem. Phys.* **121**(23), 11562 (2004).
12. M. Paterson, M. Bearpark, M. Robb, L. Blancafort and G. Worth, *Phys. Chem. Chem. Phys.* **7**(10), 2100 (2005).
13. F. Sicilia, L. Blancafort, M.J. Bearpark and M.A. Robb, *J. Chem. Theory Comput.* **4**(2), 257 (2008).
14. B. Lasorne, F. Sicilia, M.J. Bearpark, M.A. Robb, G.A. Worth and L. Blancafort, *J. Chem. Phys.* **128**(12), 124307 (2008).
15. F. Sicilia, L. Blancafort, M.J. Bearpark and M.A. Robb, *J. Phys. Chem. A* **111**(11), 2182 (2007).
16. F. Sicilia, M.J. Bearpark, L. Blancafort and M.A. Robb, *Theor. Chem. Acc.* **118**(1), 241 (2007).
17. B.H. Lengsfield and D.R. Yarkony, *Adv. Chem. Phys.* **82**, 1 (1992).
18. M. Bearpark, M. Robb and H. Schlegel, *Chem. Phys. Lett.* **223**(3), 269 (1994).
19. B.G. Levine, J.D. Coe and T.J. Martinez, *J. Phys. Chem. B* **112**(2), 405 (2008).
20. T.W. Keal, A. Koslowski and W. Thiel, *Theor. Chem. Acc.* **118**(5–6), 837 (2007).

21. G. Worth and L. Cederbaum, *Annu. Rev. Phys. Chem.* **55**, 127 (2004).
22. S. Belz, T. Grohmann and M. Leibscher, *J. Chem. Phys.* **131**(3), 034305 (2009); T. Grohmann, O. Deeb and M. Leibscher, *Chem. Phys.* **338**(2–3), 252 (2007); O. Deeb, S. Cogan and S. Zilberg, *Chem. Phys.* **325**(2), 251 (2006); M. Bearpark, L. Blancafort and M. Paterson, *Mol. Phys.* **104**(5–7), 1033 (2006).
23. M. Olivucci, I. Ragazos, F. Bernardi and M. Robb, *J. Am. Chem. Soc.* **115**(9), 3710 (1993).
24. M. Boggio-Pasqua, M. Bearpark, P. Hunt and M. Robb, *J. Am. Chem. Soc.* **124**(7), 1456 (2002).
25. P. Celani, S. Ottani, M. Olivucci, F. Bernardi and M.A. Robb, *J. Am. Chem. Soc.* **116**(22), 10141 (1994); W. Fuss, T. Schikarski, W.E. Schmid, S. Trushin and K.L. Kompa, *Chem. Phys. Lett.* **262**(6), 675 (1996); W. Fuss, P. Hering, K.L. Kompa, S. Lochbrunner, T. Schikarski, W.E. Schmid and S.A. Trushin, *Ber. Bunsenges. Phys. Chem* **101**(3), 500 (1997); M. Garavelli, P. Celani, M. Fato, M.J. Bearpark, B.R. Smith, M. Olivucci and M.A. Robb, *J. Phys. Chem. A* **101**(11), 2023 (1997); S.A. Trushin, W. Fuss, T. Schikarski, W.E. Schmid and K.L. Kompa, *J. Chem. Phys.* **106**(22), 9386 (1997); A. Hofmann and R. de Vivie-Riedle, *J. Chem. Phys.* **112**(11), 5054 (2000); A. Hofmann and R. de Vivie-Riedle, *Chem. Phys. Lett.* **346**(3–4), 299 (2001); L. Kurtz, A. Hofmann and R. de Vivie-Riedle, *J. Chem. Phys.* **114**(14), 6151 (2001); S. Zilberg and Y. Haas, *Phys. Chem. Chem. Phys.* **4**(1), 34 (2002); H. Tamura, S. Nanbu, H. Nakamura and T. Ishida, *Chem. Phys. Lett.* **401**(4–6), 487 (2005).
26. K. Kosma, S.A. Trushin, W. Fuss and W.E. Schmid, *Phys. Chem. Chem. Phys.* **11**(1), 172 (2009).
27. M. Boggio-Pasqua, M. Ravaglia, M. Bearpark, M. Garavelli and M. Robb, *J. Phys. Chem. A* **107**(50), 11139 (2003).
28. A. Migani, M. Robb and M. Olivucci, *J. Am. Chem. Soc.* **125**(9), 2804 (2003); O. Weingart, A. Migani, M. Olivucci, M. Robb, V. Buss and P. Hunt, *J. Phys. Chem. A* **108**(21), 4685 (2004).
29. I. Gomez, M. Reguero, M. Boggio-Pasqua and M.A. Robb, *J. Am. Chem. Soc.* **127**(19), 7119 (2005).
30. K.A. Zachariasse, S.I. Druzhinin, V.A. Galievsky, S. Kovalenko, T.A. Senyushkina, P. Mayer, M. Noltemeyer, M. Boggio-Pasqua and M.A. Robb, *J. Phys. Chem. A* **113**(12), 2693 (2009).
31. P. Hunt and M. Robb, *J. Am. Chem. Soc.* **127**(15), 5720 (2005).
32. G. Groenhof, M. Bouxin-Cademartory, B. Hess, S. De Visser, H. Berendsen, M. Olivucci, A. Mark and M. Robb, *J. Am. Chem. Soc.* **126**(13), 4228 (2004).
33. L.V. Schaefer, G. Groenhof, A.R. Klingen, G.M. Ullmann, M. Boggio-Pasqua, M.A. Robb and H. Grubmueller, *Angew Chem. Int. Edit.* **46**(4), 530 (2007).
34. M. Paterson, M. Robb, L. Blancafort and A. DeBellis, *J. Phys. Chem. A* **109**(33), 7527 (2005).
35. D. Asturiol, B. Lasorne, M.A. Robb and L. Blancafort, *J. Phys. Chem. A* **113**(38), 10211 (2009).
36. M. Boggio-Pasqua, M.J. Bearpark, F. Ogliaro and M.A. Robb, *J. Am. Chem. Soc.* **128**(32), 10533 (2006).

37. I. Gomez, M. Reguero and M.A. Robb, *J. Phys. Chem.* A **110**(11), 3986 (2006).
38. M. Boggio-Pasqua, M.J. Bearpark and M.A. Robb, *J. Org. Chem.* **72**(12), 4497 (2007).
39. M. Araujo, B. Lasorne, M.J. Bearpark and M.A. Robb, *J. Phys. Chem.* A **112**(33), 7489 (2008).
40. A. Nenov, P. Kolle, M. Robb and R. de Vivie-Riedle, *J. Org. Chem.* **75**, 123 (2010).
41. W.T.A.M. van der Lugt and L.J. Oosterhoff, *J. Am. Chem. Soc.* **91**, 6042 (1969); W.T.A.M. van der Lugt and L.J. Oosterhoff, *Chem. Commun.* 1235–1236 (1968).
42. M.J. Paterson, M.A. Robb, L. Blancafort and A.D. DeBellis, *J. Am. Chem. Soc.* **126**(9), 2912 (2004); A. Migani, L. Blancafort, M.A. Robb and A. D. Debellis, *J. Am. Chem. Soc.* **130**(22), 6932 (2008).
43. G. Groenhof, L.V. Schaefer, M. Boggio-Pasqua, M. Goette, H. Grubmueller and M.A. Robb, *J. Am. Chem. Soc.* **129**(21), 6812 (2007).
44. Y. Haas and S. Zilberg, *Adv. Chem. Phys.* **124**, 433 (2002); Y. Haas, S. Cogan and S. Zilberg, *Int. J. Quantum Chem.* **102**(5), 961 (2005).
45. S. Vanni, M. Garavelli and M.A. Robb, *Chem. Phys.* **347**(1–3), 46 (2008).
46. B. Dick, Y. Haas and S. Zilberg, *Chem. Phys.* **347**(1–3), 65 (2008).
47. M. Abe, Y. Ohtsuki, Y. Fujimura, Z.G. Lan and W. Domcke, *J. Chem. Phys.* **124**(22), 224316 (2006); S. Althorpe, *J. Chem. Phys.* **124**(8), 084105 (2006).
48. H.C. Longuet–Higgins, *Proc. Roy. Soc. London* A **392**, 147 (1975).
49. S.C. Althorpe, *J. Chem. Phys.* **124**(8), 084105 (2006).
50. F. Bernardi, M. Olivucci, M. Robb and G. Tonachini, *J. Am. Chem. Soc.* **114**(14), 5805 (1992).
51. D.S. Ruiz, A. Cembran, M. Garavelli, M. Olivucci and W. Fuss, *Photochem Photobiol* **76**(6), 622 (2002).
52. I. Palmer, I. Ragazos, F. Bernardi, M. Olivucci and M. Robb, *J. Am. Chem. Soc.* **115**(2), 673 (1993).
53. S. Zilberg and Y. Haas, *Chem. Eur. J.* **5**(6), 1755 (1999).
54. G. Worth and M. Robb, *Adv. Chem. Phys.* **124**, 355 (2002); M.D. Hack, A.M. Wensmann, D.G. Truhlar, M. Ben-Nun and T.J. Martinez, *J. Chem. Phys.* **115**(3), 1172 (2001); J. Quenneville, M. Ben-Nun and T.J. Martinez, *J. Photoch. Photobio.* A **144**(2–3), 229 (2001); A. Jasper, S. Nangia, C. Zhu and D. Truhlar, *Accounts Chem. Res.* **39**(2), 101 (2006); M. Barbatti, G. Granucci, M. Persico, M. Ruckenbauer, M. Vazdar, M. Eckert-Maksic and H. Lischka, *J. Photoch. Photobio.* A **190**(2–3), 228 (2007).
55. A. Sanchez-Galvez, P. Hunt, M. Robb, M. Olivucci, T. Vreven and H. Schlegel, *J. Am. Chem. Soc.* **122**(12), 2911 (2000).
56. B. Dietzek, B. Brueggemann, P. Persson and A. Yartsev, *Chem. Phys. Lett.* **455**(1–3), 13 (2008).
57. I.J. Palmer, M. Olivucci, F. Bernardi and M.A. Robb, *J. Org. Chem.* **57**(19), 5081 (1992); W. Domcke, A. Sobolewski and C. Woywod, *Chem. Phys. Lett.* **203**(2–3), 220 (1993).
58. W. Domcke, A.L. Sobolewski and C. Woywod, *Chem. Phys. Lett.* **203**(2–3), 220 (1993); H. Koppel, *Chem. Phys. Lett.* **205**(4–5), 361 (1993); I.J. Palmer,

I.N. Ragazos, F. Bernardi, M. Olivucci and M.A. Robb, *J. Am. Chem. Soc.* **115**(2), 673 (1993); A. Sobolewski, C. Woywod and W. Domcke, *J. Chem. Phys.* **98**(7), 5627 (1993); B.R. Smith, M.J. Bearpark, M.A. Robb, F. Bernardi and M. Olivucci, *Chem. Phys. Lett.* **242**(1–2), 27 (1995); G.A. Worth, *J. Photoch Photobio* A **190**(2–3), 190 (2007).

59. B. Lasorne, M.J. Bearpark, M.A. Robb and G.A. Worth, *J. Phys. Chem.* A **112**(50), 13017 (2008).
60. B. Lasorne, M.A. Robb and G.A. Worth, *Phys. Chem. Chem. Phys.* **9**(25), 3210 (2007); B. Lasorne, M.J. Bearpark, M.A. Robb and G.A. Worth, *Chem. Phys. Lett.* **432**(4–6), 604 (2006).
61. G. Groenhof, L.V. Schaefer, M. Boggio-Pasqua, H. Grubmueller and M.A. Robb, *J. Am. Chem. Soc.* **130**(11), 3250 (2008).
62. A. Strambi, P.B. Coto, L.M. Frutos, N. Ferre and M. Olivucci, *J. Am. Chem. Soc.* **130**(11), 3382 (2008); P.B. Coto, A. Strambi and M. Olivucci, *Chem. Phys.* **347**(1–3), 483 (2008); F. Santoro, A. Lami and M. Olivucci, *Theor. Chem. Acc.* **117**(5–6), 1061 (2007); A.M. Virshup, C. Punwong, T.V. Pogorelov, B.A. Lindquist, C. Ko and T.J. Martinez, *J. Phys. Chem.* B **113**(11), 3280 (2009).
63. H.R. Hudock, H.G. Levine, A.L. Thompson and T.J. Martinez, presented at the International Conference on Computational Methods in Science and Engineering, Corfu, Greece, 2007 (unpublished); M. Boggio-Pasqua, G. Groenhof, L.V. Schafer, H. Grubmuller and M.A. Robb, *J. Am. Chem. Soc.* **129**(36), 10996 (2007); G. Groenhof, L.V. Schafer, M. Boggio-Pasqua, M. Goette, H. Grubmuller and M.A. Robb, *J. Am. Chem. Soc.* **129**(21), 6812 (2007).
64. G. Villani, *J. Chem. Phys.* **128**(11), 114306 (2008); T. Rozgonyi and L. Gonzalez, *J. Phys. Chem.* A **112**(25), 5573 (2008); M. Basler, E. Gindensperger, H. D. Meyer and L.S. Cederbaum, *Chem. Phys.* **347**(1–3), 78 (2008); B.G. Levine, J.D. Coe, A.M. Virshup and T.J. Martinez, *Chem. Phys.* **347**(1–3), 3 (2008).
65. D. Geppert, P. von den Hoff and R. de Vivie-Riedle, *J. Phys. B-At Mol. Opt.* **41**(7), 074006 (2008); J. Hauer, T. Buckup and M. Motzkus, *J. Phys. Chem.* A **111**(42), 10517 (2007); P.S. Christopher, M. Shapiro and P. Brumer, *J. Chem. Phys.* **123**(6), 064313 (2005); M. Abe, Y. Ohtsuki, Y. Fujimura and W. Domcke, *J. Chem. Phys.* **123**(14), 144508 (2005); A. Muller and K. Kompa, *J. Mod. Optic.* **49**(3–4), 627 (2002).
66. L.M. Frutos, T. Andruniow, F. Santoro, N. Ferre and M. Olivucci, *Proc. Natl. Acad. Sci. U.S.A.* **104**(19), 7764 (2007); L.V. Schafer, G. Groenhof, M. Boggio-Pasqua, M.A. Robb and H. Grubmuller, *Plos. Comput. Biol.* **4**(3), 14 (2008); L.V. Schaefer, G. Groenhof, M. Boggio-Pasqua, M.A. Robb and H. Grubmueller, *Plos. Comput. Biol.* **4**(3), e1000034 (2008).
67. M.A. Robb, M. Garavelli, M. Olivucci and F. Bernardi, in *Reviews in Computational Chemistry*, **15**, 87 (2000).

Chapter 2

Efficient Excited-State Deactivation in Organic Chromophores and Biologically Relevant Molecules: Role of Electron and Proton Transfer Processes

Andrzej L. Sobolewski* and Wolfgang Domcke[†]

1. Introduction 52
2. The $\pi\sigma^*$-driven Photochemistry of Acidic
   Aromatic Systems 53
3. Excited-State Electron/Proton Transfer in
   Intra-Molecularly Hydrogen-Bonded Aromatic Systems 58
4. Excited-State Electron/Proton Transfer in
   Inter-Molecularly Hydrogen-Bonded Aromatic Systems 64
5. Chromophores with Flexible Side-Chains:
   Electron/Proton Transfer in Amino Acids and Peptides 70
6. Conclusions 77
   Acknowledgments 79
   References 79

---

*Institute of Physics, Polish Academy of Sciences, PL-02668 Warsaw, Poland.
[†]Institute of Physical and Theoretical Chemistry, Technische Universität München, D-85747 Garching, Germany.

1. Introduction

Electron and proton transfer reactions are among the most fundamental and most widespread photoinduced processes in physical and biophysical chemistry. In this chapter, we shall mainly be concerned with electron and proton transfer processes which are facilitated by intra-molecular or inter-molecular hydrogen bonds.

Hydrogen bonds are ubiquitous in chemistry and biochemistry. Examples are the structure and dynamics of liquids and molecular crystals, solvation in protic solvents, molecular recognition and supra-molecular self-organization, as well as enzymatic catalysis. While the properties of hydrogen bonds in the electronic ground state have been investigated for decades with powerful experimental techniques, such as infrared and Raman vibrational spectroscopy as well as neutron scattering, much less is known, in general, about the structure and dynamics of hydrogen bonds in excited electronic states and their role in photochemical processes. The questions of the time scale of excited-state proton transfer and the presence or absence of barriers have been disputed for a long time.[1-4] A widely studied phenomenon involving excited-state dynamics of hydrogen bonds is fluorescence quenching in intra-molecularly or inter-molecularly hydrogen-bonded complexes.[5-9] Most importantly, photosynthesis is based on a sequence of electron and proton transfer processes.[10]

One reason of our rather limited knowledge of excited-state hydrogen-bond dynamics is the extremely short time scale of these processes, which often may be beyond the limits of present-day time-resolved experiments. Another reason is the difficulty of performing accurate *ab initio* electronic-structure calculations for excited states of relatively large polyatomic molecules. Excited electronic states are open-shell, generally multi-configurational, and often subject to intricate valence-Rydberg mixing effects. Apart from the complexities of the electronic structure, chemically interesting excited-state dynamics involves large-amplitude nuclear motion such as fragmentation or isomerization. The identification of the relevant excited electronic states and the chemically relevant nuclear degrees of freedom requires the systematic exploration of truly high-dimensional electronic potential-energy surfaces over extended regions of nuclear coordinate space.

The intention of this chapter is to give an overview, at a qualitative level, of generic electron/proton-transfer reaction mechanisms in which conical intersections of the potential-energy surfaces play an essential role. The description will be in terms of minimum-energy reaction paths, reaction-path energy profiles, and a local characterization of the conical intersections. A description of the detailed electronic/nuclear dynamics at these conical intersections is beyond the scope of this chapter. While a comprehensive theoretical analysis of the nonadiabatic electronic-nuclear dynamics at conical intersections involved in electron/proton transfer certainly is of great interest, such calculations are still at a rudimentary stage or were not attempted at all for the polyatomic systems to be discussed in this chapter.

The chemical literature is replete with (sometimes controversial) discussions of processes like concerted or sequential electron and proton transfer, proton-coupled electron transfer (PCET), electron-coupled proton transfer (ECPT), hydrogen-atom transfer (HAT), etc.[11-18] These concepts are ambiguous for (at least) two reasons. First, with the exception of hydrogen-atom detachment from isolated molecules, never exactly one unit of electronic charge is transferred with a proton. Electron transfer and proton transfer are thus limiting cases, with hydrogen-atom transfer in the middle. Second, the actual time scales of electron transfer and proton transfer cannot be experimentally resolved so far. A precise definition of PCET, ECPT, HAT, concerted or sequential electron-proton transfer, etc., would require the preparation of temporarily and spatially localized electronic and protonic wave packets, as well as a prescription for the detection of these wave packets as a function of time. Lacking these precise definitions, much of the extensive literature on ECPT, PCET, HAT, etc., does not have a firm conceptual basis. Only very recently, the first steps toward an electron wave-packet description of coupled electron and proton dynamics have been undertaken.[19,20]

To avoid such ambiguities, we shall speak of "electron/proton transfer" in a general sense, rather than discriminate between different types of electron and proton transfer processes. Electron/proton transfer is defined to include both complete as well as partial transfer of an electron with a proton, independent of the relative time scale of the electron and proton motions. If the meaning is clear from the context, the expressions "proton transfer" and "hydrogen transfer" will be used synonymously.

2. The $\pi\sigma^*$-driven Photochemistry of Acidic Aromatic Systems

A decade ago, *ab initio* electronic-structure calculations for indole, pyrrole and phenol revealed the existence of optically dark and photochemically highly reactive singlet excited states, the so-called $\pi\sigma^*$ states of azoles and aromatic enoles.[21–23] The $^1\pi\sigma^*$ state in these systems had previously been classified as a 3s Rydberg state.[24–26] The calculations of scans of the electronic potential-energy (PE) surfaces along the NH or OH stretching coordinates revealed, however, the generically dissociative character of the $^1\pi\sigma^*$ states, which originates from three properties of the σ^* orbital: (i) it is localized on NH or OH groups, (ii) it is antibonding with respect to the NH or OH bonds, and (iii) upon stretching of the NH/OH bond, the 3s-type σ^* orbital shrinks to the 1s orbital of the hydrogen atom, which results in a large energy gain.[27]

PE profiles of the electronic ground state (S_0) and the lowest $^1\pi\pi^*$ and $^1\pi\sigma^*$ excited states of phenol, indole and pyrrole are shown in Fig. 1. These curves have been calculated as "relaxed scans", that is, for a given fixed value of the NH/OH bond length, the energy (of the excited state) was minimized with respect to all other internal nuclear coordinates.[27] Rigid detachment of the hydrogen atom (that is, changing only the NH/OH bond length, while keeping all other internal coordinates fixed at the ground-state equilibrium geometry) yields qualitatively similar results.[28] Over the years, a variety of electronic-structure methods, such as CASSCF/CASPT2,[27]

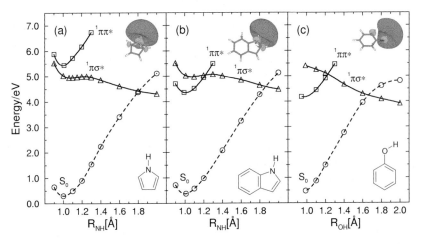

Fig. 1. PE profiles (relaxed scans) of the lowest $^1\pi\pi^*$ states (squares), the lowest $^1\pi\sigma^*$ state (triangles) and the electronic ground state (circles) as a function of the OH (phenol) or NH (indole, pyrrole) reaction coordinate. The energy of the ground state is calculated at the optimized geometries of the $^1\pi\sigma^*$ state. The symbols represent computed data. The curves are interpolations. The σ^* orbital obtained by a CASSCF calculation for the $^1\pi\sigma^*$ state of a given system in shown as an insertion.

CASSCF/MRCI,[29–31] CC2,[28] MRCI/DFT,[32] TDDFT[33] and ROKS[33] have been employed for the calculation of $^1\pi\sigma^*$ PE functions of various molecules, yielding qualitatively similar results. The generically repulsive character of the $^1\pi\sigma^*$ PE surface thus is a robust result. It is essential, however, that a basis set with sufficiently diffuse functions at least on the NH/OH group under consideration is employed, in order to account for the diffuse Rydberg character of the σ^* orbital near the ground-state equilibrium geometry.

A graphical representation of the σ^* orbital at the ground-state equilibrium geometry of phenol, indole and pyrrole is shown in Fig. 1. The large spatial extension and the antibonding character (node across the NH/OH bond) can be seen, as well as the fact that the σ^* electronic charge is localized largely outside the aromatic ring. This is reflected in an unusually large dipole moment of the $^1\pi\sigma^*$ state, typically 10 Debye at the ground-state equilibrium geometry.[27] Due to the diffuseness of the electronic wave function and the large dipole moment, the energy of the $^1\pi\sigma^*$ state is particularly sensitive to perturbations, such as substitutions at the aromatic ring or interactions with a solvent. While the Rydberg character causes a blue-shift of the $^1\pi\sigma^*$ energy in condensed phases, the large dipole moment favors a red-shift in a polar environment. Due to the subtle counterbalance of these effects, it is difficult to predict the response of the $^1\pi\sigma^*$ energy to intra-molecular or external perturbations.

The crossings of the energy profiles in Fig. 1 are allowed crossings in the planar molecules, since the S_0 state and the $^1\pi\pi^*$ states are of A' symmetry, while the $^1\pi\sigma^*$ state is of A'' symmetry. Upon out-of-plane deformation, the A' and A'' states are allowed to interact, resulting in a repulsion of the adiabatic PE surfaces. The curve crossings in Fig. 1 are thus converted into conical intersections.[22] In pyrrole,[22] as well as in imidazole,[34] the $^1\pi\sigma^*$ state is located below the lowest $^1\pi\pi^*$ state in the Franck–Condon region. Its PE surface is, apart from a shallow well in the Franck–Condon region, repulsive with respect to NH stretching, resulting in a conical intersection with the attractive S_0 surface. A survey of all symmetry-allowed A'–A'' coupling modes in pyrrole showed that the NH out-of-plane bending mode is by far the strongest coupling mode at the $^1\pi\sigma^*$–S_0 conical intersection.[35] The adiabatic S_0 and S_1 PE surfaces in the two-dimensional subspace spanned by the NH bond length r_{NH} and the NH out-of-plane bending angle θ are shown in Fig. 2 (obtained by a state-averaged CASSCF calculation with a $6\pi/7\sigma$ active space and the aug-cc-pVDZ basis set[29]). The shallow well of the $^1\pi\sigma^*$ PE surface in the Franck–Condon region and the steep cone at the intersection with the S_0 surface are clearly visible.

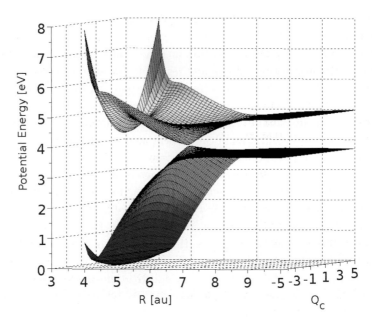

Fig. 2. Adiabatic PEs of the S_0 state and the $^1\pi\sigma^*$ state of pyrrole as a function of the NH stretching coordinate R_{NH} and the H-atom out-of-plane angle θ.

The quantum wave-packet dynamics on the nonadiabatically coupled S_1 and S_0 PE surfaces of pyrrole, assuming instantaneous vertical electronic excitation from the S_0 state, has been investigated for the two-dimensional model of Fig. 2 as well as for three-dimensional models including a second coupling mode.[35] If the $^1\pi\sigma^*$ state is prepared in its (0,0) vibrational level, the time evolution of the nuclear wave packet is determined by barrier tunneling on a picosecond time scale.[29,33] If the initially prepared state contains one quantum of the NH stretching mode, on the other hand, the energy of a significant fraction of the wave packet is above the barrier and the system dissociates on a time scale of about 20 fs.[29] As expected, most of the wave packet passes diabatically through the S_1–S_0 conical intersection, dissociating to the $^2\pi$ ground state of the pyrrolyl radical.[29]

The photoinduced detachment of fast hydrogen atoms from pyrrole was first observed by Y.T. Lee and collaborators, using photofragment translational spectroscopy at 248 nm and 193 nm excitation.[36] Temps and collaborators used ion imaging spectroscopy at 243 nm and confirmed the bimodal kinetic-energy distribution, corresponding to direct dissociation in the $^1\pi\sigma^*$ state and statistical unimolecular decay after internal conversion, respectively.[37] Ashfold and collaborators applied high-resolution photofragment translational spectroscopy (Rydberg tagging) to pyrrole and obtained high-resolution kinetic-energy spectra for a multitude of excitation wavelengths.[38] These measurements revealed that the pyrrolyl fragment (as well as phenoxyl, imidazolyl and indolyl fragments) are formed in very limited subsets of their available vibrational states, confirming the ultrafast and highly nonstatistical nature of the photodissociation process.[31,34,38] The first time-resolved detection of the nascent H-atoms was reported by Radloff, Hertel and collaborators for pyrrole.[39] The observed fast time scale of about 100 fs and a slower time scale of about 1 ps were interpreted as direct and indirect (temporarily captured at the first transition through the S_1–S_0 conical intersection) photodissociation, respectively. Very recently, Stavros and collaborators applied time-resolved velocity map ion imaging to clock the H-atom elimination in phenol and indole,[40,41] revealing that both fast and slow H-atoms appear on an ultrafast time scale (< 200 fs). A detailed discussion of these experimental results for azoles, phenols and related systems can be found in the chapter by Ashfold and co-workers (Chap. 15).

Additional insight can be gained by the preparation of the chromophore in specific vibrational levels prior to UV excitation. Computational studies for pyrrole and phenol indicate that pre-excitation in excited levels of the coupling mode(s) has a significant impact on the branching ratio of the $^2\pi$ and $^2\sigma$ product states by enhancing the adiabatic versus the diabatic reaction channel.[29,42] Experimental studies by Crim and collaborators revealed a significant increase of the fraction of slow H-atoms upon pre-excitation of the OH stretching mode of phenol,[43] resulting from the preferred formation of excited-state phenoxyl fragments. Both observations can be rationalized in terms of the nonadiabatic dynamics at the $^1\pi\sigma^*$–S_0 conical intersection.[29,42,43]

These extensive experimental investigations have established the role of photodissociated H-atoms as a novel messenger particle providing information on the complex photochemistry of isolated polyatomic molecules, in addition to photons (time- and/or energy-resolved fluorescence) and electrons (time- and/or energy-resolved photoelectron spectroscopy).

The availability of these powerful experimental tools heralds a new level of understanding of organic photochemistry and photobiology. The detection of fast H-atoms is an unequivocal spectroscopic signature of the involvement of dark $^1\pi\sigma^*$ states in the photochemical dynamics. Beyond this, the high resolution kinetic-energy distribution of the H atoms carries information on competing processes, such as out-of-plane deformation or opening of aromatic rings in $^1\pi\pi$ states, or efficient intramolecular vibrational relaxation (IVR), for example. As emphasized in several theoretical and experimental studies,[31,34,42] the comparison of H-detachment spectra in pyrrole or imidazole, in which the $^1\pi\sigma^*$ state is populated directly (via intensity borrowing from higher allowed $^1\pi\pi^*$ states) with H-detachment spectra in indole or phenol, in which the $^1\pi\sigma^*$ state is populated by a $^1\pi\pi^*$–$^1\pi\sigma^*$ conical intersection, provides deep insight into the nature of the photochemical processes, see also Chap. 15.

Being generic for OH, NH and NH_2 groups, photochemically reactive $^1\pi\sigma^*$ states also must exist in some of the most important building blocks of life, the DNA bases and the aromatic amino acids. PE profiles for hydrogen detachment via $^1\pi\sigma^*$ states have been calculated for adenine,[44–48] guanine,[49–51] xanthine[52] and uracil,[53] as well as for the protonated amino acids tyrosine (Tyr) and tryptophan (Trp).[54] Fast H-atoms arising from the $^1\pi\sigma^*$ state in adenine have been detected by Nix et al. with high kinetic-energy resolution for excitation wavelengths < 233 nm.[55] Wells et al. could demonstrate, for the first time, the participation of the amino group in the generation of fast H-atoms from adenine.[56]

Another interesting aspect of the $^1\pi\sigma^*$ photochemistry is the yield of H-atoms as a function of molecular size. Lin et al. investigated the photodissociation of a series of substituted indoles, including tryptamine and tryptophan.[57] It was observed that H-atom detachment is quenched with increasing molecular size, internal conversion becoming the major nonradiative process. A similar observation has been made by Poterya et al. for size-selected pyrrole clusters: the fraction of slow H-atoms gains in intensity with respect to the fast fraction with increasing cluster size.[58] These findings address the role of the $\pi\sigma^*$ photochemistry for the photostability of biological matter[59]: while fragmentation (H-atom loss) prevails in small and isolated chromophores, efficient internal conversion seems to be the dominant process in large molecules and in the condensed phase.[57,58]

3. Excited-State Electron/Proton Transfer in Intra-Molecularly Hydrogen-Bonded Aromatic Systems

The investigation of the photochemistry of (hetero)aromatic systems with one (or more) intra-molecular hydrogen bond has a long history in molecular spectroscopy.[1−4] The photoinduced excited-state intra-molecular proton-transfer (ESIPT) process in bifunctional molecules is of particular interest, since it is assumed to play an essential role for the functionality of so-called photostabilizers.[60,61] Photostabilizers are in wide technical use for the protection of organic polymers against degradation by the UV components of sunlight. The generally accepted mechanistic model for the function of organic photostabilizers assumes a barrierless (or nearly barrierless) proton transfer from an acidic (e.g. enol) to a basic (e.g. keto) group in the $S_1(\pi\pi^*)$ state, followed by an ultrafast ($<$ 100 fs) radiationless decay from S_1 to S_0. The reaction cycle is closed by a barrierless proton back-transfer in the electronic ground state.[1−4,60,61] Salicylic acid (SA), methyl salicylate (MS) and o-hydroxybenzaldehyde (OHBA) are among the systems for which the detailed reaction mechanisms have been explored with spectroscopic[62−65] and computational[66−69] methods.

Another paradigmatic model system for the investigation of the ESIPT process is 2-(2′-hydroxyphenyl)benzotriazole (known as Tinuvin or TIN-H) and its 5′-methylated derivative (TIN-P). These compounds are particularly efficient UV photostabilizers.[70] According to Ref. 2, more than 99.9% of the light energy absorbed by the $S_1 \leftarrow S_0$ transition of TIN-P is dissipated as heat into the condensed-phase environment. One of the remarkable features of TIN-P is the extremely rapid depopulation of the proton-transferred excited state (S_1'). It is orders of magnitude faster than the excited-state decay of related intra-molecularly H-bonded heterocycles, such as 2-(2′-hydroxyphenyl)benzoxazole (HBO) or 2-(2′-hydroxyphenyl)benzothiazole (HBT).[2,4]

In recent work, we have used the MP2 and CC2 *ab initio* methods for a systematic exploration of the ground-state and excited-state PE surfaces of TIN-H.[71] Two excited-state reaction paths have been studied: the transfer of the mobile H-atom along the intra-molecular hydrogen bond, and the torsion of the phenolic H-donor moiety against the triazole H-acceptor moiety. Two local energy minima of TIN-H have been found on the ground-state PE surface. The corresponding molecular structures are shown in Figs. 3(a) and 3(b), with indication of the most relevant bond lengths. The global minimum [Fig. 3(a)] represents the enol tautomer of TIN-H, which exhibits a strong intra-molecular hydrogen bond between the phenolic hydrogen atom and the nearest nitrogen atom of the triazole ring. The two rings are coplanar at the global energy minimum (C_s symmetry). The second local minimum on the S_0 energy surface is the open form of the keto tautomer, with a twist angle of the rings of about 150° [Fig. 3(b)]. The two

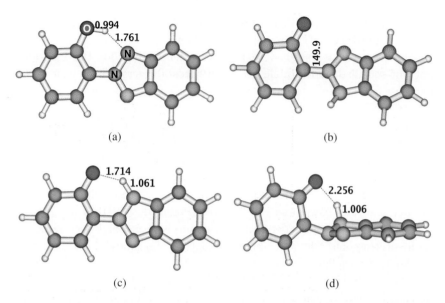

Fig. 3. Equilibrium geometries of stable structures of TIN-H in the electronic ground state (a, b) and distinguished structures of the lowest excited singlet state: the metastable hydrogen-transferred (keto) configuration (c) and the S_1–S_0 conical intersection (d). The numbers give selected bond lengths (in Å) and bond angles (in degrees). In (a), oxygen atoms (dark gray) are identified by "O", nitrogen atoms by "N"; the remaining atoms are carbon (gray) and hydrogen (light gray).

minima are separated by a barrier which is 0.24 eV (5.5 kcal/mol) higher than the local minimum.

PE profiles of the minimum-energy paths (more precisely, relaxed scans) have been calculated for hydrogen transfer, defining the OH bond length as the driving coordinate, as well as for inter-ring twisting, defining the central C-C-N-N dihedral angle as the driving coordinate. We refer to Ref. 71 for the details of the calculations. The resulting PE profiles are displayed in Fig. 4. The S_0 PE profile in Fig. 4(a) confirms that the ground-state equilibrium geometry is planar. The PE profiles for hydrogen transfer, shown in Fig. 4(b), reveal that the enol → keto H-transfer is endothermic on the S_0 surface, but exothermic on the S_1 surface, as expected for an ESIPT system.[1–4] In both cases, the PE function is barrierless. The planar proton-transferred S_1 structure [$S_1{}'$, see structure shown in Fig. 3(c)] is predicted to lie 3.2 eV above the global minimum of the S_0 surface. The fluorescence is strongly allowed ($f = 0.32$) and would have a center wavelength of about 600 nm (2.0 eV) if it could be observed in the gas phase. As Fig. 4(c) shows,

Efficient Excited-State Deactivation in Organic Chromophores 61

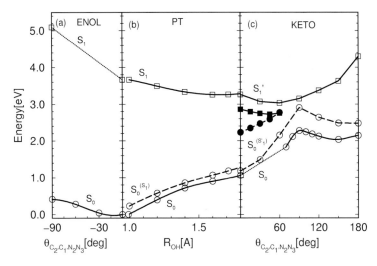

Fig. 4. PE profiles (relaxed scans) of the S_0 state (circles) and the S_1 state (squares) as a function of the torsional reaction coordinate (a, c) and the hydrogen-transfer reaction coordinate (b). Full lines: energy profiles of reaction paths determined in the same electronic state (S_0, S_1). Dashed lines: energy profiles of reaction paths determined in complementary electronic states ($S_0^{(S_1)}$, $S_0^{(S_1')}$). The symbols (circles for S_0, squares for S_1) are computed data; full, dashed, and dotted lines are interpolations. In (c), open symbols correspond to twisting without pyramidization, filled symbols correspond to the reaction path with inclusion of pyramidization at the central N-atom.

the S_1' state is unstable with respect to the torsional coordinate [open squares in Fig. 5(c)]. The S_1' energy is additionally stabilized when pyramidization at the central N-atom is allowed (filled squares). The filled circles represent the S_0 energy at these geometries. It is seen that a S_1–S_0 degeneracy is reached by about 60° torsion and by moderate pyramidization. The molecular structure of this conical intersection is shown in Fig. 3(d). As illustrated by Figs. 4(b) and 4(c), this conical intersection can be accessed from the Franck–Condon region of the $S_1(\pi\pi^*)$ state in a barrierless manner by H-transfer, followed by torsion and pyramidization. From the conical intersection, a barrierless reaction path on the S_0 surface steers the system back to the planar geometry. Finally, H-atom back-transfer on the S_0 surface restores the molecule in the original ground-state conformation.

Further insight into the electronic aspects of the photophysics of TIN-H can be obtained by the inspection of the highest occupied molecular orbital (HOMO) and the lowest unoccupied molecular orbital (LUMO) which are displayed in Fig. 5 for several relevant geometries. All frontier orbitals are

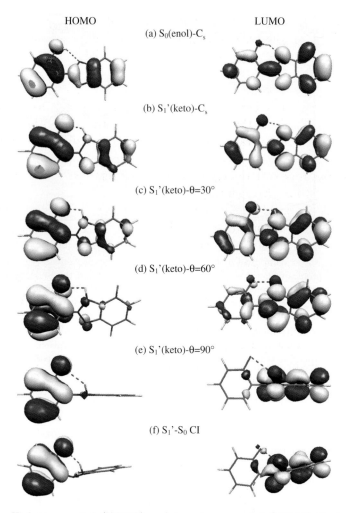

Fig. 5. Highest occupied (HOMO) and lowest unoccupied (LUMO) Hartree–Fock orbitals of TIN-H at selected geometries: ground-state equilibrium geometry (a) S_1' (keto) saddle point (b), torsional angles of 30° (c), 60° (d), 90° (e), and S_1–S_0 conical intersection (f).

of π, π^* type. It can be seen that $S_1 \leftarrow S_0$ excitation at the S_0 equilibrium geometry is accompanied by a moderate shift of electron density from the phenyl ring to the benzotriazole moiety [Fig. 5(a)]. At the planar keto conformation, the ground state is highly polar ($\mu = 4.64$ Debye) due to the preferential location of the HOMO on the phenyl ring, while the proton is attached to the triazole ring [Fig. 5(b)]. In the S_1' excited state, this polarity is reduced due to the delocalized character of the LUMO [Fig. 5(b)]. In Figs. 5(c)–(e), the HOMO and LUMO are shown as a function of the torsional angle. It is seen that the HOMO and LUMO become increasingly localized on the phenol and benzotriazole moieties, respectively, with increasing torsional angle θ. At $\theta = 90°$ [Fig. 5(e)], the half-filled orbitals are nearly completely localized on either the phenoxy and the benzotriazole part of TIN-H. At the S_1–S_0 conical intersection [Fig. 5(f)], which is reached by moderate pyramidization at the central N-atom, the electronic structure of the S_1 state represents a nearly perfect biradical. This biradical character is typical for degeneracies of the open-shell S_1 state with the closed shell S_0 state in organic systems.[72]

It should be noted that the overall stabilization of the $S_1(\pi\pi^*)$ energy by hydrogen transfer and torsion/pyramidization contributes less to the shrinking of the S_1–S_0 energy gap than the destabilization of the S_0 state, see Fig. 4. The strong preference of the S_0 state for the planar conjugated structure and the OH... N hydrogen bond are the reasons for the substantial rise in the ground-state energy, which leads to the crossing of the S_0 surface with the S_1 surface.

A generic scheme of the mechanistic function of an organic photostabilizer is shown in Fig. 6. The generic system is a covalently bonded pair of a proton donor (with heteroatom X) and a proton acceptor (with heteroatom Y). The initial excited-state process is electron/proton transfer. This reaction is barrierless and thus extremely fast, corresponding

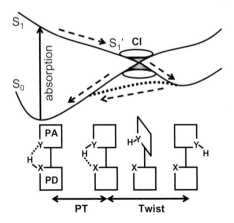

Fig. 6. PE scheme of the photophysics of a photostabilizer. PD: proton donor, PA: proton acceptor, X: proton-donating heteroatom, Y: proton-accepting heteroatom.

to "ballistic" wave-packet motion.[65,73,74] The resulting planar proton-transferred conformation (S_1') is unstable with respect to torsion (and possibly pyramidization). The torsion breaks the weak hydrogen bond of S_1' and provides a barrierless reaction path toward the S_1–S_0 conical intersection, see Fig. 6. At the S_1–S_0 conical intersection, an ultrafast nonadiabatic transition from the S_1 surface to the S_0 surface takes place. In the S_0 state, the steep torsional potential and the re-formation of the original hydrogen bond steer the system back to the original planar S_0 conformation.

While several computational studies, e.g. for SA,[66] MS,[75] OHBA,[76] and a benzotriazole model system,[77] have confirmed the qualitative reaction mechanism of Fig. 6, only indirect experimental evidence exists so far for this mechanism. The ultrafast ballistic character of the excited-state hydrogen transfer has been confirmed by femtosecond pump-probe spectroscopy with femtosecond time resolution.[65,73,74] For HBO and HBT, which exhibit comparatively long S_1' lifetimes, the "ringing" of the quasi-rigid H-donor and acceptor groups after the proton transfer has been observed.[73,74] A characteristic feature of SA, MS, OHBA, hydroxyflavones and related systems is the existence of a distinct energy threshold in the S_1 state, beyond which the fluorescence is completely quenched.[1–4,62–65] The highly efficient (sub-ps) internal conversion process for excess energies above this threshold is indicative of the directly accessible conical intersection. The suppression of the fluorescence quenching in viscous media provides indirect evidence that torsional motion or another large-amplitude motion of the donor and acceptor groups are involved in the excited-state deactivation process, supporting the qualitative scheme of Fig. 6.

4. Excited-State Electron/Proton Transfer in Inter-Molecularly Hydrogen-Bonded Aromatic Systems

The phenomenon of fluorescence quenching through inter-molecular hydrogen bonding between aromatic chromophores is a well-known phenomenon. The photophysics of aromatic hydrogen-bonded donor–acceptor pairs in various solvents has been extensively investigated by Förster, Weller, Mataga, Waluk and their co-workers.[5–9] Mataga has advocated a generic model of fluorescence quenching which emphasizes the role of charge-transfer (CT) states as promoters of proton transfer from the hydrogen donor to the hydrogen acceptor. Curve crossings of the CT state with the spectroscopic locally excited (LE) states were considered to facilitate rapid internal conversion, thus quenching the fluorescence.[8,9]

The most interesting example of hydrogen bonding between heteroaromatic chromophores is the Watson–Crick (WC) base pairing in DNA. Only two base pairs, guanine–cytosine (GC) and adenine–thymine (AT) encode the genetic information of all living (and extinct) species. While the GC and AT base pairs can exist in numerous H-bonded conformations in the gas phase, only a single conformer of GC and AT, the WC structure, occurs in DNA.

In DNA or RNA oligomers, the individual nucleic acid bases interact with each other via hydrogen bonding (base-pairing), noncovalent π–π interactions (stacking) as well as via covalent bonding to the sugar-phosphate backbone. The complex interplay of these interactions, as well as solvation, determine the photophysics of DNA. The internal interactions in DNA and external perturbations may affect the energetic ordering of the excited electronic states of the DNA bases and may also provide additional decay channels. It is therefore of scientific interest to investigate the GC and AT base pairs in isolation. This has become possible since de Vries, Kleinermanns and co-workers demonstrated that intact nucleic acid base pairs can be brought into supersonic jets by laser evaporation.[78]

The experimental investigation of the photophysical dynamics of isolated WC base pairs is complicated, however, by several factors. Most importantly, the excited-state lifetimes of the monomers are already very short.[79] Moreover, several of the many possible conformers of the base pairs are expected to be present in the jet. The investigation of the photophysics of simplified mimetic models of the WC base pairs, which are free of the complications of the actual WC base pairs, has the potential to provide useful insights. A well-known model is the 7-azaindole dimer, for which the competition of concerted double proton transfer vs. step-wise single proton transfer has been extensively investigated.[80] The 2-aminopyridine-2-pyridone H-bonded pair is another well-studied model.[81]

A particularly simple model system is the 2-aminopyridine (2AP) dimer, see Fig. 7(a). It possesses the relevant hydrogen bonds, but lacks most of the complexities of the purine and pyrimidine bases. Femtosecond time-resolved mass spectroscopy of 2AP clusters revealed an excited-state lifetime of 65 ps for the H-bonded dimer,[82] which is significantly shorter than the lifetime of the monomer (1.5 ns). Earlier *ab initio* calculations at the CASSCF/CASPT2 level had predicted a simple and generic mechanism for the efficient excited-state quenching in the 2AP dimer[83]: in addition to the intra-monomer $^1\pi\pi^*$ excited state, a state corresponding to inter-monomer $\pi \to \pi^*$ excitation exists. While this CT state is rather high in energy in the Franck–Condon region, it is strongly stabilized by the transfer of a

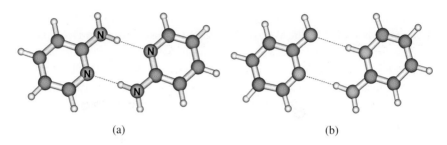

Fig. 7. (a) Ground-state equilibrium geometry of the 2-aminopyridine dimer. (b) Equilibrium geometry of the biradical after the transfer of a hydrogen atom. In (a), nitrogen atoms are identified by "N"; the remaining atoms are carbon (gray) and hydrogen (light gray).

proton across one of the NH...N hydrogen bonds ("the proton follows the electron"), see Fig. 7(b). The PE surface of the ^{1}CT state intersects the PE surfaces of both the locally-excited (^{1}LE) state as well as the ground state. This sequence of conical intersections provides the mechanism for the efficient quenching of the fluorescence of the ^{1}LE state.[83]

In 2005, de Vries and collaborators discovered an astonishing conformational sensitivity of the photophysics of GC base-pair structures in a supersonic jet.[84] The UV excitation spectra of three conformers of GC could be identified by UV-UV and UV-IR hole-burning spectroscopy and could be assigned by comparison with DFT calculations of ground-state vibrational frequencies. While the two biologically irrelevant conformers (dubbed B, C) were found to exhibit intense and sharp resonant multi-photon ionization (REMPI) spectra, the WC conformer was found to give rise to a very weak and extremely broad REMPI signal, indicating an anomalously short excited-state lifetime.[84] It has been concluded that a particularly efficient excited-state deactivation process must exist in the WC conformer of GC. It has been speculated that the rapid excited-state quenching may be essential for the prevention of destructive photochemical reactions, endowing this particular molecular structure with a unique degree of photostability.[84]

These findings of molecular-beam spectroscopy were confirmed by an alternative experiment of Schwalb and Temps, who measured time-resolved pump-probe spectra of the WC conformer of GC in a nonpolar and nonprotic solvent.[85] These measurements provided additional evidence for a pronounced shortening of the excited-state lifetime of the WC base pair in comparison with the already very short lifetime of guanine.[85]

Efficient Excited-State Deactivation in Organic Chromophores 67

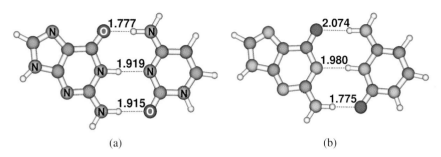

Fig. 8. (a) Ground-state equilibrium geometry of the GC Watson–Crick base pair. (b) Minimum-energy geometry of the $^1\pi\pi^*$ charge-transfer state after the transfer of the proton. The numbers denote hydrogen-bond lengths in Å. In (a), oxygen atoms (dark gray) are identified by "O", nitrogen atoms by "N"; the remaining atoms are carbon (gray) and hydrogen (light gray).

For brevity, we discuss here the photophysics of just the biologically relevant WC conformer of GC, referring to Refs. 84 and 86 for the structures, REMPI spectra and PE functions of the nonbiological conformers B and C. The ground-state equilibrium geometry of the WC form of the GC base pair is displayed in Fig. 8(a) (optimized with the CC2 method). The lengths of the three hydrogen bonds are indicated; they are a measure of the relative strengths of the hydrogen bonds. The lowest CT state in GC corresponds to the excitation of an electron from the HOMO of guanine to the LUMO of cytosine. These orbitals are shown in Figs. 9(a) and 9(b) (π orbitals) and 9(c) and 9(d) (π^* orbitals). The charge separation of the G → C charge-transfer excited state can be compensated by the transfer of one of the protons of the NH groups of guanine to cytosine (the calculation shows that the proton of the middle hydrogen bond is preferred[86]). The neutralization of the electronic charge separation results in a pronounced stabilization of the CT state. The CC2-optimized structure of the proton-transferred system, which is a biradical, is shown in Fig. 8(b). The singly occupied orbitals of the biradical are displayed in Figs. 9(e) and 9(f). It is seen that the molecular orbitals (MOs) do not change significantly upon proton transfer. They are non-overlapping already at the ground-state equilibrium geometry. In this case, the CT state is a biradical in the planar configuration, in contrast to the intra-molecularly H-bonded systems of Sec. 3, where an additional torsion is necessary to generate a biradicalic system.

The reaction path for proton transfer in the GC base pair was constructed as a linearly interpolated transit path (LITP) in internal coordinates between the ground-state equilibrium geometry [Fig. 8(a)] and the

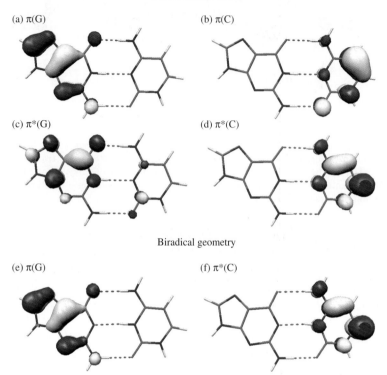

Fig. 9. Highest occupied (π) and lowest unoccupied (π^*) orbitals located on guanine (G) and cytosine (C). (a)–(d): ground-state equilibrium geometry. (e),(f): equilibrium geometry of the biradical.

equilibrium geometry of the biradical [Fig. 8(b)]. The energies of the ground state and several excited states were calculated along this path (see Ref. 86 for details). The resulting PE profiles are shown in Fig. 10. Since the energies of all states are calculated at the same geometries (i.e. the LITP geometries), the crossings in Fig. 10 are true crossings (conical intersections).

The enormous stabilization of the CT state by the transfer of the proton from G to C is eye-catching. An equally strong destabilization of the S_0 state leads to a CT–S_0 conical intersection, see Fig. 10. Since the energy surface of the ^{1}CT state crosses also the energy surfaces of the spectroscopic ^{1}LE states, it provides the mechanism for a highly efficient deactivation of the UV absorbing states. In the non-WC conformers B and C, the photoreactive CT state is higher in energy,[86] allowing the existence of long-lived vibronic

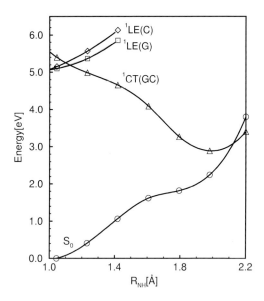

Fig. 10. PE profiles of the electronic ground state (circles), the lowest locally-excited states of guanine (squares) and cytosine (diamonds), and the guanine-to-cytosine charge-transfer state (triangles) as a function of the linearly-interpolated transit path for hydrogen transfer. The symbols are computed data; the curves are interpolations.

states, as observed experimentally.[84] Qualitatively similar computational results have been obtained for the AT base pair, predicting an extremely short lifetime of the WC conformer of AT.[87] Experimentally, the REMPI signal of the WC conformer of AT could not be detected so far.

A generic scheme of the reaction mechanism in inter-molecularly hydrogen-bonded aromatic systems is shown in Fig. 11. It consists of the PE profiles of the electronic ground state, the lowest LE singlet state and the lowest singlet state of CT character, plotted as a function of the proton-transfer coordinate. The LE–CT and CT–S_0 curve crossings become intersections when the appropriate coupling modes are included. Absorption of a UV photon leads to the (X-H...Y)* system in the ^{1}LE state. If the available vibrational excess energy in the ^{1}LE state comes close to the ^{1}LE–^{1}CT intersection, the ^{1}CT state is populated, resulting in a temporal formation of the X-H$^+$...Y$^-$ ion pair. Rapid motion of the proton on the steep PE surface of the ^{1}CT state results in the neutralization of the ion pair and the formation of the X$^\bullet$...H-Y$^\bullet$ biradical. At the ^{1}CT–S_0 conical intersection, an electron can jump from Y to X, resulting in the X$^-$...H-Y$^+$ ion pair in its electronic ground state. The back-transfer of the proton and the re-formation

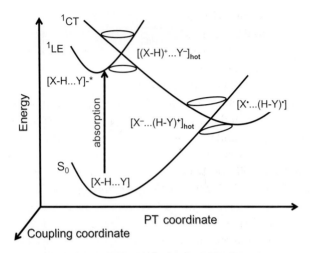

Fig. 11. Schematic view of the electron-driven proton-transfer process in intermolecularly hydrogen-bonded aromatic systems. Abbreviations: LE: locally-excited state; CT: charge-transfer state; X(Y): hydrogen-donating (hydrogen-accepting) aromatic system. The circles schematically indicate the conical intersections.

of the X-H...Y hydrogen bond close the reaction circle. Assuming that the dynamics of the much lighter electron precedes the dynamics of the proton, the reaction mechanism illustrated in Fig. 11 has been termed the "electron-driven proton transfer" (EDPT) mechanism.[59] The crucial parameter in the EDPT reaction is the height of the ^{1}LE–^{1}CT curve crossing relative to the minimum of the ^{1}LE surface. If this crossing is sufficiently low, the EDPT mechanism leads to very efficient internal conversion from the ^{1}LE to the S_0 state. It is plausible that the particularly rapid excited-state deactivation via the EDPT mechanism plays an essential role for the photostability of biopolymers such as DNA and proteins.[59]

5. Chromophores with Flexible Side-Chains: Electron/Proton Transfer in Amino Acids and Peptides

Amino acids and peptides are highly flexible molecules which exhibit a multitude of low-energy conformers. Recent spectroscopic investigations of aromatic amino acids and small peptides containing an aromatic chromophore in supersonic jets provided a wealth of information on the conformer-specific photophysics of these elementary building blocks of proteins. By the combination of laser-based UV-UV and UV-IR double resonance spectroscopy with first-principles calculations of vibrational spectra, the measured laser-induced fluorescence (LIF) or REMPI spectra of amino acids or peptides could be assigned to specific conformers.[88–92] It was found that the UV excitation spectra of the aromatic amino acids tryptophan (Trp), tyrosine (Tyr) and phenylalanine (Phe), as well as the spectra of small peptides containing aromatic chromophores, exhibit a pronounced dependence on the ground-state conformation, see, e.g. Refs. 93–96. Moreover, evidence is accumulating that many of the expected low-energy conformers of aromatic amino acids and peptides cannot be detected at all by LIF and REMPI spectroscopy, presumably due to extremely short excited-state lifetimes.[96–99] There is thus convincing experimental evidence for a decisive control of the photophysics of aromatic amino acids and peptides by the conformation of flexible side-groups.

In two recent communications, we have reported preliminary results of *ab initio* electronic-structure calculations for a dipeptide (Trp-Gly) and a tripeptide (Gly-Phe-Ala), where Gly stands for glycine and Ala for alanine.[100,101] We give here a brief account of the results for the tripeptide, emphasizing the reaction mechanisms which most likely are generic for peptides and proteins.

Gly-Phe-Ala has a multitude of possible conformations. An extensive and thorough exploration of the conformational space of Gly-Phe-Ala has been performed by Valdes *et al.*, resulting in the reliable identification of the lowest-energy conformers.[99] We consider here the conformer of lowest free energy in the family of structures which exhibit a so-called γ-turn of the peptide backbone. The ground-state equilibrium geometry (optimized at the MP2 level) of this conformer is shown in Fig. 12(a), with the two

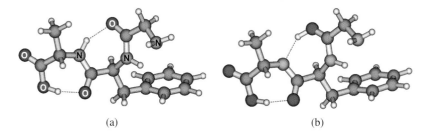

Fig. 12. (a) Ground-state equilibrium geometry of the Gly-Phe-Ala tripeptide. (b) Minimum-energy geometry of the CT state after the transfer of the proton. The dotted lines indicate hydrogen bonds. In (a), oxygen atoms (dark gray) are identified by "O", nitrogen atoms by "N"; the remaining atoms are carbon (gray) and hydrogen (light gray).

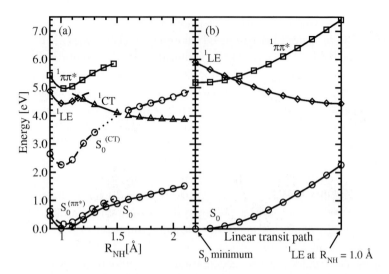

Fig. 13. (a) PE profiles of minimum-energy paths (relaxed scans) of the $^1\pi\pi^*$ state (squares), the locally-excited state (diamonds), the charge-transfer state (triangles), and the electronic ground state (circles). The $^1\pi\pi^*$, ^{1}LE and ^{1}CT energies have been determined at the respective optimized geometries. The ground-state energies designated as $S_0^{(\pi\pi^*)}$ and $S_0^{(CT)}$ have been determined at the optimized geometries of the $^1\pi\pi^*$ state and the ^{1}CT state, respectively. Symbols denote computed data; the full and dotted curves are interpolations. (b) PE profiles of the $^1\pi\pi^*$, ^{1}LE and S_0 states along the linearly-interpolated transit path between the S_0 equilibrium geometry and the ^{1}LE minimum-energy geometry an $R_{NH} = 1$ Å.

hydrogen bonds indicated. The O-H...O=C hydrogen bond involving the terminal carboxyl group is characteristic for small (uncapped) peptides. The N-H...O=C hydrogen bond, on the other hand, is a characteristic hydrogen bond in proteins, representing a so-called γ-turn (or C7 structure) of the backbone.

PE profiles (relaxed scans, calculated with the CC2 method) for electron/proton transfer along the N-H...O=C hydrogen bond are displayed in Fig. 13(a). The UV absorbing state is the $^1\pi\pi^*$ state of the phenyl ring [squares in Fig. 13(a)]. The corresponding HOMO and LUMO are shown in Fig. 14(a). It is seen that the $^1\pi\pi^*$ excitation is completely localized on the aromatic ring. The $^1\pi\pi^*$ energy minimum is, however, not the lowest minimum of the S_1 surface of the system. The lowest minimum corresponds to the excitation from the nonbonding orbital of the carbonyl group to an unoccupied orbital which is localized on the same branch of the peptide backbone [diamonds in Fig. 13(a)]. The orbitals involved in this excitation

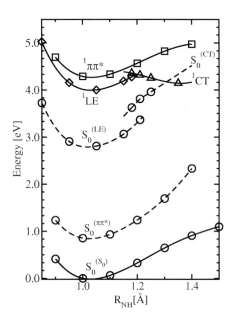

Fig. 16. PE profiles (relaxed scans) of minimum-energy paths for proton transfer in the ^{1}L$_a$($\pi\pi^*$) state (squares), the ^{1}LE state (diamonds) and the ^{1}CT state (triangles) in NATMA C. The energy profiles of the electronic ground state (circles) designated as $S_0^{(S_0)}$, $S_0^{(\pi\pi^*)}$, $S_0^{(LE)}$ and $S_0^{(CT)}$ have been computed for minimum-energy paths of the respective electronic states. The symbols represent computed data; the curves are interpolations.

The differences in the spectroscopic properties of the C5 and C7 conformers of NATMA arise from the relative location of the ^{1}LE state. In NATMA C, both the $^1\pi\pi^*$ state as well as the ^{1}LE state are lower in energy than in NATMA A, B.[104] As a result, the barrier for energy transfer from the $^1\pi\pi^*$ state to the ^{1}LE state and the ^{1}CT state is lower in NATMA C. It seems that excitation transfer to the ^{1}LE state of the peptide backbone is just possible in NATMA C, while excitation transfer cannot take place in NATMA A, B. NATMA C may be special in so far as it exhibits the spectroscopic signatures of radiationless excited-state quenching, while the chromophore excitation lifetime is still sufficiently long to allow the detection of the LIF signal.

A mechanistic scheme of the excited-state deactivation in aromatic amino acids and peptides is shown in Fig. 17. The UV absorbing state is the $^1\pi\pi^*$ state of the chromophore. The excitation energy transfer from the chromophore to an intermediate locally-excited singlet state of the peptide

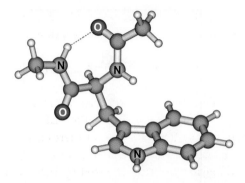

Fig. 15. Ground-state equilibrium geometry of NATMA C. The dotted line indicates the hydrogen bond of the γ-turn. Oxygen atoms (dark gray) are identified by "O", nitrogen atoms by "N"; the remaining atoms are carbon (gray) and hydrogen (light gray).

infrared fundamentals in the excited-state (S_1) infrared spectrum of the C7 conformer, while these fundamentals (with the exception of the fundamental of the indole NH group) were sharp and easily observed in the C5 conformers.[102,103]

For brevity, we restrict the discussion to the photophysics of the C7 conformer of NATMA and refer to Ref. 104 for a comparative discussion of the C5 and C7 conformers. The ground-state equilibrium structure of NATMA C (optimized with the MP2 method) is shown in Fig. 15. The γ-turn and the associated hydrogen bond are clearly visible.

The PE profiles for the minimum-energy electron/proton-transfer reaction paths (relaxed scans, calculated with the ADC(2) method[105]) along the hydrogen bond of the γ-turn are shown in Fig. 16. The electronic states involved are analogous to those discussed for Gly-Phe-Ala above. The $^1\pi\pi^*$ state (squares) is the lowest excited state of the indole chromophore (at its equilibrium geometry). The ^{1}LE (diamonds) and ^{1}CT (triangles) states are the locally-excited and charge-transfer excited states of the peptide backbone, analogous to Gly-Phe-Ala. As in the latter, a rather low barrier exists on the reaction path from the $^1\pi\pi^*$ state to the ^{1}LE state (data not shown). The curves with circles represent cuts of the multi-dimensional S_0 PE surface along the S_0 minimum-energy path and the various excited-state minimum-energy paths. It is seen that the proton transfer in the ^{1}CT state strongly stabilizes this state, while the S_0 energy rises by more than 4.0 eV. The result is a ^{1}CT–S_0 conical intersection at an NH bond length of 1.3 Å (see Fig. 16).

while the crossings of the $^1\pi\pi^*$, ^{1}LE and ^{1}CT curves in Fig. 13(a) are apparent crossings (since the corresponding reaction paths have individually been optimized).

To explore the possibility of an efficient nonadiabatic transition from the optically bright $^1\pi\pi^*$ state to the dark ^{1}LE state, the $^1\pi\pi^*$ and ^{1}LE energies have been calculated along the linearly interpolated transit path (LITP) connecting the S_0 minimum with the ^{1}LE minimum. These PE profiles are shown in Fig. 13(b). It is seen that there exists a crossing (conical intersection) only slightly above the minimum of the $^1\pi\pi^*$ PE surface.

The overall mechanistic picture suggested by these *ab initio* electronic-structure data is as follows. Absorption of a UV photon populates the $^1\pi\pi^*$ state. The $^1\pi\pi^*$ energy can be transferred to the ^{1}LE state via a low-lying conical intersection [see Fig. 13(b)]. The ^{1}LE surface is, in turn, intersected by the ^{1}CT surface [see Fig. 13(a)]. Population of the ^{1}CT state corresponds to the transfer of an electron from the N-H group to the C=O group along the intra-molecular hydrogen bond [see Fig. 14(c)]. The proton then follows the electron, which leads to a pronounced stabilization of the CT state and thus to a conical intersection with the S_0 state [see Fig. 13(a)]. In this way, the electronic excitation energy is converted, via three conical intersections, into comparatively harmless vibrational energy in the closed-shell ground state.

The experimental signature of the rather complex photophysics of this conformer of Gly-Phe-Ala is the absence of a REMPI signal,[99] which may be felt as somewhat disappointing by spectroscopists. We briefly discuss, therefore, a second example, the capped dipeptide N-acetyl tryptophan methyl amide (NATMA), for which particularly dramatic effects of the conformational flexibility of the peptide chain on the UV spectroscopy of the chromophore have been reported by Zwier and collaborators.[102,103]

The conformers of NATMA can be classified by the hydrogen-bonding motifs: C5 structures, where the H-bond closes a five-membered ring, and C7 structures, where the H-bond closes a seven-membered ring. The former structures correspond to an extended, β-sheet-type backbone, the latter correspond to γ-turns of the protein structure. Two C5 structures (NATMA A, B) and one C7 structure (NATMA C) have been identified by Dian *et al.*[102,103] The C5 and C7 structures exhibit startlingly different UV excitation spectra: while the C5 conformers possess sharp vibronically resolved UV spectra, the C7 conformer exhibits extremely broad and nearly structureless UV absorptions. In addition to the unusual UV excitation spectra of NATMA C, Dian *et al.* observed the complete absence of all CH and NH

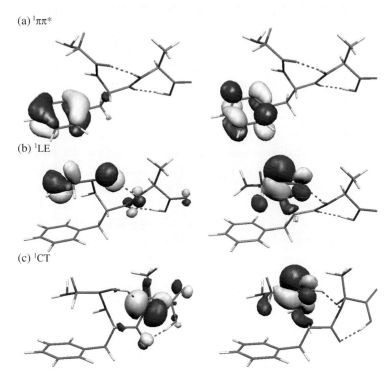

Fig. 14. Frontier Hartree–Fock molecular orbitals of Gly-Phe-Ala at the ground-state equilibrium geometry. (a) The π (left) and π^* (right) MOs of the phenyl chromophore. (b) The half-filled orbitals of the ^{1}LE state. (c) The half-filled orbitals of the ^{1}CT state.

(at the S_0 equilibrium geometry) are displayed in Fig. 14(b). This state, which we designate as locally-excited (^{1}LE) state, cannot be excited directly by light absorption due to very unfavorable Franck–Condon factors. It can be populated, however, by excitation-energy transfer from the $^1\pi\pi^*$ state of the phenyl chromophore. In addition, there exists a low-lying singlet state which involves excitation from an occupied orbital localized on the N-H hydrogen-donor group [Fig. 14(b), left] to an unoccupied orbital localized on the O=C hydrogen-acceptor group [Fig. 14(c), right]. This state can be classified as a charge-transfer state (^{1}CT) with respect to the N-H...O=C hydrogen bond. The triangles in Fig. 13(a) represent the energy profile of this state along its minimum-energy path. The PE function denoted as $S_0^{(CT)}$ in Fig. 13(a) represents the energy of the S_0 state calculated at the geometries of the minimum-energy path of the ^{1}CT state. The crossing of the ^{1}CT and the $S_0^{(CT)}$ curves is thus a true crossing (conical intersection),

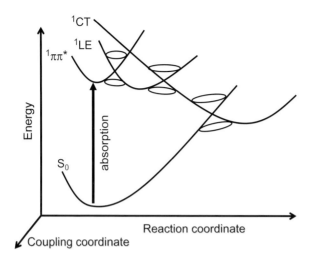

Fig. 17. Schematic view of the PE surfaces involved in the excited-state deactivation of aromatic amino acids and peptides. $^1\pi\pi^*$ is the excited-state of the chromophore; ^{1}LE and ^{1}CT are the locally-excited state and the charge-transfer excited state of the peptide backbone. The circles schematically indicate the conical intersections.

backbone (^{1}LE) is the new aspect of the radiationless decay path of peptides compared to the radiationless decay of the hydrogen-bonded DNA base pairs (Fig. 11). In both cases, an EDPT process strongly stabilizes a ^{1}CT state, leading to a biradical and a conical intersection of the biradicalic PE surface with the S_0 PE surface.

6. Conclusions

We have surveyed in this chapter generic mechanisms of excited-state deactivation via coupled electron-proton-transfer dynamics in isolated aromatic chromophores, in hydrogen-bonded pairs of chromophores, as well as in chromophores with flexible side-chains. It has been shown that excited-state reactions involving the detachment or the transfer of a hydrogen atom play a decisive role for the photophysics of these systems.

The photoinduced hydrogen-atom detachment from acidic groups of chromophores is driven by $^1\pi\sigma^*$ states with Rydberg character. The high-resolution spectroscopy (either in the energy or in the time domain) of the detached hydrogen atoms has led to a new paradigm of nonstatistical photochemistry in polyatomic molecules. The photodetached hydrogen atom has emerged as a new spectroscopic messenger particle which carries uniquely detailed information on the dynamics of conical intersections in the excited-state manifold of aromatic chromophores.[34]

In intra-molecularly hydrogen-bonded bifunctional aromatic systems, in particular in so-called photostabilizers, the ultrafast electron/proton transfer in the lowest $^1\pi\pi^*$ excited state leads to an intermediate structure which is unstable with respect to torsion about a covalent bond. The torsion, possibly accompanied by pyramidization, leads in a barrierless manner to a S_1–S_0 conical intersection and thus ultrafast internal conversion.

In inter-molecularly hydrogen-bonded aromatic pairs, in particular in the DNA base pairs, it is a highly polar CT state of $^1\pi\pi^*$ valence character which provides the driving force for the coupled electron/proton transfer: driven by the large dipole moment of the CT state, the proton follows the electron across a hydrogen bond, resulting in a pronounced energetic stabilization of a biradical and an S_1–S_0 conical intersection. After the back-transfer of the electron at the S_1–S_0 conical intersection, the proton is driven back to its original location on the S_0 PE surface, thereby closing the photophysical cycle.

In aromatic chromophores with flexible side-chains, in particular in aromatic amino acids and peptides, the hydrogen bonds of the flexible chain play an essential role for the efficient radiationless deactivation of the chromophore. After excitation transfer from the chromophore to the peptidic chain, a CT state can be populated which drives a proton across a hydrogen bond, resulting in a biradical structure and an S_1–S_0 conical intersection.

As is well known, hydrogen bonds are ubiquitous in biological matter. Their universal role in structure formation (proteins), molecular recognition (DNA) and catalysis (enzymes) is well established. Here, we have addressed the much less explored field of excited-state dynamics of hydrogen bonds. The latter provides a highly efficient mechanism for the deactivation of the potentially reactive excited states, which may be of unrivalled efficiency. The excited-state chemistry of hydrogen bonds may therefore be as essential for the existence of life as the ground-state chemistry.

Another interesting aspect of the phenomena discussed above is their potential application in molecular devices. For example, the ESIPT process along intra-molecular hydrogen bonds may be utilized for the construction of optically driven photostable molecular switches,[106] while the EDPT reaction along inter-molecular hydrogen bonds may provide a template for the design of molecular systems which can split water using solar radiation.[107]

Acknowledgments

We would like to thank our postdocs who have made substantial contributions to the results surveyed in this chapter: Zhenggang Lan, Valerie Vallet, Serhiy Perun, Luis Manuel Frutos, Shohei Yamazaki and Dorit Shemesh. The collaboration of the authors has continuously been supported by the Deutsche Forschungsgemeinschaft and the Ministry of Science and Education of Poland for more than 20 years.

References

1. S.J. Formosinho and L.G. Arnaut, *J. Photochem. Photobiol. A* **75**, 21 (1993).
2. L.G. Arnaut and S.J. Formosinho, *J. Photochem. Photobiol. A* **75**, 1 (1993).
3. S.M. Ormson and R.G. Brown, *Progr. Reaction Kin.* **19**, 45 (1994).
4. A. Douhal, F. Lahmani and A.H. Zewail, *Chem. Phys.* **207**, 477 (1996).
5. T. Förster, *Z. Elektrochem.* **54**, 531 (1950).
6. D. Rehm and A. Weller, *Isr. J. Chem.* **8**, 259 (1970).
7. N. Mataga, *Pure Appl. Chem.* **56**, 1255 (1984).
8. N. Mataga and H. Miyasaka, *Adv. Chem. Phys.* **107**, 431 (1999).
9. J. Waluk, *Acc. Chem. Res.* **36**, 832 (2003).
10. D.R. Ort and C.F. Yoccum (Eds.), *Oxygenic Photosynthesis: The Light Reactions*, Kluwer, Dordrecht, 1996.
11. R.I. Cukier and D.G. Nocera, *Annu. Rev. Phys. Chem.* **49**, 337 (1998).
12. S. Hammes-Schiffer, *Acc. Chem. Res.* **34**, 273 (2001).
13. S. Hammes-Schiffer, *ChemPhysChem* **3**, 33 (2002).
14. J.M. Mayer and I.J. Rhile, *Biochim. Biophys. Acta* **1665**, 51 (2004).
15. J.M. Mayer, *Annu. Rev. Phys. Chem.* **55**, 363 (2004).
16. M.H.V. Huynh and T.J. Meyer, *Chem. Rev.* **107**, 5004 (2007).
17. O. Tishchenko, D.G. Truhlar, A. Ceulemans and M.T. Nguyen, *J. Am. Chem. Soc.* **130**, 7000 (2008).
18. S. Hammes-Schiffer and A.V. Soudackov, *J. Phys. Chem. B* **112**, 14108 (2008).
19. U. Ushiyama and K. Takatsuka, *Angew. Chem. Int. Ed.* **46**, 587 (2007).
20. N. Nagashima and K. Takatsuka, *J. Phys. Chem. A* **113**, 15240 (2009).
21. A.L. Sobolewski and W. Domcke, *Chem. Phys. Lett.* **315**, 293 (1999).
22. A.L. Sobolewski and W. Domcke, *Chem. Phys.* **259**, 181 (2000).
23. A.L. Sobolewski and W. Domcke, *J. Phys. Chem. A* **105**, 9275 (2001).
24. L. Serrano-Andres, M. Merchan, I. Nebot-Gil, B.O. Roos and M. Fülscher, *J. Am. Chem. Soc.* **115**, 6184 (1993).
25. L. Serrano-Andres and B.O. Roos, *J. Am. Chem. Soc.* **118**, 185 (1996).
26. J. Lorentzon, P.-A. Malmqvist, M. Fülscher and B.O. Roos, *Theor. Chim. Acta* **91**, 91 (1995).

27. A.L. Sobolewski, W. Domcke, C. Dedonder-Lardeux and C. Jouvet, *Phys. Chem. Chem. Phys.* **4**, 1093 (2002).
28. A.L. Sobolewski and W. Domcke, *J. Phys. Chem. A* **111**, 11725 (2007).
29. V. Vallet, Z. Lan, S. Mahapatra, A.L. Sobolewski and W. Domcke, *J. Chem. Phys.* **123**, 144307 (2005).
30. M. Barbatti, M. Vazdar, A.J.A. Aquino, M. Eckert-Maksic and H. Lischka, *J. Chem. Phys.* **125**, 164323 (2006).
31. M.N.R. Ashfold, G.A. King, D. Murdock, M.G.D. Nix, T.A.A. Oliver and A.G. Sage, *Phys. Chem. Chem. Phys.* **12**, 1218 (2010).
32. M. Barbatti, H. Lischka, S. Salzmann and C.M. Marian, *J. Chem. Phys.* **130**, 034305 (2009).
33. I. Frank and K. Damianos, *J. Chem. Phys.* **126**, 125105 (2007).
34. M.N.R. Ashfold, B. Cronin, A.L. Devine, R.N. Dixon and M.G.D. Nix, *Science* **312**, 1637 (2006).
35. Z. Lan, A. Dupays, V. Vallet, S. Mahapatra and W. Domcke, *J. Photochem. Photobiol. A* **190**, 177 (2007).
36. D.A. Blank, S.W. North and Y.T. Lee, *Chem. Phys.* **187**, 35 (1994).
37. J. Wei, A. Kuczmann, J. Riedel, F. Renth and F. Temps, *Phys. Chem. Chem. Phys.* **5**, 315 (2003).
38. B. Cronin, M.G.D. Nix, R.H. Qadiri and M.N.R. Ashfold, *Phys. Chem. Chem. Phys.* **6**, 5031 (2004).
39. H. Lippert, H.-H. Ritze, I.V. Hertel and W. Radloff, *ChemPhysChem* **5**, 1423 (2004).
40. A. Iqbal, M.S.Y. Cheung, M.G.D. Nix and V.G. Stavros, *J. Phys. Chem. A* **113**, 8157 (2009).
41. A. Iqbal and V.G. Stavros, *J. Phys. Chem. A* **114**, 68 (2010).
42. Z. Lan, W. Domcke, V. Vallet, A.L. Sobolewski and S. Mahapatra, *J. Chem. Phys.* **122**, 224315 (2005).
43. M.L. Hause, Y.H. Yoon, A.S. Case and F.F. Crim, *J. Chem. Phys.* **128**, 104307 (2008).
44. A.L. Sobolewski and W. Domcke, *Eur. Phys. J. D* **20**, 369 (2002).
45. S. Perun, A.L. Sobolewski and W. Domcke, *Chem. Phys.* **313**, 107 (2005).
46. C.M. Marian, M. Kleinschmidt and J. Tatchen, *Chem. Phys.* **347**, 346 (2008).
47. W.C. Chung, Z. Lan, Y. Ohtsuki, N. Shimakura, W. Domcke and Y. Fujimura, *Phys. Chem. Chem. Phys.* **9**, 2075 (2007).
48. I. Conti, M. Garavelli and G. Orlandi, *J. Am. Chem. Soc.* **131**, 16108 (2009).
49. H. Chen and S. Li, *J. Chem. Phys.* **124**, 154315 (2006).
50. S. Yamazaki and W. Domcke, *J. Phys. Chem. A* **112**, 7090 (2008).
51. S. Yamazaki, W. Domcke and A.L. Sobolewski, *J. Phys. Chem. A* **112**, 11965 (2008).
52. S. Yamazaki, A.L. Sobolewski and W. Domcke, *Phys. Chem. Chem. Phys.* **11**, 10165 (2009).

53. V.B. Delchev, A.L. Sobolewski and W. Domcke, *Phys. Chem. Chem. Phys.* **12**, 5007 (2010).
54. G. Gregoire, C. Jouvet, C. Dedonder and A.L. Sobolewski, *J. Am. Chem. Soc.* **129**, 6223 (2007).
55. M.G.D. Nix, A.L. Devine, B. Cronin and M.N.R. Ashfold, *J. Chem. Phys.* **126**, 124312 (2007).
56. K.L. Wells, D.J. Hadden, M.G.D. Nix and V.G. Stavros, *J. Phys. Chem. Lett.* **1**, 993 (2010).
57. M.-F. Lin, C.M. Tzeng, Y.A. Dyakov and C.-K. Ni, *J. Chem. Phys.* **126**, 241104 (2007).
58. V. Poterya, V. Profant, M. Farnik, P. Slavicek and U. Buck, *J. Chem. Phys.* **127**, 064307 (2007).
59. A.L. Sobolewski and W. Domcke, *Europhysicsnews* **37**, 20 (2006).
60. H.J. Heller and H.R. Blattmann, *Pure Appl. Chem.* **30**, 145 (1972); **36**, 141 (1974).
61. J.-E. Otterstedt, *J. Chem. Phys.* **58**, 5716 (1973).
62. J.L. Herek, S. Pedersen, L. Banares and A.H. Zewail, *J. Chem. Phys.* **97**, 9046 (1992).
63. P.B. Bisht, H. Petek and K. Yoshihara, *J. Chem. Phys.* **103**, 5290 (1995).
64. F. Lahmani and A. Zehnacker-Rentien, *J. Phys. Chem. A* **101**, 6141 (1997).
65. K. Stock, T. Biziak and S. Lochbrunner, *Chem. Phys. Lett.* **354**, 409 (2002).
66. A.L. Sobolewski and W. Domcke, *Chem. Phys.* **232**, 257 (1998).
67. S. Scheiner, *J. Phys. Chem. A* **104**, 5898 (2000).
68. A.J.A. Aquino, H. Lischka and C. Hättig, *J. Phys. Chem. A* **109**, 3201 (2005).
69. A.L. Sobolewski and W. Domcke, *Phys. Chem. Chem. Phys.* **8**, 3410 (2006).
70. M. Wiechmann, H. Port, W. Frey, F. Laermer and T. Elsaesser, *J. Phys. Chem.* **95**, 1918 (1991).
71. A.L. Sobolewski, W. Domcke and C. Hättig, *J. Phys. Chem. A* **110**, 6301 (2006).
72. M. Klessinger and J. Michl, *Excited States and Photochemistry of Organic Molecules*, VCH, Weinheim, 1995, Chap. 4.3.
73. S. Lochbrunner, A.J. Wurzer and E. Riedle, *J. Chem. Phys.* **112**, 10699 (2000).
74. M. Barbatti, A.J.A. Aquino, H. Lischka, C. Schriever, S. Lochbrunner and E. Riedle, *Phys. Chem. Chem. Phys.* **11**, 1406 (2009).
75. J.D. Coe, B.G. Levine and T.J. Martinez, *J. Phys. Chem. A* **111**, 11302 (2007).
76. A. Migani, L. Blancafort, M.A. Robb and A.D. De Bellis, *J. Am. Chem. Soc.* **130**, 6932 (2008).
77. M.J. Paterson, M.A. Robb, L. Blancafort and A.D. De Bellis, *J. Am. Chem. Soc.* **126**, 2912 (2004).
78. E. Nir, K. Kleinermanns and M.S. de Vries, *Nature* **408**, 949 (2000).

79. E.C. Crespo-Hernandez, B. Cohen, P.M. Hare and B. Kohler, *Chem. Rev.* **104**, 1977 (2004).
80. A. Douhal, S.K. Kim and A.H. Zewail, *Nature* **378**, 260 (1995).
81. A. Müller, F. Talbot and S. Leutwyler, *J. Am. Chem. Soc.* **124**, 14486 (2002).
82. T. Schultz, E. Somoylova, W. Radloff, I.V. Hertel, A.L. Sobolewski and W. Domcke, *Science* **306**, 1765 (2004).
83. A.L. Sobolewski and W. Domcke, *Chem. Phys.* **294**, 73 (2003).
84. A. Abo-Riziq, L. Grace, E. Nir, M. Kabelac, P. Hobza and M.S. de Vries, *Proc. Natl. Acad. Sci. USA* **102**, 20 (2005).
85. N.K. Schwalb and F. Temps, *J. Am. Chem. Soc.* **129**, 9272 (2007).
86. A.L. Sobolewski, W. Domcke and C. Hättig, *Proc. Natl. Acad. Sci. USA* **102**, 17903 (2005).
87. S. Perun, A.L. Sobolewski and W. Domcke, *J. Phys. Chem. A* **110**, 9031 (2006).
88. E.G. Robertson and J.P. Simons, *Phys. Chem. Chem. Phys.* **3**, 1 (2001).
89. T.S. Zwier, *J. Phys. Chem. A* **105**, 8827 (2001).
90. E. Nir, C. Plützer, K. Kleinermanns and M.S. de Vries, *Eur. Phys. J. D* **20**, 317 (2002).
91. W. Chin, F. Piuzzi, I. Dimicoli and M. Mons, *Phys. Chem. Chem. Phys.* **8**, 1033 (2006).
92. M.S. de Vries and P. Hobza, *Annu. Rev. Phys. Chem.* **58**, 585 (2007).
93. M.J. Tubergen, J.R. Cable and D.H. Levy, *J. Chem. Phys.* **92**, 51 (1990).
94. L.C. Snoek, R.T. Kroemer, M.R. Hockridge and J.P. Simons, *Phys. Chem. Chem. Phys.* **3**, 1819 (2001).
95. Y. Inokuchi, Y. Kobayashi, T. Ito and T. Ebata, *J. Phys. Chem. A* **111**, 3209 (2007).
96. I. Hünig and K. Kleinermanns, *Phys. Chem. Chem. Phys.* **6**, 2650 (2004).
97. D. Reha, H. Valdes, J. Vondrasek, P. Hobza, A. Abo-Riziq, B. Crews and M.S. de Vries, *Chem. Eur. J.* **11**, 6083 (2005).
98. H. Valdes, D. Reha and P. Hobza, *J. Phys. Chem. B* **110**, 6385 (2006).
99. H. Valdes, V. Spiwok, J. Rezac, A.G. Abo-Riziq, M.S. de Vries and P. Hobza, *Chem. Eur. J.* **14**, 4886 (2008).
100. D. Shemesh, C. Hättig and W. Domcke, *Chem. Phys. Lett.* **482**, 38 (2009).
101. D. Shemesh, A.L. Sobolewski and W. Domcke, *J. Am. Chem. Soc.* **131**, 1374 (2009).
102. B.C. Dian, A. Longarte, S. Mercier, D.A. Evans, D.J. Wales and T.S. Zwier, *J. Chem. Phys.* **117**, 10688 (2002).
103. B.C. Dian, A. Longarte and T.S. Zwier, *J. Chem. Phys.* **118**, 2696 (2003).
104. D. Shemesh, A.L. Sobolewski and W. Domcke, *Phys. Chem. Chem. Phys.* **12**, 4899 (2010).
105. J. Schirmer, *Phys. Rev. A* **26**, 2395 (1982).
106. A.L. Sobolewski, *Phys. Chem. Chem. Phys.* **10**, 1243 (2008).
107. A.L. Sobolewski and W. Domcke, *J. Phys. Chem. A* **112**, 7311 (2008).

Chapter 3
Three-State Conical Intersections

Spiridoula Matsika[*]

1. Introduction 84
2. Noncrossing Rule 85
3. Branching and Seam Spaces 87
4. Symmetry and Three-State Conical Intersections:
   The Jahn–Teller Effect 88
5. First Accidental Three-State Conical Intersection: CH_4^+ 90
6. Locating Accidental Three-State Conical Intersections 92
7. Consequences of Three-State Conical Intersections 95
   7.1. Connectivity 95
   7.2. Geometric phase 99
   7.3. Nonadiabatic couplings 100
8. Characterizing Three-State Conical Intersections 103
9. Three-State Conical Intersections in Chemical Systems 105
   9.1. Three-state conical intersections in radicals 105
   9.2. Three-state conical intersections in closed-shell
        organic systems 108
10. Dynamics 111
11. Concluding Remarks 113
    Acknowledgment 113
    References 114

---

[*]Department of Chemistry, Temple University, Philadelphia, PA 19122, USA. E-mail: smatsika@temple.edu.

1. Introduction

The study of molecular systems using quantum mechanics is primarily based on the Born–Oppenheimer (BO) approximation.[1] Since the electrons have a smaller mass than the nuclei, they move much faster and follow the motion of the nuclei adiabatically, while the latter move on the average potential of the former. This approximation is essential to our understanding of molecular structure and dynamics. There are, however, essential processes in nature where this approximation breaks down, causing nonadiabatic phenomena. Even in this case, most often we are able to use the Born–Oppenheimer framework as the basis to develop the theory of nonadiabatic processes. Nonadiabatic processes occur when two or more potential energy surfaces (PESs) of a molecular system approach each other, making the coupling of electronic states possible through nuclear motion. In the extreme case, the PESs become degenerate and, if the degeneracy can be lifted linearly in two or more directions, conical intersections (CIs) are formed.[2,3]

Conical intersections can exist between two or more electronic states, and they can be categorized based on the number of electronic states involved in the degeneracy. In the most common case, when we use the term "conical intersections", we refer to two-state degeneracies. Conical intersections between two electronic states have been studied extensively and this book, along with its first volume,[4] is a testament to this. Theoretical developments have enabled the location of conical intersections in polyatomic molecules, showing that they are present in many cases and they can facilitate photoinitiated processes.[4–11] Experimentally, there has also been progress in detecting ultrafast processes, which often occur through CIs, providing further evidence for their importance.

Degeneracy between three electronic states has received less attention. Three-state degeneracies imposed by symmetry have been studied in the context of the Jahn–Teller problem for many decades,[12–14] but accidental three-state degeneracies in molecules without symmetry are a more recent discovery.[15,16] Since most molecular systems in nature have low or no symmetry, these accidental intersections have the potential for a greater impact on the photophysics and photochemistry of molecules, although there has not been enough work to verify or disprove this. Efficient algorithms have facilitated the location of three-state conical intersections in recent years,[16,17] and have identified the existence of such intersections in many systems, as will be discussed in later sections.

Three-state degeneracies may provide a more efficient relaxation pathway when more than one interstate transition is needed. Moreover, there are indirect effects that appear because of the three-state CIs, such as geometric phase effects and double-valued derivative couplings,[18-21] and they can affect the system's dynamics and pathways available for radiationless transitions.[22]

In this chapter, we will discuss the basic theoretical description of three-state CIs, and their effects and consequences. Examples of several systems where they have been found and analyzed will also be presented.

2. Noncrossing Rule

The first important questions to ask when discussing three-state CIs are in which cases they occur and how common they are expected to be. These questions can be addressed by invoking the noncrossing rule. Von Neumann and Wigner showed, in their seminal work in 1929,[2] that for a molecular system with N^{int} internal nuclear coordinates ($N^{int} = 3N-6$ for nonlinear molecules or $N^{int} = 3N - 5$ for linear molecules), two electronic surfaces become degenerate in a subspace of dimension $N^{int} - 2$. This can be illustrated considering a 2×2 matrix representing the electronic Hamiltonian of a system with two electronic states. We consider two intersecting adiabatic electronic states, ψ_1 and ψ_2, which are expanded in terms of two diabatic states ϕ_1 and ϕ_2,[23]

$$\psi_1 = c_{11}\phi_1 + c_{21}\phi_2, \qquad (1)$$

$$\psi_2 = c_{12}\phi_1 + c_{22}\phi_2. \qquad (2)$$

ϕ_1 and ϕ_2 are orthogonal to all the remaining electronic states and to each other. The adiabatic electronic energies are the eigenvalues of the Hamiltonian matrix:

$$\mathbf{H} = \begin{pmatrix} H_{11} & H_{12} \\ H_{21} & H_{22} \end{pmatrix}, \qquad (3)$$

where $H_{ij} = \langle \phi_i | H | \phi_j \rangle$. The eigenvalues of $\mathbf{H}$ are given by

$$E_{1,2} = \bar{H}_{12} \pm \sqrt{\Delta H_{12}^2/4 + H_{12}^2}, \qquad (4)$$

where $\bar{H}_{ij} = \frac{1}{2}(H_{ii} + H_{jj})$, and $\Delta H_{ij} = H_{ii} - H_{jj}$. The eigenfunctions are

$$\psi_1 = \cos(\alpha/2)\phi_1 + \sin(\alpha/2)\phi_2, \tag{5}$$

$$\psi_2 = -\sin(\alpha/2)\phi_1 + \cos(\alpha/2)\phi_2, \tag{6}$$

where α satisfies

$$\sin\alpha = \frac{H_{12}}{\sqrt{\Delta H_{12}^2/4 + H_{12}^2}}, \tag{7}$$

$$\cos\alpha = \frac{\Delta H_{12}}{2\sqrt{\Delta H_{12}^2/4 + H_{12}^2}}. \tag{8}$$

For the eigenvalues of this matrix to be degenerate, two conditions must be satisfied,

$$H_{11} - H_{22} = 0, \tag{9}$$

$$H_{12} = 0. \tag{10}$$

In an N^{int}-dimensional space, the two conditions are satisfied in an $N^{int}-2$ subspace. This subspace, where the states are degenerate, is called the seam space. The two-dimensional space orthogonal to it, where the degeneracy is lifted, is called the branching or $g-h$ space.[8,23]

Extending the noncrossing rule to three states being degenerate can best be understood by considering a 3×3 electronic Hamiltonian matrix instead of the 2×2 matrix. To illustrate this dimensionality rule, consider three intersecting adiabatic electronic states, ψ_1, ψ_2, ψ_3. These states are expanded in terms of three diabatic states ϕ_1, ϕ_2 and ϕ_3, which are orthogonal to all the remaining electronic states and to each other similarly to the two-state case, and the Hamiltonian matrix is

$$\mathbf{H} = \begin{pmatrix} H_{11} & H_{12} & H_{13} \\ H_{12} & H_{22} & H_{23} \\ H_{13} & H_{23} & H_{33} \end{pmatrix}, \tag{11}$$

where the matrix elements H_{ij} are defined as above. To obtain degeneracy between all three states, the following five requirements must be satisfied: (1) all off-diagonal matrix elements have to be zero, i.e. $H_{12} = H_{13} = H_{23} = 0$, leading to three requirements; (2) the diagonal matrix elements have to be equal, i.e. $H_{11} = H_{22} = H_{33}$, leading to two more requirements.

In general, for an $n \times n$ matrix, n-fold degeneracy is obtained when all diagonal elements are equal and all off-diagonal elements are zero. This can be seen by considering $\mathbf{H}$ as an $n \times n$ symmetric matrix that has n degenerate eigenvalues and $\mathbf{\Lambda}$ the diagonal matrix of eigenvalues λ_i. Then $\mathbf{U}^{-1}\mathbf{H}\mathbf{U} = \mathbf{\Lambda}$ or $\mathbf{H} = \mathbf{U}\mathbf{\Lambda}\mathbf{U}^{-1}$, where $\mathbf{U}$ is a unitary matrix. If $\mathbf{H}$ has n degenerate eigenvalues, then $\mathbf{\Lambda} = \lambda \mathbf{I}$, where $\mathbf{I}$ is the unit matrix. Therefore, $\mathbf{H} = \lambda \mathbf{U}\mathbf{U}^{-1} = \lambda \mathbf{I}$. The conditions for degeneracy then are a combination of $n - 1$ diagonal conditions and $n(n - 1)/2$ off-diagonal conditions. The total number of conditions to be satisfied is $(n - 1) + n(n - 1)/2 = (n - 1)(n + 2)/2$.

For molecules lacking any spatial symmetry and containing four or more atoms, conical intersections of three states are possible in an $N^{int} - 5$ dimensional space. One can see that the dimensionality where a three-state CI can be found is greatly reduced compared to the dimensionality of the two-state CI ($N^{int} - 2$), and this leads to the belief that three-state CIs will be extremely rare. For example, if a two-state seam is a 3D surface, the three-state CI will be just a point. Contrary to this belief, however, recent work has shown that they are much more common than believed, as will be discussed later in this chapter.

It should be noted that the dimensionalities derived above are for a nonrelativistic Hamiltonian ignoring the spin-orbit coupling. If the spin-orbit coupling is included, the dimensionality will change for a system with an odd-number of electrons.[24, 25] This is because of Kramers' degeneracy in these systems and because the spin-orbit operator leads to complex matrix elements.

3. Branching and Seam Spaces

The branching space[23] is defined as the space in which the degeneracy is lifted linearly, and for the three-state conical intersections, this space is five-dimensional. In order to understand the behavior of the energy in the branching and seam spaces, it is useful to expand the energy in a Taylor series in the neighborhood of the crossing. By subtracting H_{22} from the diagonal elements in Eq. (11), we obtain

$$\mathbf{H} = H_{22}\mathbf{I} + \begin{pmatrix} H_{11} - H_{22} & H_{12} & H_{13} \\ H_{12} & 0 & H_{23} \\ H_{13} & H_{23} & H_{33} - H_{22} \end{pmatrix}. \quad (12)$$

We assume a degeneracy at $\mathbf{R}_0$. At a nearby point $\mathbf{R} = \mathbf{R}_0 + \delta\mathbf{R}$, the matrix elements of the Hamiltonian, when expanded in a Taylor expansion to first order around the point of conical intersection $\mathbf{R}_0$, can be written as[23,26]

$$\Delta H_{ij}(\mathbf{R}) = H_{ii}(\mathbf{R}) - H_{jj}(\mathbf{R}) = 0 + \nabla(\Delta H_{ij})(\mathbf{R}_0) \cdot \delta\mathbf{R} \qquad (13)$$

$$H_{ij}(\mathbf{R}) = 0 + \nabla H_{ij}(\mathbf{R}_0) \cdot \delta\mathbf{R}. \qquad (14)$$

The requirements for a three-state conical intersection at $\mathbf{R}$ then become

$$\nabla(\Delta H_{ij}) \cdot \delta\mathbf{R} = 0 \qquad (15)$$

$$\nabla H_{ij} \cdot \delta\mathbf{R} = 0, \qquad (16)$$

so that $\delta\mathbf{R}$ must be orthogonal to the subspace spanned by the vectors $\nabla(\Delta H_{12})$, $\nabla(\Delta H_{32})$, and ∇H_{ij}, where $ij = 12, 23, 13$, for the degeneracy to remain. The degeneracy is lifted linearly, as seen in Eqs. (13) and (14), in the 5-dimensional branching space defined by these five vectors. The subspace which is orthogonal to the 5-dimensional branching space is the seam space. The five branching vectors represent nuclear motion similar to the normal modes of molecules, although they are not normal modes but in general some combination of them. Examples of the five vectors for an accidental three-state CI in a molecule are shown in Fig. 1.

4. Symmetry and Three-State Conical Intersections: The Jahn–Teller Effect

Although accidental three-state CIs are relatively new features, three-fold degeneracy due to symmetry has been known and studied for many years.[27,28] It is instructive to see how these symmetry-required CIs behave before moving to the more complicated accidental ones. Systems with tetrahedral, octahedral or icosahedral symmetry point groups $(T, T_d, T_h, O, O_h, I, I_h)$ can have three-fold degenerate states, T_1 or T_2. Vibronic effects for T_1 or T_2 are similar and the results obtained for one case can be transfered to the other. In this case all the conditions for degeneracy are satisfied by symmetry, and the dimensionality of the seam space is equal to the number of degrees of freedom that retain the high symmetry. The Jahn–Teller (JT) theorem states that these molecules will distort from the symmetrical configuration to low-symmetry configurations. In T_d and O_h point groups, there are five active JT coordinates which have e or t_2 symmetry, and the Jahn–Teller problem is defined as $T \otimes (e + t_2)$.

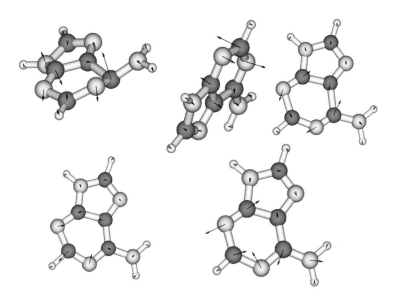

Fig. 1. Five vectors defining the branching space for an $S_1/S_2/S_3$ conical intersection in adenine. The three top vectors correspond to the coupling vectors and the two bottom ones to the energy difference gradient vectors. (Reproduced from Ref. 54 with permission.)

In the I_h point group, the Jahn–Teller active modes are combined in a five-dimensional h vibration. The first step in analyzing a JT system is concentrating on the static problem and the electronic Hamiltonian in order to find the new minima caused by the distortion. This is a complicated problem when three electronic states and five modes are involved. Usually the problem is studied first by including only the linear terms in the Hamiltonian.

Öpic and Pryce considered the $T \otimes (e+t_2)$ JT problem in an octahedral system.[27] In general, in the formulation of the JT problem, the diabatic electronic states are employed, where the nuclear kinetic energy is diagonal while the off-diagonal coupling terms are in the potential energy operator. The diabatic states are the degenerate electronic states at high symmetry. In the case of the $T \otimes (e+t_2)$ JT problem, the coupling matrix is given by[28]

$$\mathbf{W} = \begin{pmatrix} F_E(\tfrac{1}{2}Q_\theta - \tfrac{\sqrt{3}}{2}Q_\epsilon) & -F_T Q_\zeta & -F_T Q_\eta \\ -F_T Q_\zeta & F_E(\tfrac{1}{2}Q_\theta + \tfrac{\sqrt{3}}{2}Q_\epsilon) & -F_T Q_\xi \\ -F_T Q_\eta & -F_T Q_\xi & -F_E Q_\theta \end{pmatrix}. \quad (17)$$

Q_i are normal modes, where Q_θ and Q_ϵ have e symmetry and Q_η, Q_ζ, Q_ξ have t_2 symmetry. F_i are the linear vibronic constants obtained by differentiating the diagonal and off-diagonal matrix elements of the potential operator.

Further simplification is achieved if the problem is divided into two cases, which are studied separately first and then combined, as Öpic and Pryce showed.[27] The e modes are considered first in a $T \otimes e$ problem and the t_2 modes are considered separately in a $T \otimes t_2$ problem. Solving these two cases separately, one finds three minima for the $T \otimes e$ system displaced along Q_θ and Q_ϵ, and four minima for the $T \otimes t_2$ case displaced along Q_η, Q_ζ, Q_ξ. When one solves the combined $T \otimes (e + t_2)$ system, in addition to the $3 + 4 = 7$ stationary points, there are six more orthorhombic saddle points. The seven initial stationary points can be minima depending on the values of the various parameters in the Hamiltonian, but the six intermediate points can never be minima; they are higher-order saddle points. Quadratic terms have also been considered for this problem as well as the effect of the spin-orbit coupling, adding complexity to the solutions.[27-32]

In icosahedral systems, one finds JT problems with three, four or fivefold degeneracy. In the threefold degeneracy, a fivefold degenerate mode h interacts with the triply degenerate T state, giving rise to the $T \otimes h$ JT problem.[33-36] The $T \otimes h$ problem is similar to the $T \otimes (e + t)$ when the vibronic couplings to the e and t vibrations and their frequencies are the same. It can be shown that in the linear approximation, the lowest branch of the PES has a two-dimensional trough of minima points in the five-dimensional space, similar to the $E \otimes e$ "mexican hat" case, where a one-dimensional trough exists.[28]

5. First Accidental Three-State Conical Intersection: CH_4^+

Although the JT effect on T states was analyzed early on, decades passed before there was any indication of three-state degeneracy not imposed by symmetry. The first study on accidental three-state conical intersections was done for the CH_4^+ cation by Katriel and Davidson using Frost functions.[15] In a tetrahedral geometry, the ground state of CH_4^+ is a T_2 state. Therefore symmetry predicts that the ground state will be triply degenerate. Because of the Jahn–Teller effect, there is distortion that lowers the symmetry and produces six equivalent minima. The rovibronic structure of methane cation in these six minima has been studied experimentally and

Table 1. Number of coordinates for various conditions in CH_4^+, when the system is in T_d, C_{3v}, C_s and C_1 symmetry.

	T_d	C_{3v}	C_s	C_1
No. of coord. specified by symmetry	8	6	3	0
No. of degrees of freedom	1	3	6	9
No. of conditions satisfied by symmetry	5	4	2	0
No. of conditions to satisfy for degeneracy	0	1	3	5
Dimensionality of seam	1	2	3	4

Taken from Ref. 15.

theoretically, confirming the C_{2v} geometry of CH_4^+ and the presence of six minima connected by tunneling motion.[37–41]

The tetrahedral symmetry is preserved along only one degree of freedom, the totally symmetric vibration, and the dimensionality of the seam is one, since all the requirements for degeneracy are satisfied by symmetry. Katriel and Davidson, however, found additional threefold degeneracies in this system even when the tetrahedral symmetry is broken. More specifically, points of threefold degeneracy were found for CH_4^+ in T_d, C_{3v}, C_s symmetry and when no symmetry was present. Table 1, taken from the original reference, shows the dimensionality of the seam space, where degeneracy is expected to remain, in each case. As the symmetry is lowered, the number of coordinates specified by symmetry is reduced and the number of degrees of freedom which can be varied without changing the symmetry increases. For example, in T_d symmetry, there is only one coordinate that can be varied while the symmetry is retained, the CH symmetric stretch, thus the number of degrees of freedom has been reduced from nine to one. The number of degrees of freedom in intermediate symmetries, C_{3v} and C_s, is three and six, respectively. In T_d symmetry, the dimensionality of the seam space is also one, since any change that retains the T_d symmetry will retain the degeneracy. In the other extreme, if no symmetry is present, the cation has nine degrees of freedom and the dimensionality of the seam becomes $9-5=4$, since five constraints need to be satisfied. In intermediate symmetries, some of the constraints are satisfied by symmetry, and thus the seam dimensionality is equal to the number of degrees of freedom minus the number of constraints. For example, in C_{3v} symmetry, the T state transforms as $A_1 + E$, so there is still a twofold degeneracy imposed by symmetry in the E state. There is only one condition required to achieve a threefold degeneracy again, namely that the energy of state A_1 is equal to the energy of state E. The off-diagonal coupling matrix element between A_1 and E

states is zero because of their different symmetry. So the dimensionality of the three-state seam within C_{3v} symmetry is $3 - 1 = 2$. In C_s symmetry, the T state splits into $2A' + A''$ and there are three conditions to be satisfied to obtain threefold degeneracy: all diagonal matrix elements have to be equal, and the coupling between the two A' states has to vanish. So the dimensionality of the three-state seam within C_s symmetry is $6 - 3 = 3$. Katriel and Davidson were able to obtain threefold degeneracies in each of these cases, showing that this is possible even when the symmetry does not impose it.

6. Locating Accidental Three-State Conical Intersections

Besides the study of the methane cation described above,[15] until 2002 there had been no other study or evidence of accidental three-state degeneracy without the presence of a Jahn–Teller effect. Such degeneracies were considered highly unlikely to occur, and based on the dimensionality, it would be very difficult, if not impossible, to locate them without any algorithms designed to do so. Degeneracy of three states can be located using similar ideas and algorithms as have been used for two-state degeneracies, which have been found to be successful over the years.

The first to be developed, and most frequently used, algorithm for locating three-state CIs[16] uses the lagrange multiplier method developed initially for two states.[42] The basics of this algorithm are discussed below. The electronic wavefunctions are expressed as a linear combination of Configuration State Functions (CSFs), Θ_m,

$$\Phi_I(\mathbf{r};\mathbf{R}) = \sum_{m=1}^{N^{CSF}} c_m^I(\mathbf{R})\Theta_m(\mathbf{r};\mathbf{R}), \qquad (18)$$

where N^{CSF} is the number of CSFs used in the expansion, $c_m^I(\mathbf{R})$ are the expansion coefficients for state I, $\mathbf{r}$ are electronic coordinates and $\mathbf{R}$ are nuclear coordinates. The adiabatic electronic energies and wavefunctions are obtained by solving the eigenvalue equation

$$[\mathbf{H}(\mathbf{R}) - E_I(\mathbf{R})\mathbf{I}]\mathbf{c}^I(\mathbf{R}) = \mathbf{0}, \qquad (19)$$

where the Hamiltonian $\mathbf{H}$ is built in the CSF basis.

Using quasidegenerate perturbation theory[43,44] in the vicinity of a conical intersection for three states I, J, K, the electronic Hamiltonian can be written to first order in displacements $\delta\mathbf{R}$ from the point of degeneracy $\mathbf{R}^x$ as

$$\mathbf{H}^{(1)}(\mathbf{R}) = [E_J(\mathbf{R}^x) + \mathbf{g}^J \cdot \delta\mathbf{R}]\mathbf{I}$$
$$+ \begin{pmatrix} \mathbf{g}^{IJ} \cdot \delta\mathbf{R} & \mathbf{h}^{IJ} \cdot \delta\mathbf{R} & \mathbf{h}^{IK} \cdot \delta\mathbf{R} \\ \mathbf{h}^{IJ} \cdot \delta\mathbf{R} & 0 & \mathbf{h}^{JK} \cdot \delta\mathbf{R} \\ \mathbf{h}^{IK} \cdot \delta\mathbf{R} & \mathbf{h}^{JK} \cdot \delta\mathbf{R} & \mathbf{g}^{KJ} \cdot \delta\mathbf{R} \end{pmatrix}, \qquad (20)$$

where the vectors $\mathbf{g}^I, \mathbf{g}^{IJ}$ and $\mathbf{h}^{IJ}$ are the gradient of state I, the energy difference gradient between states I, J and the coupling, respectively. They are defined as

$$\mathbf{g}^I = \nabla \langle \Phi_I | \hat{H} | \Phi_I \rangle, \qquad (21)$$
$$\mathbf{g}^{IJ} = \nabla (E_I - E_J) = \nabla (\langle \Phi_I | \hat{H} | \Phi_I \rangle - \langle \Phi_J | \hat{H} | \Phi_J \rangle), \qquad (22)$$
$$\mathbf{h}^{IJ} = \langle \Phi_I | \nabla \hat{H} | \Phi_J \rangle, \qquad (23)$$

or in matrix notation in terms of the CSF coefficients the vectors:

$$\mathbf{g}^I = c^{I,\dagger}(\mathbf{R}^x) \nabla \mathbf{H}(\mathbf{R}) c^I(\mathbf{R}^x), \qquad (24)$$
$$\mathbf{g}^{IJ} = \mathbf{g}^I - \mathbf{g}^J, \qquad (25)$$
$$\mathbf{h}^{IJ} = c^{I,\dagger}(\mathbf{R}^x) \nabla \mathbf{H}(\mathbf{R}) c^J(\mathbf{R}^x), \qquad (26)$$

where ∇ denotes the gradient with respect to nuclear coordinates only. At $\mathbf{R} = \mathbf{R}^x + \delta\mathbf{R}$, the Hamiltonian in the adiabatic representation becomes

$$\mathbf{H}^{(1)}(\mathbf{R}) = (E_J(\mathbf{R}))\mathbf{I} + \begin{pmatrix} \Delta E_{IJ}(\mathbf{R}) & 0 & 0 \\ 0 & 0 & 0 \\ 0 & 0 & \Delta E_{KJ}(\mathbf{R}) \end{pmatrix}. \qquad (27)$$

In order to achieve degeneracy, a displacement $\delta\mathbf{R}$ should be taken such that

$$\Delta E_{IJ} + \mathbf{g}^{IJ} \cdot \delta\mathbf{R} = \Delta E_{JK} + \mathbf{g}^{JK} \cdot \delta\mathbf{R} = 0 \qquad (28)$$

and

$$\mathbf{h}^{IJ} \cdot \delta\mathbf{R} = \mathbf{h}^{JK} \cdot \delta\mathbf{R} = \mathbf{h}^{IK} \cdot \delta\mathbf{R} = 0 \qquad (29)$$

for I, J, K. These are the five conditions that need to be satisfied. The problem becomes a constrained minimization which can be solved using the Lagrange-constrained minimization method. Since there is an infinite number of degenerate points, one usually seeks the minimum energy point on the seam. The algorithms seek to minimize the energy based on two, five or more constraints. Alternatively, instead of minimizing the energy of one state, the average of the states involved can be minimized.[45,46] The algorithms developed for the two-state case[42] can be extended to include five constraints. A Lagrangian is built

$$L(\mathbf{R}, \boldsymbol{\xi}, \boldsymbol{\zeta}, \boldsymbol{\lambda}) = E_J + \xi_1 \Delta E_{IJ} + \xi_2 \Delta E_{KJ} + \zeta_1 H_{IJ}$$
$$+ \zeta_2 H_{JK} + \zeta_3 H_{IK} + \sum_i \lambda_i K_i, \qquad (30)$$

where K_i are additional geometrical constraints that can be added to the problem. These constraints are used when we not only are interested in the minimum energy point on the seam but also need to map out some portion of the seam.

An extremum is sought for the Lagrangian by requiring its gradient with respect to $\mathbf{R}, \boldsymbol{\xi}, \boldsymbol{\zeta}, \boldsymbol{\lambda}$ to vanish. By searching for extrema of the Lagrangian, a Newton–Raphson equation can be set up,

$$\begin{bmatrix} \mathbf{L}^{RR} & \mathbf{g}^{IJ,\dagger} & \mathbf{g}^{KJ,\dagger} & \mathbf{h}^\dagger & \mathbf{k}^\dagger \\ \mathbf{g}^{IJ} & 0 & 0 & 0 & 0 \\ \mathbf{g}^{KJ} & 0 & 0 & 0 & 0 \\ \mathbf{h} & 0 & 0 & 0 & 0 \\ \mathbf{k} & 0 & 0 & 0 & 0 \end{bmatrix} \begin{bmatrix} \delta \mathbf{R} \\ \delta \xi_1 \\ \delta \xi_2 \\ \delta \boldsymbol{\zeta} \\ \delta \boldsymbol{\lambda} \end{bmatrix} = - \begin{bmatrix} \nabla L \\ \Delta E_{IJ} \\ \Delta E_{KJ} \\ \mathbf{0} \\ \mathbf{K} \end{bmatrix}, \qquad (31)$$

which, when solved, provides the solution $\delta \mathbf{R}$. The following relations have been used:

$$L^{RR} = \nabla(\nabla L), \frac{\partial^2 L}{\partial R_i \partial \xi_1} = (g^{IJ})_i, \frac{\partial^2 L}{\partial R_i \partial \xi_2} = (g^{KJ})_i, \frac{\partial^2 L}{\partial R_i \partial \zeta_1} = (h^{IJ})_i,$$

$$\frac{\partial^2 L}{\partial R_i \partial \zeta_2} = (h^{JK})_i, \frac{\partial^2 L}{\partial R_i \partial \zeta_3} = (h^{IK})_i, \nabla \mathbf{K} = \mathbf{k}, \mathbf{h} = (\mathbf{h}^{IJ}, \mathbf{h}^{JK}, \mathbf{h}^{IK})$$

and

$$\nabla L = \nabla E_J + \xi_1 \mathbf{g}^{IJ} + \xi_2 \mathbf{g}^{KJ} + \zeta_1 \mathbf{h}^{IJ} + \zeta_2 \mathbf{h}^{JK} + \zeta_3 \mathbf{h}^{IK} + \sum_i \lambda_i \mathbf{k}_i. \qquad (32)$$

This method has been implemented using analytic gradients from multireference configuration interaction (MRCI) wavefunctions[47–50] and has been added to the COLUMBUS suite of programs.[51] It has proven to be a very efficient algorithm capable of locating three-state degeneracies in a few iterations. This algorithm has been used in most of the three-state accidental conical intersections reported to-date.

It is often desirable to have algorithms for locating conical intersections without the need of a derivative coupling vector, since these are difficult to compute in many cases. There has been work in this direction, particularly for locating the common two-state CIs. Martinez and co-workers have developed such a method and extended it to the optimization of three-state CIs.[52] The algorithm for obtaining a two-state CI minimizes an objective function

$$F_{IJ}(\mathbf{R};\sigma,\alpha) = \bar{E}_{IJ}(\mathbf{R}) + \sigma G_{IJ}(\Delta E_{IJ}(\mathbf{R});\alpha), \quad (33)$$

where $\bar{E}_{IJ}$ is the average energy of the two states I, J, ΔE_{IJ} is their energy difference, σ is a sequentially updated parameter driving the optimization towards the seam space minimum, and G_{IJ} is a penalty function used to smooth discontinuities in the gradient of the potential surface. G_{IJ} is a monotonically increasing function of the energy gap, and it has the form

$$G_{IJ}(\Delta E_{IJ};\alpha) = \frac{\Delta E_{IJ}^2}{\Delta E_{IJ} + \alpha}. \quad (34)$$

α is a user-defined smoothing parameter. Minimization of Eq. (33) corresponds to minimizing the average energy of states I and J, subject to the constraint that the gap between the states vanishes. Extending this to locating a three-state CI is achieved by minimizing the objective function

$$F_{IJK}(\mathbf{R};\sigma,\alpha) = F_{IJ}(\mathbf{R};\sigma,\alpha) + F_{IK}(\mathbf{R};\sigma,\alpha) + F_{JK}(\mathbf{R};\sigma,\alpha). \quad (35)$$

This algorithm has been used to locate three-state CIs in malonaldehyde as discussed later in this chapter.

7. Consequences of Three-State Conical Intersections

7.1. Connectivity

Displacements along the five-dimensional branching space of a three-state CI lift the degeneracy. It is possible, however, to keep two of the three states degenerate and lift the degeneracy only of the third state. This can occur in three-dimensional subspaces of the five-dimensional branching space.[17,18,20] Within the branching space, there exist two such three-dimensional degeneracy subspaces in which the degeneracy between two states remains. For example, in an $S_0/S_1/S_2$ CI, one can keep the degeneracy between the lower pair of states S_0/S_1 or between the upper pair of states S_1/S_2. These two-state CI seams can be referred to as linked seams since they have a common state between them. Alternatively, one can view this as the two linked seams intersecting to form the three-state CI.

The degeneracy spaces originating from three-state CIs were originally studied by Keating and Mead,[18] where they focused on the X_4 system. Symmetry in this system can guide the analysis. In T_d symmetry, there exists a threefold degeneracy which will lead to twofold degeneracy when the symmetry is lowered to C_{3v} or D_{2d}. Thus displacements from the T_d symmetry will cause lifting of the threefold degeneracy but this can be done in a way that the two states remain degenerate. Yarkony and co-workers[20, 53] have also examined these linked seams analytically using perturbation theory.

These two-state CI seams originating from the three-state CI have been confirmed in several systems using *ab initio* energies and wavefunctions. They were initially shown using *ab initio* MRCI energies in the allyl radical.[17, 20] In this radical, a three-state CI between states $4\,^2A$, $5\,^2A$, $6\,^2A$ was found. Pathways from a point on the three-state CI seam along the $4,5\,^2A$ seam and $5,6\,^2A$ seam were calculated. These seams, along with similar ones in pyrazolyl radical, were used to numerically test the above-mentioned perturbation theory,[20, 53] thus providing a connection between analytical results and *ab initio* calculations.

In larger polyatomic molecules, the situation starts getting more complicated since there are more than one three-state CI seams, as will be discussed in Sec. 9.2. Then different seams of two-state conical intersections originate from each of the three-state conical intersections, leading to a great number of two-state conical intersections at energies lower than the three-state seams.[54, 55] The connectivity of different two-state seams through three-state CIs has been seen in adenine and cytosine,[54, 55] and has been explored in more detail in cytosine. Cytosine has multiple seams of two-state and three-state CIs. Two three-state CIs have been found to involve the ground state $(S_0/S_1/S_2)$ and one to involve $S_1/S_2/S_3$. The $S_0/S_1/S_2$ seams are discussed here as different seams because they involve different diabatic states and the molecule has a very different structure, but we cannot exclude the possibility that they are actually connected in one seam and what we have found are just different stationary points on one seam. Figures 2 and 3 show various pathways connecting seams for cytosine. Figure 2 shows an S_0/S_1 seam starting from an $S_0/S_1/S_2$ CI point (*ci*012). The two coordinates used in the plot are

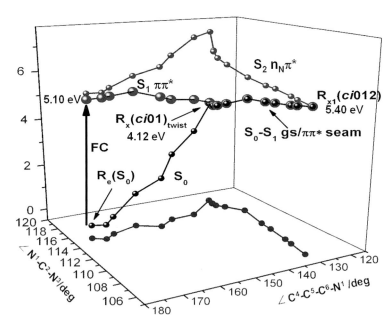

Fig. 2. S_0-S_1 seam path from cytosine's three-state CI ci012 to a two-state CI minimum $R_x(ci01)_{\text{twist}}$ and gradient pathway from $R_x(ci01)_{\text{twist}}$ to S_0 minimum. The first three singlet energies are plotted with respect to the dihedral angle $C^4C^5C^6N^1$ and the angle $N^1C^2N^3$. The Franck–Condon (FC) region is shown with a vertical arrow. The gray points on the graph floor are the projection of data onto the $E = -0.5$ eV plane. (Reproduced from Ref. 55 with permission.)

both branching coordinates for the three-state CI, but one is a seam coordinate for S_0/S_1 (N^1-C^2-N^3) and the other is a branching coordinate for S_0/S_1 (C^4-C^5-C^6-N^1). Figure 3 shows two different paths. First starting from ci012 the S_1/S_2 seam is followed while S_0 is not degenerate anymore. In the same figure, there is also a path starting from a different three-state CI, an $S_1/S_2/S_3$ CI (ci123), and following the S_1/S_2 seam. Interestingly, both paths end up at the same point which is the minimum on the S_1/S_2 seam [$R_x(ci12)'$]. A third path has also been drawn on the same figure which connects the minimum energy point on the S_1/S_2 seam to the vertical excitation. This shows clearly how all these seams are connected to each other and ultimately to the Franck–Condon region, suggesting their relevance to the photophysical behavior of cytosine.[56]

Martinez and co-workers support the conjecture that three-state intersections play a key role in extending the connectivity of intersection spaces of pairs of states, e.g. the S_1/S_0 and S_2/S_1 intersection spaces.[57] They have

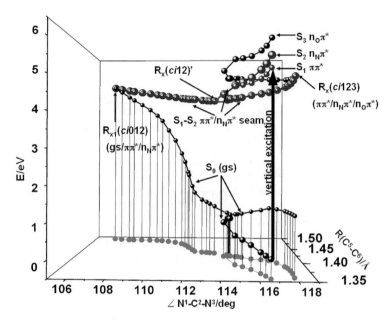

Fig. 3. S_1-S_2 seam paths from cytosine's three-state CIs $ci012$ and $ci123$, and from vertical excitation. The first three to four singlet energies are plotted with respect to the R(C^5C^6) in Å and the angle $N^1C^2N^3$ in degrees. The Franck–Condon (FC) region is shown with a vertical arrow. The gray points on the graph floor are the projection of data onto the $E = -0.5$ eV plane. (Reproduced from Ref. 55 with permission.)

used malonaldehyde to show connectivity of very different two-state seams that support this conjecture. Figure 4 shows a two-state seam S_1/S_0 that starts close to the Franck–Condon region, passes through a three-state CI and continues to reach a completely different region where the S_1/S_0 states are still degenerate. The torsional angle changes along the seam from 0° to 180°. A search for stationary points between S_1/S_0 in this case will locate two points, the HTI at torsional angle 0° and the ME-3SI at angle torsional 90°, as seen in Fig. 4, and it is not obvious that these two points are, or should be, connected. Furthermore, they seem to lead to different distortions and dynamics. Interestingly, even though the two extreme points HTI and ME-3SI have been located at 0° and 90°, dynamical studies reveal that population transfer does not occur at these points but mainly at angles in between. This points to the fact that we should always be careful when we assign too much significance to minimum energy points on a seam. Mapping seams may be a lot more informative, and three-state CIs are very helpful in connecting the different regions of conformational space.

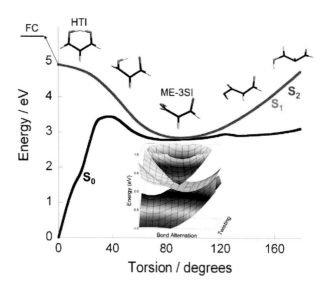

Fig. 4. Connectivity of the hydrogen transfer intersection (HTI) and minimal energy three-state intersection (ME-3SI) points. Constrained relaxation in 5° intervals from the planar to 90° twisted geometries maintains the degeneracy of S_2 with S_1 at all points along the connecting pathway. The Franck–Condon (FC) point is located at 0° torsion, slightly higher in energy than the HTI. The inset shows an approximate view of the surface topography in the vicinity of the ME-3SI. (Reproduced from Ref. 57 with permission.)

7.2. Geometric phase

It was first pointed out by Longuet–Higgins and Herzberg[58,59] that a real electronic wavefunction changes sign when traversing around a conical intersection. Mead and Truhlar[60] incorporated this geometric phase effect into the single electronic state problem and Berry generalized the theory.[61] Because of his work, this effect is often called the Berry phase effect.[61–63] Calculations in Na_3 trimer, which use the single-surface adiabatic approach, reproduce the experimental data only if, as required by theory, a geometric phase of π under pseudorotation around the equilateral configuration is imposed.[62] The geometric phase effect has in general important consequences in the vibronic levels in a JT system.[28] Effects of the geometric phase have been observed theoretically not only for bound states but also for scattering wavefunctions. For the X_3 system, it has been shown that inclusion of this effect leads to scattering changes in the sign of reactive scattering, which can alter the conclusions regarding the interference between reactive and nonreactive scattering.[64] The geometric phase effect can be cancelled when product distributions are integrated over all scattering angles.[65–67] For a more comprehensive discussion see Chapter 5 in this book.

Although the geometric phase effect in a twofold degeneracy is the most commonly studied, there has been work generalizing to three or n-fold degeneracy.[19,68] It was shown that for a threefold degeneracy of a real 3×3 Hamiltonian, the geometric phase factors $\sigma_i = e^{i\gamma_i}$ of the eigenfunctions $i = 1, 2, 3$ after traversing a circle around the degeneracy will be $\sigma_2 = 1$ and $\sigma_1 = \sigma_3 = \pm 1$. These geometric phases are related to the number of conical intersections contained in the paths.[20,21] A wavefunction i changes sign when traversing a path if the path encloses one CI (or an odd number of CIs) between state i and another state. The wavefunction will not change sign if the path encloses two or an even number of CIs involving state i. Thus, if the path encloses the two linked seams $1, 2$ and $2, 3$, wavefunctions 1 and 3 are involved in one CI each and will change sign but wavefunction 2 is involved in two CIs and will preserve its sign.

In a three-state CI beyond the geometric phase effect on the wavefunction, there is additional geometric phase effect on the derivative coupling, which becomes also double valued and changes sign when a conical intersection is encircled. This is a consequence of the wavefunction phases and will be discussed further in the next section.

7.3. Nonadiabatic couplings

The efficiency of a radiationless transition between two states depends not only on the energy difference between those states, but also on the derivative coupling $\mathbf{f}_{IJ}$ of the states. The nonadiabatic or derivative coupling between electronic states I, J is defined as

$$\mathbf{f}_{IJ} = \langle \Phi_I | \nabla \Phi_J \rangle. \tag{36}$$

In the adiabatic representation, the derivative coupling is responsible for nonadiabatic transitions between different states. The diabatic representation is formally defined by setting the derivative coupling equal to zero. In reality, the equation $\mathbf{f}_{IJ} = 0$ has no solution since there is a nonremovable part. At conical intersections, the singular part is removable, and efficient ways to transform between the diabatic and adiabatic representations can use this property.[69–71]

In a regular CI, the wavefunctions change sign if they traverse in a loop around the point of CI, but the derivative coupling does not change

Table 2. Sign of the derivative couplings when one or more CIs are enclosed within a loop.

Seams contained within loop	f_{10}	f_{20}	f_{21}
S_0-S_1 seam only	+	−	−
S_0-S_1 seam plus one S_1-S_2 seam	−	+	−
S_0-S_1 seam plus two S_1-S_2 seams	+	−	−

"+" indicates overall phase retention along the loop, and "−" overall phase inversion along the loop.

sign. This is because the derivative coupling includes a product of both functions, and, since both of them change sign, this will cancel the signs in the product. The situation is more complicated when the loop encloses more than one seam, as is the case in the close vicinity of a three-state CI. If the loop contains two CIs, where both include the same state, for example S_0/S_1 and S_1/S_2, then the wavefunction describing S_1 will change sign twice, once because of S_0/S_1 and once because of S_1/S_2. The product involved in the derivative coupling then will also change sign. In fact around the three-state CI, there are three different derivative couplings, f_{10}, f_{20} and f_{12}. Table 2 shows different cases where one or more CIs are enclosed in the loop and how this affects the sign of the derivative couplings.[20,55,72]

These sign changes were first analyzed by *ab initio* wavefunctions and derivative couplings for the pyrazolyl radical,[72] although they had been seen earlier too.[73] Figure 5 shows one component of the three derivative coupling vectors around a loop that encloses two conical intersections, one between states 1,2 and one between states 2,3. The derivative couplings f_{12} and f_{23} are double-valued (change sign around the loop) while f_{13} is not.

The sign changes in derivative couplings have also been tested in cytosine using *ab initio* MRCI wavefunctions.[55] Figure 6 shows the vicinity of a three-state conical intersection where one can see one S_0/S_1 point of degeneracy and two disconnected points of S_1/S_2 degeneracy. By making loops of different radius around the S_0/S_1 point, we can tune whether additional conical intersection points are enclosed in the loop. If the radius is very small, only the S_0/S_1 point is encircled. If the radius increases, we can encircle only one of the S_1/S_2 points or both of them. Figure 7 shows the major component f_{IJ}^θ of the *ab initio* derivative couplings for different loops around S_0/S_1 in cytosine. In Fig. 7(a), the radius is only 0.01 bohr and only the S_0/S_1 CI is enclosed. In this case the derivative coupling f_{10}^θ does not change sign. In Fig. 7(b) the radius increases to 0.03 bohr but still there is

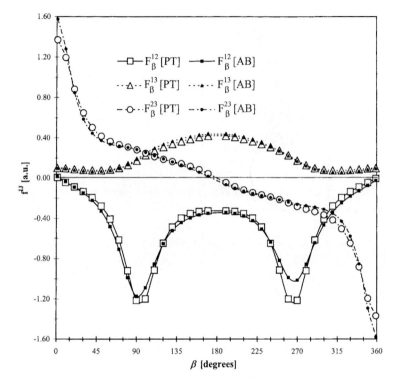

Fig. 5. β component of the derivative coupling, $f_\beta^{I,J}$, at points on a loop encircling conical intersections between states 1,2 and 2,3 on the pyrazolyl radical. Open markers represent perturbation theory results, [PT], closed markers indicate *ab initio* data, [AB]. (Reproduced from Ref. 72 with permission.) Notice that notation in the figure is switched from the text here (f_{IJ}^β vs. $f_\beta^{I,J}$).

only one CI enclosed. More interesting is Fig. 7(c) where the radius is 0.05 bohr. Now in addition to the S_0/S_1 CI, one of the S_1/S_2 CIs is enclosed. According to Table 2, f_{10}^θ will change sign and this is seen in the plot. In Fig. 7(d), the radius increases even more and both S_1/S_2 CIs are enclosed in the loop. In this case, f_{10}^θ changes sign twice so it is again single-valued in the end.

The double-valued derivative coupling is a very interesting consequence of the three-state CIs and it can be confirmed with the *ab initio* data as seen here. How this characteristic will affect the dynamical behavior around the three-state CIs is still unknown. What is already obvious is that this induced geometric effect can be used to locate conical intersections.

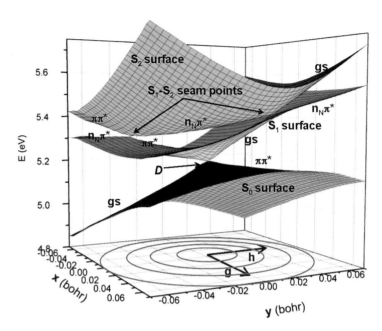

Fig. 6. S_0, S_1, and S_2 PESs in the branching space of a two-state CI for cytosine, generated from the energies calculated on loops with radii $\rho = 0.01$, 0.03, 0.05, and 0.07 bohr, shown on the floor of the plot. The dominant diabatic characters of the surfaces are shown in various regions. (Reproduced from Ref. 55 with permission.)

A further consequence of the double-valued derivative coupling is that it alters the line integral of $\mathbf{f}_{IJ}$ along a path Γ[21,74]

$$A = \int_\Gamma \mathbf{f}_{IJ} \cdot d\mathbf{R}. \tag{37}$$

The line integral of $\mathbf{f}_{IJ}$ along a path can be used to detect or verify the presence of a two-state CI between states I, J. In this case it has a value of $\pm\pi$ independent of the path. If there is no CI, the line integral is 0. However, in the presence of three-state CIs and double-valued $\mathbf{f}$, this is not true anymore. It can be shown that the line integral is not a unique function of the path.[21]

8. Characterizing Three-State Conical Intersections

Two-state conical intersections have been successfully analyzed and characterized using degenerate perturbation theory.[8,43,44] Analytic expressions

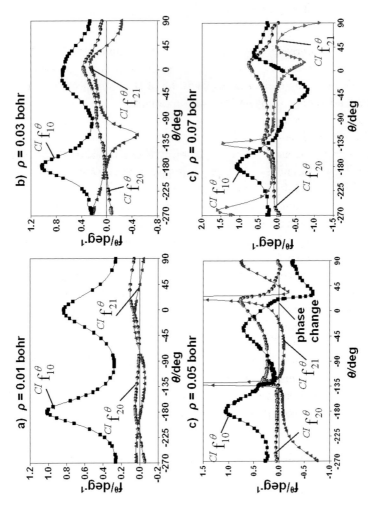

Fig. 7. Derivative couplings f_{IJ} along four branching plane loops around a point of CI. $^{CI}f^{\theta}_{I,J}$ designates the θ component of the configuration interaction *ab initio* derivative coupling. (a) $\rho = 0.01$ bohr, (b) $\rho = 0.03$ bohr, (c) $\rho = 0.05$ bohr, and (d) $\rho = 0.07$ bohr. (Adapted from Ref. 55 with permission.)

have been developed, which give both a computational advantage and pedagogical insight. For example, if only first-order terms are included in the two-state CI, one obtains the familiar form of the Hamiltonian

$$\mathbf{H} = (s_x x + s_y y)\mathbf{I} + \begin{pmatrix} gx & hy \\ hy & -gx \end{pmatrix}, \tag{38}$$

where x, y are the intersection adapted coordinates defining the branching plane.[8,23] The parameters s_x, s_y, g, h can be obtained from the gradients in Eqs. (21)–(23), and they define the topography of the conical intersection. An important advantage of using intersection-adapted coordinates is that instead of using N^{int} coordinates, only two are needed to describe the CI, or five to describe a three-state CI.

This analysis is not as easy for three-state intersections. Yarkony and co-workers applied degenerate perturbation theory through second order to characterize the vicinity of a three-state conical intersection.[20,53,72,75] They used a group homomorphism approach to make progress in this problem. Their approach produces an approximately diabatic Hamiltonian whose eigenenergies and eigenstates can accurately describe the three adiabatic potential energy surfaces, the interstate derivative couplings, and the branching and seam spaces in their full dimensionality.[72] Deriving diabatic Hamiltonians is very important for being able to do dynamics around conical intersections, since dynamics studies in the diabatic representation are easier and they do not deal with the singular derivative coupling vector. Ten parameters are needed to describe the linear terms (analogous to the four s_x, s_y, g, h parameters in the two-state CI) and over 300 parameters for the second-order terms. The accuracy of the perturbation theory can be seen in Fig. 5 where a component of the derivative coupling derived from perturbation theory is compared to *ab initio* values.

9. Three-State Conical Intersections in Chemical Systems

9.1. *Three-state conical intersections in radicals*

Since the development of the algorithms for locating accidental three-state CIs, these features have been found in many systems. The first systems where accidental three-state conical intersections were located using the search algorithms were radicals. Specifically, three-state CIs were first located between Rydberg states in the ethyl and allyl radicals.[16,17] Rydberg states are close energetically even at vertical excitations, so small distortions can lead to degeneracy. In the ethyl radical, the third, fourth and fifth excited states correspond to the $3p$ Rydberg states at vertical excitation and form a three-state CI seam with a minimum energy about 5500 cm^{-1} above vertical excitation to $5\,^2A$.[16] One can think of the $3p$ states as perturbed $3p$ atomic-like states where the perturbation of the molecule breaks the original degeneracy. Small molecular displacements can find a point where the states will become degenerate again. In the allyl radical, a seam of three-state CIs between the $\tilde{B}(^2A_1)$, $\tilde{C}(^2B_1)$, and $\tilde{D}(^2B_2)$ states (in C_{2v} symmetry) (4,5,6 ^2A states without symmetry) was located and a seam between the 3,4,5 ^2A states at higher energy. The minimum energy point on the 4,5,6 ^2A seam is similar to the equilibrium structure of the ground $\tilde{X}(^2A_2)$ state and only 1.1 eV above the $\tilde{D}$ state at its equilibrium geometry. This seam joins two two-state seams of conical intersection, the 4,5 ^2A and 5,6 ^2A conical intersection seams. The energy of the minimum energy point on the 4,5 ^2A two-state seam is only 0.15 eV above that of the $\tilde{D}$ state at its equilibrium structure.

Three-state conical intersections can affect excited state dynamics, but also ground state vibrational spectra if the ground state is involved. The latter case is particularly useful since experimental spectra may contain the signature of the three-state CIs. This is the case in five-membered ring heterocyclic radicals, such as pyrazolyl, where the ground state and two excited states cross at an energy only ca. 0.42 eV above the ground state minimum.[76] Interest in this radical is due, in part, to the general interest in polynitrogen compounds as potential high-performance rocket fuels. The ground state is involved in the degeneracy, causing the three-state CI to be affecting the spectroscopy. As was discussed earlier, if the conical intersection involves three states, the number of extrema and internal coupling modes increases. For example, the three-state, five vibrational mode, $T \otimes (e + t_2)$ Jahn–Teller problem can have 13 local extrema coupled by motion along five vibrational coordinates. Because of this, complicated vibronic spectra were expected, and observed experimentally.[77] The photoelectron spectrum of the pyrazolide anion (showing the vibronic structure of pyrazolyl radical) and its deuterium-substituted isotope have been calculated by Stanton and co-workers.[78] The spectrum was modeled using one, two or three states, and it was clear that three states are needed to reproduce the experimental peaks accurately. The vertical energies of $^2A_2$ and $^2B_1$ at the geometry of the anion are 0.327 and 0.354 eV respectively, so in these experiments, the three-state CI may not be accessed directly. Direct

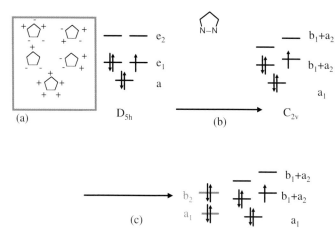

Fig. 8. Correlation between orbitals in the cyclopentadienyl and pyrazolyl radicals. (a) Orbitals in cyclopentadienyl radical in D_{5h} symmetry. (b) Introduction of two N atoms in the ring reduces the symmetry to C_{2v} and breaks the degeneracy of orbitals e_1 and e_2. (c) The two N atoms also introduce two lone pairs with symmetries a_1 and b_2. This leads to the orbitals in pyrazolyl.

access may complicate the spectra even more. The three-state CIs in pyrazolyl have also been used by Schuurman and Yarkony to develop methods to describe the vicinity of the CI.[53,72,75]

In order to understand better the origin of degenerate states in the pyrazolyl radical, we can initially consider the isoelectronic cyclopentadienyl radical. The five π orbitals in this system are shown in Fig. 8. The system has D_{5h} symmetry and the π orbitals have a, e_1, e_2 symmetries, thus the unpaired electron occupies a doubly degenerate orbital, resulting in a degenerate ground state E_1. If two nitrogen atoms substitute the CH groups, as is the case in pyrazolyl, the symmetry is reduced to C_{2v}, and the degeneracy is lifted, giving rise to two states A_2 and B_1, which will still remain very close energetically. The introduction of nitrogen atoms, besides reducing the symmetry, has the additional effect of introducing lone pair orbitals. Two orbitals with symmetry a_1 and b_2 result from the lone pair on nitrogen, with the b_2 having energy similar to the a_2 and b_1. A third state B_2 is then also energetically close to the A_2 and B_1 states. Ab initio calculations confirm the proximity of the three states. CCSD(T) adiabatic excitation energies including zero point energy corrections are 0.046 eV and 0.261 eV for $^2B_1$ and $^2B_2$, respectively.[78] Two-state CIs between all three pairs of states have been located using CCSD with energies 0.335 eV ($^2A_2/^2B_1$), 0.436 eV

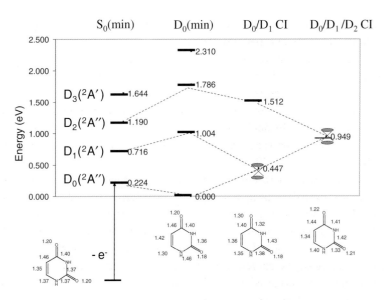

Fig. 9. Energies (in eV) and geometries (in Å) of the important points (S_0-minimum, D_0-minimum, D_0/D_1 CI, $D_0/D_1/D_2$ CI) in the uracil cation. (modified from Ref. 79)

($^2B_2/^2B_1$) and 0.292 eV ($^2B_2/^2A_2$).[78] The three-state CI was found using MRCI methods only 0.425 eV above the minimum, slightly higher than the adiabatic excitation energies.[76]

Two- and three-state CIs can be found in radical cations produced by ionizing a neutral molecule. CIs in radical cations may have an effect on time-resolved photoelectron spectra of the neutral systems, making interpretation of the signals more complicated. These spectra are often used to examine the dynamics and nonadiabatic effects in excited states of a neutral molecule, but nonadiabatic effects of the cationic molecule may also be important. Three-state CIs have been found in the uracil radical cation connecting the first three states.[79] The minimum energy point on the seam is 0.9 eV above the ground state minimum for this cation, as seen in Fig. 9. It is also connected to a two-state CI which is 0.45 eV above the ground state minimum. The geometry remains planar in all of these structures, and the deformations involve bond length changes.

9.2. Three-state conical intersections in closed-shell organic systems

Conical intersections in radicals should be anticipated since the electronic states are often not too far separated even at equilibrium geometries. In neutral closed shell molecules, however, the ground state is usually several eV below any excited states and it is more difficult for degeneracy to occur between the ground state and excited states. Such degeneracies do occur, usually requiring considerable distortion of the molecule. Two-state CIs between the ground state and an excited state are very important for relaxation of excited states and they have become a common mechanism in describing photophysical and photochemical processes. Three-state CIs have also been located recently and we present a brief summary indicating their frequency and significance.

In recent years, there have been many theoretical studies investigating the role of two-state CIs in the radiationless decay of nucleobases. Many seams of two-state CIs have been found in all the natural nucleobases, which are believed to facilitate the short excited state lifetimes and ultrafast radiationless decay. The involvement of three-state conical intersections in the photophysics and radiationless decay processes of nucleobases has also been investigated.[54,55,80] Three-state conical intersections have been located for the pyrimidine bases, uracil and cytosine, and for the purine base, adenine. Figure 10 shows the energies of the three-state conical

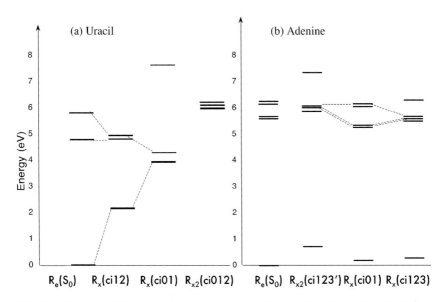

Fig. 10. Diagram of the energy levels at the conical intersections in uracil and adenine. (a) The S_0, S_1, S_2 states of uracil at equilibrium geometry $R_e(S_0)$ and at conical intersections $ci12$, $ci01$, and $ci012$. (b) S_0, S_1, S_2, S_3, S_4 states of adenine at equilibrium geometry $R_e(S_0)$ and at conical intersections $ci123'$, $ci01$ and $ci123$. (Reproduced from Ref. 54 with permission.)

intersections compared with the vertical excitations in uracil and adenine. In uracil, a three-state degeneracy between the S_0, S_1, and S_2 states has been located 6.2 eV above the ground state minimum energy.[54] This energy is 0.4 eV higher than the vertical excitation to S_2 and at least 1.3 eV higher than the two-state conical intersections found previously. In adenine, two different three-state degeneracies between the S_1, S_2, and S_3 states have been located at energies close to the vertical excitation energies.[54] The distortions are also not significant, with adenine retaining its planar structure. The energetics and distortions of these three-state conical intersections suggest that they can play a role in the radiationless decay pathways in adenine.

Blancafort and Robb[80] have reported a three-state CI in cytosine between the first three singlet states, optimized at the CASSCF level. A more detailed study of three-state CIs in cytosine and its analog 5-methyl-2-pyrimidinone was done using MRCI methods.[55] The potential energy surfaces for each of these molecules contain three different three-state seams: two different seams involving the ground state and S_1 and S_2, and a seam involving S_1, S_2, S_3 states. The first $S_0/S_1/S_2$ seam in cytosine involves the closed shell ground state plus $\pi\pi^*$ and $n_N\pi^*$ states, and the second seam involves the closed shell ground state plus $\pi\pi^*$ and $n_O\pi^*$ states. We describe them as different seams because of the different character of states involved. There has not been an attempt to connect them directly. The $S_1/S_2/S_3$ seam (labeled $ci123$) involves $\pi\pi^*$, $n_O\pi^*$ and $n_N\pi^*$ states. The minimum energy of the first $S_0/S_1/S_2$ seam (labeled $ci012$) is 5.3 eV while the minimum energy of the second is much higher, 6.6 eV, using the higher-level MRCI treatment in that study. The minimum energy point of the $S_1/S_2/S_3$ seam was calculated to be at 4.35 eV. Since vertical excitations in cytosine are 4.37 eV, 4.42 eV and 5.07 eV at the same level of theory, the $S_1/S_2/S_3$ seam may be relevant and important in its photophysical behavior. Even the $S_0/S_1/S_2$ seam with a minimum at 5.3 eV may be relevant. This point can be further examined by studying the branching of these three-state seams to two-state seams. Two-state seam paths from these intersections were calculated and they were shown to be connected to energetically low two-state CIs that participate in radiationless decay. Figures 2 and 3 show some of these pathways which were also discussed in Sec. 7.1. Figure 2 shows $ci012$ being connected to a two-state S_0/S_1 seam that facilitates radiationless transition to the ground state. Figure 3 shows that $ci012$ and $ci123$ are actually connected through a two-state seam S_1/S_2, ($ci12'$). This two-state seam is further connected to the Franck–Condon region, indirectly showing how the three-state seams are also connected to the FC region.

Malonaldehyde is another system where a three-state CI has been implicated in its intra-molecular proton transfer.[22,81] This is the simplest system where excited-state intra-molecular proton transfer is observed. Femtosecond experiments on related molecules have found that proton transfer occurs in less than 100 fs after excitation to the bright $\pi\pi^*$ state. The bright state in malonaldehyde is the S_2 state. Studies involving conical intersection searches and molecular dynamics reveal that there are two conical intersections in malonaldehyde that enable this process. There is an S_2/S_1 conical intersection energetically close to the S_2 vertical excitation, which involves H migration from one oxygen to the other. There is also a three-state CI between $S_2/S_1/S_0$ which is lower in energy and involves torsion around the C=C bond (Figs. 4 and 11). Consequently, there are two primary decay channels after photoexcitation to the bright S_2 state: (1) in-plane evolution, leading to reversible hydrogen exchange accompanied by quenching to the S_1 state through the S_2/S_1 conical intersection, and (2) torsion about the C=C bond, leading to opening of the chelate ring, which shuts off hydrogen atom transfer and leads to efficient quenching to both S_1 and S_0 through the three-state $S_2/S_1/S_0$ conical intersection.

The possibility of the three-state CI in malonaldehyde can be anticipated if one considers the behavior of the different states as the C=C bond twists. This causes a destabilization of the ground state and a strong stabilization of the $\pi\pi^*$ S_2 state, while the $n\pi^*$ S_1 state is not affected much. Thus there may be a point where all these three states can meet. Of course this is not a guarantee that a true three-state CI exists but a plausible explanation for its existence.

In summary, these results show that three-state conical intersections are quite common, and they can complicate the potential energy surfaces of molecules. The most relevant question then becomes whether they are accessible during a photoinitiated event and how the nuclear motion is affected by them.

10. Dynamics

The only study so far where nuclear dynamics has been studied around accidental three-state CIs is a study that applied the *ab initio* multiple spawning (AIMS) method for the study on malonaldehyde.[22,57,81] For a

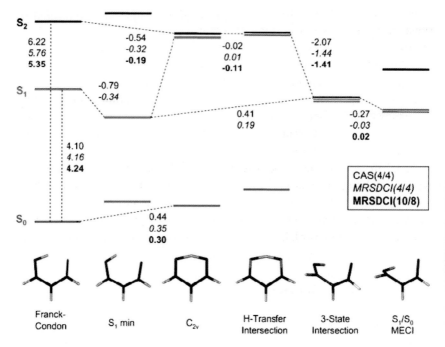

Fig. 11. Energy differences (in eV) between important geometries of malonaldehyde, obtained using SA-3-CAS(4/4), SA-3-CAS(4/4)*SDCI (italics), and SA-3-CAS(10/8)*-SDCI (bold). Geometries are determined by optimization with SA-3-CAS(4/4). S_2 excitation leads to an energetically favorable route to rotamerization that may or may not be accompanied by hydrogen transfer. The S_1-transfer barrier is quite large, and approach to the 3SI from the minimum on this state is unfavorable. Energy orderings are not significantly affected by the addition of dynamic correlation. Intersections are shown with thin lines and degenerate states slightly displaced for visual clarity. (Reproduced from Ref. 81 with permission.)

more comprehensive discussion on AIMS see Chapter 9 in this book. The electronic structure of this molecule was described in the previous section. The three-state CI region was entered about 150 fs after initial excitation. The most obvious question is whether direct quenching from S_2 to S_0 is observed, facilitated by the three-state CI. However, this is not seen in this study. This suggests that population decays sequentially from S_2 to S_1 and then from S_1 to S_0 despite the proximity to the three-state CI. Once the three-state CI region is entered, the average S_2/S_0 energy gap remains near 1 eV for the remainder of the simulation. This indicates that S_2 trajectories entering the region never fully exit, because population can be trapped on S_1, as seen in Fig. 4. In summary, the dynamics showed preference for a sequential population decay through the two two-state conical intersections rather than a one-step decay directly from the three-state conical intersection.[22] Further studies showed that the minimum energy point of the three-state CI is not the whole story, and radiationless decay occurs stepwise (S_2-S_1-S_0) in the vicinity of the three-state seam.[57] The fact that a minimum energy point on a seam is not the only important point and may be insufficient to discuss nonadiabatic events has been seen and discussed previously in two-state seams.[82–84] Although the direct effects of the three-state CI on the dynamics of malonaldehyde are small, indirectly the three-state CI brings together both the S_2/S_1 and the S_1/S_0 seams in a larger region, enabling efficiency of the stepwise deactivation. This is expected to be general around three-state CIs. The details of the dynamics around three-state CIs in general will depend on the topography of the PESs in each case and much more work is needed before we can draw general conclusions.

11. Concluding Remarks

The studies of conical intersections have been extended in recent years beyond the common case of two intersecting states to include three intersecting states. Although three-state conical intersections had been known and studied in detail in the symmetry-required Jahn–Teller problem, the recent work makes considerable progress in the more general case where there is no symmetry present to simplify the situation. Algorithms have been developed to locate accidental three-state CIs and they have enabled the location of these CIs in a large number of polyatomic systems, either radicals or neutral systems. The frequency with which three-state CIs appear suggests that they may play an important role in nonadiabatic dynamics, but more work is needed to confirm this suggestion. Substantial work in the characterization of three-state CIs and nuclear dynamics involving them should be expected in the coming years in order to reach the same level of understanding we currently have for the two-state CIs.

Acknowledgment

The author thanks the National Science Foundation (CHE-0911474) and Department of Energy (DE-FG02-08ER15983) for financial support.

References

1. M. Born and R. Oppenheimer, *Ann. Phys.* **84**, 457 (1927).
2. J. von Neumann and E.P. Wigner, *Physik. Z.* **30**, 467 (1929).
3. E. Teller, *J. Phys. Chem.* **41**, 109 (1937).
4. W. Domcke, D.R. Yarkony and H. Köppel (Eds.), *Conical Intersections: Electronic Structure, Dynamics & Spectroscopy* (World Scientific, Singapore, 2004).
5. D.R. Yarkony, Electronic structure aspects of nonadiabatic processes, in *Modern Electronic Structure Theory Part I*, edited by D.R. Yarkony, Advance Series in Physical Chemistry (World Scientific, Singapore, 1995), pp. 642–721.
6. M. Klessinger and J. Michl, *Excited States and Photochemistry of Organic Molecules* (VCH Publishers, Inc., New York, 1995).
7. M.A. Robb, M. Garavelli, M. Olivucci and F. Bernardi, A computational strategy for organic photochemistry, in *Reviews in Computational Chemistry*, volume 15, edited by K.B. Lipkowitz and D.B. Boyd (Wiley-VCH, New York, 2000), pp. 87–146.
8. D.R. Yarkony, *J. Phys. Chem. A* **105**, 6277 (2001).
9. G.A. Worth and L.S. Cederbaum, *Annu. Rev. Phys. Chem.* **55**, 127158 (2004).
10. S. Matsika, Conical intersections in molecular systems, in *Reviews in Computational Chemistry*, volume 23, edited by K.B. Lipkowitz and T.R. Cundari (Wiley-VCH, New Jersey, 2007), pp. 83–124.
11. B.G. Levine and T.J. Martinez, *Annu. Rev. Phys. Chem.* **58**, 613 (2007).
12. I.B. Bersuker, *The Jahn–Teller Effect and Vibronic Interactions in Modern Chemistry* (Plenum Press, New York, 1984).
13. R. Englman, *The Jahn–Teller Effect in Molecules and Crystals* (Wiley-Interscience, New York, 1972).
14. I.B. Bersuker and V.Z. Polinger, *Vibronic Interactions in Molecules and Crystals*, volume 49 (Springer-Verlag, Berlin Heidelberg, 1989).
15. J. Katriel and E.R. Davidson, *Chem. Phys. Lett.* **76**, 259 (1980).
16. S. Matsika and D.R. Yarkony, *J. Chem. Phys.* **117**, 6907 (2002).
17. S. Matsika and D.R. Yarkony, *J. Chem. Soc.* **125**, 10672 (2003).
18. S.P. Keating and C.A. Mead, *J. Chem. Phys.* **82**, 5102 (1985).
19. D.E. Manolopoulos and M.S. Child, *Phys. Rev. Lett.* **82**, 2223 (1999).
20. S. Han and D.R. Yarkony, *J. Chem. Phys.* **119**, 11562 (2003).
21. S. Han and D.R. Yarkony, *J. Chem. Phys.* **119**, 5058 (2003).
22. J.D. Coe and T.J. Martinez, *J. Am. Chem. Soc.* **127**, 4560 (2005).
23. G.J. Atchity, S.S. Xantheas and K. Ruedenberg, *J. Chem. Phys.* **95**, 1862 (1991).
24. C.A. Mead, *J. Chem. Phys.* **70**, 2276 (1979).
25. S. Matsika and D.R. Yarkony, *J. Chem. Phys.* **115**, 2038 (2001).
26. D.R. Yarkony, Conical intersections: Their description and consequences, in Ref. 4, pages 41–128 (World Scientific, Singapore, 2004).

27. U. Öpik and M.H.L. Pryce, *Proc. R. Soc. London Ser. A* **238**, 425 (1957).
28. I.B. Bersuker, *The Jahn–Teller Effect* (Cambridge University Press, Cambridge UK, 2006).
29. I.B. Bersuker and V.Z. Polinger, *Phys. Lett. A* **44**, 495 (1973).
30. S. Muramatsu and T. Iida, *J. Phys. Chem. Solids* **31**, 2209 (1970).
31. M. Bacci, M.P. Fontana, M. Cetica and G. Viliani, *Phys. Rev. B* **12**, 5907 (1975).
32. M. Bacci, M.P. Fontana, A. Ranfagni and G. Viliani, *Phys. Lett. A* **50**, 405 (1975).
33. V.P. Khlopin, V.Z. Polinger and I.B. Bersuker, *Theor. Chim. Acta* **48**, 87 (1978).
34. C.C. Chancey and M.C.M. O'Brien, *The Jahn-Teller Effect in C_{60} and Other Icosahedral Complexes* (Princeton University Press, Princeton, New Jersey, 1997).
35. A. Ceulemans and P.W. Fowler, *J. Chem. Phys.* **93**, 1221 (1990).
36. J.L. Dunn and A.C. Bates, *Phys. Rev. B* **52**, 5996 (1995).
37. L.B. Knight, Jr., J. Steadman, D. Feller and E.R. Davidson, *J. Am. Chem. Soc.* **106**, 3700 (1984).
38. W. Meyer, *J. Chem. Phys.* **58**, 1017 (1973).
39. M.N. Paddon-Row, D.J. Fox, J.A. Pople, K.N. Houk and D.W. Pratt, *J. Am. Chem. Soc.* **107**, 7696 (1985).
40. R. Signorell and F. Merkt, *Faraday Discuss* **115**, 205 (2000).
41. H.J. Wörner, R. van der Veen and F. Merkt, *Phys. Rev. Lett.* **97**, 173003 (2006).
42. M.R. Manaa and D.R. Yarkony, *J. Chem. Phys.* **99**, 5251 (1993).
43. C.A. Mead, *J. Chem. Phys.* **78**, 807 (1983).
44. D.R. Yarkony, *J. Phys. Chem. A* **101**, 4263 (1997).
45. D.R. Yarkony, *Faraday Discuss.* **127**, 325 (2004).
46. D.R. Yarkony, *J. Phys. Chem. A* **108**, 3200 (2004).
47. B.H. Lengsfield and D.R. Yarkony, *Adv. Chem. Phys.* **82**, 1 (1992).
48. H. Lischka, M. Dallos, P.G. Szalay, D.R. Yarkony and R. Shepard, *J. Chem. Phys.* **120**, 7322 (2004).
49. M. Dallos, H. Lischka, R. Shepard, D.R. Yarkony and P.G. Szalay, *J. Chem. Phys.* **120**, 7330 (2004).
50. D.R. Yarkony, Determination of potential energy surface intersections and derivative couplings in the adiabatic representation, in Ref. 4, pages 129–174 (World Scientific, Singapore, 2004).
51. H. Lischka et al., *Phys. Chem. Chem. Phys.* **3**, 664 (2001).
52. B.G. Levine, J.D. Coe and T.J. Martinez, *J. Phys. Chem. B* **112**, 405 (2008).
53. M.S. Schuurman and D.R. Yarkony, *J. Phys. Chem. B* **110**, 19031 (2006).
54. S. Matsika, *J. Phys. Chem. A* **109**, 7538 (2005).
55. K.A. Kistler and S. Matsika, *J. Chem. Phys.* **128**, 215102 (2008).
56. K.A. Kistler and S. Matsika, *J. Phys. Chem. A* **111**, 2650 (2007).
57. J.D. Coe, M.T. Ong, B.G. Levine and T.J. Martinez, *J. Phys. Chem. A* **112**, 12559 (2008).

58. H.C. Longuet-Higgins, U. Opik, M.H.L. Pryce and R.A. Sack, *Proc. R. Soc. London Ser. A* **244**, 1 (1958).
59. G. Herzberg and H.C. Longuet-Higgins, *Discuss. Faraday Soc.* **35**, 77 (1963).
60. C.A. Mead and D.G. Truhlar, *J. Chem. Phys.* **70**, 2284 (1979).
61. M.V. Berry, *Proc. R. Soc. London Ser. A* **392**, 45 (1984).
62. H. von Busch et al., *Phys. Rev. Lett.* **81**, 4584 (1984).
63. G. Delacretàz, E.R. Grant, R.L.K. Whetton, L. Wöste and J.W. Zwanziger, *Phys. Rev. Lett.* **56**, 2598 (1986).
64. C.A. Mead, *J. Chem. Phys.* **72**, 3839 (1980).
65. B.K. Kendrick, *J. Chem. Phys.* **112**, 5679 (2000).
66. B.K. Kendrick, *J. Phys. Chem. A* **107**, 6739 (2003).
67. J.C. Juanes-Marcos, S.C. Althorpe and E. Wrede, *Phys. Rev. Lett.* **309**, 1227 (2005).
68. J. Samuel and A. Dhar, *Phys. Rev. Lett.* **87**, 260401 (2001).
69. H. Köppel, Diabatic representation: Methods for the construction of diabatic electronic states, in Ref. 4, page 175 (World Scientific, Singapore, 2004).
70. K. Ruedenberg and G.J. Atchity, *J. Chem. Phys.* **110**, 3799 (1993).
71. D.R. Yarkony, *J. Chem. Phys.* **112**, 2111 (2000).
72. M.S. Schuurman and D.R. Yarkony, *J. Chem. Phys.* **124**, 124109 (2006).
73. G. Halász, A. Vibók, A.M. Mebel and M. Baer, *J. Chem. Phys.* **118**, 3052 (2003).
74. M. Baer and R. Englman, *Chem. Phys. Lett.* **335**, 85 (2001).
75. M.S. Schuurman and D.R. Yarkony, *J. Chem. Phys.* **124**, 244103 (2006).
76. S. Matsika and D.R. Yarkony, *J. Am. Chem. Soc.* **125**, 12428 (2003).
77. A.J. Gianola, T. Ichino, S. Kato, V.M. Bierbaum and W.C. Lineberger, *J. Phys. Chem. A* **110**, 8457 (2006).
78. T. Ichino, A.J. Gianola, W.C. Lineberger and J.F. Stanton, *J. Chem. Phys.* **125**, 084312 (2006).
79. S. Matsika, *Chem. Phys.* **349**, 356 (2008).
80. L. Blancafort and M.A. Robb, *J. Phys. Chem. A* **108**, 10609 (2004).
81. J.D. Coe and T.J. Martinez, *J. Phys. Chem. A* **110**, 618 (2006).
82. M. Boggio-Pasqua, M.J. Bearpark, P.A. Hunt and M.A. Robb, *J. Am. Chem. Soc.* **124**, 1456 (2002).
83. M. Boggio-Pasqua, M. Ravaglia, M.J. Bearpark, M. Garavelli and M.A. Robb, *J. Phys. Chem. A* **107**, 11139 (2003).
84. M. Boggio-Pasqua, M.J. Bearpark, F. Ogliaro and M.A. Robb, *J. Am. Chem. Soc.* **128**, 10533 (2006).

Chapter 4

Spin-Orbit Vibronic Coupling in Jahn–Teller Systems

Leonid V. Poluyanov* and Wolfgang Domcke[†]

1. Introduction 117
2. Spin-Free and Relativistically Generalized
   Jahn–Teller Selection Rules 120
3. JT and SO Coupling in Trigonal Systems 123
4. Relativistic Jahn–Teller Coupling in Tetrahedral Systems 129
5. Jahn–Teller and Spin-Orbit Coupling in High-Spin
   States of Trigonal Systems 137
6. Examples: Transition-Metal Trifluorides and Group V
   Tetrahedral Cations 141
7. Conclusions 151
   Acknowledgments 151
   References 152

1. Introduction

The photochemistry of chromophores containing heavy elements is a theoretically and computationally little explored field. Experimentally, the photoinduced decomposition of transition-metal carbonyls, for example,

---

*Institute of Chemical Physics, Russian Academy of Sciences, Chernogolovka, Moscow 142432, Russian Federation.
[†]Department of Chemistry, Technische Universität München, D-85747 Garching, Germany.

has been of considerable interest in femtochemistry since many years.[1-4] Transition-metal photochemistry has also served as a testbed for the demonstration of laser control of chemical reactions.[5,6] Complexes of ruthenium, iridium or other transition metals with aromatic molecules are important chromophores for the photosensitation of photovoltaic cells.[7,8] The photophysics of many of these complexes has been investigated with femtosecond spectroscopy.[9-13] Most of these metal-organic complexes exhibit high symmetry (e.g. tetrahedral or octahedral symmetry) and are therefore prone to Jahn–Teller distortions in open-shell electronic states. The high density of excited electronic states in the visible or UV region of the spectrum and the multiple crossings and nonadiabatic couplings of potential-energy (PE) surfaces of the same or different spin-multiplicities render the photochemistry of transition-metal complexes considerably more challenging for theory and computation than the photochemistry of molecules built from first-row atoms.

The relevance of spin-orbit (SO) coupling for the dimensionality and topology of seams of intersections of PE surfaces of polyatomic molecules has been discussed by Mead.[14,15] Matsika and Yarkony comprehensively analyzed the role of SO coupling at conical intersections of nonrelativistic PE surfaces in a general context.[16-20]

The concepts developed by the pioneers of vibronic-coupling theory for the description of the Jahn–Teller (JT) effect in molecules and crystals have played a paradigmatic role for the understanding of general conical intersections in the nonrelativistic limit. The concepts of a diabatic electronic basis, singular nonadiabatic coupling elements and the geometric phase carried by the adiabatic electronic wave functions, for example, have first been elaborated for the simplest JT systems.[21-23] In an analogous manner, the investigation of the interplay of JT coupling and SO coupling in highly symmetric complexes containing heavy elements may provide the frame for the theoretical description and computational treatment of general conical intersections in the presence of strong SO coupling.

The basic concepts of vibronic coupling in molecules, clusters and crystals in the nonrelativistic limit, including the Renner[24-26] and JT[27-31] effects as special cases, can be summarized as follows:

(a) Representation of the (nonrelativistic) electronic Hamiltonian in a basis of diabatic electronic states.
(b) Expansion of the electronic Hamiltonian in powers of normal-mode displacements at the reference geometry.

According to Eq. (8a), the SO interaction lifts the four-fold degeneracy of the ^2E state. According to Eq. (8b), the SO-split components $E_{1/2}$, $E_{3/2}$ are coupled by B_1, B_2 and E modes in first order. The JT activity of the E modes must arise from the SO operator, since the E modes are not JT active in the electrostatic Hamiltonian, see Eq. (7). The same selection rules apply in C_{4v} and D_{4h}. The relativistic JT forces are thus complementary to the electrostatic JT forces in tetragonal symmetry.

Let us finally consider tetrahedral systems as examples of highly symmetric systems. The spin-free JT selection rules in the group T_d are[27]

$$[E^2] = A + E, \tag{9a}$$

$$[T^2_{1,2}] = A + E + T_2. \tag{9b}$$

Vibrational modes transform as A, E or T_2 in T_d. In electronic states of E symmetry, only E modes are JT active [Eq. (9a)]. In electronic states of T_1 or T_2 symmetry, the T_2 modes as well as the E modes are JT active [Eq. (9b)].

$^2T_1$, $^2T_2$ and ^2E states transform as follows in T'_d:

$$^2T_1 = T_1 \times E_{1/2} = G_{3/2} + E_{1/2}, \tag{10a}$$

$$^2T_2 = T_2 \times E_{1/2} = G_{3/2} + E_{5/2}, \tag{10b}$$

$$^2E = E \times E_{1/2} = G_{3/2}, \tag{10c}$$

where $E_{1/2}$, $E_{5/2}$, $G_{3/2}$ are the three double-valued representations of T'_d.[41] $E_{1/2}$ and $E_{5/2}$ are two-dimensional irreducible representations, while $G_{3/2}$ is a four-dimensional irreducible representation. According to Eqs. (10a) and (10b), a $^2T_2$ ($^2T_1$) nonrelativistic electronic manifold splits into $G_{3/2}$ and $E_{5/2}$ ($G_{3/2}$ and $E_{1/2}$) irreducible representations. The six-fold degeneracy of a $^2T_2$ ($^2T_1$) state is thus partially lifted by the SO coupling, resulting in a four-fold degenerate $G_{3/2}$ ($G_{3/2}$) level and a two-fold degenerate $E_{5/2}$ ($E_{1/2}$) level. The four-fold degeneracy of a nonrelativistic ^2E state, on the other hand, is not lifted by the SO coupling at the tetrahedral reference geometry. As a consequence of time-reversal symmetry, the two-fold degeneracy of the $E_{1/2}$ and $E_{5/2}$ levels cannot be lifted by any intra-molecular interaction,[42] while the energy levels of the $G_{3/2}$ manifold can split into two two-fold degenerate levels.

for a given C_n of the point group G, the solution of Eq. (4) defines the corresponding SU2 matrix U_n and thus the symmetry operator Z_n of the spin double group G'.[38,39] Given the set of symmetry operators $\{Z_n\}$ and the group multiplication table, the classes and the irreducible representations can be constructed in the usual manner.[37]

Let us consider C_{3v} as one of the simplest non-Abelian point groups. The spin-free JT selection rule in the C_{3v} point group is[27]

$$[E^2] = A + E, \tag{5}$$

where $[\Gamma^2]$ denotes the symmetrized square of the irreducible representation Γ.[37] In electronic states of E symmetry, vibrational modes of E symmetry are JT active in first-order according to Eq. (5).

The relativistically generalized selection rules in the spin double group C'_{3v} are (for a single unpaired electron)[40]

$$^2E = E \times E_{1/2} = E_{1/2} + E_{3/2}, \tag{6a}$$

$$E_{1/2} \times E_{3/2} = 2E. \tag{6b}$$

The spin function of a single electron transforms as $E_{1/2}$ in C'_{3v}. According to Eq. (6a), the four-fold degenerate ^2E state splits into two-fold degenerate $E_{1/2}$ and $E_{3/2}$ states by SO coupling. According to Eq. (6b), vibrational modes of E symmetry can couple the $E_{1/2}$ and $E_{3/2}$ SO-split states. Since the E modes already are JT active in the electrostatic Hamiltonian, no new JT couplings arise from the SO operator. In C_{3v} (and D_{3h}) systems, the SO operator thus lifts the degeneracy of the nonrelativistic ^2E state in zeroth order (in normal-mode displacements), while the JT forces are of electrostatic origin.

Interestingly, a different situation is encountered in tetragonal groups (D_{2d}, C_{4v}, D_{4h}). In D_{2d}, the spin-free JT selection rule is[27]

$$[E^2] = A_1 + B_1 + B_2. \tag{7}$$

It is well known that B_1 and B_2 modes are JT active (in first order) in tetragonal symmetry.[28-31] The relativistically generalized selection rules in D'_{2d} are (for a single unpaired electron)

$$^2E = E \times E_{1/2} = E_{1/2} + E_{3/2}, \tag{8a}$$

$$E_{1/2} \times E_{3/2} = B_1 + B_2 + E. \tag{8b}$$

(b) Expansion of the Breit–Pauli operator in powers of normal-mode displacements at the reference geometry.

(c) Use of symmetry selection rules to determine the nonvanishing matrix elements.

The use of nonrelativistic basis functions in (a) requires that the SO interaction can be considered as a relatively weak perturbation of the nonrelativistic Hamiltonian, which typically is the case for second- and third-row atoms and transition metals. For systems with heavier atoms, two-component relativistic electronic basis functions should be employed.

In Sec. 3 of this chapter, we outline the systematic treatment of SO coupling for the simplest and most common JT system, the $E \times E$ JT effect in spin-$\frac{1}{2}$ states of trigonal systems. This example serves to explain the application of the group-theoretical tools. It is demonstrated how the well-known $^2E \times E$ Hamiltonian with SO coupling is obtained by a systematic derivation from first principles.

In Sec. 4, the interplay of electrostatic and relativistic JT coupling is discussed for orbitally degenerate spin-$\frac{1}{2}$ states in tetrahedral systems. In Sec. 5, the effects arising in states of high spin multiplicity are discussed for the example of M-fold spin-degenerate E states in trigonal systems. In Sec. 6, we discuss evidence for the relevance of SO splittings and relativistic JT forces on the basis of *ab initio* calculations for specific systems.

2. Spin-Free and Relativistically Generalized Jahn–Teller Selection Rules

The symmetry group of the Hamiltonian including electron spin and SO coupling is the so-called spin double group of the respective point group.[36,37] The elements of the spin double group can be constructed by determining the complete set of generalized symmetry operators of spin-$\frac{1}{2}$ systems,

$$Z_n = C_n U_n^\dagger, \qquad (3)$$

where the C_n are the symmetry operators of the point group and the U_n are SU2 matrices acting on the spin functions α and β. Requiring the invariance of the BP operator,

$$Z_n H_{SO} Z_n^{-1} = H_{SO} \qquad (4)$$

(c) Use of symmetry selection rules for the determination of the nonvanishing matrix elements.

Diabatic electronic states (an example of which are the "crude adiabatic states" of Longuet–Higgins[23]) are defined as slowly varying functions of the nuclear geometry in the vicinity of the reference geometry.[32–34] The final vibronic-coupling Hamiltonian is obtained by adding the nuclear kinetic-energy operator which is assumed to be diagonal in the diabatic representation.

SO coupling is the most obvious relativistic effect in molecular electronic structure. SO coupling lifts, in general, orbital and spin degeneracies of electronic states in open-shell systems. Since SO coupling scales approximately with the fourth power of the nuclear charge Z, it is essential to take SO coupling effects into account in systems containing heavy elements. In a nonempirical treatment, SO coupling is described by the Breit–Pauli (BP) operator, which results from the reduction of the four-component Dirac equation to two-component form in the so-called Pauli approximation.[35] In this approximation, the electronic Hamiltonian can be written as the sum of the electrostatic Hamiltonian H_{es} and the SO operator H_{SO}

$$H = H_{es} + H_{SO}. \tag{1}$$

H_{es} may be chosen, for example, as the restricted open-shell Hartree–Fock Hamiltonian of the many-electron system.

While the BP operator was known since the late 1920s, the description of SO coupling in the vast literature on the spectroscopy of impurity centers in crystals has been based on empirical atom-like SO operators of the type

$$H_{SO} = \lambda \mathbf{L} \cdot \mathbf{S}, \tag{2}$$

where $\mathbf{L}$ and $\mathbf{S}$ are total orbital and spin angular momenta and λ is a phenomenological SO coupling constant. The approximation (2) eliminates any dependence of the SO operator on the nuclear coordinates. While the empirical ansatz (2) may be appropriate for a partially occupied inner shell of an impurity atom in a rigid crystal environment, a more accurate description of SO coupling is necessary for molecules, atomic clusters and multi-center transition-metal complexes.

In a systematic vibronic-coupling theory including SO coupling effects, the SO operator should be treated in exactly the same manner as the electrostatic Hamiltonian, that is,

(a) Representation of the Breit–Pauli operator in a basis of (nonrelativistic) diabatic electronic states.

The JT selection rules in the T'_d spin double group are[43]

$$E_{1/2} \times G_{3/2} = E + T_1 + T_2, \qquad (11a)$$

$$E_{5/2} \times G_{3/2} = E + T_1 + T_2, \qquad (11b)$$

$$\{G^2_{3/2}\} = A_1 + E + T_2. \qquad (11c)$$

Here $\{\Gamma^2\}$ denotes the antisymmetrized square of Γ.[37]

Equation (11c) reveals that the four-fold degeneracy of the $G_{3/2}$ manifold can be lifted in first order by vibrational modes of E or T_2 symmetry. For a ^2E state, which transforms as $G_{3/2}$ in the T'_d double group, the T_2 mode is not JT active in the nonrelativistic limit, see Eq. (9a). It follows that the JT activity of the T_2 mode according to Eq. (11c) must arise from the SO operator.

The JT selection rules in cubic symmetry (O_h, O'_h) are the same as Eqs. (9)–(11) apart from the additional inversion symmetry.

The treatment of electronic states of higher spin multiplicity in odd-electron systems is a straightforward extension of the above analysis, since the representations of quartet, sextet, etc., spin states can be decomposed into the irreducible representations with the character table of the spin double group.

3. JT and SO Coupling in Trigonal Systems

In this section, we outline the systematic derivation of the relativistically generalized JT Hamiltonian for the simplest example of the JT effect, that is, a single unpaired electron in the field of three identical nuclei which form an equilateral triangle (D_{3h} symmetry).

For the purpose of symmetry analysis, the electrostatic Hamiltonian can be written as (in atomic units)

$$H_{es} = -\frac{1}{2}\nabla^2 - q\Phi(\mathbf{r}), \qquad (12)$$

where

$$\Phi(\mathbf{r}) = \sum_{k=1}^{3} \frac{1}{r_k} \qquad (13)$$

and

$$r_k = |\mathbf{r} - \mathbf{R}_k|. \qquad (14)$$

Here **r** is the radius vector of the single unpaired electron, $\mathbf{R}_k, k = 1, 2, 3$, denote the positions of the nuclei, and q is the effective charge of the three identical nuclei.

The Breit–Pauli Hamiltonian of this system is[44]

$$H_{SO} = -ig_e \beta_e^2 \, q\mathbf{S} \cdot \sum_{k=1}^{3} \frac{1}{r_k^3} (\mathbf{r}_k \times \nabla), \tag{15}$$

where

$$\mathbf{S} = \frac{1}{2}(\mathbf{i}\sigma_x + \mathbf{j}\sigma_y + \mathbf{k}\sigma_z), \tag{16}$$

$\sigma_x, \sigma_y, \sigma_z$ are the Pauli spin matrices,

$$\beta_e = \frac{e\hbar}{2m_e c} \tag{17}$$

is the Bohr magneton, $g_e = 2.0023$ is the g-factor of the electron, and **i**, **j**, **k** are the Cartesian unit vectors.

It is seen that the Breit–Pauli operator has the structure of Eq. (2) for each atomic center, but depends explicitly on the distances r_k of the unpaired electron from the atomic centers, defined in Eq. (14). While the magnetic interaction energy is $\sim r_k^{-2}$ and thus of shorter range than the electrostatic interaction, it can nevertheless result in a nonnegligible dependence of the SO operator on the nuclear coordinates. This effect is neglected when the empirical SO operator [Eq. (2)] is employed.

It is useful for the symmetry analysis to write the Breit–Pauli operator [Eq. (15)] in determinantal form

$$H_{SO} = \frac{1}{2} i g_e \beta_e^2 \, q \begin{vmatrix} \sigma_x & \sigma_y & \sigma_z \\ \Phi_x & \Phi_y & \Phi_z \\ \dfrac{\partial}{\partial x} & \dfrac{\partial}{\partial y} & \dfrac{\partial}{\partial z} \end{vmatrix}, \tag{18}$$

where Φ is given by Eq. (13) and

$$\Phi_x = \frac{\partial \Phi}{\partial x}, \quad \text{etc.} \tag{19}$$

From Eq. (18) it is obvious that the BP operator is completely determined by the effective electrostatic potential $\Phi(\mathbf{r})$ for the unpaired electron. Since Φ depends on the nuclear coordinates, so does H_{SO}.

H_{SO} is time-reversal invariant. The time-reversal operator for a single electron is the antiunitary operator (up to an arbitrary phase factor)[42]

$$\tau = -i\sigma_y \hat{cc} = \begin{pmatrix} 0 & -1 \\ 1 & 0 \end{pmatrix} \hat{cc}, \qquad (20)$$

where $\hat{cc}$ denotes the operation of complex conjugation. The full symmetry group $\widetilde{G}$ of H_{SO} of Eq. (15) is thus

$$\widetilde{G} = D'_{3h} \otimes (1, \tau), \qquad (21)$$

of order 48. The operations of D'_{3h} commute with τ.

A pair of electronic basis functions transforming as x and y in D_{3h} is

$$\psi_x = 6^{-1/2}[2\chi(\mathbf{r}_1) - \chi(\mathbf{r}_2) - \chi(\mathbf{r}_3)], \qquad (22a)$$
$$\psi_y = 2^{-1/2}[\chi(\mathbf{r}_2) - \chi(\mathbf{r}_3)], \qquad (22b)$$

where the $\chi(\mathbf{r}_k)$ are atom-centered basis functions. Introducing the spin of the electron, we have four nonrelativistic spin-orbital basis functions

$$\begin{aligned} \psi_x^+ &= \psi_x \alpha, \\ \psi_y^+ &= \psi_y \alpha, \\ \psi_x^- &= \psi_x \beta, \\ \psi_y^- &= \psi_y \beta, \end{aligned} \qquad (23)$$

where $\alpha(\beta)$ represent the spin projection $\frac{1}{2}(-\frac{1}{2})$ of the electron. The time-reversal operator τ acts on these spin orbitals as follows:

$$\begin{aligned} \tau \psi_x^+ &= \psi_x^- \hat{cc}, & \tau \psi_x^- &= -\psi_x^+ \hat{cc}, \\ \tau \psi_y^+ &= \psi_y^- \hat{cc}, & \tau \psi_y^- &= -\psi_y^+ \hat{cc}. \end{aligned} \qquad (24)$$

Note that $\tau \psi_x^+$, etc., is still an operator, acting on the nuclear part of the wave function.

The representation of the operator τ is thus the 4×4 matrix

$$\tau = \begin{pmatrix} 0 & 0 & 1 & 0 \\ 0 & 0 & 0 & 1 \\ -1 & 0 & 0 & 0 \\ 0 & -1 & 0 & 0 \end{pmatrix} \hat{cc}. \qquad (25)$$

Note that $\tau^2 = -1_4$, as is required for an odd-electron system.[42]

The vibrational displacements are described in terms of dimensionless normal coordinates Q_x, Q_y of the degenerate vibrational mode of E symmetry. The electrostatic Hamiltonian is expanded at the reference geometry in powers of Q_x, Q_y up to second order

$$H_{es} = H_0 + H_x Q_x + H_y Q_y + \frac{1}{2} H_{xx} Q_x^2 + \frac{1}{2} H_{yy} Q_y^2 + H_{xy} Q_x Q_y, \qquad (26)$$

where

$$H_0 = H_{es}(0),$$
$$H_x = \left(\frac{\partial H_{es}}{\partial Q_x} \right)_0, \qquad (27)$$
$$H_{xy} = \left(\frac{\partial^2 H_{es}}{\partial Q_x \partial Q_y} \right)_0, \quad \text{etc.}$$

H_0 transforms totally symmetric, $H_x(H_y)$ transforms as $Q_x(Q_y)$, H_{xy} transforms as $Q_x Q_y$, etc.

The electrostatic vibronic matrix is obtained by taking matrix elements of the Hamiltonian [Eq. (26)] with the electronic wave functions ψ_x, ψ_y. The well-known result is[28–31]

$$H_{es} = \frac{1}{2} \omega (Q_x^2 + Q_y^2) 1_2$$
$$+ \begin{pmatrix} \kappa Q_x + \frac{1}{2} g(Q_x^2 - Q_y^2) & \kappa Q_y - g Q_x Q_y \\ \kappa Q_y - g Q_x Q_y & -\kappa Q_x - \frac{1}{2} g(Q_x^2 - Q_y^2) \end{pmatrix}, \qquad (28)$$

where 1_2 denotes the 2×2 unit matrix, ω is the vibrational frequency of the E mode and $\kappa(g)$ denotes the linear (quadratic) JT coupling constant.

To obtain the SO vibronic matrix, H_{SO} is expanded in analogy to Eq. (26)

$$H_{SO} = h_0 + h_x Q_x + h_y Q_y + \cdots \qquad (29)$$

Assuming that the SO coupling is weak compared to the electrostatic interactions, we terminate the expansion after the first order. The individual SO operators in Eq. (29) can be written as

$$\begin{aligned} h_0 &= h^x \sigma_x + h^y \sigma_y + h^z \sigma_z, \\ h_x &= h_x^x \sigma_x + h_x^y \sigma_y + h_x^z \sigma_z, \\ h_y &= h_y^x \sigma_x + h_y^y \sigma_y + h_y^z \sigma_z, \end{aligned} \qquad (30)$$

with

$$h^x = ig_e \beta_e^2 q \left(\frac{\partial \Phi}{\partial y} \frac{\partial}{\partial z} - \frac{\partial \Phi}{\partial z} \frac{\partial}{\partial y} \right),$$

$$h^y = ig_e \beta_e^2 q \left(\frac{\partial \Phi}{\partial z} \frac{\partial}{\partial x} - \frac{\partial \Phi}{\partial x} \frac{\partial}{\partial z} \right), \quad (31)$$

$$h^z = ig_e \beta_e^2 q \left(\frac{\partial \Phi}{\partial x} \frac{\partial}{\partial y} - \frac{\partial \Phi}{\partial y} \frac{\partial}{\partial x} \right),$$

and

$$h_x^x = \left(\frac{\partial h_x}{\partial Q_x} \right)_0,$$

$$h_y^x = \left(\frac{\partial h_x}{\partial Q_y} \right)_0, \quad \text{etc.} \quad (32)$$

It is straightforward to calculate the matrix elements of H_{SO} with the basis functions Eq. (23). The result is

$$H_{SO} = i \begin{pmatrix} 0 & \Delta_z & 0 & \Delta_x - i\Delta_y \\ -\Delta_z & 0 & -\Delta_x + i\Delta_y & 0 \\ 0 & \Delta_x + i\Delta_y & 0 & -\Delta_z \\ -\Delta_x - i\Delta_y & 0 & \Delta_z & 0 \end{pmatrix}, \quad (33)$$

where Δ_x, Δ_y, Δ_z are real constants. It can easily be verified that H_{SO} of Eq. (33) commutes with the time-reversal operator of Eq. (25).

When transformed to complex-valued spatial electronic basis functions

$$\psi_\pm = \frac{1}{\sqrt{2}} (\psi_x \pm i\psi_y) \quad (34)$$

and expressed in terms of complex-valued normal-mode displacements

$$Q_\pm = \rho e^{\pm i\phi} = Q_x \pm iQ_y, \quad (35)$$

H_{es} takes the more familiar form[28–31]

$$H_{es} = \frac{1}{2} \omega \rho^2 1_2 + \begin{pmatrix} 0 & X \\ X^* & 0 \end{pmatrix} \quad (36)$$

with

$$X = \kappa \rho e^{i\phi} + \frac{1}{2} g \rho^2 e^{-2i\phi}. \quad (37)$$

H_{SO} of Eq. (33) becomes

$$H_{SO} = \begin{pmatrix} \Delta_z & 0 & \Delta_x - i\Delta_y & 0 \\ 0 & -\Delta_z & 0 & -\Delta_x + i\Delta_y \\ \Delta_x + i\Delta_y & 0 & -\Delta_z & 0 \\ 0 & -\Delta_x - i\Delta_y & 0 & \Delta_z \end{pmatrix}. \quad (38)$$

The SO vibronic matrix [Eq. (38)], which does not depend on the nuclear geometry (within first order in ρ), can be transformed to diagonal form by a unitary 4×4 matrix S, yielding[45]

$$S^\dagger H_{SO} S = \begin{pmatrix} \Delta & 0 & 0 & 0 \\ 0 & -\Delta & 0 & 0 \\ 0 & 0 & -\Delta & 0 \\ 0 & 0 & 0 & \Delta \end{pmatrix} \quad (39)$$

with

$$\Delta = \sqrt{\Delta_x^2 + \Delta_y^2 + \Delta_z^2}. \quad (40)$$

The electrostatic vibronic matrix is invariant with respect to S. The final form of the $^2E \times E$ JT Hamiltonian is thus

$$H = \left(T_N + \frac{1}{2}\omega\rho^2\right) 1_4 + \begin{pmatrix} \Delta & X & 0 & 0 \\ X^* & -\Delta & 0 & 0 \\ 0 & 0 & -\Delta & X \\ 0 & 0 & X^* & \Delta \end{pmatrix}, \quad (41)$$

where 1_4 is the 4×4 unit matrix, T_N is the nuclear kinetic-energy operator and ω is the vibrational frequency of a nondegenerate reference state. Equation (41) agrees with previous results, which have been derived in a more heuristic manner.[46–48] The derivation of the SO matrix [Eq. (39)] from the microscopic BP operator has also been discussed by Marenich and Boggs[49] and Yarkony and collaborators.[50] The adiabatic electronic potential-energy surfaces, that is, the eigenvalues of $(H - T_N 1_4)$, are doubly degenerate (Kramers degeneracy). The adiabatic electronic wave functions carry nontrivial geometric phases which depend on the radius of the loop of integration.[46–48]

It should be noted that the SO operator is nondiagonal in the diabatic spin-orbital electronic basis which usually is employed to set up the $E \times E$ JT Hamiltonian, see Eq. (33). The (usually *ad hoc* assumed) diagonal form of H_{SO} is obtained by the unitary transformation S which mixes spatial orbitals and spin functions of the electron. In this transformed basis, the electronic spin projection is thus no longer a good quantum number.

4. Relativistic Jahn–Teller Coupling in Tetrahedral Systems

The spin-free Hamiltonian for linear and quadratic $T_2 \times T_2$ JT coupling in tetrahedral systems is well established and can be found in many reviews and monographs.[28–31]

Symmetry-adapted atomic displacements transforming as x, y and z in the T_d point group are

$$s_x = \frac{1}{\sqrt{2}}(R_{12} - R_{34}), \tag{42a}$$

$$s_y = \frac{1}{\sqrt{2}}(R_{14} - R_{23}), \tag{42b}$$

$$s_z = \frac{1}{\sqrt{2}}(R_{13} - R_{24}), \tag{42c}$$

where

$$R_{km} = |\mathbf{R}_k - \mathbf{R}_m| \tag{43}$$

and the $\mathbf{R}_k$ are the radius vectors from the origin to the four corners of the tetrahedron. Dimensionless normal coordinates Q_x, Q_y, Q_z of a suitable nondegenerate reference state can be obtained by multiplication with the appropriate inverse L-matrix which depends on the atomic masses and the harmonic force constants of the reference state.[51]

Molecular orbitals transforming according to the T_2 representation of T_d can be constructed as linear combinations of atomic p orbitals on the four atoms

$$\psi_x(\mathbf{r}) = \frac{1}{2}(\phi_x^{(1)} + \phi_x^{(2)} + \phi_x^{(3)} + \phi_x^{(4)}), \tag{44a}$$

$$\psi_y(\mathbf{r}) = \frac{1}{2}(\phi_y^{(1)} + \phi_y^{(2)} + \phi_y^{(3)} + \phi_y^{(4)}), \tag{44b}$$

$$\psi_z(\mathbf{r}) = \frac{1}{2}(\phi_z^{(1)} + \phi_z^{(2)} + \phi_z^{(3)} + \phi_z^{(4)}). \tag{44c}$$

Expansion of the electrostatic Hamiltonian

$$H_{es} = -\frac{1}{2}\nabla^2 - \Phi(\mathbf{r}), \tag{45a}$$

$$\Phi(\mathbf{r}) = \sum_{k=1}^{4} \frac{q}{r_k}, \tag{45b}$$

where $r_k = |\mathbf{r} - \mathbf{R}_k|$ and q is the effective charge on the four identical nuclei, up to first order in $\mathbf{Q}$ and calculation of matrix elements with the basis functions [Eq. (44)] yields the spin-free $T_2 \times T_2$ JT Hamiltonian[28–31] (we suppress the quadratic electrostatic JT coupling for simplicity and brevity)

$$H_{es}(T_2 \times T_2) = (T_N + \frac{1}{2}\omega R^2)1_3 + H_{es}^{(1)}(T_2 \times T_2), \tag{46a}$$

$$H_{es}^{(1)}(T_2 \times T_2) = a\begin{pmatrix} 0 & Q_z & Q_y \\ Q_z & 0 & Q_x \\ Q_y & Q_x & 0 \end{pmatrix}, \tag{46b}$$

where

$$R = \sqrt{Q_x^2 + Q_y^2 + Q_z^2}, \tag{46c}$$

1_3 denotes the 3×3 unit matrix, ω is the harmonic vibrational frequency of the T_2 mode and a is the first-order electrostatic JT coupling constant. Discussions of the adiabatic electronic potential-energy surfaces (the eigenvalues of $H_{es}(T_2 \times T_2) - T_N 1_3$) can be found in the literature.[28–31]

H_{SO} defined by Eq. (18) is expanded in a Taylor series in Q_x, Q_y, Q_z in analogy to Eqs. (29)–(32). Assuming that SO coupling is weak compared to the electrostatic interactions, the expansion is terminated after the first order. The calculation of the matrix elements with the spin-orbital basis functions

$$\psi_x^+ = \psi_x(\mathbf{r})\alpha \tag{47a}$$

$$\psi_y^+ = \psi_y(\mathbf{r})\alpha \tag{47b}$$

$$\psi_z^+ = \psi_z(\mathbf{r})\alpha \tag{47c}$$

$$\psi_z^- = \psi_z(\mathbf{r})\beta \tag{47d}$$

$$\psi_y^- = \psi_y(\mathbf{r})\beta \tag{47e}$$

$$\psi_x^- = \psi_x(\mathbf{r})\beta \tag{47f}$$

yields the SO-induced $^2T_2 \times T_2$ JT Hamiltonian (see Ref. 39 for details)

$$H_{SO}(^2T_2 \times T_2) = H_{SO}^{(0)}(^2T_2) + H_{SO}^{(1)}(^2T_2 \times T_2), \tag{48a}$$

$$H_{SO}^{(0)}(^2T_2) = \Delta \begin{pmatrix} 0 & i & 0 & -1 & 0 & 0 \\ -i & 0 & 0 & i & 0 & 0 \\ 0 & 0 & 0 & 0 & -i & 1 \\ -1 & -i & 0 & 0 & 0 & 0 \\ 0 & 0 & i & 0 & 0 & i \\ 0 & 0 & 1 & 0 & -i & 0 \end{pmatrix}, \tag{48b}$$

$$H_{SO}^{(1)}(^2T_2 \times T_2) = \alpha \begin{pmatrix} 0 & 0 & -iQ_x & -iQ_z & Q_+ & 0 \\ 0 & 0 & iQ_y & Q_z & 0 & -Q_+ \\ iQ_x & -iQ_y & 0 & 0 & -Q_z & iQ_z \\ iQ_z & Q_z & 0 & 0 & iQ_y & -iQ_x \\ Q_- & 0 & -Q_z & -iQ_y & 0 & 0 \\ 0 & -Q_- & -iQ_z & iQ_x & 0 & 0 \end{pmatrix}, \tag{48c}$$

where

$$Q_\pm = Q_x \pm iQ_y \tag{48d}$$

and Δ and α are real constants which represent the zeroth-order SO splitting of the $^2T_2$ state and the first-order relativistic $^2T_2 \times T_2$ JT coupling, respectively.

This analysis reveals that the T_2 mode is JT active in first order both in the electrostatic Hamiltonian [Eq. (46b)] as well as via the SO operator [Eq. (48c)]. The electrostatic (a) and relativistic (α) coupling constants are real, but can be positive or negative. There may thus be constructive or destructive interference of the electrostatic and relativistic JT couplings.

The hermitian matrix $H_{SO}^{(0)}(^2T_2)$ of Eq. (48b) can be transformed to diagonal form by a unitary transformation U

$$\widetilde{H}_{SO}^{(0)}(^2T_2) = U^\dagger H_{SO}^{(0)}(^2T_2)U. \tag{49}$$

In the transformed basis

$$\widetilde{\psi} = U^\dagger \psi, \tag{50}$$

where ψ is the vector of electronic basis functions defined in Eq. (47), the SO vibronic matrix takes the form

$$\widetilde{H}_{SO}^{(0)}(^2T_2) = \text{diag}(-\Delta, -\Delta, -\Delta, -\Delta, 2\Delta, 2\Delta). \tag{51}$$

In agreement with the group-theoretical result [Eq. (10b)], the zeroth-order SO coupling splits the 6-fold degenerate $^2T_2$ manifold into a doubly degenerate manifold ($E_{5/2}$) and a quadruply degenerate manifold ($G_{3/2}$).

Rewriting the first-order electrostatic vibronic matrix [Eq. (46b)] in the spin-orbital basis (47) (which means doubling of the 3×3 matrix to a 6×6 matrix) and transformation of $H_{es}^{(1)} + H_{SO}^{(1)}$ to the SO adapted basis

$$\widetilde{H}_{es}^{(1)}(^2T_2 \times T_2) = U^\dagger H_{es}^{(1)}(T_2 \times T_2)U \qquad (52a)$$

$$\widetilde{H}_{SO}^{(1)}(^2T_2 \times T_2) = U^\dagger H_{SO}^{(1)}(^2T_2 \times T_2)U \qquad (52b)$$

yields

$$\widetilde{H}(^2T_2 \times T_2) = \left(T_N + \frac{1}{2}\omega R^2\right)1_6 + \widetilde{H}^{(0)}(^2T_2)$$

$$+ \widetilde{H}_{es}^{(1)}(^2T_2 \times T_2) + \widetilde{H}_{SO}^{(1)}(^2T_2 \times T_2) = \left(T_N + \frac{1}{2}\omega R^2\right)1_6$$

$$+ \begin{pmatrix} -\Delta & -i\widetilde{a}_1 Q_+ & 0 & i\widetilde{a}_1 Q_z & 0 & i\widetilde{a}_2 Q_- \\ i\widetilde{a}_1 Q_- & -\Delta & -i\widetilde{a}_1 Q_z & 0 & -\frac{i}{\sqrt{3}}\widetilde{a}_2 Q_- & \frac{i}{\sqrt{3}}\widetilde{a}_2 Q_z \\ 0 & i\widetilde{a}_1 Q_z & -\Delta & i\widetilde{a}_1 Q_- & -i\widetilde{a}_2 Q_+ & 0 \\ -i\widetilde{a}_1 Q_z & 0 & -i\widetilde{a}_1 Q_+ & -\Delta & -i\frac{2}{\sqrt{3}}\widetilde{a}_2 Q_z & -\frac{i}{\sqrt{3}}\widetilde{a}_2 Q_+ \\ 0 & \frac{i}{\sqrt{3}}\widetilde{a}_2 Q_+ & i\widetilde{a}_2 Q_- & i\frac{2}{\sqrt{3}}\widetilde{a}_2 Q_z & 2\Delta & 0 \\ -i\widetilde{a}_2 Q_+ & -\frac{i}{\sqrt{3}}\widetilde{a}_2 Q_z & 0 & \frac{i}{\sqrt{3}}\widetilde{a}_2 Q_- & 0 & 2\Delta \end{pmatrix},$$

(53)

where

$$\widetilde{a}_1 = \frac{1}{\sqrt{3}}(a + 2\alpha), \qquad (54a)$$

$$\widetilde{a}_2 = \frac{1}{\sqrt{2}}(a - \alpha). \qquad (54b)$$

It is seen that the effective first-order JT coupling parameters $\widetilde{a}_1$, $\widetilde{a}_2$ have an electrostatic (a) and a relativistic (α) contribution.

For sufficiently large SO splitting Δ, the $G_{3/2}$ manifold (with eigenvalue $-\Delta$) can be considered to be approximately decoupled from the $E_{5/2}$ manifold (with eigenvalue 2Δ). In this approximation, the $^2T_2 \times T_2$ JT

Hamiltonian is reduced to a 4×4 matrix

$$\widetilde{H}^{(1)}\left(G_{3/2} \times T_2\right) = \left(T_N + \frac{1}{2}\omega R^2 - \Delta\right) 1_4$$

$$+ i\widetilde{a}_1 \begin{pmatrix} 0 & -Q_+ & 0 & Q_z \\ Q_- & 0 & -Q_z & 0 \\ 0 & Q_z & 0 & Q_- \\ -Q_z & 0 & -Q_+ & 0 \end{pmatrix}. \qquad (55)$$

The eigenvalues of the vibronic matrix (55) are

$$V_{1,2} = -\Delta + \frac{1}{2}\omega R^2 - |\widetilde{a}_1|R, \qquad (56a)$$

$$V_{3,4} = -\Delta + \frac{1}{2}\omega R^2 + |\widetilde{a}_1|R, \qquad (56b)$$

where R defined in Eq. (46c) and $\widetilde{a}_1$ is given by Eq. (54a).

The adiabatic potentials (56) are doubly degenerate (Kramers degeneracy) and represent a "Mexican Hat" in four-dimensional space (the energy as a function of three nuclear coordinates). While the $G_{3/2} \times T_2$ JT Hamiltonian [Eq. (55)] of a ^2T state in tetrahedral symmetry has been given previously,[52] the above analysis reveals explicitly that the electrostatic potential as well as the SO operator contribute to the linear JT coupling.

It can be shown that the adiabatic electronic wave functions of the $G_{3/2} \times T_2$ JT Hamiltonian [Eq. (55)] carry a nontrivial geometric phase.[39] The geometric phase is defined by the contour integral[53]

$$\gamma_n(C) = i \oint_C d\mathbf{Q}\, \langle U(\mathbf{Q}) | \nabla_Q U(\mathbf{Q}) \rangle, \qquad (57)$$

where $U(\mathbf{Q})$ is a single-valued adiabatic wave function which depends parametrically on the nuclear coordinate vector $\mathbf{Q}$. Evaluation of the contour integral in an arbitrary plane through origin

$$aQ_x + bQ_y + cQ_z = 0 \qquad (58)$$

yields[39]

$$\gamma_1 = \gamma_3 = -\gamma_2 = -\gamma_4 = \frac{\pi c}{\sqrt{a^2 + b^2 + c^2}}. \qquad (59)$$

It is seen that the geometric phases depend on the orientation of the plane spanned by the integration contour.

The $^2T_2 \times E$ JT Hamiltonian with inclusion of relativistic forces can be derived in the same manner. Symmetry-adapted coordinates transforming according to the E representation in T_d are given by the following linear combinations of the edges R_{km} of the tetrahedron

$$s_a = \frac{1}{2}(R_{13} + R_{24} - R_{14} - R_{23}), \tag{60a}$$

$$s_b = \frac{1}{2\sqrt{3}}(2R_{12} + 2R_{43} - R_{13} - R_{24} - R_{14} - R_{23}). \tag{60b}$$

Dimensionless normal coordinates q_a, q_b can be obtained by multiplication with the appropriate inverse L-matrix.[51]

In the SO-adapted basis [defined by Eqs. (49) and (50)] the $^2T_2 \times E$ Hamiltonian up to first order in q_a, q_b takes the form[39]

$$\widetilde{H}(^2T_2 \times E) = \left(T_N + \frac{1}{2}\omega\rho^2\right)1_6$$

$$+ \begin{pmatrix} -\Delta + \widetilde{c}_1 q_- & 0 & 0 & \widetilde{c}_1 q^+ & -\widetilde{c}_2 q_- & 0 \\ 0 & -\Delta - \widetilde{c}_1 q_- & -\widetilde{c}_1 q^+ & 0 & 0 & \widetilde{c}_2 q^+ \\ 0 & -\widetilde{c}_1 q^+ & -\Delta + \widetilde{c}_1 q_- & 0 & 0 & \widetilde{c}_2 q_- \\ \widetilde{c}_1 q^+ & 0 & 0 & -\Delta - \widetilde{c}_1 q_- & \widetilde{c}_2 q^+ & 0 \\ -\widetilde{c}_2 q_- & 0 & 0 & \widetilde{c}_2 q^+ & 2\Delta & 0 \\ 0 & \widetilde{c}_2 q^+ & \widetilde{c}_2 q_- & 0 & 0 & 2\Delta \end{pmatrix}, \tag{61}$$

where

$$\widetilde{c}_1 = \frac{c}{2} - \gamma, \tag{62a}$$

$$\widetilde{c}_2 = \frac{1}{\sqrt{2}}(c + \gamma), \tag{62b}$$

and

$$q_\pm = \frac{\sqrt{3}}{2}q_a \pm \frac{1}{2}q_b, \tag{63a}$$

$$q^\pm = \frac{1}{2}q_a \pm \frac{\sqrt{3}}{2}q_b, \tag{63b}$$

$$\rho = \sqrt{q_a^2 + q_b^2}. \tag{63c}$$

Here, c and γ are the linear electrostatic and relativistic $T_2 \times E$ JT coupling constants, respectively.

If Δ is sufficiently large, the $E_{5/2}$ manifold can be approximately decoupled from the $G_{3/2}$ manifold. The 4×4 JT Hamiltonian of the $G_{3/2}$ manifold becomes

$$\widetilde{H}^{(1)}(G_{3/2} \times E) = \left(T_N + \frac{1}{2}\omega\rho^2 - \Delta\right) 1_4$$

$$+ \widetilde{c}_1 \begin{pmatrix} q_- & 0 & 0 & q^+ \\ 0 & -q_- & -q^+ & 0 \\ 0 & -q^+ & q_- & 0 \\ q^+ & 0 & 0 & -q_- \end{pmatrix}. \quad (64)$$

The adiabatic potential-energy surfaces of the Hamiltonian (64) are

$$V_{1,2} = -\Delta + \frac{1}{2}\omega\rho^2 - |\widetilde{c}_1|\rho, \quad (65a)$$

$$V_{3,4} = -\Delta + \frac{1}{2}\omega\rho^2 + |\widetilde{c}_1|\rho. \quad (65b)$$

Equation (65) represents a (doubly-degenerate) "Mexican Hat" in three-dimensional space (the energy as a function of two nuclear coordinates). The adiabatic eigenvector matrix of the Hamiltonian (64) is real-valued, which implies that the geometric phases (57) are zero (since $\gamma_n(C)$ must be real).

In a ^2E state of tetrahedral systems, as discussed in Sec. 2, E modes are JT active through the electrostatic forces, while T_2 modes are JT active through the relativistic forces. While the electrostatic $E \times E$ JT Hamiltonian of tetrahedral systems is well known,[28–31] the linear relativistic $^2E \times T_2$ Hamiltonian has been derived only recently.[54]

Two degenerate molecular orbitals transforming as the E representation can be obtained with the projection-operator technique

$$\psi_a(\mathbf{r}) = 2^{-\frac{3}{2}}(-\phi_y^{(1)} + \phi_z^{(1)} + \phi_y^{(2)} - \phi_z^{(2)}$$
$$+ \phi_y^{(3)} + \phi_z^{(3)} - \phi_y^{(4)} - \phi_z^{(4)}), \quad (66a)$$

$$\psi_b(\mathbf{r}) = 2^{-\frac{3}{2}} 3^{-\frac{1}{2}}(2\phi_x^{(1)} + 2\phi_x^{(2)} - 2\phi_x^{(3)} - 2\phi_x^{(4)} - \phi_y^{(1)} + \phi_y^{(2)}$$
$$+ \phi_y^{(3)} - \phi_y^{(4)} - \phi_z^{(1)} + \phi_z^{(2)} - \phi_z^{(3)} + \phi_z^{(4)}), \quad (66b)$$

where the $\phi_x^{(k)}$, $\phi_y^{(k)}$, $\phi_z^{(k)}$, $k = 1, 2, 3, 4$, are atomic p orbitals. In analogy to the above analysis for $^2T_2$ states, we evaluate matrix elements of H_{SO} in the basis of spin orbitals

$$\psi_a^+ = \psi_a(\mathbf{r})\alpha, \tag{67a}$$

$$\psi_b^+ = \psi_b(\mathbf{r})\alpha, \tag{67b}$$

$$\psi_a^- = \psi_a(\mathbf{r})\beta, \tag{67c}$$

$$\psi_b^- = \psi_b(\mathbf{r})\beta. \tag{67d}$$

The SO coupling matrix in first order of T_2 normal-mode displacements is given by[54]

$$H_{SO} = i\epsilon \begin{pmatrix} 0 & Q_z & Q_x - iQ_y & 0 \\ -Q_z & 0 & 0 & -Q_x + iQ_y \\ -Q_x - iQ_y & 0 & 0 & Q_z \\ 0 & Q_x + iQ_y & -Q_z & 0 \end{pmatrix}, \tag{68}$$

where ϵ is a real constant. The diagonal elements of H_{SO} are zero according to the group-theoretical selection rule (10c). The elements of the antidiagonal of the matrix (68) are zero as a consequence of time-reversal symmetry.

Including the nuclear kinetic-energy operator T_N and the harmonic potential of the reference state, the total $^2E \times T_2$ Hamiltonian up to first order in T_2 normal modes given by

$$H = \left(T_N + \frac{1}{2}\omega R^2\right) 1_4$$

$$+ \epsilon \begin{pmatrix} 0 & iQ_z & i(Q_x - iQ_y) & 0 \\ -iQ_z & 0 & 0 & -i(Q_x - iQ_y) \\ -i(Q_x + iQ_y) & 0 & 0 & iQ_z \\ 0 & i(Q_x + iQ_y) & -iQ_z & 0 \end{pmatrix}.$$
$$\tag{69}$$

The adiabatic electronic potentials of the Hamiltonian (69) are

$$V_{1,2} = \frac{1}{2}\omega R^2 - |\epsilon|R, \tag{70a}$$

$$V_{3,4} = \frac{1}{2}\omega R^2 + |\epsilon|R. \tag{70b}$$

They represent a (doubly-degenerate) "Mexican Hat" in four-dimensional space. The geometric phases of the adiabatic eigenvectors have been determined in Ref. 54. They are identical with those of the $G_{3/2} \times T_2$ Hamiltonian of the T_2 state, Eq. (59).

The JT Hamiltonians obtained in cubic symmetry (O_h, O'_h) are identical with those of tetrahedral systems, with the exception of possible additional selection rules due to inversion symmetry.

5. Jahn–Teller and Spin-Orbit Coupling in High-Spin States of Trigonal Systems

In transition-metal complexes as well as in rare-earth impurity centers in crystals, one often encounters orbitally degenerate states (E or T) with high spin multiplicity (triplets, quartets, quintets, etc.). It is therefore of interest to develop microscopically founded JT Hamiltonians including SO coupling for E and T electronic states with higher than two-fold spin multiplicity. In this chapter we discuss the derivation of the SO–JT Hamiltonian for triplet electronic states in trigonal symmetry as an example of higher spin multiplicity. The derivations for quartet, quintet, etc. states are completely analogous.

Let us consider two unpaired electrons in a molecular system with trigonal symmetry, e.g. a pyramidal molecule with C_{3v} point-group symmetry. The electrostatic Hamiltonian is given by Eqs. (12)–(14). For two unpaired electrons, we have to include the two-electron terms of the BP operator. The complete BP operator reads:[55]

$$H_{SO} = \sum_k H_{SO}^{(k)} + \sum_{k<l} H_{SO}^{(kl)}, \tag{71a}$$

$$H_{SO}^{(k)} = -ig\,\beta^2 \mathbf{S}_k \cdot \sum_{n=1}^{3} \frac{\widetilde{q}}{r_{kn}^3}(\mathbf{r}_{kn} \times \nabla_k), \tag{71b}$$

$$H_{SO}^{(kl)} = \frac{ig\beta^2}{r_{kl}^3}[\mathbf{S}_k \cdot [\mathbf{r}_{kl} \times (\nabla_k - 2\nabla_l)] + \mathbf{S}_l \cdot [\mathbf{r}_{lk} \times (\nabla_l - 2\nabla_k)]], \tag{71c}$$

with

$$\mathbf{S}_k = \frac{1}{2}(\mathbf{i}\sigma_x^{(k)} + \mathbf{j}\sigma_y^{(k)} + \mathbf{k}\sigma_z^{(k)}). \tag{72}$$

Here $\sigma_x^{(k)}, \sigma_y^{(k)}, \sigma_z^{(k)}$ are the Pauli spin matrices acting on the spin eigenstates of the electron k. It should be noted that H_{SO} is a two-electron operator in the electronic coordinate space, but is a one-electron operator in spin space.

The generalization of the time-reversal operator for several unpaired electrons is

$$\hat{T} = \prod_k \begin{pmatrix} 0 & -1 \\ 1 & 0 \end{pmatrix}_k \hat{cc}, \tag{73}$$

where the product is over unpaired electrons.

Let us consider two unpaired electrons, where one electron occupies one of the orbitals $\psi_\pm$ of an E state [Eq. (34)] and the second electron occupies a spatially nondegenerate orbital ψ_A. Denoting spin-orbitals by $p\sigma, q\sigma$, where $p, q = \psi_A, \psi_\pm$ and $\sigma = \alpha, \beta$, and Slater determinants by $|p\sigma q\sigma\rangle$, spin-adapted nonrelativistic two-electron basis functions for a ^3E state are

$$\begin{aligned}
\Psi_\pm^{(1)} &= |\psi_A \alpha \psi_\pm \alpha\rangle, \\
\Psi_\pm^{(0)} &= \frac{1}{\sqrt{2}}[|\psi_A \alpha \psi_\pm \beta\rangle + |\psi_A \beta \psi_\pm \alpha\rangle], \\
\Psi_\pm^{(-1)} &= |\psi_A \beta \psi_\pm \beta\rangle.
\end{aligned} \tag{74}$$

These functions are orthonormal eigenfunctions of S^2 (with eigenvalue 2) and S_z (with eigenvalues $\pm 1, 0$). Being linear combinations of Slater determinants, they are antisymmetric with respect to electron exchange.

The basis functions (74) can be rewritten as follows

$$\begin{aligned}
\Psi_\pm^{(1)} &= \Phi_\pm(\mathbf{r}_1, \mathbf{r}_2)\alpha_1 \alpha_2, \\
\Psi_\pm^{(0)} &= \frac{1}{\sqrt{2}}\Phi_\pm(\mathbf{r}_1, \mathbf{r}_2)(\alpha_1 \beta_2 + \alpha_2 \beta_1), \\
\Psi_\pm^{(-1)} &= \Phi_\pm(\mathbf{r}_1, \mathbf{r}_2)\beta_1 \beta_2,
\end{aligned} \tag{75}$$

where

$$\Phi_\pm = \frac{1}{\sqrt{2}}[\psi_A(\mathbf{r}_1)\psi_\pm(\mathbf{r}_2) - \psi_A(\mathbf{r}_2)\psi_\pm(\mathbf{r}_1)]. \tag{76}$$

These two-electron basis functions span the 6-dimensional Hilbert space of the ^3E state. They transform under C_3 as

$$C_3 \Psi_\pm^{(m)} = e^{\pm \frac{2i\pi}{3}} \Psi_\pm^{(m)}, \tag{77a}$$

$$C_3^2 \Psi_\pm^{(m)} = e^{\mp \frac{2i\pi}{3}} \Psi_\pm^{(m)}, \quad m = 1, 0, -1. \tag{77b}$$

The action of the time-reversal operator on these basis functions is

$$\hat{T}\Psi_{\pm}^{(1)} = \Psi_{\mp}^{(-1)}\hat{c}c,$$
$$\hat{T}\Psi_{\pm}^{(0)} = -\Psi_{\mp}^{(0)}\hat{c}c, \quad (78)$$
$$\hat{T}\Psi_{\pm}^{(-1)} = \Psi_{\mp}^{(1)}\hat{c}c.$$

The representation of $\hat{T}$ in this basis is thus a 6×6 matrix which can be written as

$$\hat{T} = \begin{pmatrix} 0 & 0 & 1 \\ 0 & -1 & 0 \\ 1 & 0 & 0 \end{pmatrix} \otimes \sigma_x \hat{c}c. \quad (79)$$

Note that $\hat{T}^2 = 1$, as is required for a two-electron system.

Using the six diabatic two-electron basis functions of Eq. (75), the electronic Hamiltonian can be written as a 6×6 vibronic matrix. To derive the matrix elements, H_{es} and H_{SO} are expanded in a Taylor series in powers of the normal-mode displacement ρ up to second and first order, respectively. The derivation of the 6×6 spin-vibronic matrix is described in Ref. 56. The result is

$$H = H_{es} + H_{SO}, \quad (80a)$$

$$H_{es} = \left(T_N + \frac{1}{2}\omega\rho^2\right) 1_6 + 1_3 \otimes \gamma, \quad (80b)$$

$$H_{SO} = A_3 \otimes \sigma_z, \quad (80c)$$

where 1_3 and 1_6 denote the three-dimensional and six-dimensional unit matrix, respectively,

$$\gamma = \begin{pmatrix} 0 & X \\ X^* & 0 \end{pmatrix}, \quad (80d)$$

$$A_3 = \begin{pmatrix} \Delta_z & \delta e^{i\chi} & 0 \\ \delta e^{-i\chi} & 0 & \delta e^{i\chi} \\ 0 & \delta e^{-i\chi} & -\Delta_z \end{pmatrix}, \quad (80e)$$

$$\delta = \sqrt{\Delta_x^2 + \Delta_y^2}, \quad (80f)$$

$$\chi = \tan^{-1}\left(\frac{\Delta_y}{\Delta_x}\right), \quad (80g)$$

and X is defined in Eq. (37). $\Delta_x, \Delta_y, \Delta_z$ are real-valued matrix elements of the SO operator. The symbol $\otimes$ means that each element of A_3 is to be multiplied by the Pauli matrix σ_z.

The matrix (80c) can be transformed to diagonal form by a unitary matrix C. Since C is independent of the nuclear coordinates ρ, ϕ, this unitary transformation defines an alternative diabatic electronic basis. The electrostatic part of the Hamiltonian is invariant under this transformation. The final $^3E \times E$ Hamiltonian is

$$H = \left(T_N + \frac{1}{2}\omega\rho^2\right)\mathbf{1}_6 + \begin{pmatrix} -\Delta & X & 0 & 0 & 0 & 0 \\ X^* & \Delta & 0 & 0 & 0 & 0 \\ 0 & 0 & 0 & X & 0 & 0 \\ 0 & 0 & X^* & 0 & 0 & 0 \\ 0 & 0 & 0 & 0 & \Delta & X \\ 0 & 0 & 0 & 0 & X^* & -\Delta \end{pmatrix}, \quad (81)$$

where

$$\Delta = \sqrt{\Delta_z^2 + 2\delta^2}. \quad (82)$$

The upper-left and lower-right 2×2 matrices represent the familiar $^2E \times E$ JT Hamiltonian with SO splitting Δ, cf. Eq. (41). The central 2×2 block, on the other hand, represents the $E \times E$ JT Hamiltonian without SO coupling, cf. Eq. (36).

The adiabatic potentials are, after a suitable reordering of the eigenvalues,

$$V_1 = V_2 = \frac{1}{2}\omega\rho^2 - \sqrt{\Delta^2 + |q|^2}, \quad (83a)$$

$$V_3 = \frac{1}{2}\omega\rho^2 - |q|, \quad (83b)$$

$$V_4 = \frac{1}{2}\omega\rho^2 + |q|, \quad (83c)$$

$$V_5 = V_6 = \frac{1}{2}\omega\rho^2 + \sqrt{\Delta^2 + |q|^2}. \quad (83d)$$

In the case of linear JT coupling ($g = 0$), we have $|q| = \kappa\rho$ and the adiabatic potentials are independent of the azimuthal angle χ.

The lowest (V_1, V_2) and the uppermost (V_5, V_6) potentials are doubly degenerate in the approximation of an isolated ^3E state. They exhibit the well-known quenching of the JT coupling by the SO splitting.[28–31] V_3 and V_4, on the other hand, represent an $E \times E$ JT effect which is strictly unaffected by the SO coupling.

The extension of this analysis to quartet and quintet states can be found in Ref. 56. In systems with an odd number of electrons (and thus even spin multiplicity M), Kramers degeneracy applies and the two-fold spatial degeneracy of the $^M E$ state is lifted by the SO coupling. In systems with an even number of electrons (and thus odd spin multiplicity M), there exist $M-1$ pairs of adiabatic potentials, in which the JT effect is quenched by the SO coupling, while one pair exhibits a JT effect which is strictly unaffected by the SO splitting.[56] This implies that non-Born–Oppenheimer effects tend to zero in systems with even spin multiplicity in the limit of large SO coupling, while pronounced non-Born–Oppenheimer effects remain in one pair of adiabatic potentials in systems with odd spin multiplicity. The nonadiabatic effects in triplet, quintet, etc. E states are thus fundamentally different from those in doublet, quartet, etc. E states in trigonal systems.

JT coupling in quartet states has recently been experimentally observed and theoretically analyzed by Hauser *et al.* in alkali trimers on the surface of helium droplets.[57] The vibronic spectra have been accurately reproduced by calculations of vibronic energies and intensities based on *ab inito* calculated JT and SO coupling parameters.[57]

6. Examples: Transition-Metal Trifluorides and Group V Tetrahedral Cations

Transition-metal (TM) halogenides form crystals of octahedral structure.[58,59] In the gas phase, transition-metal trifluorides exhibit D_{3h} structure.[60] In MF_3 molecules, orbitally degenerate states (E', E'') with high spin multiplicities ($M = 3, 4, 5, \ldots$) are very common. For a few of these states, calculations have indicated strong JT coupling involving the degenerate bending mode.[60,61] SO coupling has been ignored in these calculations. Gas-phase TM trifluorides have been investigated with electron diffraction[61] and vibrational spectroscopy.[62] To the knowledge of the authors, no gas-phase electronic spectra of MF$_3$ molecules have been reported.

A systematic investigation of the JT effects in the electronic spectra of TM trifluorides is in progress in the laboratories of the authors.[63] Many $^4E'$, $^4E''$, $^5E'$, $^5E''$ states of TM trifluorides exhibit moderate to strong JT couplings, involving the degenerate bending and stretching modes. The SO splittings of the $^ME'$, $^ME''$ states are of the order of 100–300 cm^{-1} for the first row of the TMs and of the same order of magnitude as the bending vibrational frequencies. Naturally, the SO splittings are larger for the second and third rows of the TMs. Nevertheless, the use of nonrelativistic electronic basis functions is appropriate for TM compounds.

For the sake of illustration, let us consider two examples: the lowest $^5E'$ state of MnF$_3$ and the lowest $^5E'$ state of CoF$_3$. For brevity, only the (stronger) JT coupling by the degenerate bending mode is considered. The electronic-structure calculations have been performed with the complete-active-space self-consistent-field (CASSCF) method using the MOLPRO suite of programs. The computational details can be found in Ref. 63. The vibrational frquency of the E mode has been obtained from the second derivative of the mean value of the *ab initio* energies. The calculated spectroscopic data for the ^5E states of MnF$_3$ and CoF$_3$ are collected in Table 1.

The adiabatic potentials as a function of the bending coordinate Q_x of the lowest $^5E'$ state of MnF$_3$, calculated without (a) and with (b) SO coupling are displayed in Fig. 1. Figure 1(b) illustrates the generic SO-splitting pattern of a quintet E state which has been discussed in the preceding section. The calculated value of the SO parameter Δ is 145 cm^{-1}. It can be seen that there exists a pair of PE functions which are unaffected by the SO coupling, as pointed out in Sec. 5.

Vibronic spectra, corresponding to transitions from a nondegenerate reference state to the JT active states, have been calculated by the

Table 1. Zeroth-order SO splitting (Δ), harmonic bending vibrational frequency (ω) and linear (κ) and quadratic (g) JT coupling constants for the $^5E'$ states of MnF$_3$ and CoF$_3$.

molecule	Δ(cm^{-1})	ω(cm^{-1})	κ(cm^{-1})	g(cm^{-1})
MnF$_3$	145.4	186.8	786.7	41.5
CoF$_3$	292.9	191.2	545.9	27.4

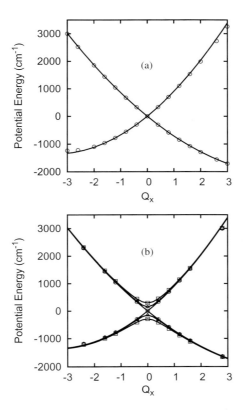

Fig. 1. Nonrelativistic (a) and relativistic (b) *ab initio* potential-energy of the $^5E'$ state of MnF$_3$ as a function of the coordinate Q_x of E symmetry.

diagonalization of the Hamiltonian [Eq. (81)] in a basis set of two-dimensional harmonic-oscillator functions. The calculated vibronic spectrum of the $^5E'$ state of MnF$_3$ is shown in Fig. 2 without (a) and with (b) inclusion of SO coupling. The spectrum of Fig. 2(b) consists of a superposition of three JT spectra with SO splittings $0, \Delta, 2\Delta$. The double-hump envelope of the spectrum in Fig. 2(a) is characteristic for a strong linear $E \times E$ JT effect.[64] The irregular vibronic line position and intensities in Fig. 2(a) are the result of a rather strong quadratic JT coupling. The inclusion of SO coupling [Fig. 2(b)] leads to a drastic increase of the line density. The upper hump of the spectral envelope is reduced in intensity and shifted to higher energies by the SO coupling. It should be noted that the spectrum

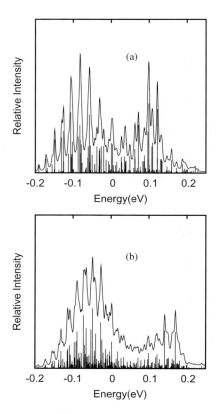

Fig. 2. Vibronic spectrum of the $^5E'$ state of MnF$_3$ without (a) and with (b) inclusion of SO coupling. The envelope represents the convolution of the stick spectrum with a Lorentzian function of 5 meV FWHM.

of Fig. 2(b) contains exactly one replica of the electrostatic JT spectrum in Fig. 2(a).

The adiabatic potentials of the lowest $^5E'$ state of CoF$_3$ are shown in Fig. 3 without (a) and with (b) inclusion of SO coupling. As expected, the SO splitting is larger in CoF$_3$ ($\Delta = 293\,\text{cm}^{-1}$) than in MnF$_3$. The corresponding electronic spectra are shown in Fig. 4 without (a) and with (b) inclusion of SO coupling. The rather strong SO coupling ($\Delta/\omega = 1.6$) has a pronounced effect on the vibronic spectrum. The intensity of the upper spectral hump in Fig. 4(b) is strongly reduced. The separated clumps of lines at about 0.15 eV and 0.20 eV in Fig. 4(b) correspond to the lowest

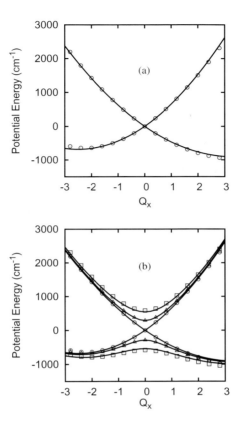

Fig. 3. Nonrelativistic (a) and relativistic (b) *ab initio* potential-energy of the $^5E'$ state of CoF$_3$ as a function of the coordinate Q_x of E symmetry.

energy levels of the two upper adiabatic surfaces, which are perturbed by nonadiabatic coupling with the lower adiabatic PE surfaces.

The *ab inito* electronic-structure calculations for MF$_3$ molecules confirm that the SO coupling is represented by a constant (the SO splitting at the D_{3h} reference geometry) to a very good approximation. There is no need to take corrections (quadratic in ρ) into account. In tetrahedral (and octahedral) systems, the situation is quite different, as explained in Sec. 4. Here the SO coupling gives rise to relativistic JT forces which interfere constructively or destructively with the electrostatic JT forces.

The vibronic spectra of the cluster cations P$_4^+$, As$_4^+$, Sb$_4^+$, Bi$_4^+$ have been observed by photoelectron spectroscopy of the tetrahedral clusters

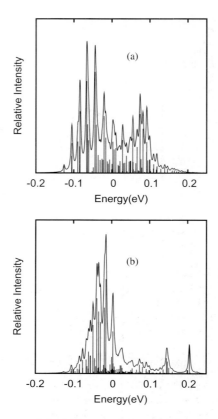

Fig. 4. Vibronic spectrum of the $^5E'$ state of CoF$_3$ without (a) and with (b) inclusion of SO coupling. The envelope represents the convolution of the stick spectrum with a Lorentzian function of 5 meV FWHM.

P$_4$, As$_4$, Sb$_4$, Bi$_4$.[65,66] It has been found that they exhibit very pronounced electrostatic $E \times E$ and $T_2 \times T_2$ JT effects.[65–67]

To determine the magnitude of the relativistic JT coupling parameters, we have performed relativistic *ab initio* electronic-structure calculations for the ^2E ground state and the $^2T_2$ first excited electronic state of the cluster cations P$_4^+$, As$_4^+$, Sb$_4^+$, Bi$_4^+$. The wavefunction of the electrostatic Hamiltonian has been constructed in the state-averaged CASSCF approximation, employing the cc-pVTZ basis set. Matrix elements of the BP Hamiltonian have been computed with the CASSCF wave functions. For P$_4^+$, all-electron calculations have been performed, while relativistic small-core pseudopotentials have been employed for the heavier systems.

All *ab initio* calculations have been performed with the MOLPRO suite of programs. Vibrational frequencies of the T_2 and E modes have been obtained as second derivatives of the mean energy $(V_{1,2} + V_{3,4})/2$ for ^2E states and $(V_{1,2} + V_{3,4} + V_{5,6})/3$ for $^2T_2$ states, respectively. The relativistic JT coupling parameters have been determined by fitting the eigenvalues of the $^2E \times (E + T_2)$ and $^2T_2 \times (E + T_2)$ vibronic matrices to the *ab initio* data.

To visualize the effects of the nonrelativistic and relativistic JT couplings, the *ab initio* adiabatic PE surfaces of the ^2E and $^2T_2$ states of Sb_4^+ are shown in Fig. 5 as a function of the symmetry coordinate s_z of T_2 symmetry (s_x, s_y and s_z are equivalent) without (a) and with (b) inclusion of SO coupling. Figure 5(a) illustrates the very strong nonrelativistic $T_2 \times T_2$ JT coupling in the $^2T_2$ state and the absence of a nonrelativistic $E \times T_2$ coupling in the ^2E state. Figure 5(b) illustrates the zeroth-order SO splitting of the T_2 state and the existence of a relativistic linear $E \times T_2$ JT coupling in the ^2E state, which partially lifts the four-fold degeneracy of this state. Figure 6 displays the *ab initio* adiabatic PE surfaces of the ^2E and $^2T_2$ states of Sb_4^+ as a function of the symmetry coordinate s_b of E symmetry. Figure 6(a) illustrates the very pronounced nonrelativistic $E \times E$ JT coupling in the ^2E state, while the $T_2 \times E$ JT coupling is weak. Figure 6(b) illustrates the absence of a quenching of the strong electrostatic JT effect in the ^2E state by the SO coupling.

Table 2 (for the $^2T_2$ state) and Table 3 (for the ^2E state) contain the zeroth-order SO splitting Δ, the vibrational frequencies ω and the dimensionless linear relativistic JT coupling parameters. The tables include the linear electrostatic $T_2 \times T_2$ (a/ω_{T_2}), $T_2 \times E$ (c/ω_E) and $E \times E$ (b/ω_E) JT coupling parameters for comparison. It should be mentioned that the electrostatic $E \times E$ and $T_2 \times T_2$ JT effects in the P_4^+, As_4^+ and Sb_4^+ cluster cations are among the strongest JT effects known in nature.[65,67]

The zeroth-order SO splitting Δ of the $^2T_2$ state exhibits the expected increase with the atomic number Z of the atoms, see Table 2. While the magnitudes of the electrostatic $T_2 \times T_2$ (a/ω_{T_2}) and $T_2 \times E$ (c/ω_E) JT coupling parameters fluctuate throughout the series, the relativistic JT coupling parameters increase monotonously with Z. For Bi_4^+, the relativistic $T_2 \times T_2$ JT coupling is of the same order of magnitude as the electrostatic JT coupling and represents a truly pronounced JT effect ($|a|/\omega_{T_2} \approx 2$). For the ^2E ground state, the electrostatic $E \times E$ JT coupling (b/ω_E) is likewise roughly constant throughout the series, while

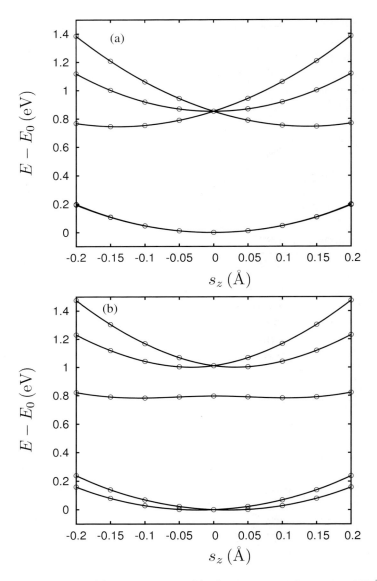

Fig. 5. Nonrelativistic (a) and relativistic (b) *ab initio* potential-energies of Sb_4^+ as a function of the coordinate s_z of T_2 symmetry.

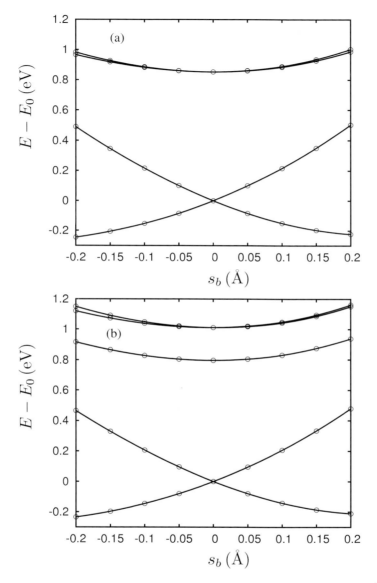

Fig. 6. Nonrelativistic (a) and relativistic (b) *ab initio* potential-energies of Sb_4^+ as a function of the coordinate s_b of E symmetry.

Table 2. Zeroth-order SO splitting Δ, vibrational frequencies (ω_{T_2}, ω_E), electrostatic (a/ω_{T_2}, c/ω_E) and relativistic (γ/ω_E, α/ω_{T_2}) linear JT coupling parameters for the $^2T_2$ state of the group V tetramers. For the T_2 and E modes of P_4^+ and the E mode of As_4^+, the relativistic coupling parameters are too small to be reliably determined by the least-squares fit.

$T_2 \times (E + T_2)$	P_4^+	As_4^+	Sb_4^+	Bi_4^+
Δ (meV)	−5.18	−25.07	−71.45	−264.30
ω_E (meV)	41.01	17.61	14.90	10.49
ω_{T_2} (meV)	51.88	22.72	19.97	14.21
$\dfrac{c}{\omega_E}$	−0.51	0.68	0.11	0.55
$\dfrac{a}{\omega_{T_2}}$	2.78	4.11	3.19	3.49
$\dfrac{\gamma}{\omega_E}$	—	—	0.21	0.54
$\dfrac{\alpha}{\omega_{T_2}}$	—	−0.43	−0.96	−2.01

Table 3. Vibrational frequencies (ω_{T_2}, ω_E), electrostatic (b/ω_E) and relativistic (ϵ/ω_{T_2}) linear JT coupling parameters for the ^2E state of the group V tetramers. For P_4^+, the relativistic coupling parameter (ε/ω_{T_2}) is too small to be reliably determined by the least-squares fit.

$E \times (E + T_2)$	P_4^+	As_4^+	Sb_4^+	Bi_4^+
ω_E (meV)	40.57	18.16	15.00	10.65
ω_{T_2} (meV)	46.46	18.07	17.99	12.89
$\dfrac{b}{\omega_E}$	4.86	6.29	5.68	5.98
$\dfrac{\varepsilon}{\omega_{T_2}}$	—	0.30	0.48	1.18

the magnitude of the relativistic $^2E \times T_2$ JT coupling ($|\epsilon|/\omega_{T_2}$) increases with Z (see Table 3). It should be noted that the T_2 mode is not JT active in the ^2E state in the nonrelativistic approximation. The relativistic forces, however, result in a distortion $\Delta R = |\epsilon|/\omega_{T_2}$ along the T_2 radial coordinate R which is substantial for the heavier group V cluster cations (see Table 3).

7. Conclusions

The intention of this chapter has been the outline of a systematic generalization of JT theory to include the SO operator at the same level of approximation as the electrostatic Hamiltonian in the spin-free JT theory. The BP operator is expanded in a Taylor series in normal-mode displacements at the reference geometry and matrix elements are taken with nonrelativistic diabatic spin-orbital basis functions, taking account of the symmetry selection rules (spin double-group symmetry and time-reversal symmetry).

While the mutual quenching of SO splitting and linear JT coupling is very well known (often referred to as the Ham effect),[28–31] this phenomenon is just a special case of SO vibronic coupling, arising in spin-double groups which lack irreducible representations with dimension larger than two. In tetrahedral and cubic systems, on the other hand, JT forces of relativistic origin exist and may either enhance or weaken the electrostatic JT forces. Making use of existing electronic structure codes, the vibronic-coupling parameters arising from the SO operator have been calculated for a few exemplary systems. The *ab initio* calculations indicate that the relativistic JT forces are comparable with or can even exceed the electrostatic JT forces in systems with very heavy elements.[68]

As discussed in the introduction, the theory of nonadiabatic dynamics at conical intersections needs to be generalized by a systematic inclusion of SO coupling effects. Such developments are prerequisite for a first-principles theoretical understanding of the photophysics and photochemistry of complexes containing heavy elements. This problem has been addressed by Matsika and Yarkony from the point of view of *ab initio* computational chemistry.[16–20] In this chapter, we have emphasized the group-theoretical aspects in systems of high symmetry. It is suggested that the relativistically generalized JT theory can provide important guidelines for the development and computational realization of a comprehensive theory of nonadiabatic dynamics in systems with strong SO coupling.

Acknowledgments

The collaboration of the authors has been supported for many years by travel grants of the Deutsche Forschungsgemeinschaft. The authors would like to thank Sabyashachi Mishra, Padmabati Mondal, Daniel Opalka and Ilias Sioutis for their contributions to this research.

References

1. R.L. Jackson, *Acc. Chem. Res.* **25**, 581 (1992).
2. S.K. Kim, S. Pedersen and A.H. Zewail, *Chem. Phys. Lett.* **233**, 500 (1995).
3. L. Banares, T. Baumert, M. Bergt, B. Kiefer and G. Gerber, *Chem. Phys. Lett.* **267**, 141 (1997).
4. S.A. Trushin, W. Fuss, W.E. Schmid and K.L. Kompa, *J. Phys. Chem. A* **102**, 4129 (1998).
5. A. Assion, T. Baumert, M. Bergt, T. Brixner, B. Kiefer, V. Seyfried, M. Strehle and G. Gerber, *Science* **282**, 919 (1998).
6. T. Brixner, N.H. Damrauer and G. Gerber, *Adv. At. Mol. Opt. Phys.* **46**, 1 (2001).
7. B. Oregan and M. Grätzel, *Nature* **353**, 737 (1991).
8. M. Grätzel, *Nature* **414**, 338 (2001).
9. G.C. Walker, P.F. Barbara, S.K. Doorn, Y. Dong and J.T. Hupp, *J. Phys. Chem.* **95**, 5712 (1991).
10. S.K. Doorn, R.B. Dyer, P.O Stoutland and W.H. Woodruff, *J. Am. Chem. Soc.* **115**, 6398 (1993).
11. G. Benko, J. Kallioinen, J.E.I. Korppi-Tommola, A.P Yartsev and V. Sundstrom, *J. Am. Chem. Soc.* **124**, 489 (2002).
12. A. Cannizzo, F. van Mourik, W. Gawelda, G. Zgrablic, C. Bressler and M. Chergui, *Angew. Chem. Int. Ed.* **45**, 3174 (2007).
13. G.J. Hedley, A. Rusekas and I.D.W. Samuel, *J. Phys. Chem. A* **113**, 2 (2008).
14. C.A. Mead, *J. Chem. Phys.* **70**, 2276 (1979).
15. C.A. Mead, *Chem. Phys.* **49**, 33 (1980).
16. S. Matsika and D.R. Yarkony, *J. Chem. Phys.* **115**, 2038 (2001).
17. S. Matsika and D.R. Yarkony, *J. Chem. Phys.* **115**, 5066 (2001).
18. S. Matsika and D.R. Yarkony, *J. Chem. Phys.* **116**, 2825 (2002).
19. S. Matsika and D.R. Yarkony, *J. Phys. Chem. B* **106**, 8108 (2002).
20. S. Matsika and D.R. Yarkony, *Adv. Chem. Phys.* **124**, 557 (2002).
21. W. Moffitt and A.D. Liehr, *Phys. Rev.* **106**, 1195 (1957).
22. A.D. Liehr, *J. Phys. Chem.* **67**, 389 (1963).
23. H.C. Longuet-Higgins, in *Advances in Spectroscopy*, Vol. 2, Ed. H.W. Thompson (Interscience, New York, 1961), p. 429.
24. E. Renner, *Z. Physik* **92**, 172 (1934).
25. J.A. Pople and H.C. Longuet-Higgins, *Mol. Phys.* **1**, 372 (1958).
26. T. Barrow, R.N. Dixon and G. Duxbury, *Mol. Phys.* **27**, 1217 (1974).
27. H.A. Jahn and E. Teller, *Proc. Roy. Soc. (London) A* **161**, 220 (1937).
28. M.D. Sturge, *Solid State Phys.* **20**, 91 (1967).
29. R. Englman, *The Jahn-Teller Effect* (Wiley, New York, 1972).
30. I.B. Bersuker and V.Z. Polinger, *Vibronic Interactions in Molecules and Crystals*, Springer, Heidelberg, 1989.
31. I.B. Bersuker, *The Jahn-Teller Effect*, Cambridge University Press, Cambridge, 2006.
32. W. Lichten, *Phys. Rev.* **131**, 229 (1963).

33. F.T. Smith, *Phys. Rev.* **179**, 111 (1969).
34. V. Sidis, *Adv. At. Mol. Phys.* **26**, 161 (1989).
35. A. Wolf, M. Reiher and B.A. Heß, in *Relativistic Electronic-Structure Theory. Part 1. Fundamentals*, Ed. P. Schwerdtfeger, Chap. 11 (Elsevier, Amsterdam, 2002).
36. L.D. Landau and E.M. Lifshitz, *Quantum Mechanics*, Nauka, Moscow, 1974.
37. M. Hamermesh, *Group Theory and its Application to Physical Problems*, Addison-Wesley, Reading, 1962.
38. L.V. Poluyanov and W. Domcke, in *The Jahn-Teller Effect. Fundamentals and Implications for Physical Chemistry*, Eds. H. Köppel, D.R. Yarkony and H. Barentzen (Springer, Heidelberg, 2009), p. 77.
39. L.V. Poluyanov and W. Domcke, *Chem. Phys.* **374**, 86 (2010).
40. We use the symbols $E_{1/2}$, $E_{5/2}$, etc. of Landau and Lifshitz and Hamermesh for the double-valued representations of spin double groups.
41. In the solid-state literature, the representations $E_{1/2}$, $E_{5/2}$, $G_{3/2}$ of tetrahedral and cubic groups are denoted as Γ_6, Γ_7, Γ_8, following H. A. Bethe, *Ann. Phys.* **395**, 133 (1929).
42. E. Wigner, *Group Theory*, Academic Press, New York, 1959.
43. H.A. Jahn, *Proc. Roy. Soc. (London) A* **164**, 117 (1938).
44. H.A. Bethe and E.E. Salpeter, *Quantum Mechanics for One- and Two-Electron Atoms*, Springer, Berlin, 1957.
45. W. Domcke, S. Mishra and L.V. Poluyanov, *Chem. Phys.* **322**, 405 (2006).
46. A. Stone, *Proc. Roy. Soc. (London) A* **351**, 141 (1976).
47. H. Koizumi and S. Sugano, *J. Chem. Phys.* **102**, 4472 (1995).
48. J. Schön and H. Köppel, *J. Chem. Phys.* **108**, 1503 (1998).
49. A.V. Marenich and J.E. Boggs, *J. Phys. Chem.* **108**, 10594 (2004).
50. M.S. Schuurman, D.E. Weinberg and D.R. Yarkony, *J. Chem. Phys.* **127**, 104309 (2007).
51. E.B. Wilson, J.C. Decius and P.C. Cross, *Molecular Vibrations*, McGraw-Hill, New York, 1955.
52. W. Moffitt and W. Thorson, *Phys. Rev.* **108**, 1251 (1957).
53. M.V. Berry, *Proc. Roy. Soc. (London) A* **392**, 45 (1984).
54. L.V. Poluyanov and W. Domcke, *J. Chem. Phys.* **129**, 224102 (2008).
55. B.A. Heß and C.M. Marian, in *Computational Molecular Spectroscopy*, Eds. P. Jensen and P.R. Bunker (Wiley, New York, 2000), p. 152.
56. L.V. Poluyanov and W. Domcke, *Chem. Phys.* **352**, 125 (2008).
57. G. Auböck, J. Nagl, C. Callegari and W.E. Ernst, *J. Chem. Phys.* **129**, 114501 (2008).
58. R. Colton and J.H. Canterford, *Halides of the First Row Transition Metals*, Wiley, New York, 1971.
59. A.F. Wells, *Structural Inorganic Chemistry*, Clarendon Press, Oxford, 1975.
60. J.H. Yates and R.M. Pitzer, *J. Chem. Phys.* **70**, 4049 (1979).
61. M. Hargittai, B. Reffy, M. Kolonits, C.J. Marsden and J.-L. Heully, *J. Am. Chem. Soc.* **119**, 9042 (1997).
62. V.N. Bukhmarina, A.Y. Gerasimov, Y.B. Predtechenskii and V.G. Shklyarik, *Opt. Spectrosc. (USSR)* **65**, 518 (1988).

63. P. Mondal, D. Opalka, L.V. Poluyanov and W. Domcke, to be published.
64. H.C. Longuet-Higgins, U. Öpik, M.H.L. Price and R.A. Sack, *Proc. Roy. Soc. (London) A* **244**, 1 (1958).
65. L.-S. Wang, B. Niu, Y.T. Lee, D.A. Shirley, E. Ghelichkhani and E.R. Grant, *J. Chem. Phys.* **93**, 6318 (1990).
66. L.-S. Wang, B. Niu, Y.T. Lee, D.A. Shirley, E. Ghelichkhani and E.R. Grant, *J. Chem. Phys.* **93**, 6327 (1990).
67. R. Meiswinkel and H. Köppel, *Chem. Phys. Lett.* **201**, 449 (1993).
68. D. Opalka, M. Segado, L.V. Poluyanov and W. Domcke, *Phys. Rev. A* **81**, 042501 (2010).

Chapter 5

Symmetry Analysis of Geometric-Phase Effects in Quantum Dynamics

Stuart C. Althorpe[*]

1. Introduction 156
2. Time-Independent Description of Single-Surface Dynamics . . . 157
   2.1. Topology and encirclement 157
   2.2. Symmetry approach 159
   2.3. Path-integral approach 162
   2.4. Relation between symmetry and path-integral approaches 166
   2.5. Distinction between scattering and bound states 169
3. Time-Dependent Description of Single-Surface Dynamics 172
   3.1. Symmetry approach 172
   3.2. Path-integral approach 174
4. Time-Dependent Description of Dynamics on Two Coupled Surfaces 176
   4.1. Symmetry approach 176
   4.2. Time-ordered-product path-integral approach 178
   4.3. Behaviour of Feynman paths at CI seam 181
5. Numerical Applications 182

[*]Department of Chemistry, University of Cambridge, Lensfield Road, Cambridge, CB2 1EW, UK.

5.1. Cancellation of GP effects in the
       hydrogen-exchange reaction 183
5.2. Geometric phase effects in two-surface
       population transfer 187
6. Conclusions 191
   Acknowledgments 192
   References 192

1. Introduction

One of the defining properties of a conical intersection (CI) is that the adiabatic electronic wave function changes sign upon following a closed loop around the CI. This property is referred to as the geometric phase (GP), or sometimes the Berry phase.[1] The GP was first identified in molecular systems by Herzberg and Longuet-Higgins,[2,3] who showed that the GP is a direct consequence of the first-order lifting of the degeneracy about the CI. Because the total (electronic + nuclear) wave function must be single-valued, the GP produces a corresponding change in the boundary condition of the nuclear wave function.[4-8] When one discusses GP effects on the dynamics, one is therefore addressing how the properties of the nuclear wave function are affected by this change in boundary condition.

The effect of the GP on bound state, time-independent wave functions has been known for a long time. It is easy to visualise the effect by considering a particle on a ring, with wave function $\exp(im\phi)$. In place of the usual continuity boundary condition, which requires that m be an integer, the GP boundary condition requires that m be a half-integer. As a result, the GP boundary condition changes the nodal structure of the wave function, and produces a shift in the energy levels. Similar changes in nodal structures and shifts in energies have been calculated and observed in the spectra of Jahn–Teller molecules.[9-12]

The effect of the GP on other nuclear wave functions is, however, more complicated. In time-independent scattering wave functions, or time-dependent wave packets (for both scattering and bound systems), the wave function can be generated by Feynman paths that loop a finite number of times around the CI. The effect of the GP is to change the relative sign of the odd and even-looping paths, which, in general, describe different types of dynamics. This effect of the GP was understood in the early 1970s,[13-16] where it was demonstrated in model two-dimensional calculations of the

Aharonov-Bohm effect (which is effectively a two-slit experiment with GP boundary conditions). However, these simple results were not taken up at the time by the chemical physics community; as a result, the effect of the GP on time-independent scattering wave functions and on time-dependent wave functions (of any kind) remained something of a mystery.

Here we review work we have done over the past five years[17-24] to adapt and apply the theory just described to molecular systems. This work was inspired by a long running puzzle in the prototype hydrogen-exchange reaction[25-34] but the results apply generally (and rigorously) to all dynamical systems that have a conical intersection (except for time-independent bound states, which are something of a special case, as already mentioned). It turns out that one does not need to use path-integral theory if one does not want to: one can apply symmetry arguments in a double space, which make it mathematically trivial to extract (rigorously) the contributions from even- and odd-looping Feynman paths from the nuclear wave function. This technique allowed us to solve the hydrogen-exchange puzzle just mentioned, and more generally turns out to be a useful tool for analysing nuclear dynamics at conical intersections. In particular, the technique allows one to estimate the effect of interference between even- and odd-looping paths on population transfer.

This chapter is structured as follows: Section 2 describes the application of the symmetry and equivalent path-integral approaches to time-independent scattering wave functions, in which the dynamics is confined to the lower of two conically intersecting potential energy surfaces. Section 3 describes the equivalent application in the time domain. Section 4 generalises Sec. 3, to cover time-dependent dynamics on both coupled surfaces. Section 5 summarises the results of applications to the hydrogen-exchange reaction (5.1) and to time-dependent dynamics on two surfaces (5.2). Section 6 concludes the chapter.

2. Time-Independent Description of Single-Surface Dynamics

2.1. *Topology and encirclement*

Let us consider a system with N nuclear degrees of freedom, which possesses one CI seam of dimension $N - 2$. The topology of the nuclear space can then be represented schematically as shown in Fig. 1.[18] The line at the centre represents every CI point in the seam. Each circular cut through the

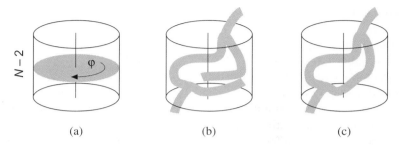

Fig. 1. (a) Diagram illustrating the topology of the N-dimensional nuclear coordinate space of a reactive system with a conical intersection (CI). The vertical line represents the $(N-2)$-dimensional space occupied by the CI seam. The grey disc represents a 2D branching-space cut through one point on the seam, with the angle ϕ describing internal rotation around the CI. (b) A nuclear wave function which wraps around the CI, but is not a torus, and thus exhibits only trivial GP effects. (c) A torus-shaped nuclear wave function encircling the CI. [Reprinted with permission from Ref. 18. Copyright 2006, American Institute of Physics.]

cylinder represents the two degrees of freedom in the nuclear "branching space", in which the adiabatic potential energy surfaces have the familiar double-cone shape, centred about the CI point. We assume that the seam line extends throughout the entire region of energetically accessible nuclear coordinates space. We also assume that the system is confined to the lower (adiabatic) electronic state, because it has insufficient energy to approach the region of strong coupling with the upper state close to the CI. The CI seam line is therefore surrounded by a tube of inaccessible coordinate space.

We then define an internal coordinate ϕ such that $\phi = 0 \to 2\pi$ denotes a path which has described one complete loop around the CI in the nuclear branching space. Other than this, we need not specify further details about ϕ. It is sufficient that ϕ permits us to count how many times a closed loop has wound around the CI. Using this definition of ϕ, we can express the effect of the GP on the adiabatic ground state electronic wave function $\Phi(\phi)$ and the nuclear wave function $\Psi(\phi)$ as

$$\Phi(\phi + 2n\pi) = (-1)^n \Phi(\phi), \tag{1}$$

$$\Psi(\phi + 2n\pi) = (-1)^n \Psi(\phi). \tag{2}$$

The dependence on the other $N-1$ nuclear degrees of freedom has been suppressed.

The effects of the GP are therefore the differences between the nuclear dynamics described by the wave function

$$\Psi^{[G]}(\phi) = (-1)^n \Psi^{[G]}(\phi + 2n\pi), \tag{3}$$

which correctly includes the GP boundary condition, and the wave function

$$\Psi^{[N]}(\phi) = \Psi^{[N]}(\phi + 2n\pi), \tag{4}$$

which ignores it (and is therefore physically incorrect). It is well known in the literature that the GP will produce a non-trivial effect on the dynamics only when the time-independent wave function $\Psi^{[G]}(\phi)$ *encircles* the CI. Otherwise the effect is simply a change in the phase of $\Psi^{[N]}(\phi)$, which has no effect on any observables. Hence we seek to explain how the dynamics described by $\Psi^{[G]}(\phi)$ differs from the dynamics described by $\Psi^{[N]}(\phi)$, when these wave functions encircle the CI. (Note that the requirement for encirclement applies to the time-*independent* wave function; it is not necessary for the corresponding time-*dependent* wave function to encircle the CI at any instant in time, as we discuss in Sec. 3.)

It is worth clarifying what is meant by encirclement. The notion of taking a cut through nuclear coordinate space in order to see whether $\Psi^{[G]}(\phi)$ encircles the CI in this cut is not useful. Figure 1(b) shows a nuclear wave function which, if a certain choice of nuclear coordinates were used, could easily be made to 'encircle' the CI if a suitable 2D cut were taken. However, this particular wave function would *not* show non-trivial GP effects, because it does not encircle the CI: it has unconnected 'ends'. For non-trivial GP effects to appear, $|\Psi^{[G]}(\phi)|^2$ must have the form of a *torus* in the nuclear coordinate space, as shown in Fig. 1(c). If one were to take a series of branching-space cuts through this wave function, none of them would encircle the CI, and hence one might get the mistaken impression that this wave function would only show a trivial phase change upon inclusion of the GP boundary condition. However, the wave function of Fig. 1(c) would definitely show strong, non-trivial GP effects. There are various ways in which one can prove this and we will mention one below. It is important to emphasise that it is $|\Psi^{[G]}(\phi)|^2$ which has the form of a torus, and not the wave function $\Psi^{[G]}(\phi)$.

2.2. *Symmetry approach*

To explain the effect of the GP on the nuclear dynamics [i.e. to explain the difference between the dynamics described by an encircling $\Psi^{[G]}(\phi)$ and an encircling $\Psi_N(\phi)$], we need to compare the topology of $\Psi^{[G]}(\phi)$ with the topology of $\Psi^{[N]}(\phi)$. In Sec. 2.3, we review how this can be done using the homotopy of the Feynman paths[13–16] that contribute to these wave functions. But first, to demonstrate the simplicity of the problem, we use the diagrammatic approach developed in Refs. 17 and 18.

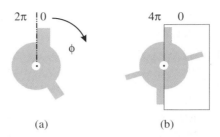

Fig. 2. (a) Schematic picture of the potential surface of a reactive system, indicating that there is an energetically accessible (grey) 'tube' through the potential surface, permitting encirclement of the CI (dot at centre). The 'arms' are the reagent and product channels. (b) The same surface, represented in the $0 \to 4\pi$ cover space. The rectangle represents a $0 \to 2\pi$ sector which can be cut out of the double space so as to map back onto the single space, where the $\phi = 0$ and $\phi = 2\pi$ 'edges' are joined together at the cut line (chains). [Reprinted with permission from Ref. 18. Copyright 2006, American Institute of Physics.]

We represent the internal coordinate space occupied by the nuclear wave function as shown in Fig. 2. The grey area represents the energetically accessible region of the potential energy surface; the conical intersection is the point at the centre; the 'arms' represent the reactant entrance and product exit channels. To simplify the discussion, we place a restriction on ϕ (which will be relaxed later), which is that ϕ tends to a constant value as the system moves down the entrance or exit channel towards an asymptotic separation of the reactants or products. This places no restriction on the generality of the diagram, other than that the conical intersection should be located in the 'strong-interaction region' of the potential energy surface, where all the nuclei are close together. Note that, although we have restricted the number of product channels to one, the diagram is immediately generalisable to systems with multiple product channels. We also assume that the reaction is bimolecular (leaving unimolecular reactions until Sec. 2.4), which means that it is initiated at the asymptotic limit of the reactant channel, at the value of ϕ which is reached in this limit. We will define this to be $\phi = 0$.

Figure 2(b) represents the potential surface of the same system, mapped onto the double-cover space.[18] The latter is obtained simply by 'unwinding' the encirclement angle ϕ, from $0 \to 2\pi$ to $0 \to 4\pi$, such that two (internal) rotations around the CI are represented as one in the page. The potential is therefore symmetric under the operation $\hat{R}_{2\pi}$, defined as an internal rotation by 2π in the double space. To map back onto the single space, one cuts out a 2π-wide sector from the double space. This is taken to be the $0 \to 2\pi$

sector in Fig. 2(b), but any 2π-wide sector would be acceptable. Which particular sector has been taken is represented by a cut line in the single space, so in Fig. 2(b) the cut line passes between $\phi = 0$ and 2π. Since the single space is the physical space, any observable obtained from the total (electronic + nuclear) wave function in this space must be independent of the position of the cut line.

To construct a diagrammatic representation of the wave function, we start in the double space, as shown in Fig. 3(a). The arrow at the top indicates that the incoming boundary condition is applied here, and the arrows at each of the other channels indicate outgoing boundary conditions. Note that we are treating the second appearance of the reactant channel (at $\phi = 2\pi$) as though it were a product channel, and are treating the second appearance of the product channel (in the $2\pi \to 4\pi$ sector) as though it were physically distinct from the first appearance of this channel (which is indicated by the use of wavy lines). As a result, the wave function $\Psi^{[e]}$ is neither symmetric nor antisymmetric under $\phi \to \phi + 2\pi$, which means it cannot be mapped back onto the physical space independently of the position of the cut line. In other words $\Psi^{[e]}$ is the wave function of a completely artificial system.

To construct wave functions which can be mapped back onto the physical space, one needs to take symmetric and antisymmetric linear combinations of $\Psi^{[e]}(\phi)$ and $\Psi^{[o]}(\phi) = \Psi^{[e]}(\phi + 2\pi)$, and these are illustrated in Fig. 3(b). It is then clear that these functions can be mapped onto the physical space [Fig. 3(c)], and that they correspond to $\Psi^{[N]}$ and $\Psi^{[G]}$

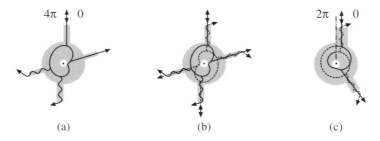

Fig. 3. (a) The unsymmetrised nuclear wave function $\Psi^{[e]}$ (solid line) in the double space. The arrows indicate the application of incoming and outgoing scattering boundary conditions. (b) The symmetrised linear combinations of $\Psi^{[e]}$ (solid) and $\Psi^{[o]}$ (dashed), which yield $\Psi^{[N/G]} = 1/\sqrt{2}\{\Psi^{[e]} \pm \Psi^{[o]}\}$. (c) The same functions mapped back onto the single space. [Reprinted with permission from Ref. 18. Copyright 2006, American Institute of Physics.]

respectively. Thus we may write,

$$\Psi^{[G]} = \frac{1}{\sqrt{2}}\{\Psi^{[e]} - \Psi^{[o]}\},$$

$$\Psi^{[N]} = \frac{1}{\sqrt{2}}\{\Psi^{[e]} + \Psi^{[o]}\}. \tag{5}$$

This equation is the main result needed to explain the effect of the GP on the nuclear dynamics of a chemical reaction. Clearly the sole effect of the GP is to change the relative sign of $\Psi^{[e]}$ and $\Psi^{[o]}$. Within each of these functions the dynamics is completely unaffected by the GP. We emphasise that, despite remaining unnoticed for so long in the chemical physics community, Eq. (5) is exact.

If we can compute $\Psi^{[G]}$ and $\Psi^{[N]}$ numerically (as described below), it is therefore trivial to extract $\Psi^{[e]}$ and $\Psi^{[o]}$ by evaluating

$$\Psi^{[e]} = \frac{1}{\sqrt{2}}\{\Psi^{[N]} + \Psi^{[G]}\},$$

$$\Psi^{[o]} = \frac{1}{\sqrt{2}}\{\Psi^{[N]} - \Psi^{[G]}\}. \tag{6}$$

Once one has extracted $\Psi^{[e]}$ and $\Psi^{[o]}$, an explanation of the GP effect on the nuclear dynamics will follow immediately. The dynamics in $\Psi^{[e]}$ is decoupled from the dynamics in $\Psi^{[o]}$, and thus any observable will show GP effects only if the corresponding operator depends on $\Psi^{[e]}$ and $\Psi^{[o]}$ in a region of space where these functions overlap. In a non-encircling nuclear wave function, $\Psi^{[e]}$ and $\Psi^{[o]}$ never overlap, and this gives us a diagrammatic proof (Fig. 4) of the well-known result that a non-encircling wave function shows no non-trivial GP effects.

2.3. *Path-integral approach*

Here, we explain the path-integral approach of Refs. 13–16 which, as we will show in Sec. 2.4, is equivalent to the wavefunction-based theory just described. This approach was developed originally to treat the Aharonov–Bohm system, in which an electron encircles but does not touch a magnetic solenoid. The vector potential of the solenoid has an effect which is exactly equivalent to the application of the GP boundary condition, and scattering boundary conditions are applied at long range. The Aharonov–Bohm system is therefore exactly analogous to a nuclear wave function in a reactive system which encircles a CI.

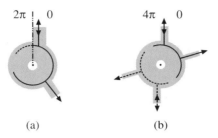

Fig. 4. (a) Single-space and (b) double-space representations of $\Psi^{[e]}$ (solid) and $\Psi^{[o]}$ (dashed) for a system which does not encircle the CI. [Reprinted with permission from Ref. 18. Copyright 2006, American Institute of Physics.]

Only a few basic concepts of path integrals are required to understand this theory. We review these here in a heuristic manner, beginning with the celebrated result of Feynman,[35] which is that the time-evolution operator or Kernel, $K = \exp(-i\hat{H}t/\hbar)$, can be constructed using

$$K(\mathbf{x}, \mathbf{x_0}|t) = \int \mathcal{D}\mathbf{x}(t) e^{iS(\mathbf{x}, \mathbf{x_0}|t)/\hbar}. \qquad (7)$$

Here $\mathcal{D}\mathbf{x}(t)$ represents the sum over all possible paths connecting the points $\mathbf{x}$ and $\mathbf{x_0}$ in the time interval t, and S is the classical action evaluated along each of these individual paths. It is useful to point out two properties of this expression: (a) the overall sign of the Kernel is arbitrary, because S is only defined up to an overall constant (because S is the time-integral over the Lagrangian, and the latter is only defined up to a total derivative in t^{36}); (b) each path has equal weight, so the relative contribution of a given path to the sum is determined by the extent to which it is cancelled out by its immediate neighbours.

Any prediction expressed in the language of path integrals must have an equivalent formulation in the language of wave functions. Point (a) is equivalent to saying that a wave function is only specified up to an overall phase factor. Point (b) can be thought of as saying that, when computing $K(\mathbf{x}, \mathbf{x_0}|t)$, all possible paths between $\mathbf{x}$ and $\mathbf{x_0}$ in time t are 'coupled'. If we start with one particular path between $\mathbf{x}$ and $\mathbf{x_0}$, then we need to know all of its immediate neighbours, in order to assess the extent to which this path is cancelled out by them. These neighbouring paths are obtained by all possible tiny distortions that can be applied to the first path. We then need to know all of the immediate neighbours of each of the latter paths (in order to assess the extent to which each of these is cancelled out), and then we need to find out the immediate neighbours of the new paths, and so on. In other words, if we start with one particular path between **x** and **x₀**, then this path is 'coupled' (in the sense just described) to all the other paths into which it can be continuously deformed. This is equivalent to saying that one cannot compute accurately just part of a wave function; one must compute all of it, since all parts of the function are coupled by the Hamiltonian operator.

In their work on the Aharonov–Bohm system, Schulman, deWitt and others[13–16] found that points (a) and (b) must be modified when applying path-integral theory in a *multiply-connected* space. The term 'multiply-connected' simply means that the space contains an inaccessible region or obstacle, which 'gets in the way', such that a given path between **x** and **x₀** cannot be continuously deformed into all other possible paths. It can only be deformed into a subset of such paths. This subset defines a *homotopy class*: paths that belong to different homotopy classes are called different *homotopes*. The concept that there exist different classes of paths, such that a path that belongs to one class cannot be continuously deformed into a path that belongs to another, is called *homotopy*.

The nuclear coordinate space shown in Fig. 1 is a multiply-connected space, because there is an energetically inaccessible 'tube' of space surrounding the CI seam. An explanation of the homotopy of such a space will be found in any elementary text on topology.[37] Let us take first a system with only two nuclear degrees of freedom, so that the CI is just a point at the centre of the branching space, and there are no other degrees of freedom. The homotopy of a given path within this space is simply the number of entire loops it follows around the CI. We can thus classify each homotopic class according to a winding number n, as defined in Fig. 5. Note that the sign of n indicates the *sense* of the path, and that it is useful to adopt the convention that even n refer to paths that make an even number

Fig. 5. Examples of Feynman paths belonging to different homotopy classes, illustrating how the winding number n is defined. [Reprinted with permission from Ref. 18. Copyright 2006, American Institute of Physics.]

of clockwise loops or an odd number of counterclockwise loops; and odd n vice versa. It is easy to prove that the set of all these homotopic classes forms an infinite group (which is called the 'Fundamental Group' of a circle[37]).

The same classification into winding numbers can be used in a system with N nuclear degrees of freedom, in which the CI seam is an $(N-2)$-dimensional hyperline as in Fig. 1. For example, if we take $N=3$, then the seam is a line; the homotopy of this system is just the same as for the $N=2$ system, since the number of loops made around the line can be represented by a winding number defined exactly as for the 2D case. Although it is difficult to visualise, the generalisation continues to all higher N in the same way, so that one can always classify a path by its winding number around the $(N-2)$-dimensional CI hyperline. In many systems, each class of paths designated by the winding number n will in fact include more than one homotopy class, because it will be possible to further classify the paths according to their winding about other energetically inaccessible regions in the potential surface (which may exist in addition to the tube around the CI seam). However, to understand the GP, we do not need to consider these classes, and so, for shorthand, we will use the terms 'homotopy class' and 'winding number' interchangeably.

Retracing the argument used to justify point (b) above, it is clear that, in a multiply-connected space, a given path is only coupled to those paths into which it can be continuously deformed. By definition, these are all the paths that belong to the same homotopy class. Paths belonging to different homotopy classes are thus decoupled from one another.[13–16] For a reactive system with a CI which has the space of Fig. 1, this means that a path with a given winding number n is coupled to all paths with the same n, but is decoupled from paths with different n. As a result, the Kernel separates into[13–16]

$$K(\mathbf{x},\mathbf{x_0}|t) = \sum_{n=-\infty}^{\infty} e^{in\beta} K^{[n]}(\mathbf{x},\mathbf{x_0}|t), \quad (8)$$

where

$$K^{[n]}(\mathbf{x},\mathbf{x_0}|t) = \int \mathcal{D}_n \mathbf{x}(t) e^{iS(\mathbf{x},\mathbf{x_0}|t)/\hbar} \quad (9)$$

and $\mathcal{D}_n \mathbf{x}(t)$ denotes the sum over all paths linking $\mathbf{x_0}$ to $\mathbf{x}$ that have winding number n.

Each $K^{[n]}$ in Eq. (8) has a different overall phase, which arises because a different Lagrangian can be used for each value of n. However, there is a strong constraint on these phases, which arises because the set of Kernels

$K^{[n]}$ must form an irreducible representation of the Fundamental Group of the circle. As a result, the phases have the form $e^{in\beta}$ [given in Eq. (8)], so that there is only one parameter β that can be varied. To determine possible values of β, let us consider the operation $\phi \to \phi + 2\pi$ on $K^{[n]}$. This operation is equivalent to rotating the end points of all the paths around the CI by 2π, thus increasing the winding number of each path from n to $n+1$. As a result, $K^{[n]} \to K^{[n+1]}$, which means that, overall, $K \to \exp(i\beta)K$. In other words, specifying β is equivalent to specifying the $\phi \to \phi + 2\pi$ boundary condition which is to be satisfied by the Kernel. Thus, we can obtain Kernels corresponding to GP and non-GP boundary conditions by choosing $\beta = \pi$ and $\beta = 0$, which gives

$$K^{[G]}(\mathbf{x},\mathbf{x_0}|t) = \frac{1}{\sqrt{2}}\{K^{[e]}(\mathbf{x},\mathbf{x_0}|t) - K^{[o]}(\mathbf{x},\mathbf{x_0}|t)\},$$

$$K^{[N]}(\mathbf{x},\mathbf{x_0}|t) = \frac{1}{\sqrt{2}}\{K^{[e]}(\mathbf{x},\mathbf{x_0}|t) + K^{[o]}(\mathbf{x},\mathbf{x_0}|t)\}, \qquad (10)$$

where $K^{[e]} = \sum K^{[n]}$, with the sum running over all even n, and $K^{[o]}$ is similarly defined for odd n.

To put this result in context, one should imagine a crude semiclassical calculation, in which one propagates Newtonian trajectories, each of which is given a phase $\exp(iS/\hbar)$. One could implement the GP boundary condition by counting the number of loops n made by each trajectory around the CI, and adding an extra $n\pi$ to the associated phase. To our knowledge, no such calculation has been reported, almost certainly because it would be difficult to disentangle genuine GP effects from errors in the approximation. However, Eq. (10) tells us that such an intuitive approach can be applied to the Feynman paths, and thus implemented rigorously, without approximation.

2.4. *Relation between symmetry and path-integral approaches*

The separation of the Feynman paths in Eq. (10) is equivalent to the splitting of the wave function into $\Psi^{[e]}$ and $\Psi^{[o]}$ in Eq. (6). To demonstrate this, we connect the Kernel to the wave function using Ref. 38,

$$\Psi(\mathbf{x}) = \frac{1}{A(E)} \int d\mathbf{x_0} \int_0^\infty dt\, e^{iEt/\hbar} K(\mathbf{x},\mathbf{x_0}|t)\chi(\mathbf{x_0}), \qquad (11)$$

where $\chi(\mathbf{x_0})$ is an initial wave packet, which contains a spread of energies $A(E)$. At time $t = 0$, $\chi(\mathbf{x_0})$ is localised in the reactant channel, at a sufficiently large reactant separation that the interaction potential can be neglected. The function $\Psi(\mathbf{x})$ given by Eq. (11) is the time-independent wave function, with incoming boundary conditions in the reactant channel, as represented schematically in Fig. 3. It follows immediately from Eq. (11) that $K^{[e]}$ generates $\Psi_e(\phi)$, and $K^{[o]}$ generates $\Psi^{[o]}(\phi)$. Hence unwinding the nuclear wave function according to Eq. (6) is equivalent to separating the even n Feynman paths [which are contained in $\Psi^{[e]}(\phi)$] from the odd n paths [which are contained in $\Psi^{[o]}(\phi)$].

Some care must be taken when applying the Feynman interpretation to Eq. (6), as the interpretation must be consistent with the position of the cut line (used to map from the double to the single space). For example, Figs. 6(a) and 6(b) show two different choices of cut line. It is clear that the relative sign of $\Psi^{[e]}(\phi)$ and $\Psi^{[o]}(\phi)$, and hence all the GP effects, are independent of the position of the cut line. However, the overall phase of $\Psi^{[G]}(\phi)$ does depend on the cut line. This phase is important, because it must cancel out a corresponding phase in the electronic wave function $\Phi(\phi)$, to give a total wave function $\Psi^{[G]}(\phi)\Phi(\phi)$ which is independent of the position of the cut line. Because of this, one needs to define the winding number n with respect to the cut line and *not* with respect to the ($\phi = 0$) point at which the nuclear wave function enters the encirclement region. Thus in

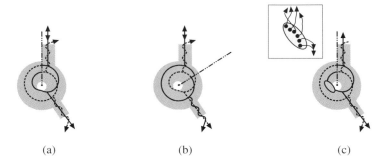

Fig. 6. Diagram showing how the winding number n of the Feynman paths should be defined with respect to the cut line. In (a) the cut line (chains) is placed between $\phi = -\epsilon$ and $2\pi - \epsilon$; in (b) between (approx) $\phi = \pi/4$ and $-7\pi/4$. In (c) the wave function describes a unimolecular reaction, in which the initial state occupies the (grey shaded) area shown. Feynman paths originate from all points within this area (inset); their winding number n is defined with respect to the common cut line. [Reprinted with permission from Ref. 18. Copyright 2006, American Institute of Physics.]

Fig. 6(b), a path which starts at $\phi = 0$ and terminates at $\phi = \pi/2$, and has made no loops around the CI, is classified as an $n = 0$ path. However, a path which starts at $\phi = 0$ and terminates just short of the cut line at, say, $\phi = \pi/6$, and has also made no loops around the CI, is an $n = -1$ path. A path which enters at $\phi = 0$ and makes one clockwise loop around the CI will be an $n = 0$ path if it terminates just short of the cut line, and will only become an $n = 1$ path once it has passed the cut line; and so on. By classifying the Feynman paths with respect to the cut line in this way, we ensure that the overall phase of $\Psi^{[G]}(\phi)$ has the correct dependence on ϕ needed to cancel the corresponding dependence of the phase of $\Phi(\phi)$.

In Fig. 3, we placed the cut line between $\phi = -\epsilon$ and $\phi = 2\pi - \epsilon$ (where ϵ is an arbitrarily small number). This will often be the most convenient choice of cut line, because n then describes exactly the number of complete loops that the system has made around the CI since entering the encirclement region. Thus the paths that scatter inelastically will each have described an (internal) rotation of exactly $\phi = 2n\pi$.

However, it will not always be possible to fix the cut line at the same value of ϕ as the entry points, for the reason that the system does not enter the encirclement region at one unique value of ϕ. Up till now, we have assumed (see Sec. 2.1) that the reaction is bimolecular, that it can only encircle the CI when the nuclei are all close together, and that the reactants and products are distinguishable. These conditions are what are required to guarantee that the system starts at one unique value of ϕ (which we have taken to be $\phi = 0$). We can now relax these conditions, and consider unimolecular reactions, and reactions that can encircle the CI at large separations of the reactants or products.

We can represent a unimolecular reaction using the diagram of Fig. 6(c). The grey blob indicates the initial state of the system. For example, it could be the Frank–Condon region accessed in a photodissociation experiment.[39] All the Feynman paths that contribute to the nuclear wave function will originate in the initial state. Hence the paths will have a spread of start points, distributed over the range of ϕ for which the initial state is non-negligible. Clearly, the symmetry argument of Sec. 2.2 applies immediately to this system, so we may unwind the wave function, and extract $\Psi^{[e]}$ and $\Psi^{[o]}$ using Eq. (6). We can then interpret these functions as containing the even n and odd n Feynman paths respectively, where n is defined with respect to a fixed cut line. Note that, when we discuss, say, the even n Feynman paths, we are not referring to paths that all necessarily complete an even number of loops around the CI, since the paths may have started on different sides of the cut line (if the latter passes through the initial state), or on different sides of the end point. The reader may verify that, in either of these cases, the even n paths will contain a mixture of paths that have looped an even and an odd number of times around the CI.

Hence, when applied to a unimolecular reaction, Eq. (6) does not give such a neat separation into even- and odd-looping Feynman paths. However, the separation that it does give (into even and odd n, each of which contains a mixture of even-looping and odd-looping paths) is the one that is necessary to explain the effect of the GP, since these are the two contributions to Ψ whose relative sign is changed by the GP. Clearly, if we were to compute directly the Kernels, we could then separate out the odd-looping and even-looping paths, because we would know the starting point of each path. In the wave function, however, we do not know the starting points of the individual paths, nor do we need to in order to explain the effect of the GP.

Similar arguments to those just given apply to bimolecular reactions in which the CI can be encircled when the reactants are still well separated from one another. For such systems, one cannot define ϕ such that it tends to a unique value as the system travels out along the reactant channel. The incoming boundary condition must then be applied across a range of ϕ, which is analogous to the range of ϕ contained in the initial state of the unimolecular reaction. Applying Eq. (6) will then separate out the even n and odd n paths with respect to a fixed cut line, which paths may contain a mixture of odd-looping and even-looping paths (as in the unimolecular case).

2.5. *Distinction between scattering and bound states*

This chapter focuses on reactive systems, in which the nuclear wave function satisfies scattering boundary conditions, applied at the asymptotic limits of the reactant and product channels. Here we discuss briefly how such wave functions differ from bound state wave functions, when subjected to the analysis described in the preceding sections.

Let us attempt to apply the procedure of Sec. 2.2 to a bound state wave function. This is illustrated schematically in Fig. 7. It is clear immediately that we cannot construct an unsymmetric $\Psi^{[e]}$ in the double space, because each bound state eigenfunction must be an irreducible representation of the double-space symmetry group. Thus a bound state function in the double

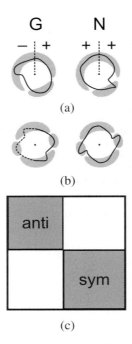

Fig. 7. Relation between $\Psi^{[G]}$ and $\Psi^{[N]}$ for a bound state system. The functions in the single space (a) can be mapped onto the double space (b) where they have opposite symmetries with respect to rotation around the CI by 2π, and therefore correspond to different symmetry blocks of the double-space hamiltonian matrix (c). [Reprinted with permission from Ref. 18. Copyright 2006, American Institute of Physics.]

space is necessarily symmetric or antisymmetric under $\hat{R}_{2\pi}$, and is thus either a $\Psi^{[G]}$ or a $\Psi^{[N]}$ function. For a $\Psi^{[G]}$ function, we have $\Psi^{[N]} = 0$ (since $\Psi^{[G]}$ and $\Psi^{[N]}$ cannot form a degenerate pair) which implies [from Eq. (6)] that

$$\Psi^{[e]} = -\Psi^{[o]} = \Psi^{[G]}. \tag{12}$$

Similarly, for a $\Psi^{[N]}$ function,

$$\Psi^{[e]} = \Psi^{[o]} = \Psi^{[N]}. \tag{13}$$

In other words, if we map a bound state wave function onto the double space using Eq. (6), we simply duplicate the function, because the contribution from the even n Feynman paths is exactly equal to (or equal and opposite to) the contribution from the odd n paths.

An encircling reactive wave function is thus topologically different from an encircling bound state wave function. This is why, in Sec. 2.1, we said that, when the wave function encircles the CI, it is $|\Psi(\phi)|^2$ which has the form of a torus, rather than $\Psi(\phi)$. A reactive wave function $\Psi(\phi)$ is not a torus — it is essentially a coil, since it can be unwound. A bound state function $\Psi^{[G]}(\phi)$, on the other hand, is a torus (with a twist), because if one imagines calculating it by propagating a function around the CI, then the two ends of the function must match. In a reactive wave function, no such matching is ever required, and we may loop each end of the function around the CI as many times as we please, before allowing each end to pass out through one of the entrance or exit channels (where it is then matched to the asymptotic scattering functions). We therefore suggest that the term 'encirclement' often used in the GP literature should be qualified as 'weak encirclement', when the system is reactive, and 'strong encirclement' when it is bound.

Mapping onto the double space will therefore reveal nothing new about the effect of the GP on a bound state system, since this is purely a boundary-condition effect. However, it does give us an alternative representation of the GP and non-GP wave functions, which may sometimes be clearer than the equivalent single-space representation (in which one deduces the symmetry Ψ from the total wave function $\Phi\Psi$). For example, in the double space, it is very clear that the GP will cause all the bound states to be doubly-degenerate when the (single-space) molecular symmetry group is isomorphic with C_{2v} (because the double-space group is then isomorphic with C_{4v}). Similarly, the double-space picture is analogous to a double-well system, with periodic boundary conditions, and this may also sometimes be useful in rationalising the effect of the GP on the spectrum.

The distinction between reactive and bound state wave functions obviously becomes less clear-cut when one considers very long-lived reactive resonances (see Ref. 18). Also, we emphasise that this distinction only applies to time-*independent* wave functions. For a time-dependent wave function (evaluated at a finite time t), the Feynman paths that survive in the Kernel describe a finite number of loops around the CI, irrespective of whether the system is bound or scattering. Time-independent bound-state wave functions are thus a special case; they are the only nuclear wave functions in which the Feynman paths can loop an infinite number of times around the CI, and for which the decomposition into $\Psi^{[e]}$ and $\Psi^{[o]}$ simply regenerates the original $\Psi^{[G]}$ and $\Psi^{[N]}$.

3. Time-Dependent Description of Single-Surface Dynamics

The various expressions obtained in Sec. 2 can be converted into forms that apply to time-dependent wave functions, by taking Fourier transforms. However, it is simpler and more instructive to apply the approach directly in the time-domain, as follows.[23]

3.1. *Symmetry approach*

Let us write the complete time-dependent wave packet for a two-surface system as

$$\Psi^{[\Lambda]}(\phi,t) = \chi_1^{[\Lambda]}(\phi,t)\Phi_1(\phi) + \chi_2^{[\Lambda]}(\phi,t)\Phi_2(\phi), \qquad (14)$$

where $\Phi_1(\phi)$ and $\Phi_2(\phi)$ are the ground and excited (adiabatic) electronic states, and $\chi_1^{[\Lambda]}(\phi,t)$ and $\chi_2^{[\Lambda]}(\phi,t)$ are the coupled nuclear wave functions associated with each of these states. Following Sec. 2, we suppress the dependence of the wave function on all nuclear degrees of freedom, except for the encirclement angle ϕ; the superscript $[\Lambda]$ denotes whether the wave function satisfies the geometric phase (G) boundary condition

$$\chi_\alpha^{[G]}(\phi + 2n\pi, t) = (-1)^n \chi_\alpha^{[G]}(\phi, t), \qquad (15)$$

(where $\alpha = 1$ or 2) or the continuity (non-geometric phase, N) boundary condition

$$\chi_\alpha^{[N]}(\phi + 2n\pi, t) = \chi_\alpha^{[N]}(\phi, t). \qquad (16)$$

We assume that the initial $t = 0$ wave function $\chi_\alpha^{[\Lambda]}(\phi, 0)$ is a wave packet which is sufficiently localised not to encircle the CI. This condition is satisfied in most of the femtosecond pump-probe experiments used to probe quantum dynamics at conical intersections. Figure 8 gives single- and double-space representations of the wave packet $\chi_\alpha^{[\Lambda]}(\phi, t)$, analogous to those used to represent the time-independent wave function in Figs. 2–6; i.e. one can map from the double to the single space by cutting out a 2π-wide sector from the double space, which we take as the interval $\phi = 0 \to 2\pi$. Each representation shows the packet at the initial time $t = 0$, when it is localised, and at some later time t, when it has spread.

Let us consider the initial wave packet $\chi_\alpha^{[\Lambda]}(\phi, 0)$ in Fig. 8 (where Λ can be either G or N). In the single (physical) space, this packet is assumed to

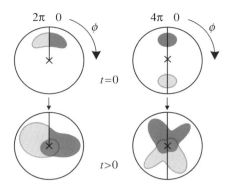

Fig. 8. Schematic diagram showing the single-space (left) and double-space (right) representations of a time-dependent wave packet $\chi_\alpha^{[\Lambda]}(\phi,t)$ ($\Lambda = G$ or N) at initial time $t = 0$, and at a later time t. The dark and light-shaded areas represent the wave packets $\chi_\alpha^{[e]}(\phi,t)$ and $\chi_\alpha^{[o]}(\phi,t)$. [Reprinted with permission from Ref. 23. Copyright 2008, American Institute of Physics.]

consist of one localised piece, as shown in Fig. 8. Equations (15) and (16) then tell us that $\chi_\alpha^{[\Lambda]}(\phi,0)$ must consist of two localised pieces in the double space, which, in anticipation of what follows, we write as $\chi_\alpha^{[e]}(\phi,0)/\sqrt{2}$ and $\chi_\alpha^{[o]}(\phi,0)/\sqrt{2}$. We then have

$$\chi_\alpha^{[G]}(\phi,0) = \frac{1}{\sqrt{2}}\{\chi_\alpha^{[e]}(\phi,0) - \chi_\alpha^{[o]}(\phi,0)\},$$

$$\chi_\alpha^{[N]}(\phi,0) = \frac{1}{\sqrt{2}}\{\chi_\alpha^{[e]}(\phi,0) + \chi_\alpha^{[o]}(\phi,0)\}, \qquad (17)$$

where

$$\chi_\alpha^{[o]}(\phi,0) = \chi_\alpha^{[e]}(\phi + 2\pi, 0). \qquad (18)$$

The pieces $\chi_\alpha^{[e]}(\phi,0)/\sqrt{2}$ and $\chi_\alpha^{[o]}(\phi,0)/\sqrt{2}$ are shaded dark and light in Fig. 8, which can be regarded as a schematic representation of either $\chi_\alpha^{[G]}(\phi,t)$ (when the dark and light-shaded packets are assumed to have opposite signs) or $\chi_\alpha^{[N]}(\phi,t)$ (when they are assumed to have the same sign). For convenience later on, we have defined ϕ in Fig. 1 such that the $0 \to 2\pi$ cut line passes through $\chi_\alpha^{[e]}(\phi,0)$ and $\chi_\alpha^{[o]}(\phi,0)$. In the single space, therefore, $\chi_\alpha^{[\Lambda]}(\phi,0)$ consists of two pieces joined together at the cut line: to the right of the cut line, $\sqrt{2}\chi_\alpha^{[\Lambda]}(\phi,0)$ is equal to the $0 \to 2\pi$ piece of $\chi_\alpha^{[e]}(\phi,0)$; to the left, it is equal to the $0 \to 2\pi$ piece of $\pm\chi_\alpha^{[o]}(\phi,0)$.

Let us now consider $\chi_\alpha^{[G]}(\phi,t)$ and $\chi_\alpha^{[N]}(\phi,t)$ at some later time t. From Eqs. (17) and (18), and from the linearity of the time-evolution operator, it follows that

$$\chi_\alpha^{[G]}(\phi,t) = \frac{1}{\sqrt{2}}\{\chi_\alpha^{[e]}(\phi,t) - \chi_\alpha^{[o]}(\phi,t)\},$$

$$\chi_\alpha^{[N]}(\phi,t) = \frac{1}{\sqrt{2}}\{\chi_\alpha^{[e]}(\phi,t) + \chi_\alpha^{[o]}(\phi,t)\}, \qquad (19)$$

and

$$\chi_\alpha^{[o]}(\phi,t) = \chi_\alpha^{[e]}(\phi + 2\pi, t). \qquad (20)$$

These expressions give us a simple interpretation of the effect of the GP. They show that the wave packet $\chi_\alpha^{[\Lambda]}(\phi,t)$ is a superposition of two components $\chi_\alpha^{[e]}(\phi,t)$ and $\chi_\alpha^{[o]}(\phi,t)$, and that the sole effect of the GP is to change their relative sign. When represented in the single (i.e. physical) space, the components $\chi_\alpha^{[e]}(\phi,t)$ and $\chi_\alpha^{[o]}(\phi,t)$ are twisted together inside $\chi_\alpha^{[\Lambda]}(\phi,t)$ as shown schematically in Fig. 8, and their dynamics are coupled by the time-evolution operator. When represented in the double space, however, $\chi_\alpha^{[e]}(\phi,t)$ and $\chi_\alpha^{[o]}(\phi,t)$ map onto two separate wave packets. The dynamics of one packet is mirrored in that of the other through the symmetry relation of Eq. (20), but the dynamics of the two packets are completely decoupled. Equation (19) is in fact Eq. (5), re-expressed in the time domain. In the single space, the components $\chi_\alpha^{[e]}(\phi,t)$ and $\chi_\alpha^{[o]}(\phi,t)$ correspond to the contributions from even and odd-looping Feynman paths, as we shall clarify in Sec. 3.2. Non-trivial GP effects will arise whenever $\chi_\alpha^{[e]}(\phi,t)$ and $\chi_\alpha^{[o]}(\phi,t)$ overlap, which is equivalent to requiring that $\chi_\alpha^{[G]}(\phi,t)$ follows a path that encircles the CI. This in turn is equivalent to requiring that the time-independent wave function encircles the CI.

3.2. *Path-integral approach*

It is straightforward[23] to link the time-dependent wave functions $\chi_\alpha^{[e]}(\phi,t)$ and $\chi_\alpha^{[o]}(\phi,t)$ to the corresponding Feynman paths, as was done for $\Psi^{[e]}$ and $\Psi^{[o]}$ in Sec. 2. We write the wave packet $\chi_1^{[\Lambda]}(\phi,t)$ in terms of the action of the kernel (time-evolution operator) on an initial wave packet

$$\chi_1^{[\Lambda]}(\mathbf{x},t) = \int d\mathbf{x}_0 K^{[\Lambda]}(\mathbf{x},\mathbf{x}_0|t)\chi_1^{[N]}(\mathbf{x}_0,0), \qquad (21)$$

where the N nuclear degrees of freedom (which include the encirclement angle ϕ) are now included explicitly as the vector $\mathbf{x}$. Note that the same initial wave packet $\chi_1^{[N]}(\mathbf{x_0}, 0)$ is used to generate both $\chi_1^{[G]}(\mathbf{x}, t)$ and $\chi_1^{[N]}(\mathbf{x}, t)$, since the effects of the GP boundary condition and cut line are included in the kernel; i.e. the effect of $K^{[G]}(\mathbf{x}, \mathbf{x_0}|0)$ on $\chi_1^{[N]}(\mathbf{x_0}, 0)$ is to switch the sign of the $\phi < 0$ part, thus converting this function into $\chi_1^{[G]}(\mathbf{x}, 0)$. We will often drop the Λ superscript in what follows, leaving it to be understood that $K(\mathbf{x}, \mathbf{x_0}|t)$ represents either $K^{[G]}(\mathbf{x}, \mathbf{x_0}|t)$ or $K^{[N]}(\mathbf{x}, \mathbf{x_0}|t)$.

From Eq. (21), it then follows that $K^{[e]}(\mathbf{x}, \mathbf{x_0}|t)$ generates $\chi_1^{[e]}(\phi, t)$ and $K^{[o]}(\mathbf{x}, \mathbf{x_0}|t)$ generates $\chi_1^{[o]}(\phi, t)$, provided one uses a definition of n which treats paths with different $\mathbf{x}$ and $\mathbf{x_0}$ consistently. We will use the definition illustrated in Fig. 9(a), which shows a single-space plot of the initial wave packet $\chi_1^{[\Lambda]}(\phi, 0)$, together with examples of Feynman paths with various n. To determine the value of n for a given path, one begins at its starting point $\mathbf{x_0}$ and sets $n = 0$ if $\mathbf{x_0}$ is located to the right of the cut line, or $n = -1$ if $\mathbf{x_0}$ is located to the left. One then traces the path through to its end point $\mathbf{x}$, increasing (or decreasing) n by 1 every time the path crosses the cut line in a clockwise (or counterclockwise) direction. When the paths are plotted in the double space [see Fig. 9(b)], only the paths with even (odd) values of n that originate in $\chi_1^{[e]}(\phi, 0)$ [$\chi_1^{[o]}(\phi, 0)$] have end-points in

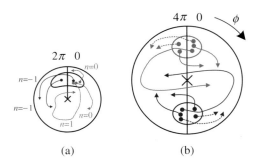

Fig. 9. (a) Single-space representation illustrating how the winding number n of a Feynman path is defined with respect to the initial wave packet and the cut-line. (b) Double-space representation showing that the even-n paths originating from $\chi_1^{[e]}(\phi, t)$ (solid light grey lines) and the odd-n paths from $\chi_1^{[o]}(\phi, t)$ (dashed dark grey lines) terminate in the $0 \to 2\pi$ sector that maps onto the single (physical) space. The odd-n paths from $\chi_1^{[e]}(\phi, t)$ (dashed light grey lines) and the even-n paths from $\chi_1^{[o]}(\phi, t)$ (solid dark grey lines) terminate in the $2\pi \to 4\pi$ sector. [Reprinted with permission from Ref. 23. Copyright 2008, American Institute of Physics.]

the $0 \to 2\pi$ region of space that maps onto the single-space. Hence, in the single-space

$$\chi_1^{[e]}(\mathbf{x},t) = \int d\mathbf{x}_0 K^{[e]}(\mathbf{x},\mathbf{x}_0|t)\chi_1^{[N]}(\mathbf{x}_0,0),$$

$$\chi_1^{[o]}(\mathbf{x},t) = \int d\mathbf{x}_0 K^{[o]}(\mathbf{x},\mathbf{x}_0|t)\chi_1^{[N]}(\mathbf{x}_0,0). \quad (22)$$

In words, $\chi_1^{[e]}(\phi,0)$ [$\chi_1^{[o]}(\phi,0)$] is the result of applying a kernel which contains only the even-n (odd-n) Feynman paths.

4. Time-Dependent Description of Dynamics on Two Coupled Surfaces

The decomposition of the wave function into even- and odd-looping Feynman paths can be generalised to treat systems in which there is sufficient energy to access both the potential surfaces coupled by CI. Such systems are most often studied using femto-second resolved pump-probe spectroscopy, and hence we extend the time-dependent approaches of Sec. 3.

4.1. *Symmetry approach*

To extend the symmetry approach of Sec. 3.1 to treat dynamics on two coupled potential energy surfaces, we have simply to replace the wave packet $\chi_1^{[\Lambda]}(\phi,t)$ by the coupled wave packets $\{\chi_1^{[\Lambda]}(\phi,t), \chi_2^{[\Lambda]}(\phi,t)\}$.[23]

At first sight, one might object that such an approach would be problematic, since a two-surface system has in general enough energy to access points along the CI seam. The angle ϕ is undefined at such points, and hence the concept of a single-to-double space mapping breaks down. However, the definition of the adiabatic electronic states also breaks down at the CI seam (for the same reason, that ϕ is undefined), but this does not invalidate the widely used concepts of adiabatic states and conical intersections. The CI seam occupies a region of space corresponding to a set of points of measure zero. Hence the properties of the wave function along the seam have no effect on its properties elsewhere (i.e. everywhere). For the same reason, our inability to define a single-to-double space mapping at points along the CI seam has no effect on the validity of the mapping elsewhere.

We may therefore map the coupled wave packets $\{\chi_1^{[\Lambda]}(\phi,t), \chi_2^{[\Lambda]}(\phi,t)\}$ from the single to the double space. We can apply every step in the approach of Sec. 3.1, with the only change being that there are now two coupled functions where before there was one. Hence, Eq. (19) applies, with $\alpha = 1, 2$, and the coupled pairs of packets $\{\chi_1^{[G]}(\phi,t), \chi_2^{[G]}(\phi,t)\}$ and $\{\chi_1^{[N]}(\phi,t), \chi_2^{[N]}(\phi,t)\}$ are found to be superpositions of two sets of coupled packets, $\{\chi_1^{[e]}(\phi,t), \chi_2^{[e]}(\phi,t)\}$ and $\{\chi_1^{[o]}(\phi,t), \chi_2^{[o]}(\phi,t)\}$. In the double space, the dynamics of the set $\{\chi_1^{[e]}(\phi,t), \chi_2^{[e]}(\phi,t)\}$ is completely decoupled from that of $\{\chi_1^{[o]}(\phi,t), \chi_2^{[o]}(\phi,t)\}$, but mirrors it through the symmetry relation of Eq. (20). The functions $\chi_1^{[e]}(\phi,t)$ and $\chi_2^{[e]}(\phi,t)$ [and similarly $\chi_1^{[o]}(\phi,t)$ and $\chi_2^{[o]}(\phi,t)$] remain coupled to one another, through the derivative coupling terms.

In the single-space, the functions $\chi_\alpha^{[e]}(\phi,t)$ and $\chi_\alpha^{[o]}(\phi,t)$ will in general overlap one another. They will also twist round the CI to join at the cut line, which ensures that $\chi_\alpha^{[\Lambda]}(\phi,t)$ satisfies Eq. (15) or (16). This overlapping, twisting and joining is shown schematically in Fig. 8. In many two-surface systems, however, the dynamics is such that this description simplifies. Figure 10 is a schematic diagram of a system whose dynamics is such that the wave packet has split into two parts. One part has become trapped and cannot reach the region of space surrounding the CI; the other part has passed through this region. Since $\chi_\alpha^{[e]}(\phi,t)$ and $\chi_\alpha^{[o]}(\phi,t)$ overlap in only the latter part of the wave packet, this is the only part which is affected by the GP. The type of dynamics illustrated in Fig. 10 is found in many conically intersecting systems, and we discuss two simple examples in Sec. 5.2.

The most important property of a CI is its ability to transfer population between the two adiabatic surfaces. Equation (19) shows that the extent of transfer between surfaces is affected by the sign of the interference between $\chi_\alpha^{[e]}(\phi,t)$ and $\chi_\alpha^{[o]}(\phi,t)$, if the overlap integral between these functions is significant. Hence the separation into $\chi_\alpha^{[e]}(\phi,t)$ and $\chi_\alpha^{[o]}(\phi,t)$ reveals the effect of the GP on population transfer.

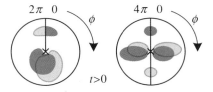

Fig. 10. Schematic showing the simplification in the structure of the upper-surface wave packet $\chi_2^{[G]}(\phi,t)$ that often occurs in a typical pump-probe experiment. The dynamics has caused the wave packet to split into two separate localised pieces. [Reprinted with permission from Ref. 23. Copyright 2008, American Institute of Physics.]

Finally, we point out that, if we can apply the mapping approach just described to a time-evolving wave packet in the adiabatic representation, then we can also apply it in the diabatic representation. However, since the main purpose of the approach is interpretation of the dynamics, we expect that it will find most use in the adiabatic representation.

4.2. *Time-ordered-product path-integral approach*

If we attempt to extend the path-integral analysis of Sec. 3.2 to a two-surface system, we encounter the usual difficulties of applying path-integral theory to a system with a discrete degree of freedom. We cannot associate a classical action S with α, and hence we cannot write down a simple sum over paths in the form of Eq. (7). Even if we could write such a sum, it is doubtful whether we could classify the paths in it according to their winding number n, since n is defined only for paths that can be continuously deformed into one another (with their ends held fixed). Paths that hop between two discrete states ($\alpha = 1, 2$) would seem not to have this property.

Fortunately, there are alternative approaches to constructing path integrals when dealing with systems with discrete degrees of freedom. The first of these is to write the path integral as a time-ordered product, which was first done for non-adiabatic quantum dynamics by Pechukas.[40] The second approach is to map the discrete degree of freedom α onto continuous degrees of freedom, and an elegant mapping for non-adiabatic quantum dynamics has been developed by Thoss, Miller and Stock.[41] We will base our analysis on the first approach, although an equivalent result can probably be obtained using the second.

We begin[23] by writing down the two-surface analogy to Eq. (21):

$$\chi^{[\Lambda]}_{\alpha_P}(\mathbf{x}_P, t) = \sum_{\alpha_0=1,2} \int d\mathbf{x}_0 \ K^{[\Lambda]}_{\alpha_P\alpha_0}(\mathbf{x}_P, \mathbf{x}_0|t)\chi^{[N]}_{\alpha_0}(\mathbf{x}_0, 0), \qquad (23)$$

where, in anticipation of the next step, we use the labels α_P and α_0 in place of the label α, and the vectors $\mathbf{x}_P$ and $\mathbf{x}_0$ in place of the N-dimensional coordinate vector $\mathbf{x}$. The label Λ will often be dropped in what follows (but should be understood).

Starting from Eq. (23), we employ the commonly used technique of splitting the time t into P equally spaced intervals $\Delta = t/P$. We can then split the propagator into a sequence of P propagators to obtain

$$K_{\alpha_P\alpha_0}(\mathbf{x}_P, \mathbf{x}_0|t) = \sum_{\alpha_{P-1}} \cdots \sum_{\alpha_2} \sum_{\alpha_1} K_{\alpha_P\ldots\alpha_1\alpha_0}(\mathbf{x}_P, \mathbf{x}_0|t), \qquad (24)$$

where

$$K_{\alpha_P\ldots\alpha_1\alpha_0}(\mathbf{x}_P,\mathbf{x}_0|t) = \int d\mathbf{x}_{P-1}\ldots\int d\mathbf{x}_2\int d\mathbf{x}_1\ K_{\alpha_P\alpha_{P-1}}(\mathbf{x}_P,\mathbf{x}_{P-1}|\Delta)\ldots$$
$$\ldots K_{\alpha_2\alpha_1}(\mathbf{x}_2,\mathbf{x}_1|\Delta)K_{\alpha_1\alpha_0}(\mathbf{x}_1,\mathbf{x}_0|\Delta). \qquad (25)$$

This time-ordered product form of the propagator allows us to develop a Feynman-path interpretation of $\chi_\alpha^{[e]}(\phi,t)$ and $\chi_\alpha^{[o]}(\phi,t)$. Let us consider first the special case that the system is confined to the lower adiabatic surface. In this case, Eq. (24) contains just one term, with $\alpha_0 = \alpha_1 = \cdots = \alpha_P = 1$, and Eq. (25) is equivalent to Eq. (7) in the limit that $P \to \infty$, because the multi-dimensional integral in Eq. (25) can be evaluated by summing all the paths obtained by tracing through a given sequence of points $\mathbf{x}_0 \to \mathbf{x}_1 \to \cdots \to \mathbf{x}_P$. Each of these paths has an associated phase, which is the sum of the individual phases $s_{i\,i+1} = \ln[K_{11}(\mathbf{x}_{i+1},\mathbf{x}_i|\Delta)]$ contributed by each $\mathbf{x}_i \to \mathbf{x}_{i+1}$ segment of the path. It is easy to show that $s_{i\,i+1}$ tends to the classical action associated with the $\mathbf{x}_i \to \mathbf{x}_{i+1}$ segment of the path as $P \to \infty$, and hence that the total phase is equal to the classical action $S(\mathbf{x},\mathbf{x}_0|t)$ associated with the path.

Let us consider now the general case, in which Eq. (24) includes 2^P terms, each describing a different sequence of hops $\alpha_0 \to \alpha_1 \to \cdots \to \alpha_P$ between the two surfaces, and let us pick out just one of the terms $K_{\alpha_P\ldots\alpha_1\alpha_0}(\mathbf{x}_P,\mathbf{x}_0|t)$, corresponding to one particular sequence of hops. We can convert this term into a path integral

$$K_{\alpha_P\ldots\alpha_1\alpha_0}(\mathbf{x}_P,\mathbf{x}_0|t) = \int \mathcal{D}\mathbf{x}(t)\ \exp[iS_{\alpha_P\ldots\alpha_1\alpha_0}(\mathbf{x}_P,\mathbf{x}_0|t)/\hbar] \qquad (26)$$

for the same reason that we could do this for the one-surface case [namely that the multi-dimensional integral over $(\mathbf{x}_1,\mathbf{x}_1\cdots\mathbf{x}_{P-1})$ describes the sum over all paths between $\mathbf{x}_0$ and $\mathbf{x}_P$]. The only difference between Eq. (26) and Eq. (7) is in the form of the action $S_{\alpha_P\ldots\alpha_1\alpha_0}(\mathbf{x}_P,\mathbf{x}_0|t)$ associated with each path $\mathbf{x}_0 \to \mathbf{x}_1 \to \cdots \to \mathbf{x}_P$. As in the one-surface case, this action is the sum of the individual actions $s_{i\,i+1}$ associated with the segments $\mathbf{x}_i \to \mathbf{x}_{i+1}$ that make up the total path. However, the form of $s_{i\,i+1}$ associated with each path segment depends on the values of α_i and α_{i+1}. When $\alpha_i = \alpha_{i+1}$, $s_{i\,i+1}$ is the classical action for motion on surface α_i (where $\alpha_i = 1$ or 2); when $\alpha_i \neq \alpha_{i+1}$, $s_{i\,i+1}$ is the action obtained by taking the $P \to \infty$ limit of the off-diagonal $\alpha_i \neq \alpha_{i+1}$ elements of $K_{\alpha_i\alpha_{i+1}}(\mathbf{x}_i,\mathbf{x}_{i+1}|\Delta)$.

Hence the paths in Eq. (26) pass continuously through the space $\mathbf{x}$, following the sequences of points $\mathbf{x}_0 \to \mathbf{x}_1 \to \cdots \to \mathbf{x}_P$. They do not 'hop' between the two surfaces. Instead, it is the form of the action that 'hops', according to the (predetermined) sequence $\alpha_0 \to \alpha_1 \to \cdots \to \alpha_P$. The only topological difference between the paths in Eq. (26) and the paths in the one-surface system of Eq. (7) is therefore that the former are free to pass through points along the CI seam. However, such paths make no contribution to the integral, since, as mentioned earlier, the CI seam occupies a set of points of measure zero. The paths in Eq. (26) may therefore be classified by a winding number n, defined exactly as in the one-surface integrals of Eq. (9). We may therefore split up the Kernel $K_{\alpha_P \ldots \alpha_1 \alpha_0}(\mathbf{x}_P, \mathbf{x}_0|t)$ into,

$$K^{[\Lambda]}_{\alpha_P \ldots \alpha_1 \alpha_0}(\mathbf{x}_P, \mathbf{x}_0|t) = \sum_n e^{in\beta} K^{[n]}_{\alpha_P \ldots \alpha_1 \alpha_0}(\mathbf{x}_P, \mathbf{x}_0|t), \qquad (27)$$

where $K^{[n]}_{\alpha_P \ldots \alpha_1 \alpha_0}(\mathbf{x}_P, \mathbf{x}_0|t)$ includes only those paths with winding number n, and the phase-angle β determines whether $\Lambda = G$ ($\beta = \pi$) or N ($\beta = 0$).

Clearly, we can apply Eq. (27) to every term in the sum in Eq. (24). Thus we can write

$$K^{[\Lambda]}_{\alpha_P \alpha_0}(\mathbf{x}_P, \mathbf{x}_0|t) = \sum_n e^{in\beta} K^{[n]}_{\alpha_P \alpha_0}(\mathbf{x}_P, \mathbf{x}_0|t), \qquad (28)$$

where

$$K^{[n]}_{\alpha_P \alpha_0}(\mathbf{x}_P, \mathbf{x}_0|t) = \sum_{\alpha_{P-1}} \cdots \sum_{\alpha_2} \sum_{\alpha_1} K^{[n]}_{\alpha_P \ldots \alpha_1 \alpha_0}(\mathbf{x}_P, \mathbf{x}_0|t). \qquad (29)$$

We can also collect together all the even-n terms into $K^{[e]}_{\alpha_P \alpha_0}(\mathbf{x}_P, \mathbf{x}_0|t)/\sqrt{2}$, and the odd-$n$ into $K^{[o]}_{\alpha_P \alpha_0}(\mathbf{x}_P, \mathbf{x}_0|t)/\sqrt{2}$, to obtain

$$K^{[G]}_{\alpha_P \alpha_0}(\mathbf{x}_P, \mathbf{x}_0|t) = \frac{1}{\sqrt{2}} \left\{ K^{[e]}_{\alpha_P \alpha_0}(\mathbf{x}_P, \mathbf{x}_0|t) - K^{[o]}_{\alpha_P \alpha_0}(\mathbf{x}_P, \mathbf{x}_0|t) \right\},$$

$$K^{[N]}_{\alpha_P \alpha_0}(\mathbf{x}_P, \mathbf{x}_0|t) = \frac{1}{\sqrt{2}} \left\{ K^{[e]}_{\alpha_P \alpha_0}(\mathbf{x}_P, \mathbf{x}_0|t) + K^{[o]}_{\alpha_P \alpha_0}(\mathbf{x}_P, \mathbf{x}_0|t) \right\}. \qquad (30)$$

If we define the winding number n exactly as in Fig. 3, then the relations

$$\chi^{[e]}_\alpha(\mathbf{x}_P, t) = \sum_{\alpha_0} \int d\mathbf{x}_0 \; K^{[e]}_{\alpha_P \alpha_0}(\mathbf{x}_P, \mathbf{x}_0|t) \chi^{[N]}_{\alpha_0}(\mathbf{x}_0, 0)$$

$$\chi^{[o]}_\alpha(\mathbf{x}_P, t) = \sum_{\alpha_0} \int d\mathbf{x}_0 \; K^{[o]}_{\alpha_P \alpha_0}(\mathbf{x}_P, \mathbf{x}_0|t) \chi^{[N]}_{\alpha_0}(\mathbf{x}_0, 0) \qquad (31)$$

hold in the single (i.e. physical) space. This equation is the desired Feynman-path interpretation of the wave packets $\chi^{[e]}_\alpha(\mathbf{x}, t)$ and $\chi^{[o]}_\alpha(\mathbf{x}, t)$. In words, it states that the wave packets $\chi_\alpha^{[e]}(\mathbf{x},t)$ ($\chi_\alpha^{[o]}(\mathbf{x},t)$) result from the application of a propagator that contains only even-looping (odd-looping) Feynman paths when the Kernel is expressed as a sum of time-ordered products.

4.3. Behaviour of Feynman paths at CI seam

As mentioned in Sec. 4.2, we do not need to consider paths that pass precisely through the CI seam (since it spans a set of points of measure zero). However, we do need to consider paths that pass through a very small region of space surrounding the CI seam. These paths have well-defined winding numbers n, and will in general make a non-zero contribution to the path integral. We discuss the properties of such paths here.[23]

Some insight into the nature of paths close to the CI can be obtained by considering the form of the hamiltonian in this region. A general expression for the hamiltonian can be obtained using standard procedures,[42,43] starting with the assumption that it can be approximated very accurately by a (quasi-)diabatic[43,44] representation of the form

$$\hat{\mathbf{H}}_{\text{dia}} = -\frac{\hbar^2}{2M}\begin{pmatrix} \nabla^2 & 0 \\ 0 & \nabla^2 \end{pmatrix} + \begin{pmatrix} W_{11}(\mathbf{x}) & W_{12}(\mathbf{x}) \\ W_{12}(\mathbf{x}) & W_{22}(\mathbf{x}) \end{pmatrix}, \quad (32)$$

where ∇ is the extended gradient operator in the N-dimensional space of nuclear coordinates $\mathbf{x}$, which have been scaled so that the same mass M is associated with each. The hamiltonian $\hat{\mathbf{H}}_{\text{dia}}$ is related to the adiabatic hamiltonian $\hat{\mathbf{H}}_{\text{adia}}$ through the mixing angle,

$$\Theta(\mathbf{x}) = \frac{1}{2}\arctan\frac{2W_{12}(\mathbf{x})}{W_{22}(\mathbf{x}) - W_{11}(\mathbf{x})}. \quad (33)$$

In the limit that the system approaches the CI, $\Theta(\mathbf{x}) \to \frac{1}{2}\phi(\mathbf{x})$ [where $\phi(\mathbf{x})$ is the encirclement angle defined above], and hence

$$\hat{\mathbf{H}}_{\text{adia}} \to -\frac{\hbar^2}{2M}\begin{pmatrix} \nabla^2 - G(\mathbf{x}) & 2\mathbf{F}(\mathbf{x})\cdot\nabla \\ -2\mathbf{F}(\mathbf{x})\cdot\nabla & \nabla^2 - G(\mathbf{x}) \end{pmatrix} + \begin{pmatrix} V_1(\mathbf{x}) & 0 \\ 0 & V_2(\mathbf{x}) \end{pmatrix}, \quad (34)$$

where $\mathbf{F}(\mathbf{x}) = \frac{1}{2}\nabla\phi(\mathbf{x})$ and $G(\mathbf{x}) = \frac{1}{4}\{\nabla\phi(\mathbf{x})\}\cdot\{\nabla\phi(\mathbf{x})\}$.

Both $\mathbf{F}(\mathbf{x})$ and $G(\mathbf{x})$ are positive and singular at the CI point. Hence it is tempting to think that the non-hopping paths experience $G(\mathbf{x})$ as a strongly repulsive potential 'spike', which stops them from passing through the region of space close to the CI; the hopping paths do not experience the 'spike', and are thus free to pass through this region. However, this picture is only partially true, since it implies that there exist non-hopping paths which reflect from $G(\mathbf{x})$, which survive the sum over paths, and which therefore contribute a reflected component to the wave packet. Clearly this cannot happen, since the equivalent diabatic description of the dynamics does not allow for reflection from the CI.

To obtain a clearer picture of the nature of the paths close to the CI, we need to consider the form of the $P \to \infty$ propagator $K_{\alpha_{i+1}\alpha_i}(\mathbf{x}_{i+1}, \mathbf{x}_i|\Delta)$ in the short-time limit, in which it can be factored into a potential energy part and a kinetic energy part $[K_0]_{\alpha_{i+1}\alpha_i}(\mathbf{x}_{i+1}, \mathbf{x}_i|\Delta)$. We are concerned only with the latter, since the potential part is diagonal in α. We can write out $[K_0]_{\alpha_{i+1}\alpha_i}(\mathbf{x}_{i+1}, \mathbf{x}_i|\Delta)$ by transforming to the diabatic representation in which it is diagonal in α, and then transforming back again, to yield

$$\mathbf{K}_0(\mathbf{x}_{i+1}, \mathbf{x}_i|\Delta) = K_0(\mathbf{x}_{i+1}, \mathbf{x}_i|\Delta)$$
$$\times \begin{pmatrix} \cos[\Theta(\mathbf{x}_{i+1}) - \Theta(\mathbf{x}_i)] & -\sin[\Theta(\mathbf{x}_{i+1}) - \Theta(\mathbf{x}_i)] \\ \sin[\Theta(\mathbf{x}_{i+1}) - \Theta(\mathbf{x}_i)] & \cos[\Theta(\mathbf{x}_{i+1}) - \Theta(\mathbf{x}_i)] \end{pmatrix},$$
(35)

where $K_0(\mathbf{x}_{i+1}, \mathbf{x}_i|\Delta)$ is the free particle propagator.

The hopping and the non-hopping path segments between $\mathbf{x}_i$ and $\mathbf{x}_{i+1}$ are thus generated by the same free particle kernel $K_0(\mathbf{x}_{i+1}, \mathbf{x}_i|\Delta)$, and are then weighted by a factor which depends on the difference in the mixing angle between $\mathbf{x}_i$ and $\mathbf{x}_{i+1}$. When the system is far away from the CI, $\Theta(\mathbf{x})$ changes slowly with $\mathbf{x}$, and the weight factors are thus approximately unity for non-hopping path segments, and zero for hopping segments. On moving closer to the CI, the weight factors for the hopping paths increase at the expense of the non-hopping paths. A path segment that passes through a small region of space enclosing the CI satisfies $\Theta(\mathbf{x}_{i+1}) - \Theta(\mathbf{x}_i) \simeq \pm\pi/2$, and hence its weight factor is approximately zero if the segment does not hop, and unity if it does. In other words, non-hopping paths that attempt to pass through a small region enclosing the CI are simply removed from the sum; there is no reflected component.

5. Numerical Applications

Here we give examples to illustrate the application of the above theory. These include an application of the time-independent theory of Sec. 2 to the hydrogen-exchange reaction (Sec. 5.1), and an application of the time-dependent theory of Sec. 4 to a reduced dimensionality model of pyrrole (Sec. 5.2).

5.1. Cancellation of GP effects in the hydrogen-exchange reaction

The hydrogen-exchange reaction $H + H_2 \to H_2 + H$ is a benchmark in reactive scattering, since it is one of the few reactions for which ab initio theory can yield quantitative agreement with experiment.[51] In particular, theoretical state-to-state differential cross sections [measuring the angular distribution of the $H_2(v', j')$ product with respect to the initial $H + H_2(v, j)$ approach vector] agree quantitatively with experimental measurements. It is well known that H_3 is an $E \times e$ Jahn–Teller system, with a CI seam running through all equilateral triangle geometries of the three nuclei.[2, 52–54] A longstanding puzzle[25–33] has been why GP effects are not seen in the cross sections of this reaction. Calculations by Kendrick showed that GP effects are present in the reaction probablities for this reaction, but that they cancel out in the cross sections. We will refer to this as the 'cancellation puzzle'. The following two subsections explain how the theory of Sec. 2 was used to solve the puzzle, and explain why the GP effects cancelled in the cross sections.

5.1.1. Reaction paths and potential energy surface

Figure 11 shows a schematic representation of the $H + H_2$ potential energy surface,[19] plotted using the hyperspherical coordinate scheme of Kuppermann.[45] We will treat the three hydrogen nuclei as distinguishable particles, ignoring the requirement that the nuclear wave function be antisymmetric under exchange of two ^1H nuclei. The exchange symmetry is easy to incorporate by taking appropriate linear combinations of the unsymmetrised (distinguishable particle) wave functions.[18] Hence we consider here only the effects of the GP on the unsymmetrised wave functions. These are the effects caused by reaction paths that encircle the CI, which give rise to the 'cancellation puzzle'.

Hence we consider wave functions in which the reaction starts at an asymptotic separation of one uniquely specified arrangement of the atoms $(A + BC)$, and analyze the cross sections produced by reactive scattering into one of the product channels $(AC + B)$. It is well known[46, 47] that the dominant $H + H_2$ reaction path passes over one transition state (1-TS),

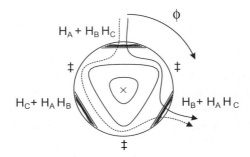

Fig. 11. Schematic representation of the 1-TS (solid) and 2-TS (dashed) reaction paths in the reaction $H_A + H_B H_C \to H_A H_C + H_B$. The H_3 potential energy surface is represented using the hyperspherical coordinate system of Kuppermann,[45] in which the equilateral-triangle geometry of the CI is in the centre (×), and the linear transition states (‡) are on the perimeter of the circle; the hyperradius $\rho = 3.9$ a.u. The angle ϕ is the internal angular coordinate which describes motion around the CI. [Reprinted with permission from Ref. 19. Copyright 2007, American Institute of Physics.]

as illustrated schematically in Fig. 11. Since GP effects are found in the reaction probabilities at sufficiently high energies[17,31–34] ($>1.8\,\text{eV}$ above the potential minimum), the wave function must encircle the CI at these energies, and thus also contain reaction paths that pass over two transition states (2-TS). For the AC + B products, the 1-TS paths make less than one full revolution, loop in a clockwise sense around the CI (see Fig. 11), and are assigned a winding number $n = 0$ (following the convention of Sec. 2.3). The 2-TS paths also make less than one full revolution, but loop in an anticlockwise sense, and are assigned $n = -1$.[18] This means that the 1-TS (Feynman) paths are contained in $\Psi^{[e]}$ and the 2-TS paths in $\Psi^{[o]}$. In principle, there are also paths with higher winding numbers present in both $\Psi^{[e]}$ and $\Psi^{[o]}$, but these can be ignored (since reaction paths passing over three or more transition states are highly unlikely in $H + H_2$). In this chapter, we will therefore use the terms '$\Psi^{[e]}$ paths' and '1-TS paths' (and '$\Psi^{[o]}$ paths' and '2-TS paths') interchangeably.

5.1.2. *Cancellation of GP effects in state-to-state cross sections*

The main experimental observable that we need to consider is the state-to-state differential cross section (DCS), which can be written

$$\frac{d\sigma^{[\Lambda]}_{n' \leftarrow n}}{d\Omega}(\theta, E) = \frac{1}{2j+1} \left| f^{[\Lambda]}_{n' \leftarrow n}(\theta, E) \right|^2, \tag{36}$$

where n and n' denote the quantum states (v, j) and (v', j') of the reactants and products, θ is the centre-of-mass scattering angle between the velocity vectors of the reactants and products, E is the total energy, $d\Omega$ is the element of solid angle swept out by θ,[48] and $f_{n'\leftarrow n}^{[\Lambda]}(\theta, E)$ is the state-to-state scattering amplitude, obtained by projecting the wave function Ψ onto the product quantum state (v', j'), in the asymptotic limit of infinite separation of the products. The wave function Ψ is expanded in terms of partial waves, each corresponding to a particular value of the total angular momentum quantum number J (which is a good quantum number). The set of projections of the partial waves onto the product quantum states, in the limit of asymptotic separation of the products, is referred to as the S-matrix, $S_{n'\leftarrow n}^{[\Lambda]}(J, E)$. One can construct $f_{n'\leftarrow n}^{[\Lambda]}(\theta, E)$ from the S-matrix using

$$f_{n'\leftarrow n}^{[\Lambda]}(\theta, E) = \frac{1}{2ik_{vj}} \sum_J F(J)(2J+1) d_{\kappa'\kappa}^J(\pi - \theta) S_{n'\leftarrow n}^{[\Lambda]}(J, E), \quad (37)$$

where $d_{\kappa'\kappa}^J(\pi - \theta)$ is a reduced Wigner rotation matrix,[49] and κ, κ' are the projections of j and j' on the initial and final velocity vectors. There is a rough correspondence between the classical impact parameter b (i.e. the perpendicular distance of the colliding reactants from the centre-of-mass approach vector) and the quantum number J, such that $\hbar J \sim b \times$ velocity. We insert a filter $F(J)$ into Eq. (37), which is set either to unity, to give the full cross section, or to a value which smoothly includes only a range of low values of J ($J = 0 \to 6$), to give the low-impact parameter cross sections (corresponding approximately to head-on collisions). Another important observable is the state-to-state integral cross section (ICS)

$$\sigma_{n'\leftarrow n}^{[\Lambda]}(E) = \frac{2\pi}{2j+1} \int_0^\pi |f_{n'\leftarrow n}^{[\Lambda]}(\theta, E)|^2 \sin\theta \, d\theta, \quad (38)$$

which is the total amount of product scattering into final quantum states (v', j') at energy E.

The 'cancellation puzzle' is that GP effects are present in the state-to-state reaction probabilities

$$P_{n'\leftarrow n}^{[\Lambda]}(J, E) = \left| S_{n'\leftarrow n}^{[\Lambda]}(J, E) \right|^2, \quad (39)$$

but cancel completely in the state-to-state ICS and low-impact parameter (i.e. $J = 0 \to 6$) DCS. Small GP effects do survive as oscillations in the complete state-to-state DCS, but then cancel out very efficiently on integrating over θ to yield the integral cross sections. To explain these cancellations we need to map contributions from the 1-TS and 2-TS reaction pathways onto the DCS.

In general, it is difficult to map contributions from different reaction paths onto the DCS. However, Eq. (6) tells us that, in a reaction with a CI, one can easily map the contributions from the e and o (Feynman) paths onto the DCS. Since Eq. (6) applies to the entire wave function, we can apply it to the asymptotic limit of the scattering wave function, and thus to $S^{[G]}(E)$ and $S^{[N]}(E)$, to obtain

$$S^{[e]}(E) = \frac{1}{\sqrt{2}}\left\{S^{[N]}(E) + S^{[G]}(E)\right\},$$

$$S^{[o]}(E) = \frac{1}{\sqrt{2}}\left\{S^{[N]}(E) - S^{[G]}(E)\right\}. \qquad (40)$$

These equations are all that we need to explain the effect of the GP on scattering cross sections such as the DCS and ICS. They allow us to compute separate e and o cross sections using Eq. (37), which show the scattering produced by the e and o reaction paths in isolation. They tell us that we can only expect GP effects if $f^{[e]}_{n'\leftarrow n}(\theta, E)$ and $f^{[o]}_{n'\leftarrow n}(\theta, E)$ overlap.

In the case of the $H + H_2$ reaction, Eq. (40) specialises to

$$S^{[1-TS]}(E) = S^{[e]}(E),$$

$$S^{[2-TS]}(E) = S^{[o]}(E). \qquad (41)$$

Hence, simply by adding and subtracting the computed $S^{[G]}(E)$ and $S^{[N]}(E)$, we can identify the contributions from the 1-TS and 2-TS reaction paths in the DCS and ICS, and thus explain the effects of the GP on the $H + H_2$ reaction.

Application of this technique[17,19] to the low-impact parameter cross-sections yields the results of Fig. 12, which demonstrate that the reason the GP effects cancel in these cross sections is simply that the 1-TS and 2-TS paths scatter their products into different regions of space. The 1-TS paths are predominantly backward-scattered, and the 2-TS forward-scattered (at these low impact parameters). When applied to the full DCS, we obtain[17,19] the results of Fig. 13. These show that the 1-TS and 2-TS paths now scatter into overlapping regions of space, which is to be expected since these cross sections manifest small GP effects. The reason these effects cancel so efficiently in the ICS is revealed by plotting the phases of the amplitudes

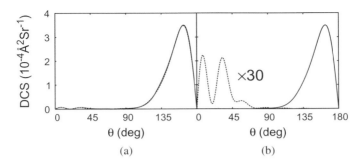

Fig. 12. Low impact parameter DCS at 2.3 eV, for $H + H_2(v = 1, j = 0) \rightarrow H_2(v' = 2, j' = 5, \kappa' = 0) + H$, describing the scattering of (a) $\Psi^{[G]}$ (solid lines) and $\Psi^{[N]}$ (dashed lines), and (b) $\Psi^{[e]}$ (solid lines) and $\Psi^{[o]}$ (dashed lines). [Reprinted with permission from Ref. 19. Copyright 2007, American Institute of Physics.]

[Fig. 13(c)]. The 1-TS and 2-TS phases vary in opposite senses as a function of θ, and hence the resulting interference term varies rapidly with θ, and thus cancels out efficiently on integration. The reason the phases vary as they do is that the 1-TS paths scatter predominantly into the nearside[55] hemisphere, and the 2-TS paths into the farside hemisphere [as defined in Fig. 13(d)]. The different scattering dynamics reflect differences in the the 1-TS and 2-TS reaction mechanisms: the 1-TS paths correspond to a recoil mechanism; the 2-TS paths to an insertion mechanism.[19]

5.2. *Geometric phase effects in two-surface population transfer*

5.2.1. *Wave packet calculations on $E \times e$ Jahn–Teller model*

Here we illustrate the time-dependent, two-surface treatment of Sec. 4 with a calculation[23] on a simple 2-dimensional model of an $E \times e$ Jahn–Teller system. The hamiltonian has the form[6]

$$\hat{H} = \hat{T}_R + \hat{T}_\phi + \begin{pmatrix} \kappa R^2/2 - kR & 0 \\ 0 & \kappa R^2/2 + kR \end{pmatrix} \quad (42)$$

with

$$\hat{T}_R = -\frac{\hbar^2}{2mR}\frac{\partial}{\partial R}R\frac{\partial}{\partial R}\begin{pmatrix} 1 & 0 \\ 0 & 1 \end{pmatrix} \quad (43)$$

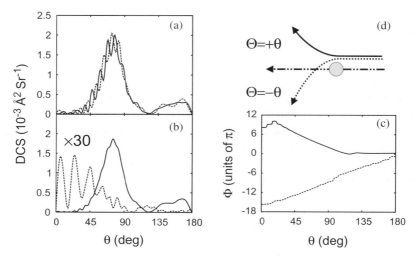

Fig. 13. Full DCS (i.e. including all impact parameters), for $H + H_2(v = 1, j = 0) \to H_2(v' = 2, j' = 5, \kappa' = 0) + H$, describing the scattering of (a) $\Psi^{[G]}$ (solid lines) and $\Psi^{[N]}$ (dashed lines), and (b) $\Psi^{[e]}$ (solid lines) and $\Psi^{[o]}$ (dashed lines). (c) The phases of the corresponding e and o scattering amplitudes. (d) Diagram illustrating 'nearside' and 'farside' scattering. The arrow (chains) represents the initial approach direction of the reagents in centre-of-mass frame; the grey rectangle represents the spread of impact parameters in the initial plane wave. [Reprinted with permission from Ref. 19. Copyright 2007, American Institute of Physics.]

and

$$\hat{\mathbf{T}}_\phi = -\frac{\hbar^2}{2mR^2} \begin{pmatrix} \partial^2/\partial\phi^2 - 1/4 & -\partial/\partial\phi \\ \partial/\partial\phi & \partial^2/\partial\phi^2 - 1/4 \end{pmatrix}. \quad (44)$$

The polar coordinates (R, ϕ) are defined such that the CI point is at the origin, and ϕ is the encirclement angle. The constants k and κ control the steepness of the CI and the depth of the 'Mexican hat' minimum in the ground state adiabatic surface; m is the mass. The values taken by these parameters (in atomic units) were $k = 0.017$, $\kappa = 6 \times 10^{-3}$ and $m = 1000$.

As explained in Secs. 3 and 4, the even- and odd-looping Feynman paths can be separated by computing $\chi_\alpha^{[G]}(\phi, t)$ and $\chi_\alpha^{[N]}(\phi, t)$, and then adding and subtracting these functions to extract $\chi_\alpha^{[e]}(\phi, t)$ and $\chi_\alpha^{[o]}(\phi, t)$, according to Eq. (19). It was straightforward to propagate $\chi_\alpha^{[G]}(\phi, t)$ and $\chi_\alpha^{[N]}(\phi, t)$ numerically for the hamiltonian of Eq. (42). To simplify the calculations, we propagated $\chi_\alpha^{[G]}(\phi, t)$ in the diabatic representation, and $\chi_\alpha^{[N]}(\phi, t)$ in the adiabatic representation, and applied single-valued boundary conditions in each case. Numerical wave packet propagation in the adiabatic representation is potentially difficult, because the off-diagonal coupling terms become singular as the system approaches the CI seam. We avoided the singularity by representing the wave packet on an equally spaced grid, which was chosen such that no grid point coincided with the CI at $R = 0$. We found that the same grid spacing (0.06 a.u.) was sufficient to yield results converged to better than 1% in both the adiabatic and diabatic calculations. A similar grid technique is used regularly in (single-surface) wave packet calculations in Jacobi coordinates,[50] where the centrifugal potential also becomes singular at $R = 0$. In both calculations, the initial wave packet was taken to be of the form

$$\chi_1(x, y|0) = 0$$

$$\chi_2(x, y|0) = \frac{1}{\sqrt{\sigma_x \sigma_y \pi}} e^{-(x-x_0)^2/(2\sigma_x^2)-(y-y_0)^2/(2\sigma_y^2)+ik_x x+ik_y y} \quad (45)$$

with $(x_0, \sigma_x, k_x) = (0, 0.3, 0)$, and $(y_0, \sigma_y, k_y) = (1.7, 0.3, -7.0)$ atomic units.

Snapshots of the resulting $|\chi_\alpha^{[e]}(\phi,t)|^2$ and $|\chi_\alpha^{[o]}(\phi,t)|^2$ are plotted in Fig. 14. In this simple example, the system is unable to loop completely around the CI, and hence $\chi_\alpha^{[e]}(\phi,t)$ and $\chi_\alpha^{[o]}(\phi,t)$ correspond respectively to the $n = 0$ and $n = -1$ paths only; i.e. to paths that make less than one complete loop around the CI in the clockwise and counterclockwise senses. At 8 fs, the system has completed its transfer from the upper to the lower state, and the parts of the wave packet that have been transferred to the lower state retain a 'memory' of the sense in which the paths looped around the CI on the upper state, previous to the transfer. In a more complex, less symmetric, system, these paths could involve very different reaction mechanisms, and the separation into $\chi_\alpha^{[e]}(\phi,t)$ and $\chi_\alpha^{[o]}(\phi,t)$ would therefore reveal the contribution made by each mechanism to the relaxation through the CI.

5.2.2. *Effect of GP on population transfer*

The ability to separate the wave packet into $\chi_\alpha^{[e]}(\phi,t)$ and $\chi_\alpha^{[o]}(\phi,t)$ allows one to investigate and explain the effect of the GP on population transfer between the two surfaces. Figure 15 shows the population transfer versus time for the Jahn–Teller system of Fig. 14, obtained from $\chi_\alpha^{[G]}(\phi,t)$ and $\chi_\alpha^{[N]}(\phi,t)$. The GP increases the terminal population ratio of the lower

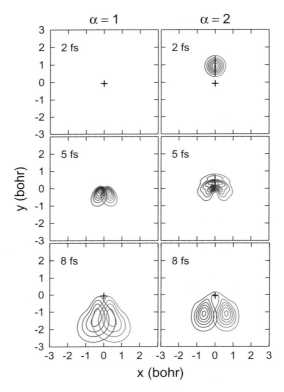

Fig. 14. Snapshots of the time-evolution of $|\chi_\alpha^{[e]}(\phi,t)|^2$ (light grey) and $|\chi_\alpha^{[o]}(\phi,t)|^2$ (dark grey) for the simple $E \times e$ Jahn–Teller model of Eq. (42). [Reprinted with permission from Ref. 23. Copyright 2008, American Institute of Physics.]

to the upper state from 1.25:1 to 1.93:1, by producing constructive interference between $\chi_1^{[e]}(\phi,t)$ and $\chi_1^{[o]}(\phi,t)$. (The interference is constructive because the non-adiabatic coupling produces opposite signs in $\chi_1^{[e]}(\phi,t)$ and $\chi_1^{[o]}(\phi,t)$, which are then cancelled out by the GP.)

Effects such as the above are significant, and would be detectable in a femtosecond pump-probe experiment measuring quantum yield versus time. However, the GP will have a large effect on the population transfer only when the overlap integral between $\chi_\alpha^{[e]}(\phi,t)$ and $\chi_\alpha^{[o]}(\phi,t)$ is significant. In many systems this integral will become very small as a result of phase averaging. For example, Fig. 15(c) shows the effect of the GP on population transfer obtained using a 2-dimensional model of the $^1B_1 - S_0$ intersection

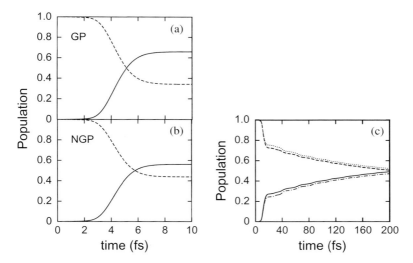

Fig. 15. Adiabatic-state population versus time, starting with the initial wave packet in the upper state, computed for (a) the simple $E \times e$ Jahn-Teller model of Eq. (42); (b) the same, but with omission of the GP; (c) the Pyrrole $^1B_1 - S_0$ model[56] both with (solid and dashed lines) and without (chained and dotted lines) inclusion of the GP. [Reprinted with permission from Ref. 23. Copyright 2008, American Institute of Physics.]

in pyrrole due to Vallet et al.[56] The two degrees of freedom in this model are a radial distance r associated with stretching of the NH hydrogen atom, and an angular degree of freedom γ associated with bending of the NH bond. The moment of inertia associated with the latter is about 40,000 a.u. The initial wave packet was a Gaussian, chosen to approximate the position and initial momentum of the photoexcited initial wave packet of Ref. 56 (the packet was centred at $r = 4.2$ bohr, with average momentum 3.0 a.u. in the positive r direction). This gave the initial wave packet an excess energy of >1 eV with respect to the CI point on the potential surface. Figure 15(c) shows that these conditions give a de Broglie wavelength that is sufficiently short for the overlap integral between $\chi_\alpha^{[e]}(\phi, t)$ and $\chi_\alpha^{[o]}(\phi, t)$ to average to a very small value. Hence the GP has no appreciable effect on population transfer in this system.

6. Conclusions

The central theme running through this chapter has been a simple one, that the nuclear wave function at a CI can be decomposed, rigorously, into contributions from Feynman paths that loop an even and an odd number of times about the CI. This decomposition does not require a semiclassical approximation in order to extract trajectories from the wave function. Rather, it can be applied to the exact nuclear wave function, without approximation, by exploiting the symmetry of the system in a double space.

For a scattering system, or for a bound state system evolving over a finite time, the even- and odd-looping components correspond to different dynamics. The decomposition technique therefore allows one to extract separate wave functions that describe each of these processes independently. Extracting these functions allows one to interpret and visualise the dynamics, and also to explain how the GP (which changes the relative sign of the interference between these components) affects experimental observables.

An important question that arises from this work is whether interference between even- and odd-looping paths in the immediate vicinity of a CI influences the amount of population transfer. This effect is easy to compute in simple systems (as we showed above). In many of the systems typically studied in pump-probe experiments, it is unlikely that such interference will be important, because the topography of the potential surfaces is such that the system accelerates through the CI and thus has a short de Broglie wavelength, causing interference effects to average out. However, in systems with sloped conical intersections, or that access the CI through tunnelling, it is likely that such interference will influence considerably the amount of population transfer. Such effects may also be important in the condensed phase on account of the rapidity of the relaxation through the CI.

Acknowledgments

It is a pleasure to thank those of my co-workers who have contributed to the work reviewed here, most notably, J.C. Juanes-Marcos, F. Bouakline, T. Stecher, E. Wrede and M.S. Holt. This work was funded by two grants from the UK Engineering and Physical Sciences Research Council.

References

1. M.V. Berry, *Proc. Roy. Soc. A* **392**, 45 (1984).
2. G. Herzberg and H.C. Longuet-Higgins, *Discuss. Faraday Soc.* **35**, 77 (1963).
3. H.C. Longuet-Higgins, *Proc. Roy. Soc. A* **344**, 147 (1975).

4. C.A. Mead and D.G. Truhlar, *J. Chem. Phys.* **70**, 2284 (1979).
5. C.A. Mead, *Rev. Mod. Phys.* **64**, 51 (1992).
6. M.S. Child, *Adv. Chem. Phys.* **124**, 1 (2002).
7. J. Schön and H. Köppel, *J. Chem. Phys.* **103**, 9292 (1995).
8. J. Schön and H. Köppel, *J. Chem. Phys.* **108**, 1503 (1998).
9. B.K. Kendrick, *Phys. Rev. Lett.* **79**, 2431 (1997).
10. B.E. Applegate, T.A. Barckholtz and T.A. Miller, *Chem. Soc. Rev.* **32**, 38 (2003).
11. C.A. Mead, *Chem. Phys.* **49**, 23 (1980).
12. D. Babikov, B.K. Kendrick, P. Zhang and K. Morokuma, *J. Chem. Phys.* **122**, 044315 (2005).
13. L.S. Schulman, *Phys. Rev.* **176**, 1558 (1968).
14. L.S. Schulman, *J. Math. Phys.* **12**, 304 (1971).
15. L.S. Schulman, *Techniques and Applications of Path Integration* (Wiley, New York, 1981).
16. M.G.G. Laidlaw and C.M. Morette DeWitt, *Phys. Rev. D* **3**, 1375 (1971).
17. J.C. Juanes-Marcos, S.C. Althorpe and E. Wrede, *Science* **309**, 1227 (2005).
18. S.C. Althorpe, *J. Chem. Phys.* **124**, 084105 (2006).
19. J.C. Juanes-Marcos, S.C. Althorpe and E. Wrede, *J. Chem. Phys.* **126**, 044317 (2007).
20. F. Bouakline, S.C. Althorpe and D. Peláez Ruiz, *J. Chem. Phys.* **128**, 124322 (2008).
21. J.C. Juanes-Marcos, A.J. Varandas and S.C. Althorpe, *J. Chem. Phys.* **128**, 211101 (2008).
22. S.C. Althorpe, J.C. Juanes-Marcos and E. Wrede, *Adv. Chem. Phys.* **138**, 1 (2008).
23. S.C. Althorpe, T. Stecher and F. Bouakline, *J. Chem. Phys.* **129**, 214117 (2008).
24. F. Bouakline, B. Lepetit, S.C. Althorpe and A. Kuppermann, in *The Jahn–Teller Effect, Springer Series in Chemical Physics vol. 97*, ed. H. Köppel, D.R. Yarkony and H. Barentzen (Springer, New York, 2010).
25. C.A. Mead, *J. Chem. Phys.* **72**, 3839 (1980).
26. B. Lepetit and A. Kuppermann, *Chem. Phys. Lett.* **166**, 581 (1990).
27. Y.M. Wu, A. Kuppermann and B. Lepetit, *Chem. Phys. Lett.* **186**, 319 (1991).
28. A. Kuppermann and Y.-S.M. Wu, *Chem. Phys. Lett.* **205**, 577 (1993).
29. A. Kuppermann and Y.-S.M. Wu, *Chem. Phys. Lett.* **241**, 229 (1995).
30. A. Kuppermann and Y.-S.M. Wu, *Chem. Phys. Lett.* **349**, 537 (2001).
31. B.K. Kendrick, *J. Chem. Phys.* **112**, 5679 (2000).
32. B.K. Kendrick, *J. Phys. Chem. A* **107**, 6739 (2003).
33. B.K. Kendrick, *J. Chem. Phys.* **118**, 10502 (2003).
34. J.C. Juanes-Marcos and S.C. Althorpe, *J. Chem. Phys.* **122**, 204324 (2005).
35. R.P. Feynman and A.R. Hibbs, *Quantum Mechanics and Path Integrals* (McGraw-Hill, New York, 1965).
36. H. Goldstein, *Classical Mechanics* (Addison-Wesley, Reading MA, 1980).

37. J. Stillwell, *Classical Topology and Combinatorial Group Theory* (Springer, New York, 1993).
38. D.J. Kouri and D.K. Hoffman, *Few. Body Systems* **18**, 203 (1995).
39. R. Schinke, *Photodissociation Dynamics* (Cambridge University Press, Cambridge, 1993).
40. P. Pechukas, *Phys. Rev.* **181**, 174 (1969).
41. M. Thoss, W.H. Miller and G. Stock, *J. Chem. Phys.* **112**, 10282 (2000).
42. G.A. Worth and L.S. Cederbaum, *Ann. Rev. Phys. Chem.* **55**, 127 (2004).
43. H. Köppel, *Faraday Discuss.* **127**, 35 (2004).
44. C.A. Mead and D.G. Truhlar, *J. Chem. Phys.* **77**, 6090 (1982).
45. A. Kuppermann, *Chem. Phys. Lett.* **32**, 374 (1975).
46. F. Fernández-Alonso and R.N. Zare, *Annu. Rev. Phys. Chem.* **53**, 67 (2002).
47. F.J. Aoiz, L. Bañares and V.J. Herrero, *Int. Rev. Phys. Chem.* **24**, 119 (2005).
48. M.S. Child, *Molecular Collision Theory* (Academic, London, 1974).
49. R.N. Zare, *Angular Momentum* (Wiley, New York, 1988).
50. S.C. Althorpe, *J. Chem. Phys.* **114**, 1601 (2001).
51. S.C. Althorpe and D.C. Clary, *Annu. Rev. Phys. Chem.* **54**, 493 (2003).
52. A.J.C. Varandas, F.B. Brown, C.A. Mead, D.G. Truhlar and N.C. Blais, *J. Chem. Phys.* **86**, 6258 (1987).
53. A. I. Boothroyd, W. J. Keogh, P. G. Martin, and M. R. Peterson, *J. Chem. Phys.* **104**, 7139 (1996).
54. S. Mahapatra, H. Köppel, and L.S. Cederbaum, *J. Phys. Chem. A* **105**, 2321 (2001).
55. A.J. Dobbyn, P. McCabe, J.N.L. Connor and J.F. Castillo, *PCCP* **1**, 1115 (1999).
56. V. Vallet, Z. Lan, S. Mahapatra, A.L. Sobolewski and W. Domcke, *J. Chem. Phys.* **123**, 144307 (2005).

Part II

Dynamics at Conical Intersections

Chapter 6

Conical Intersections in Electron Photodetachment Spectroscopy: Theory and Applications

Michael S. Schuurman[*] and David R. Yarkony[†]

1. Introduction 198
   1.1. Electron spectroscopies 198
   1.2. Scope of this work 199
2. Vibronic Coupling Model 200
   2.1. Vibronic wave functions 201
   2.2. Spectral intensity distribution function 208
   2.3. Hamiltonians 212
   2.4. Computational issues 213
3. Applications 222
   3.1. Overview: Conical intersections
        not required by symmetry 222
   3.2. Azolyls: Electronic structure 223
   3.3. Pyrrolyl: Detailed computations 226
   3.4. Isopropoxy 234
4. Conclusions 240

---

[*]Steacie Institute for Molecular Sciences, National Research Council Canada, Ottawa, Ontario K1A0R6, Canada.
[†]Department of Chemistry, Johns Hopkins University, Baltimore, MD 21218, USA.

Acknowledgments 241
Appendices 241
A. Time Reversal Adapted Basis 241
B. Electron Scattering and Electronic Transition Moments . . . 242
References 243

1. Introduction

1.1. *Electron spectroscopies*

In this chapter we are primarily concerned with electron photodetachment spectroscopies in which electrons are detached from negative ions, thereby accessing electronic states of the neutral coupled by conical intersections. We will not go into the relative advantages of these spectroscopies, which include anion photoelectron spectroscopy (anion PES),[1,2] anion zero electron kinetic energy[3,4] (anion ZEKE) spectroscopy and anion slow electron velocity-map imaging[5] (anion SEVI) spectroscopy, either with respect to each other or to photon spectroscopies. In this regard, we refer the reader to the recent review by Neumark.[6] However, several fundamental advantages are worth emphasizing. Electron based spectroscopies, although ultimately electric dipole driven processes, are "universal", in that there are no dark states[7] as in photon spectroscopies. Furthermore, the use of negative ions facilitates mass selection, which is invaluable in studies of clusters and solvation.[8–10]

With respect to spectral analysis, the identification of the electronic state from which a spectral feature originates is facilitated by determining the anisotropy parameter,[11] obtained via the measurement of the photoelectron intensity as a function of the angle between the electric vector and the detection axis. Photoelectron angular distributions yield even more detailed information regarding the electronic states involved in the detachment process, but are correspondingly more difficult to determine under general circumstances.[11–14]

While this work focuses on the frequency domain, it should be noted that time resolved photoelectron spectroscopy (TR-PES) is another extremely powerful tool in current use. This class of techniques was recently reviewed by Stolow.[7] Experiments in the time domain provide a valuable complement to frequency domain methods in the study of electronically nonadiabatic processes. Time resolved spectroscopic techniques allow for the indirect observation of the evolution of nuclear motion on coupled electronic potential energy surfaces, yielding information that was once exclusively the domain of theoretical simulation.

1.2. Scope of this work

The effect of nonadiabatic interactions on electron photodetachment spectra, or alternatively, what can be learned about nonadiabatic interactions from electron photodetachment spectra, is the general subject of this work. It is now generally appreciated that conical intersections play a key role in nonadiabatic processes.[15–17] Conical intersections fall into two categories: those required by symmetry, denoted here as Jahn–Teller conical intersections,[18, 19] and those which are not, denoted generally as accidental conical intersections.[20] Accidental conical intersections in turn exist in two forms: symmetry-allowed intersections, in which two states carrying distinct irreducible representations of a spatial point group intersect conically, and same symmetry intersections, where point group symmetry plays no role or is absent.[20] It is the appreciation of the prevalence of same symmetry conical intersections that has led to the current interest in this once arcane topographical feature.

This chapter describes the simulation of photoelectron spectra involving electronic states strongly coupled by accidental and same symmetry conical intersections using the vibronic (i.e. vibrational electronic) coupling model,[21] originally introduced by Cederbaum, Köppel and Domcke[22] more than a quarter of a century ago. In the intervening period, the original model has seen a myriad of extensions and applications, and has been the subject of numerous reviews,[22] including an excellent review in this series in 2004.[21] The purpose of this article is to describe work that has been done since that review to extend the computational utility of the vibronic coupling model and some applications that have exploited these new capabilities. We will review work done in other venues but the detailed discussion will emphasize work done by the current authors and our collaborators. In this regard we note that there are in general two distinct approaches to the simulation of electron photodetachment spectra: the time-independent approach and the time-dependent approach.[21] Considerable progress in the implementation of the time-dependent formulation has been made in the context of the multiconfigurational time-dependent Hartree (MCTDH) approach.[23] Recent progress in direct dynamics techniques,[24] including the multiple spawning[25] and the variational Gaussian wave packet approaches[26] should also be noted. However, in this chapter we focus on the time-independent method which is nicely summarized in Ref. 27. In many cases, time-independent methods permit an analysis of individual lines in the spectrum. Complementarily, time-dependent approaches, in addition to their relative computational efficiency, permit the observation of wave packet evolution in the vicinity of conical intersections.

2. Vibronic Coupling Model

We begin with a review requisite concepts for the computation of the spectral intensity distribution function,

$$I(E) = 2\pi \sum_f |A_f^i|^2 \delta(E - E_f), \qquad (1a)$$

$$A_f^i = \langle f|\mu|i\rangle, \qquad (1b)$$

using the vibronic coupling model. In Eq. (1b), $|i\rangle$ denotes the fixed initial state, $|f\rangle$ denotes a final state, and μ is a transition moment operator. The initial state, $|i\rangle$, is assumed to be well described by a single electronic potential energy surface, on which vibrational motion is harmonic. The final state $|f\rangle$ is a complicated vibronically coupled state. Extensions to coupled initial states have also been discussed.[28] In the present formulation, we consider a molecule with $N^{\text{int}} = 3N^{\text{atom}} - 6$ internal degrees of freedom and neglect molecular rotation. While the discussion that follows is phrased in the language of negative ion photoelectron spectroscopy, the concepts and algorithms discussed are applicable to other electron and even photon spectroscopies.

The presentation of the theoretical framework is organized as follows. We begin by describing the wave function prior to and after photodetachment. The description of these wave functions provides for a concise, but precise, statement of the level of treatment and the approximations inherent in our multimode vibronic coupling approach. We will consider two cases: molecular states in which the spin-orbit interaction must be included, and those for which it can be neglected. Since the spin-orbit interaction is a relativistic effect, we refer to attributes determined with the spin-orbit interaction included (ignored) as relativistic (nonrelativistic) attributes. Elegant analyses of the spin-orbit interaction in the presence of conical and Renner–Teller intersections have been reported by Domcke and Poluyanov[29–31] and are reviewed in this volume.

The final state is presented in both the relativistic and nonrelativistic formulations to explain how the spin-orbit interaction is included. These representations for the final state $|f\rangle$ are then used to organize the discussion of secular equations used in their determination. The relevant

Hamiltonians for each level of treatment are described and the origin of the Ham reduction effect[32,33] in the relativistic case is explained. We then turn to the evaluation of the spectral intensity distribution function using a pedagogical approach based on Ref. 34. The evaluation of the relativistic spectral intensity distribution function addresses issues encountered working in a time-reversal adapted electronic basis[35,36] and demonstrates the origin of spin-orbit induced intensity borrowing.

The final portion of the theoretical section deals with three computational issues. The first is central to the vibronic coupling model: the construction of the nonrelativistic coupled electronic state quasi diabatic Hamiltonian that reproduces the electronic structure aspects of the problem. This issue is germane to both time-dependent and time-independent approaches. There are several procedures for determining $\mathbf{H}^d$ currently in use, which include (i) a "diabatization by ansatz" approach in which energies (only) are matched using least squares fitting,[37–39] (ii) a diabatization by maximizing overlap approach,[40] and (iii) approaches using analytic energy gradients.[21,27] The advantages of the approach reviewed in this work are that (i) it provides an accurate representation of the region around a point of conical intersection, (ii) it can be made more accurate in a systematic manner, and (iii) given the way it incorporates derivative couplings into the fitting procedure, it is maximally diabatic in a least squares sense.

The two remaining issues that will be discussed are specific to the time-independent approach and address the solution of the nonrelativistic vibronic secular equation. In particular we describe (i) an approach for reducing the size of the Hamilonian matrix used in the vibronic Schrödinger equation by careful choice of the basis used to expand $|f\rangle$ and (ii) an open-ended fine-grained parallel Lanczos based algorithm for solving that secular problem.

2.1. *Vibronic wave functions*

The vibronic wave functions are expressed as a sum of products of an electronic term $\mathbf{\Psi}_\alpha^e(\mathbf{r}^{N^{el}}; \mathbf{q})$ and a vibrational term, $\Theta_\alpha^f(\mathbf{q}, t)$, where $\mathbf{r}^{N^{el}}$ denotes the coordinates of the N^{el} electrons and $\mathbf{q}$ is a set of N^{int} internal coordinates.[41] Here, e can take the values a or d depending on whether $\mathbf{\Psi}_\alpha^e(\mathbf{r}^{N^{el}}; \mathbf{q})$ denotes an adiabatic or quasi-diabatic electronic state, respectively. The possible time dependence of the vibrational functions is indicated by t. In a time-independent approach, t is eliminated. The adjective quasi is used to indicate that, in general, rigorous diabatic electronic states do not exist.[42–44] Having said this, we will drop the modifier quasi for the remainder of this work.

2.1.1. *The initial state*

The initial state, which in the case of many negative ion photoelectron spectroscopies is the cold anionic ground electronic state, is written as

$$|i\rangle = \Psi_0^a(\mathbf{r}^{N^{el}+1}; \mathbf{Q}')\Theta_\mathbf{I}^0(\mathbf{Q}'), \qquad (2)$$

where

$$\Theta_\mathbf{I}^0(\mathbf{Q}') = \theta_\mathbf{I}^0(\mathbf{Q}') = \prod_{j=1}^{N^{int}} \chi_{I_j}^{0,j}(\mathbf{Q}'_j). \qquad (3)$$

This representation of the initial state requires some explanation. Firstly, note that the electronic state Ψ_0^a is an adiabatic electronic state for a molecule with $N^{el}+1$ electrons, whose coordinates are denoted $\mathbf{r}^{N^{el}+1}$. This electronic state has an equilibrium geometry denoted $\mathbf{q}^{a,\min}$. The $\mathbf{Q}'$ are the normal modes of the corresponding harmonic potential and are given as a linear transformation of the $\mathbf{q}$ as

$$\mathbf{Q}' = \mathbf{L}'(\mathbf{q} - \mathbf{q}^{a,\min}), \qquad (4)$$

while $\chi_m^{0,j}$ denotes the mth harmonic oscillator function corresponding to the jth mode of the anion harmonic potential. Owing to the assumption of a harmonic potential for the anion, its (low-lying) vibrational levels are represented by an N^{int} dimensional vector $\mathbf{I} = (I_1, I_2, \ldots, I_{N^{int}})$, where I_m indicates the number of phonons in the mth mode, and $\mathbf{I} = \mathbf{0}$, is the ground vibrational state. The corresponding vibrational wave function is given by $\theta_\mathbf{I}^0(\mathbf{Q}')$, the multimode product in Eq. (3).

2.1.2. *The final state: Nonadiabatic Ansatz*

The final state is a vibronic level of the neutral molecule, which has N^{el} electrons and one electron photodetached. The final state wave function is much more challenging to determine (and describe). Firstly, there is the issue of the coordinates. In the time-independent approach, a set of normal coordinates $\mathbf{Q}$, are employed to describe the final state vibronic levels:

$$\mathbf{Q} = \mathbf{L}(\mathbf{q} - \mathbf{q}^0). \qquad (5)$$

However, there is no universal choice for either $\mathbf{L}$ or $\mathbf{q}^0$. In fact, in the applications discussed in Sec. 3, two different choices are made. We simply observe here that it is quite common to choose $\mathbf{Q} = \mathbf{Q}'$, whereas in our work $\mathbf{Q} \ne \mathbf{Q}'$ with $\mathbf{L} \ne \mathbf{L}'$ and $\mathbf{q}^0 \ne \mathbf{q}^{a,\min}$. The implications of this are discussed in Sec. 2.2.1.

Unlike the anion wave function, for which a single electronic state suffices, the wave function for the neutral species must be expanded as a sum of vibronic products, that is

$$|f\rangle = \sum_{\alpha=1}^{N^{\text{state}}} \Psi_\alpha^d(\mathbf{r}^{N^{\text{el}}}; \mathbf{Q}) \Theta_\alpha^f(\mathbf{Q}, t), \qquad (6)$$

where Ψ_α^d are diabatic electronic states. Note that these are N^{el}, rather than $N^{\text{el}} + 1$, electron wave functions. Therefore, although the wavefunction can provide a correct description of the neutral or photodetached species, it cannot describe the photodissociation process, that is the matrix element $\mu^{f,i} = \langle f|\mu|i\rangle$, since the outgoing electron is ignored. In order to incorporate the outgoing electron, and thus enable the determination of $\mu^{f,i}$ in the vibronic coupling model, the wave function is assumed to be unaltered by the photodetached electron in a simplification termed the sudden approximation.[34] This approximation allows us to determine the final state, $|f\rangle$, without first determining the transition moment, which may be a formidable computational challenge in itself.

2.1.2.1. Time-independent nonrelativistic wave functions

In the time-independent approach, the vibrational wave functions $\Theta_\alpha^f(\mathbf{Q})$ in Eq. (6), are expanded as a sum of the multimode products, that is

$$\Theta_\alpha^f(\mathbf{Q}) = \sum_{\mathbf{m}} d_{\alpha,\mathbf{m}}^f \theta_{\mathbf{m}}^\alpha(\mathbf{Q}) = \sum_{\mathbf{m}} d_{\alpha,\mathbf{m}}^f \prod_{i=1}^{N^{\text{int}}} \chi_{m_i}^{\alpha,i}(Q_i), \quad 0 \le m_i < M_i, \qquad (7)$$

so that

$$\Psi_f^{T,nr}(\mathbf{r}^{N^{\text{el}}}, \mathbf{Q}) = \sum_{\mathbf{m},\alpha} \Psi_\alpha^d(\mathbf{r}^{N^{\text{el}}}; \mathbf{Q}) d_{\alpha,\mathbf{m}}^f \prod_{i=1}^{N^{\text{int}}} \chi_{m_i}^{\alpha,i}(Q_i). \qquad (8)$$

The same multimode basis is used for each electronic state, an approximation that is essential to the efficiency of the computational approach discussed in Sec. 2.4. Note that in Eq. (7) there are $N^{\text{vib}} = \prod_{i=1}^{N^{\text{int}}} M_i$ terms. While strongly dependent on the system under study, N^{vib} is routinely in the tens to hundreds of millions and can often exceed 10^9. The special techniques required to treat such large expansions are discussed in Sec. 2.4.

2.1.2.2. Wave functions including the spin-orbit interaction

Because of the large size of the expansion in Eq. (8) and the potential need for complex arithmetic in treating the spin-orbit interaction, the relativistic eigenstates $\Psi_l^{T,so}$ are expanded in the nonrelativistic eigenbasis, $\Psi_f^{T,nr}$. In order to explain the form of these eigenstates, the spin quantum numbers of $\Psi_\alpha^d(\mathbf{r}^{N^{el}};\mathbf{Q})$ must be specified. Since this review is concerned exclusively with photodetachment from singlet states, the Ψ_α^d are doublets which we denote Ψ_α^{d,M_s} where $M_s = \pm\frac{1}{2}$. It is convenient to work in the time reversal adapted electronic basis.[35, 36]

$$\Psi_\alpha^{d,\pm} = \frac{\pm 1}{\sqrt{2}}(\Psi_\alpha^{d,1/2} \pm i\Psi_\alpha^{d,-1/2}). \qquad (9)$$

The time-reversal basis simplifies the treatment of the spin-orbit interaction and is briefly reviewed in Appendix A. Replacing Ψ_α^d in Eq. (8) with $\Psi_\alpha^{d,\pm}$, the eigenstates of $H^{T,nr}$ can be written as

$$\Psi_f^{T,nr,p}(\mathbf{r}^{N^{el}},\mathbf{Q}) = \sum_{\mathbf{m},\alpha} \Psi_\alpha^{d,p}(\mathbf{r}^{N^{el}};\mathbf{Q})d_{\alpha,\mathbf{m}}^f \prod_{i=1}^{N^{int}} \chi_{m_i}^{\alpha,i}(Q_i), \quad p=\pm. \qquad (10)$$

Since the nonrelativistic energies are independent of p, these states come in degenerate pairs, called Kramers doublets.[45] Using the lowest $2N^{eig}$ nonrelativistic eigenstates as the basis, the eigenstates of total relativistic Hamiltonian with energy $E_m^{T,so}$ are given by

$$\Psi_m^{T,so} = \sum_{\substack{p=\pm \\ k=1-N^{eig}}} c_{p,k}^m \Psi_k^{T,nr,p}. \qquad (11)$$

Since $\hat{T}\Psi_k^{T,nr,+} = \Psi_k^{T,nr,-}$, the Kramers degenerate pair $(\Psi_m^{T,so}, \hat{T}\Psi_m^{T,so})$, where $\hat{T}$ is the time reversal operator, is contained in this expansion.

2.1.3. Determination of time-independent wave functions

2.1.3.1. Nonrelativistic Schrödinger equation

In the nonrelativistic case, the $N^{state}N^{vib}$ dimensional eigenvector $d_{\alpha,\mathbf{m}}^k$ and the associated energy $E_k^{T,nr}$ are determined from the nonrelativistic nuclear-electronic Schrödinger equation

$$(H^{T,nr} - E_f^{T,nr})\Psi_f^{T,nr} = 0, \qquad (12)$$

where

$$H^{T,nr} = T^{nuc} + H^0(\mathbf{r}; \mathbf{Q}). \quad (13)$$

In the above, H^0 is the Coulomb or nonrelativistic Hamiltonian and T^{nuc} is the nuclear kinetic energy operator.

The working form of Eq. (12) is obtained by inserting Eq. (8) into Eq. (12), then multiplying by $\Psi_\alpha^d(\mathbf{r}^{N^{el}}; \mathbf{Q})\theta_{\mathbf{m}}^\alpha(\mathbf{Q})$ and integrating successively with respect to $\mathbf{r}$ and $\mathbf{Q}$ to get

$$\begin{pmatrix} \mathbf{H}_{1;1}^{vib,nr} & \mathbf{H}_{1;2}^{vib,nr} & \cdots & \mathbf{H}_{1;N^{state}}^{vib,nr} \\ \mathbf{H}_{2;1}^{vib,nr} & \mathbf{H}_{2;2}^{vib,nr} & \cdots & \mathbf{H}_{2;N^{state}}^{vib,nr} \\ \vdots & & & \vdots \\ \mathbf{H}_{N^{state};1}^{vib,nr} & \mathbf{H}_{N^{state};2}^{vib,nr} & \cdots & \mathbf{H}_{N^{state};N^{state}}^{vib,nr} \end{pmatrix} \begin{pmatrix} \mathbf{d}_1^f \\ \mathbf{d}_2^f \\ \vdots \\ \mathbf{d}_{N^{state}}^f \end{pmatrix}$$
$$= E_k^{T,nr} \begin{pmatrix} \mathbf{d}_1^f \\ \mathbf{d}_2^f \\ \vdots \\ \mathbf{d}_{N^{state}}^f \end{pmatrix}, \quad (14)$$

where we have noted that the $\Psi_\alpha^d(\mathbf{r}^{N^{el}}; \mathbf{Q})$ are diabatic, and defined

$$H_{\gamma,\mathbf{m};\gamma'\mathbf{m}}^{vib,nr} = \langle \theta_\mathbf{m}^\gamma(\mathbf{Q}) | H_{\gamma,\gamma'}^{d,0}(\mathbf{Q}) | \theta_{\mathbf{m}'}^{\gamma'}(\mathbf{Q}) \rangle_\mathbf{Q}, \quad (15)$$

and

$$H_{\alpha,\beta}^{d,0}(\mathbf{Q}) = \langle \Psi_\alpha^d(\mathbf{r}^{N^{el}}; \mathbf{Q}) | H^0 | \Psi_\beta^d(\mathbf{r}^{N^{el}}; \mathbf{Q}) \rangle_\mathbf{r}. \quad (16)$$

In Eq. (14) each $\mathbf{H}_{i;j}^{vib,nr}$ is an $N^{vib} \times N^{vib}$ matrix. The construction of $H_{\gamma,\gamma'}^{d,0}(\mathbf{Q})$ from *ab initio* electronic structure data is a significant issue in the spectral simulation and is discussed at length in Sec. 2.4.1. When no confusion will take place, $\mathbf{H}^{d,0}$ will be abbreviated $\mathbf{H}^d$. The diagonalization of $\mathbf{H}_{i;j}^{vib,nr}$ is a computational challenge, one frequently tackled employing a Lanczos algorithm.[46,47] An open-end fine grained parallel implementation[47] is described in Sec. 2.4.

2.1.3.2. *Schrödinger equation including the spin-orbit interaction*

The total nuclear-electronic Hamiltonian including the spin-orbit interaction, $H^{T,so}$, is

$$H^{T,so} = T^{nuc} + H^0(\mathbf{r}; \mathbf{Q}) + H^{so}(\mathbf{r}; \mathbf{Q}), \quad (17)$$

where H^{so} is the spin-orbit operator in the Breit–Pauli approximation[48]

$$H^{so}(\mathbf{r};\mathbf{Q}) = \sum_{i=1}^{N^{el}} \mathbf{h}^{so,1-2}(\mathbf{r}_i,\mathbf{Q}) \cdot \mathbf{s}_i, \qquad (18)$$

and $\mathbf{h}^{so,1-2}$ includes the one (spin-orbit) and two electron (spin–same orbit and spin–other orbit) parts of the Breit–Pauli spin-orbit operator.[48] The relativistic wave functions in Eq. (12) satisfy the relativistic electronic-nuclear Schrödinger equation,

$$(H^{T,so} - E_m^{T,so})\Psi_m^{T,so} = 0. \qquad (19)$$

To determine $\Psi_m^{T,so}$, the matrix elements

$$\begin{aligned}H^{T,so}_{kz;k',z'} &= \langle \Psi_k^{T,nr,z}|H^{T,so}|\Psi_{k'}^{T,nr,z'}\rangle \\ &= \delta_{k,k'}\delta_{z,z'}E_k^{T,nr} + \langle \Psi_k^{T,nr,z}|H^{so}|\Psi_{k'}^{T,nr,z'}\rangle\end{aligned} \qquad (20)$$

are required. Using the definition of $\Psi_k^{T,nr,z}(\mathbf{r},\mathbf{Q})$ in Eq. (10), we have

$$\begin{aligned}\langle \Psi_f^{T,nr,z}|H^{so}|\Psi_l^{T,nr,z'}\rangle &= \sum_{\substack{\mathbf{m},\gamma \\ \mathbf{m}',\gamma'}} \langle d^f_{\gamma,\mathbf{m}}\theta^\gamma_\mathbf{m}|H^{e,so}_{\gamma,z;\gamma',z'}|d^l_{\gamma',\mathbf{m}'}\theta^{\gamma'}_{\mathbf{m}'}\rangle_\mathbf{Q} \\ &= \sum_{\substack{\mathbf{m},\gamma \\ \mathbf{m}',\gamma'}} d^f_{\gamma,\mathbf{m}}\bar{H}^{so}_{\mathbf{m},\gamma,z;\mathbf{m}'\gamma',z'}d^l_{\gamma',\mathbf{m}'},\end{aligned} \qquad (21)$$

where

$$\bar{H}^{so}_{\mathbf{m},\gamma,z;\mathbf{m}',\gamma',z'} \equiv \langle \theta^\gamma_\mathbf{m}(\mathbf{Q})|H^{e,so}_{\gamma,z;\gamma',z'}(\mathbf{Q})|\theta^{\gamma'}_{\mathbf{m}'}(\mathbf{Q})\rangle_\mathbf{Q} \qquad (22a)$$

and

$$H^{e,so}_{\gamma,z;\gamma',z'}(Q) \equiv \langle \Psi_\gamma^{d,z}(\mathbf{r}^{N^{el}};\mathbf{Q})|H^{so}|\Psi_{\gamma'}^{d,z'}(\mathbf{r}^{N^{el}};\mathbf{Q})\rangle_\mathbf{r}. \qquad (22b)$$

To evaluate $H^{e,so}_{\gamma,z;\gamma',z'}(\mathbf{Q})$ we assume $N^{\text{state}} = 2$ but the results are readily extended to larger values of N^{state}. Using the Slater–Condon rules, with the basis functions ordered as $\Psi_\alpha^{d,+}, \Psi_\beta^{d,+}, \Psi_\alpha^{d,-}, \Psi_\beta^{d,-}$, the $H^{e,so}_{\gamma,z;\gamma',z'}(\mathbf{Q})$ are given by

$$\mathbf{H}^{e,so} = \begin{pmatrix} 0 & iH_Y^{rso} & 0 & iH_Z^{rso} + H_X^{rso} \\ -iH_Y^{rso} & 0 & -iH_Z^{rso} - H_X^{rso} & 0 \\ 0 & iH_Z^{rso} - H_X^{rso} & 0 & -iH_Y^{rso} \\ -iH_Z^{rso} + H_X^{rso} & 0 & iH_Y^{rso} & 0 \end{pmatrix}, \quad (23)$$

where

$$H_\lambda^{rso}(\mathbf{Q}) = i \left\langle \Psi_\alpha^{d,1/2} \left| \sum_{i=1}^{N^{el}} h_\lambda^{1-2}(\mathbf{r}_i, \mathbf{Q}) s_z(i) \right| \Psi_\beta^{d,1/2} \right\rangle_\mathbf{r}, \quad \lambda = X, Y, Z. \quad (24)$$

Note that Eq. (23) involves only the approximation that interactions with other nonrelativistic electronic wave states can be neglected. The fact that only one combination of the M_s values is required in Eq. (24) is a consequence of the Wigner–Eckart theorem.[48–50] Using Eq. (23), Eq. (21) reduces to

$$\langle \Psi_f^{T,nr,z} | H^{so} | \Psi_l^{T,nr,z'} \rangle$$
$$= \sum_{\mathbf{m},\mathbf{m}'} (d_{1,\mathbf{m}}^f d_{2,\mathbf{m}'}^l - d_{2,\mathbf{m}}^f d_{1,\mathbf{m}'}^l)[iH_{Y,\mathbf{m},\mathbf{m}'}^{so}]$$
$$\equiv i(H_Y^{A,so})_{f,l} \quad \text{for } z = z' = + \quad (25a)$$
$$= \sum_{\mathbf{m},\mathbf{m}'} (d_{1,\mathbf{m}}^f d_{2,\mathbf{m}'}^l - d_{2,\mathbf{m}}^f d_{1,\mathbf{m}'}^l)[iH_{Z,\mathbf{m},\mathbf{m}'}^{so} + H_{X,\mathbf{m},\mathbf{m}'}^{so}]$$
$$\equiv (iH_Z^{A,so} + H_X^{A,so})_{f,l} \quad \text{for } z = +, z' = - \quad (25b)$$

and the $z = z' = -$ and $z = -, z' = +$ results are obtained from

$$\langle \Psi_k^{T,nr,-}(\mathbf{r},\mathbf{Q}) | H^{so} | \Psi_l^{T,nr,-}(\mathbf{r},\mathbf{Q}) \rangle$$
$$= \langle \Psi_k^{T,nr,+}(\mathbf{r},\mathbf{Q}) | H^{so} | \Psi_l^{T,nr,+}(\mathbf{r},\mathbf{Q}) \rangle^*, \quad (26a)$$
$$\langle \Psi_k^{T,nr,-}(\mathbf{r},\mathbf{Q}) | H^{so} | \Psi_l^{T,nr,+}(\mathbf{r},\mathbf{Q}) \rangle$$
$$= -\langle \Psi_K^{T,nr,+}(\mathbf{r},\mathbf{Q}) | H^{so} | \Psi_l^{T,nr,-}(\mathbf{r},\mathbf{Q}) \rangle^*, \quad (26b)$$

which is a consequence of the use of a time-reversal adapted electronic basis as reflected in Eq. (23) [and see also Appendix A].[35,36] Equations (25) and (26) are particularly illuminating in the limit that $H^{e,so}_{\alpha \cdot z; \alpha', z'}(\mathbf{Q})$ is independent of $\mathbf{Q}$. In that case, Eqs. (21), or (25), (26) reduce to

$$\langle \Psi^{T,nr,z}_k(\mathbf{r},\mathbf{Q})|H^{so}|\Psi^{T,nr,z'}_l(\mathbf{r},\mathbf{Q})\rangle = O_{k,l} H^{e,so}_{\alpha,z;\beta z'}, \qquad (27a)$$

where

$$O_{f,l} = \sum_m (d^f_{1,m} d^l_{2,m} - d^f_{2,m} d^l_{1,m}). \qquad (27b)$$

Since $|O_{k,l}|$ is less than 1, it has the effect of reducing the net spin-orbit interaction [Eq. (27a)] and hence is referred to as a generalized Ham reduction factor.[32,33]

Using Eqs. (11), (20), (25a) and (25b), Eq. (19) becomes:

$$\begin{pmatrix} \mathbf{E}^{T,nr} + i\mathbf{H}^{A,so}_Y & i\mathbf{H}^{A,so}_Z + \mathbf{H}^{A,so}_X \\ i\mathbf{H}^{A,so}_Z - \mathbf{H}^{A,so}_X & \mathbf{E}^{T,nr} - i\mathbf{H}^{A,so}_Y \end{pmatrix} \begin{pmatrix} \mathbf{c}^m_+ \\ \mathbf{c}^m_- \end{pmatrix} = E^{T,so}_m \begin{pmatrix} \mathbf{c}^m_+ \\ \mathbf{c}^m_- \end{pmatrix}; \qquad (28)$$

here $E^{T,nr}_{k,l} = \delta_{k,l} E^{T,nr}_k$. In practice, Eqs. (27a) and (27b) are used to simplify the $N^{eig} \times N^{eig}$ antisymmetric matrices $\mathbf{H}^{A,so}_W$. Since the matrix on the left hand side of Eq. (28) is complex hermitian it is diagonalized by the standard technique of separating the real and imaginary parts, such that

$$\begin{pmatrix} \mathbf{E}^{T,nr} & \mathbf{H}^{A,so}_X & -\mathbf{H}^{A,so}_Y & -\mathbf{H}^{A,so}_Z \\ -\mathbf{H}^{A,so}_X & \mathbf{E}^{T,nr} & -\mathbf{H}^{A,so}_Z & \mathbf{H}^{A,so}_Y \\ \mathbf{H}^{A,so}_Y & \mathbf{H}^{A,so}_Z & \mathbf{E}^{T,nr} & \mathbf{H}^{A,so}_X \\ \mathbf{H}^{A,so}_Z & -\mathbf{H}^{A,so}_Y & -\mathbf{H}^{A,so}_X & \mathbf{E}^{T,nr} \end{pmatrix} \begin{pmatrix} \mathbf{c}^{m,R}_+ \\ \mathbf{c}^{m,R}_- \\ \mathbf{c}^{m,I}_+ \\ \mathbf{c}^{m,I}_- \end{pmatrix} = E^{T,so}_m \begin{pmatrix} \mathbf{c}^{m,R}_+ \\ \mathbf{c}^{m,R}_- \\ \mathbf{c}^{m,I}_+ \\ \mathbf{c}^{m,I}_- \end{pmatrix}, \qquad (29)$$

where $\mathbf{c}^m_\pm = \mathbf{c}^{m,R}_\pm + i\mathbf{c}^{m,I}_\pm$.

2.2. *Spectral intensity distribution function*

This section evaluates the working expressions for the spectral intensity distribution functions, $I^{nr}(E)$ and $I^{so}(E)$. Two key issues are addressed: (i) the nature of the transition moments to the diabatic states of the neutral and (ii) the intensity borrowing induced by the spin-orbit interaction. The derivation of $I^{nr}(E)$ presented here follows that of Ref. 34. As noted previously, the photodetached electron must be explicitly considered in order to determine $I(E)$. The outgoing electron is assumed to be in an orbital ϕ_f^c where c denotes the fact that this orbital describes a continuum electron and f denotes the vibronic channel. While even this simplified analysis is beyond the scope of this review, Appendix B suggests how this question might be addressed.

2.2.1. *Nonrelativistic formulation*

Using $\Psi_\gamma^{d,\pm 1/2}$ and the continuum orbital ϕ_f^c, we can construct the $N^{\mathrm{el}}+1$ electron spin singlet as

$$\Psi_\gamma^{d,S} = \frac{1}{\sqrt{2}} A[\Psi_\gamma^{d,-1/2}(r^{N^{\mathrm{el}}};\mathbf{Q})\phi_f^c(\mathbf{r}_{N^{\mathrm{el}}+1})\alpha$$
$$- \Psi_\gamma^{d,1/2}(r^{N^{\mathrm{el}}};\mathbf{Q})\phi_f^c(\mathbf{r}_{N^{\mathrm{el}}+1})\beta], \quad (30)$$

where A is the $N^{\mathrm{el}}+1$ electron antisymmetrizer and $\phi_f^c(\mathbf{r}_{N^{\mathrm{el}}+1})$ is a one-electron function of the $N^{\mathrm{el}}+1$ electron. In this case,

$$\mu^{nr,f,I} = \left\langle \sum_\gamma \frac{1}{\sqrt{2}} A \begin{bmatrix} \Psi_\gamma^{d,-1/2}(\mathbf{r}^{N^{\mathrm{el}}};\mathbf{Q})\phi_f^c(\mathbf{r}_{N^{\mathrm{el}}+1})\alpha - \\ \Psi_\gamma^{d,+1/2}(\mathbf{r}^{N^{\mathrm{el}}};\mathbf{Q})\phi_f^c(\mathbf{r}_{N^{\mathrm{el}}+1})\beta \end{bmatrix} \right.$$
$$\left. \times \Theta_\gamma^f(\mathbf{Q})|\mu|\Psi_0^a(\mathbf{r}^{N^{\mathrm{el}}+1};\mathbf{Q}')\Theta_I^0(\mathbf{Q}') \right\rangle_{\mathbf{r},\mathbf{Q}}. \quad (31)$$

The diabatic state transition moment, $\mu^{\gamma,0}(\mathbf{Q})$ is defined as:

$$\mu^{\gamma,0}(\mathbf{Q}) = \left\langle \frac{1}{\sqrt{2}} A[\Psi_\gamma^{d,-1/2}(\mathbf{r}^{N^{\mathrm{el}}};\mathbf{Q})\phi_f^c(\mathbf{r}_{N^{\mathrm{el}}+1})\alpha \right.$$
$$\left. - \Psi_\gamma^{d,+1/2}(\mathbf{r}^{N^{\mathrm{el}}};\mathbf{Q})\phi_f^c(\mathbf{r}_{N^{\mathrm{el}}+1})\beta]|\mu|\Psi_0^a(\mathbf{r}^{N^{\mathrm{el}}+1};\mathbf{Q}') \right\rangle_{\mathbf{r}}. \quad (32)$$

The diabatic state dependence of $\mu^{\gamma,0}(\mathbf{Q})$ is attributed entirely to the diabatic electronic state, since ϕ_f^c depends on the vibronic level only. Using Eqs. (3) and (6) the nonrelativistic spectral distribution function reduces to

$$\mu^{nr,f,\mathbf{I}} = \sum_\gamma \langle \Theta_\gamma^f(\mathbf{Q})|\mu^{\gamma,0}|\Theta_I^0(\mathbf{Q}')\rangle$$
$$\approx \sum_\gamma \mu^{\gamma,0} \langle \Theta_\gamma^f(\mathbf{Q})|\Theta_I^0(\mathbf{Q}')\rangle$$
$$= \sum_{\mathbf{m},\gamma} d_{\gamma,\mathbf{m}}^f[\mu^{\gamma,0} o(\mathbf{m},\mathbf{I})] = \mathbf{s}^\dagger \mathbf{d}, \quad (33)$$

where, the approximate equality in Eq. (33) assumes $\mu^{\alpha,0}(\mathbf{Q})$ is independent of $\mathbf{Q}$,

$$o(\mathbf{m}, \mathbf{I}) = \left\langle \prod_{i=1}^{N^{\text{int}}} \chi_{m_i}^{\alpha,i}(Q_i) \middle| \prod_{j=1}^{N^{\text{int}}} \chi_{I_j}^{0,j}(Q_j') \right\rangle, \tag{34}$$

and

$$s_{\alpha,\mathbf{m}} = \mu^{\alpha,0} o(\mathbf{m}, \mathbf{I}). \tag{35}$$

The vector of Franck–Condon overlap factors $o(\mathbf{m}, \mathbf{I})$ has length N^{vib}. Although N^{vib} is large, efficient algorithms for evaluating $o(\mathbf{m}, \mathbf{I})$ based on well-known recursion relations[51–54] have been reported.[55]

The evaluation of $\mathbf{s}^\dagger \mathbf{d}^f$, the dot product of two vectors of dimension $N^{\text{state}} N^{\text{vib}}$, is intimately related to the diagonalization procedure used to solve Eq. (14), as discussed in Ref. 21 and Sec. 2.4.3. The transition intensities, $\mathbf{s}^\dagger \mathbf{d}^f$, are readily determined if $\mathbf{s}$ is used as the seed or initial guess vector in the Lanczos procedure used to determine $\mathbf{d}^f$.

2.2.2. *Inclusion of spin-orbit coupling*

The derivation of the spectral intensity distribution including the spin-orbit interaction proceeds in a fashion similar to that in the nonrelativistic case, except that some care must be taken in dealing with the time reversal adapted electronic basis. Starting with Eq. (11) and using the sudden approximation, we construct two $N^{\text{el}} + 1$ electron wave functions

$$\Psi_m^{T,so,\alpha} = \sum_{\substack{p = \pm \\ k = 1 - N^{\text{eig}}}} c_{p,k}^m \Psi_k^{T,nr,p}(\phi_m^c \alpha) \quad \text{and}$$

$$\Psi_m^{T,so,\beta} = \sum_{\substack{p = \pm \\ k = 1 - N^{\text{eig}}}} c_{p,k}^m \Psi_k^{T,nr,p}(\phi_m^c \beta), \tag{36}$$

where the parenthesis indicates that when the continuum orbital is included, $N^{\text{el}} + 1$ antisymmetrized products must be formed. These linear combinations can be re-expressed in terms of spin-eigenfunctions by forming

$$\Psi_m^{T,so,\pm} = \frac{1}{\sqrt{2}}(\Psi_m^{T,so,\alpha} \pm i\Psi_m^{T,so,\beta}). \tag{37}$$

Since the $c_{p,k}^m$ are independent of ϕ_m^c and its spin, the electronic part of the wave functions in Eq. (37) is given below (suppressing the N^{el} superscript), for $p = +$, terms in the sum in Eq. (36):

$$\frac{1}{\sqrt{2}}(\Psi_\gamma^{d,+}(\phi_m^c\alpha) \pm i\Psi_\gamma^{d,+}(\phi_m^c\beta))$$
$$= \frac{1}{2}(\Psi_\gamma^{d,1/2}(\phi_m^c\alpha) + i\Psi_\gamma^{d,-1/2}(\phi_m^c\alpha) \pm i(\Psi_\gamma^{d,1/2}(\phi_m^c\beta) + i\Psi_\gamma^{d,-1/2}(\phi_m^c\beta))). \tag{38a}$$

and for $p = -$, terms in the sum in Eq. (36):

$$\frac{1}{\sqrt{2}}(\Psi_\gamma^{d,-}(\phi_m^c\alpha) \pm i\Psi_\gamma^{d,-}(\phi_m^c\beta))$$
$$= \frac{1}{2}(\Psi_\gamma^{d,1/2}(\phi_m^c\alpha) - i\Psi_\gamma^{d,-1/2}(\phi_m^c\alpha) \pm i(\Psi_\gamma^{d,1/2}(\phi_m^c\beta) - i\Psi_\gamma^{d,-1/2}(\phi_m^c\beta))). \tag{38b}$$

It can be seen from the form of $\Psi_\alpha^{d,S}$ in Eq. (30) that Eq. (38a) contains one singlet term which occurs for the $-i$ combination in Eq. (37), while Eq. (38b) contains one singlet term which occurs for the $+i$ combination in Eq. (37). Thus each of the wave functions in Eq. (37) makes an independent contribution to the line intensity

$$\sqrt{2}\mu^{so,m,\mathbf{I},+}$$
$$= \left\langle \sum_{\substack{p = \pm \\ f = 1 - N^{\text{eig}}}} c_{p,f}^m(\Psi_f^{T,nr,p}(\phi_m^c\alpha) + i\Psi_f^{T,nr,p}(\phi_m^c\beta)) |\mu| \Psi_0^a \Theta_\mathbf{I}^0 \right\rangle$$
$$= \left\langle \sum_{f=1-N^{\text{eig}}} c_{-,f}^m(\Psi_f^{T,nr,-}(\phi_m^c\alpha) + i\Psi_f^{T,nr,-}(\phi_m^c\beta)) |\mu| \Psi_0^a \Theta_\mathbf{I}^0 \right\rangle \tag{39a}$$

and

$$\sqrt{2}\mu^{so,m,\mathbf{I},-}$$
$$= \left\langle \sum_{\substack{p = \pm \\ f = 1 - N^{\text{eig}}}} c_{p,f}^m(\Psi_f^{T,nr,p}(\phi_m^c\alpha) - i\Psi_f^{T,nr,p}(\phi_m^c\beta)) |\mu| \Psi_0^a \Theta_\mathbf{I}^0 \right\rangle$$
$$= \left\langle \sum_{f=1-N^{\text{eig}}} c_{+,f}^m(\Psi_f^{T,nr,+}(\phi_m^c\alpha) - i\Psi_f^{T,nr,+}(\phi_m^c\beta)) |\mu| \Psi_0^a \Theta_\mathbf{I}^0 \right\rangle. \tag{39b}$$

Then the line intensity is given by

$$2|A^{so}_{m,\mathbf{I}}| = |\mu^{so,m,\mathbf{I},+}|^2 + |\mu^{so,m,\mathbf{I},-}|^2. \quad (40)$$

To be specific, we insert Eqs. (2), (6) and (7) into Eq. (39a) giving:

$$\mu^{so,m,\mathbf{I},+} = \Bigg\langle \sum_{f=1}^{N^{\text{eig}}} c^m_{-,f} \sum_{\mathbf{s},\lambda} d^f_{\lambda,\mathbf{s}} \left[\prod_{i=1}^{N^{\text{int}}} \chi^{(n),i}_{s_i}(Q_i) \right] (\Psi^{d,-}_\lambda (\phi^c_m \alpha)$$

$$+ i\Psi^{d,-}_\lambda (\phi^c_m \beta))|\mu|\Psi^a_0 \prod_{j=1}^{N^{\text{int}}} \chi^{0,j}_{I_j}(Q'_i) \Bigg\rangle. \quad (41a)$$

Using Eqs. (32)–(35) and assuming the electronic transition moment is independent of $\mathbf{Q}$, this becomes

$$\mu^{so,m,\mathbf{I},+} = \sum_{f=1}^{N^{\text{eig}}} c^m_{-,f} \sum_{\mathbf{m}',\gamma} d^f_{\gamma,\mathbf{m}'} \mu^{\gamma,0} o(\mathbf{m}',\mathbf{I}) = \sum_{f=1}^{N^{\text{eig}}} c^m_{-,f} \mu^{nr,f,\mathbf{I}}. \quad (41b)$$

Similarly,

$$\mu^{so,m,\mathbf{I},-} = \sum_{f=1-N^{\text{eig}}} c^m_{+,f} \mu^{nr,f,\mathbf{I}}. \quad (41c)$$

Note that in the nonrelativistic limit, $c^m_{p,f} = \delta_{f,m}\delta_{p,+}$ or $c^m_{p,f} = \delta_{f,m}\delta_{p,-}$, so Eq. (40) becomes equivalent to Eq. (33). Equations (41b) and (41c) evince the intensity borrowing in the relativistic case.

2.3. Hamiltonians

2.3.1. Coulomb or nonrelativistic diabatic Hamiltonian, $\mathbf{H}^d$

In the vibronic coupling approximation, the nonrelativistic diabatic Hamiltonian matrix $\mathbf{H}^d$ in Eq. (16), has the form

$$H^{d,0}_{\alpha,\beta}(\mathbf{Q}) = \langle \Psi^d_\alpha(\mathbf{r}^{N^{\text{el}}}; \mathbf{Q})|H^0|\Psi^d_\beta(\mathbf{r}^{N^{\text{el}}}; \mathbf{Q}) \rangle_r$$

$$= E^0_\alpha(Q^0)\delta_{\alpha,\beta} + \sum_i V^{(1),\alpha,\beta}_i Q_i + \frac{1}{2} \sum_{i,j} V^{(2),\alpha,\beta}_{i,j} Q_i Q_j + \cdots \quad (42)$$

Here we have explicitly written $\mathbf{H}^{d,0}$ through second order. The determination of the unknown coefficients is discussed in Sec. 2.4.1. This quadratic vibronic coupling Hamiltonian is the workhorse of the field, although Hamiltonians with higher order terms have been reported.[38,39,56] In this chapter we will review a study which addresses the limits of utility of this quadratic approach and how that limit can be achieved.[57] Note that $\mathbf{H}^d$ is unchanged if the time reversal adapted electronic wave functions in Eq. (10), $\Psi_\alpha^{d,\pm}$, are used, that is

$$\langle \Psi_\alpha^{d,z} | H^0 | \Psi_\beta^{d,z'} \rangle_\mathbf{r} = H_{\alpha,\beta}^{d,0}(\mathbf{Q}) \delta_{z,z'}, \tag{43}$$

where $z, z' = \pm$. The electronic Schrödinger equation corresponding to $\mathbf{H}^d$ is

$$(\mathbf{H}^d(\mathbf{Q}) - E_J^0(\mathbf{Q})\mathbf{I})\mathbf{e}^J(\mathbf{Q}) = \mathbf{0}. \tag{44}$$

2.3.2. *Kinetic energy operator*

As Eqs. (14) and (18) show, construction of $H^{T,nr}$ or $H^{T,so}$ requires determination of the kinetic energy operator. This is generally addressed by working in a normal mode basis of a reference Hamiltonian. Thus

$$H^{T,nr} = T^{nuc} + H^0 = (T^{nuc} + V^{\text{ref}}) + (H^0 - V^{\text{ref}}) \equiv H^{T,\text{ref}} + \Delta H. \tag{45a}$$

The contributions of $H^{T,\text{ref}}$ will only appear in the diagonal elements of $H^{T,nr}$ and are given by the harmonic oscillator energy expression:

$$\langle \theta_\mathbf{m}^\alpha | H^{T,\text{ref}} | \theta_\mathbf{m}^\alpha \rangle = \sum_{i=1}^{N^{\text{int}}} \omega_i^\alpha (m_i + 1/2). \tag{45b}$$

The choice of which normal coordinates, and thus the form of $H^{T,\text{ref}}$, to employ in a vibronic coupling computation will be discussed in further detail in the following section.

2.4. *Computational issues*

In this section we review three computational issues relevant to the simulation of photoelectron spectra. Firstly we consider the construction of diabatic Hamiltonians from electronic structure data. This key issue is relevant to both the time-dependent and time-independent formulations of the multimode vibronic coupling model. Then we turn to two issues which have enabled us to extend considerably the range of applicability of the time-independent approach: (i) the reduction of the size of the vibronic expansion in Eq. (7) by flexible choice of the $\chi_m^{\alpha,i}(Q_i)$ and (ii) the development of an open-ended fine-grained parallel Lanczos solver to treat Eq. (14).

2.4.1. Construction of diabatic Hamiltonians for bound states

The working equations are obtained by differentiating the electronic Schrödinger equation shown in Eq. (44) and inserting the gradient of $\mathbf{H}^d$ from Eq. (42). The resulting equations, based on energy gradients and derivative couplings, through which the second-, and potentially higher, order coefficients may be obtained, are given by:

$$\bar{\bar{M}}_k^{I,J}(\mathbf{q}^{(n)}) = \sum_{\substack{1 < \alpha, \beta \leq N^{\text{state}} \\ 1 \leq l \leq N^{\text{int}}}} [e_\alpha^I(\mathbf{q}^{(n)})e_\beta^J(\mathbf{q}^{(n)})q_l^{(n)}]V_{k,l}^{(2),\alpha,\beta}, \qquad (46a)$$

where

$$\bar{M}_k^{I,J}(\mathbf{q}^{(n)}) = M_k^{I,J}(\mathbf{q}^{(n)}) - \sum_{\alpha,\beta=1}^{N^{\text{state}}} e_\alpha^I(\mathbf{q}^{(n)})e_\beta^J(\mathbf{q}^{(n)})V_k^{(1),\alpha,\beta}, \qquad (46b)$$

$$M_k^{I,J} = (E_J^0 - E_I^0)f_k^{I,J}; \quad M_k^{J,J} = \frac{\partial E_J^0}{\partial q_k}; \qquad (46c)$$

for $1 \leq n \leq N^{\text{point}}, 1 \leq k \leq N^{\text{int}}$, and $1 \leq I, J \leq N^{\text{state}}$. In these equations, the diagonal $\mathbf{M}^{I,I}$ are the energy gradient of state I, while the $\mathbf{M}^{I,J}$, $I \neq J$ are the energy difference scaled derivative couplings, referred to as the interstate coupling gradients, with $\mathbf{f}^{I,J}$ the derivative coupling of adiabatic states I and J. The $\mathbf{q}^{(n)}$ denote the set of selected nuclear configurations. To these equations we (may choose to) add the energy equations

$$\bar{\bar{M}}_0^{I,I}(\mathbf{q}^{(n)}) = \frac{1}{2}\sum_{k,l=1}^{N^{\text{int}}}\sum_{\alpha,\beta=1}^{N^{\text{state}}} [e_\alpha^I(\mathbf{q}^{(n)})e_\beta^J(\mathbf{q}^{(n)})q_k^{(n)}q_l^{(n)}]V_{k,l}^{(2),\alpha,\beta}, \qquad (47a)$$

where

$$\bar{M}_0^{I,I}(\mathbf{q}^{(n)}) = M_0^{I,I}(\mathbf{q}^{(n)}) - E_I^0(\mathbf{q}^0)$$

$$- \sum_{k=1}^{N^{\text{int}}}\sum_{\alpha,\beta=1}^{N^{\text{state}}} e_\alpha^I(\mathbf{q}^{(n)})e_\beta^I(\mathbf{q}^{(n)})q_k^{(n)}V_k^{(1),\alpha,\beta} \qquad (47b)$$

and $M_0^{I,I}(\mathbf{q}) = E_I^0(\mathbf{q})$. The $M_k^{I,J}$ are obtained from *ab initio* MRCI wave functions. The first order coefficients, $V_k^{(1),\alpha,\beta}$, are determined exactly using energy gradient and interstate coupling gradients at the origin of the expansion and thus appear on the right-hand side of Eq. (46b) as known quantities. As a practical matter, the $\mathbf{q}^{(n)}$ may be expressed in displacements in natural internal coordinates,[41] or in intersection adapted coordinates[58] (a particular linear combination of the natural internal coordinates, see Sec. 3.3.3.) if the origin of the expansion is a conical intersection. The coefficients obtained in these cases can be transformed to a normal coordinate basis for use in Eq. (8) using the transformations shown Eqs. (4) and (5).

Taking all the equations represented by Eqs. (46a) and (46b) and casting them as a matrix equation, one obtains,

$$\mathbf{W}\mathbf{v} = \bar{\mathbf{m}}, \qquad (48)$$

where $\mathbf{v}$ is a vector denoting the unique elements of $\mathbf{V}$, $\mathbf{W}$ is constructed from the terms in the square brackets in Eqs. (46a) and (47a), and $\bar{\mathbf{m}}$ is a vector containing all the values of $\bar{M}^{I,J}$. Neglecting any potential numerical issues, if the number of coefficients on the left-hand side of Eq. (48) are exactly equal to the number of equations, then the *ab initio* data employed in the fit will be exactly reproduced by the resultant $\mathbf{H}^d$.

For typical values of N^{point}, however, there are many more equations than unknowns, which results in Eq. (48) being over determined. Furthermore, the length of $\mathbf{v}$, and thus the number of unique non-zero coefficients may be reduced via the enforcement of vibronic point group symmetry (resulting in numerous coefficients being identically zero), or by only including the symmetry unique coefficients (in which a subset of coefficients can be represented as linear combinations of other coefficients). This latter case is realized when degenerate point group symmetry is present, which results in numerous symmetry related coefficients for e or t symmetry vibrational modes. Requiring that these over determined equations be solved in a least squares sense yields the pseudo normal equations[59]

$$\mathbf{W}^\dagger \mathbf{W} \mathbf{v} = \mathbf{W}^\dagger \bar{\mathbf{m}}. \qquad (49)$$

Equation (49) has the form of a set of normal equations,[60] but is deemed "pseudo" since these equations must be solved self-consistently given that $\mathbf{W}$ includes contributions from the $\mathbf{e}^J$, which are in turn determined using the $V_{k,l}^{(2),\alpha,\beta}$.

The unique aspect of Eq. (49) comes from Eq. (46a) for $I \neq J$ which requires that $\mathbf{H}^d$ reproduce an interstate coupling gradient. Consequently, Eq. (49) yields an $\mathbf{H}^d$ that is as diabatic as possible, in a least squares sense. In Sec. 3.2. we show computationally that the derivative couplings are in general well reproduced by $\mathbf{H}^d$ and are particularly well reproduced near a conical intersection. Since Eq. (49) is a least squares system of equations, $\mathbf{H}^d$ can be straightforwardly and systematically improved through the inclusion of additional data points in regions of interest on the potential energy surfaces, or by including weights in Eq. (49).

2.4.2. *Optimal bases and Franck–Condon overlaps*

The choice of $\mathbf{Q}$ in Eq. (5) significantly impacts the computational effort required to simulate a photoelectron spectrum. The predominant approach[21,27] has been to employ the normal modes at the minimum of the initial state, which in the case of a photodetachment spectrum, results in the ω_i in Eq. (50) and the $\mathbf{Q'}, \mathbf{Q}$ in Eqs. (4) and (5) corresponding to the harmonic frequencies and normal coordinates, respectively, of the anion.

The impetus for this approach lies in the significant formal simplifications that are realized with this choice of basis. For, in this case, the determination of $|i\rangle$ is trivial and the Franck–Condon integrals in Eq. (34) are simple δ-functions, $o(\mathbf{m}, \mathbf{I}) = \delta_{\mathbf{m}, \mathbf{I}}$. In some instances, for example the study of short time nuclear dynamics in the Franck–Condon region, this choice of basis may be desirable for physical reasons. In the time-independent case, however, the trade-off for this formal simplification is that the neutral states are expanded in a sub-optimal basis, particularly if the relevant minima on the neutral state manifold are significantly different from the minimum energy structure of the anion. Employing such an "anion biased" basis results in slower rates of convergence of the vibronic levels with respect to basis set size, necessitating larger basis set expansions.

To address this issue, recent studies have examined the utility of employing basis sets tailored to the vibronic states of interest in the neutral species.[55] The price one pays for this change of basis is that the initial anion state is no longer straightforward to determine, but rather, must be expanded in a "neutral biased" vibronic basis, necessitating the computation of N^{vib} Franck–Condon overlap integrals, as shown in Eq. (34). This additional complication is justified given that it is computationally more efficient to describe the single anion state using a neutral biased basis, than to determine the multitude of neutral states using an anion biased basis. In addition, these integrals need only to be computed a single time for the generation of the seed vector, shown in Eq. (33).

The evaluation of the $o(\mathbf{m}, \mathbf{I})$ employing generating function techniques has been discussed by several authors.[61–64] However, since the sizes of the vibronic expansions have the potential to grow to truly large dimensions

($> 10^9$), it is imperative that the determination of these terms be as efficient as possible. The most computationally economical approach involves the use of well-known recursion relations.[51, 52, 54, 65, 66]

The harmonic oscillator basis functions have the form:

$$\chi_{m_i}^{\alpha,i}(Q_i) = \frac{1}{\sqrt{2^{m_i} m_i!}} \left(\frac{\omega_i}{\pi}\right)^{1/4} H_{m_i}(\sqrt{\omega_i} Q_i) e^{-\omega_i Q_i^2 / 2}, \quad (50)$$

where ω_i are the harmonic frequencies (in atomic units), and the H_n are Hermite polynomials. The two sets of functions are related by a linear transformation that converts one set of normal mode coordinates to another:

$$\mathbf{Q} = \mathbf{T}\mathbf{Q}' + \mathbf{d}, \quad (51)$$

with ω and ω' the harmonic frequencies associated with normal modes $\mathbf{Q}$ and $\mathbf{Q}'$, respectively. Using this definition, it is possible to derive (see Ref. 55) a recursion relation, for $\mathbf{I} = 0$

$$C(\mathbf{m}, m_i + 1) = 2(\mathbf{b}^\dagger \bar{\mathbf{A}})_i C(\mathbf{m})$$

$$- 2 \sum_{j=1}^{N^{\text{int}}} (\mathbf{I} - \tilde{\mathbf{A}})_{i,j} C(\mathbf{m}, m_j - 1) m_j, \quad i = 1 - N^{\text{int}}, \quad (52\text{a})$$

$$o(\mathbf{m}, \mathbf{0}) = \tilde{G} \left[\prod_i^{N^{\text{int}}} \frac{1}{\sqrt{2^{m_i} m_i!}} \right] C(\mathbf{m}), \quad (52\text{b})$$

where $C(\mathbf{0}) = 1$, $C(\mathbf{m})$ is a coefficient for the vibronic basis function $\mathbf{m}$, and $C(\mathbf{m}, m_i + 1)$ is the $C(\mathbf{m})$ with the ith index increased by 1. In Eqs. (52a) and (52b) $\mathbf{I}$ is an $N^{\text{int}} \times N^{\text{int}}$ unit matrix and $\tilde{G}$ is a scalar constant given as

$$\tilde{G} = \sqrt{\frac{\det \mathbf{T} \prod_i (\omega_i \omega_i')^{1/2}}{\det \mathbf{A}}} \exp\left[\sum_i -\frac{\omega_i'}{2} d_i^2\right] \exp[\mathbf{b}^\dagger \mathbf{A}^{-1} \mathbf{b}], \quad (53)$$

and the matrices $\mathbf{A}$, $\tilde{\mathbf{A}}$, and $\bar{\mathbf{A}}$ and vector $\mathbf{b}$, are defined by

$$A_{i,j'} = \frac{\omega_j}{2} \delta_{i,j'} + \sum_l \frac{\omega_l'}{2} T_{l,i} T_{l,j};$$

$$\tilde{A}_{i,j} = \sqrt{\omega_i} A_{i,j}^{-1} \sqrt{\omega_j}; \quad \bar{A}_{i,j} = A_{i,j}^{-1} \sqrt{\omega_j}; \quad (54\text{a})$$

$$b_j = \sum_i d_i \frac{\omega_i'}{2} T_{i,j}. \quad (54\text{b})$$

Most importantly, with regards to computational efficiency, Eq. (52a) demonstrates that each of the $o(\mathbf{m}, \mathbf{0})$ may be determined at a cost of only $N^{\text{int}} + 1$ multiplies. In order to make efficient use of the recursion relation in Eq. (52a), each of the $C(\mathbf{m}, m_i - 1)$ in the summation must be readily available. Thus, the terms $C(\mathbf{m})$ are determined in batches for ascending values of $k(\mathbf{m}) = \sum_{i=1}^{N^{\text{int}}} m_i$, ensuring that each $C(\mathbf{m}, m_i - 1)$ required to compute $C(\mathbf{m})$ has already been calculated from the recursion relation. In this way, each element of the seed vector for the Lanczos procedure, $s_{\alpha, \mathbf{m}}$ in Eq. (35), is determined only once.

The different computational approaches to evaluating these integrals using the recursion relations has been discussed previously.[53,67,68] While the above discussion has emphasized determining a set of Franck–Condon overlap integrals relative to a ground vibrational state, denoted $o(\mathbf{m}, \mathbf{0})$, a more general solution would involve a second recursion to iterate the initial state as well in order to compute $o(\mathbf{m}, \mathbf{m}')$. Storage and indexing requirements for the determination of an arbitrary Franck–Condon integral between two vibronic states generally necessitates the use of a binary tree algorithm. In this approach, the integrals required to evaluate the recursion relation are located using pointer arithmetic to traverse a binary tree, as opposed to explicitly computing an index in order to retrieve the desired value from a table. Algorithms for implementing these techniques may be found in the literature.[53,67,68]

2.4.3. *Lanczos procedure: Open-ended fine grained parallel approach*

Owing to the large dimension of $\mathbf{H}^{\text{vib,nr}}$, the iterative Lanczos diagonalization technique,[46,69–71] which only requires that the matrix-vector product be computed, is the method of choice for determining the eigenvalues of $\mathbf{H}^{\text{vib,nr}}$. In this approach, the eigenvalues of the large vibronic Hamiltonian matrix are determined by diagonalizing a smaller, tri-diagonal matrix, $\mathbf{T}$, constructed such that the eigenspectrum of this matrix approximates that of $\mathbf{H}^{\text{vib,nr}}$. The elements that compose $\mathbf{T}$ are determined at each Lanczos iteration. While subtle variations on the algorithm exist,[69–72] the ith step of the Lanczos algorithm may given by:

1. $\mathbf{p}_i = \mathbf{s}_{i-1} / \|\mathbf{s}_{i-1}\|$,
2. $\mathbf{s}_i = \mathbf{H}\mathbf{p}_i$,
3. $\alpha_i = \mathbf{p}_i^\dagger \mathbf{s}_i$,
4. $\mathbf{s}_i = \mathbf{s}_i - \alpha_i \mathbf{p}_i - \beta_{i-1} \mathbf{p}_{i-1}$,
5. $\beta_i = \|\mathbf{s}_i\|$,

where $\mathbf{p}_i$ is the ith Lanczos vector, the elements of the matrix $\mathbf{T}$ are given by $T_{i,i} = \alpha_i$, $T_{i,i+1} = T_{i+1,i} = \beta_i$, and the initial seed vector, $\mathbf{s}_0$, is appropriately chosen to reproduce the line intensities.[21,28] See also comments following Eq. (35). Since $\mathbf{T}$ is only of dimension $N^{\text{iter}} \times N^{\text{iter}}$, where N^{iter} is the number of Lanczos iterations, it may be diagonalized using standard techniques. In general, N^{iter} is on the order of 10^3, with both the size of $\mathbf{H}^{\text{vib,nr}}$ and the number of eigenvalues desired influencing the number of iterations performed.

While the reader is referred to other sources for a more detailed discussion of the specifics of the algorithm,[21,70,71] a couple characteristic properties of the method should be noted.

Firstly, while in exact arithmetic each Lanczos vector $\mathbf{p}_i$ is orthogonal to $\mathbf{p}_j$, $1 \leq j \leq i-1$, the orthogonality of the Lanczos vectors deteriorates as the number iterations increase due round-off error. If one does not periodically pause to re-orthogonalize the Lanczos vectors, "ghost" roots, which are spurious duplicates of converged eigenvalues, will appear in the eigenspectrum of $\mathbf{T}$. If these roots are recognized as being specious, little harm is done. However, if computational resources are sufficient, periodic reorthogonalization of the Lanczos vectors ensure that effort is not wasted on the convergence of these extra roots, and thus more unique eigenvalues can be determined for a given number of iterations. Futhermore, if Ham reduction factors, shown in Eq. (43b), are desired for the subsequent determination of the spin-orbit spectrum, a rigorously orthogonal Lanczos space is required.

Secondly, the algorithm as presented above will converge the extremal eigenvalues of $\mathbf{H}^{\text{vib,nr}}$ first and then proceed to move "inwards" to larger and smaller values.[69–71] Generally, one is primarily concerned with the smallest eigenvalues $\mathbf{H}^{\text{vib,nr}}$, up to a given threshold. Thus, approximately half the converged eigenvalues determined from the Lanczos algorithm are not of much use in spectral simulations. This potential shortcoming is addressed with the spectral transformation formulation[73] of the Lanczos algorithm and with recent advances in filter diagonalization approaches.[74]

From a cursory examination of the above algorithm, it is readily apparent that the majority of the computational effort, from a floating point operation perspective, is focused on the evaluation of the matrix-vector product $\mathbf{H}\mathbf{p}_i$. Furthermore, the recursion relation requires the previous two N^{vib} length Lanczos vectors in order to iterate. As the vibronic expansions increase to the order of 10^8 multimode basis functions, this creates an additional storage issue that requires attention.

Fortunately, one of the defining characteristics of these vibronic Hamiltonian matrices is that they are very sparse. In particular, the number of non-zero matrix elements scales only linearly with the total dimension of the vibronic Hamiltonian. In fact, it is possible to enumerate the number of non-zero elements in Hamiltonian matrix. Taking for example the fully quadratic vibronic coupling model discussed in this chapter, if the total dimension of $\mathbf{H}^{\text{vib,nr}}$ is given by $D^H = N^{\text{state}} N^{\text{vib}}$, then the number of non-zero elements in $\mathbf{H}^{\text{vib,nr}}$ is given by:

$$N_{\text{nz}} = \frac{N^{\text{state}} + 1}{2} \left(1 + 2\sum_i^{N^{\text{int}}} \frac{(M_i - 1)}{M_i} + 2\sum_i^{N^{\text{int}}} \frac{(M_i - 2 + \delta_{M_i,1})}{M_i} + 4\sum_{i>j}^{N^{\text{int}}} \frac{(M_i - 1)(M_j - 1)}{M_i M_j} \right) D^H \quad (55a)$$

$$= \frac{N^{\text{state}} + 1}{2} \left(1 + 4[N^{\text{int}}] - 2\sum_i^{N^{\text{int}}} \frac{3 - \delta_{M_i,1}}{M_i} - 4\sum_{i>j}^{N^{\text{int}}} \frac{M_i + M_j - 1}{M_i M_j} \right) D^H, \quad (55b)$$

where M_i is the number of basis functions in ith mode, and a quantity in square brackets is evaluated as $[x] = \frac{x(x+1)}{2}$. The first term in the parentheses in Eq. (55a), the scalar 1, counts the diagonal elements of $\mathbf{H}^{\text{vib,nr}}$, which are computed via contributions from the kinetic energy, diagonal second-order (i.e. matrix elements of the form $\langle \chi_{m_j}^i | x_i^2 | \chi_{m_j}^i \rangle$), and constant terms. The three remaining terms enumerate the number of first-order, one-index second-order, and two-index second-order terms (given by the matrix elements $\langle \chi_{m_j}^i | x_i | \chi_{m_j \pm 1}^i \rangle$, $\langle \chi_{m_j}^i | x_i^2 | \chi_{m_j \pm 2}^i \rangle$, and $\langle \chi_{m_j}^i \chi_{m_{j'}}^{i'} | x_i x_{i'} | \chi_{m_j \pm 1}^i \chi_{m_{j'} \pm 1}^{i'} \rangle$), respectively. As the basis approaches the complete basis limit (i.e. $M_i \to \infty$, $i = 1 - N^{\text{int}}$), the slope of the scaling curve, N_{nz}/D^H, is a maximum given by $(N^{\text{state}} + 1)(1 + 4[N^{\text{int}}])/2$. However, since the signs of the third and fourth terms in Eq. (55b) are negative, the linear prefactor is generally significantly less than this value for typical basis set sizes (i.e. $M_i \approx 10^1$, $i = 1 - N^{\text{int}}$). Furthermore, if symmetry is present and enforced, many of the first- and second-order coefficients will be zero, which reduces the number of terms in the above summations.

Given both the computational demands of the problem, but also the favorable scaling of the number of non-zero elements as the dimension of the vibronic Hamiltonian increases, an efficient parallel Lanczos solver has been developed for use in vibronic coupling computations. Firstly, the parallel environment enables Lanczos vectors, $\mathbf{p}_i$, to be divided up among many processors, thus reducing the memory storage requirement per processor by a factor of N^{proc}. This aspect of the problem was addressed by employing the Global Arrays Toolkit.[75]

The initial step in the Lanczos procedure is the determination of the seed vector, which in this case corresponds to the ground vibrational state of the initial electronic state [see Eq. (35)]. Employing the recursion relation shown in Eq. (52), the $C(\mathbf{m})$ are computed in order of increasing value of the phonon counter $\kappa(\mathbf{m})$. Since it is assumed that accessing $C(\mathbf{m})$ that are located in non-local sections of $\mathbf{s}_0$ is relatively expensive, each process initially determines only a contribution to a given $C(\mathbf{m}, \mathbf{m}_i + 1)$ using locally available $C(\mathbf{m})$ and $C(\mathbf{m}, \mathbf{m}_j - 1)$. Following the determination of the contributions to all the $C(\mathbf{m})$ with a given value of $\kappa(\mathbf{m})$, the algorithm pauses to sum all the contributions and distribute the $C(\mathbf{m})$ to the appropriate vector locations in $\mathbf{s}_0$ before incrementing $\kappa(\mathbf{m})$. The number of $C(\mathbf{m})$ terms for a particular value of the phonon counter is given by the binomial coefficient $\binom{N^{\text{int}}+\kappa}{\kappa}$.

As stated above, once the Lanczos iterations begin, the majority of the computational effort is focused on evaluating the product of the vibronic Hamiltonian with the current Lanczos vector. To take full advantage of the sparsity of $\mathbf{H}^{\text{vib,nr}}$, the algorithm pre-processes the Hamiltonian by identifying the location of the non-zero matrix elements. This may be accomplished without explicitly storing the indices of the elements, which itself would present a significant storage issue. Rather, given a basis set indexing scheme that enumerates the vibronic basis functions, $i = 1 - N^{\text{vib}}$, where the index of vibronic basis function $\mathbf{m}$ is given by

$$index(\mathbf{m}) = \sum_{i=1}^{N^{\text{int}}} \left[m_i \prod_{j=i+1}^{N^{\text{int}}} M_j \right], \quad (56)$$

one may determine the regular pattern with which the non-zero elements arise as a set of offsets and strides through $\mathbf{H}^{\text{vib,nr}}$. Each potential coefficient in $\mathbf{V}^{(1),\alpha,\beta}$, $\mathbf{V}^{(2),\alpha,\beta}$, $\mathbf{V}^{(3),\alpha,\beta}$, etc. will have an associated indexing pattern that must be determined before the Lanczos iterations begin. Once this pre-processing is complete, each matrix-vector multiplication involves simply looping over the non-zero elements for each non-zero potential term in $\mathbf{V}^{(1),\alpha,\beta}$ and $\mathbf{V}^{(2),\alpha,\beta}$. Initial applications of this algorithm[47,76] demonstrated that vibronic expansions where $D^H > 10^9$ could now be readily treated.

3. Applications

All nonrelativistic electronic structure calculations discussed in this work employed the COLUMBUS suite of electronic structure codes.[77,78]

3.1. Overview: Conical intersections not required by symmetry

In the preceding five years, there have been a large number of experimental and theoretical studies of nonadiabatic effects in electron photodetachment spectra, including Refs. 27, 30, 79–86. While this section will focus on electron detachment from negative ions, theoretical studies of electron detachment from neutrals producing the spectrum of the corresponding cation have also been reported including studies of fluorobenzene,[85] cyclopropane,[80,87] pentatetraene,[81,88] and fluoromethane.[89] The pentatetraene and fluorobenzene studies considered five coupled electronic states and used the MCTDH method to compute the photoelectron spectrum. The pentatetraene work emphasized the advantages of the MCTDH method when both a large number of internal degrees of freedom (there 21) and a broad spectral range must be considered.

This section will describe recent work on the negative ion photoelectron spectra for two classes of molecules — the azolides and the alkoxides — revealing the vibronic structure of the azolyls, five-member carbon-nitrogen heterocycles of the form $(CH)_{5-m}N_m$, $m = 1 - 5$, and the alkoxy, R-O, radicals. There has been considerable recent interest[27,76,79,90–93] in the azolyls, attributable at least in part to their potential role in energetic materials. Experimental,[94–101] and computational[37,47,102,103] interest in the alkoxy radicals reflects their role in the combustion of hydrocarbon fuels, as well as in atmospheric and interstellar chemistry.[100,104,105]

Here we focus on the enticing theoretical challenges presented by these radicals. The common thread for these molecules is that they are substitutional derivatives of the classic Jahn–Teller molecules, cyclopentadienyl (C_5H_5) and methoxy (CH_3O). The ground electronic state of the cyclopentadienyl radical is a $^2E_1''$ state,[56,86,106] arising from five electrons in five π-orbitals. For methoxy the ground state is a ^2E state and its spectrum is impacted by both the seam of symmetry-required conical intersections,[107] and the spin-orbit interaction.

Conical intersections in low symmetry species which correlate with degenerate states in higher symmetry analogues provide an example of the quasi Jahn–Teller effect.[22] However, substitution of key substituents can alter the topography near a conical intersection in non-uniform ways. For example, in the case of the alkoxy radicals which are obtained by replacing the hydrogen atoms of methoxy, (CH_3O), with methyl or larger alkyl groups, the low-lying $^2A'$ and $^2A''$ states, correlate with methoxy's ^2E ground state. In this class of molecules, the quasi Jahn–Teller stabilization energy is small and the relevant extrema can be found in close proximity to the conical intersection. Thus, non-adiabatic effects are expected to impact the minima on both the $^2A'$ and $^2A''$ states in an approximately equivalent manner. Our alkoxy study will focus on the isopropoxy radical and the effect of the spin-orbit interaction on a low-lying accidental conical intersection.

The azolyls present a more diverse range of energetics. We discuss three azolyls for which detailed theoretical analyses have been reported: pyrrolyl, imidazolyl, and pyrazolyl. Pyrrolyl is discussed in detail. In these molecules which have C_{2v} symmetry, the $^2E''$ ground state of cyclopentadienyl is split into $^2A_2$ and $^2B_1$ electronic states.

3.2. Azolyls: Electronic structure

In imidazolyl, $(CH)_3N_2$, where the nitrogen atoms are nonadjacent, the ground state minimum has $^2B_1$ symmetry. The first excited state has $^2A_2$ symmetry and is approximately 0.836 eV above the ground state including differential zero point energy effects.[27] A second excited state, of $^2B_2$ symmetry, is higher at 0.958 eV, again including zero point effects.[27]

In pyrrolyl, $(CH)_4N$, the ground state minimum has $^2A_2$ symmetry and is approximately 0.5 eV lower in energy than the lowest energy stationary point on the $^2B_1$ surface.[79] This $^2B_1$ extremum is in fact a saddle point, and is much closer, geometrically and energetically, to the minimum energy point on the $^2A_2 -^2B_1$ seam of conical intersection than is the $^2A_2$ extremum, which is found in a relatively distant region of nuclear coordinate space.[57]

In pyrazolyl, $(CH)_3N_2$, with the nitrogens adjacent, the ground state is again of $^2A_2$ symmetry. The first excited state has $^2B_1$ symmetry but is only 0.046 eV above the ground state at the CCSD(T) level including differential zero point vibrational energies.[27] The lowest energy point on the $^2A_2 - ^2B_1$ seam of conical intersection is only 0.294 eV above the ground state minimum.[76] A second excited state, with $^2B_2$ symmetry, is higher, but still only 0.261 eV above the ground state at the CCSD(T) level including zero point effects.[27] This state, like the $^2B_1$ state in pyrrolyl, is a saddle point. The existence of three low-lying states is reflected in (or is a consequence of) the existence of a low-lying conical intersection of three electronic states only 0.454 eV[76] (0.426 eV[93]) above the $^2A_2$ minimum. Three state intersections are the subject of a chapter in this volume.

From these observations we conclude that the affects of conical intersections on the observed photoelectron spectra will vary dramatically among these azolyl isomers.

3.2.1. *Imidazolyl*

The photoelectron spectrum of imidazolide, revealing the vibronic structure of imidazolyl, has been measured and analyzed by the Lineberger[108] and Stanton[27] groups. The measured spectrum spans approximately 0.5 eV from the ground state band origin. The theoretical analyses conclude that the measured spectrum is well described by a model in which the neutral imidazolyl is represented by a single adiabatic $\tilde{X}^2B_1$ ground state. This is understandable in terms of the above discussion since the minimum of the excited state, which is below the lowest two state conical intersection, is ~0.83 eV above the spectral threshold.

3.2.2. *Pyrazolyl*

Pyrazolyl is the most challenging of the azolyls studied to date. The photoelectron spectra of pyrazolide-h_3 and its fully deuterated analog pyrazolide-d_3, revealing the vibronic structure of pyrazolyl-h_3 and pyrazolyl-d_3 respectively, have been measured by Lineberger and co-workers.[27,90] The measured spectra span approximately 0.4 eV from the ground state band origin. Whereas nonadiabatic effects were found to be negligible for imidazolyl, quite the opposite is expected for pyrazolyl, since in this case there are two low-lying excited states within ~0.3 eV of the ground state and a three-state intersection within ~0.45 eV of the ground state minimum.

Adiabatic simulations involving the $\tilde{X}^2A_2$ and $\tilde{A}^2B_1$ states for pyrazolide-h_3 (Ref. 76) and the $\tilde{X}^2A_2$, $\tilde{A}^2B_1$ and $\tilde{B}^2B_2$ states for pyrazolide-d_3, (Ref. 27) failed to reproduce the measured spectra. Of particular interest for the discussion below, is the significant over estimation of the intensity of the lines in the region greater than 0.2 eV above threshold.

In a series of careful calculations by Stanton and co-workers,[27] using a variant of the time-independent vibronic coupling approach described in this chapter, it was shown that proper simulation of the measured photoelectron spectra of the pyrazolides required inclusion of the $^2A_2$, $^2B_1$ and $^2B_2$ states. This is not unexpected given the electronic structure data noted above. More difficult to anticipate was the fact that the pyrazolide-h_3 photoelectron spectrum was much more challenging to simulate than was the pyrazolide-d_3 photoelectron spectrum. Indeed, Stanton et al. observed that the pyrazolide-h_3 photoelectron spectrum was not computable given the state-of-the-art time-independent methods available at that time.[27] Subsequently we were able to compute the pyrazolide-h_3 photoelectron spectrum using the neutral biased basis approach and open-ended Lanczos algorithm described in this chapter.[76] The reader is referred to Ref. 76 for the details of that treatment.

3.2.3. *Pyrrolyl*

The photodetachment spectrum of pyrrolide reveals the vibronic structure of pyrrolyl. The pyrrolyl radical occupies a middle position between the largely adiabatic (over the measured spectral range) imidazolyl radical and the completely nonadiabatic pyrazolyl radical. The photoelectron spectrum has been measured[91,109] and was initially analyzed using an adiabatic model by Lineberger's group,[91] in a study denoted *GIHKBL* below. It was subsequently studied by Domcke and co-workers,[79] in a study denoted *MLWD* below, using a variant of the time-independent methodology described in this work. In *GIHKBL*, an analysis based on an adiabatic representation of the $^2A_2$ and $^2B_1$ states predicted that spectral features attributable to both states should be observed in the measured spectrum. However, only features attributable to the $^2A_2$ state were observed. In the spectral region where features attributable to the $^2B_1$ state were expected to be seen, only a diffuse continuum was observed. As discussed by *GIHKBL* and *MLWD*, at least a partial explanation for this observation is that the proximity of the $^2B_1$ state extremum to the $^2A_2 - ^2B_1$ conical intersection seam in pyrrolyl leads to a dispersion of the intensity of transitions due to the $^2B_1$ state in the photoelectron spectrum (PES) of pyrrolide.

As noted by *MLWD*, a second factor may be relevant to the diminished contribution from the $^2B_1$ state to the spectrum: the strength of the bound-free electronic transition moments, Eq. (32). This same issue has been raised in analyses of photodetachment spectra of CH_3CC^- (Refs. 110, 111) and NO_3^-.[112]

In this review we use recent studies of the pyrrolide photoelectron spectrum in Refs. 57, 113 to consider two issues. For pyrrolyl, the region of strong nonadiabaticity is distinct from the region of the lowest energy minimum. This makes the simulation of its PES challenging since a correct description of both regions is required if the low-lying vibrational levels and the onset of nonadiabaticity are to be accurately described. A fundamental question is whether the quadratic $\mathbf{H}^d$, the most commonly employed model for quantitative computations, has sufficient flexibility to describe this situation. The second issue to be addressed is the (diabatic) state dependence of the electronic transition moment.

3.3. Pyrrolyl: Detailed computations

3.3.1. Electronic structure treatment

The geometric structure of the pyrrolyl radical is given in Fig. 1. The electronic structure data used to construct $\mathbf{H}^d$ was obtained from multireference configuration interaction (MRCI) wave functions using orbitals computed employing a state-averaged multiconfiguration self-consistent field (SA-MCSCF) procedure.[114] The SA-MCSCF treatment averaged two states with equal weights and employed a five electron in five orbital complete active space (CAS) expansion, comprising the five π orbitals from the CH and N moieties. The correlation-consistent cc-pVTZ basis set[115] was employed on nitrogen and carbon while a polarized double zeta (DZP) basis

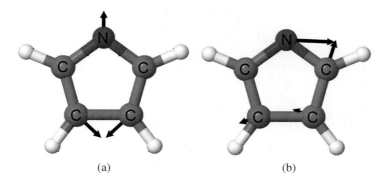

Fig. 1. The mass-weighted $\mathbf{g}$ and $\mathbf{h}$ vectors at the minimum energy intersection of the 1^2A and 2^2A states of the pyrrolyl radical. Dark (light) gray atoms are nitrogen (carbon). Hydrogens are white. Redrawn using data from Ref. 113 with permission.

was used on the hydrogens.[116] Dynamic correlation was included at the second order configuration interaction level, with interacting space restrictions enforced. The resulting MRCI expansion consisted of 108.5 million configuration state functions (CSFs).

The anion wave functions were computed employing an aug-cc-pVTZ atomic basis[117] on the carbons and nitrogen atoms, and the DZP basis set[116] on the hydrogens. The molecular orbitals were determined from a single configuration SCF procedure. Dynamic correlation was included at the single and double excitation CI level. The CI expansion consists of 27.5 million CSFs.

3.3.2. *Construction and accuracy of* $\mathbf{H}^d$

We begin with an analysis of the limits of accuracy for the fully quadatic $\mathbf{H}^d$ and the implication of those limitations for spectral simulations. Below we designate four extrema: $\mathbf{q}^{mex}$, the minimum energy crossing point; $\mathbf{q}^{min,a}$, the minimum on the anion potential energy surface; $\mathbf{q}^{min}$, the minimum on the lowest potential energy surface; and $\mathbf{q}^{ts}$, the first-order saddle point near the minimum energy crossing. See Fig. 2 for an illustration.

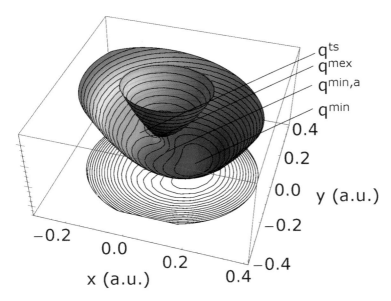

Fig. 2. Electronic energies as a function of branching plane coordinates of the minimum energy intersection of the 1^2A and 2^2A states of the pyrrolyl radical. Redrawn using data from Ref. 113 with permission.

Since the matrix elements of $\mathbf{H}^d$ are polynomials in the nuclear coordinates, most procedures for constructing $\mathbf{H}^d$ use data around a single point. That point is usually taken as $\mathbf{q}^{\min,a}$ for reasons discussed in Refs. 55 and 118 (and in Sec. 2.4.2) although $\mathbf{q}^{\min}$ is also used.[27] Our experience indicates that when states strongly coupled by conical intersections are involved, $\mathbf{q}^{\mathrm{mex}}$ is the preferred choice. This is a consequence of the accurate description of the singularity at the conical intersection by the $\mathbf{V}^{(1),\alpha,\beta}$ terms in Eq. (42). This choice works nicely near $\mathbf{q}^{\mathrm{mex}}$ (and $\mathbf{q}^{ts}$) However, it can be less than optimal near $\mathbf{q}^{\min}$. Below we illustrate the issues associated with the choice of origin, and how potential problems can be overcome by using geometrically diverse data in the context of our pseudo normal equations approach. Details of this study can be found in Ref. 57. Stanton's vertical and adiabatic approach[27] for constructing $\mathbf{H}^d$ is an alternative approach to addressing these issues.

3.3.2.1. Nascent $\mathbf{H}^d$

The diabatic Hamiltonian determined using only data in the vicinity of $\mathbf{q}^{\mathrm{mex}}$, also known as the nascent $\mathbf{H}^d$, will be subsequently denoted $\mathbf{H}^{d,W=0}$. The nascent $\mathbf{H}^d$ employed in this work was constructed from Eq. (49) using a data set, denoted D^{mex}, which consists of energy gradients and derivative couplings at $N^{\mathrm{point}} = 21$ points in an $N^{\mathrm{int}} = 21$ dimensional sphere of radius 0.01 surrounding $\mathbf{q}^{\mathrm{mex}}$. To this data set, data at the single point $\mathbf{q}^{\min,a}$ was also added.

Firstly, we consider how well the nascent $\mathbf{H}^d$ reproduces the *ab initio* determined structures of the four extrema noted above. Since $\mathbf{q}^{\mathrm{mex}}$ was chosen as the origin of $\mathbf{H}^{d,W=0}$, the *ab initio* and $\mathbf{H}^{d,W=0}$ results for the coordinates of $\mathbf{q}^{\mathrm{mex}}$ are essentially in perfect agreement. However, $\mathbf{H}^{d,W=0}$ also works quite well near the neighboring point $\mathbf{q}^{ts}$. The performance of $\mathbf{H}^{d,W=0}$ near the more distant $\mathbf{q}^{\min}$ is less satisfactory, with errors as large as of 0.01–0.02 Å in the predicted bond lengths. The energies at these points exhibit similar trends.

Since the ultimate goal of these calculations is to determine a photodetachment spectrum, the accuracy of the $\mathbf{H}^{d,W=0}$ determined harmonic frequencies at $\mathbf{q}^{ts}$ and $\mathbf{q}^{\min}$ is germane. Figure 3 reports the root mean square (RMS) error in the $\mathbf{H}^{d,W=0}$ derived harmonic frequencies at $\mathbf{q}^{\min}$ and $\mathbf{q}^{ts}$. As expected, the $\mathbf{H}^{d,W=0}$ derived frequencies are significantly better at $\mathbf{q}^{ts}$ than at $\mathbf{q}^{\min}$.

These deficiencies in $\mathbf{H}^{d,W=0}$ at $\mathbf{q}^{\min}$ do not represent a fundamental limitation of the fully quadratic $\mathbf{H}^d$ as we will now demonstrate.

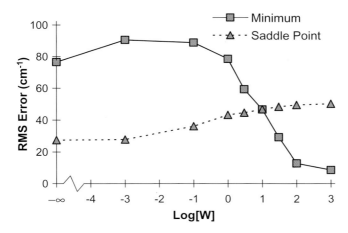

Fig. 3. The RMS error between the *ab initio* and $\mathbf{H}^{d,W}$ determined harmonic frequencies at the ground state minimum, $\mathbf{q}^{\min}$ (squares markers), and transition state, $\mathbf{q}^{ts}$ (diamond markers), of the pyrrolyl radical. Redrawn using data from Ref. 57 with permission.

3.3.2.2. Improving the accuracy and extending the domain of utility of $\mathbf{H}^d$

There are two interrelated ways to improve the performance of $\mathbf{H}^{d,W=0}$ in the vicinity of $\mathbf{q}^{\min}$: (i) explicitly include data from nuclear configurations in that region in Eq. (49), and (ii) adjust the weight (or number of repetitions for integer weights) of that data in those equations. To study these approaches, a data set $D^{\min}$ consisting of energy gradients and derivative couplings at $N^{\text{point}} = 21$ points, in an $N^{\text{int}} = 21$ dimensional sphere of radius 0.005 surrounding $\mathbf{q}^{\min}$ was determined and employed to form a composite data set, D^m comprising D^{mex} and $D^{\min}$ with a weight $W = m$, formally (for m an integer) $D^m = m(D^{\min}) \cup D^{\text{mex}}$. The $\mathbf{H}^d$ obtained from D^m is denoted $\mathbf{H}^{d,W=m}$.

Increasing m is found to improve the performance of $\mathbf{H}^{d,W=m}$ with regard to structures and energetics near $\mathbf{q}^{\min}$ while inducing only modest adverse effects on the description of the vicinity of $\mathbf{q}^{ts}$. To evince this consider, Fig. 3 which reports the RMS error of the $\mathbf{H}^{d,W=m}$ derived harmonic frequencies. Here increasing m improves the performance of $\mathbf{H}^{d,W=m}$ near $\mathbf{q}^{\min}$ while inducing limited, but systematically adverse, effects on the description of the vicinity of $\mathbf{q}^{ts}$. Note that Fig. 3 shows that the frequencies at $\mathbf{q}^{\min}$ improve dramatically until $W = m \sim 100$ and then improve only incrementally. Similarly the degradation of the frequencies at $\mathbf{q}^{ts}$ is most significant when $m \leq 1$. Analogous results are found for the energy gradient and interstate coupling gradients.[57] For example for $m \lesssim 0.001$, significant errors in the energy gradients near $\mathbf{q}^{\min,a}$ for both states are found. However for $m > 0.01$ these large errors are eliminated. This reflects the observation that data near $\mathbf{q}^{\text{mex}}$ alone does not provide a viable description of the region of $\mathbf{q}^{\min}$.

These comparisons illustrate both the versatility of the fully quadratic vibronic coupling model and the ability of the normal equations approach to exploit that flexibility. However, the results also indicate that the quadratic model is not without its limitations since (significant) improvement in one region comes at the expense of (here limited) diminution of the performance in another region. Limitations in the fully quadratic model can be reduced by including higher order terms in $\mathbf{H}^d$.[38,84]

The present analysis also indicates that within the fully quadratic model there is an optimal value for W, which is not necessarily the largest value. For W beyond this optimal value, little improvement in the $\mathbf{H}^d$ description of the *ab initio* data is observed. In general, it would be recommended that the sensitivity of the resultant spectral simulation to the chosen value of W, in this case $W = 100$, be investigated. With this in mind, we next assess the dependence of the simulated PES on this parameter.

3.3.2.3. Sensitivity of simulated photoelectron spectrum to $H^{d,W}$

Here we consider how the changes in $\mathbf{H}^d$ caused by changes in the domain of fitting points and the weight of those points in the normal equations are reflected in the simulated spectrum. The spectral data, spectral intensity distribution functions, are reported as spectral envelopes, employing a comparatively narrow $20\,\text{cm}^{-1}$ Gaussian convolution. The spectra are converged with respect to the neutral biased basis, the origin for which is $\mathbf{q}^{\min}$. The origin for $\mathbf{H}^d$ is $\mathbf{q}^{\text{mex}}$. In the nonadiabatic simulation, $\chi_m^{\alpha,j}$ are the normal modes of the Hessian corresponding to the $\mathbf{H}^d$ harmonic frequencies at $\mathbf{q}^{\min}$.

The anion is described by Eq. (2) with its equilibrium geometry, $\mathbf{q}^{\min,a}$ and $\chi_0^{0,j}$ obtained from *ab initio* Hessian used to compute the anion frequencies. Since neither the origins, nor frequencies of the harmonic oscillator bases are the same for the anion and neutral, the Franck–Condon overlaps, $s_{0,m}$ defined in Eq. (34) were calculated. $D = \sum_{\mathbf{m}} |o(\mathbf{m},0)|^2 = 0.97$. Based on our previous experience with ethoxy[47] and pyrazolyl,[76] the remaining contribution to D is likely due to the C-H stretches which do not contribute to this spectrum.

The origin for the simulated pyrrolyl spectrum is taken as the experimentally determined ionization potential of the anion, 2.145 eV from *GIHKBL*. The peak heights were uniformly scaled so that the first peak height agrees with that of *GIHKBL*.

Figure 4 reports the PES constructed from $\mathbf{H}^{d,W}$ for $W = 0, 1, 100, 1000$. Comparing the $W = 0, 1, 100$ plates with the $W = 1000$ plate demonstrates the significant changes in going from $W = 0$ to $W = 1$, and $W = 1$ to $W = 100$ with the much smaller changes in going from $W = 100$ to $W = 1000$. The changes are particularly pronounced for the low electron binding energy (eBE) portion of the spectrum. Comparing the $W = 100$ results to the $W = 1000$ results evinces changes in the individual peaks. However, despite these differences, the spectral envelope is reasonably stable as a function of W. The high energy portion of the spectrum, eBE > 2.6 eV, is particularly stable in this regard.

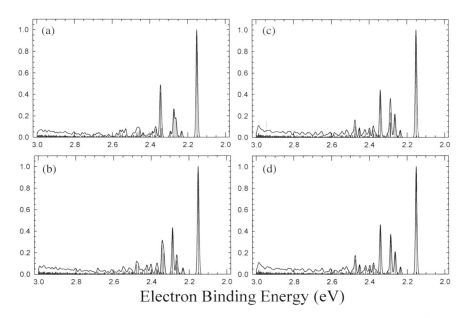

Fig. 4. The simulated photoelectron spectra of the pyrrolide ion. The spectra were determined employing $W = 0$ (a), 1 (b), 100 (c), 1000 (d). Redrawn using data from Ref. 57 with permission.

3.3.3. *Photoelectron spectrum of pyrrolyl*

In this section the PES of pyrrolide-h_4 based on $\mathbf{H}^{d,W=100}$ is discussed. The sharp spectral lines of the time-independent calculations are convoluted with a Gaussian of width 15 meV ($\sim 121\,\mathrm{cm}^{-1}$) to reflect the instrumental resolution. The simulations will be compared with the experimental spectrum of *GIHKBL* presented in Fig. 5(a). In addition to the origin labeled *a*, *GIHKBL* identify three lines at 925 ± 65, 1012 ± 25, and $1464 \pm 20\,\mathrm{cm}^{-1}$ denoted *b*, *c* and *d* respectively in Fig. 5(a). The focus of this discussion is the transition from adiabatic to nonadiabatic behavior and the inferences that can be drawn concerning the electronic transition moments in Eq. (32).

Figure 5(b) presents the nonadiabatic simulation for the pyrrolide-h_4 PES. The low energy region eBE < 2.5 eV, attributed to the $^2A_2$ state, is well reproduced, with the exception of the peak b', the low intensity shoulder to red of peak b.

For electron binding energies greater than 2.9 eV, the well-resolved spectrum characteristic of a long-lived excited state is absent and in its stead a broad continuum, quite similar to that reported by *MLWD*, is observed. When the nonadiabatic PES is obtained for $(\mu^{^2A_2,0}, \mu^{^2B_1,0}) = (1,1)$, so that $r = \mu^{^2A_2,0}/\mu^{^2B_1,0} = 1$, where $\mu^{J,0}$, $J = \,^2A_2$ and $^2B_1$, are the transition moments to the indicated diabatic states, this broad continuum, although somewhat reduced in intensity compared to *MLWD*, is appreciably higher in intensity than that found in the experimentally measured spectrum. The observed lower intensities could be attributable to near threshold effects on the photodetachment cross section.[79] However, we believe that this discrepancy is due to differences in the transition moments for the production of pyrrolyl in its $^2A_2$ and $^2B_1$ states by electron detachment from pyrrolide, as we now explain.

Figure 5(b) reports nonadiabatic spectrum with $(\mu^{^2A_2,0}, \mu^{^2B_1,0}) = (2,1)$, $r = 2$. It was found that relative intensity of peaks a, b and c is largely independent of r. However, the intensity of the broad continuum and that of peak b' are quite sensitive to r. The intensity of peak b' is found to be a quadratic function of r indicating that it is a b_2 vibrational level of the $^2A_2$ diabat that borrows intensity from an a_1 vibrational level of the $^2B_1$ diabat. Decreasing r too much eliminates the shoulder on peak *B* which is contraindicated by the measured spectrum. The ratio $r = 2$ appears to strike the best balance between preserving peak b' and reducing the intensity of the broad continuum. The peak labeled b'

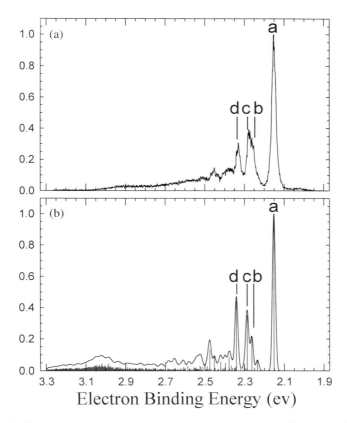

Fig. 5. (a) Experimental photoelectron spectrum of the pyrrolide ion, taken from Ref. 91. Lines labeled a, b, c, d were assigned in Ref. 91. (b) Nonadiabatic simulation of the photoelectron spectrum of the pyrrolide ion where the ratio of the transition dipole matrix elements is given by $\mu^{2\,A_2,0}/\mu^{2\,B_1,0} = 2$. Redrawn using data from Ref. 113 with permission.

should be readily discernable using high resolution photoelectron detection techniques.

A recurring issue in the simulation of vibronic spectra involving coupled electronic states is the determination of the relative magnitudes of the transition dipole moments, $\mu^{I,0}$, for each of the diabatic states. Computational techniques to accurately determine these transition moments are currently lacking.[21] When such techniques do become available, the results presented here, based on accurate treatment of the nonadiabatic effects, will provide valuable benchmarks.

3.4. *Isopropoxy*

In this section, we describe a simulation of the isopropoxide-h_7 anion PES, which reveals the vibronic structure of isopropoxy radical. This spectrum has been reported by Lineberger's group,[96] and is denoted *L-PES* below. The vibronic structure of the isopropoxy radical has also been probed via dispersed fluorescence spectroscopy, by Miller's group,[100] denoted *M-DFS* below. Figure 6 gives the molecular structure of this radical and indicates the atom labeling used in this section.

In isopropoxy, the degenerate ground state of methoxy is split into the nondegenerate $\tilde{X}^2A$ and $\tilde{A}^2A$ states. The presented simulation provided[119] the first theoretical determination of the $\tilde{A} - \tilde{X}$ splitting in isopropoxy which properly accounts for nonadiabatic and zero point energy effects as well as the spin-orbit interaction. We will explain that the nominal $\tilde{A} - \tilde{X}$ splitting is largely a consequence of the spin-orbit interaction.

3.4.1. *Electronic structure treatment*

The electronic structure data used to construct $\mathbf{H}^d$ was obtained from MRCI wave functions using orbitals obtained from SA-MCSCF wave functions. The SA-MCSCF treatment averaged two states with equal weights and used wave functions obtained from a five electron in four orbital complete active space expansion. The atomic orbital basis was composed

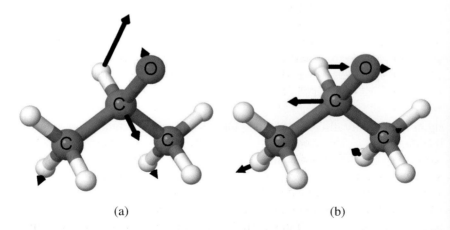

Fig. 6. The mass-weighted **g** (a) and **h** (b) vectors at the minimum energy intersection of the 1^2A and 2^2A states of the isopropoxy radical. Redrawn using data from Ref. 119 with permission.

of correlation-consistent basis sets,[115] with cc-pVTZ bases on the carbon and oxygen atoms and a cc-pVDZ basis on the hydrogen atoms. Dynamic correlation was included at the second order configuration interaction level, with the generalized interacting space restrictions included. The resulting MRCI expansion consists of 26 million CSFs.

The spin-orbit coupling between the 1^2A and 2^2A adiabatic states was determined within the Breit–Pauli approximation[48] including all one- and two- electron terms, using a computer code based on the methodology described in Ref. 120. The molecular orbitals were determined using the DZP basis set and the above noted active space and SA-MCSCF procedure. The corresponding second-order MRCI expansion comprises $\sim$1.3 million CSFs and includes core–core correlation.

The anion wave functions were computed using an atomic orbital basis comprising aug-cc-pVTZ basis set,[116] on the carbons and oxygen, and the cc-pVDZ basis set on the hydrogen atoms. A single reference single and double excitation CI treatment of electron correlation, comprising 3.5 million CSFs, was used.

3.4.2. *An unexpected conical topography*

As in the case of pyrrolyl, the key points on the ground state potential energy surface include $\mathbf{q}^{\min,a}(^1A')$, the minimum energy structure of the anion ground state; $\mathbf{q}^{\mathrm{mex}}(1^2A' - 2^2A'')$, the minimum energy point on the 1^2A-2^2A seam of accidental conical intersection; $\mathbf{q}^{\min}(1^2A')$, the minimum on the ground state potential energy surface; and $\mathbf{q}^{ts}(1^2A'')$, a first-order saddle point, also on the lowest energy adiabatic state. All these points also possess C_s symmetry, and the electronic state designations at these points reflect this fact. However, there exists an additional symmetry related pair of minima, denoted $\mathbf{q}^{\min'}(1^2A)$ and $\mathbf{q}^{\min''}(1^2A)$, as well as two symmetry related saddle points, $\mathbf{q}^{ts'}(1^2A)$ and $\mathbf{q}^{ts''}(1^2A)$ which connect $\mathbf{q}^{\min}(1^2A')$ with $\mathbf{q}^{\min'}$ and $\mathbf{q}^{\min''}$. Both these sets of structures are of C_1 symmetry, which is again reflected by the electronic state notation. These lower symmetry pair of structures are the lowest energy minima on the ground state potential energy surface. The quasi Jahn–Teller stabilization energy of this system is given by $\mathrm{E}^{\mathrm{qJTS}} = E_1^0(\mathbf{q}^{\mathrm{mex}}(^2A'' - {}^2A')) - E_1^0(\mathbf{q}^{\min''}(^2A)) = 193.8\,\mathrm{cm}^{-1}$. It is a bending of the HCO bond angle that is primarily responsible for the quasi Jahn–Teller stabilization. The six extrema and the conical intersection at $\mathbf{q}^{\mathrm{mex}}$ are indicated on a contour plot, as shown in Fig. 7, which reports the energies $E_1^0(x,y)$ with (x,y) in the branching plane (see

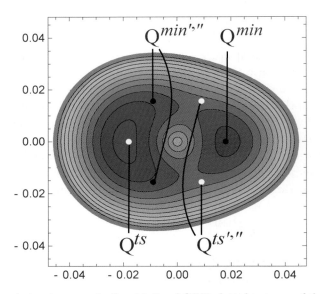

Fig. 7. The electronic energy in the vicinity of $\mathbf{Q}^{\text{mex}}$ plotted in terms of the branching plane coordinates x and y presented as contour plot. The location of the three minima and three saddle points that would be present in C_{3v} symmetry are denoted $\mathbf{Q}^{\text{min}}$ and $\mathbf{Q}^{ts}$, respectively. Redrawn using data from Ref. 119 with permission.

below) of $\mathbf{q}^{\text{mex}}$. While the branching plane section of the potential energy surface captures the essential character of the topography near the conical intersection, it misses the detailed structure required to fully evince the six extrema.

In order to better understand the existence of the three minima and three saddle points, it is useful to express these extrema in terms of intersection adapted coordinates.[58] Intersection adapted coordinates, $\mathbf{Q}$, are an orthogonal transformation of a set of internal coordinates, $\mathbf{q}$, with a conical intersection serving as the origin of the coordinate system. Of primary concern are the $\mathbf{g}$ direction (denoted the $\mathbf{x}$ coordinate, which in this case transforms as a') and the $\mathbf{h}$ direction (denoted the $\mathbf{y}$ coordinate, which transforms as a''). These two coordinates, which are pictured in Figs. 6(a) and 6(b) respectively, lift the degeneracy at the conical intersection at first order. The $\mathbf{x}$ and $\mathbf{y}$ coordinates define the branching[58] or $g-h$[20] plane. It is the explicit identification of the branching plane that gives intersection adapted coordinates their conceptual value. The relative magnitudes of the gradients that define these directions indicate the degree of asymmetry in the cone, with $\|\mathbf{g}\| = \|\mathbf{h}\|$ for a symmetry required Jahn–Teller intersection. In the case of isopropoxy, one finds that $\|\mathbf{g}\| = 0.0211 \approx \|\mathbf{h}\| = 0.0210$, which suggests that this intersection is closely analogous to what one would expect to find in the C_{3v} intersection of the methoxy radical. While the true extrema are not located in the branching plane,[119] they each have significant projections onto it. Replacing the x, y coordinates with the polar coordinates, ρ and θ, where $x = \rho \cos\theta$, $y = \rho \sin\theta$, the projection (ρ, θ) onto the branching plane of $\mathbf{q}^{\min}$, $\mathbf{q}^{\min'}$, $\mathbf{q}^{\min''}$ are given by $(0.079, 0°)$, $(0.084, 125.8°)$, $(0.084, -125.8°)$ while $\mathbf{q}^{ts}$, $\mathbf{q}^{ts'}$, $\mathbf{q}^{ts''}$ are located at $(0.065, 180°)$, $(0.057, 57.1°)$, $(0.057, -57.1°)$, respectively.

The following analysis serves to explain the locus of the extrema. For a Jahn–Teller ($E \times \varepsilon$) system,[18,121] described by the coordinates ρ and θ, the energy of the two adiabatic electronic states through second order in displacements from the origin, a conical intersection, is given by:

$$E_{\pm}^0 = b\rho^2 \pm \rho g \sqrt{1 - 2(\rho/\rho_0)\cos 3\theta + (\rho/\rho_0)^2}, \qquad (57)$$

where $\rho_0 = g/a$, g is the linear constant, $g = \|\mathbf{g}\| = \|\mathbf{h}\|$, and a and b are the quadratic constants. When the quadratic coupling constant a is small, the term in the square root in Eq. (57) approaches 1 and $E_{\pm}$ is independent of θ. However, when quadratic coupling is large, and one defines the coordinate system such that $\rho_0 < 0$, then at a given value of ρ, E_{-}^0 will exhibit three minima located at $\theta = 0°, 120°$, and $-120°$, and three maxima at $\theta = 180°, 60°$, and $-60°$. These restricted extrema become true (local) minima and saddle points when the ρ dependence is taken into account. On the basis of the polar coordinates for $(\mathbf{q}^{\min}, \mathbf{q}^{\min'}, \mathbf{q}^{\min''})$ and $(\mathbf{q}^{ts}, \mathbf{q}^{ts'}, \mathbf{q}^{ts''})$ noted above, the branching plane projections of the three minima and three saddle points observed on the ground state potential energy surface for isopropoxy are seen to correspond quite closely to the locus of the extrema in a quadratic $E \times \varepsilon$ Jahn–Teller system.

3.4.3. *Photoelectron spectrum of isopropoxide*

In each simulation, the anion wave function required to determine $I^{nr}(E)$ or $I^{so}(E)$ uses the wave function form in Eq. (2) with the anion equilibrium geometry and $\chi_0^{\alpha,j}$ obtained from the *ab initio* calculations described above.

Figure 8(a) presents a nonrelativistic, nonadiabatic simulation of the PES of isopropoxide-h_7 appropriate for comparison with the measured spectrum of *L-PES*, which is reproduced in Fig. 8(b). Given the near C_{3v} symmetry behavior observed in the electronic structure discussed above, the transition dipole moments employed were $\mu^{1,0} = \mu^{2,0}$. The location of the origin band for the simulated nonrelativistic isopropoxide-h_7 spectrum is taken as the experimentally determined ionization potential of the

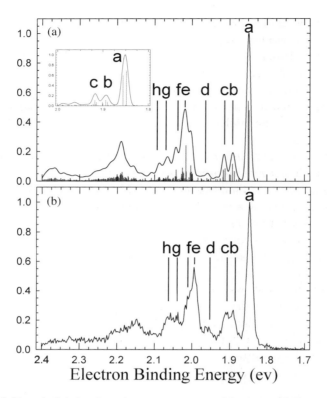

Fig. 8. (a) Nonrelativistic photoelectron spectrum of isopropoxide-h_7, employing a 10-meV convolution. Inset in panel (a) is the simulated relativistic photoelectron spectrum of isopropoxide-h_7 for the region near the vibronic band origins. (b) Experimental photoelectron spectrum of isopropoxide-h_7 from L-PES. Redrawn using data from Ref. 119 with permission.

anion, 1.847 eV from L-PES. The peak heights were uniformly scaled so that the first peak height agrees with that of L-PES. To obtain a simulation that most closely mirrors the experimental spectral resolution, the lines of the time-independent calculation are convoluted with a Gaussian of width 10 meV ($\sim$80 cm^{-1}). The converged vibronic basis, which comprises $N^{\text{state}}N^{\text{vib}} = 0.5$ billion basis functions and for which $D = 0.96$. The remaining contribution to D is likely due to the C-H stretches which do not contribute to this spectrum.

The nonadiabatic, nonrelativistic simulated spectrum and the measured spectrum of L-PES are in quite good agreement, with the possible exception of the small feature in the simulation at electron binding energy (eBE) of

~2.4 eV, which is not found in the PES of *L-PES*. This feature is not energetically precluded from the experimental spectrum, as the available photon energy is 3.4 eV.[96] The spectral features at 1.9 (labeled b and c), 1.95 (labeled d), 2.0–2.1 (labeled e, f, g, h) and ~2.15 eV on the other hand are well reproduced. The observation in *L-PES* of an absence of distinctive β (anisotropy) parameters for any of these peaks is consistent with our prediction of approximately ^2E behavior for the strongly coupled 1,2 ^2A electronic states.

Perhaps the most interesting feature observed in Fig. 8(a) is the apparent single peak at threshold. A closer examination of this region of the simulation reveals that this peak is in fact two transitions separated by ~17.7 cm^{-1}. Taking a nonadiabatic perspective, the second peak could potentially be identified as the first excited vibronic level in the lower sheet of the Mexican hat, i.e. an excited pseudo rotation[122] state. It is this feature which *M-DFS* denotes as the $\tilde{A} - \tilde{X}$ splitting. While the computed splitting is much smaller than the $\tilde{A} - \tilde{X}$ splitting reported by *M-DFS*, analysis of the eigenvectors shows that these two vibronic states are largely the zero phonon states of the two diabats. A small separation is perhaps not entirely unexpected, since in Sec. 3.4.2. it was argued that the two electronic states in question behave approximately like the components of a ^2E state. The described situation is precisely what would happen for a true ^2E state, except that in the case of a true ^2E state the degeneracy would be exact.[18,123]

This analysis suggests that, as in the case of a true ^2E state, the splitting of these two states would be increased by including the effects of the spin-orbit interaction. To this end $\mathbf{H}^{\mathrm{rso}}$ [Eq. (24)] was determined at $\mathbf{q}^{\mathrm{mex}}$ and found to be $\mathbf{H}^{\mathrm{rso}} = (50.3, 0.0, 37.3)$ cm^{-1}, with $\|\mathbf{H}^{\mathrm{rso}}\| = 62.6$ cm^{-1}. This computed magnitude for the spin-orbit coupling is in good accord both with the spin-orbit coupling determined previously for ethoxy using a similar methodology, $\mathbf{H}^{\mathrm{rso}} = (6.32, 0.0, 64.7)$, $\|\mathbf{H}^{\mathrm{rso}}\| = 65.0$ cm^{-1},[103] as well as a high level *ab initio* calculation for methoxy,[37] which found $\mathbf{H}^{\mathrm{rso}} = (0, 0, 67)$ cm^{-1}. When this value of $\mathbf{H}^{\mathrm{rso}}$ (taken as geometry independent) is used in Eq. (29), the splitting of the origin bands, the nominal $\tilde{A} - \tilde{X}$ splitting, is increased significantly, as shown in the inset of Fig. 8(a), to 60.6 cm^{-1}, in excellent agreement with the experimental value of 68 cm^{-1} reported in *M-DFS*. Note that these spin-orbit effects depend only on $\|\mathbf{H}^{\mathrm{rso}}\|$, and not on the individual components of $\mathbf{H}^{\mathrm{rso}}$. Significantly the vibronic Ham reduction effect[32,33] has reduced the maximum possible spin-orbit induced splitting of ~120 cm^{-1} by approximately a factor of 2, suggesting that the good agreement between the computed and measured nominal $\tilde{A} - \tilde{X}$ splitting is more than just fortuitous.

Finally, we note that the $\tilde{A} - \tilde{X}$ splitting would not have been visible in the *L-PES* experiment owing to the resolution available in that experiment. This is illustrated in the inset to Fig. 8(a) in which the relativistic PES is convoluted with a 80 cm^{-1} FWHM Gaussian to simulate the *L-PES* resolution. Only a single peak is evident at threshold in this inset.

3.4.4. *Implications*

This analysis demonstrates that the quantity denoted as the $\tilde{A}-\tilde{X}$ splitting is better thought of as the spin-orbit splitting of a nearly degenerate ground state. It is therefore no surprise that a similar result has been obtained for methoxy. There the spin-orbit induced ground state splitting was computed to be 68 cm^{-1}, giving again a Ham reduction (of the above noted spin-orbit interaction) of $\sim$2.[37] This computed value of the spin-orbit splitting in methoxy is consistent with measured values of 61.8 (Ref. 124) 64 (Ref. 125, 126) and 63 (Ref. 94) cm^{-1}.

4. Conclusions

Since the review of this field in 2004,[21] considerable progress has been made in the area of electron photodetachment spectroscopy. On the experimental side, the introduction of new time-resolved methods[7] and the high resolution SEVI method[6] allows for a richer comparison to computational simulations. Theoretical approaches for simulating such spectra in systems for which nonadiabatic effects are important have also advanced considerably. These include new techniques to describe the effects of the spin-orbit interaction, which have been discussed in this chapter as well as elsewhere in this volume. In the time-dependent approach, advances in the solution of the time-dependent Schrödinger equation using the multiconfiguration time-dependent Hartree approach,[127] and direct dynamics formulations[24, 26, 128] have been significant. The range of systems that can be treated is further increased by new approaches for separating system and bath modes.[129] Current methods that extend the accuracy of the quasi-diabatic Hamiltonians that describe the electronic structure aspects of photoelectron spectral simulations include (i) the "diabatization by ansatz" approach which includes terms through fourth and even higher orders in favorable circumstances,[38, 39, 84] (ii) the analytic evaluation of coupling terms using the equations of motion coupled cluster approach,[130] and (iii) the pseudo-normal equations method discussed at length in this work.[59,131] Working within the time-independent approach, we have discussed methods that reduce the size of the vibronic expansion by employing flexible choice of origin and basis functions. This methodology is made computationally tractable thanks to the development of efficient algorithms for the evaluation of large numbers of Franck–Condon overlaps.[55] A significant reduction in the time to solution has been achieved by a complementary algorithm which enables the solution of vibronic Schrödinger equation using a fine-grained parallel version of the Lanczos diagonalization routine.[47]

Given the power of electron photodetachment spectroscopies, one can anticipate significant applications and methodological advances in the years to come. One area of importance noted in this chapter is the first principles determination of photodissociation cross sections in systems where conical intersections play an essential role.

Acknowledgments

D. R. Y. is pleased to acknowledge the support of NSF grant CHE 0513952 during the preparation of this manuscript. The authors thank Joseph Dillon and Xiaolei Zhu for providing the raw data for Figures 1–4, 5(b), 6, 7, 8(a), and Carl Lineberger for the raw data used in Figs. 5(a) and 8(b).

Appendices

A. Time Reversal Adapted Basis

This appendix summarizes some results for time reversal adapted bases taken principally from Ref. 36. Let $\hat{T}$ denote the time reversal operator.[45] Then for doublet states

$$\hat{T}\Psi_\alpha^{d,+} = \hat{T}(\Psi_\alpha^{d,1/2} + i\Psi_\alpha^{d,-1/2})/\sqrt{2} = -(\Psi_\alpha^{d,1/2} - i\Psi_\alpha^{d,-1/2})/\sqrt{2} \equiv \Psi_\alpha^{d,-}. \tag{A1a}$$

So

$$\hat{T}\Psi_\alpha^{d,-} = -\Psi_\alpha^{d,+}. \tag{A1b}$$

From these relations, the fact that $\hat{T}$ commutes with H and is antiunitary, that is

$$\langle \hat{T}f|\hat{T}g\rangle = \langle f|g\rangle^*; \tag{A2}$$

we readily derive[36]

$$\langle \Psi_\alpha^{d,+}|H|\Psi_\alpha^{d,-}\rangle = 0, \quad \text{(A3a)}$$
$$\langle \Psi_\alpha^{d,+}|H|\Psi_\beta^{d,+}\rangle = \langle \Psi_\alpha^{d,-}|H|\Psi_\beta^{d,-}\rangle^*, \quad \text{(A3b)}$$

and

$$\langle \Psi_\alpha^{d,+}|H|\Psi_\beta^{d,-}\rangle = -\langle \Psi_\alpha^{d,-}|H|\Psi_\beta^{d,+}\rangle^*. \quad \text{(A3c)}$$

Equation (A3a) gives the 0 in the (1,3) and (2,4) matrix elements in Eq. (23). The (1,2) and (3,4) matrix elements of Eq. (23) are related by Eq. (A3b) and the (1,4) and (3,2) matrix elements of Eq. (23) are related by Eq. (A3c).

In the isopropoxy radical considered in Sec. 3, the spin-orbit interaction is evaluated at the minimum energy point of conical intersection, a nuclear configuration with C_s symmetry. The diabatic wave functions transform as $^2A'$ and $^2A''$. Using Eq. (23) and the fact that, assuming that the symmetry plane is the xz plane, $H_Y^{rso} = 0$ by symmetry [see Eq. (24)], only the pairs $(\Psi_{2_{A'}}^{d,+}, \Psi_{2_{A''}}^{d,-})$ and $(\Psi_{2_{A'}}^{d,-}, \Psi_{2_{A''}}^{d,+})$ are coupled by H^{so}. This is consistent with the observation that these pairs carry distinct double-valued irreducible representations of the C_s double group.[132] This group theoretical result is readily deduced from the observation that spinors (α, β) satisfy $\sigma^{xz}\binom{\alpha}{\beta} = \binom{\beta}{-\alpha}$ (Ref. 133).

B. Electron Scattering and Electronic Transition Moments

The evaluation of the electronic transition moment integral in Eq. (32) is a complex problem in electron scattering whose evaluation is beyond the scope of the present review. In this appendix we limit ourselves to a brief discussion of the relevant issues. Since the measured spectrum is that of the neutral molecule, it is reasonable to use the sudden approximation[34] in which the orbital for the outgoing electron, ϕ_f^c, is assumed to not alter the molecular core, but is shaped by the molecular environment. Furthermore, it is also important to observe that ϕ_f^c need not be orthogonal to the orbitals in $\mathbf{\Psi}_\alpha^d$.[134,135]

When nonadiabatic effects are negligible, it is possible to exploit the Born–Oppenheimer separation of electronic and nuclear motion and determine ϕ_f^c as a parametric function of the internal coordinates $\mathbf{Q}$, which is subsequently vibrationally averaged. It might even be possible to perform the scattering calculation required to determine the ϕ_f^c at a single representative geometry. The situation for states strongly coupled by conical intersections is much more complicated since there is no representative geometry in the region of strong coupling. Thus, considering the nonrelativistic case, the wave function for a particular final state f is given by

$$|f(\phi_f^c)\rangle = \sum_{\gamma=1}^{N^{\text{state}}} \Theta_\gamma^f(\mathbf{Q})[\Psi_\gamma^{d,-1/2}(\mathbf{r}^{N^{\text{el}}};\mathbf{Q})(\phi_f^c(\mathbf{r}_{N^{\text{el}}+1})\alpha)$$
$$- \Psi_\gamma^{d,+1/2}(\mathbf{r}^{N^{\text{el}}};\mathbf{Q})(\phi_f^c(\mathbf{r}_{N^{\text{el}}+1})\beta)]. \tag{B1}$$

Since we are using the sudden approximation, only the ϕ_f^c are to be determined. Note that there is a distinct orbital for each channel and that the channel index refers to a vibronic level rather than an electronic state at a fixed $\mathbf{Q}$, as is usually the case when nonadiabatic effects can be neglected. At a total energy E the multichannel wave function

$$\Psi^{N^{\text{el}}+1,E} = \sum_f \left| f(\phi_f^c) \right\rangle \tag{B2}$$

satisfies the projected Schrödinger equation[134]

$$\left\langle \sum_f f(\delta\phi_f^c) | (T^{\text{nuc}} + H^0(\mathbf{r}^{N^{\text{el}}+1},\mathbf{Q}) - E) | \Psi^{N^{\text{el}}+1,E} \right\rangle = 0, \tag{B3}$$

where the integration is over the N^{int} nuclear coordinates and $N^{\text{el}}+1$ electrons. Although qualitatively correct, even this discussion is over simplified. See Refs. 136 and 137. Solution of this problem is an active area of research.

References

1. B. Brehm, M.A. Gusinow and J.L. Hall, *Phys. Rev. Lett.* **19**, 737 (1967).
2. P.G. Wenthold and W.C. Lineberger, *Acc. Chem. Res.* **32**, 597 (1999).
3. K. Muller-Dethlefs, M. Sander and E.W. Schlag, *Chem. Phys. Lett.* **112**, 291 (1984).
4. T.N. Kitsopoulos, I.M. Waller, J.G. Loeser and D.M. Neumark, *Chem. Phys. Lett.* **159**, 300 (1989).
5. A. Osterwalder, M.J. Nee, J. Zhou and D.M. Neumark, *J. Chem. Phys.* **121**, 6317 (2004).
6. D.M. Neumark, *J. Phys. Chem. A* **112**, 13287 (2008).

7. A. Stolow and J.G. Underwood, in *Adv. Chem. Phys.*, edited by S.A. Rice (John Wiley and Sons, New York, 2008), Vol. 139.
8. J.V. Coe, S.T. Arnold, J. Eaton, G.H. Lee and K.H. Bowen, *J. Chem. Phys.* **125**, 014315 (2006).
9. V. Dribinski, J. Barbera, Joshua P. Martin, A. Svendsen, M.A. Thompson, R. Parson and W.C. Lineberger, *J. Chem. Phys.* **125**, 133405 (2006).
10. M.A. Sobhy, J.U. Reveles, U. Gupta, S.N. Khanna and A.W. Castleman, *J. Chem. Phys.* **130**, 054304 (2009).
11. K.L. Reid, *Annu. Rev. Phys. Chem.* **54**, 397 (2003).
12. D. Dill, *J. Chem. Phys.* **54**, 397 (1976).
13. P. Hockett, M. Staniforth, K.L. Reid and D. Townsend, *Phys. Rev. Lett.* **102**, 253002 (2009).
14. C.Z. Bisgaard, O.J. Clarkin, G. Wu, A.M. Lee, O. Gessner, C.C. Hayden and A. Stolow, *Science* **323**, 1464 (2009).
15. D.R. Yarkony, *Rev. Mod. Phys.* **68**, 985 (1996).
16. F. Bernardi, M. Olivucci and M.A. Robb, *J. Photochemistry and Photobiology A: Chemistry* **105**, 365 (1997).
17. M. Klessinger and J. Michl, *Excited States and the Photochemistry of Organic Molecules* (VCH, New York, 1995).
18. I. Bersuker, *The Jahn–Teller Effect* (Cambridge University Press, Cambridge, 2006).
19. *The Jahn–Teller Effect. Fundamentals and Implications for Physics and Chemistry*, edited by H. Köppel, D.R. Yarkony, and H. Barentzen (Spinger-Verlag, Heidelberg, 2010).
20. D.R. Yarkony, *Acc. Chem. Res.* **31**, 511 (1998).
21. H. Köppel, W. Domcke and L.S. Cederbaum, in *Conical Intersections: Electronic Structure, Dynamics and Spectroscopy*, edited by W. Domcke, D.R. Yarkony and H. Köppel (World Scientific Publishing, Singapore, 2004), Vol. 15, p. 323.
22. H. Köppel, W. Domcke and L.S. Cederbaum, *Adv. Chem. Phys.* **57**, 59 (1984).
23. M. Beck, A. Jackle, G.A. Worth and H.-D. Meyer, *Phys. Rep.* **324**, 1 (2000).
24. G.A. Worth, M.A. Robb and B. Lasorne, *Mol. Phys.* **106**, 2077 (2008).
25. J. Quenneville, M. Ben-Nun and T.J. Martinez, *J. Photochem. and Photobio.* **144**, 229 (2001).
26. G.A. Worth, M.A. Robb and I. Burghardt, *Farad. Discuss.* **127**, 307 (2004).
27. T. Ichino, A.J. Gianola, W.C. Lineberger and J.F. Stanton, *J. Chem. Phys.* **125**, 084312 (2006).
28. J.F. Stanton, *J. Chem. Phys.* **126**, 134309 (2007).
29. S. Mishra, L.V. Poluyanov and W. Domcke, *J. Chem. Phys.* **126**, 134312 (2007).
30. S. Mishra, V. Vallet, L.V. Poluyanov and W. Domcke, *J. Chem. Phys.* **124**, 044317 (2006).
31. L.V. Poluyanov and W. Domcke, *J. Chem. Phys.* **129**, 224102 (7 pages) (2008).

32. F.S. Ham, *Phys. Rev.* **138**, A1727 (1965).
33. F.S. Ham, *Phys. Rev.* **166**, 307 (1968).
34. A.B. Trofimov, H. Köppel and J. Schirmer, *J. Chem. Phys.* **109**, 1025 (1998).
35. D.G. Truhlar, C.A. Mead and M.A. Brandt, *Adv. Chem. Phys.* **33**, 295 (1975).
36. C.A. Mead, *J. Chem. Phys.* **70**, 2276 (1979).
37. A.V. Marenich and J.E. Boggs, *J. Chem. Phys.* **122**, 024308 (2005).
38. A. Viel and W. Eisfeld, *J. Chem. Phys.* **120**, 4603 (2004).
39. W. Eisfeld and A. Viel, *J. Chem. Phys.* **122**, 204317 (2005).
40. M. Nooijen, *Int. J. Quantum Chem.* **95**, 768 (2003).
41. G. Fogarasi, X. Zhou, P.W. Taylor and P. Pulay, *J. Amer. Chem. Soc.* **114**, 8191 (1992).
42. M. Baer, *Mol. Phys.* **40**, 1011 (1980).
43. C.A. Mead and D.G. Truhlar, *J. Chem. Phys.* **77**, 6090 (1982).
44. M. Baer, *Phys. Rep.* **358**, 75 (2002).
45. M. Tinkham, *Group Theory and Quantum Mechanics* (McGraw-Hill, New York, 1964).
46. W. Domcke, H. Köppel and L. Cederbaum, *Mol. Phys.* **43**, 851 (1981).
47. M.S. Schuurman, R.A. Young and D.R. Yarkony, *Chem. Phys.* **347**, 57 (2008).
48. S.R. Langhoff and C.W. Kern, in *Modern Theoretical Chemistry*, edited by H.F. Schaefer (Plenum, New York, 1977), Vol. 4, p. 381.
49. R. McWeeny, *J. Chem. Phys.* **42**, 1717 (1965).
50. T.R. Furlani and H.F. King, *J. Chem. Phys.* **82**, 5577 (1985).
51. E.V. Doktorov, I.A. Malkin and V.I. Man'ko, *J. Mol. Spec.* **56**, 1 (1975).
52. E.V. Doktorov, I.A. Malkin and V.I. Man'ko, *J. Mol. Spec.* **64**, 302 (1977).
53. A. Hazra and M. Nooijen, *Int. J. Quantum Chem.* **95**, 643 (2003).
54. H. Kupka and P.H. Cribb, *J. Chem. Phys.* **85**, 1303 (1986).
55. M.S. Schuurman and D.R. Yarkony, *J. Chem. Phys.* **128**, 044119 (9 pages) (2008).
56. H.J. Wörner and F. Merkt, *J. Chem. Phys.* **127**, 034303 (16 pages) (2007).
57. X. Zhu and D.R. Yarkony, *J. Chem. Phys.* **130**, 234108 (11 pages) (2009).
58. G.J. Atchity, S.S. Xantheas and K. Ruedenberg, *J. Chem. Phys.* **95**, 1862 (1991).
59. B.N. Papas, M.S. Schuurman and D.R. Yarkony, *J. Chem. Phys.* **129**, 124104 (10 pages) (2008).
60. S.J. Leon, *Linear Algebra with Applications* (Prentice-Hall, Upper Saddle River, NJ, 2002).
61. A. Warshel, (preprint communicated by author).
62. A. Warshel and P. Dauber, *J. Chem. Phys.* **66**, 5477 (1977).
63. A. Warshel and M. Karplus, *Chem. Phys. Lett.* **17**, 7 (1972).
64. D.J. Tannor and E.J. Heller, *J. Chem. Phys.* **77**, 202 (1982).
65. T.E. Sharp and H.M. Rosenstock, *J. Chem. Phys.* **41**, 3453 (1964).
66. L.S. Cederbaum and W. Domcke, *J. Chem. Phys.* **64**, 603 (1976).

67. D. Gruner and P. Brumer, *Chem. Phys. Lett.* **138**, 310 (1987).
68. P.T. Ruhoff and M.A. Ratner, *Int. J. Quant. Chem.* **77**, 383 (2000).
69. C. Lanczos, *J. Res. Natl. Bur. Stand.* **49**, 33 (1952).
70. J. Cullum and R. Willoughby, *Lanczos Algorithms for Large Symmetric Eigenvalue Problems* (Birkhauser, Boston, 1985).
71. H. Simon, *Math. Comp.* **42**, 115 (1984).
72. W. Domcke, H. Köppel and L. Cederbaum, *Mol. Phys.* **43**, 851 (1981).
73. T. Erricsson and A. Ruhe, *J. Math. Comp.* **35**, 1251 (1980).
74. D. Neuhauser, *J. Chem. Phys.* **93**, 2611 (1990).
75. J. Nieplocha, R.J. Harrison and R.J. Littlefield, *J. Supercomput.* **10**, 169 (1996).
76. M.S. Schuurman and D.R. Yarkony, *J. Chem. Phys.* **129**, 064304 (14 pages) (2008).
77. H. Lischka, M. Dallos, P. Szalay, D.R. Yarkony and R. Shepard, *J. Chem. Phys.* **120**, 7322 (2004).
78. H. Lischka, R. Shepard, I. Shavitt, R. Pitzer, M. Dallos, T. Müller, P.G.Szalay, F.B. Brown, R. Alhrichs, H.J. Böhm, A. Chang, D.C. Comeau, R. Gdanitz, H. Dachsel, C. Erhard, M. Ernzerhof, P. Höchtl, S. Irle, G. Kedziora, T. Kovar, V. Parasuk, M. Pepper, P. Scharf, H. Schiffer, M. Schindler, M. Schüler and J.-G. Zhao, COLUMBUS, An ab initio Electronic Structure Program (2003).
79. A. Motzke, Z. Lan, C. Woywod and W. Domcke, *Chem. Phys.* **329**, 50 (2006).
80. T.S. Venkatesan, S. Mahapatra, L.S. Cederbaum and H. Köppel, *J. Mol. Struc.* **838**, 100 (2007).
81. A. Markmann, G.A. Worth, S. Mahapatra, H.-D. Meyer, H. Köppel and L.S. Cederbaum, *J. Chem. Phys.* **123**, 204310 (9 pages) (2005).
82. I. Baldea, J. Franz and H. Koppel, *J. Mol. Struc.* **838**, 94 (2007).
83. I. Baldea, J. Franz and H. Köppel, *Chem. Phys.* **329**, 65 (2006).
84. S. Faraji, H. Koppel, W. Eisfeld and S. Mahapatra, *Chem. Phys.* **347**(1–3), 110 (2008).
85. E. Gindensperger, I. Baldea, J. Franz and H. Koppel, *Chem. Phys.* **338**, 207 (2007).
86. T. Ichino, S.W. Wren, K.M. Vogelhuber, A.J. Gianola, W.C. Lineberger and J.F. Stanton, *J. Chem. Phys.* **129**, 084310 (19 pages) (2008).
87. T.S. Venkatesan, S. Mahapatra, L.S. Cederbaum and H. Köppel, *J. Phys. Chem. A* **108**, 2256 (2004).
88. A. Markmann, G.A. Worth and L.S. Cederbaum, *J. Chem. Phys.* **122**, 144320 (15 pages) (2005).
89. S. Mahapatra, V. Vallet, C. Woywod, H. Köppel and W. Domcke, *Chem. Phys.* **304**, 17 (2004).
90. A.J. Gianola, T. Ichino, S. Kato, V.M. Bierbaum and W.C. Lineberger, *J. Phys. Chem. A* **110**, 8457 (2006).
91. A.J. Gianola, T. Ichino, R.L. Hoenigman, S. Kato, V.M. Bierbaum and W.C. Lineberger, *J. Phys. Chem. A* **108**, 10326 (2004).

92. T. Ichino, D.H. Andrews, G.J. Rathbone, F. Misaiozu, R.M.D. Calvi, S.W. Wren, S. Kato, V.M. Bierbaum and W.C. Lineberger, *J. Phys. Chem. B* **112**, 545 (2008).
93. S. Matsika and D.R. Yarkony, *J. Amer. Chem. Soc.* **125**, 12428 (2003).
94. S.C. Foster, P. Misra, T.-Y.D. Lin, C.P. Damo, C.C. Carter and T.A. Miller, *J. Phys. Chem.* **92**, 5914 (1988).
95. X. Zhu, M.M. Kamal and P. Misra, *Pure Appl. Opt.* **5**, 1021 (1996).
96. T.M. Ramond, G.E. Davico, R.L. Schwartz and W.C. Lineberger, *J. Chem. Phys.* **112**, 1158 (2000).
97. C.C. Carter, J.R. Atwell, S. Gopalakrishnan and T.A. Miller, *J. Phys. Chem. A* **104**(40), 9165 (2000).
98. C.C. Carter, S. Gopalakrishnan, J.R. Atwell and T.A. Miller, *J. Phys. Chem. A* **105**, 2925 (2001).
99. S. Gopalakrishnan, L. Zu and T. Miller, *J. Phys. Chem. A* **107**, 5189 (2003).
100. J. Jin, I. Sioutis, G. Tarczay, S. Gopalakrishnan, A. Bezanat and T.A. Miller, *J. Chem. Phys.* **121**, 11780 (2004).
101. S. Gopalakrishnan, L. Zu and T.A. Miller, *Chem. Phys. Lett.* **380**, 749 (2003).
102. G. Tarczay, S. Gopalakrishnan and T.A. Miller, *J. Mol. Spec.* **220**, 276 (2003).
103. R.A. Young Jr. and D.R. Yarkony, *J. Chem. Phys.* **125**, 234301 (14 pages) (2006).
104. R. Atkinson, *Int. J. Chem. Kinet.* **29**, 99 (1997).
105. M.E. Jenkin and G.D. Hayman, *Atmos. Environ.* **33**, 1275 (1999).
106. B.E. Applegate, T.A. Miller and T.A. Barckholtz, *J. Chem. Phys.* **114**(11), 4855 (2001).
107. B.E. Applegate, T.A. Barckholtz and T.A. Miller, *Chem. Soc. Rev.* **32**, 38 (2003).
108. A.J. Gianola, T. Ichino, R.L. Hoenigman, S. Kato, V.M. Bierbaum and W.C. Lineberger, *J. Phys. Chem. A* **109**, 11504 (2005).
109. J.H. Richardson, L.M. Stephenson and J.I. Brauman, *J. Am. Chem. Soc.* **97**, 1160 (1975).
110. J. Zhou, E. Garand, W. Eisfeld and D.M. Neumark, *J. Chem. Phys.* **127**, 034304 (7 pages) (2007).
111. B.N. Papas, M.S. Schuurman and D.R. Yarkony, *J. Chem. Phys.* **130**, 064306 (12 pages) (2009).
112. A. Weaver, D.W. Arnold, S.E. Bradforth and D.M. Neumark, *J. Chem. Phys.* **94**, 1740 (1991).
113. X. Zhu and D.R. Yarkony, *J. Phys. Chem. C* **114**, 5312 (2010).
114. B.H. Lengsfield and D.R. Yarkony, in *State-Selected and State to State Ion-Molecule Reaction Dynamics: Part 2 Theory*, edited by M. Baer and C.-Y. Ng (John Wiley and Sons, New York, 1992), Vol. 82, p. 1.
115. T.H. Dunning Jr., *J. Chem. Phys.* **90**, 1007 (1989).

116. T.H. Dunning and P.J. Hay, in *Modern Theoretical Chemistry*, edited by H.F. Schaefer (Plenum, New York, 1976), Vol. 3.
117. R.A. Kendall, T.H. Dunning and R.J. Harrison, *J. Chem. Phys.* **96**, 6796 (1992).
118. J. Schmidt-Klügmann, H. Köppel, S. Schmatz and P. Botschwina, *Chem. Phys. Lett.* **369**, 21 (2003).
119. J.J. Dillon and D.R. Yarkony, *J. Chem. Phys.* **130**, 154312 (11 pages) (2009).
120. D.R. Yarkony, *Int. Rev. Phys. Chem.* **11**, 195 (1992).
121. R. Englman, *The Jahn–Teller Effect in Molecules and Crystals* (Wiley-Interscience, New York, 1972).
122. R.S. Berry, *J. Chem. Phys.* **32**, 933 (1960).
123. M.S. Schuurman, D.E. Weinberg and D.R. Yarkony, *J. Chem. Phys.* **127**, 104309 (12 pages) (2007).
124. S.D. Brossard, P.G. Carrick, E.L. Chappell, S.C. Hulegaard and P.C. Engelking, *J. Chem. Phys.* **84**, 2459 (1986).
125. P. Misra, X. Zhu, C.-Y. Hsueh and J.B. Halpern, *Chem. Phys.* **178**, 377 (1993).
126. Y.-Y. Lee, G.-H. Wann and Y.-P. Lee, *J. Chem. Phys.* **99**, 9465 (1993).
127. G.A. Worth, H.-D. Meyer, H. Köppel, L.S. Cederbaum and I. Burghardt, *Int. Rev. Phys. Chem.* **27**, 569 (2008).
128. M. Ben-Nun and T.J. Martinez, *J. Phys. Chem. A* **104**, 5161 (2000).
129. G. Worth, H. Meyer and L. Cederbaum, *J. Chem. Phys.* **109**, 3518 (1998).
130. T. Ichino, J. Gauss and J.F. Stanton, *J. Chem. Phys.* **130**, 174105 (2009).
131. M.S. Schuurman and D.R. Yarkony, *J. Chem. Phys.* **127**, 094104 (9 pages) (2007).
132. M. Hamermesh, *Group Theory and Its Application to Physical Problems* (Addison-Wesley, Reading MA, 1962).
133. P.J. Hay, T.H. Dunning and R.C. Raffenetti, *J. Chem. Phys.* **65**, 2679 (1976).
134. G. Bandarage and R.R. Lucchese, *Phys. Rev. A* **47**, 1989 (1993).
135. T.M. Rescigno, B.H. Lengsfield III and C.W. McCurdy, in *Modern Electronic Structure Theory*, edited by D.R. Yarkony (World Scientific Publishing, Singapore, 1995), Vol. 2, p. 501.
136. S. Han and D.R. Yarkony, *J. Chem. Phys.* **133**, 194107 (2010).
137. S. Han and D.R. Yarkony, *J. Chem. Phys.* **134**, 134110 (2011).

Chapter 7

Multistate Vibronic Dynamics and Multiple Conical Intersections

Shirin Faraji, Susana Gómez-Carrasco and Horst Köppel*

1. Introduction 249
2. Methodological Framework 251
   2.1. The multimode vibronic coupling approach 251
   2.2. The concept of regularized diabatic states 254
   2.3. Electronic structure calculations 258
   2.4. Quantum dynamical calculations 261
3. Illustrative Examples 263
   3.1. Formaldehyde 263
   3.2. Pyrrole 273
   3.3. Fluorinated benzene cations 281
4. Concluding Remarks 293
   Acknowledgments 294
   References 295

1. Introduction

Nuclear motion following photoexcitation of even small polyatomic molecules is often subject to strong nonadiabatic coupling effects and ceases

---

Theoretische Chemie, Universität Heidelberg, INF 229, D-69120 Heidelberg, Germany.
*E-mail: Horst.Koeppel@pci.uni-heidelberg.de

to be confined — even approximately — to a single potential energy surface (PES). Rather, transitions between different surfaces occur on the same time scale as the molecular vibrations and need to be treated on the same footing.[1,2] Strong nonadiabatic coupling gives rise to a wide range of phenomena in spectroscopy, collision dynamics and elementary photophysical and photochemical processes. In the past one to two decades, conical intersections (CoIns) have been established as the key topography signalling the presence of strong nonadiabatic couplings,[3-13] as is also testified by the various contributions in the present book.

Quite naturally, the first step in extending the description of the dynamics beyond the adiabatic approximation consists in considering *two* strongly coupled electronic states. The ansatz for the molecular wavefunction then amounts to a sum of two products rather than a single one. Indeed, for low excitation energies, when the density of electronic states is moderate, primarily two states will approach energetically or become degenerate, while the others tend to be further away, and to interact only weakly or not at all with the states under consideration. On the other hand, for higher excitation energies the density of states will increase and *several* become close and be mutually coupled. The importance of more than two states interacting strongly even increases when not only simultaneous interactions are considered, but also consecutive ones, affecting a given dynamical process in the course of time.

It is the purpose of the present chapter to survey some of our theoretical work on the strong nonadiabatic interaction of more than two electronic states, be it simultaneous or consecutive, as just described. In the light of the earlier findings for two-state cases, this will be typically associated with multiple CoIns of the underlying PESs. These multiple intersections can be triple intersections, i.e. the degeneracy of three adiabatic PESs at a point (or set of points) in nuclear coordinate space. They may also consist in a set of intersections which are arranged such that the nuclear motion undergoes an ultrafast sequence of sub-*ps* transitions through major parts of, or the whole, set. We may also encounter the situation of intersecting seams of CoIns,[14] where we recall that intersections do not occur at isolated points in nuclear coordinate space (dimensionality N) but rather in subspaces of dimension $N - 2$, called seams.

A natural scenario for three or more intersecting PESs are molecules characterized by non-Abelian point groups with symmetry-induced spatial degeneracies of electronic states.[3,15] Then, the interaction of a degenerate state with even a nondegenerate state gives rise to a 3-state coupling situation and three intersecting surfaces. With different types of vibrational modes (coupling and tuning[4]) consecutive intersections may arise which turn out (in full dimensional coordinate space) to be cuts through intersecting seams of degeneracies. When two degenerate states interact strongly, the topography of the PESs will be correspondingly more complex. However, also for Abelian point groups similar situations may arise. For the simplicity of nomenclature, we will call all cases with more than two interacting states "multistate" systems and also tacitly assume that interactions or couplings are always meant to be strong ones.

The remainder of this chapter is organized as follows. In Sec. 2 we give an overview over the theoretical methodology use to describe multistate vibronic interactions. This comprises, in particular, the multimode vibronic coupling (MMVC) approach, developed long ago in the Heidelberg group,[4] which lends itself naturally to a treatment of multistate coupling situations. Its modification to cover also general adiabatic PESs, the concept of regularized diabatic states, has so far been applied mostly to two-state problems.[16, 17] Recent advances, however, deal with more than two coupled PESs and are thus also addressed here. Further, computational aspects addressed below include the *ab initio* determination of coupling constants and the solution of the nuclear dynamics by quantal methods, namely, integration of the time-dependent Schrödinger equation. In Sec. 3 we present selected examples to demonstrate some technical aspects of the application, but especially relevant phenomena and implications. Since the aforementioned scenario with degenerate electronic states has been exposed quite comprehensively in a recent review paper,[15] we focus here on systems without spatial degeneracies (i.e. Abelian point groups), namely formaldehyde, pyrrole and fluorinated benzene cations. Finally, Sec. 4 concludes.

2. Methodological Framework

2.1. *The multimode vibronic coupling approach*

To analyse vibronic coupling phenomena, i.e. those which go beyond the well-known adiabatic or Born–Oppenheimer approximation,[1, 2] it is often advantageous to employ a diabatic electronic representation,[18–21] as also done here. Contrary to the usual adiabatic electronic basis, the off-diagonal matrix elements which lead to a coupling between different electronic states derive from the potential energy part, rather than from the nuclear kinetic energy. The diabatic electronic functions are generally smooth functions of the nuclear coordinates, even at degeneracies of PESs where the adiabatic electronic wave functions become discontinuous.[18–21] The electronic matrix elements in the diabatic basis may therefore be expanded as a Taylor series about a reference nuclear configuration, and the series be suitably truncated to best fit the *ab initio* computed electronic energies.

In typical applications we consider a photoexcitation or photoionization process where the nuclear kinetic energy T_N and potential energy operators V_0 relate to the initial electronic state (usually the ground state), described here in the harmonic approximation. Therefore, the kinetic energy and potential energy operators of the electronic ground state take the simple form

$$T_N = -\frac{1}{2}\sum_i \hbar\omega_i \frac{\partial^2}{\partial Q_i^2}, \tag{1}$$

$$V_0 = \frac{1}{2}\sum_i \hbar\omega_i Q_i^2. \tag{2}$$

The minimum energy configuration $\mathbf{Q} = \mathbf{0}$ of V_0 in this expansion is also chosen as the reference configuration $\mathbf{Q_0}$, used in the construction of the diabatic basis out of the adiabatic one (which means that the adiabatic and diabatic states are identical at $\mathbf{Q_0} = \mathbf{0}$). General expressions for the total potential energy matrix $\mathbf{W}(\mathbf{Q})$ are derived by decomposing its matrix elements $W^{tot}_{\alpha\beta}(\mathbf{Q})$ into the above term $V_0(\mathbf{Q})$ describing the initial electronic state prior to the optical transition, and the changes $W_{\alpha\beta}(\mathbf{Q})$ induced by the latter. Here, α and β label the electronic states of the system. Therefore, we have

$$W^{tot}_{\alpha\beta}(\mathbf{Q}) = V_0(\mathbf{Q})\delta_{\alpha\beta} + W_{\alpha\beta}(\mathbf{Q}), \tag{3}$$

$$W_{\alpha\alpha}(\mathbf{Q}) = E_\alpha + \sum_i \kappa_i^\alpha Q_i + \sum_{ij} g_{ij}^{(\alpha)} Q_i Q_j + \cdots, \tag{4}$$

$$W_{\alpha\beta}(\mathbf{Q}) = \sum_i \lambda_i^{(\alpha\beta)} Q_i + \cdots (\alpha \neq \beta), \tag{5}$$

where, for example,

$$\kappa_i^{(\alpha)} = \left.\frac{\partial \Delta V_\alpha(\mathbf{Q})}{\partial Q_i}\right|_{\mathbf{Q}=\mathbf{0}}, \tag{6}$$

$$g_{ii}^{(\alpha)} = \left.\frac{\partial^2 \Delta V_\alpha(\mathbf{Q})}{2\partial Q_i^2}\right|_{\mathbf{Q}=\mathbf{0}}, \tag{7}$$

$$\lambda_i^{(\alpha\beta)} = \left.\frac{\partial W_{\alpha\beta}(\mathbf{Q})}{\partial Q_i}\right|_{\mathbf{Q}=\mathbf{0}}. \tag{8}$$

The energies E_α which appear in the diagonal part are constants given by $W_{\alpha\alpha}(\mathbf{0})$. The latter quantities have the meaning of vertical excitation or ionization energies, referring to the center of the Franck–Condon (FC) zone, $\mathbf{Q} = \mathbf{0}$ (boldface denotes the vector of all coordinates). Because we take the diabatic and adiabatic basis states to coincide at this geometry, the E_α have no counterpart in the off-diagonal elements of Eq. (3). The quantities $\kappa_i^{(\alpha)}$ and $\lambda_i^{(\alpha\beta)}$ are referred to as first-order *intrastate* and *interstate* electron-vibrational coupling constants, respectively. The second-order intrastate coupling constants $g_{ij}^{(\alpha)}$ are responsible for frequency changes and (for $i \neq j$) for the Duschinsky rotation[22] of the normal modes in the excited state. V_α and V_β are the adiabatic PESs and $\Delta V_\alpha = V_\alpha - V_0$. Truncating the series after the first-order terms defines the *linear vibronic coupling* model (LVC), while including second-order terms leads to the — as a short-hand notation — *quadratic vibronic coupling* model (QVC), and so forth. The full vibronic coupling (VC) Hamiltonian is finally obtained by adding the nuclear kinetic energy T_N to Eq. (3)

$$H = (T_N + V_0(\mathbf{Q}))\mathbf{1} + \mathbf{W}(\mathbf{Q}), \tag{9}$$

where $\mathbf{1}$ denotes the $N \times N$ unit matrix, for N interacting states under consideration. We emphasize that the diagonal form of T_N represents an additional model assumption because strictly diabatic electronic states do not exist in general,[18,23] but only approximately diabatic (quasi-diabatic) states can be obtained. Corrections are nevertheless expected to be small if the quasi-diabatic basis is constructed properly, and the prefix 'quasi' is dropped in the following for notational simplicity.

In applying the LVC Hamiltonian to the subset of electronic states one has to take into account important symmetry selection rules[4] which impose important restrictions on the modes appearing in the various summations of Eqs. (4) and (5). These are relevant, in particular, for the linear coupling terms, for which they read

$$\Gamma_\alpha \otimes \Gamma_i \otimes \Gamma_\beta \supset \Gamma_A. \tag{10}$$

Explicitly, a given vibrational mode with symmetry Γ_i can couple electronic states with symmetries Γ_α and Γ_β in first order only if the direct product on the left-hand side of Eq. (10) comprises the totally symmetric irreducible representation Γ_A of the point group in question. The generalization to the second-order terms should be apparent, though it is less restrictive.

The above Hamiltonian lends itself immediately to the treatment of a general number of interacting electronic states. For apparent reasons it has been applied mostly to two intersecting PESs in early applications. Here we focus on more recent work with three or more strongly interacting states. This also affects the determination of the various coupling constants, as is described in Sec. 2.3. below.

2.2. *The concept of regularized diabatic states*

The MMVC scheme can be extended by adopting more general functional forms of the diabatic potentials and coupling elements, which has also been termed "diabatization by ansatz" in the literature.[24] Nevertheless, sometimes it is desirable to construct diabatic states directly — at a given point in nuclear coordinate space — out of the adiabatic states, as is discussed now.

Direct methods for diabatization have been divided into three groups, depending on the type of information the diabatic states are built from Ref. 25: derivative-based methods, property-based methods and energy-based methods, in a decreasing order regarding the required computational effort. Derivative-based methods[18,19] involve the calculation of the nonadiabatic derivative couplings and require very accurate electronic wavefunctions, as is viable for very small polyatomics, but becomes very tedious in the rest of the cases. Property-based methods[26,27] also require the calculation of the wavefunction to evaluate the matrix elements of some property, for example, the dipole moment, which changes smoothly with the configuration space. Finally, the energy-based methods[28] represent conceptually the simplest approximation because they do not require any other information than that contained in the adiabatic PESs $V_i(\mathbf{Q})$ alone.

The concept of *regularized diabatic states* belongs to the third class of methods and the main idea behind it is that the singular coupling terms can be determined from the behaviour of the adiabatic PESs in the close vicinity of the CoIns. In particular, it has been shown[4,16] that only the leading terms in a Taylor series expansion of the PES around an intersection are responsible for the singularity of the coupling terms and, therefore, these leading terms can be used to define an adiabatic-to-quasidiabatic transformation angle. The other residual coupling terms are assumed to be small and, therefore, neglected. This idea underlies the VC approach just explained in the previous section that allows the treatment of complex systems using model Hamiltonians.

The method was initially proposed by Thiel and Köppel[28] and tested succesfully for a $E \times e$ Jahn–Teller situation. Later, it was generalized to any symmetry-allowed CoIn[16] and, subsequently, extended to treat general, i.e. not symmetry-allowed, CoIns.[17] Let us consider initially a two-state problem, although it will be generalized to more states later on. Assuming that diabatic states exist for our specific problem, the general diabatic Hamiltonian can be written as:

$$\mathscr{H}_{\text{dia}} = (T_N + \Sigma(\mathbf{Q}))\mathbf{1} + \mathbf{W}(\mathbf{Q}), \qquad (11)$$

with a *traceless* matrix $\mathbf{W}(\mathbf{Q})$. The transformation to the adiabatic representation through the unitary transformation matrix $\mathbf{S}$ leads to

$$\mathbf{V} = \begin{pmatrix} V_1 & 0 \\ 0 & V_2 \end{pmatrix} = \Sigma(\mathbf{Q})\mathbf{1} + \mathbf{S}^\dagger \mathbf{W}(\mathbf{Q})\mathbf{S} = \Sigma(\mathbf{Q})\mathbf{1} + \begin{pmatrix} \Delta V & 0 \\ 0 & -\Delta V \end{pmatrix}, \qquad (12)$$

with $\Sigma = (V_1 + V_2)/2$ and $\Delta V = (V_1 - V_2)/2$.

Expanding the elements of the diabatic potential energy matrix $\mathbf{W}$ in a Taylor series expansion (see previous section) around the point of degeneracy $\mathbf{Q_0}$, i.e.

$$\mathbf{W}(\mathbf{Q}) = \mathbf{W}^{(1)}(\mathbf{Q}) + \mathbf{W}^{(2)}(\mathbf{Q}) + \cdots \qquad (13)$$

It has been shown[16] that the singular part of the derivative coupling terms can be removed by a unitary transformation through the adiabatic-to-diabatic mixing angle $\theta^{(1)}$, defined by the leading term $\mathbf{W}^{(1)}$ of Eq. (13). That is,

$$\mathbf{S}^{\dagger(1)} \mathbf{W}^{(1)}(\mathbf{Q}) \mathbf{S}^{(1)} = \begin{pmatrix} \Delta V^{(1)} & 0 \\ 0 & -\Delta V^{(1)} \end{pmatrix} \quad \text{with } \mathbf{S}^{(1)} = \begin{pmatrix} c\,\theta^{(1)} & -s\,\theta^{(1)} \\ s\,\theta^{(1)} & c\,\theta^{(1)} \end{pmatrix}, \qquad (14)$$

where c and s stand for cos and sin, respectively, and $\Delta V^{(1)}$ represents the half-difference of the eigenvalues of the first-order diabatic potential energy matrix $\mathbf{W}^{(1)}$, analogous to ΔV given above.

By applying the first-order transformation matrix $\mathbf{S}^{(1)}$ to the *general* adiabatic potential energy matrix $\mathbf{V}$ of Eq. (12), we can thus remove the singular derivative couplings associated with the latter and define the regularized diabatic potential energy matrix, $\mathbf{W}^{\text{reg}}$, as

$$\mathbf{W}^{\text{reg}} = \mathbf{S}^{(1)}\mathbf{V}\mathbf{S}^{(1)\dagger} = \Sigma\mathbf{1} + \frac{\Delta V}{\Delta V^{(1)}}\mathbf{W}^{(1)}, \qquad (15)$$

which is our working equation.

Finally, the corresponding regularized diabatic Hamiltonian is obtained by adding the (diagonal) contribution of the kinetic energy operator as

$$\mathscr{H}^{\text{reg}} = \hat{T}_N \mathbf{1} + \mathbf{W}^{\text{reg}}. \qquad (16)$$

$\mathscr{H}^{\text{reg}}$ represents the VC Hamiltonian in the scheme of regularized diabatic states.

For the case of two relevant vibrational modes, the first-order diabatic matrix $\mathbf{W}^{(1)}$ that enters in the Eq. (15) can be written as

$$\mathbf{W}^{(1)} = \begin{pmatrix} \sum k_i Q_i & \sum \lambda_i Q_i \\ \sum \lambda_i Q_i & -\sum k_i Q_i \end{pmatrix}. \qquad (17)$$

From the equation above it follows that $\Delta V^{(1)} = [(\sum k_i Q_i)^2 + (\sum \lambda_i Q_i)^2]^{1/2}$. Here and below we deal with *not symmetry-allowed CoIns* that occur at some general point in a two-dimensional coordinate space. The coordinates will be denoted as Q_1 and Q_2 and both participate in all matrix elements of $\mathbf{W}^{(1)}$. There are two coupling constants for each mode, k_i and $\lambda_i (i = 1, 2)$, and we now indicate briefly their determination from electronic structure data for the PESs (a more detailed description can be found in Ref. 17). This is accomplished by noting that the squared half-difference of the adiabatic PESs in the vicinity of the CoIn can be described by the following expression

$$(\Delta V)^2 = f_{11} Q_1^2 + f_{22} Q_2^2 + 2 f_{12} Q_1 Q_2, \qquad (18)$$

and the "difference force constants" $f_{ij}(i, j = 1, 2)$ by determined by a least-squares fit to the ab initio data. On the other hand, we can write the difference of eigenvalues $\Delta V^{(1)}$ of the first-order coupling matrix $\mathbf{W}^{(1)}$ as follows

$$(\Delta V^{(1)})^2 = (k_1^2 + \lambda_1^2) Q_1^2 + (k_2^2 + \lambda_2^2) Q_2^2 + 2(k_1 k_2 + \lambda_1 \lambda_2) Q_1 Q_2. \qquad (19)$$

Comparing the two Eqs. (18) and (19) shows that the constants k_i and $\lambda_i (i = 1, 2)$ needed to define the first-order coupling matrix $\mathbf{W}^{(1)}$ can be obtained directly from the $f_{ij}(i, j = 1, 2)$. Although this procedure is not unique, it only leaves open a constant orthogonal transformation which does not affect the final results and can be chosen to fulfill suitable constraints.[17]

Let us now turn to the case of three or more intersecting adiabatic surfaces and restrict attention again to two relevant degrees of freedom. In principle one may think of using the adiabatic-to-diabatic transformation of the $N \times N$ matrix of Eq. (4) as a "whole" and apply it to the diagonal matrix of (N) adiabatic PESs V_i. While this is technically straightforward in principle, due to the model character of the LVC (or QVC) coupling scheme this may not be able to correctly reproduce the complicated set of interconnected CoIns that may exist in the electronic manifold. Therefore one should address the various CoIns separately.

To be more specific, we focus on the case $n = 3$. There are a number of important special cases, like the interaction of a doubly degenerate state with a nondegenerate one, where a symmetry is present and one of the three PESs is not affected in shape by the couplings.[15] While all three component states are interacting nonadiabatically, the shape of the PESs is like in a two-state problem. Therefore, the diabatization procedure for two coupled states discussed above and in earlier work may also be applied for such three-state systems.

A somewhat more general three-state situation is encountered when no symmetries are present, but the different CoIns are separated in space. Then the above procedure for the two-state case may be applied to each intersection separately and the transformation matrices simply be multiplied. For example, let us consider the following adiabatic 3-state situation

$$\mathbf{V}_{3\times 3} = \begin{pmatrix} V_1 & 0 & 0 \\ 0 & V_2 & 0 \\ 0 & 0 & V_3 \end{pmatrix}, \qquad (20)$$

where the system consecutively undergoes a nonadiabatic transition at a CoIn between the potential surfaces V_1 and V_2 and at one between the PESs V_2 and V_3. For each intersection, a diabatization is performed along the lines described above. Let the resulting mixing angles be denoted by θ_{12} and θ_{23}, where the subscript refers to the pair of intersecting surfaces and the superscript (1), denoting the first order in the expansion, has been dropped for notational simplicity. The individual diabatic-to-adiabatic transformation matrices

$$\mathbf{S}_{12} = \begin{pmatrix} c_{12} & -s_{12} & 0 \\ s_{12} & c_{12} & 0 \\ 0 & 0 & 1 \end{pmatrix} \quad \text{and} \quad \mathbf{S}_{23} = \begin{pmatrix} 1 & 0 & 0 \\ 0 & c_{23} & -s_{23} \\ 0 & s_{23} & c_{23} \end{pmatrix}, \qquad (21)$$

(c_{ij} and s_{ij} stand for $\cos \theta_{ij}$ and $\sin \theta_{ij}$ respectively), and are then multiplied to give the whole transformation matrix:

$$\mathbf{S_{123}} = \mathbf{S_{12}} \cdot \mathbf{S_{23}} = \begin{pmatrix} c_{12} & -s_{12}\, c_{23} & s_{12}\, s_{23} \\ s_{12} & c_{12}\, c_{23} & -c_{12}\, s_{23} \\ 0 & s_{23} & c_{23} \end{pmatrix}. \qquad (22)$$

Finally the diabatic potential energy matrix is obtained as

$$\mathbf{W_{3\times 3}} = \mathbf{S_{123}}\ \mathbf{V_{3\times 3}}\, \mathbf{S_{123}}^{\dagger} \qquad (23)$$

in the usual way. The mixing angles could be expressed in terms of the various potential energy matrix elements as above. However, contrary to the two-state case, lengthy expressions result which do not offer a simplification compared to the implicit equations above. Therefore, the above results are the working equations to be used in corresponding calculations such as on formaldehyde below.

2.3. *Electronic structure calculations*

In order to determine the various system parameters entering the Hamiltonian, Eq. (3), and provide a solid basis for the subsequent dynamical simulations, *ab initio* electronic structure calculations are performed. These rely generally on basis sets of double-zeta or triple zeta quality. Electron correlation is included at the CCSD (EOM-CCSD for excitation or ionization energies) or CASSCF/MRCI level of theory. Alternative approaches are MP2 computations for the electronic ground state, used mostly in earlier work, and the outer-valence Greens function (OVGF) formalism for ionization potentials, employed here for 1,2,3-trifluoro benzene. Since all these schemes have been amply described in the literature, we do not go into more details here.

As a first step in the calculation, a ground state geometry optimization is performed, followed by a frequency analysis. The former step also provides the geometrical parameters needed for computing the vertical excitation or ionization energies. The latter characterizes the (harmonic) PES V_0 in Eq. (2) and defines the normal coordinates for the subsequent treatment.

The determination of the various coupling constants can proceed by directly evaluating the elements of Eq. (4) as matrix elements of derivatives of the molecular Hamiltonian at the reference geometry ($\mathbf{Q_0} = \mathbf{0}$) with the electronic wavefunctions (see, for example, Ref. 29). By a suitable averaging over relevant nuclear configurations the accuracy of this procedure may be considerably enhanced.[24] Alternatively, as done in the present work, the coupling constants may be deduced from the adiabatic PESs. In practice several variants emerge which are compared here briefly with respect to computational ease and accuracy.

In the simplest case only "local" information is required, characterized by the immediate vicinity of the expansion point $\mathbf{Q} = \mathbf{0}$ in Eq. (4). For the totally symmetric modes, the corresponding expressions have already been given above as Eqs. (6) and (7). That is, the first-order (second-order) coupling constant for mode i in state α is the corresponding derivative of the corresponding PES along the coordinate Q_i.

For a non-totally symmetric mode u and two interacting states an alternative expression to the above Eq. (8) can be given as follows. The two-state coupling matrix can be written as

$$W_{\text{eff}}^{\alpha\beta}(Q_i) = \begin{pmatrix} E_\alpha & \lambda_i^{\alpha\beta} Q_i \\ \lambda_i^{\alpha\beta} Q_i & E_\beta \end{pmatrix}, \quad (24)$$

from which one gets the well-known hyperbolic shape of the square of the difference $\Delta V_{\alpha\beta}$ of the PESs V_α and V_β

$$\Delta V_{\alpha\beta} = \sqrt{(E_\alpha - E_\beta)^2 + 4(\lambda_i^{\alpha\beta} Q_i)^2}. \quad (25)$$

From this one easily deduces[4]

$$\lambda_i^{(\alpha\beta)} = \sqrt{\frac{1}{8} \frac{\partial^2 (V_\alpha - V_\beta)^2}{\partial Q_i^2}}\bigg|_{\mathbf{Q}=\mathbf{0}}. \quad (26)$$

For more than two coupled states the adiabatic PES can normally not be given in closed form and the above "local" approach using first and second derivatives is no longer viable. Then an appropriate coupling matrix is set up, like

$$W_{\text{eff}}^{\alpha\beta\gamma}(Q_i) = \begin{pmatrix} E_\alpha & 0 & \lambda_i^{\alpha\gamma} Q_i \\ 0 & E_\beta & \lambda_i^{\beta\gamma} Q_i \\ \lambda_i^{\alpha\gamma} Q_i & \lambda_i^{\beta\gamma} Q_i & E_\gamma \end{pmatrix}, \quad (27)$$

and its eigenvalues are determined numerically for a suitable set of finite displacements. These are then fitted to the corresponding *ab initio* data, typically in a least-squares sense.

This latter procedure amounts to a more general, reduced-dimensionality approach to determine the coupling constants. In this scheme the least-squares fit is applied to *any* vibrational mode, given a suitable set of finite displacements along the coordinate of that mode. Apparently, the least squares-fit may be applied also to totally-symmetric modes, and to non-totally symmetric modes coupling two states only. Thus, for example, also higher-order coupling constants can be determined together with the first-order (and second-order) constants. The advantage of this procedure is that it allows to test the accuracy of the coupling scheme in Eq. (4), by comparing its predictions with the *ab initio* data. This is important especially when larger-amplitude displacements are of relevance. In the earlier "local" approach, on the other hand, its applicability is assumed to hold and not controlled.

It should be emphasized that the effort of this procedure increases only linearly with the number of vibrational degrees of freedom [or possibly quadratic, if inter-mode coupling constants are to be included in Eq. (4)]. The scaling behavior is thus the same as in the local approach, only the prefactor differs. This is a huge saving over the general case, because for the number of degrees of freedom of interest (10–30), full-dimensional electronic structure calculations would not be feasible owing to the exponential growth of the number of data points with the number of degrees of freedom (although promising recent developments should be mentioned here[30]).

The last, third scheme would consist in performing a least-squares fit of the model PESs to a global set of electronic-structure data points. Formally this is the most general and rigorous way to determine the coupling constants, but at the expense of an enormous numerical effort (see last paragraph). In practice, suitable subspaces could be selected, thus making the *ab inito* calculations numerically tractable. Indeed, in some cases appropriate linear combinations of normal mode displacements have been selected to determine relative signs and higher-order coupling constants.[31–33] Normally, however, the reduced-dimensionality approach utilizing finite displacements along the coordinates of the individual vibrational modes, is considered as the best compromise between effort and accuracy in applications of the MMVC approach. It underlies the studies on the fluorobenzene cations and singlet-excited pyrrole reported below. Within the concept of regularized diabatic states, on the other hand, general PESs are of relevance, and suitable global or semi-global, data are to be provided, typically for a smaller number of degrees of freedom than for the MMVC approach.

2.4. Quantum dynamical calculations

The dynamical calculations performed in this work and reported below rely on fully quantal, time-dependent methods, namely wavepacket propagation techniques. For the applications of the MMVC scheme the multiconfiguration time-dependent Hartree (MCTDH) method is a highly powerful, nearly ideally suited tool to integrate the time-dependent Schrödinger equation and is therefore used in this work. The MCTDH method[34–40] uses a time development of the wavefunction expanded in a basis of sets of variationally optimized time-dependent functions called single-particle functions (SPFs). The MCTDH equations of motion are obtained from the Dirac–Frenkel variational principle. By virtue of this optimization the length of this expansion can be much smaller than in standard integration schemes (so-called MCTDH contraction effect). The efficiency is even enhanced by two important additional features: each of the coordinates used in the integration scheme can comprise several physical coordinates Q_i. Furthermore, for vibronically coupled systems the wave function is written as a sum of several wavefunctions, one for each electronic state

$$\Psi(t) = \sum_\alpha^{n_s} \Psi_\alpha(t)|\alpha\rangle, \tag{28}$$

where n_s is the number of electronic states. The SPFs may then be optimized separately for each electronic state, and therefore fewer coefficients are needed in the wavefunction expansion. Both choices are employed in this work and, in combination, lead to an MCTDH contraction effect of about 6 orders of magnitude in typical applications.

We emphasize that the form, Eq. (4), of the VC Hamiltonian represents a sum of low-order products of the Q_i which is exactly the form that makes the application of the MCTDH algorithm efficient. — In applications of the regularization scheme, on the other hand, general expressions for the potential energy operator are to be dealt with, but high efficiency is not a severe aspect because of the smaller number of degrees of freedom treated.

Given the time-dependent wavepacket (28) various time-dependent and time-independent quantities can be computed directly or indirectly. Two quantities easily extracted from $\Psi(t)$ are the electronic populations, $P_\alpha(t)$, and reduced densities $\rho_\alpha(Q_i, t)$ for the electronic state α:

$$P_\alpha(t) = \langle \Psi_\alpha(t)|\Psi_\alpha(t)\rangle, \tag{29}$$

$$\rho_\alpha(Q_i, t) = \int \Psi_\alpha^*(t)\Psi_\alpha(t) \prod_{l \neq i} dQ_l. \tag{30}$$

From the definition of the Hamiltonian (4), the populations refer to *diabatic* electronic states. Adiabatic populations have also been obtained, originally for two-state cases, subsequently also for three-state situations.[41]

Of special interest is the spectral intensity distribution $P(E)$ for an optical transition from the ground state, charaterized by the potential V_0 of Eq. (3), to the interacting manifold. This can be computed within Fermi's golden rule, as the Fourier transform of the time autocorrelation function $C(t)$.

$$P(E) \propto \int e^{iEt} C(t) dt, \tag{31}$$

$$C(t) = \langle \Psi(0) | \Psi(t) \rangle = \langle \mathbf{0} | \boldsymbol{\tau}^\dagger e^{-i\mathbf{H}t} \boldsymbol{\tau} | \mathbf{0} \rangle. \tag{32}$$

In Eq. (32), **H** is the vibronic Hamiltonian of Eqs. (9) or (16), $|\mathbf{0}\rangle$ is the vibrational ground state of the initial electronic state with PES V_0, and

$$\boldsymbol{\tau}^\dagger = (\tau_1, \tau_2, \ldots, \tau_{n_s}) \tag{33}$$

is the vector of individual transition matrix element τ_α between the initial state and the final electronic states labeled by α. The autocorrelation function $C(t)$ measures the overlap between the time-evolving wave-packet and the initial one, generated by the optical transition. The scalar product involving the vector $\boldsymbol{\tau}$ of transition matrix elements implies a summation over various partial spectra, each being proportional to $|\tau_\alpha|^2$ (different final electronic states).

Working with real symmetric Hamiltonians and a real initial state as done here, allows to reduce the propagation time by a factor of two according to[42, 43]

$$C(t) = \langle \Psi(t/2)^* | \Psi(t/2) \rangle. \tag{34}$$

Due to the finite propagation time T of the wavepackets, the Fourier transformation causes artifacts known as the Gibbs phenomenon.[44] In order to reduce this effect, the autocorrelation function is first multiplied by a damping function $\cos^2(\pi t/2T)$.[37, 45] Furthermore, to simulate the experimental line broadening, the autocorrelation functions will be damped by an additional multiplication with a Gaussian function $\exp[-(t/\tau_d)^2]$, where τ_d is the damping parameter. This multiplication is equivalent to a convolution of the spectrum with a Gaussian with a full width at half maximum (FWHM) of $4(\ln 2)^{1/2}/\tau_d$. The convolution thus simulates the resolution of the spectrometer used in experiments, plus intrinsic line broadening effects.

3. Illustrative Examples

In this section we present some examples treated by us in past and current work along the lines described above. These serve to highlight methodological aspects as well as phenomena associated with multistate interactions and multiple CoIns. While the work on fluorinated benzene cations has been detailed in several original publications, the results on the other systems covered (formaldehyde and pyrrole) are partly preliminary, but nevertheless included to arrive at a more complete picture of the important features. As stated in the introduction, another class of systems treated in the past are molecules with degenerate electronic states, interacting with other (degenerate or nondegenerate) states. These 3- and 4-state coupling cases have been covered rather comprehensively in a recent review paper.[15] They are therefore not further treated here, where we address (with a single exception) only Abelian molecular point groups without symmetry-induced spatial degeneracies.

3.1. *Formaldehyde*

3.1.1. *General; electronic structure calculations*

Formaldehyde, H_2CO, is the simplest carbonyl compound and is also one of the polyatomic organic molecules relevant to atmospheric chemistry and to the interstellar medium.[46] The presence of a π system and the lone pairs on the oxygen give rise to both n, π^* and π, π^* transitions, which are characteristics of this kind of compounds[47,48] and are responsible for important photochemical reactions. Many of the theoretical studies on formaldehyde[49-60] have focused on the vertical excitations. Grein *et al.*[54,57,61] provided also information on the perturbations of the $^1A_1(\pi, \pi^*)$ and $^1B_1(\sigma, \pi^*)$ valence states with the surrounding Rydberg states. Recently, high-level MRCI and multireference averaged quadratic coupled clusters (MR-AQCC) calculations based on complete active space self consistent field (CASSCF) wavefunctions have been presented by Lischka *et al.*[59,60] along with full geometry optimizations for several electronic states. They have also analyzed the CoIn occurring between the 1 $^1B_1(\sigma, \pi^*)$ and 2 $^1A_1(\pi, \pi^*)$ electronic states, coupled through the out-of-plane bending motion. The electronic spectrum of formaldehyde has been also analyzed experimentally.[62-67] Optical[62,64,68-70] and electron impact techniques[71-73] have been used to record the vacuum-ultraviolet (VUV) absorption region from approximately 7 to 11 eV.

Despite of the large number of studies on formaldehyde, there are still several issues not well understood. For example, in the absorption spectrum of this system, the electric dipole forbidden transition $^1A_2(n, \pi^*) \leftarrow X\ ^1A_1$ is observed and well studied; however, the $^1A_1(\pi, \pi^*) \leftarrow X^1A_1$ transition, that is allowed and should be one of the most intense ones given its large oscillator strength, has not been yet experimentally detected. Likewise, the experimental assignment of the $^1B_1(\sigma, \pi^*) \leftarrow X\ ^1A_1$ electronic transition remains still uncertain.[54–56,59,74,75] The reason for these difficulties is associated with the interaction between the valence and Rydberg states mainly along the r_{CO} stretching coordinate.[57,58] From the dynamical point of view, the photodissociation of formaldehyde has been also extensively studied both theoretically[76–85] and experimentally,[86–91] partly because of the important role that formaldehyde plays in atmospheric chemistry.[92] However, the dynamical studies have mainly focused on the photodissociation involving the lowest-lying electronic states: the ground (S_0), the $1\ ^1A_2$ (S_1) and the $1\ ^3A_2$ (T_1) electronic states. There are virtually no studies on the dynamics of high-lying electronic states.

In this work we present a coupled surface photodynamics study along the r_{CO} stretching coordinate and the HCO angle, involving five interacting (singlet) electronic states in the 7–10 eV excitation energy range.[93] The ground and excited electronic states of formaldedyde have been calculated using an equation-of-motion coupled cluster method with single and double excitations (EOM-CCSD) as implemented in the MOLPRO suite of programs.[94] The triple zeta (VTZ) valence-type basis set of Dunning[95] has been used for the Oxygen, Hydrogen and Carbon atoms. In order to have a balanced description of the valence and Rydberg states of formaldehyde the aforementioned functions were supplemented by a quadruply-augmented double-zeta basis set of diffuse functions placed in the center of masses of the molecule. Altogether, the basis set contains a total of 124 contracted functions. In order to describe properly the potential energy curves for long CO distances, additional state averaged complete active space (SA-CASSCF) calculations followed by an internally contracted multireference configuration interaction (icMRCI) calculations were performed up to $r_{CO} = 20$ a.u. For more details of the calculations, see Ref. 93.

All calculations have been performed in C_{2v} symmetry, varying the r_{CO} and the HCO angle valence coordinates. The C_2 axis is aligned with the z-axis and the x-axis taken perpendicular to the molecular plane. The valence electronic configuration of the ground state can be described as:

$$(a_1\sigma(CO))^2(a_1\sigma(CH))^2(b_2\sigma(CH))^2(a_1\sigma n(CO))^2(b_1\pi(CO))^2(b_2 n(O))^2.$$

For our purposes, the most important orbitals are the HOMO, HOMO-1 and HOMO-2, denoted in short as n, π and σ orbitals, respectively. Rydberg MO's will be labeled in an atomic-like notation (nl) as, ns, np, nd, as usual.

Valence and Rydberg excited electronic states have been calculated up to an energy threshold slightly above $10\,\text{eV}$.[93] The nature of the wavefunction (such as valence or Rydberg) has been confirmed by inspecting the expectation value $\langle x^2 \rangle$, where x is the out-of-plane direction. A large number of Rydberg states interspersed among the valence states has been obtained. This constitutes a substantial challenge for the accuracy of the calculation because of the need to treat a relatively large number of states simultaneously. The comparison with previous calculations and experimental data regarding the vertical excitation energies is quite good,[93] although our results for the B_2 states appear in general shifted towards higher energies ($\approx 0.10\,\text{eV}$ for most of the states). From the analysis of the oscillator strengths, it follows that the main contribution to the absorption spectrum comes from the A_1 and B_2 electronic states. Therefore we confine our attention to these states in the following subsections.

The 2D electronic structure calculations as a function of the r_{CO} bond distance and the HCO angle have been perfomed using the following grid

$$r_{CO} = [1.52 - 3.40]\quad \text{a.u. (220 points)},$$
$$\widehat{HCO} = [100 - 140]\quad \text{deg. (21 points)},$$

where the rest of the coordinates have been kept fixed at the equilibrium values of the electronic ground state. Potential energy curves along the other degrees of freedom, that is the HCH out-of-plane motion, symmetric and antisymmetric CH stretching and the rocking motion were also calculated, and are analyzed in Ref. 93.

3.1.2. *Potential energy surfaces and diabatic Hamiltonian*

Figure 1 displays several potential energy curves for the $^1A_1$ (the two panels on the left) and $^1B_2$ (the two panels on the right) symmetries, keeping the rest of the coordinates fixed to their values at the equilibrium geometry of the ground state. The calculated ground-state optimized equilibrium

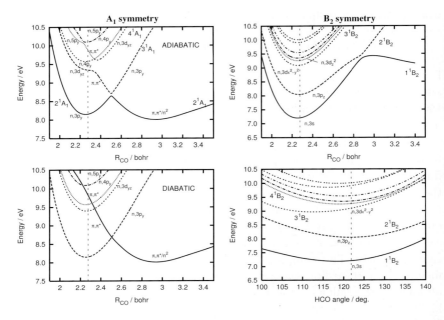

Fig. 1. 1D potential energy curves of formaldehyde. The two panels on the left correspond to the adiabatic (top) and diabatic (bottom) curves of A_1 symmetry along the CO bond length. The two panel on the right correspond to the B_2 symmetry along the CO bond length (top) and HCO angle (bottom). For the B_2 symmetry, only adiabatics curves are displayed.

geometry of formaldehyde is r_{CO} = 2.2766 a.u., r_{CH} = 2.0800 a.u. and $\widehat{HCO}$ = 121.7°, in good agreement with the experimental values r_{CO} = 2.2733 a.u., r_{CH} = 2.0806 a.u. and $\widehat{HCO}$ = 121.9°.[96] The vertical dashed line in Fig. 1 indicates the equilibrium geometry of the ground electronic state, i.e. the FC point. For the B_2 symmetry, the adiabatic potential energy curves are shown along the r_{CO} stretching coordinate (top) and along the HCO angle (bottom). All the B_2 electronic states have a Rydberg character. No avoided crossings are present among the electronic states, except for the one between the 1 $^1B_2$ and the 2 $^1B_2$ states, which is located at a larger distance and a higher energy than the corresponding minima.

In contrast to the B_2 states, the A_1 states show a very complex shape due to several avoided crossings, involving either Rydberg and valence states. In particular, the $^1A_1$ (π, π^*) valence state undergoes several avoided crossings with all the Rydberg states. In order to account for nonadiabatic effects on the dynamics of formaldehyde along the r_{CO} coordinate, a 5×5-states diabatic model has been used for the five $^1A_1$ electronic states shown in the top left panel in Fig. 1. The diabatic Hamiltonian reads as[a]

$$\mathscr{H} = T_N + \mathbf{W}, \tag{35}$$

where T_N is the kinetic energy operator and $\mathbf{W}$ is the 5×5 diabatic potential energy matrix given by

$$\mathbf{W} \equiv \mathbf{W}_{5\times 5} = \begin{pmatrix} W_{11} & W_{12} & W_{13} & 0 & W_{15} \\ & W_{22} & 0 & 0 & 0 \\ & & W_{33} & 0 & 0 \\ & & & W_{44} & 0 \\ & & & & W_{55} \end{pmatrix}. \tag{36}$$

The diagonal elements of $\mathbf{W}_{5\times 5}$ correspond to the diabatic surfaces and they were estimated graphically connecting the electronic energy points with the same character. The diabatic surfaces are displayed in the bottom panel on the left side of Fig. 1, so that $W_{11} = \pi, \pi^*$, $W_{22} = n, 3p_y$, $W_{33} = n, 3d_{zy}$, $W_{44} = n, 4p_y$ and $W_{55} = n, 5p_y$. In contrast to the adiabatic potential energy curves, the diabatic ones are smooth functions of the CO coordinate. In the simplest (1D) model, the W_{ij} ($i \neq j$) coupling terms are one half of the smallest energy difference at the avoided crossing region with values $W_{12} = 0.034$ eV, $W_{13} = 0.169$ eV and $W_{15} = 0.026$ eV. Note that the $W_{44} = n, 4p_y$ electronic state is uncoupled from the rest of states in this simplified treatment. The π, π^* valence electronic state has a minimum located at ≈ 2.95 a.u., in contrast to the A_1 Rydberg states whose minima are close to that of the ground state.

In order to describe more correctly the dynamical behaviour of a system, 2D PESs including the r_{CO} bond coordinate and the HCO angle have been calculated. When doing this, the avoided crossings displayed in Fig. 1 along the r_{CO} turn into CoIns.

Table 1 shows the energy and the location of the CoIns ocurring between the $^1A_1$ electronic states. The regularization method explained in Sec. 2.2 has been applied for the diabatization of the CoIns. The interaction involving the 3^1A_1, 4^1A_1 and 5^1A_1 electronic states (see energy range from 9.0–9.75 eV in the top-left panel of Fig. 1) has been treated in two steps: (1) the $4^1A_1/5^1A_1$ CoIn has been diabatized first, yielding two

---

[a]Note that, contrary to the other equations of this chapter, the matrix $\mathbf{W}$ includes here the ground state potential energy $V_0(\mathbf{Q})$.

Table 1. Energy and geometry location of the CoIns between the $^1A_1$ electronic states.

CoIn	Energy/eV	r_{CO}/a.u.	HCO angle/deg.
$2\,^1A_1/3\,^1A_1$	8.91	2.49	128.4
$3\,^1A_1/^1A_1^{4,5}$	9.46	2.34	117.0
$4\,^1A_1/5\,^1A_1$	9.63	2.33	121.7
$5\,^1A_1/6\,^1A_1$	10.56	2.31	104.0

diabatic states, namely, the state with $n, 4p_y$ character and another diabatic state with a $\pi, \pi^*/n, 3d_{yz}$ mixed character. The latter will be denoted as $^1A_1^{4,5}$. (2) After that, the $3\,^1A_1/^1A_1^{4,5}$ interaction has been subsequently treated.

In order to exemplify the procedure of regularization, we present in Fig. 2 the results obtained for the case of the $4\,^1A_1/5\,^1A_1$ CoIn (or V_4/V_5 as denoted in the figure). The axes represent the displacement coordinates: $\Delta y = y - y_0$, for $y = r_{CO}$ and $\Delta x = (x - x_0) \times r_{CH}^{(0)}$ for $x = \widehat{HCO}$, with $r_{CH}^{(0)}$ being the value of the CH bond length in the equilibrium. x_0 and y_0 are the corresponding values at the CoIn given in Table 1. The intersection is located at the crossing of the vertical and horizontal lines displayed in each panel. The top panels show the adiabatic PESs V_4 and V_5 [see Eq. (15)], which provide all the information needed to build the diabatic states. The left-middle panel shows the half-difference of the adiabatic surfaces, ΔV, in solid lines along with the half-difference of the eigenvalues $\Delta V^{(1)}$ of the first-order diabatic matrix, plotted in dashed lines. $\Delta V^{(1)}$ is obtained by a least-squares fitting of $(\Delta V)^2$ in the vicinity of the CoIn. ΔV and $\Delta V^{(1)}$ enter in the numerator and denominator of Eq. (15), respectively. Likewise, the coupling constants that enter in the diagonal and off-diagonal elements of $\mathbf{W}^{(1)}$ are also deduced from the diagonalization of the matrix of force constants f_{ij}, obtained from the fitting of $(\Delta V)^2$ (Eq. 18).[17] Finally, the right-middle panel displays the off-diagonal term $(W_{45}^{(1)}*(\Delta V/\Delta V^{(1)}))$ of the total regularized diabatic matrix given by Eq. (15), and both panels of the bottom represent the regularized diabatic surfaces, W_{44} and W_{55}. The same procedure has been followed for the other CoIns. Note that since the ellipse in Fig. 2 does not match the half-difference of the adiabatic surfaces in the bottom-right quadrant, the corresponding regularized diabatic surfaces W_{44} and W_{55} show a slight bump in that region. Second-order terms could be added in the fitting to avoid these structures.

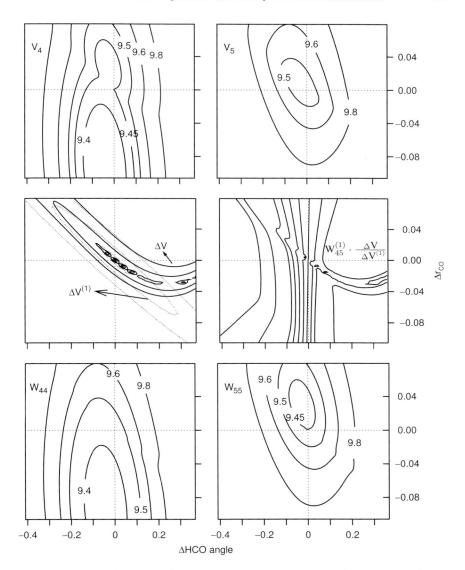

Fig. 2. Adiabatic (V_4 and V_5) and the corresponding diabatic (W_{44} and W_{55}) two-dimensional potential energy surfaces as a function of the displacement coordinates, ΔHCO angle and Δr_{CO} bond length. The coupling is displayed in the middle-right panel and the half of the adiabatic (ΔV) and the first order ($\Delta V^{(1)}$) energy differences are shown in the middle-left panel using solid and dashed lines, respectively. Distances and (scaled) angles are given in a.u. and rad*a.u., respectively. Energies are in eV. For more information, see text.

3.1.3. *VUV absorption spectrum*

Dynamics calculations for the r_{CO} stretching coordinate and for the HCO angle have been performed on the A_1 and B_2 PESs using a wave packet method implemented in the MCTDH Heidelberg package of programs.[34–40] Initially, the ground state wavefunction (for the r_{CO} or the HCO angle) is generated by an energy relaxation method of an initial guess wavefunction on the ground electronic state of formaldehyde, $^1A_1$. The photodynamics of the system is then studied placing the ground state wave function on the excited PESs and following its evolution. Regarding the set of primitive basis functions, a fast Fourier transform representation was used for describing both coordinates and, after a study of convergence, 1024 equally-spaced grid points within the range [1.89–9.44] a.u. and [100–140] deg. were included for the r_{CO} stretching and HCO angle, respectively. A complex absorbing potential (CAP) of the form $-i\mathscr{A} = \eta(x - x_0)^\beta$ was used for the π, π^* electronic state along the $x = r_{CO}$ coordinate, where η is the strength parameter, β is the order of the CAP and x_0 is the place where the CAP starts acting. The parameters were optimized to $\eta = 3.32$ a.u, $x_0 = 8.94$ a.u. and $\beta = 3$. The propagation time was $300\,fs$.

The electronic absorption spectrum of formaldehyde has been obtained by Eq. (31) and the result is presented in Fig. 3 (lower panel). For comparison, the experimental result from Ref. 70 is included in the upper panel of the figure. The calculated total spectrum can be divided clearly into three regions: at low energies, we find the transition to the $1\,^1B_2(n, 3s)$ Rydberg state, whose calculated excitation energy is slightly larger than the experimental one (7.20 vs. 7.09 eV). The structure of this band is found to be associated to a vibrational excitation in the HCO angle, although the experimental vibrational spacing is smaller than the calculated one. The calculated peak is less intense than the experimental one since the calculation understimates the oscillator strength for the correspoding transition. The next region located between 145 to 160 nm corresponds also to transitions to Rydbergs states: the peak at higher wavenumbers ($\approx$154 nm) corresponds to a transition to the $2\,^1B_2(n, 3p_z)$ Rydberg state and is shown in the bottom panel. At $\approx$152 nm we find the vibrational progression in the HCO angle, belonging to the transition to the $2\,^1A_1(n, 3p_y)$ electronic state. The relative energetic separation between both states, $2\,^1B_2$ and $2\,^1A_1$, is smaller than the experimental one since, as commented above, most of the $^1B_2$ state appears shifted to higher energies. The third region

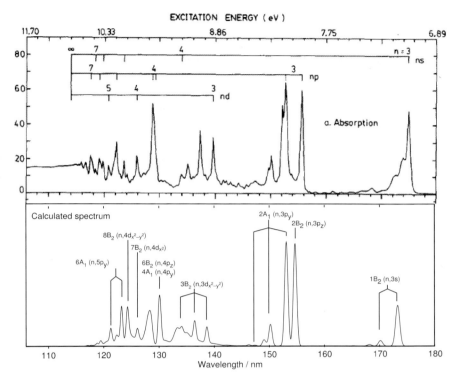

Fig. 3. Experimental and calculated absorption spectrum of formaldehyde. The experimental spectrum is taken from Fig. 1 of Ref. 70.

is the most striking part of the spectrum, since it corresponds to the region of strong interaction. The most noticeable feature is the intensity redistribution of the π, π^* state due to the interaction with the "surrounding" Rydberg states, showing a vibronic structure that becomes denser and irregular at higher energies. The two peaks at 139.55 and 137.20 nm in the experimental spectrum are also found theoretically. They are associated experimentally to 3d transitions.[70] At higher energies we find a rather intense peak that theoretically corresponds to the $4\,^1A_1(n, 4p_y)$ transition, in good agreement with the experimental assignment although the oscillator strength estimated experimentally is larger than the calculated one. Similarly, the peak assigned theoretically as a transition to the $6\,^1A_1(n, 5p_y)$ electronic state is also in good agreement with the experimental assignment.

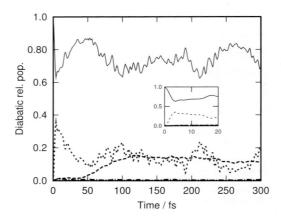

Fig. 4. Relative diabatic populations along the propagation time for the five A_1 potential energy curves. The initial wave packet is placed in the π, π^* state. The inset displays a detail at short propagation times. For the line codings, see Fig. 1.

3.1.4. *Time-dependent electronic populations*

A deeper insight into the nonadiabatic effects on the dynamics of the system can be obtained by analyzing time-dependent quantities such as the electronic populations, shown in Fig. 4 up to a propagation time of 300 fs. Figure 4 displays the probability of the wave packet to be located in any of the five $^1A_1$ diabatic electronic states after excitation to the π, π^* state. The inset shows a detail at short propagation times.

The dynamics of the system can be described as follows: the system, after initial photoexcitation, undergoes a $\pi, \pi^* \to n, 3d_{yz}$ nonadiabatic transition of the order of 5 fs (see inset of Fig. 4). The transfer is very fast owing to the proximity of the crossing to the FC-point. Approximately 40% of the wave packet is transferred initially between these two states. The part of the wave packet transferred to the $n, 3d_{yz}$ state evolves on this state and subsequently reaches again the interaction region, transferring some population back to the π, π^* electronic state. At ≈ 30 fs the system reaches the second interaction region between the π, π^* and $n, 3p_y$ states, populating slightly the latter owing to the smaller W_{12} coupling constant compared to W_{13}. After that, the remaining part of the system keeps evolving on the π, π^* state, bouncing back at the outer turning point and reaching again the intersection $\pi, \pi^*/n, 3d_{yz}$. In general, the dynamics can be mainly seen as the population transfer among the π, π^* and $n, 3d_{yz}$ states, with the corresponding 'transfer' periods of ≈ 100 fs and ≈ 10 fs on the π, π^* and on the $n, 3d_{yz}$ states, respectively. The $n, 3p_y$ state remains slightly populated

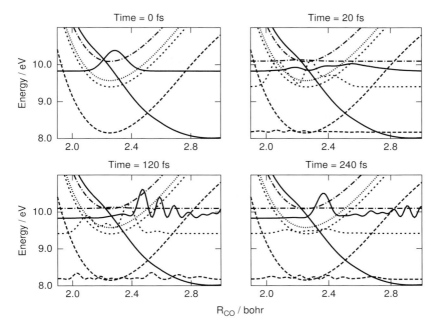

Fig. 5. Diabatic wavepackets on the different diabatic A_1 electronic states as a function of the propagation time (in fs) after excitation from the ground electronic state to the π, π^* state. For the line codings, see Fig. 1.

along the propagation. On the other hand, the population transfer to the highest state is practically not visible under these conditions. Finally, the dynamics of the system can be also visualized in Fig. 5 where the diabatic nuclear densities are depicted along with the corresponding potential energy curves for different propagation times. This confirms the characterization of the system dynamics based above on the electronic populations.

3.2. *Pyrrole*

Heteroaromatic systems, such as phenol, indole, pyrrole and their derivatives,[97–107] are often considered as simplified models for the investigation of the photophysical and photochemical properties of important biomolecules, such as DNA bases or aromatic amino acids. Previous experimental observations and theoretical calculations have revealed that the near-UV photolysis of these heteroaromatic molecules essentially involves ultrafast radiationless decay processes governed by nonadiabatic transitions at CoIns. Short excited-state lifetimes may provide a mechanism to protect the building blocks of life against photochemical damage, by providing a pathway for rapid internal conversion to the ground state.[108]

Since pyrrole and its derivatives play a prominent role in the synthesis of biologically active compounds, it has been under intense experimental and theoretical studies towards the understanding of its photochemical properties,[109] largely focusing on its UV absorption spectrum and its photodynamics. The near-UV photolysis of pyrrole essentially involves the four lowest excited singlet states, which are of $^1\pi\sigma^*$, $^1\pi\sigma^*$, $^1\pi\pi^*$ character and $^1\pi\pi^*$, in order of increasing vertical excitation energy.[102] A summary of more recent *ab initio* studies, and a comparison with the most similar present states is shown in Table 2. The first and the second excited singlet states correspond to excitations from the valence π orbitals of pyrrole to the σ^*/Rydberg-type 3s orbital. These electronic transitions possess very small oscillator strengths.[100] The third and the fourth excited singlet states, which correspond to valence excitations from π to π^* orbitals, are the UV absorbing states and are responsible for the intense band around 6 eV in the absorption spectrum. The $^1\pi\pi^*$ states are bound as a function of the NH stretching coordinate and the $^1\pi\sigma^*$ states are repulsive and therefore

Table 2. Comparison of the present results with previous Rydberg and valence state calculations reported in the more recent literature. f refers to the oscillator strengths (F means dipole forbidden).

		S_0	S_1	S_2	S_3	S_4
Roos et al.[a]	Symmetry	A_1	A_2	A_1	B_2	B_1
		S_0	$\pi\sigma^*(3s)$	$\pi\pi^*$	$\pi\pi^*$	$\pi\sigma^*(3p_y)$
	E_{vert}(CASPT2)		5.22	5.82	5.87	5.87
	f		F	0.036	0.209	0.026
Vallet et al.[b]	Symmetry	A_1	A_2	B_1	B_2	A_1
		S_0	$\pi\sigma^*$	$\pi\sigma^*$	$\pi\pi^*$	$\pi\pi^*$
	E_{vert}(CASSCF/MRCI)		4.45	5.3		
Barbatti et al.[c]	Symmetry	A_1	A_2	B_1	B_2	A_1
		S_0	$\pi\sigma^*$	$\pi\sigma^*$	$\pi - 3P_x$	$\pi\pi^*$
	E_{vert}(MRCI)		5.09	5.86	5.94	6.39
Present work	Symmetry	A_1	A_2	B_1	A_1	B_2
		S_0	$\pi\sigma^*$	$\pi\sigma^*$	$\pi\pi^*$	$\pi\pi^*$
	E_{vert}(MRCI)		5.34	6.30	6.53	6.78
	f		F	8.7×10^{-6}	0.002	0.228

[a]Ref. 99, [b]Ref. 105, [c]Ref. 106

can predissociate the $^1\pi\pi^*$ and $^1n\pi^*$ states. These $^1\pi\sigma^*$ states are optically forbidden and thus not spectroscopically detectable.[110,111]

Theoretical studies done so far on pyrrole,[99,100,105,106,112–115] mainly focused on model PESs of reduced dimensionality constructed on the basis of accurate *ab initio* electronic structure calculations. The model includes the NH stretching coordinate plus one or two additional coupling modes for the separate treatments of the two ($^1\pi\sigma^* - S_0$) CoIns. Time-dependent quantum wave-packet propagation has been used to explore the ultrafast internal conversion and photodissociation dynamics of pyrrole which works essentially fine under the reduced dimensionality model. The latter model reveals that for both CoIns the nonadiabatic dynamics is governed by the strongest coupling mode to a good approximation, and the branching ratio for different photodissociation channels depends on the initial excitation of the strong coupling mode.[113,114]

We have analyzed the multistate multimode dynamics of pyrrole by applying the MMVC model.[116] The quantum dynamical calculations are done using the MCTDH method which allows us to consider four electronic states ($^1\pi\sigma^*$, $^1\pi\sigma^*$, $^1\pi\pi^*$ and $^1\pi\pi^*$) and up to 12 vibrational degrees of freedom. The goal is to shed more light on its low quantum yield fluorescence after UV absorption as well as its photodynamics.

3.2.1. *Vibronic coupling Hamiltonian for pyrrole*

In the present work, we are dealing with the four lowest excited singlet states of pyrrole which play a central role in its photoinduced dynamics.[97,99–102,105,112] Their C_{2v} symmetry assignments and energies are given in Table 2 (present work). The 24 vibrational modes of pyrrole consist of 17 planar and 7 out-of-plane modes according to the following symmetry species

$$\Gamma_{vib} = 9A_1 + 3A_2 + 4B_1 + 8B_2. \tag{37}$$

The NH stretching coordinate r_{NH} (Q_{24}) of A_1 symmetry is the reaction coordinate for the hydrogen abstraction reaction. The selection of the relevant modes of the reduced dimensionality models can be performed with the help of symmetry selection rules for the linear (κ and λ) contributions in applying the VC Hamiltonian, Eqs. (4) and (5), one has the following general symmetry selection rule for the linear (κ and λ) contributions:

$$\Gamma_\alpha \otimes \Gamma_\beta \supset \Gamma_i. \tag{38}$$

The most important modes which have been used in the subsequent treatment of the VC in pyrrole are collected in Table 3. In order to set up

Table 3. Frequencies and coupling constants entering the Hamiltonian (39) for pyrrole. All quantities are in eV.

		Freq.	$A_2(\pi\sigma^*)$	$B_1(\pi\sigma^*)$	$A_1(\pi\pi^*)$	$B_2(\pi\pi^*)$
A_1	ν_9	0.1129	0.0723	−0.1027	−0.0408	0.0254
	ν_{10}	0.1292	−0.0441	−0.1696	−0.0334	−0.0030
	ν_{12}	0.1366	0.1992	−0.0343	−0.0665	0.1377
	ν_{13}	0.1468	−0.0466	0.0239	−0.2124	−0.0336
	ν_{16}	0.1795	0.1710	−0.1299	0.1571	0.1883
	ν_{24}	0.4844	0.1811	0.2194	−0.0091	0.0379
B_2	ν_8	0.1106	0.0860		0.0636	
	ν_{14}	0.1490	0.1815		...	
	ν_{17}	0.1841	0.2118		0.0364	
A_2	ν_2	0.0759	0.0		0.0181	
	ν_7	0.1050	0.0		0.0227	

the working Hamiltonian, further simplifications are used. These regard in particular the values of the couplings and/or the energetic location of the minimum of the intersection seams between the various electronic states. (Apparently, when the minimum of a seam of intersections is too high in energy with respect to our energy range of 5–7 eV, the intersection will not play a significant role.) The interactions of the $\pi\sigma^*$ states with the lowest $\pi\pi^*$ state are suppressed because the ab initio data reveal negligible couplings between the corresponding electronic states, either by virtue of small coupling constants or high-energy minima of the corresponding CoIn seams. The 4×4 vibronic Hamiltonian matrix for the description of the four lowest S_1–S_4 singlet excited states of pyrrole then reads as follows:

$$\mathbf{H} = (T_N + V_0)\,\mathbf{1} + \mathbf{W}, \tag{39}$$

$$\mathbf{W} = \begin{pmatrix} E_1 + \boldsymbol{\kappa}^{(1)}\mathbf{Q} + \mathbf{g}^{(1)}\mathbf{Q}^2 & \sum_{j \in B_2} \lambda_j^{(1,2)} Q_j & 0 & 0 \\ \sum_{j \in B_2} \lambda_j^{(1,2)} Q_j & E_2 + \boldsymbol{\kappa}^{(2)}\mathbf{Q} + \mathbf{g}^{(2)}\mathbf{Q}^2 & 0 & \sum_{j \in A_2} \lambda_j^{(2,4)} Q_j \\ 0 & 0 & E_3 + \boldsymbol{\kappa}^{(3)}\mathbf{Q} + \mathbf{g}^{(3)}\mathbf{Q}^2 & \sum_{j \in B_2} \lambda_j^{(3,4)} Q_j \\ 0 & \sum_{j \in A_2} \lambda_j^{(2,4)} Q_j & \sum_{j \in B_2} \lambda_j^{(3,4)} Q_j & E_4 + \boldsymbol{\kappa}^{(4)}\mathbf{Q} + \mathbf{g}^{(4)}\mathbf{Q}^2 \end{pmatrix}.$$

Here

$$\kappa^{(\alpha)}\mathbf{Q} = \sum_{i\in\Gamma_i}\kappa_i^{(\alpha)}Q_i \quad \text{and} \quad \mathbf{g}^{(\alpha)}\mathbf{Q}^2 = \sum_{i\in\Gamma_i}g_{ii}^{(\alpha)}Q_i^2 \qquad (40)$$

with Γ_i being the irreducible representation to which the mode i belongs. The vertical excitation potentials E_α refer to the excited electronic states ($\alpha = S_1, S_2, S_3, S_4$) and all the Q_i represent dimensionless normal coordinates of the electronic ground state of neutral pyrrole. The other quantities entering Eq. (39) are defined in Eqs. (4) and (5). As was emphasized before, the nonadiabatic dynamics described by the above VC Hamiltonian is essentially controlled by the energies of the minima of the various diabatic surfaces as well as of the various CoIn seams.[4,117] One should note that the determination of the seam minima in the presence of quadratic couplings represents a nontrivial extension of the LVC model, because this equation becomes nonlinear.[118]

3.2.2. *Electronic structure calculations and coupling constants*

The various coupling parameters entering the VC Hamiltonian, Eq. (39), are obtained by performing *ab initio* electronic structure calculations to provide a solid basis for the dynamical calculations. The previous *ab initio* quantum theoretical investigations[97,99,100,102,114,119–121] have demonstrated the underlying difficulties in calculating vertical excited states of pyrrole. One is strong valence-Rydberg mixing. Calculations based only on the valence basis are obviously inadequate and this is a major reason for the failure of early semiempirical studies and early *ab initio* studies. Second, the roles of electron correlations are crucial. Accurate and reliable results are obtained only by sophisticated electron-correlation methods for ground and excited states. Third, dynamic polarization of the σ electrons is strong for some electronic excitations. However, there are still many inconsistencies among the results of these recent high-level theoretical studies with large basis sets.

The *ab initio* calculations for the potential energy curves are done in collaboration with Hans Lischka and colleagues.[116] The ground state geometry optimization and vibrational frequency analysis of pyrrole was performed by using the density functional theory with the B3LYP functional and the 6-311+G(d)' basis set. The complete active space (CAS) used for vertical excitation calculations consisted of five orbitals and four electrons. The reference configurations for the MRCI were constructed within the CAS(4,5) by allowing single and double excitations from the two π orbitals into the two π^* orbitals and the σ^* orbital (MR-CISD). The state averaging is performed over five states (SA-5): ground state, two $\pi\pi^*$ states and two $\pi\sigma^*$ states (see above). The coupling constants were extracted by a least squares fitting procedure, considering various displacements along a given normal mode coordinate, as described in Sec. 2.3 above.

The *ab initio* calculations reveal that not all vibrational modes of pyrrole play a significant role in the VC mechanism within the QVC scheme because of small coupling constants. Table 3 collects the linear coupling constants as well as the harmonic frequencies of the most important vibrational modes. As revealed there, we are dealing with 11 nonseparable degrees of freedom, which still represents a formidable numerical problem. The coupling constants of Table 3 imply a rich variety of CoIns between the PESs of the $S_1 - S_4$ states; several of them are low in energy (see Table 4) and thus accessible to the nuclear motion following photoexcitation. It has been pointed out[118, 122] that quadratic couplings often play a crucial role in the energetic lowering of several seams of CoIn. Thus, in the present study we have also computed the quadratic coupling terms for totally symmetric modes using the least-squares fitting mentioned above (Sec. 2.3.).

3.2.3. *Potential energy surfaces and conical intersections*

The sets of coupling constants and the Hamiltonian, Eq. (39), define the high dimensional PESs of the lowest four electronic states of pyrrole. Representative cuts through the various PESs of pyrrole are presented in Fig. 6. The curves represent the potential energies along a straight line from the origin $\mathbf{Q} = \mathbf{0}$ to the minimum energy of intersection between the S_1 and

Table 4. Summary of important electronic energies, for the interacting states of pyrrole using *ab initio* vertical excitation energies. The diagonal values represent the minima of the diabatic potential energies, off-diagonal entries are minima of the corresponding intersection seams. The numbers on the left side display the results for the quadratic vibronic coupling scheme including all the totally symmetric modes, while those on the right side display the results when only those modes are considered which are included in the dynamical calculations.

$$\begin{pmatrix} & S_1 & S_2 & S_3 & S_4 \\ S_1 & 4.99 & 6.03 & 6.40 & 9.24 \\ S_2 & & 5.99 & 6.25 & 6.55 \\ S_3 & & & 6.19 & 6.54 \\ S_4 & & & & 6.53 \end{pmatrix} \quad \begin{pmatrix} & S_1 & S_2 & S_3 & S_4 \\ S_1(A_2) & 5.01 & 6.04 & 6.49 & 9.28 \\ S_2(B_1) & & 6.01 & 6.29 & 6.58 \\ S_3(A_1) & & & 6.25 & 6.57 \\ S_4(B_2) & & & & 6.56 \end{pmatrix}$$

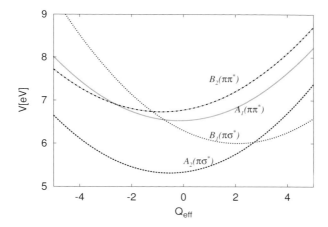

Fig. 6. Representative cut through the PESs of pyrrole. The effective coordinate connects the center of the Franck–Condon zone to the minimum of the intersection seam between the S_1 and S_2 states.

S_2 electronic states. The value of the effective coordinate Q_{eff} denotes the distance from the origin $\mathbf{Q} = \mathbf{0}$ (center of FC zone) along this line. Figure 6 shows a whole set of interstate curve crossings which interconnect all four lowest excited states. These all represent a point on a seam of CoIns and are expected to give rise to a rich vibronic dynamics which will be explored in the following subsection. The curve crossings other than the S_1-S_2 crossing do not occur at their minimal energy in the figure since the same "cut" (effective coordinate) is used in the drawing for all electronic states. There are substantial vibronic interactions within the S_1-S_4 electronic manifold.

Although the multidimensional PESs for the totally symmetric modes are harmonic oscillators, we emphasize that (pronounced) anharmonicity of the adiabatic PESs comes into play as soon as non-totally symmetric modes are included.[4] The minima of the diabatic PESs can be determined by retaining only the totally symmetric modes. Table 4 contains a full comparison of the minimum energy curve crossings for all pairs of electronic states within the QVC scheme. The diagonal entries refer to the diabatic minima of the various PESs and the off-diagonal values are minima of the corresponding intersection seams. The numbers may be compared with the analogous data from the right panel considering only the modes which are included in the dynamical calculations. There are various low-energy curve crossings within the S_1-S_4 excited state manifolds except the minimum energy intersection between S_1 and S_4 which is high in energy. Comparisons of the resulting potential energy curves of the QVC coupling model with the *ab initio* data for many modes (not shown) illustrate how well the quadratic coupling approach reproduces the *ab initio* data. However this is not always the case and the deviations of the model potential energy curves and *ab initio* data points are clearly visible for some modes. Thus, the QVC scheme, is not able to reproduce all *ab initio* data very well. This points to the importance of higher-order coupling terms. Further improvements of the present potential energy curves are being undertaken[116] but are beyond the scope of the present chapter.

3.2.4. *Time-dependent electronic populations*

We now turn to the time-dependent (diabatic) electronic populations resulting from our theoretical approach. Figure 7 represents the electronic populations for a state-specific preparation of the initial wave-packet in the highest-energy $^1B_2(\pi\pi^*)$ state. Almost all four states become populated to a moderate extent, owing to the high energy of the wave-packet and the rather low energies of the CoIns seams discussed above. After the initial decay, the $^1B_2(\pi\pi^*)$ state becomes less likely populated than the $^1A_1(\pi\pi^*)$ state owing to its higher energy. One can clearly see the fast population

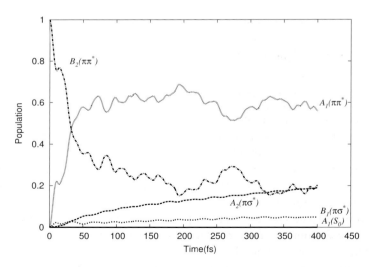

Fig. 7. Electronic population dynamics of pyrrole for state-specific preparation of the initial wave-packet in the S_4 valence excited state.

transfer between these two states occurring on a time scale of around 30 fs. Despite of the fact that the minimum of the intersection seam between the $^1A_1(\pi\pi^*)$ and $^1B_1(\pi\sigma^*)$ states is lower than the $^1B_2(\pi\pi^*)-^1B_1(\pi\sigma^*)$ seam minimum (see Table 4), the population transfer to the $^1B_1(\pi\sigma^*)$ state is more efficient from $^1B_2(\pi\pi^*)$ due to the zero coupling between the $^1A_1(\pi\pi^*)$ and $^1B_1(\pi\sigma^*)$ states (see Table 3). In fact, the $^1B_1(\pi\sigma^*)$ state is weakly populated on a time scale of less than 10 fs, but its population is immediately transferred to the $^1A_2(\pi\sigma^*)$ state. This can be rationalized by the corresponding S_1-S_2 low-energy intersection seam (see Table 4) and the strong coupling between these two states (see Table 3). Additional calculations with the initial wavepacket located in the $^1B_1(\pi\sigma^*)$ state (not shown) indeed feature a two-state population dynamics for joint ultrafast motion in the two lowest excited electronic states $^1B_1(\pi\sigma^*)-^1A_2(\pi\sigma^*)$. One can conclude that the population transfer from the bright valence state $^1B_2(\pi\pi^*)$ to the dark Rydberg states $(\pi\sigma^*)$ is incomplete ($\sim 20\%$). However, the population of the $^1A_2(\pi\sigma^*)$ state shows a gradual increase for longer propagation time. The deactivation of photoexcited pyrrole and other related compounds through ultrafast internal conversion from the $\pi\sigma^*$ states to the S_0 ground state proposed by Sobolewski et al.[101, 110–112] is not treated in our current investigation within the QVC scheme. However, our current coupling scheme provides a mechanism for populating the initial states used by these authors in their calculations. The present results thus help to explain the lack of detectable fluorescence in pyrrole; one requires to go beyond the QVC coupling approach for these complex molecular systems to fully interpret the experimental data.

3.3. *Fluorinated benzene cations*

Apart from systematic studies and individual examples, it is of considerable interest to have available a set of related molecules which can serve as a means to vary one or several system parameters and, thus, establish their impact on the vibronic interactions in general and on the nonadiabatic coupling effects in particular. The benzene and benzenoid radical cations are prototype organic species of fundamental importance, and represent important showcases in this respect. Their electronic structure, spectroscopy, and dynamics have received great attention in the literature over the past, including nonadiabatic interactions in their elementary photophysical and photochemical processes. For example, the parent benzene radical cation Bz^+,[31, 32, 123–130] the sym-trifluoro and hexafluoro derivatives as well as their chlorinated counterparts or the deuterated isotopomers possess degenerate electronic states, and the multimode dynamical Jahn–Teller effect has been studied intensely over the past, both experimentally and theoretically.[118,131–145] The fluorobenzene derivatives are of systematic interest for at least two different reasons. (1) The reduction of symmetry by incomplete fluorination leads to a disappearance of the Jahn–Teller effect present in the parent cation. (2) A specific, more *chemical* effect of fluorination consists in the energetic increase of the lowest σ-type electronic states of the radical cations. Even in the parent cation Bz^+, the interactions between distinctly different states play a role[31,32,125] and have been made responsible for the absence of detectable emission in Bz^+[146,147]: the internal conversion processes which are caused by nonadiabatic vibronic interactions are so fast that fluorescence cannot compete.[31,32] Similar observations have been made for several derivatives,[131] such as mono- and di-fluorobenzene cations. This motivated us to analyze not only the multimode dynamical Jahn–Teller and pseudo Jahn–Teller effect in the unsubstituted species, but also the changes that occur upon fluorination and accompanying reduction in symmetry.[118,122,148] All four difluoro- and the 1, 2, 3-trifluoro derivatives are selected to that end.

3.3.1. *Vibronic coupling Hamiltonian for the fluorobenzene cations*

We are focusing on the five lowest electronic states of the fluorobenzene cations (component states in case of the parent cation, Bz^+). These states lie, for all six cations, in the ionization energy range from 9 to 14 eV and give rise to the low energy band systems of the experimental photoelectron spectra (PE).[149] Important features of these states, such as the vertical ionization potentials, the nature of some of the orbitals out of which ionization takes place, and, in particular, their correlation among the various species considered, are depicted in Fig. 8. The labeling of the species follows an obvious notation, Bz^+, F-Bz^+, 1, 2, 1, 3, 1, 4 and 1, 2, 3 where the charge state (+1) of the di- and trifluoro isomers has been suppressed for notational simplicity.

By using the QVC model (augmented by purely quadratic couplings only for totally symmetric modes), the symmetry-selection rule, Eq. (10), can be directly applied to deduce the vibronic Hamiltonian matrices for the description of the five lowest $\tilde{X}$–$\tilde{D}$ doublet states of these fluorobenzene cations. We shall not write down all five matrices here, but rather

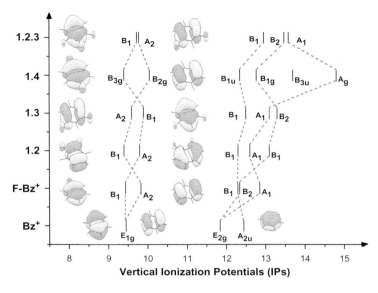

Fig. 8. Correlation between the lowest ionization potentials of the benzene, mono-, di- and trifluoro derivatives according to the adjusted IPs. The Hartree–Fock canonical orbitals are included for the E_{1g} and the E_{1g}-derived states (and also for the A_{2u} state, see text for details). Since these orbitals, and the corresponding ones for the A_{2u}-derived states are π orbitals, these states are also referred to as π-type states. For the analogous reason, the E_{2g} and E_{2g}-derived states are also termed σ-type states.

provide the basic features regarding their QVC Hamiltonian. The general form of the QVC potential energy matrix, $\mathbf{W}_{\text{fluoro}}$, for the above-mentioned fluorobenzene cations is depicted below:

$$\mathbf{H} = (T_N + V_0)\mathbf{1} + \mathbf{W}_{\text{fluoro}}, \tag{41}$$

$$\mathbf{W}_{fluoro} = \begin{pmatrix} E_X + \boldsymbol{\kappa}^{(X)}\mathbf{Q} \\ + \mathbf{g}^{(X)}\mathbf{Q}^2 & \boldsymbol{\lambda}^{(XA)}\mathbf{Q} & 0 & 0 & 0 \\ \boldsymbol{\lambda}^{(XA)}\mathbf{Q} & E_A + \boldsymbol{\kappa}^{(A)}\mathbf{Q} \\ + \mathbf{g}^{(A)}\mathbf{Q}^2 & 0 & \boldsymbol{\lambda}^{(AC)}\mathbf{Q} & 0 \\ 0 & 0 & E_B + \boldsymbol{\kappa}^{(B)}\mathbf{Q} \\ + \mathbf{g}^{(B)}\mathbf{Q}^2 & \boldsymbol{\lambda}^{(BC)}\mathbf{Q} & \boldsymbol{\lambda}^{(BD)}\mathbf{Q} \\ 0 & \boldsymbol{\lambda}^{(AC)}\mathbf{Q} & \boldsymbol{\lambda}^{(BC)}\mathbf{Q} & E_C + \boldsymbol{\kappa}^{(C)}\mathbf{Q} \\ + \mathbf{g}^{(C)}\mathbf{Q}^2 & \boldsymbol{\lambda}^{(CD)}\mathbf{Q} \\ 0 & 0 & \boldsymbol{\lambda}^{(BD)}\mathbf{Q} & \boldsymbol{\lambda}^{(CD)}\mathbf{Q} & E_D + \boldsymbol{\kappa}^{(D)}\mathbf{Q} \\ + \mathbf{g}^{(D)}\mathbf{Q}^2 \end{pmatrix},$$

where the quantities $\boldsymbol{\kappa}^{(\alpha)}\mathbf{Q}$ and $\mathbf{g}^{(\alpha)}\mathbf{Q}^2$ are given by Eq. (40), and $\boldsymbol{\lambda}^{(\alpha\beta)}$ by:

$$\boldsymbol{\lambda}^{(\alpha\beta)}\mathbf{Q} = \sum_{i \in \Gamma_i} \lambda_i^{(\alpha\beta)} Q_i. \tag{42}$$

Some of the off-diagonal entries are put to zero, because the subsequent electronic structure calculations reveal only negligible interactions between the corresponding electronic states. The details about the construction of the Hamiltonian matrices, as well as their explicit form, can be found in Refs. 118 as well as 122 and 148 for the mono- and di-fluorobenzene cations, respectively.

3.3.2. *Electronic structure calculations and coupling constants*

In order to determine the various coupling parameters entering the Hamiltonian Eq. (41), *ab initio* electronic structure calculations have been carried out with the TZ2P one-particle basis set. This basis consists of the triple zeta set of Dunning[150] augmented by polarization functions as given in Ref. 151 and 152. The coupled-clusters singles and doubles (CCSD) method has been employed for ground state geometry optimization and vibrational frequency analysis in ACES electronic structure package.[153] The ground state structural parameters thus obtained agree very well with available literature data.[154–157] Ionization potentials and ionic state energies have been determined by means of the EOMIP-CCSD method[158,159] as implemented in the development version of the ACES program system.[153]

We briefly discuss the key quantities, the vertical ionization potentials (IPs) and coupling constants, and come back to Fig. 8. To be precise, the IPs displayed there do not represent the pure ab initio data but adjusted values, obtained in order to better reproduce the band centres of the various PE spectral bands.[148] We do not give more details here but only emphasize that the difference to the *ab initio* data is typically 0.2–0.3 eV only, and that the adjusted IPs are used in all subsequent calculations reported below. The underlying molecular orbitals, included for the lowest IPs for all species in the figure, show a characteristic behavior. Whereas for benzene one can see the familiar components of the degenerate HOMO of E_{1g} symmetry, for all the fluoro derivatives this degeneracy is necessarily lifted, although the key features remain similar for all cases studied. For example, the nodal characters of the two lowest electronic states of fluoro derivatives are seen to reflect clearly the well-known shape of the two components of the doubly degenerate HOMO (symmetry E_{1g}) of benzene. The nodal properties of the B_1 HOMO of F-Bz$^+$ are shared, for example, by the A_2, B_1, B_{3g} and B_1 orbitals of 1,2-, 1,3-, 1,4-difluorobenzene and 1,2,3-trifluorobenzene cations as indicated in Fig. 8. The energetic ordering of the components of the same symmetry changes, which can be attributed to the different number (and strength) of the C-F antibonding interactions.[122,148]

More important proves to be the systematic increase of the σ state with increasing fluorination. This holds in absolute energy as well as in relation to the second π-type IP, which corresponds to the state of A_{2u} symmetry in the benzene radical cation and the higher one of B_1 symmetry in the fluoro derivatives (B_{1u} symmetry for the 1,4-difluoro isomer). This energetic increase is known in the literature as perfluoro-effect,[160] and seen here to lead to an interchange of the energetic ordering of the E_{2g} and A_{2u} derived ionization processes. This will be seen below to play a crucial role for the nonadiabatic interactions in the cations and their change upon fluorination. This will be discussed further below in relation to their fluorescence dynamics (see Sec. 3.3.6.).

An overview of important vibrational and VC constants is presented in Ref. 161. Out of the many coupling constants computed,[118,122,148] only a few first-order couplings which are large and correspond to vibrational modes that can be correlated between the various isomers, were retained. These comprise mode 1, denoting the totally symmetric C-C stretching mode of Bz$^+$ and the modes 6a–8a and 6b–8b, deriving from the doubly degenerate E_{2g} modes 6–8 of Bz$^+$. The similarity of the vibrational frequencies and also coupling constants throughout the series is to be noted.[161]

3.3.3. *Potential energy surfaces and conical intersections*

The sets of coupling constants together with the Hamiltonian, Eq. (41), define the high-dimensional PESs of the lowest five electronic states of the various cations treated. Typically 6–8 totally symmetric modes and 8–10 non-totally symmetric modes are found to have non-negligible coupling constants in the C_{2v} systems; in the case of higher symmetry these numbers decrease, e.g. to 3 relevant totally symmetric modes for the 1,4-difluoro isomer. The multidimensional PESs thus defined imply a rich variety of different CoIns in the various cations. To better visualize the situation, we present in Fig. 9 representative cuts through the PESs of the benzene cation (upper panel) as well as the 1,2-difluorobenzene derivative (lower panel). A linear combination of the normal coordinates of the Jahn–Teller active modes $\nu_6-\nu_8$ is chosen for the benzene cation, and of totally symmetric modes for the 1,2-difluoro benzene cation. Both are defined to minimize the energy of the CoIn between the $\tilde{A}$ and $\tilde{C}$ states of the difluoro derivative, and between the $\tilde{X}$ and $\tilde{B}$ states of the parent cation. For the

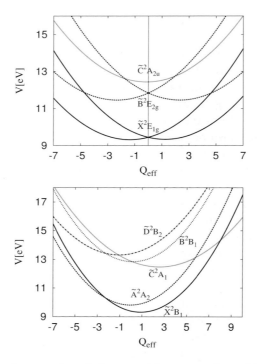

Fig. 9. Representative cuts through the potential energy surfaces of Bz$^+$ (upper panel) and its 1,2-difluoro derivative, 2F-Bz$^+$ (lower panel). The upper panel shows the results for the linear vibronic coupling model, while in the lower one the quadratic coupling terms are also included. In both panels the effective coordinate connects the centre of the Franck–Condon zone to the minimum of the intersection seam between the $\tilde{A}$ and $\tilde{C}$ states of 2F-Bz$^+$, and between the $\tilde{X}$ and $\tilde{B}$ states of the parent cation.

parent cation one identifies a low-energy inter-state curve crossing which is mediated by the multimode Jahn–Teller effect in the two degenerate electronic states.[31,32] The latter is reflected by the symmetric crossing between the two lowest ($^2E_{1g}$) potential energy curves in upper panel of Fig. 9 which actually represents a cut through the multidimensional Jahn–Teller split PESs in this state. These are the well-known Mexican hat[3] PESs of the $E \otimes e$ Jahn–Teller effect. They are recovered also from the $^2E_{2g}$ state curves in the figure.

For Bz$^+$, the two component states are degenerate by symmetry, and the slopes of the PESs are necessarily oppositely equal at the origin $\mathbf{Q} = \mathbf{0}$. Figure 9 illustrates two main trends in the series of molecules. First, by the asymmetric substitution the degeneracy is lifted and the slopes in question are no longer equal in modulus and opposite in sign at $\mathbf{Q} = \mathbf{0}$ (see lower panel of Fig. 9). Thus, the Jahn–Teller effect in the parent cation 'disappears' in the fluoro derivatives; nevertheless, the deviations from the high-symmetry case Bz^+ are seen to be minor. The shapes of their lowest two PESs still resemble those of the parent cation, regarding the opposite slopes, the rather small energetic splitting at the origin $\mathbf{Q} = \mathbf{0}$, and the presence of a low-energy CoIn of relevance to the nuclear motion (in the $\tilde{A}$ state). Therefore this has also been termed a replica of the Jahn–Teller intersection in Bz^+.[162] This topological, or more "physical" effect is complemented by the second, more "chemical" effect, caused by the energetic increase of the second π-type IP by fluorination. We note that the adiabatic ordering of these states (i.e. at their respective minima) is identical to their vertical ordering discussed in relation to Fig. 8. This trend, already mentioned in relation to Fig. 8 above, is specially related to the substituents (F atoms) and manifests itself in a growing separation of the $\tilde{X}-\tilde{A}$ and the $\tilde{B}-\tilde{C}-\tilde{D}$ sets of states. While the effect is rather moderate for F-Bz^+ and in Fig. 9, it increases upon increasing fluorination and thus leads to a higher energy of the corresponding intersection, see Table 5. These two trends, caused by the substitution in general and fluorination in particular, will provide useful guidelines in the discussion of the dynamical results in Secs. 3.3.4 and 3.3.5. Finally we point out again that the results for the inter-set crossings depend crucially on the inclusion of the quadratic coupling constants for the totally symmetric modes. The latter lower them energetically, thus making them accessible to the nuclear motion. They are included in the results of the present subsection and also in all dynamical calculations on the fluoro derivatives reported below.

3.3.4. *Photoelectron spectra*

Figure 10 shows the calculated PE spectrum of 1, 2-difluorobenzene compared to the experimental recording of Ref. 149. This is representative of similar results for the mono- and difluoro derivatives published earlier,[118,122,148] and may also be compared to Fig. 9. The theoretical spectra are presented for two different resolutions: the upper traces correspond to a Lorentzian line width FWHM (full width at half maximum) = 66.6 meV for a better comparison with experiment, while for the lower ones we have used a higher resolution (FWHM = 13.3 meV) to reveal more vibronic structure in the spectral envelopes. Also, the latter spectra have been decomposed into $\tilde{X}$ and $\tilde{A}$ bands for clarity.

Table 5. Summary of important electronic energies, for the interacting states of the fluorobenzene radical cations including the quadratic coupling terms and considering only the modes which are included in the dynamical calculations. The diagonal values represent the minima of the diabatic potential energies, off-diagonal entries are minima of the corresponding intersection seams.

Benzene

	$\tilde{X}$	$\tilde{X}$	$\tilde{B}$	$\tilde{B}$	$\tilde{C}$
$\tilde{X}$	9.27	9.27	11.58	11.58	...
$\tilde{X}$	9.27	9.27	11.58	11.58	...
$\tilde{B}$			11.42	11.42	12.27
$\tilde{B}$			11.42	11.42	12.27
$\tilde{C}$					12.25

mono-fluorobenzene

	$\tilde{X}$	$\tilde{A}$	$\tilde{B}$	$\tilde{C}$	$\tilde{D}$
$\tilde{X}$	9.22	9.69	>16	12.84	>14
$\tilde{A}$		9.69	>15	12.29	>14
$\tilde{B}$			12.22	12.24	12.45
$\tilde{C}$				11.91	12.58
$\tilde{D}$					12.43

1,2-difluorobenzene

	$\tilde{X}$	$\tilde{A}$	$\tilde{B}$	$\tilde{C}$	$\tilde{D}$
$\tilde{X}$	9.15	9.61	>16	>13	>13
$\tilde{A}$		9.61	>16	12.70	>13
$\tilde{B}$			12.12	12.28	12.60
$\tilde{C}$				12.16	12.76
$\tilde{D}$					12.57

1,3-difluorobenzene

	$\tilde{X}$	$\tilde{A}$	$\tilde{B}$	$\tilde{C}$	$\tilde{D}$
$\tilde{X}$	9.35	9.70	>16	>14	>13
$\tilde{A}$		9.69	>16	13.67	>13
$\tilde{B}$			12.32	12.67	12.89
$\tilde{C}$				12.64	13.04
$\tilde{D}$					12.88

1,4-difluorobenzene

	$\tilde{X}$	$\tilde{A}$	$\tilde{B}$	$\tilde{C}$	$\tilde{D}$
$\tilde{X}$	9.11	9.92	>16	>14	>16
$\tilde{A}$		9.88	>16	13.09	>16
$\tilde{B}$			12.17	12.39	14.61
$\tilde{C}$				12.31	13.46
$\tilde{D}$					13.43

1,2,3-fluorobenzene

	$\tilde{X}$	$\tilde{A}$	$\tilde{B}$	$\tilde{C}$	$\tilde{D}$
$\tilde{X}$	9.48	9.62	>16	>16	>16
$\tilde{A}$		9.53	>16	15.10	>16
$\tilde{B}$			12.81	13.17	13.30
$\tilde{C}$				13.17	13.40
$\tilde{D}$					13.30

The PE spectra are seen to consist of two distinct groups of bands, representing the $\tilde{X}-\tilde{A}$ and $\tilde{B}-\tilde{C}-\tilde{D}$ electronic states. The nonadiabatic coupling effects manifest themselves as irregularities in the spectral structures of Fig. 10. Their vibronic structure is revealed more clearly by the lower drawing which corresponds to a higher resolution. This gives an impression of the highly complex, irregular and dense underlying line structure; it is not fully resolved even here, because the resolution is still too limited except for the low-energy spectral regimes.

The $\tilde{X}-\tilde{A}$ group of bands is indeed rather regular for low vibronic energies ($\tilde{X}$-state), but becomes increasingly irregular in the energy region of the $\tilde{A}$-state, that is, for energies above the $\tilde{X}-\tilde{A}$ seam of conical intersections. Their energetic minima exhibit significant changes between the systems considered. According to Table 5, the numbers are 9.61, 9.70, 9.92 and 9.62 eV for the 1,2-, 1,3-, 1,4-difluorobenzene and 1,2,3-trifluorobenzene cations. They may be compared to the $\tilde{X}$-state energetic minima of 9.15, 9.35, 9.11 and 9.48 eV, in the same series. Thus, in the 1,4-difluorobenzene

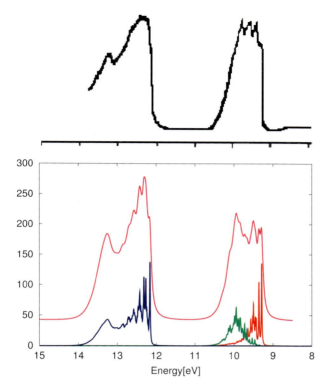

Fig. 10. Comparison of theoretical (lower panel) and experimental (upper panel)[149] photoelectron spectra of 1, 2-difluorobenzene. The linewidths of the theoretical spectra are FWHM = 66.6 meV (upper curve) and 13.3 meV (lower curve). In the higher-resolution theoretical spectrum, the $\tilde{X}$–$\tilde{A}$ electronic bands are drawn separately.

cation there is a particularly large "adiabatic" $\tilde{X}$-state energy range, whereas in the 1, 2, 3-trifluorobenzene cation it is particularly small. This follows from the near-degeneracy of the $\tilde{X}$- and $\tilde{A}$-states and renders this system in the 1, 2, 3-trifluorobenzene cation a particularly close replica[163] of the JT effect in the parent system Bz$^+$. For the other isomers, like that considered in the present figure, a somewhat intermediate situation prevails. — The lowest-energy range in the higher group of $\tilde{B}$–$\tilde{C}$–$\tilde{D}$-bands is also characterized by a rather well-resolved and regular structure. Here the energies are again slightly below that of the conical intersection (12.28 eV according to Table 5). The complexity rapidly increases for higher energy, soon exceeding that of the CoIn. Under low-to-moderate resolution a diffuse spectral profile results, because the highly irregular and very dense individual spectral lines cannot be resolved any more. This is a typical consequence of CoIns between the various PESs[4] and generalized here to multistate coupling situations.

3.3.5. *Time-dependent electronic populations*

In order to get further insight into the multistate dynamics in these systems, the time-dependent (diabatic) electronic populations have been systematically computed, for all the fluoro-derivatives in question, and for all possible initial electronic states.[118, 122, 148] To avoid an excessive number of drawings, we confine ourselves here to the results of the wavepacket located initially in the $\tilde{B}$ (emissive state) electronic state. This is motivated by the relevance of this state to the fluorescence dynamics, see next subsection. The results are presented in Fig. 11. As for the parent cation Bz^+,[32] we see a rich population dynamics proceeding on the *fs* time scale. Generally all five states become populated to a significant or moderate extent owing to the relatively high initial energy of the wave packet. The $\tilde{D}$ state always becomes least likely populated as expected from its high energy. Its decrease

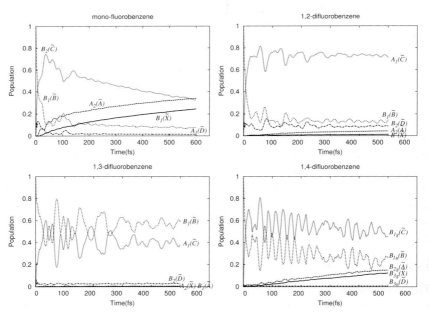

Fig. 11. Electronic population dynamics of fluorobenzene isomers for initial preparation of each cation in the emissive electronic state ($\tilde{B}$ state).

in the series 1, 2-, 1, 3- and 1, 4-difluorobenzene cations can be rationalized by density of states arguments and the increase of the diabatic minimum of this state in the same series (see Table 5). The diabatic minima for the $\tilde{B}$ and $\tilde{C}$ states also help to understand their different populations, especially if a coherent superposition of the electronic states is chosen as initial wavepacket.[148,161]

Of further interest is the transfer of population from the $\tilde{B}-\tilde{C}-\tilde{D}$ to the $\tilde{X}-\tilde{A}$ group of states. As pointed out above, these two sets of states are far apart energetically in the center of the FC zone, but nevertheless interconnected through one CoIn (between the $\tilde{A}-\tilde{C}$ states) which is low (mono, 1, 2- and 1, 4-isomers), or moderately or very high (1, 3- and 1, 2, 3-isomers) in energy. This energetic trend, seen in Table 5, is again reflected in the population curves of Fig. 11. There the combined $\tilde{X}/\tilde{A}$ population after $\sim 500\,fs$ propagation time amounts to only $\sim 1\%$ for 1, 3-difluorobenzene, but to $\sim 60\%$, $\sim 5\%$ and $\sim 27\%$ for the mono, 1, 2- and 1, 4-difluorobenzene cations, respectively. The results for the 1, 2, 3-trifluorobenzene cation reveal no population transfer to the $\tilde{X}-\tilde{A}$ set.[163] The minimum energies of the intersection seams are 13.67 eV, 12.29, 12.70, 13.09 eV and 15.10 eV in the same series. Thus the difference between the 1, 3- and 1, 2, 3-isomers, on one hand, and the mono and 1, 4- isomers, on the other hand, is well reflected by these energetic data (while the situation with the 1, 2-isomer is somewhat less clear). The average populations stay fairly constant after $\sim 100\,fs$ in the case of 1, 2- and 1, 3-isomers, but show a gradual decrease ($\tilde{B}/\tilde{C}$) and increase ($\tilde{X}/\tilde{A}$) for the mono and 1, 4-difluorobenzene cations. The reason for this difference remains unclear at present, as is the case for the different oscillatory or fluctuating time dependences of the various populations. Apparently, the underlying complex and multidimensional dynamics still awaits a more detailed analysis and understanding. Some of this was explored recently for the parent cation, Bz^+.[41] The general trends of the electronic populations, and their relations to the respective energetic quantities, remain the same also for other choices of the initial wavepacket,[161] and also for the purely *ab initio* vertical IPs reported in Ref. 148.

3.3.6. *Relation to the fluorescence yield*

The behavior of the time-dependent electronic populations has important consequences for the fluorescence of these radical cations. The parent cation Bz^+, as well as the monofluoro-derivative, are non-emitting species, with an upper limit for the fluorescence quantum yield of $10^{-4}-10^{-5}$.[131,146,147]

Given typical radiative lifetimes of 10^{-8} sec, these low quantum yields imply a subpicosecond timescale for the radiationless deactivation of the electronically excited radical cations. Increasing fluorination, however, changes the situation, and for at least threefold fluorination of the parent cation there is clear emission.[123, 131–139] The difluorobenzene cations represent a transitional regime, and only one of them, the 1,3-isomer, has been found to emit weakly.[131]

For the parent system Bz^+ itself, a detailed mechanism could be established[31, 32] in terms of the multimode dynamical Jahn–Teller effect in the $\tilde{X}$ and $\tilde{B}$ electronic states, which leads to a low-energy CoIn between the corresponding PESs (see Fig. 9). Also it has been conjectured[164] that the stabilization of the $e_{2g}(\sigma)$ orbital by fluorination leads to an increase of the corresponding ionization potential and a corresponding increase of the (minimum energy of) CoIn, thus weakening the vibronic interactions and rendering the excited states long lived to make emission eventually (i.e., for a sufficient degree of fluorination) observable.

This earlier conjecture is fully confirmed, regarding the general trends upon fluorination, by the present mechanism and results.[122, 148] The radiationless deactivation in the Bz^+ is not a direct one (from the state where dipole-allowed transitions are possible, the $\tilde{C}$ state, to the ground state) but involves the $\tilde{B}$ state as an intermediate.[31, 148] Already for the monofluoro derivative, the two IPs deriving from the σ orbital of benzene (the $\tilde{C}$ and $\tilde{D}$ states of the monofluoro benzene cation) are sufficiently high in energy so that their energetic ordering with the π-type IP is interchanged.[118, 141, 144, 149] For the three difluoro isomers and 1,2,3-trifluorobenzene, the shifts in energy are more pronounced (see Fig. 8). Correspondingly, already for the monofluoro derivative the $\tilde{B}$–$\tilde{X}$ internal conversion, competing with the strongly dipole-allowed transition, is substantially slower than in the parent cation.[118, 144] In the difluoro isomers and the 1,2,3-trifluorobenzene cation this decay is further slowed down owing to the higher-energy vertical IPs and CoIns as discussed above. Although we cannot make quantitative predictions, the present electronic populations allow to draw important conclusions on the different emission properties of these 6 systems. As seen from Fig. 11 and stated above, the internal conversion to the $\tilde{X} + \tilde{A}$ states is indeed slowest, and inefficient also on an absolute scale, for the 1,3- and 1,2,3-isomers. We find it intriguing that emission has indeed been observed for these species, but not for the others. For the $F-Bz^+$ and 1,4-isomer, on the other hand the $\tilde{X} + \tilde{A}$ populations keep increasing after 500 fs and may be expected to dominate after several ps. This behavior is expected to suppress fluorescence, in accord with the experimental results. Only for the $1,2$-isomer the situation is somewhat less clear. However, other modes, not included in the present treatment, may further enhance the $\tilde{X} + \tilde{A}$ populations and thus be consistent with the absence of fluorescence in the $1,2$-isomer.

Additional evidence comes from the consideration of the dipole transition matrix elements. It is well known that the transition from the σ-type state of the benzene cation to the ground state is dipole forbidden.[31,147] For the fluoro derivatives the molecular symmetry is reduced and the selection rules are relaxed.[148] One finds that there is always one component of the σ-type $(\tilde{C} + \tilde{D})$ states that has a finite dipole matrix element for transitions to one component of the lower π-type $(\tilde{X} + \tilde{A})$ states (at least for the C_{2v} molecular point group). Nevertheless, EOM-CCSD/TZ2P calculations clearly show that the corresponding transition dipole matrix elements are smaller by 2–3 orders of magnitude than those for the $\tilde{B}-(\tilde{X} + \tilde{A})$ transition, corresponding to the dipole-allowed transition in the case of Bz$^+$.[148] Thus, the π-type electronic state is the 'emitting' state also for the fluorobenzene cations. Its combined $\tilde{X} + \tilde{A}$ oscillator strength is almost the same for all five systems. Comparing again the various populations of Fig. 11 we find that the $\tilde{B}$ state of $1,3$-difluorobenzene indeed remains more populated after 300–$500\,fs$ (probability ~ 0.6) than the $\tilde{B}$ state of $1,2$-, $1,4$-difluorobenzene and F-Bz$^+$ (probability ~ 0.15, ~ 0.25 and ~ 0.08, respectively). For the $1,2,3$-isomer, the $\tilde{B}$ state stays populated even more than for the $1,3$-difluorobenzene cation (probability 0.9–0.95).[163] Similar results are obtained for broadband excitation of the $\tilde{B}-\tilde{C}-\tilde{D}$ electronic states.[161] This confirms that the $1,2,3$-trifluorobenzene cation is an emissive species, and also underlines the similarity of the $1,3$-isomer with the systems with three or more fluorine atoms (in agreement with the observations.[131]) It demonstrates that the internal conversion mechanism considered here, namely multiple CoIns involving one of the σ-type electronic states of benzene and its fluoro derivatives, is of key importance to the fluorescence dynamics in this family of compounds.

4. Concluding Remarks

In this chapter we have emphasized the importance of more than two strongly coupled electronic states, with multiply intersecting PES. The calculations relied on the MMVC scheme, supplemented by suitable

27. G. Hirsch, R.J. Buenker and C. Petrongolo, *Mol. Phys.* **70**, 835 (1990).
28. A. Thiel and H. Köppel, *J. Chem. Phys.* **110**, 9371 (1999).
29. I.B. Bersuker, N.B. Balabanov, D. Pekker and J.E. Boggs, *J. Chem. Phys.* **117**, 10478 (2002).
30. C.R. Evenhuis and M.A. Collins, *J. Chem. Phys.* **121**, 2525 (2004).
31. M. Döscher, H. Köppel and P.G. Szalay, *J. Chem. Phys.* **117**, 2645 (2002).
32. H. Köppel, M. Döscher, I. Bâldea, H.D. Meyer and P.G. Szalay, *J. Chem. Phys.* **117**, 2657 (2002).
33. S. Faraji, H. Köppel, W. Eisfeld and S. Mahapatra, *Chem. Phys.* **347**, 110 (2008).
34. H.D. Meyer, U. Manthe and L.S. Cederbaum, *Chem. Phys. Lett.* **165**, 73 (1990).
35. U. Manthe, H.D. Meyer and L.S. Cederbaum, *J. Chem. Phys.* **97**, 3199 (1992).
36. G.A. Worth, H.D. Meyer and L.S. Cederbaum, *J. Chem. Phys.* **105**, 4412 (1996).
37. M. Beck, A. Jäckle, G.A. Worth and H.D. Meyer, *Phys. Rep.* **324**, 1 (2000).
38. H.D. Meyer and G.A. Worth, *Theor. Chem. Acc.* **109**, 251 (2003).
39. H.D. Meyer, F. Gatti and G.A. Worth, (Eds.), *Multidimensional Quantum Dynamics: MCTDH Theory and Applications* (Wiley-VCH, Weinheim, 2009).
40. G.A. Worth, H.D. Meyer and L.S. Cederbaum, *J. Chem. Phys.* **109**, 3518 (1998).
41. H. Köppel and I. Bâldea, *J. Chem. Phys.* **124**, 064101 (2006).
42. U. Manthe, H.D. Meyer and L.S. Cederbaum, *J. Chem. Phys.* **97**, 9062 (1992).
43. V. Engel, *Chem. Phys. Lett.* **189**, 76 (1992).
44. A.J. Jerri, *The Gibbs Phenomenon in Fourier Analysis, Splines and Wavelet Approximations* (Kluwer, New York, 1998).
45. A. Raab, G.A. Worth, H.D. Meyer and L.S. Cederbaum, *J. Chem. Phys.* **110**, 936 (1999).
46. L.E. Snyder, D. Buhl, B. Zuckerman and P. Palmer, *Phys. Rev. Lett.* **22**, 679 (1969).
47. J.G. Calvert and J.N. Pitts, in *Photochemistry* (Wiley, New York, 1966).
48. M.B. Robin, in *Higher Excited States of Polyatomic Molecules: Vol. III* (Academic Press, New York, 1985).
49. J.L. Whitten and M.J. Hackmeyer, *J. Chem. Phys.* **51**, 5584 (1969).
50. R.J. Buenker and S.D. Peyerimhoff, *J. Chem. Phys.* **53**, 1368 (1970).
51. S.D. Peyerimhoff, R.J. Buenker, W.K. Kammer and H. Hsu, *Chem. Phys. Lett.* **8**, 129 (1971).
52. L.B. Harding and W.A. Goddard, *J. Am. Chem. Soc.* **99**, 677 (1977).
53. G. Fitzgerald and H.F. Schaefer, *J. Chem. Phys.* **83**, 1162 (1985).
54. M.R.J. Hachey, P.J. Bruna and F. Grein, *J. Phys. Chem.* **99**, 8050 (1995).
55. M. Merchan and B.O. Roos, *Theor. Chem. Acc.* **92**, 227 (1995).
56. S.R. Gwaltney and R.J. Barlett, *Chem. Phys. Lett.* **241**, 26 (1995).

dominate after several *ps*. This behavior is expected to suppress fluorescence, in accord with the experimental results. Only for the 1, 2-isomer the situation is somewhat less clear. However, other modes, not included in the present treatment, may further enhance the $\tilde{X} + \tilde{A}$ populations and thus be consistent with the absence of fluorescence in the 1, 2-isomer.

Additional evidence comes from the consideration of the dipole transition matrix elements. It is well known that the transition from the σ-type state of the benzene cation to the ground state is dipole forbidden.[31, 147] For the fluoro derivatives the molecular symmetry is reduced and the selection rules are relaxed.[148] One finds that there is always one component of the σ-type ($\tilde{C} + \tilde{D}$) states that has a finite dipole matrix element for transitions to one component of the lower π-type ($\tilde{X} + \tilde{A}$) states (at least for the C_{2v} molecular point group). Nevertheless, EOM-CCSD/TZ2P calculations clearly show that the corresponding transition dipole matrix elements are smaller by 2–3 orders of magnitude than those for the $\tilde{B}-(\tilde{X} + \tilde{A})$ transition, corresponding to the dipole-allowed transition in the case of Bz$^+$.[148] Thus, the π-type electronic state is the 'emitting' state also for the fluorobenzene cations. Its combined $\tilde{X} + \tilde{A}$ oscillator strength is almost the same for all five systems. Comparing again the various populations of Fig. 11 we find that the $\tilde{B}$ state of 1, 3-difluorobenzene indeed remains more populated after 300–500 *fs* (probability $\sim$0.6) than the $\tilde{B}$ state of 1, 2-, 1, 4-difluorobenzene and F-Bz$^+$ (probability $\sim$0.15, $\sim$0.25 and $\sim$0.08, respectively). For the 1, 2, 3-isomer, the $\tilde{B}$ state stays populated even more than for the 1, 3-difluorobenzene cation (probability 0.9–0.95).[163] Similar results are obtained for broadband excitation of the $\tilde{B}-\tilde{C}-\tilde{D}$ electronic states.[161] This confirms that the 1, 2, 3-trifluorobenzene cation is an emissive species, and also underlines the similarity of the 1, 3-isomer with the systems with three or more fluorine atoms (in agreement with the observations.[131]) It demonstrates that the internal conversion mechanism considered here, namely multiple CoIns involving one of the σ-type electronic states of benzene and its fluoro derivatives, is of key importance to the fluorescence dynamics in this family of compounds.

4. Concluding Remarks

In this chapter we have emphasized the importance of more than two strongly coupled electronic states, with multiply intersecting PES. The calculations relied on the MMVC scheme, supplemented by suitable extensions such as the concept of regularized diabatic states. A similar account for Jahn–Teller systems with degenerate electronic states and their mutual interactions (also with nondegenerate ones) has been given recently.[15] Therefore we focused here on Abelian point groups with only nondegenerate electronic states. Among the most important findings for the systems covered, formaldehyde, pyrrole and fluorinated benzene radical cations, we summarize as follows.

In formaldehyde, the strong interactions of the $\pi - \pi^*$ valence excited state with nearby Rydberg states lead to irregularities in the VUV absorption spectrum and the disappearance of clear signatures of this state in the intensity distribution. However, the time-dependent population of this state stays remarkably large following photoexcitation: while the system undergoes a femtosecond transition to nearby Rydberg states, the population is "transferred back" to the initial state after a somewhat longer time. In pyrrole, the (partly preliminary) calculations reveal a mechanism for populating the S_1 $\pi - \sigma^*$ excited state which is considered important for its photochemistry, but not optically bright due to dipole selection rules. Rather, the strongly dipole-allowed transition to the higher-energy $\pi - \pi^*$ state is followed by a sub-ps sequence of S_4-S_1 transitions with a corresponding, though incomplete, transfer of population. In the mono-, di- and trifluoro benzene cations, the multistate interactions are further characterized by a systematic energetic increase of one set (σ-type) of states which leads to a corresponding increase of the CoIn between the different sets. This gradually weakens the multistate interactions for increasing fluorination and turns out essential for understanding the emission properties of the whole family of compounds.

Beyond the individual examples treated, the above phenomena are considered important features of multistate interactions. Further developments could consist in considering their effects also for photochemical processes. While in its full generality the quantal treatment of the problem will face considerable difficulties, relevant simpler cases may consist in the photochemical rearrangement taking place only in the lowest of the interacting electronic states. This may be a promising line of future work.

Acknowledgments

The authors are indebted to H. Lischka, Th. Müller, M. Vazdar and M. Eckert-Maksic for fruitful collaboration and providing the *ab initio* data on pyrrole. They are grateful to E. Gindensperger for stimulating cooperation on the monofluoro benzene cation. S. Gómez-Carrasco and S. Faraji acknowledge financial support by the Alexander von Humboldt Foundation and the Deutsche Forschungsgemeinschaft through Graduiertenkolleg 850.

References

1. M. Born and R. Oppenheimer, *Ann. Physik* **84**, 457 (1927).
2. M. Born and K. Huang, *Dynamical Theory of Crystal Lattices* (Oxford University Press, New York, 1954).
3. I.B. Bersuker, *The Jahn–Teller Effect* (University Press, Cambridge, 2006).
4. H. Köppel, W. Domcke and L.S. Cederbaum, *Adv. Chem. Phys.* **57**, 59 (1984).
5. M. Klessinger and J. Michl, *Excited States and Photochemistry of Organic Molecules* (VCH Publishers, New York, 1995).
6. T.A. Barckholtz and T. Miller, *Int. Rev. Phys. Chem.* **17**, 435 (1998).
7. D.R. Yarkony, *Rev. Mod. Phys.* **68**, 985 (1996).
8. D.R. Yarkony, *Acc. Chem. Res.* **31**, 511 (1998).
9. F. Bernardi, M. Olivucci and M. Robb, *Chem. Soc. Rev.* **25**, 321 (1996).
10. M. Robb, F. Bernardi and M. Olivucci, *Pure Appl. Chem.* **67**, 783 (1995).
11. W. Domcke, D.R. Yarkony and H. Köppel (Eds.), *Conical Intersections: Electronic Structure, Dynamics and Spectroscopy* (Word Scientific, Singapore, 2004).
12. G.A. Worth and L.S. Cederbaum, *Annu. Rev. Phys. Chem.* **55**, 127 (2004).
13. W. Domcke and G. Stock, *Adv. Chem. Phys.* **100**, 1 (1997).
14. D.R. Yarkony, *J. Phys. Chem. A* **105**, 2642 (2001).
15. H. Köppel, L.S. Cederbaum and S. Mahapatra, in *Handbook of High-resolution Spectroscopy* (John Wiley and Sons, Ltd., 2011).
16. H. Köppel, J. Gronki and S. Mahapatra, *J. Chem. Phys.* **115**, 2377 (2001).
17. H. Köppel and B. Schubert, *Mol. Phys.* **104**, 1069 (2006).
18. M. Baer, *Chem. Phys. Lett.* **35**, 112 (1975).
19. F.T. Smith, *Phys. Rev.* **179**, 111 (1969).
20. T.F.O. Malley, *Adv. At. Mol. Phys.* **7**, 223 (1971).
21. T. Pacher, L.S. Cederbaum and H. Köppel, *Adv. Chem. Phys.* **84**, 293 (1993).
22. F. Duschinsky, *Acta Physicochim. URSS* **7**, 551 (1937).
23. C.A. Mead and D.G. Truhlar, *J. Chem. Phys.* **77**, 6090 (1982).
24. M.S. Schuurman and D.R. Yarkony, Conical intersections in electron photodetachment spectroscopy: Theory and applications, in W. Domcke, D.R. Yarkony and H. Köppel (Eds.), *Conical Intersections: Theory, Computation and Experiment* (Word Scientific, Singapore, 2011).
25. H. Köppel, Jahn–Teller and Pseudo-Jahn–Teller intersections: Spectroscopy and vibronic dynamics, in Ref. 11, pp. 429–472.
26. H. Werner and W. Meyer, *J. Chem. Phys.* **74**, 5802 (1981).

27. G. Hirsch, R.J. Buenker and C. Petrongolo, *Mol. Phys.* **70**, 835 (1990).
28. A. Thiel and H. Köppel, *J. Chem. Phys.* **110**, 9371 (1999).
29. I.B. Bersuker, N.B. Balabanov, D. Pekker and J.E. Boggs, *J. Chem. Phys.* **117**, 10478 (2002).
30. C.R. Evenhuis and M.A. Collins, *J. Chem. Phys.* **121**, 2525 (2004).
31. M. Döscher, H. Köppel and P.G. Szalay, *J. Chem. Phys.* **117**, 2645 (2002).
32. H. Köppel, M. Döscher, I. Bâldea, H.D. Meyer and P.G. Szalay, *J. Chem. Phys.* **117**, 2657 (2002).
33. S. Faraji, H. Köppel, W. Eisfeld and S. Mahapatra, *Chem. Phys.* **347**, 110 (2008).
34. H.D. Meyer, U. Manthe and L.S. Cederbaum, *Chem. Phys. Lett.* **165**, 73 (1990).
35. U. Manthe, H.D. Meyer and L.S. Cederbaum, *J. Chem. Phys.* **97**, 3199 (1992).
36. G.A. Worth, H.D. Meyer and L.S. Cederbaum, *J. Chem. Phys.* **105**, 4412 (1996).
37. M. Beck, A. Jäckle, G.A. Worth and H.D. Meyer, *Phys. Rep.* **324**, 1 (2000).
38. H.D. Meyer and G.A. Worth, *Theor. Chem. Acc.* **109**, 251 (2003).
39. H.D. Meyer, F. Gatti and G.A. Worth, (Eds.), *Multidimensional Quantum Dynamics: MCTDH Theory and Applications* (Wiley-VCH, Weinheim, 2009).
40. G.A. Worth, H.D. Meyer and L.S. Cederbaum, *J. Chem. Phys.* **109**, 3518 (1998).
41. H. Köppel and I. Bâldea, *J. Chem. Phys.* **124**, 064101 (2006).
42. U. Manthe, H.D. Meyer and L.S. Cederbaum, *J. Chem. Phys.* **97**, 9062 (1992).
43. V. Engel, *Chem. Phys. Lett.* **189**, 76 (1992).
44. A.J. Jerri, *The Gibbs Phenomenon in Fourier Analysis, Splines and Wavelet Approximations* (Kluwer, New York, 1998).
45. A. Raab, G.A. Worth, H.D. Meyer and L.S. Cederbaum, *J. Chem. Phys.* **110**, 936 (1999).
46. L.E. Snyder, D. Buhl, B. Zuckerman and P. Palmer, *Phys. Rev. Lett.* **22**, 679 (1969).
47. J.G. Calvert and J.N. Pitts, in *Photochemistry* (Wiley, New York, 1966).
48. M.B. Robin, in *Higher Excited States of Polyatomic Molecules: Vol. III* (Academic Press, New York, 1985).
49. J.L. Whitten and M.J. Hackmeyer, *J. Chem. Phys.* **51**, 5584 (1969).
50. R.J. Buenker and S.D. Peyerimhoff, *J. Chem. Phys.* **53**, 1368 (1970).
51. S.D. Peyerimhoff, R.J. Buenker, W.K. Kammer and H. Hsu, *Chem. Phys. Lett.* **8**, 129 (1971).
52. L.B. Harding and W.A. Goddard, *J. Am. Chem. Soc.* **99**, 677 (1977).
53. G. Fitzgerald and H.F. Schaefer, *J. Chem. Phys.* **83**, 1162 (1985).
54. M.R.J. Hachey, P.J. Bruna and F. Grein, *J. Phys. Chem.* **99**, 8050 (1995).
55. M. Merchan and B.O. Roos, *Theor. Chem. Acc.* **92**, 227 (1995).
56. S.R. Gwaltney and R.J. Barlett, *Chem. Phys. Lett.* **241**, 26 (1995).

57. F. Grein and M.R.J. Hachey, *Int. J. Quant. Chem: Quant. Chem. Symp.* **30**, 1661 (1996).
58. M. Perić, F. Grein and M.R.J. Hachey, *J. Chem. Phys.* **113**, 9011 (2000).
59. T. Müller and H. Lischka, *Theor. Chem. Acc.* **106**, 369 (2001).
60. M. Dallos, T. Müller, H. Lischka and R. Shepard, *J. Chem. Phys.* **114**, 746 (2001).
61. M.R.J. Hachey, P.J. Bruna and F. Grein, *J. Mol. Spec.* **176**, 375 (1996).
62. J.E. Metall, E.P. Gentiu, M. Krauss and D. Neumann, *J. Chem. Phys.* **55**, 5471 (1971).
63. D.C. Moule and A.D. Walsh, *Chem. Rev.* **75**, 67 (1975).
64. C.R. Lessard and D.C. Moule, *J. Chem. Phys.* **66**, 3908 (1977).
65. C.B. Moore and J.C. Weisshaar, *Annu. Rev. Phys. Chem.* **34**, 525 (1983).
66. D.J. Clouthier and D.A. Ramsay, *Annu. Rev. Phys. Chem.* **34**, 31 (1983).
67. D.J. Clouthier and D.C. Moule, *Topics in Current Chemistry* (Springer-Verlag, Berlin; Vol. 150, p. 167, 1989).
68. W.C. Price, *J. Chem. Phys.* **3**, 256 (1935).
69. P. Brint, J.P. Connerade, C. Mayhew and K. Sommer, *J. Chem. Soc. Faraday Trans. 2* **81**, 1643 (1985).
70. M. Suto, X. Wang and L.C. Lee, *J. Chem. Phys.* **85**, 4228 (1986).
71. M.J. Weiss, C.E. Kuyat and S.J. Mielczraek, *J. Chem. Phys.* **54**, 4147 (1971).
72. A.J. Chutjian, *J. Chem. Phys.* **61**, 4279 (1974).
73. S. Taylor, D.G. Wilden and J. Comer, *J. Chem. Phys.* **70**, 291 (1982).
74. S.R. Langhoff, S.T. Elbert, C.F. Jackels and E.R. Davidson, *Chem. Phys. Lett.* **29**, 247 (1974).
75. B. Niu, D.A. Shirley, Y. Bai and E. Daymo, *Chem. Phys. Lett.* **201**, 212 (1993).
76. R. Schinke, *J. Chem. Phys.* **84**, 1487 (1986).
77. Y.-T. Chang and W.H. Miller, *J. Phys. Chem.* **94**, 5884 (1990).
78. X. Zhang, J.L. Rheinecker and J.M. Bowman, *J. Phys. Chem.* **122**, 114313 (2005).
79. H.M. Yin, S.H. Kable, X. Zhang and J.M. Bowman, *Science* **311**, 1443 (2006).
80. J.D. Farnum, X. Zhang and J.M. Bowman, *J. Chem. Phys.* **126**, 134305 (2007).
81. M. Araujo, B. Lasorne, M.J. Bearpark and M.A. Robb, *J. Phys. Chem. A* **112**, 7489 (2008).
82. M. Araujo, B. Lasorne, A.L. Magalhaes, G.A. Worth, M.J. Bearpark and M.A. Robb, *J. Chem. Phys.* **131**, 144301 (2009).
83. H.-M. Yin, S.J. Rowling, A. Büll and S.H. Kable, *J. Chem. Phys.* **127**, 064302 (2007).
84. P. Zhang, S. Maeda, K. Morokuma and B.J. Braams, *J. Chem. Phys.* **130**, 114304 (2009).
85. B.C. Shepler, E. Epifanovsky, P. Zhang, J.M. Bowman, A.I. Krylov and K. Morokuma, *J. Phys. Chem. A* **112**, 13267 (2008).

86. K.L. Carleton, T.J. Butenhoff and C.B. Moore, *J. Chem. Phys.* **93**, 3907 (1990).
87. A.C. Terentis and S.H. Kable, *Chem. Phys. Lett.* **258**, 626 (1996).
88. W.S. Hopkins, H.-P. Loock, B. Cronin, M.G.D. Nix, A.L. Devine, R.N. Dixon and M.N.R. Ashfold, *J. Chem. Phys.* **127**, 064301 (2007).
89. M.-C. Chuang, M.F. Foltz and C.B. Moore, *J. Chem. Phys.* **87**, 3855 (1987).
90. L.R. Valachovic, M.F. Tuchler, M. Dulligan, T.D. Georget, M. Zyrianov, A. Kolessov, H. Reisler and C. Wittig, *J. Chem. Phys.* **112**, 2752 (2000).
91. J.B. Simonsen, N. Rusteika, M.S. Johnson and T. Solling, *Phys. Chem. Chem. Phys.* **10**, 674 (2008).
92. B.J. Finlayson-Pitts and J.N. Pitts, in *Atmospheric Chemistry* (Wiley, New York, 1986).
93. S. Gómez-Carrasco, Th. Müller and H. Köppel, *J. Phys. Chem. A*, **114**, 11436 (2010).
94. MOLPRO is a package of ab initio programs written by H.-J. Werner and P.J. Knowles, with contributions from R.D. Amos, A. Berning, D.L. Cooper, M.J.O. Deegan, A.J. Dobbyn, F. Eckert, C. Hampel, G. Hetzer, T. Leininger, R. Lindh, A.W. Lloyd, W. Meyer, M.E. Mura, A. Nicklaß, P. Palmieri, K. Peterson, R. Pitzer, P. Pulay, G. Rauhut, M. Schütz, H. Stoll, A.J. Stone and T. Thorsteinsson.
95. T.H. Dunning, Jr., *J. Chem. Phys.* **90**, 1007 (1989).
96. K. Takagi and T. Oka, *J. Phys. Soc. Jpn.* **18**, 1174 (1977).
97. I.C.W.M.H. Palmer and M.F. Guest, *Chem. Phys.* **238**, 179 (1998).
98. D.A. Blank, S.W. North and Y.T. Lee, *Chem. Phys.* **187**, 35 (1994).
99. B.O. Roos, P. Malmqvist, V. Molina, L. Serrano-Andres and M. Merchan, *J. Chem. Phys.* **116**, 7526 (2002).
100. O. Christiansen, J. Gauss, J.F. Stanson and P. Jørgensen, *J. Chem. Phys.* **111**, 525 (1999).
101. A.L. Sobolewski, W. Domcke, C. Dedonder and C. Jouvet, *Phys. Chem. Chem. Phys.* **4**, 1093 (2002).
102. P. Celani and H.J. Werner, *J. Chem. Phys.* **119**, 5044 (2003).
103. H. Lippert, H.H. Ritze, I.V. Hertel and W. Radloff, *Chem. Phys. Chem.* **5**, 1423 (2004).
104. J. Wei, J. Riedel, A. Kuczmann, F. Renth and F. Temps, *J. Chem. Soc. Faraday Discuss.* **127**, 267 (2004).
105. V. Vallet, Z. Lan, S. Mahapatra, A.L. Sobolewski and W. Domcke, *J. Chem. Phys.* **123**, 144307 (2005).
106. M. Barbatti, M. Vazdar, A.J.A. Aquino, M. Eckert-Maksic and H. Lischka, *J. Chem. Phys.* **125**, 164323 (2006).
107. B. Cronin, M.G.D. Nix, R.H. Qadiri, M.N.R. Ashfold, *Phys. Chem. Chem. Phys.* **6**, 5031 (2004).
108. A. Broo, *J. Phys. Chem. A* **102**, 526 (1998).
109. R.A. Jones, E.C. Taylor and A. Weissberger, *The Chemistry of Heterocyclic Compounds* (Wiley, New York, 1990).
110. A.L. Sobolewski and W. Domcke, *Chem. Phys.* **259**, 181 (2000).

111. A.L. Sobolewski and W. Domcke, *Chem. Phys. Lett.* **321**, 479 (2000).
112. V. Vallet, Z. Lan, S. Mahapatra, A.L. Sobolewski and W. Domcke, *J. Chem. Soc. Faraday Discuss.* **127**, 283 (2004).
113. Z. Lan, A. Dupays, V. Vallet, S. Mahapatra, A.L. Sobolewski and W. Domcke, *J. Photochem. A* **190**, 177 (2007).
114. Z. Lan and W. Domcke, *Chem. Phys.* **350**, 125 (2008).
115. L. Serrano-Andrés, M. Merchán, I. Nebot-Gil, B.O. Roos and M. Fülscher, *J. Am. Chem. Soc.* **115**, 6184 (1993).
116. S. Faraji, M. Vazdar, S. Reddy, M. Eckert-Maksic, H. Lischka and H. Köppel, *J. Chem. Phys.*, submitted (2011).
117. H. Köppel and W. Domcke, in *Encyclopedia of Computational Chemistry*, P. von Ragué Schleyer (Ed.) (Wiley, New York, 1998), p. 3166.
118. E. Gindensperger, I. Bâldea, J. Franz and H. Köppel, *Chem. Phys.* **338**, 207 (2007).
119. H. Nakatsuji, O. Kitao and T. Yonezawa, *J. Chem. Phys.* **83**, 723 (1985).
120. H. Nakano, T. Tsuneda, T. Hashimoto and K. Hirao, *J. Chem. Phys.* **104**, 2312 (1996).
121. A.B. Trofimov and J. Schirmer, *Chem. Phys.* **214**, 153 (1996).
122. S. Faraji and H. Köppel, *J. Chem. Phys.* **129**, 074310 (2008).
123. T. Miller and V.E. Bondybey, in *Molecular Ions: Spectroscopy, Structure and Chemistry* (North-Holland Publ. Company, Amsterdam, 1983), p. 201.
124. J. Eiding, R. Schneider, W. Domcke, H. Köppel and W. von Niessen, *Chem. Phys. Lett.* **177**, 3 (1991).
125. H. Köppel, L.S. Cederbaum and W. Domcke, *J. Chem. Phys.* **89**, 2023 (1988).
126. K. Müller-Dethlefs and J.B. Peel, *J. Chem. Phys.* **111**, 10550 (1999).
127. M. Döscher and H. Köppel, *Chem. Phys.* **225**, 93 (1997).
128. P.M. Johnson, *J. Chem. Phys.* **117**, 9991 (2002).
129. P.M. Johnson, *J. Chem. Phys.* **117**, 10001 (2002).
130. B.E. Applegate and T.A. Miller, *J. Chem. Phys.* **117**, 10654 (2002).
131. M. Allan, J.P. Maier and O. Marthaler, *Chem. Phys.* **26**, 131 (1977).
132. C. Cossart-Magos, D. Cossart and S. Leach, *Mol. Phys.* **37**, 793 (1979).
133. C. Cossart-Magos, D. Cossart and S. Leach, *Chem. Phys.* **41**, 375 (1979).
134. V.E. Bondybey, T.J. Sears, J.H. English and T.A. Miller, *J. Chem. Phys.* **73**, 2063 (1980).
135. T.J. Sears, T.A. Miller and V.E. Bondybey, *J. Am. Chem. Soc.* **103**, 326 (1981).
136. C. Cossart-Magos, D. Cossart, S. Leach, J.P. Maier and L. Misev, *J. Chem. Phys.* **78**, 3673 (1983).
137. G. Dujardin and S. Leach, *J. Chem. Phys.* **79**, 658 (1983).
138. D. Klapstein, S. Leutwyler and J.P. Maier, *Mol. Phys.* **51**, 413 (1984).
139. D. Winkoun, D. Champoulard, G. Dujardin and S. Leach, *Can. J. Phys.* **62**, 1361 (1984).
140. K. Walter, K. Schern and U. Boesl, *J. Phys. Chem.* **95**, 1188 (1991).
141. R. Anand, J.E. LeClaire and P.M. Johnson, *J. Phys. Chem. A* **1999**, 2618 (1999).

142. C.H. Kwon, H.L. Kim and M.S. Kim, *J. Chem. Phys.* **116**, 10361 (2002).
143. I. Bâldea, J. Franz and H. Köppel, *J. Mol. Struct.* **838**, 94 (2007).
144. I. Bâldea, J. Franz, P. Szalay and H. Köppel, *Chem. Phys.* **329**, 65 (2006).
145. V.P. Vysotsky, G.E. Salnikov and L.N. Shchegoleva, *Int. J. Quantum Chem. S* **100**, 469 (2004).
146. J.P. Maier, *Kinetics of Ion-Molecule Reactions* (Plenum Press, New York, 1979).
147. O. Braitbart, E. Castellucci, G. Dujardin and S. Leach, *J. Phys. Chem.* **87**, 4799 (1983).
148. S. Faraji, H.D. Meyer and H. Köppel, *J. Chem. Phys.* **129**, 074311 (2008).
149. G. Bieri, L. Asbrink and W. von Niessen, *J. Electron Spectrosc. Relat. Phenom.* **23**, 281 (1981).
150. T.H. Dunning, *J. Chem. Phys.* **55**, 716 (1971).
151. P. Szalay, J.F. Stanton and R.J. Bartlett, *Chem. Phys. Lett.* **193**, 573 (1992).
152. O. Christiansen, J. Stanton and J. Gauss, *J. Chem. Phys.* **108**, 3987 (1998).
153. J.F. Stanton, J. Gauss, J.D. Watts *et al.*, ACES II Mainz-Austin-Budapest version; integral packages: MOLECULE (J. Almlöf and P.R. Taylor); PROPS (P.R. Taylor); ABACUS (T. Helgaker, H.J. Aa, Jensen, P. Jørgensen and J. Olsen); current version see http://www.aces2.de
154. C.H. Kwon, H.L. Kim and M.S. Kim, *J. Chem. Phys.* **118**, 6327 (2003).
155. E.J.H.V. Schaick, H.J. Geise, F.C. Mijilhoff and G. Renes, *J. Mol. Struct.* **16**, 389 (1973).
156. G.J.D. Otter, J. Gerritsen and C. MacLean, *J. Mol. Struct.* **16**, 379 (1973).
157. A. Domenicano, G. Schultz and I. Harigittai, *J. Mol. Struct.* **78**, 97 (1982).
158. M. Nooijen and J.G. Snijders, *Int. J. Quantum Chem. S* **26**, 55 (1992).
159. J.F. Stanton and J. Gauss, *J. Chem. Phys.* **101**, 8938 (1994).
160. C. Brundle, M. Robin and N. Kuebler, *J. Am. Chem. Soc.* **94**, 1466 (1972).
161. S. Faraji, E. Gindensperger and H. Köppel, *Springer Series in Chemical Physics* **97**, 239 (2009).
162. T.S. Venkatesan, S. Mahapatra, H. Köppel and L.S. Cederbaum, *J. Mol. Struct.* **838**, 100 (2007).
163. S. Faraji and H. Köppel, *J. Chem. Phys.*, submitted (2011).
164. H. Köppel, *Chem. Phys. Lett.* **205**, 361 (1993).

Chapter 8

Conical Intersections Coupled to an Environment

Irene Burghardt[*,**], Keith H. Hughes[†], Rocco Martinazzo,[‡]
Hiroyuki Tamura[§], Etienne Gindensperger[¶],
Horst Köppel[‖] and Lorenz S. Cederbaum[‖]

1. Introduction 302
2. Multimode System-Bath Hamiltonian 306
   2.1. System-bath Hamiltonian 307
   2.2. Spectral densities 309
3. Effective Modes at a Conical Intersection 311
   3.1. Definition of effective-mode subspace 312
   3.2. Hierarchical structure of the bath 313
   3.3. Alternative sets of effective modes 314
   3.4. Generalization to more than two electronic states 316

---

[*]Département de Chimie, Ecole Normale Supérieure, 24 rue Lhomond, F–75231 Paris cedex 05, France. Present address: Institute of Physical and Theoretical Chemistry, Goethe University Frankfurt, Max-von-Laue-Str. 7, 60438 Frankfurt/Main, Germany
[†]School of Chemistry, Bangor University, Bangor, Gwynedd LL57 2UW, UK.
[‡]Department of Physical Chemistry and Electrochemistry, University of Milan, Via Golgi 19, 20122 Milan, Italy.
[§]Advanced Institute for Materials Research, Tohoku University, 2-1-1 Katahira Aobaku, Sendai, Japan.
[¶]Laboratoire de Chimie Quantique, Université de Strasbourg, CNRS UMR 7177, 4 rue Blaise Pascal, 67000 Strasbourg, France.
[‖]Theoretische Chemie, Universität Heidelberg, Im Neuenheimer Feld 229, 69120 Heidelberg, Germany.
[**]Corresponding author: burghardt@theochem.uni-frankfurt.de

4.	Effective-Plus-Residual Bath Models	317
	4.1. Model 1: Secondary bath coupled to primary effective modes	318
	4.2. Model 2: Mori-type chain	319
	4.3. Model 3: Mori-type chain with Markovian closure	320
	4.4. Dynamical evolution and moment conservation rules	321
	4.5. Example: Ultrafast photophysics of semiconductor polymer junctions	323
5.	Hierarchical Approximations of the Spectral Density	324
	5.1. Spectral densities in the transformed representation	325
	5.2. Tuning mode bath	326
	5.3. General correlated bath	330
6.	Discussion and Conclusions	331
	Acknowledgment	334
	Appendices	335
	A. Spectral Densities at a Conical Intersection	335
	B. Hierarchy of Spectral Densities for a Tuning-Mode Bath	337
	C. Hierarchy of Spectral Densities for a General Bath	341
	References	343

1. Introduction

Many photochemical processes of interest occur in an environment, e.g. in a solvent, cluster, or in biological environments like proteins. Since conical intersection (CoIn) topologies play a ubiquitous role in such processes, the question naturally arises as to what is the role of environmental effects — do these tend to catalyze or impede the nonadiabatic transfer processes at conical intersections? On the ultrafast time scale that is characteristic of processes at CoIn's, the environment is usually neither static nor rapidly fluctuating (Markovian limit). Instead, photoexcitation of the subsystem, or chromophore, entails a nonequilibrium response of the environment that is interleaved with the subsystem evolution, thus generating a dynamical evolution in the high-dimensional system-plus-environment space. This situation corresponds to a markedly non-Markovian case. Furthermore, since the topology and dynamics associated with conical intersections are of considerable complexity already for isolated polyatomic species, it is not clear *a priori* how to systematically construct a "reduced dynamics" to include environmental effects.

Some guidance is provided by existing theories of system-environment interactions. For example, solvent-induced energy gap fluctuations can be dominant and their effect can be formulated in terms of a corresponding macro-variable, here an energy gap coordinate.[1-4] This is the case, in particular, if chromophore-solvent interactions are electrostatic and translate to a Marcus-type collective coordinate.[3-6] Further, the high-dimensional nature of the environment is expected to lead to dissipation and decoherence. Given that quantum coherence plays a prominent role in the ultrafast dynamical processes at conical intersections, a rapid quenching of coherence by the environment could substantially modify the transfer efficiency.

Various examples illustrate that environmental effects can indeed be very pronounced. In the Green Fluorescent Protein (GFP) chromophore, the S_1-S_0 transfer rate increases dramatically in a solvent.[7,8] Likewise, in the Photoactive Yellow Protein (PYP) chromophore, the S_1-S_0 dynamics is known to depend in a highly sensitive fashion on the environment's electrostatic effects and hydrogen bonding properties.[9-11] In both cases, the conical intersections in question involve charge transfer, and one would therefore expect that the polar/polarizable solvent (or protein) environment couples strongly to the chromophore's excited-state evolution.[12] In other cases, though, experiments carried out in situations where the electrostatic coupling could be dominant, seem to suggest that intramolecular factors are decisive. For example, retinal and related photoswitches in solution are remarkably insensitive to the solvent's dielectric properties.[13,14] In addition, viscosity does not necessarily play an important role even in cases where large aromatic groups are displaced, as in the retinal analog described in Ref. 14. Furthermore, coherent vibrational dynamics has been observed experimentally for these systems, suggesting that environment-induced decoherence[15,16] sets in with a considerable time delay. The interpretation of many of these observations is currently still open.

Very similar issues arise for polyatomic species where a limited number of modes can often be identified as "system" modes that couple strongly to the electronic subsystem, while the remaining modes constitute an intramolecular bath.[17-20] Likewise, in spatially extended systems like semiconducting polymers[21-23] or carbon nanotubes,[24,25] a subset of modes can be identified that dominate the vibronic coupling (typically high-frequency carbon-carbon stretch modes) while other modes (in particular, low frequency torsional or breathing modes) act as a residual bath that can, however, significantly influence the dynamics.[26,27]

Over the past decade, various theoretical approaches and simulation techniques have been developed and applied that are able to account for envionmental effects at CoIn's. Broadly, two types of approaches can be distinguished. First, system-bath models, often used in conjunction with vibronic coupling model Hamiltonians.[28, 29] These models have the advantage that bath-induced effects like relaxation and decoherence can be systematically identified and expressed in terms of relevant parameters of a system-bath model. However, these models often rely on the validity of typical approximations like the Markovian limit (i.e. fast fluctuations). The second type of approach relates to an explicit treatment of the system-environment supermolecular system. Several realizations of this approach exist, ranging from multidimensional vibronic coupling models in conjunction with accurate quantum dynamical calculations,[19, 20, 30] to quantum mechanics–molecular mechanics (QM/MM) calculations combined with mixed quantum-classical dynamics simulations.[8–10, 14] Many of the relevant methods are based on hybrid approaches, both for the electronic structure side and for the dynamics side. Regarding non-adiabatic dynamics simulation techniques, methods range from explicit multidimensional quantum dynamics to *on-the-fly* techniques that are either trajectory-based[9, 10] or rely on time-evolving Gaussian wavepackets.[7, 8, 31, 32]

The picture that emerges from many of the experimental and theoretical studies is that (i) specific interactions between the chromophore and the local environment (first solvent shell, or nearest-neighbor layer in a chromophore-protein system) can play a dominant role, (ii) the dynamical nonequilibrium evolution of the environment cannot be neglected, (iii) conventional system-bath theories using the Markovian approximation are not generally appropriate, (iv) collective environmental modes, i.e. generalized solvent coordinates, could be useful in describing the environment's dynamical effects.

In this chapter, we focus on an *explicit but reduced-dimensional* representation of the environment. Our starting point is a multimode vibronic coupling model[17] which describes the coupling of a high-dimensional or infinite-dimensional bath to the electronic subsystem, assuming an interaction which is linear in the bath coordinates. Similarly to the spin-boson model,[2, 33] this system-bath model is of considerable generality. The model employed here is tailored to the conical intersection topology and includes both diagonal and off-diagonal couplings of the bath modes to the electronic subsystem, i.e. bath-induced fluctuations affect both the energy gap and the electronic coupling. Furthermore, since the environment's modes do not generally conform to the symmetry of the subsystem (chromophore), they may couple simultaneously diagonally and off-diagonally, giving rise to correlated fluctuations.[34] Using this model, we seek to extract a set of *effective environmental modes* which are generated by suitable coordinate transformations.[35,36] The modes in question are collective coordinates which are chosen in such a way that they describe (i) the effects of the system-environment interaction on short time scales, and (ii) the effect of the environment on the conical intersection topology, which is found to be closely connected to (i),[37] and further provide (iii) a systematic procedure by which chains of effective environmental modes are generated[26,38,39] which unravel the dynamics as a function of time. The effective modes in question can be interpreted as generalized Brownian oscillator modes, and the effective-mode chains are related to Mori chains known from statistical mechanics.[40–42] In keeping with this picture, the irreversible nature of the environment is maintained by adding a Markovian closure to a truncated chain representation.[43,44]

Following the analysis of Refs. 35–37, three effective modes can be defined which completely capture the short-time dynamics at a two-state conical intersection, and by extension, $n_{el}(n_{el}+1)/2$ such modes can be defined for n_{el} electronic states. A truncated Hamiltonian containing only these modes correctly reproduces the moments of the exact propagator up to the third order.[36] In the transformed Hamiltonian, the remaining modes are coupled bilinearly to the effective modes and among each other. If further transformations are introduced so as to cast the couplings within the residual subspace into a band-diagonal form, a Mori-type chain is obtained. Truncation of this chain at the nth order conserves the moments of the propagator up to the $(2n+3)$rd order.[46,47]

Building upon this analysis, it can be shown that starting from a given environmental spectral density, a *hierarchy of approximate spectral densities* can be constructed which are expressed in terms of chains comprising a limited number of effective modes.[43–45] The remaining modes are approximated in terms of the Markovian closure referred to above. This family of spectral densities are coarse-grained realizations of the true spectral density, which guarantee an accurate representation of the overall system-plus-bath dynamics up to increasing times. This development, which has recently been demonstrated for the simpler case of a spin-boson type system,[43,44] is addressed below in the more general case of a conical intersection.[34,45]

The system-bath dynamics at each level of approximation is treated by an explicit dynamics in the system subspace augmented by the effective environmental modes, in conjunction with a Markovian master equation[2,6,49,50] acting on the last member of the effective-mode chain. The effective-mode decomposition *de facto* leads to a modified system-bath partitioning. The picture developed here therefore interpolates between an explicit representation of the environment and a reduced-dynamics type description. The non-Markovian aspect of the dynamics is emphasized and cast in terms of a set of effective environmental modes. Importantly, the Mori-chain procedure generates a unique series of approximations which successively unravel the non-Markovian dynamics.

In Sec. 2, the multimode system-bath Hamiltonian is introduced that is the basis of the following discussion. Section 3 details the effective-mode transformations that are the key ingredient in obtaining hierarchical representations of the system-bath Hamiltonian, and Sec. 4 addresses several realizations of such system-bath models. In Sec. 5, we construct a series of approximate spectral densities which reproduce the effects of the environment over increasing time intervals. Finally, Sec. 6 summarizes and concludes.

2. Multimode System-Bath Hamiltonian

In the following, we consider a model Hamiltonian describing multimode processes at a conical intersection.[17] A system-bath perspective is adopted, where the "system" part contains the electronic subsystem, along with a certain number of nuclear modes which couple strongly to the electronic subsystem. The "bath" part is composed of a — potentially very large — number of nuclear modes which also couple to the electronic system. For certain geometries of the combined system and bath coordinates, a degeneracy arises which corresponds to a CoIn point (in two dimensions) and more generally to an $(N-2)$-dimensional intersection space (in N dimensions).[17,51-54] In general, we will assume that the system part by itself features a conical intersection. However, the analysis also includes situations where a conical intersection is generated by the interaction with the environment, along with the limiting case where all nuclear modes are part of the bath subspace.

For convenience, we focus below on the specific case of an electronic two-level system, but generalizations to more than two states are straightforward. An application to a three-level system will be addressed in Sec. 3.4 and Sec. 4.5.

2.1. System-bath Hamiltonian

In accordance with the above, we consider a system-bath type Hamiltonian,

$$\hat{H} = \hat{H}_S + \hat{H}_{SB} + \hat{H}_B \qquad (1)$$

with the system part[35–37,55]

$$\hat{H}_S = \hat{V}_\Delta + \sum_{i=1}^{N_S} \frac{\omega_{S,i}}{2}(\hat{p}_{S,i}^2 + \hat{x}_{S,i}^2)\hat{1} + \hat{V}_S(\hat{\boldsymbol{x}}_S), \qquad (2)$$

where $\hat{V}_\Delta = -\Delta\,\hat{\sigma}_z$ gives the electronic splitting, with $\hat{\sigma}_z = |1\rangle\langle 1| - |2\rangle\langle 2|$ the operator representation of the Pauli matrix, and $\hat{p}_i = (\hbar/i)\,\partial/\partial x_i$. We use mass and frequency weighted coordinates throughout. The potential part $\hat{V}_S$ represents anharmonicities in the subsystem Hamiltonian as well as the coupling of the system modes to the electronic subsystem and is of the form,

$$\hat{V}_S(\hat{\boldsymbol{x}}_S) = \hat{v}_1(\hat{x}_{S,1}, \ldots, \hat{x}_{S,N_S})\,\hat{1} + \hat{v}_z(\hat{x}_{S,1}, \ldots, \hat{x}_{S,N_S})\,\hat{\sigma}_z \\ + \hat{v}_x(\hat{x}_{S,1}, \ldots, \hat{x}_{S,N_S})\,\hat{\sigma}_x, \qquad (3)$$

where $\hat{\sigma}_x = |1\rangle\langle 2| + |2\rangle\langle 1|$. This form of the potential, in conjunction with the diagonal form of the kinetic energy, corresponds to a (quasi-)diabatic representation.[17,53,54,56]

A particular instance is given by a linearized form at the CoIn, i.e. the linear vibronic coupling (LVC) model,[17,53,54,56]

$$\hat{V}_S(\hat{\boldsymbol{x}}_S) = \sum_{i=1}^{N_S} \hat{V}_{S,i}(\hat{x}_{S,i}) \qquad (4)$$

with

$$\hat{V}_{S,i}(\hat{x}_{S,i}) = \kappa_{S,i}^{(+)}\,\hat{x}_{S,i}\,\hat{1} + \kappa_{S,i}^{(-)}\,\hat{x}_{S,i}\,\hat{\sigma}_z + \lambda_{S,i}\,\hat{x}_{S,i}\,\hat{\sigma}_x, \qquad (5)$$

or equivalently,

$$\hat{V}_{S,i}(\hat{x}_{S,i}) = \kappa_{S,i}^{(1)}\,\hat{x}_{S,i}\,\hat{\sigma}_{11} + \kappa_{S,i}^{(2)}\,\hat{x}_{S,i}\,\hat{\sigma}_{22} + \lambda_{S,i}\,\hat{x}_{S,i}\,(\hat{\sigma}_{12} + \hat{\sigma}_{21}), \qquad (6)$$

where $\hat{\sigma}_{nm} = |n\rangle\langle m|$ and $\kappa_{S,i}^{(\pm)} = 1/2(\kappa_{S,i}^{(1)} \pm \kappa_{S,i}^{(2)})$. By a linear expansion around the conical intersection, the LVC model accounts for the removable part of the nonadiabatic coupling.[57–60] This model can be augmented so as to yield a correct, global representation of the adiabatic surfaces away from the conical intersection geometry, by the construction of so-called regularized diabatic states.[58,61,62]

In Eqs. (5) and (6), the ith nuclear mode can couple both to $\hat{\sigma}_z$ (diagonally) and to $\hat{\sigma}_x$ (off-diagonally). If the system is characterized by symmetry — i.e. in the case of symmetry-allowed conical intersections[57–60] — the modes which couple diagonally (totally symmetric, tuning modes) are distinct from those which couple off-diagonally (non-totally symmetric, coupling modes). The basic, two-dimensional conical intersection topology is represented by the combination of one coupling mode and one tuning mode.

Further, the bath Hamiltonian $\hat{H}_B$ of Eq. (1) represents the zeroth-order Hamiltonian for N_B environmental modes,

$$\hat{H}_B = \sum_{i=1}^{N_B} \frac{\omega_{B,i}}{2} (\hat{p}_{B,i}^2 + \hat{x}_{B,i}^2)\hat{1}. \tag{7}$$

The system-bath interaction $\hat{H}_{SB}$ corresponds to the electronic-nuclear interaction involving all bath modes, which is of the same form as the linear vibronic coupling potential of Eq. (5),

$$\hat{H}_{SB} = \sum_{i=1}^{N_B} \left(\kappa_{B,i}^{(+)} \hat{x}_{B,i} \hat{1} + \kappa_{B,i}^{(-)} \hat{x}_{B,i} \hat{\sigma}_z + \lambda_{B,i} \hat{x}_{B,i} \hat{\sigma}_x \right). \tag{8}$$

Here, it is assumed that there is no direct vibration-vibration coupling among the N_S system modes and the N_B bath modes, but the coupling acts entirely *via* the electronic subsystem. The interaction Hamiltonian therefore corresponds to a generalized spin-boson model.[2,33,63] However, the interaction is more complex than that in the standard spin-boson case, since several subsystem operators are involved in Eq. (8) while the conventional spin-boson Hamiltonian only includes a system-bath interaction term proportional to $\hat{\sigma}_z$. In the case of an electronic n_{el}-level system, the most general form of $\hat{H}_{SB}$ involves $n_{el}(n_{el}+1)/2$ coupling terms.[35,36]

The first term on the r.h.s. of Eq. (8), involving the $\kappa_{B,i}^{(+)}$ parameters, is formally included even though it is not a system-bath coupling in a strict sense. Its effect is a mode-specific shift of the bath oscillators, independent of the electronic state. This term could therefore alternatively be included in $\hat{H}_B$.

2.2. Spectral densities

If the frequency distribution of the bath modes is dense, it is natural to characterize the influence of the bath on the subsystem in terms of a spectral density, or its discretized representation.[2,63] In the case where the bath modes couple only to one of the subsystem operators, for instance $\hat{H}_{SB} = \sum_{i=1}^{N_B} \kappa_{B,i}^{(-)} \hat{x}_{B,i} \hat{\sigma}_z$, the definition of the spectral density corresponds to the form known for the spin-boson Hamiltonian,

$$J(\omega) = \pi \sum_{i=1}^{N_B} \kappa_{B,i}^{(-)2} \delta(\omega - \omega_{B,i}). \quad (9)$$

This spectral density characterizes a bath that induces energy gap fluctuations in the subsystem.

For the more general form of the system-bath coupling Eq. (8) where the bath modes couple both diagonally ($\sim \hat{\sigma}_z$) and off-diagonally ($\sim \hat{\sigma}_x$), spectral densities are defined component-wise. Furthermore, if the bath modes couple simultaneously to several subsystem operators, we will refer to a *correlated bath*.[34] The subsystem variables then do not experience independent fluctuations, and this is reflected in the definition of the spectral densities which involve *cross-correlation* contributions. For example, the spectral density component,

$$J_{zx}(\omega) = \pi \sum_{i=1}^{N_B} \kappa_{B,i}^{(-)} \lambda_{B,i} \delta(\omega - \omega_{B,i}), \quad (10)$$

contains non-zero contributions from modes which couple simultaneously to $\hat{\sigma}_z$ and $\hat{\sigma}_x$.

There are various procedures to systematically construct the relevant spectral densities in the general case. One of these procedures expresses the bath-induced relaxation properties of the subsystem in the Heisenberg picture.[64] The spectral densities can then be defined as follows in terms of the bath-induced portion of the Heisenberg evolution, described by the operator $\hat{\mathbf{L}}_B$,[43,64]

$$J_{ij}(\omega) = \lim_{\epsilon \to 0^+} \text{Im}\, \hat{L}_{B,ij}(z) \Big|_{z=\omega+i\epsilon}, \quad (11)$$

where $\hat{L}_{B,ij}(z)$ determines the dissipative evolution of the subsystem Heisenberg operators in a Fourier/Laplace transformed representation. Equation (11) corresponds to the definition of the spectral density as the imaginary part of the dynamic susceptibility.[1,64] For a two-level system characterized by operators $\hat{\boldsymbol{\sigma}} = \{\hat{\sigma}_x, \hat{\sigma}_y, \hat{\sigma}_z\}$, we have for the Heisenberg evolution of the subsystem operators,

$$(z - \hat{\mathbf{L}}_B)\,\hat{\boldsymbol{\sigma}}(z) = \hat{\mathbf{L}}_S \hat{\boldsymbol{\sigma}}(z) + i\hat{\boldsymbol{\sigma}}_0. \tag{12}$$

A detailed description of the steps leading to Eqs. (11) and (12) is given in Appendix A and Ref. 34. The resulting spectral density matrix, following from the Hamiltonian Eqs. (1)–(8), is given as

$$\boldsymbol{J}(\omega) = \begin{pmatrix} J_{xx}(\omega) & J_{xy}(\omega) & J_{xz}(\omega) \\ J_{yx}(\omega) & J_{yy}(\omega) & J_{yz}(\omega) \\ J_{zx}(\omega) & J_{zy}(\omega) & J_{zz}(\omega) \end{pmatrix}$$

$$= \pi \sum_{n=1}^{N_B} \begin{pmatrix} \kappa_n^{(-)2} & i\kappa_n^{(-)}\kappa_n^{(+)} & -\kappa_n^{(-)}\lambda_n \\ -i\kappa_n^{(-)}\kappa_n^{(+)} & (\kappa_n^{(-)2} + \lambda_n^2) & i\lambda_n\kappa_n^{(+)} \\ -\kappa_n^{(-)}\lambda_n & -i\lambda_n\kappa_n^{(+)} & \lambda_n^2 \end{pmatrix} \delta(\omega - \omega_{B,n}). \tag{13}$$

As anticipated, cross-correlation contributions arise which provide a second-order coupling between the subsystem operators. Note that some of the cross-correlation contributions are imaginary; these involve the $\{\kappa_n^{(+)}\}$ couplings of Eq. (8) which are not genuine system-bath couplings as discussed above.

One of the key questions that we attempt to answer is the following: Given the spectral density (or spectral density matrix) of the environment, can one extract the relevant frequency components that determine the dynamical behavior of the subsystem as a function of time? In particular, we seek to determine those components that determine the short-time behavior which is of particular importance for the dynamics at conical intersections. Anticipating the results described in Sec. 5, it can be shown that a given spectral density can be approximated by a sequence of simpler spectral densities, which successively resolve the system-bath dynamics in time. Thus, a coarse-graining is introduced in the frequency domain, building upon the observation that the evolution on the shortest time scale will be determined by *few collective modes*. Even at the simplest level of approximation, which would reduce a highly structured, multipeaked spectral density to an analytical form constructed from few effective-mode frequencies, the short-time dynamics can be shown to be described *exactly*. Overall, a rigorous procedure can be formulated, within the restrictions of the LVC model, by which a high-dimensional environment can be reduced to an effective environment composed of few collective modes that are active on ultrafast time scales.

In order to construct the approximate spectral densities in question, we employ an effective-mode representation of the environment as detailed in the following (Secs. 3 and 4). This will in turn be used to derive a hierarchy of modes, and an associated hierarchy of approximate spectral densities (Sec. 5), which are based upon lower-dimensional approximants to the initial N_B-dimensional representation of the bath in Eq. (1).

3. Effective Modes at a Conical Intersection

The LVC model that is employed for the system-bath coupling of Eq. (8) allows one to introduce coordinate transformations by which a set of effective, or collective modes are extracted that act as generalized reaction coordinates for the dynamics. As shown in Refs. 35–37, $n_{\text{eff}} = n_{\text{el}}(n_{\text{el}} + 1)/2$ such coordinates can be defined for an electronic n_{el}-state system, in such a way that the short-time dynamics is completely described in terms of these effective coordinates. Thus, three effective modes are introduced for an electronic two-level system, six effective modes for a three-level system etc., for an arbitrary number of phonon modes that couple to the electronic subsystem according to the LVC model. In order to capture the dynamics on longer time scales, chains of such effective modes can be introduced.[26,39,46] These transformations, which are summarized below, will be shown to yield a unique perspective on the environmental effects at a conical intersection.

The effective-mode transformation is conceptually related to early work by Toyozawa and Inoue[65] on the identification of "interaction modes" in Jahn–Teller systems, and further, to work by O'Brien and others[66–69] on the construction of "cluster modes". Our recent results reported in Refs. 35–37 represent a generalization beyond the Jahn–Teller case, to generic conical intersection situations described by the LVC Hamiltonian Eq. (8) which requires consideration of *three* effective modes.

For convenience, the following discussion focuses on the case of an electronic two-level system. The generalization to more than two states is straightforward and is addressed in more detail in Sec. 3.4.

3.1. Definition of effective-mode subspace

Following the analysis of Refs. 35–37, we note that the bath modes produce cumulative effects by their coupling to the electronic subsystem. For an electronic two-level system, the interaction Hamiltonian Eq. (8) can be formally re-written in terms of a set of *three collective bath modes* $(\hat{X}_{B,+}, \hat{X}_{B,-}, \hat{X}_{B,\Lambda})$,

$$\hat{H}_{SB} = \hat{X}_{B,+}\hat{1} + \hat{X}_{B,-}\hat{\sigma}_z + \hat{X}_{B,\Lambda}\hat{\sigma}_x, \tag{14}$$

defined as

$$\hat{X}_{B,+} = \sum_{i=1}^{N_B} \kappa_{B,i}^{(+)} \hat{x}_{B,i}, \quad \hat{X}_{B,-} = \sum_{i=1}^{N_B} \kappa_{B,i}^{(-)} \hat{x}_{B,i}, \quad \hat{X}_{B,\Lambda} = \sum_{i=1}^{N_B} \lambda_{B,i} \hat{x}_{B,i}, \tag{15}$$

which reflect the collective shift $(\hat{X}_{B,+})$, tuning $(\hat{X}_{B,-})$, and coupling $(\hat{X}_{B,\Lambda})$ effects induced by the bath. These modes entirely define the environment's coupling to the electronic subsystem. The modes $(\hat{X}_{B,-}, \hat{X}_{B,\Lambda})$ span the projection of the branching plane[51,52,70] onto the environment subspace, i.e. they define the directions within the environment subspace along which the degeneracy at the conical intersection is lifted.

However, the modes of Eq. (15) are not of immediate use since they are not generally orthogonal on the space defined by the original coordinates $\{\hat{x}_{B,i}\}$. When orthogonalizing the modes of Eq. (15),[35–37] one obtains a set of coordinates $(\hat{X}_{B,1}, \hat{X}_{B,2}, \hat{X}_{B,3})$ in terms of which the dynamical problem can be reformulated. With these new, orthogonal modes, the most general form of the interaction Hamiltonian reads as follows,

$$\hat{H}_{SB} = \sum_{i=1}^{n_{\text{eff}}=3} \left(K_{B,i}^{(+)} \hat{X}_{B,i}\hat{1} + K_{B,i}^{(-)} \hat{X}_{B,i}\hat{\sigma}_z + \Lambda_{B,i} \hat{X}_{B,i}\hat{\sigma}_x \right). \tag{16}$$

This expression is formally the same as the one of the original system-bath coupling, Eq. (8), except that the system-bath interaction is entirely absorbed by the three effective modes. Depending on the orthogonalization procedure, different couplings can result, as discussed further in Sec. 3.3.

The introduction of the set of effective modes $(\hat{X}_{B,1}, \hat{X}_{B,2}, \hat{X}_{B,3})$ is the first step in defining an overall orthogonal transformation which leaves the subsystem coordinates $\{\hat{x}_S\}$ unaffected while transforming the bath coordinates,

$$(\hat{\boldsymbol{X}}_B \pm i\hat{\boldsymbol{P}}_B) = \boldsymbol{T}(\hat{\boldsymbol{x}}_B \pm i\hat{\boldsymbol{p}}_B). \tag{17}$$

As a result, one obtains the bath Hamiltonian in the following form,

$$\hat{H}_B = \sum_{i=1}^{N_B} \frac{\Omega_{B,i}}{2}(\hat{P}_{B,i}^2 + \hat{X}_{B,i}^2)\hat{1} + \sum_{i,j=1,j>i}^{N_B} d_{ij}\left(\hat{P}_{B,i}\hat{P}_{B,j} + \hat{X}_{B,i}\hat{X}_{B,j}\right)\hat{1}, \tag{18}$$

where bilinear coupling terms now appear in the bath subspace. The new frequencies $\Omega_{B,i}$ and couplings d_{ij} result from the coordinate transformation introduced above, such that $\Omega_{B,i} = \sum_{j=1}^{N_B} \omega_{B,j} t_{ji}^2$ and $d_{ij} = \sum_{k=1}^{N_B} \omega_{B,k} t_{ki} t_{kj}$, where t_{ji} are the elements of the transformation matrix $\boldsymbol{T}$.

The transformed bath Hamiltonian $\hat{H}_B$ of Eq. (18) and the transformed interaction part $\hat{H}_{SB}$ of Eq. (16) define the new system-bath Hamiltonian. The subsystem part, comprising the electronic subspace and possibly a subset of strongly coupled vibrational modes, has remained unchanged.

3.2. Hierarchical structure of the bath

In the new coordinates, the bath Hamiltonian takes a hierarchical form: The three effective modes $(\hat{X}_{B,1}, \hat{X}_{B,2}, \hat{X}_{B,3})$ couple directly to the electronic subsystem, while the remaining (residual) $(N_B - 3)$ bath modes couple in turn to the effective modes. This chain of interactions is illustrated in Fig. 1. The new bath Hamiltonian $\hat{H}_B$ of Eq. (18) can thus be split as follows:

$$\hat{H}_B = \hat{H}_B^{\text{eff}} + \hat{H}_B^{\text{eff-res}} + \hat{H}_B^{\text{res}}, \tag{19}$$

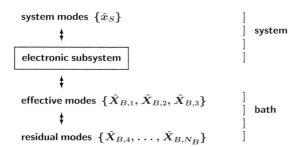

Fig. 1. Schematic illustration of the effective-mode construction (here, for an electronic two-level system), by which a subset of effective bath modes carry all vibronic coupling effects.

with the effective (eff) 3-mode bath portion

$$\hat{H}_B^{\text{eff}} = \sum_{i=1}^{3} \frac{\Omega_{B,i}}{2}(\hat{P}_{B,i}^2 + \hat{X}_{B,i}^2)\hat{1} + \sum_{i,j=1}^{3} d_{ij}\left(\hat{P}_{B,i}\hat{P}_{B,j} + \hat{X}_{B,i}\hat{X}_{B,j}\right)\hat{1}, \quad (20)$$

the effective-residual (eff-res) mode interaction

$$\hat{H}_B^{\text{eff-res}} = \sum_{i=1}^{3}\sum_{j=4}^{N_B} d_{ij}\left(\hat{P}_{B,i}\hat{P}_{B,j} + \hat{X}_{B,i}\hat{X}_{B,j}\right)\hat{1}, \quad (21)$$

and a definition analogous to Eq. (20) for the residual (res) Hamiltonian $\hat{H}_B^{\text{res}}$ comprising the $(N_B - 3)$ residual bath modes.

From this hierarchical structure, various approximation schemes emerge, which are discussed in detail in Sec. 4. As shown in Refs. 35–37 and 55, even the simplest level of approximation, by which all terms except for $\hat{H}_B^{\text{eff}}$ are disregarded,

$$\hat{H}' = \hat{H}_S + \hat{H}_{SB} + \hat{H}_B^{\text{eff}} \quad (22)$$

can be meaningful, since the effective modes entirely determine the short-time dynamical behavior. For example, it has been shown for a model of the butatriene cation comprising two strongly coupled system modes and 20 more weakly coupled intramolecular bath modes, that the dynamics over the first tens of femtoseconds can be described accurately by truncating the intramolecular bath at the level of the first three modes.[35,55] However, coherent artifacts are expected to appear beyond the shortest time scale, since the multimode nature of the bath has been disregarded.

3.3. *Alternative sets of effective modes*

The definition of the effective modes is not unique. The coordinate set $(\hat{X}_{B,1}, \hat{X}_{B,2}, \hat{X}_{B,3})$ is a member of a *manifold* of coordinate triples which are interrelated by orthogonal transformations within the effective-mode subspace.[37] Two choices are of particular relevance: (i) First, a definition of the new coordinates which eliminates the bilinear couplings $\{d_{ij}\}$ within the effective-mode subspace, and creates a diagonal form of the kinetic and potential energy in $\hat{H}_B^{\text{eff}}$.[36] (ii) Second, a definition leading to *topology-adapted* vectors, two of which span the branching plane.[37] These two choices will now be briefly discussed.

By the first choice,[35,36] the bilinear couplings of Eq. (20) are eliminated for the primary effective modes, i.e. $d_{ij} = 0$, $i,j = 1,\ldots,n_{\text{eff}}$. This form leads to mathematical simplicity in the effective-mode subspace, such that the truncated effective-mode Hamiltonian H' of Eq. (22), does not exhibit any bilinear coupling terms since $\hat{H}_B^{\text{eff}} = \sum_{i=1}^{3} \Omega_{B,i}/2(\hat{P}_{B,i}^2 + \hat{X}_{B,i}^2)\hat{1}$. As in the original Hamiltonian, no direct couplings occur between the effective bath modes, and all interactions are absorbed into the vibronic coupling.

The second choice is motivated by topological considerations and constrains two of the effective modes to lie in the branching plane.[37,46] Starting from the vectors $(\hat{X}_{B,+}, \hat{X}_{B,-}, \hat{X}_{B,\Lambda})$ of Eqs. (14) and (15), the orthogonalization procedure is carried out in such a way that the vectors $(\hat{X}_{B,-}, \hat{X}_{B,\Lambda})$ which lie in the branching plane are orthogonalized first, so as to yield the effective modes $(\hat{X}_{B,1}, \hat{X}_{B,2})$. Following this, the third mode $\hat{X}_{B,3}$ is constructed which lies in the intersection space. The diabatic branching plane modes generated from the LVC model are closely related to the gradient difference (**g**) and nonadiabatic coupling (**h**) vectors, and the $\hat{X}_{B,3}$ vector is in turn related to the average gradient (**s**) vector;[51,52,71,72] see Ref. 37 for a detailed discussion of this point. This topology-adapted construction has the advantage that the effect of the environment on the local CoIn topology can be immediately expressed in terms of the effective modes. In fact, the (**g**, **h**, **s**) vectors can be decomposed into system vs. environment contributions where the latter are calculated directly from the $(\hat{X}_{B,1}, \hat{X}_{B,2}, \hat{X}_{B,3})$ modes.

The two prescriptions described above for generating the $(\hat{X}_{B,1}, \hat{X}_{B,2}, \hat{X}_{B,3})$ effective modes are not compatible, i.e. the topological construction entails the presence of bilinear coupling terms. Independently of this choice — and other possible choices that could be motivated by physical considerations — the essence of the effective-mode construction remains unaffected: that is, all possible orthogonal effective-mode combinations guarantee that *the short-time dynamics is accurately captured by the primary effective modes, in the absence of the residual modes*. This is the case since the truncated Hamiltonian H' of Eq. (22) reproduces the first few moments (or cumulants) of the Hamiltonian exactly.[36] Section 4.4 gives further details on this aspect.

Finally, a comment is in order regarding the expansion point used for the effective-mode construction. The conventional choice made in the applications shown here, as well as those discussed in Refs. 27, 35, 37, 46 and 55, is the use of the Franck–Condon (FC) geometry as reference geometry.

This is consistent with the goal of describing the short-time dynamics accurately, and is expected to give a good representation of the dynamics at the conical intersection if the spatial separation between the FC and CoIn geometries is small. If this is not the case, but an initial condition for the dynamics can be defined close to the intersection (possibly extrapolated from the real dynamics taking one from the FC geometry to the CoIn), an appropriate expansion point could be a location on the CoIn seam of the subsystem. Unless the system and bath are represented by an LVC model, one then has to account for the fact that the intersection seam is curved,[73,74] such that the composition of the branching plane vectors varies as a function of the seam location. Following, e.g. Ref. 73, optimized locations on the seam can be identified, usually corresponding to extrema along the seam.

3.4. *Generalization to more than two electronic states*

A generalization of the effective-mode construction to three or more electronic states is straightforward, using a set of $n_{\text{eff}} = n_{\text{el}}(n_{\text{el}}+1)/2$ effective modes as mentioned above. In Refs. 27 and 38, we have thus employed effective mode transformations for a three-state LVC model, representing the ultrafast dissociation of an exciton state at a semiconductor polymer interface. Besides the exciton (XT) state, this model involves an interfacial charge transfer (CT) state and an intermediate (IS) state. While further comments on the model and dynamics are deferred to Sec. 4.5, we cite here the general form of the transformed Hamiltonian which generalizes the system-bath interaction Eq. (16) and otherwise exhibits the same structure as before,

$$\hat{H} = \hat{H}_S + \hat{H}_{SB} + \hat{H}_B, \tag{23}$$

$$\hat{H}_{SB} = \sum_{n,m>n}^{n_{\text{el}}=3} \sum_{i=1}^{n_{\text{eff}}=6} K_i^{nm} \hat{X}_{B,i} \left(|n\rangle\langle m| + |m\rangle\langle n| \right), \tag{24}$$

$$\hat{H}_B = \sum_{i=1}^{N_B} \frac{\Omega_{B,i}}{2}(\hat{P}_{B,i}^2 + \hat{X}_{B,i}^2)\hat{1} + \sum_{i,j=1,j>i}^{N_B} d_{ij}\left(\hat{P}_{B,i}\hat{P}_{B,j} + \hat{X}_{B,i}\hat{X}_{B,j}\right)\hat{1}. \tag{25}$$

In this case, the system Hamiltonian $\hat{H}_S$ is restricted to the electronic subsystem while all vibrational modes are included in the bath part, using

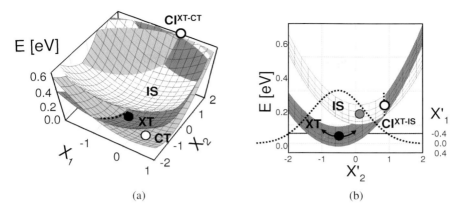

Fig. 2. For the three-state LVC model of Refs. 27 and 38, involving a photogenerated excitonic (XT) state, an interfacial charge transfer (CT) state, and an intermediate state (IS), two different branching plane projections of the coupled diabatic PESs are shown: (a) XT, CT, and IS PESs projected onto the XT-CT branching plane and (b) XT and IS PESs projected onto the XT-IS branching plane. In (b), the width of the initial wavepacket (localized at the FC point) and the trajectory of the wavepacket center are also indicated. The two branching space projections are associated with two different sets of effective modes, with respective $(\hat{X}_{B,1}, \hat{X}_{B,2})$ vectors spanning either of the two branching planes. Reprinted with permission from Ref. 38.

the LVC model. In Refs. 27 and 38, the effective-mode construction was chosen such that two of the six effective modes correspond to topology-adapted modes that span the branching plane of a two-state intersection between a selected pair of electronic states. Figure 2 illustrates two alternative choices involving different pairs of electronic states, i.e. XT/CT or XT/IS; these two choices are not compatible with each other due to the orthogonality constraint. Other construction schemes could be envisaged, e.g. one could choose modes that are adapted to the five-dimensional branching space of the three-state intersection[75,76] that occurs in the model (even though this intersection might not be directly involved in the dynamics). Further applications to three and more electronic states can be found in Ref. 77.

4. Effective-Plus-Residual Bath Models

The effective vs. residual bath partitioning is the key piece in the treatment of the electron-vibration coupling. Beyond this, various representations can be chosen for the residual bath subspace, which play an important role in practice in view of reducing the dimensionality of the system-bath problem. Three possible realizations are now outlined. These realizations correspond to different orthogonal transformations, resulting in different intra-chain couplings.

4.1. Model 1: Secondary bath coupled to primary effective modes

In this case, the bilinear coupling matrix $\{d_{ij}\}$ is diagonalized in the subspace of the residual bath modes $\{4, \ldots, N_B\}$ such that only couplings $\{d_{ij}\}$, between the primary effective modes $\hat{X}_{B,i}$, $i = 1, \ldots, 3$, and the secondary bath modes $\hat{X}_{B,i}$, $i = 4, \ldots, N_B$, occur, i.e.

$$\hat{H}_B = \sum_{i=1}^{N_B} \frac{\Omega_{B,i}}{2}(\hat{P}_{B,i}^2 + \hat{X}_{B,i}^2)\hat{1} + \sum_{i=1}^{3}\sum_{j=4}^{N_B} d_{ij}(\hat{P}_{B,i}\hat{P}_{B,j} + \hat{X}_{B,i}\hat{X}_{B,j})\hat{1}$$
$$\equiv (\hat{\boldsymbol{X}}_B + i\hat{\boldsymbol{P}}_B)^T \mathbf{d}\,(\hat{\boldsymbol{X}}_B - i\hat{\boldsymbol{P}}_B), \tag{26}$$

where the diagonal elements of the **d** matrix correspond to the transformed frequencies. The corresponding coupling pattern is illustrated in Fig. 3.

Within this model, the secondary modes act as an unstructured bath with respect to the primary effective modes. In the simplest case, their influence can be approximated in terms of the Markovian limit, such that

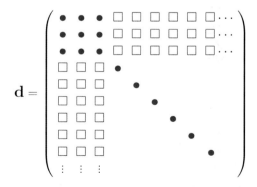

Fig. 3. Schematic illustration of the coupling pattern of Model 1: couplings within the effective-mode sub-block as well as diagonal terms (transformed frequencies $\Omega_{B,i}$) are indicated by bullets, while couplings between the effective-mode vs. residual-mode blocks are indicated by open squares.

the $\{d_{ij}\}$ conform to the spectral density

$$J_i(\omega) = \pi \sum_{j=1}^{N_B-3} d_{ij}^2 \delta(\omega - \omega_j) = \eta\omega \qquad (27)$$

with the friction coefficient η. The primary bath modes $\{\hat{X}_{B,i}\}$ then play the role of Brownian oscillator modes.[1]

This level of approximation can be appropriate if the coupling between the electronic subspace and the primary effective modes is stronger than the effective-residual bath coupling. This can be the case for typical CoIn situations, see our studies of Ref. 55.

4.2. Model 2: Mori-type chain

In this scheme, the bilinear coupling matrix is cast into a band-diagonal form[46,78]

$$\hat{H}_B = \hat{H}_B^{\text{eff}} + \hat{H}_B^{\text{eff-res}} + \sum_{l=1}^{n} \hat{H}_{B,\text{res}}^{(l)} \qquad (28)$$

with the effective-residual space coupling

$$\hat{H}_B^{\text{eff-res}} = \sum_{i=1}^{3} \sum_{j=4}^{i+3} d_{ij} \left(\hat{P}_{B,i}\hat{P}_{B,j} + \hat{X}_{B,i}\hat{X}_{B,j} \right) \hat{1} \qquad (29)$$

and the residual bath Hamiltonian

$$\hat{H}_{B,\text{res}}^{(l)} = \sum_{i=3l+1}^{3l+3} \frac{\Omega_{B,i}}{2} (\hat{P}_{B,i}^2 + \hat{X}_{B,i}^2)\hat{1}$$

$$+ \sum_{i=3l+1}^{3l+3} \sum_{j=i+1}^{i+3} d_{ij} \left(\hat{P}_{B,i}\hat{P}_{B,j} + \hat{X}_{B,i}\hat{X}_{B,j} \right) \hat{1}; \qquad (30)$$

see the schematic representation of Fig. 4. We have referred to this variant as a hierarchical electron-phonon (HEP) model.[46] This representation is closely related to a Mori chain representation.[40–42] A very similar approach was suggested in Ref. 47 where a block-diagonal form of the coupling matrix is constructed.

As more (triples of) modes are added in the chain construction, the dynamics is reproduced accurately over increasing intervals of time. The

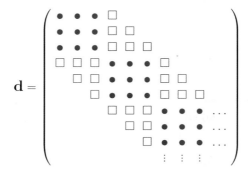

Fig. 4. Schematic illustration of the coupling pattern in Model 2, involving a band-diagonal structure of the **d** matrix. Couplings within each lth sub-block corresponding to effective-mode triples $(\hat{X}_{B,1+l}, \hat{X}_{B,2+l}, \hat{X}_{B,3+l})$ are indicated by bullets, while couplings between the sub-blocks are indicated by open squares.

underlying analysis in terms of Hamiltonian moments is briefly addressed below (Sec. 4.4). This model is most appropriate in cases where the bilinear couplings within the chain are non-negligible as compared with the vibronic coupling of $\hat{H}_{SB}$.

4.3. Model 3: Mori-type chain with Markovian closure

This variant uses the Mori-type construction of Model 2 but the hierarchy of modes is now terminated at a chosen order M, with a dissipative closure acting on the end of the chain (Fig. 5).[34,43–45] The bath Hamiltonian

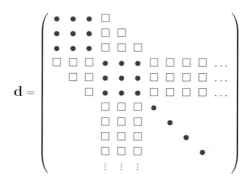

Fig. 5. Schematic illustration of the coupling pattern in Model 3. Here, the hierarchy of Model 2 is terminated at a chosen order M (here, $M = 2$ for illustration). The remaining modes couple to the Mth sub-block, and a Markovian approximation is subsequently introduced for these modes.

takes the form

$$\hat{H}_B = \hat{H}_B^{\text{eff}} + \hat{H}_B^{\text{eff-res}} + \sum_{l=1}^{M} \hat{H}_{B,\text{res}}^{(l)} + \hat{H}_{B,\text{diss}}^{(M)}, \quad (31)$$

where $\hat{H}_{\text{diss}}^{(M)}$ corresponds to a residual bath composed of modes $\{M+1, \ldots, N_B\}$ which are all coupled to the Mth mode of the chain,

$$\hat{H}_{\text{diss}}^{(M)} = \sum_{i=M+1}^{N_B} \frac{\Omega_{B,i}}{2}(\hat{P}_{B,i}^2 + \hat{X}_{B,i}^2)\hat{1} + \sum_{i=M+1}^{N_B} d_{M\,i}(\hat{P}_{B,M}\hat{P}_{B,i} + \hat{X}_{B,M}\hat{X}_{B,i})\hat{1} \quad (32)$$

The Brownian oscillator model of Sec. 4.1 represents a special case of this construction with $M = 1$. Again, the distribution of residual bath modes may be approximated in terms of an Ohmic spectral density, in which case the picture of a Mori-type chain with Markovian truncation at the Mth order arises. In practice, the $\hat{H}_{\text{diss}}^{(M)}$ portion can be treated either by Markovian master equations, or else by an explicit representation of the Ohmic bath (see the example presented below).[43,44]

As demonstrated in Ref. 79, the model in fact provides a *rigorous* description of non-Markovian dynamics, since the residual bath can be shown to converge towards a Markovian form under very general conditions. Model 3 is the cornerstone of the description in terms of approximate spectral densities, to be detailed in Sec. 5 below.

4.4. *Dynamical evolution and moment conservation rules*

The key advantage of the transformations addressed above is that the effective modes unravel the multimode dynamics as a function of time. On the shortest time scale, the dynamics is entirely captured by the primary modes. As time proceeds, the excitation propagates into the residual bath. The hierarchical representations of Model 2 and Model 3 construct a sequential picture of this process, very similarly to Mori theory.[40–42] This scenario leads to truncation schemes which define reduced-dimensional representations that are *exact* over certain time scales.

A formal analysis can be carried out in terms of moment or cumulant expansions of the propagator, showing that the moments (cumulants) are exactly reproduced up to a certain order by employing a truncated Hamiltonian comprising a limited number of effective modes. In Refs. 36, 46 and 47, such an analysis has been carried out for the representation of the propagator in the initial wavefunction state, yielding the autocorrelation function $C(t, t_0) = \langle \psi(t_0) | \hat{U}(t, t_0) | \psi(t_0) \rangle = \langle \psi(t_0) | \psi(t) \rangle$. An analogous moment analysis can be performed for the reduced propagator, where the bath has been integrated out.[80] The result of the analysis of Refs. 46 and 47 for Model 2 (Mori chain) is that truncation of the chain at a given order n — i.e. $(3 + 3n)$ modes for a two-state system, or $(n_{\text{eff}} + n_{\text{eff}} \times n)$ modes for n_{el} states with $n_{\text{eff}} = n_{\text{el}}(n_{\text{el}} + 1)/2$ — exactly reproduces the first $(2n + 3)$th order moments (cumulants) of the total Hamiltonian. A prerequisite of the above analysis is the initial condition of the bath, which is taken to correspond to the ground state at zero temperature.

As mentioned above, the primary effective modes conserve the first three moments, which can already account for an accurate dynamics over a time scale which determines the passage from the Franck–Condon point through a conical intersection region. As demonstrated by the example presented in the next section (Fig. 6), a low-order truncation often gives rather good results over reasonably long time scales. However, counter-examples can be given; for example, our recent study of Refs. 26 and 46 demonstrates a case where the primary effective modes do not give a qualitatively correct picture of the dynamics. The coupling strength between the different orders of the hierarchy, in conjunction with the relevant dynamical mechanisms, eventually determine the quality of the approximation. Furthermore, the initial condition and the choice of the expansion point for

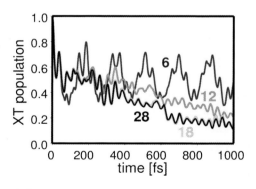

Fig. 6. Evolution of the XT state population as a function of time, for comparative calculations according to Model 2, involving truncation at successive orders $M = 1$ (6 modes), $M = 2$ (12 modes), and $M = 3$ (18 modes), along with the 28-mode full-dimensional reference calculation. Reproduced from Ref. 27. Copyright 2008 by the American Physical Society.

the effective-mode development play an important role, see the discussion in Sec. 3.3.

4.5. *Example: Ultrafast photophysics of semiconductor polymer junctions*

We have recently employed Models 2 and 3 to describe the ultrafast exciton decay at semiconductor polymer interfaces.[26,27,38,46] As briefly explained in Sec. 3.4, the subsystem is purely electronic in this case, and all vibrational modes were included as bath modes in an LVC description. Typically, the model comprises a 20–30 mode phonon distribution, composed of a high-frequency branch corresponding to carbon-carbon stretch modes and a low-frequency branch corresponding to ring-torsional modes. Both two- and three-state models were constructed, based upon a semiempirical lattice model of the polymer system.[23] The system does not exhibit any particular symmetry, such that all coupling parameters are non-zero.

A transformed version of the three-state Hamiltonian corresponding to Model 1 has been addressed in Sec. 3.4, and we focus here on Models 2 and 3. The three-state Hamiltonian corresponding to Model 2 reads as follows in matrix form:

$$\hat{\boldsymbol{H}}_B^{(n)}(\hat{X}_{B,1},\ldots,\hat{X}_{B,6n+6}) = \hat{\boldsymbol{H}}_{\text{eff}}(\hat{X}_{B,1},\ldots,\hat{X}_{B,6})$$
$$+ \sum_{l=1}^{n} \hat{\boldsymbol{H}}_{\text{res}}^{(l)}(\hat{X}_{B,6l+1},\ldots,\hat{X}_{B,6l+6}), \quad (33)$$

with the following effective Hamiltonian part, which comprises the effective-mode portion of the Hamiltonian of Eqs. (23)–(25),

$$\hat{\boldsymbol{H}}_{\text{eff}} = \sum_{i=1}^{6} \frac{\Omega_{B,i}}{2}(\hat{P}_{B,i}^2 + \hat{X}_{B,i}^2)\mathbf{1}$$
$$+ \sum_{i=1}^{6} \begin{pmatrix} (K_{B,i}+D_{B,i})\hat{X}_{B,i} & \Lambda_{B,i}^{(12)}\hat{X}_{B,i} & \Lambda_{B,i}^{(13)}\hat{X}_{B,i} \\ \Lambda_{B,i}^{(12)}\hat{X}_{B,i} & (K_{B,i}-D_{B,i})\hat{X}_{B,i} & \Lambda_{B,i}^{(23)}\hat{X}_{B,i} \\ \Lambda_{B,i}^{(13)}\hat{X}_{B,i} & \Lambda_{B,i}^{(23)}\hat{X}_{B,i} & K_{B,i}^{(3)}\hat{X}_{B,i} \end{pmatrix}$$
$$+ \sum_{i=1}^{6}\sum_{j=i+1}^{6} d_{ij}\left(\hat{P}_{B,i}\hat{P}_{B,j} + \hat{X}_{B,i}\hat{X}_{B,j}\right)\mathbf{1} + \boldsymbol{C}, \quad (34)$$

where **C** is a constant matrix and six effective modes were constructed according to a topology-adapted representation (see Sec. 3.3).[27,38] Each lth-order residual term also comprises six modes,

$$\hat{\boldsymbol{H}}_{\text{res}}^{(l)} = \sum_{i=6l+1}^{6l+6} \frac{\Omega_{B,i}}{2}(\hat{P}_{B,i}^2 + \hat{X}_{B,i}^2)\mathbf{1}$$
$$+ \sum_{i=6l+1}^{6l+6} \sum_{j=i-6}^{i-1} d_{ij}\left(\hat{P}_{B,i}\hat{P}_{B,j} + \hat{X}_{B,i}\hat{X}_{B,j}\right)\mathbf{1}. \qquad (35)$$

As explained in Sec. 3.4, the dynamical process of interest corresponds to the transition from a photogenerated excitonic (XT) state to an interfacial charge transfer (CT) state, possibly involving indirect transitions XT → IS → CT via an intermediate state (IS).[27,38] These states exhibit multiple intersections, as illustrated in Fig. 2 which depicts the projection of the multidimensional potential surfaces on the XT-CT vs. XT-IS branching plane.

An analysis based upon the HEP hierarchy of models 2 and 3 shows that the primary effective modes $(\hat{X}_{B,1}, \hat{X}_{B,2}, \hat{X}_{B,3})$, or $(\hat{X}_{B,1}, \ldots, \hat{X}_{B,6})$ in the case of a three-state model, are exclusively of high-frequency type since the relevant carbon-carbon stretch modes dominate the electron-phonon coupling. However, low-frequency torsional modes appear at the second order of the hierarchy and turn out to have an important effect on the dynamics. Indeed, the resonant interplay between the two types of modes is an important ingredient in mediating the XT state decay.[26,44,46] Figure 6 illustrates the convergence of the HEP hierarchy with increasing orders. Quantum dynamical calculations were carried out using the multiconfiguration time-depedent Hartree (MCTDH) method.[20,81–84]

5. Hierarchical Approximations of the Spectral Density

In light of the above transformation schemes, we now return to the characterization of the environment in terms of spectral densities, as introduced in Sec. 2.2. Using the series of approximate bath realizations that are generated from Model 3 of the preceding section, i.e. a truncated Mori-chain representation with Markovian closure, we identify the spectral density that each of these realizations is associated with. The family of bath spectral densities that are thus generated define coarse-grained representations of the bath, which lead to an accurate system-bath dynamics over increasing times as the order of the Mori chain is increased. In the following, we explicitly formulate these spectral densities in terms of continued-fraction expressions. The derivation is analogous to our recent study of Refs. 43 and 44, but includes a generalization to the set of spectral densities defined in Sec. 2.2. A detailed description is given in Ref. 34.

The overall picture that emerges from the present discussion is the following: using the transformation techniques described in the previous section, a set of relevant effective modes are extracted that accurately define the system-bath dynamics over increasing time scales. In the frequency domain, the successive reduced-dimensional models define a family of approximate environmental spectral densities. These spectral densities are constructed from a limited number of effective environmental modes in conjunction with a Markovian closure acting on the last member of the effective-mode chain. These reduced-dimensional dynamical problems can be treated non-perturbatively, using an explicit dynamical treatment for the subsystem plus effective mode space, along with a Markovian approximation for the remaining space.

5.1. *Spectral densities in the transformed representation*

As a result of the transformation from the original Hamiltonian Eqs. (1)–(8) to the effective-mode Hamiltonian Eqs. (16)–(18), the spectral densities introduced in Sec. 2.2 have to be re-written in terms of the transformed quantities. At each order generated by the effective-mode development including Markovian closure (Model 3), the Heisenberg evolution of the subsystem operators can be described by an Mth-order reduced-dimensional approximant,

$$(z - \hat{\mathbf{L}}_B^{(M)})\,\hat{\boldsymbol{\sigma}}(z) = \hat{\mathbf{L}}_S \hat{\boldsymbol{\sigma}}(z) + i\hat{\boldsymbol{\sigma}}_0 \tag{36}$$

and an associated effective Mth-order spectral density,

$$\mathbf{J}_{\text{eff}}^{(M)}(\omega) = \lim_{\epsilon \to 0^+} \operatorname{Im} \hat{\mathbf{L}}_B^{(M)}(z) \bigg|_{z=\omega+i\epsilon}. \tag{37}$$

Equations (36) and (37) yield Eqs. (11) and (12) if the full dimensionality is taken into account. As shown in Appendix B, the reduced Mth order quantities can be expressed using continued-fraction techniques. For clarity, we consider first the special case of a tuning-mode bath (i.e. a bath inducing energy gap fluctuations) and then turn to the multivariate problem defined by the full system-bath Hamiltonian Eq. (16).

5.2. Tuning mode bath

In the case of a tuning mode bath, the system-bath interaction is characterized by a single spectral density, which corresponds to Eq. (9) in the pre-transformed representation. A single effective mode appears at each order of the Mori chain construction, such that the chain takes a tridiagonal form.

5.2.1. Spectral densities

As shown in Refs. 43 and 44 and summarized in Appendix B, a series of Mth-order approximate spectral densities can be defined in terms of the following continued-fraction expression,

$$\hat{L}_B^{(M)}(z) = \hat{L}_{B,+}^{(M)}(z) + \hat{L}_{B,-}^{(M)}(z)$$

with

$$\hat{L}_{B,\pm}^{(M)}(z) = \frac{-\kappa_{\text{eff}}^{(-)2}}{\Omega_1 \mp z - \Xi_\pm^{(M)}}$$

$$= -\cfrac{\kappa_{\text{eff}}^{(-)2}}{\Omega_1 \mp z - \cfrac{d_{1,2}^2}{\Omega_2 \mp z - \cdots \cfrac{d_{M-2,M-1}^2}{\Omega_{M-1} \mp z - \cfrac{d_{M-1,M}^2}{\Omega_M \mp z + i2\eta z}}}}, \quad (38)$$

where $\kappa_{\text{eff}}^{(-)}$ denotes the coupling of the first effective mode to the electronic subsystem, and $\Xi_\pm^{(M)}$ subsumes the relaxation of the effective mode under the effect of the coupling to the chain of residual bath modes, with a Markovian closure at the Mth order. According to Eq. (37), the spectral density $J_{\text{eff}}^{(M)}(\omega)$ is related to the imaginary part of $\hat{L}_B^{(M)}(z)$,

$$J_{\text{eff}}^{(M)}(\omega) = J_{\text{eff},+}^{(M)}(\omega) + J_{\text{eff},-}^{(M)}(\omega)$$
$$= \lim_{\epsilon \to 0^+} \{\text{Im}\hat{L}_{B,+}^{(M)}(\omega + i\epsilon) + \text{Im}\hat{L}_{B,-}^{(M)}(\omega + i\epsilon)\}. \quad (39)$$

At the first order of the hierarchy, $M = 1$, we obtain the explicit expression[85]

$$J_{\text{eff},\pm}^{(1)}(\omega) = \frac{2\pi\eta\omega\kappa_{\text{eff}}^{(-)2}}{(\Omega_1 \mp \omega)^2 + (2\eta\omega)^2}, \quad (40)$$

which corresponds to a result first obtained by Garg et al.[64] (however, mass-weighted coordinates were used in Refs. 43, 44 and 64, which modifies the form of the spectral density). This level of approximation represents a simple Brownian oscillator model, with a central environmental mode and a Markovian damping exerted by the remaining modes.

An analytical expression can also be found at the second order,[85]

$$J^{(2)}_{\text{eff},\pm}(\omega)$$
$$= \frac{2\pi \kappa_{\text{eff}}^{(-)2} d_{12}^2 \eta \omega}{\{(\Omega_1 \mp \omega)[\Omega_2 \mp \omega(1-i2\eta)] - d_{12}^2\}\{(\Omega_1 \mp \omega)[\Omega_2 \mp \omega(1+i2\eta)] - d_{12}^2\}}, \quad (41)$$

while the third and higher orders are most conveniently generated numerically from the expressions of Eqs. (38) and (39).

The spectral densities of Eqs. (40) and (41) provide the lowest-order approximants for a tuning mode bath exhibiting a spectral density of *arbitrary complexity*. All parameters are determined from the initial spectral density, here Eq. (9), using the transformations described in Sec. 4. In the following section, these lowest-order approximations will be demonstrated for a typical example.

5.2.2. *Example: S_2-S_1 CoIn in pyrazine coupled to a tuning mode bath*

We illustrate the spectral density construction for a tuning mode bath that is coupled to a 4-mode subsystem model of the S_2-S_1 conical intersection in pyrazine, described according to the second-order vibronic coupling Hamiltonian of Raab et al.[19] In the present model,[45] a linear vibronic coupling approximation is thus only made for the bath part. A continuous reference spectral density is constructed by a Lorentzian convolution procedure[45] from an $N_B = 20$ tuning mode distribution obtained by Krempl et al.[18] as a weighted random ensemble, see Fig. 7, panel (a). Following the continued-fraction construction described above, a series of approximate spectral densities $J^{(M)}_{\text{eff}}(\omega)$ are then generated, the lowest orders of which are shown in Fig. 7, panel (b). The Mth order spectral densities are re-discretized, here again for $N_B = 20$ bath modes.

Wavepacket calculations at $T = 0$ K were carried out for the combined 4-mode subsystem plus 20-mode bath,[45] using the MCTDH method.[81–84] The explicit representation of all bath modes is not a necessity (and, in fact,

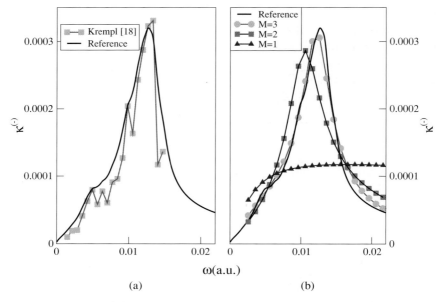

Fig. 7. Couplings $\{\kappa_i^{(-)}\}$ associated with the 20-mode intramolecular bath described in Sec. 5.2.2. Panel (a) shows the coupling parameters used by Krempl et al.[18] for a 20-mode tuning mode distribution adapted to the S_2-S_1 CoIn in pyrazine (squares), along with the continuous reference spectral density which was obtained from these data by the fitting procedure described in Ref. 45. Panel (b) illustrates the discrete couplings associated with the spectral densities $J_{\text{eff}}^{(M)}(\omega)$, $M = 1, \ldots, 3$ which were constructed by the procedure described in Sec. 5.2.1. As detailed in Ref. 45, mass-weighted coordinates rather than mass-and-frequency weighted coordinates were employed. Note that the lowest-order approximants $J_{\text{eff}}^{(M)}$, $M = 1, 2$, are not centered on the reference spectral density, even though their construction ensures that they reproduce the short-time dynamics. (In this sense, the $J_{\text{eff}}^{(M)}$'s cannot simply be understood as coarse-grained versions of the actual spectral density.) From $M = 3$ onwards, the approximants coincide more closely with the center of the reference spectral density and tend to converge. (Adapted from Ref. 45.)

the general method is designed so as to treat only the effective modes explicitly); however, an explicit wavepacket dynamics for all modes is convenient to demonstrate the convergence of the procedure for a zero-temperature system. In Refs. 43 and 44, we have shown that explicit calculations for high-dimensional system-plus-bath wavefunctions are in excellent agreement with reduced density matrix calculations.

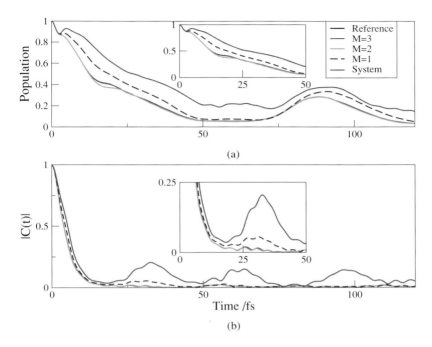

Fig. 8. (a) Time dependence of the S_2 state (diabatic) population for successive 4+20-mode models based on the bath spectral densities $J_{\text{eff}}^{(M)}$, $M = 1, 2, 3$, as well as the reference spectral density illustrated in Fig. 7. The 4-mode system dynamics is also shown for comparison (slowly decaying trace). (b) Time dependence of the wavepacket autocorrelation function (absolute value), $|C(t)| = |\langle \psi(0) | \psi(t) \rangle|$. At the level of the $M = 3$ approximation, the dynamics is indistinguishable from the reference dynamics, even though the corresponding spectral density is not fully converged yet in the frequency domain, see Fig. 7. (Adapted from Ref. 45.)

Initial conditions for the wavepacket calculations correspond to the Franck–Condon geometry, and the effective-mode expansion is defined with respect to this reference geometry. Figure 8 shows the time-dependent diabatic S_2 populations and autocorrelation functions $|C(t)| = |\langle \psi(0) | \psi(t) \rangle|$ generated from the successive spectral density approximants $J_{\text{eff}}^{(M)}(\omega)$, $M = 1, \ldots, 3$. All orders agree over the shortest time scale (~ 5 fs), and the orders $M = 2, 3$ are found to be very close over the complete observation interval. The $M = 3$ result is virtually indistinguishable from the result obtained for the reference spectral density and can be considered converged.

To summarize, the general procedure that is proposed is as follows: Starting from an arbitrary spectral density, successive approximations $J_{\text{eff}}^{(M)}$ are obtained in continued-fraction form according to Eqs. (37) and (38). These Mth-order models provide approximate representations of the non-Markovian system-environment dynamics, and become accurate over increasing time intervals as M increases. For complicated, structured environments, this implies that *the subsystem explores the details of the spectral density as time evolves.*

Importantly, the use of the Markovian closure of Model 3, which underlies the present spectral density construction, implies that no artifacts appear due to the truncation of the chain. The excitation thus cannot "propagate back" along the chain, and irreversible dynamics results.

5.3. General correlated bath

In the case of a correlated bath, a full set of auto-correlation and cross-correlation spectral densities are required. As before, we focus on an electronic two-level system, whose spectral density matrix in the most general form is given in Eq. (13). As detailed in Appendix C, the Mth order spectral density components are obtained according to Eq. (37) as the imaginary part of the operator

$$\hat{\mathbf{L}}_B^{(M)} = \begin{pmatrix} L_{xx}^{(M)} & L_{xy}^{(M)} & L_{xz}^{(M)} \\ L_{yx}^{(M)} & L_{yy}^{(M)} & L_{yz}^{(M)} \\ L_{zx}^{(M)} & L_{zy}^{(M)} & L_{zz}^{(M)} \end{pmatrix} \equiv \hat{\mathbf{L}}_{B,+}^{(M)} + \hat{\mathbf{L}}_{B,-}^{(M)} \qquad (42)$$

with components

$$\hat{\mathbf{L}}_{B,\pm}^{(M)}(z)$$

$$= \sum_{n=1}^{n_{\text{eff}}} \frac{-2}{z \mp \Omega_{1n} \mp \Xi_{1n,\pm}^{(M)}} \begin{pmatrix} K_n^{(-)2} & iK_n^{(-)}K_n^{(+)} & -K_n^{(-)}\Lambda_n \\ -iK_n^{(-)}K_n^{(+)} & (K_n^{(-)2} + \Lambda_n^2) & i\Lambda_n K_n^{(+)} \\ -K_n^{(-)}\Lambda_n & -i\Lambda_n K_n^{(+)} & \Lambda_n^2 \end{pmatrix}$$

$$\equiv \hat{\mathbf{L}}_{B,\pm}^{R(M)} + i\hat{\mathbf{L}}_{B,\pm}^{I(M)}, \qquad (43)$$

where the $\hat{\mathbf{L}}_B^{I(M)}$ part accounts for imaginary cross-correlation contributions that are associated with the $\{K_n^{(+)}\}$ parameters, analogously to Eq. (A9).

The corresponding spectral density matrix $\boldsymbol{J}_{\text{eff}}^{(M)}(\omega)$ is given according to Eq. (37), evaluated for $\hat{\mathbf{L}}_B^{R(M)}$ and $\hat{\mathbf{L}}_B^{I(M)}$, respectively.

Comparison of Eq. (43) with the matrix expression Eq. (A9) obtained from the pre-transformed Hamiltonian leads to the conclusion that all terms appearing in the $\hat{\mathbf{L}}_B^{(M)}$ matrix are formally identical to the pre-transformed version. This follows from the formal similarity of the pre- vs. post-transformed interaction Hamiltonians Eqs. (8) and (16). Differently from Eq. (A9), though, the summation of Eq. (43) involves only the first-order effective modes. All effects of the many-particle residual bath are absorbed into the terms $\{\Xi_\pm^{(M)}\}$ which correspond to a generalized version of the continued fraction representation of Eq. (38). Further details are provided in Appendix C.

The usefulness of the present procedure follows precisely from the fact that all approximations are shifted into the $\{\Xi_\pm^{(M)}\}$ part while the system-bath coupling pattern only necessitates information about the low-dimensional effective-mode subspace. Equations (42) and (43) together with the definition of the spectral density Eq. (37) thus give a complete characterization of the Mth order approximation to an arbitrary spectral density at a two-state CoIn. Cross-correlation contributions arise naturally, since the effective modes couple to several subsystem variables by construction.

Again, generalizations to more than two electronic states can be obtained along the same lines. For example, our recent calculations on semiconducting polymer systems[26, 27, 38, 46] as briefly summarized in Sec. 4.5, could be analyzed in terms of the spectral densities derived from Eqs. (42) and (43) and their analog for three electronic states. Work in this direction is currently in progress.

6. Discussion and Conclusions

Environmental effects on the dynamics at conical intersections often fall into a short-time regime where inertial, coherent effects dominate and the many-particle dissipative dynamics has not yet set in. This generally precludes the use of standard system-bath approaches and necessitates an explicit dynamical treatment of the combined subsystem-plus-environment supermolecular system. The powerful QM/MM based simulation techniques that have been developed over recent years for the explicit simulation of photochemical processes in chromophore–solvent and chromophore–protein complexes have made great strides in this direction.[8–10, 14] Even so, the need for complementary reduced-dimensional models and dynamical interpretations persists. In the present chapter, we have attempted to provide such a complementary perspective.

Starting from an LVC model representing the system–environment interaction, the present approach identifies a set of $n_{el}(n_{el}+1)/2$ effective modes which describe the collective environmental effects on the short-time non-adiabatic dynamics.[35,36] Importantly, these modes describe the effects exerted by the environment both on the CoIn topology and dynamics. Beyond the identification of the set of modes that predominate on the shortest time scale, further transformations are introduced by which chains of residual modes are created that successively unravel the dynamics as a function of time. As shown above, this construction has proven useful, e.g. in the analysis of photoinduced dynamics in extended systems like semiconducting polymers.[26,27] For this type of systems, where distinct high-frequency vs. low-frequency phonon branches exist, one can further envisage alternative transformations by which effective modes are assigned to each phonon branch separately.[86,87]

As detailed in Sec. 3.3, the effective modes constructed here can be related to the characteristic topological vectors spanning the branching space, i.e. the gradient difference and non-adiabatic coupling vectors, as well as the average gradient vector.[51,52] This relation reflects that the CoIn topology is determined at the leading order by the linear (LVC) contributions. By contrast, effects which appear at the quadratic and higher orders of the Hamiltonian are not accounted for within the environmental subspace. These effects include, for example, the notion of gateway modes which couple the branching plane and seam space,[88] and the identification of photoactive modes introduced in Ref. 48. In the latter case, a scheme is suggested by which modes which are relevant for the dynamics are selected according to their second-order contributions to the energy gap at the Franck–Condon geometry. This information is not contained in our construction for the environmental subspace, but could be used in a complementary way to provide a rationale for the system-bath partitioning employed in the present model.

Despite the limitations of the LVC model, our approach should capture the dominant environmental influences in many system-bath type situations. As the photophysics of more complex systems like semiconducting polymers and other nanostructured materials is addressed, model Hamiltonians comprising the linear vibronic coupling contributions are a natural choice, which is acceptable as long as large-amplitude motions do not play an important role. Regarding the identification of relevant environmental modes, the effective-mode approach brings the clear advantage that no selection needs to be made among the environmental modes in the first place — i.e. all bath modes are introduced democratically in the transformation scheme, and a reduction is carried out only at the level of the transformed, collective modes.

Based upon the construction of chains of effective modes, a systematic approximation procedure for the environment can be formulated in terms of a series of coarse-grained spectral densities.[43, 44] These spectral densities are generated from successive orders of a truncated chain model with Markovian closure. Analytical expressions can be given in terms of Mori-type continued fractions. Assuming that an — *a priori* arbitrarily complicated — reference spectral density can be obtained independently, e.g. from experiments or classical molecular dynamics simulations, one can thus (i) extract those features of the spectral density that determine the interaction with the subsystem on successive time scales, and (ii) carry out reduced-dimensional simulations which are in exact agreement with the complete system-bath dynamics up to a certain time. Typically, quantum dynamical simulations will be carried out for the subsystem degrees of freedom augmented by the environment's effective modes, while the remaining modes are treated by a master equation.[44]

While the above analysis in terms of spectral densities has been carried out before for a spin-boson system,[43, 44] we have here extended our previous treatment so as to account for multiple system-environment coupling mechanisms available at a CoIn.[34] This entails the possibility of cross-correlated fluctuations, due to the simultaneous coupling of environmental modes to several subsystem variables (e.g. $\hat{\sigma}_z$ and $\hat{\sigma}_x$). This complexity of the system-environment coupling is a reflection of the fact that the environmental modes can be involved in the same vibronic coupling pathways that are open to the intramolecular modes as well. Furthermore, the environmental modes are not generally subject to the symmetry constraints of the intramolecular modes, such that symmetry-breaking interactions and the cross-correlation effects which were just mentioned are frequent occurrences.

Even though the effective-mode construction can be applied to various types of quantum subsystems involving linear interactions with a multimode environment, the approach is especially tailored to the ultrafast dynamics at conical intersections. Thus, the Mori-chain type expansions described here are expected to converge rapidly. In view of the markedly non-Markovian character of the dynamics, the dissipative many-body effects of the environment act with a delay. For the same reason, one may conjecture that environment-induced decoherence effects are less pronounced than usually expected,[15, 16] similarly to our observations for other types of nonadiabatic dynamics.[44]

In many relevant cases, diagonal interactions which give rise to energy gap fluctuations dominate. The tuning-mode model that was addressed in Sec. 5.2 is adapted to this case. Solute–solvent interactions can often be mapped upon such a model as well, such that the picture of a solvent coordinate[3, 4, 12] can be accommodated within the present class of models, even if the actual microscopic interactions cannot be described at the level of a harmonic oscillator bath.

We expect the effective-mode models described here to be versatile tools that can predict general trends and can be used in conjunction with microscopic information provided from other sources, i.e. spectral densities, energy gap correlation functions, and possibly cross-correlation functions. Further, model parametrizations could be provided by QM/MM type simulations, and the model-based dynamics could be employed to analyze the wealth of microscopic information provided by such simulations. Such complementary strategies would bridge the gap between system-bath theory approaches and explicit multidimensional simulations for ultrafast photochemical processes in various types of environments.

Note Added in Proof

Since submission of this chapter, a recent application of the effective-mode approach to elementary processes in organic photovoltaics has been published [H. Tamura, I. Burghardt, and N. Tsukada, *J. Phys. Chem. C*, 115, 10205 (2011)]. Further, the time-domain convergence of the effective-mode chain representation has been analyzed in a Langevin equation setting [R. Martinazzo, K. H. Hughes, and I. Burghardt, *Phys. Rev. E*, 84, 030102(R) (2011)].

Acknowledgment

We thank Eric Bittner and Andrey Pereverzev for their contributions to the applications described in Sec. 4.5. Financial support by the Centre National de la Recherche Scientifique (CNRS) and the Agence Nationale de la Recherche ANR-NT05-3-42315 project, as well as the Deutsche Forschungsgemeinschaft (DFG) is gratefully acknowledged.

Appendices

A. Spectral Densities at a Conical Intersection

Following Ref. 34, we describe here the derivation of the spectral density matrix Eq. (13). Starting from the original, pre-transformation Hamiltonian Eq. (1), we seek to obtain coupled equations for the subsystem operators $\hat{\boldsymbol{\sigma}} = \{\hat{\sigma}_x, \hat{\sigma}_y, \hat{\sigma}_z\}$ in the Heisenberg representation, in the following form, cf. Eq. (12),

$$(z - \hat{\mathbf{L}}_B)\,\hat{\boldsymbol{\sigma}}(z) = \hat{\mathbf{L}}_S \hat{\boldsymbol{\sigma}}(z) + i\hat{\boldsymbol{\sigma}}_0, \tag{A1}$$

where the Fourier–Laplace transformed operators are referred to,

$$\hat{\sigma}_n(z) = \mathcal{L}\{\hat{\sigma}_n(t)\} = \int_0^\infty dt\,\hat{\sigma}_n(t)\exp(izt) \tag{A2}$$

with $z = \omega + i\epsilon$, Im $z > 0$. In Eq. (A1), $\hat{\boldsymbol{\sigma}}_0 = \hat{\boldsymbol{\sigma}}(t=0)$ is the initial condition and $\hat{\mathbf{L}}_S$ represents the evolution within the subsystem space (i.e. in the simplest case, evolution under the pure electronic subsystem operators), while $\hat{\mathbf{L}}_B$ represents the evolution induced by the interaction with the bath. Note that Eq. (A1) has been obtained by integrating out the bath.

We now derive an explicit expression for $\hat{\mathbf{L}}_B$ and the associated spectral density that was given in Eq. (13) of the main text,

$$\boldsymbol{J}(\omega) = -\lim_{\epsilon \to 0^+} \text{Im }(z - \hat{\mathbf{L}}_B(z))\Big|_{z=\omega+i\epsilon} = \lim_{\epsilon \to 0^+} \text{Im}\hat{\mathbf{L}}_B(z)\Big|_{z=\omega+i\epsilon}. \tag{A3}$$

Our starting point is the operator equation of motion in the Heisenberg picture, $\dot{\hat{O}} = i[\hat{H}, \hat{O}]$ for the subsystem operators $\{\hat{\sigma}_x, \hat{\sigma}_y, \hat{\sigma}_z\}$ and the bath operators $\hat{x}_n = 1/2^{1/2}(\hat{a}_n + \hat{a}_n^\dagger)$. For convenience, we use an interaction picture with respect to the subsystem Hamiltonian,

$$\dot{\hat{\sigma}}_n^I = i\,[\hat{H}_{SB}(t), \hat{\sigma}_n^I]; \quad \hat{H}_{SB}(t) = \exp(i\hat{H}_S t)\hat{H}_{SB}\exp(-i\hat{H}_S t), \tag{A4}$$

while the bath oscillators are expressed in the usual Heisenberg representation. Using the Hamiltonian of Eq. (1) and the Fourier/Laplace-transform Eq. (A2), the equations of motion for the subsystem operators read as follows,[34]

$$-iz\hat{\sigma}_x^I(z) - \hat{\sigma}_{x,0}^I = -2^{1/2}\sum_n \kappa_n^{(-)}(\hat{a}_n(z) + \hat{a}_n^\dagger(z)) * \hat{\sigma}_y^I(z),$$

$$-iz\hat{\sigma}_y^I(z) - \hat{\sigma}_{y,0}^I$$
$$= 2^{1/2}\sum_n \left(\kappa_n^{(-)}(\hat{a}_n(z) + \hat{a}_n^\dagger(z)) * \hat{\sigma}_x^I(z) - \lambda_n(\hat{a}_n(z) + \hat{a}_n^\dagger(z)) * \hat{\sigma}_z^I(z)\right),$$

$$-iz\hat{\sigma}_z^I(z) - \hat{\sigma}_{z,0}^I = 2^{1/2}\sum_n \lambda_n(\hat{a}_n(z) + \hat{a}_n^\dagger(z)) * \hat{\sigma}_y^I(z), \quad (A5)$$

where the $*$ symbol denotes the frequency-domain convolution and the equations of motion were obtained using the commutation relations $[\hat{\sigma}_i, \hat{\sigma}_j] = 2i\sum_k \epsilon_{ijk}\hat{\sigma}_k$ for the Pauli matrices, with the Levi–Civita symbol ϵ_{ijk}.

The equations for the bath oscillators are in turn given as

$$-iz\hat{a}_n(z) - \hat{a}_{n,0} = -i\omega_n\hat{a}_n(z) - \frac{i}{2^{1/2}}\left(\kappa_n^{(+)} + \kappa_n^{(-)}\hat{\sigma}_z(z) + \lambda_n\hat{\sigma}_x(z)\right),$$

$$-iz\hat{a}_n^\dagger(z) - \hat{a}_{n,0}^\dagger = i\omega_n\hat{a}_n(z) + \frac{i}{2^{1/2}}\left(\kappa_n^{(+)} + \kappa_n^{(-)}\hat{\sigma}_z(z) + \lambda_n\hat{\sigma}_x(z)\right), \quad (A6)$$

where $\hat{a}_{n,0} = \hat{a}_n(t=0)$ and $\hat{a}_{n,0}^\dagger = \hat{a}_n^\dagger(t=0)$. From the last two relations, we obtain

$$\hat{a}_n(z) + \hat{a}_n^\dagger(z) = \frac{1}{2^{1/2}}\frac{2\omega_n}{z^2 - \omega_n^2}(\kappa_n^{(+)} + \kappa_n^{(-)}\hat{\sigma}_z(z) + \lambda_n\hat{\sigma}_x(z)) + \hat{s}_n(z) \quad (A7)$$

with the initial value (source) term $\hat{s}_n(z)$,

$$\hat{s}_n(z) = \frac{\hat{a}_{n,0}}{z - \omega_n} + \frac{\hat{a}_{n,0}^\dagger}{z + \omega_n}. \quad (A8)$$

The strategy to be followed in view of obtaining Eq. (A1) is to eliminate the bath oscillators by inserting Eq. (A7) into Eq. (A5). As detailed in Ref. 34, the convolution form on the r.h.s. of Eq. (A5) will be approximated by assuming $\hat{\sigma}_n^I(z-z') \sim \hat{\sigma}_{n,0}^I\delta(z-z')$ for the subsystem operators in the interaction frame. (Equivalently, it is assumed that these interaction frame operators are constant in the time domain, $\hat{\sigma}_n^I(t) \sim \hat{\sigma}_{n,0}^I$.) This amounts to a weak-coupling approximation, which is justified since the spectral densities in question can be obtained within a second-order approximation with respect to the system-bath coupling.[89] With this approximation, the product forms of subsystem operators can then be simplified using the relations $\hat{\sigma}_z\hat{\sigma}_y = -i\hat{\sigma}_x$, $\hat{\sigma}_x\hat{\sigma}_y = i\hat{\sigma}_z$, and $\hat{\sigma}_z\hat{\sigma}_x = i\hat{\sigma}_y$.

These steps eventually lead to the second-order equation Eq. (A1) for the subsystem operators, with the following matrix form for $\hat{\mathbf{L}}_B(z)$:

$$\hat{\mathbf{L}}_B(z) = \sum_{n=1}^{N_B} \frac{-2\omega_n}{(z^2 - \omega_n^2)} \begin{pmatrix} \kappa_n^{(-)2} & i\kappa_n^{(-)}\kappa_n^{(+)} & -\kappa_n^{(-)}\lambda_n \\ -i\kappa_n^{(-)}\kappa_n^{(+)} & (\kappa_n^{(-)2} + \lambda_n^2) & i\lambda_n\kappa_n^{(+)} \\ -\kappa_n^{(-)}\lambda_n & -i\lambda_n\kappa_n^{(+)} & \lambda_n^2 \end{pmatrix} + \hat{\mathbf{S}}_B$$

$$\equiv \hat{\mathbf{L}}_B^R + i\hat{\mathbf{L}}_B^I + \hat{\mathbf{S}}_B, \tag{A9}$$

where the components $\hat{\mathbf{L}}_B^R$ and $\hat{\mathbf{L}}_B^I$ are identified according to the real vs. imaginary-valued entries in the parameter matrix, and $\hat{\mathbf{S}}_B$ denotes an initial value term related to the source term of Eq. (A8).

The next and final step is the derivation of the spectral densities Eq. (13) from Eq. (A9). To this end, we take the limit of a continuous distribution of bath modes,

$$\sum_{n=1}^{N} \frac{2\omega_n}{(z^2 - \omega_n^2)} \longrightarrow \int_{-\infty}^{\infty} d\omega_B \frac{2\omega_B}{(z^2 - \omega_B^2)}. \tag{A10}$$

Using $z = \omega + i\epsilon$, the integral can be written as follows for small ϵ,

$$\int_{-\infty}^{\infty} d\omega_B \frac{2\omega_B}{(z^2 - \omega_B^2)} \xrightarrow{\epsilon \to 0+} \mathcal{P} \int d\omega_B \frac{2\omega_B}{(\omega^2 - \omega_B^2)} - \pi i\, \delta(\omega - \omega_B), \tag{A11}$$

where the Cauchy principal-value integral modifies the frequency and the imaginary part gives rise to dissipation. Taking the imaginary parts of the components of $\hat{\mathbf{L}}_B$ according to Eq. (11) (where $\hat{\mathbf{L}}_B^R$ and $\hat{\mathbf{L}}_B^I$ of Eq. (A9) are taken separately) results in the spectral densities of Eq. (13).

B. Hierarchy of Spectral Densities for a Tuning-Mode Bath

Similarly to Appendix A, we consider the concerted evolution of the electronic subsystem variables and bath oscillators, using a Fourier/Laplace transform representation. The electronic subsystem variables are expressed in the interaction representation with respect to the subsystem Hamiltonian (denoted $\hat{\sigma}^I$), while the bath oscillators are expressed in the usual Heisenberg representation. Furthermore, we now refer to the effective-mode, chain-type representation of the bath oscillators, following the development of Sec. 3. In order to simplify the discussion, we focus here upon a tuning mode bath, such that only the coupling terms $\{\kappa_n^{(-)}\}$ are non-zero.

Differently from Appendix A, we now replace the bosonic creation and annihilation operators by their classical c-number analogs,[85]

$$\alpha_n = \frac{1}{2^{1/2}}(x_n + ip_n),$$
$$\alpha_n^* = \frac{1}{2^{1/2}}(x_n - ip_n), \quad (B1)$$

noting that α_n is the eigenvalue of the annihilation operator when acting on the corresponding coherent state, $\hat{a}_n|\alpha_n\rangle = \alpha_n|\alpha_n\rangle$, and likewise, $\langle\alpha_n|\hat{a}_n^\dagger = \langle\alpha_n|\alpha_n^*$. This strategy has the advantage of avoiding the introduction of inverse operators in the continued-fraction treatment that will be detailed below. Even though inverse operators can be defined despite the singular character of the boson operators in question,[90] the classical treatment is more straightforward. Further, the quantum-classical level of treatment adopted here is analogous to our previous developments of Refs. 43 and 44. We emphasize that the classical approximation for the bath degrees of freedom only concerns the construction of the spectral densities, and does not impose any approximation regarding the dynamical treatment of the bath.

For the quantum subsystem, we employ the operators $\hat{\sigma}_\pm^I = 1/2(\hat{\sigma}_x^I \pm i\hat{\sigma}_y^I)$ which are found to evolve independently under the coupling to the bath. Using again the Fourier/Laplace transform of Eq. (A2), we obtain the following quantum-classical Heisenberg evolution for the subsystem operators,[85]

$$-iz\hat{\sigma}_\pm^I(z) - \hat{\sigma}_{\pm,0}^I = \pm i2^{1/2}\kappa_{\text{eff}}^{(-)}(\alpha_1(z) + \alpha_1^*(z)) * \hat{\sigma}_\pm^I(z),$$
$$-iz\hat{\sigma}_z^I(z) - \hat{\sigma}_{z,0}^I = 0, \quad (B2)$$

where $\hat{\sigma}_{\pm,0}^I = \hat{\sigma}_\pm^I(t=0)$ and we used $[\hat{\sigma}_z^I, \hat{\sigma}_\pm^I] = \pm 2\hat{\sigma}_\pm^I$. In the following, we will disregard the second equation of Eqs. (B2) since $\hat{\sigma}_z^I$ is not coupled to the bath.

The equations for the bath variables read as follows:[85]

$$-iz\alpha_1(z) = -i\Omega_1\alpha_1(z) - \frac{i}{2^{1/2}}\kappa_{\text{eff}}^{(-)}\hat{\sigma}_z^I(z) - id_{1,2}\alpha_2(z),$$
$$-iz\alpha_2(z) = -i\Omega_2\alpha_2(z) - id_{1,2}\alpha_1(z) - id_{2,3}\alpha_3(z),$$
$$\vdots \quad \vdots$$

$$-iz\alpha_M(z) = -i\Omega_M\alpha_M(z) - id_{M-1,M}\alpha_{M-1}(z) - i\sum_{n=M+1}^{N} d_{M,n}\alpha_n(z),$$

$$-iz\alpha_{M+1}(z) = -i\Omega_{M+1}\alpha_{M+1}(z) - id_{M,M+1}\alpha_M(z),$$

$$\vdots \quad \vdots$$

$$-iz\alpha_{N_B}(z) = -i\Omega_N\alpha_N(z) - id_{M,N}\alpha_M(z), \tag{B3}$$

and a corresponding set of equations are obtained for the α_n^* quantities, using $\dot{\alpha}_n = -i\omega_n\alpha_n$ and $\dot{\alpha}_n^* = i\omega_n\alpha_n^*$ from Hamilton's equations. Further, $\alpha_n(t=0) = 0$ was assumed as the initial condition. The structure of the above equations corresponds to the transformed Hamiltonian of Model 3 of Sec. 4.3 where the bath modes $\{\alpha_1, \ldots, \alpha_M\}$ form a chain structure while the modes $\{\alpha_{M+1}, \ldots, \alpha_N\}$ are all coupled to α_M, and a Markovian approximation will be employed for these modes.

From the first equation of Eq. (B3), an expression for $\alpha_1(z)$ is obtained, which is to be inserted into Eq. (B2),

$$\alpha_1(z) = \frac{2^{-1/2}\kappa_{\text{eff}}^{(-)}\hat{\sigma}_z^I(z)}{z - \Omega_1 - d_{1,2}\frac{\alpha_2(z)}{\alpha_1(z)}}. \tag{B4}$$

An analogous equation for $\alpha_1^*(z)$ is obtained, see Eq. (B6) below.

Focusing on the $\hat{\sigma}_\pm^I$ subsystem operators, we now combine Eq. (B4) with the first equation of Eq. (B2), and re-write the latter by analogy with Eq. (A1) and Eq. (12),

$$(z - \hat{L}_B(z))\hat{\sigma}_\pm^I(z) = i\hat{\sigma}_{\pm,0}^I, \tag{B5}$$

where

$$\hat{L}_B(z) = -\frac{\kappa_{\text{eff}}^{(-)}}{2^{1/2}}(\alpha_1(z) + \alpha_1^*(z))$$

$$= -\kappa_{\text{eff}}^{(-)2}\left(\frac{1}{z - \Omega_1 - d_{1,2}\frac{\alpha_2(z)}{\alpha_1(z)}} + \frac{1}{z + \Omega_1 + d_{1,2}\frac{\alpha_2^*(z)}{\alpha_1^*(z)}}\right)$$

$$\equiv \hat{L}_{B,+}(z) + \hat{L}_{B,-}(z). \tag{B6}$$

To obtain Eqs. (B5) and (B6), we again used a weak-coupling approximation by which the convolution form Eq. (B2) is replaced by a local-in-frequency product yielding $\hat{\sigma}_z^I(z)\hat{\sigma}_\pm(z) = \pm\hat{\sigma}_\pm^I(z)$ (see Appendix A for further comments on this approximation).

To continue, an equation for $\alpha_2(z)/\alpha_1(z)$ is required, to be inserted into Eq. (B6). This equation is obtained from the third equation of Eq. (B3), and similarly for the following orders. A continued fraction pattern thus develops.[43,44]

When the Mth member is reached, the hierarchy terminates as follows:

$$\frac{\alpha_M(z)}{\alpha_{M-1}(z)} = \frac{d_{M-1,M}}{z - \Omega_M - \sum_{n=M+1}^{N_B} \frac{d_{M,n}^2}{z-\Omega_n}}, \tag{B7}$$

where we used $\alpha_n(z) = d_{M,n}\alpha_M(z)/(z-\Omega_n)$ from the last $N-M$ equations of Eq. (B3).

In the limit where the last $N-M$ modes conform to an Ohmic bath, we can replace

$$\sum_{n=M+1}^{N_B} \frac{d_{M,n}^2}{z - \Omega_n} \longrightarrow -i2\eta z, \tag{B8}$$

where η is the friction coefficient from a Langevin treatment. Analogous expressions can again be obtained for the α_n^* variables.

With a Markovian closure at the Mth order, we thus obtain $\hat{L}_B^{(M)}$ as a continued fraction of order M,[85]

$$\hat{L}_B^{(M)}(z) = \hat{L}_{B,+}^{(M)}(z) + \hat{L}_{B,-}^{(M)}(z)$$

with

$$\hat{L}_{B,\pm}^{(M)}(z) = -\cfrac{\kappa_{\text{eff}}^{(-)2}}{\Omega_1 \mp z - \cfrac{d_{1,2}^2}{\Omega_2 \mp z - \cdots \cfrac{d_{M-2,M-1}^2}{\Omega_{M-1} \mp z - \cfrac{d_{M-1,M}^2}{\Omega_M \mp z + i2\eta z}}}}$$

$$\equiv -\frac{\kappa_{\text{eff}}^{(-)2}|}{|\Omega_1 \mp z} - \frac{d_{1,2}^2|}{|\Omega_2 \mp z} - \cdots \frac{d_{M-2,M-1}^2|}{|\Omega_{M-1} \mp z} - \frac{d_{M-1,M}^2|}{|\Omega_M \mp z + i2\eta z};$$
(B9)

see Eq. (38) of the main text. In the last part of Eq. (B9), the continued-fraction notation of Pringsheim[91] was used.

By this procedure, a family of Mth order spectral densities is generated, based on subsets of effective modes. In accordance with the moment conservation rules of Refs. 36, 46 and 47, successive Mth order truncation schemes allow for an accurate description of the system-bath dynamics over increasing time intervals.

C. Hierarchy of Spectral Densities for a General Bath

We now consider a generalization of the derivation of Appendix B, which is applicable to the case where several effective modes appear at each level of the chain structure.[34] We thus obtain coupled matrix equations for the subsystem operators $\hat{\boldsymbol{\sigma}}^I = \{\hat{\sigma}_x^I, \hat{\sigma}_y^I, \hat{\sigma}_z^I\}$, which will again be of the form Eq. (A1), but now involving a hierarchy of Mth order approximations to the bath propagator,

$$(z - \hat{\mathbf{L}}_B^{(M)})\,\hat{\boldsymbol{\sigma}}(z) = \hat{\mathbf{L}}_S\hat{\boldsymbol{\sigma}}(z) + i\hat{\boldsymbol{\sigma}}_0. \tag{C1}$$

Our starting point is a generalization of Eqs. (B2) and (B3) of Appendix B, for the subsystem operators in the interaction representation,

$$z\hat{\boldsymbol{\sigma}}^I(z) - i\hat{\boldsymbol{\sigma}}_0^I = \mathbf{C}\,(\boldsymbol{\alpha}_1(z) + \boldsymbol{\alpha}_1^*(z))^T * \hat{\boldsymbol{\sigma}}^I(z), \tag{C2}$$

and the hierarchy of effective modes

$$z\boldsymbol{\alpha}_1(z) = \boldsymbol{\Omega}_1\boldsymbol{\alpha}_1(z) + \mathbf{d}^{11}\boldsymbol{\alpha}_1(z) + \mathbf{C}\,\hat{\boldsymbol{\sigma}}^I(z) + \mathbf{d}^{12}\boldsymbol{\alpha}_2(z), \tag{C3}$$

$$z\boldsymbol{\alpha}_2(z) = \boldsymbol{\Omega}_2\boldsymbol{\alpha}_2(z) + \mathbf{d}^{22}\boldsymbol{\alpha}_2(z) + \mathbf{d}^{12}\boldsymbol{\alpha}_1(z) + \mathbf{d}^{23}\boldsymbol{\alpha}_3(z), \tag{C4}$$

$$\vdots = \vdots$$

$$z\boldsymbol{\alpha}_M(z) = \boldsymbol{\Omega}_M\boldsymbol{\alpha}_M(z) + \mathbf{d}^{MM}\boldsymbol{\alpha}_M(z) + \mathbf{d}^{M-1M}\boldsymbol{\alpha}_{M-1}(z) - 2i\eta z\boldsymbol{\alpha}_M(z). \tag{C5}$$

Here, the electron-phonon coupling matrix $\mathbf{C}$ subsumes the couplings of the transformed interaction Hamiltonian Eq. (16), and $\boldsymbol{\alpha}_1$ denotes the set of primary effective modes, $\boldsymbol{\alpha}_2$ denotes the set of secondary effective modes, etc. The matrices $\mathbf{d}^{nm}$ are $n_{\text{eff}} \times n_{\text{eff}}$ (here, 3×3) sub-blocks of the bilinear coupling matrix. Note that the $\mathbf{d}^{nn}$ matrices are purely off-diagonal since they complement the diagonal $\boldsymbol{\Omega}_n$ matrices. Finally, a Markovian closure is introduced at the order M, as explained in Appendix B.

As before, the aim is to obtain equations of the form Eq. (C1) and Eq. (12) by inserting a solution for $\boldsymbol{\alpha}_1(z)$ from Eq. (C3) into Eq. (C2) for $\hat{\boldsymbol{\sigma}}^I(z)$. As compared with Appendix B, the coupling to the residual bath now induces a more complicated dynamics for the primary effective modes contained in $\boldsymbol{\alpha}_1(z)$. As a result, Eq. (C3) yields component-wise

$$\alpha_{1n}(z) = \frac{2}{z - \Omega_{1n} - \Xi_{1n}}\left(K_n^{(+)} + K_n^{(-)}\hat{\sigma}_z^I(z) + \Lambda_{1n}\hat{\sigma}_x^I(z)\right), \tag{C6}$$

with $n = 1, \ldots, n_{\text{eff}}$, and the Ξ_{1n} term reflects the coupling among the effective modes and the coupling to the residual bath,

$$\Xi_{1n} = \frac{(\mathbf{d}^{11}\boldsymbol{\alpha}_1(z))_n + (\mathbf{d}^{12}\boldsymbol{\alpha}_2(z))_n}{\alpha_{1n}(z)}. \tag{C7}$$

The expression Eq. (C7) can be developed further by inserting the formal solution for $\boldsymbol{\alpha}_2(z)$ from Eq. (C4),

$$\alpha_{2n}(z) = \frac{(\mathbf{d}^{12}\boldsymbol{\alpha}_1(z))_n}{z - \Omega_{2n} - \dfrac{(\mathbf{d}^{22}\boldsymbol{\alpha}_2(z))_n + (\mathbf{d}^{23}\boldsymbol{\alpha}_3(z))_n}{\alpha_{2n}(z)}}. \tag{C8}$$

A continued fraction pattern thus again emerges, similarly to the analysis of Appendix B. A closure is obtained at the Mth order according to Eq. (C5). For example, for $M = 2$, Eq. (C8) is replaced with

$$\alpha_{2n}(z) = \frac{(\mathbf{d}^{12}\boldsymbol{\alpha}_1(z))_n}{z - \Omega_{2n} - \dfrac{(\mathbf{d}^{22}\boldsymbol{\alpha}_2(z))_n - 2iz(\boldsymbol{\eta}\boldsymbol{\alpha}_2(z))_n}{\alpha_{2n}(z)}}. \tag{C9}$$

With these ingredients, we finally obtain a form that is equivalent to Eq. (C1),

$$(z - \hat{\mathbf{L}}_B^{(M)}(z))\hat{\boldsymbol{\sigma}}^I(z) = i\hat{\boldsymbol{\sigma}}_0, \tag{C10}$$

with the post-transformed Mth order $\hat{\mathbf{L}}_B^{(M)}$ matrix,

$$\hat{\mathbf{L}}_B^{(M)} = \hat{\mathbf{L}}_{B,+}^{(M)} + \hat{\mathbf{L}}_{B,-}^{(M)}, \tag{C11}$$

where the $\hat{\mathbf{L}}_{B,\pm}^{(M)}$ parts again refer to the $\boldsymbol{\alpha}$ vs. $\boldsymbol{\alpha}^*$ components as in Eq. (B6) and can be identified as follows,[34]

$$\hat{\mathbf{L}}_{B,\pm}^{(M)}(z)$$

$$= \sum_{n=1}^{n_{\text{eff}}} \frac{-2}{z \mp \Omega_{1n} \mp \Xi_{1n,\pm}^{(M)}} \begin{pmatrix} K_n^{(-)2} & iK_n^{(-)}K_n^{(+)} & -K_n^{(-)}\Lambda_n \\ -iK_n^{(-)}K_n^{(+)} & (K_n^{(-)2} + \Lambda_n^2) & i\Lambda_n K_n^{(+)} \\ -K_n^{(-)}\Lambda_n & -i\Lambda_n K_n^{(+)} & \Lambda_n^2 \end{pmatrix}$$

$$\equiv \hat{\mathbf{L}}_{B,\pm}^{R(M)} + i\hat{\mathbf{L}}_{B,\pm}^{I(M)}. \tag{C12}$$

Here, the $\hat{\mathbf{L}}_B^{I(M)}$ part accounts for imaginary cross-correlation contributions that are associated with the $\{K_n^{(+)}\}$ parameters, analogously to Eq. (A9). The associated Mth order spectral densities are again given as follows,

$$J_{\text{eff},ij}^{(M)}(\omega) = \lim_{\epsilon \to 0^+} -\text{Im}\,(z - \hat{L}_{B,ij}^{(M)}(z))\Big|_{z=\omega+i\epsilon} = \lim_{\epsilon \to 0^+} \text{Im}\hat{L}_{B,ij}^{(M)}(z)\Big|_{z=\omega+i\epsilon},$$
(C13)

evaluated for $\hat{\mathbf{L}}_B^{R(M)}$ and $\hat{\mathbf{L}}_B^{I(M)}$, respectively.

Overall, the analysis of Appendix B carries over to the multivariate, correlated case which necessitates introducing a matrix of spectral density components. Importantly, the formal structure of the pre- vs. post-transformed evolution operators and spectral densities remains essentially the same, as is clear from the comparison between Eq. (A9) and Eq. (C12).

References

1. S. Mukamel, *Principles of Nonlinear Optical Spectroscopy* (Oxford University Press, New York/Oxford, 1995).
2. U. Weiss, *Quantum Dissipative Systems* (World Scientific Singapore, 1999).
3. A. Warshel and R.M. Weiss, *J. Am. Chem. Soc.* **102**, 6218 (1980).
4. I. Benjamin, P. Barbara, B. Gertner and J.T. Hynes, *J. Phys. Chem.* **99**, 7557 (1995).
5. R.A. Marcus and N. Sutin, *Biochim. Biophys. Acta* **811**, 265 (1985).
6. A. Nitzan, *Chemical Dynamics in Condensed Phases* (Oxford University Press, New York/Oxford, 2006).
7. A. Toniolo, S. Olsen, L. Manohar and T.J. Martínez, *Faraday Discuss. Chem. Soc.* **127**, 149 (2004).
8. A.M. Virshup, C. Punwong, T.V. Pogorelov, B.A. Lindquist, C. Ko and T.J. Martínez, *J. Phys. Chem. B* **113**, 3280 (2009).
9. G. Groenhof, M. Bouxin-Cademartory, B. Hess, S.P. de Visser, H.J.C. Berendsen, M. Olivucci, A.E. Mark and M.A. Robb, *J. Am. Chem. Soc.* **126**, 4228 (2004).
10. M. Boggio-Pasqua, M.A. Robb and G. Groenhof, *J. Am. Chem. Soc.* **131**, 13580 (2009).
11. E.V. Gromov, I. Burghardt, H. Köppel and L.S. Cederbaum, *J. Am. Chem. Soc.* **129**, 6798 (2007).
12. I. Burghardt and J.T. Hynes, *J. Phys. Chem. A* **110**, 11411 (2006).
13. G. Zgrablić, K. Voïtchovsky, M. Kindermann, S. Haake and M. Chergui, *Biophys. J.* **88**, 2779 (2005).
14. A. Sinicropi, E. Martin, M. Ryazantsev, J. Helbing, J. Briand, D. Sharma, J. Léonard, S. Haake, A. Cannizzo, M. Chergui, V. Zanirato, S. Fusi,

F. Santoro, R. Basosi, N. Ferré and M. Olivucci, *Proc. Natl. Acad. Sci. USA* **105**, 17642 (2007).
15. O.V. Prezhdo and P.J. Rossky, *Phys. Rev. Lett.* **81**, 5294 (1998).
16. M. Schlosshauer, *Decoherence* (Springer, Berlin, 2007).
17. H. Köppel, W. Domcke and L.S. Cederbaum, *Adv. Chem. Phys.* **57**, 59 (1984).
18. S. Krempl, M. Winterstetter, H. Plöhn and W. Domcke, *J. Chem. Phys.* **100**, 926 (1994).
19. A. Raab, G.A. Worth, H.-D. Meyer and L.S. Cederbaum, *J. Chem. Phys.* **110**, 936 (1999).
20. G.A. Worth, H.-D. Meyer, H. Köppel, L.S. Cederbaum and I. Burghardt, *Int. Rev. Phys. Chem.* **27**, 569 (2008).
21. J.L. Brédas, D. Beljonne, V. Cropceanu and J. Cornil, *Chem. Rev.* **104**, 4971 (2004).
22. S. Tretiak, A. Saxena, R.L. Martin and A.R. Bishop, *Phys. Rev. Lett.* **89**, 097402 (2002).
23. I. Burghardt, E.R. Bittner, H. Tamura, A. Pereverzev and J. Ramon, in *Energy Transfer Dynamics in Biomaterial Systems*, edited by I. Burghardt, V. May, D.A. Micha and E.R. Bittner (Springer Chemical Physics Series, Heidelberg/Berlin, Vol. **93**, 2009), p. 183.
24. L. Lüer, C. Gadermaier, J. Crochet, T. Hertel, D. Brida and G. Lanzani, *Phys. Rev. Lett.* **102**, 127401 (2009).
25. B.F. Habenicht, H. Kamisaka, K. Yamashita and O.V. Prezhdo, *Nano Lett.* **7**, 3260 (2007).
26. H. Tamura, E.R. Bittner and I. Burghardt, *J. Chem. Phys.* **126**, 021103 (2007).
27. H. Tamura, J. Ramon, E.R. Bittner and I. Burghardt, *Phys. Rev. Lett.* **100**, 107402 (2008).
28. A. Kühl and W. Domcke, *J. Chem. Phys.* **116**, 263 (2002).
29. D. Gelman, G. Katz, R. Kosloff and M.A. Ratner, *J. Chem. Phys.* **123**, 134112 (2005).
30. I. Burghardt, K. Giri and G.A. Worth, *J. Chem. Phys.* **129**, 174104 (2008).
31. B.G. Levine and T.J. Martínez, *Annu. Rev. Phys. Chem.* **58**, 613 (2007).
32. G.A. Worth, M.A. Robb and I. Burghardt, *Faraday Discuss.* **127**, 307 (2004).
33. A.J. Leggett, S. Chkravarty, A.T. Dorsey, M.P.A. Fisher, A. Garg and W. Zwerger, *Rev. Mod. Phys.* **59**, 1 (1987).
34. K.H. Hughes and I. Burghardt, "Effective-mode representation of non-Markovian dynamics: A hierarchical approximation of the spectral density. III. Application to conical intersections", *J. Chem. Phys.*, submitted (2011).
35. L.S. Cederbaum, E. Gindensperger and I. Burghardt, *Phys. Rev. Lett.* **94**, 113003 (2005).
36. E. Gindensperger, I. Burghardt and L.S. Cederbaum, *J. Chem. Phys.* **124**, 144103 (2006).
37. I. Burghardt, E. Gindensperger and L.S. Cederbaum, *Mol. Phys.* **104**, 1081 (2006).

38. H. Tamura, J. Ramon, E.R. Bittner and I. Burghardt, *J. Phys. Chem. B* **112**, 495 (2008).
39. E. Gindensperger, H. Köppel and L.S. Cederbaum, *J. Chem. Phys.* **126**, 034106 (2007).
40. H. Mori, *Prog. Theor. Phys.* **34**, 399 (1965).
41. M. Dupuis, *Prog. Theor. Phys.* **37**, 502 (1967).
42. P. Grigolini and G.P. Parravicini, *Phys. Rev. B* **25**, 5180 (1982).
43. K.H. Hughes, C.D. Christ and I. Burghardt, *J. Chem. Phys.* **131**, 024109 (2009).
44. K.H. Hughes, C.D. Christ and I. Burghardt, *J. Chem. Phys.* **131**, 124108 (2009).
45. R. Martinazzo, K.H. Hughes, F. Martelli and I. Burghardt, *Chem. Phys.*, **377**, 21 (2010).
46. H. Tamura, E.R. Bittner and I. Burghardt, *J. Chem. Phys.* **127**, 034706 (2007).
47. E. Gindensperger and L.S. Cederbaum, *J. Chem. Phys.* **127**, 124107 (2007).
48. B. Lasorne, F. Sicilia, M.J. Bearpark, M.A. Robb, G.A. Worth and L. Blancafort, *J. Chem. Phys.* **128**, 124307 (2008).
49. A.O. Caldeira and A.J. Leggett, *Phys. Rev. A* **31**, 1059 (1985).
50. H. Breuer and F. Petruccione, *The Theory of Open Quantum Systems* (Oxford University Press, Oxford, 2002).
51. D.R. Yarkony, *Rev. Mod. Phys.* **68**, 985 (1996).
52. D.R. Yarkony, *Acc. Chem. Res.* **31**, 511 (1998).
53. G.A. Worth and L.S. Cederbaum, *Ann. Rev. Phys. Chem.* **55**, 127 (2004).
54. H. Köppel, W. Domcke and L.S. Cederbaum, in *Conical Intersections: Electronic Structure, Dynamics and Spectroscopy*, edited by W. Domcke, D.R. Yarkony and H. Köppel (World Scientific, New Jersey, Vol. **15**, 2004), p. 323.
55. E. Gindensperger, I. Burghardt and L.S. Cederbaum, *J. Chem. Phys.* **124**, 144104 (2006).
56. H. Köppel and W. Domcke, in *Encyclopedia in Computational Chemistry* (Wiley, New York, 1998), p. 3166.
57. T. Pacher, H. Köppel and L.S. Cederbaum, *Adv. Chem. Phys.* **84**, 293 (1993).
58. H. Köppel, J. Gronki and S. Mahapatra, *J. Chem. Phys.* **115**, 2377 (2001).
59. D.R. Yarkony, *J. Chem. Phys.* **112**, 2111 (2000).
60. B.K. Kendrick, C.A. Mead and D.G. Truhlar, *Chem. Phys.* **277**, 31 (2002).
61. H. Köppel and A. Thiel, *J. Chem. Phys.* **110**, 9371 (1999).
62. H. Köppel, *Faraday Discuss. Chem. Soc.* **127**, 35 (2004).
63. N.G. van Kampen, *Stochastic Processes in Physics and Chemistry* (North-Holland, Amsterdam, 1992).
64. A. Garg, J.N. Onuchic and V. Ambegaokar, *J. Chem. Phys.* **83**, 4491 (1985).
65. Y. Toyozawa and M. Inoue, *J. Phys. Soc. Jpn.* **21**, 1663 (1966).
66. M.C.M. O'Brien, *J. Phys. C* **5**, 2045 (1971).
67. R. Englman and B. Halperin, *Ann. Phys.* **3**, 453 (1978).
68. E. Haller, L.S. Cederbaum and W. Domcke, *Mol. Phys.* **41**, 1291 (1980).
69. L.S. Cederbaum, E. Haller and W. Domcke, *Solid State Comm.* **35**, 879 (1980).

70. G.J. Atchity, S.S. Xantheas and K. Ruedenberg, *J. Chem. Phys.* **95**, 1862 (1991).
71. D.R. Yarkony, *J. Phys. Chem. A* **101**, 4263 (1997).
72. D.R. Yarkony, *J. Chem. Phys.* **114**, 2601 (2001).
73. M.J. Paterson, M.J. Bearpark, M.A. Robb and L. Blancafort, *J. Chem. Phys.* **121**, 11562 (2004).
74. D.R. Yarkony, *J. Chem. Phys.* **123**, 204101 (2005).
75. S. Matsika and D.R. Yarkony, *J. Chem. Phys.* **117**, 6907 (2002).
76. J.D. Coe, M.T. Ong, B.G. Levine and T.J. Martínez, *J. Phys. Chem. A* **112**, 12559 (2008).
77. E. Gindensperger and H. Köppel, "Effective-mode approach for multi-state quantum dynamics in the fluorobenzene cation", *J. Chem. Phys.*, submitted (2011).
78. I. Burghardt and H. Tamura, in *Dynamics of Open Quantum Systems*, edited by K.H. Hughes (CCP6, Daresbury Laboratory, 2006), p. 53.
79. R. Martinazzo, B. Vacchini, K.H. Hughes and I. Burghardt, *J. Chem. Phys.*, **134**, 011101 (2011).
80. I. Burghardt, in *Quantum Dynamics of Complex Molecular Systems*, edited by I. Burghardt and D.A. Micha (Springer Chemical Physics Series, Vol. **83**, Heidelberg/Berlin, 2007), p. 135.
81. H.-D. Meyer, U. Manthe and L.S. Cederbaum, *Chem. Phys. Lett.* **165**, 73 (1990).
82. U. Manthe, H.-D. Meyer and L.S. Cederbaum, *J. Chem. Phys.* **97**, 3199 (1992).
83. M.H. Beck, A. Jäckle, G.A. Worth and H.D. Meyer, *Phys. Rep.* **324**, 1 (2000).
84. G.A. Worth, M.H. Beck, A. Jäckle and H. Meyer, "The MCTDH Package" (Version 8.2, 2000); H.-D. Meyer (Version 8.3, 2002). See http://www.pci.uni-heidelberg.de/tc/usr/mctdh/.
85. I. Burghardt and K.H. Hughes, "Non-Markovian reduced dynamics based upon a hierarchical effective-mode representation", *J. Chem. Phys.*, submitted (2011).
86. A. Pereverzev, I. Burghardt and E.R. Bittner, *J. Chem. Phys.* **131**, 034104 (2009).
87. B. Halperin and R. Englman, *Phys. Rev. B* **9**, 2264 (1974).
88. D.R. Yarkony, *J. Chem. Phys.* **123**, 134106 (2005).
89. Note that this approximation is not necessary if the subsystem corresponds to a boson operator.[43,64,92]
90. C.L. Mehta and A.K. Roy, *Phys. Rev. A* **46**, 1565 (1992).
91. A. Cuyt, V. Petersen, B. Verdonk, H. Waadeland and W.B. Jones, *Handbook of Continued Fractions for Special Functions* (Springer, Berlin, 2008).
92. M. Wubs, L.G. Suttorp and A. Lagendijk, *Phys. Rev. A* **70**, 053823 (2004).

in interfacing quantum mechanical dynamics with "on-the-fly" solution of the electronic Schrödinger equation is the seeming contradiction between the global nature of quantum dynamics and the locality of quantum chemistry. Traditional quantum chemistry begins with the Born–Oppenheimer approximation and solves the electronic wavefunctions at a single configuration of the nuclei. Thus, one can easily obtain PESs and their gradients at a single point. However, quantum dynamics requires knowledge of the PESs over a wider set of configurations at each time step. In principle, the PESs are required for all possible geometries, although the situation is often much less severe in practice. Note that there is no such difficulty when using AIMD with classical mechanics because trajectories are perfectly localized in phase space and one only requires the calculation of the gradient of the PES at a single molecular geometry in each time step.

The AIMS method introduces a quasi-localization through the choice of a nuclear basis set which is local in phase space — namely coherent states or Gaussian wavepackets.[49–51] While integrals of the basis functions over configuration space are still required, one can expect that these will be well approximated by low-order Taylor expansions or sparse grid quadratures, provided that the basis functions are well localized. Furthermore, the coherent states are by some measures the most "classical-like" quantum objects and thus the time evolution of the phase space centers can be chosen to follow Hamilton's equations. In this case, it is natural to denote the nuclear basis functions as "trajectory basis functions" (TBFs) since they are closely related to classical trajectories. Another ingredient in the AIMS method is the "spawning" which refers to the adaptive expansion of the basis set. The number of basis functions used to describe the time-evolving nuclear wavefunction may need to change as a consequence of the requirements of the dynamics and AIMS allows for this through spawning. In principle, one could identify the needed basis functions at the beginning of a simulation, but in practice it is difficult to predict what basis functions will be needed *a priori*. By spawning, i.e. adaptively increasing the basis set during the simulation, we circumvent the need for foreknowledge of what regions of phase space will be covered during the dynamical evolution.

2.2. *Full multiple spawning dynamics — wavefunction ansatz*

The dynamical method underlying AIMS is referred to as "full multiple spawning" or FMS. As discussed above, the FMS method has been designed for use in the AIMD context, where *ab initio* electronic structure theory is solved simultaneously with the dynamical evolution to obtain PESs, gradients, and any other related information that might be needed. However, the FMS method is equally valid when analytical PESs and nonadiabatic couplings are available. We first discuss the FMS method, independent of the origin of the PESs.

The FMS method is a basis set technique and therefore is guaranteed to converge to the exact result for sufficiently large basis sets. Of course, this guarantee is only in principle and the more relevant question is the practical convergence behavior in cases where a finite (and often very small in the case of AIMS, because of the expense associated with "on-the-fly" evaluation of the PESs) basis set is used. Indeed, much of the motivation for the adaptive basis set increase associated with "spawning" is to ensure that the basis set spans the region relevant to the nuclear dynamics while keeping the basis set as small as possible. The FMS wavefunction (nuclear and electronic) is written as:[33,40–43,52,53]

$$\Psi(\mathbf{r}, \mathbf{R}, t) = \sum_{I}^{N_{el}} \chi_I(\mathbf{R}, t) \, \phi_I(\mathbf{r}; \mathbf{R}), \qquad (1)$$

where $\mathbf{R}$ and $\mathbf{r}$ denote nuclear and electronic coordinates, respectively. The total wavefunction is written as a sum over nuclear wavefunctions associated with each of the N_{el} electronic states. Each of the nuclear wavefunctions corresponding to a given electronic state is itself expanded as a sum over TBFs:

$$\chi_I(\mathbf{R}; t) = \sum_{m=1}^{N_I(t)} c_m^I(t) \chi_m^I(\mathbf{R}; \bar{\mathbf{R}}_m^I, \bar{\mathbf{P}}_m^I, \gamma_m^I, \alpha_m^I), \qquad (2)$$

where N_I is the number of TBFs associated with the Ith electronic state and this number will change during the simulation as new TBFs are spawned. Each individual TBF χ_m^I corresponds to a multidimensional product of one-dimensional Gaussians $\chi_{m\rho}^I$,

$$\chi_m^I(\mathbf{R}; \bar{\mathbf{R}}_m^I, \bar{\mathbf{P}}_m^I, \gamma_m^I, \alpha_m^I) = e^{i\gamma_m^I(t)} \prod_{\rho=1}^{N_{DOF}} \chi_{m\rho}^I(R; \bar{R}_{m\rho}^I, \bar{P}_{m\rho}^I, \alpha_{m\rho}^I), \qquad (3)$$

where N_{DOF} is the total number of degrees of freedom and each one-dimensional frozen Gaussian[54–66] is given as

$$\chi_{m\rho}^I(R; \bar{R}_{m\rho}^I, \bar{P}_{m\rho}^I, \alpha_{m\rho}^I) = \left(\frac{2\alpha_{m\rho}^I}{\pi}\right)^{1/4} e^{-\alpha_{m\rho}^I(R-\bar{R}_{m\rho}^I)^2 + i\bar{P}_{m\rho}^I(R-\bar{R}_{m\rho}^I)}. \qquad (4)$$

Thus, each of the TBFs is parameterized by its phase space center whose positions and momenta are given by the $\bar{\mathbf{R}}_m^I$ and $\bar{\mathbf{P}}_m^I$ vectors, respectively. For numerical convenience, each TBF also carries a phase γ_m^I, although this could just as well be absorbed in the complex amplitude $c_m^I(t)$. Defining γ_m^I as a semiclassical phase for reference amounts to an interaction picture and allows the use of larger time steps during the integration of the equations of motion. Finally, each TBF also has an associated width α_m^I which is time-independent and usually also independent of electronic state. This parameter determines the position space uncertainty of the Gaussian basis functions. Although there are clear guidelines for the choice of this parameter based on physical considerations, the more important point is that the results (e.g. probabilities of nonadiabatic transitions) are quite insensitive to the precise value over a wide range.[42,67,68] If the dynamics are carried out in normal mode coordinates, the physically reasonable choice for the width[41,49,51] should be related to the force constant in each mode, i.e. $\alpha \approx m\omega/2$ in a one-dimensional problem. In practice, one often prefers to use Cartesian coordinates and we recently outlined a procedure for defining the widths in this case.[67]

Classical Hamiltonian equations of motion are used to propagate the phase space centers of all the TBFs (using forces corresponding to the associated electronic state). At the same time, the nuclear Schrödinger equation within the basis set of TBFs is solved at each time step to calculate the complex amplitude $c_m^I(t)$ associated with each TBF:

$$\dot{c}_l^I(t) = -i \sum_{J,K}^{N_{el}} \sum_{m}^{N_J(t)} \sum_{n}^{N_K(t)} (\mathbf{S}^{-1})_{Il,Jm}(H_{Jm,Kn} - i\dot{\vec{S}}_{Jm,Kn})c_n^K(t), \quad (5)$$

where the overlap, Hamiltonian, and right-acting derivative matrix elements are defined as:

$$S_{Il,Jm} = \langle \chi_l^I \phi_I | \chi_m^J \phi_J \rangle_{R,r},$$
$$H_{Il,Jm} = \langle \chi_l^I \phi_I | \hat{H} | \chi_m^J \phi_J \rangle_{R,r} = \langle \chi_l^I \phi_I | \hat{T} | \chi_m^J \phi_J \rangle_{R,r} + \langle \chi_l^I \phi_I | \hat{V} | \chi_m^J \phi_J \rangle_{R,r},$$
$$\dot{\vec{S}}_{Il,Jm} = \langle \chi_l^I \phi_I | \frac{\partial}{\partial t} \chi_m^J \phi_J \rangle_{R,r}. \quad (6)$$

The matrix elements of the Hamiltonian operator contain both kinetic and potential (with respect to the nuclear degrees of freedom) parts. For matrix elements which are diagonal in the electronic index, the kinetic energy integrals can be done analytically since the basis functions are of Gaussian form. For both off-diagonal (in the electronic state index) kinetic energy matrix elements and potential energy matrix elements, one usually requires a numerical approximation. Numerical quadrature is one option to evaluate the required integrals over the PES and the localized nature of the TBFs can be used to minimize the number of quadrature points needed for this. However, a simpler approach is to use saddle-point approximations (SPA) which are motivated by the localized nature of the basis functions. This amounts to a Taylor expansion of the operator about the centroid of the product of the TBFs involved in the matrix element. In AIMS, we usually truncate this Taylor expansion at the first term (the "zeroth order" SPA) and thus,

$$\langle \chi_m^I | V_{IJ} | \chi_n^J \rangle_R \approx V_{IJ}\left(\mathbf{R}_{\text{centroid}}^{IJ}\right) \langle \chi_m^I | \chi_n^J \rangle_R, \qquad (7)$$

where $\mathbf{R}_{\text{centroid}}^{IJ}$ is the coordinate space location of the centroid of the product of TBFs $|\chi_m^I\rangle$ and $|\chi_n^J\rangle$. We have integrated over the electronic coordinates to define the potential energy V_{IJ}:

$$V_{IJ}(\mathbf{R}) = \langle \phi_I | \hat{H}_{\text{electronic}}(\mathbf{R}, \mathbf{r}) | \phi_J \rangle_r, \qquad (8)$$

where the electronic Hamiltonian operator is the usual Born–Oppenheimer Hamiltonian when the adiabatic representation is used. Higher-order SP approximations may also be defined. For example, the first-order SPA is given as:

$$\langle \chi_m^I | V_{IJ} | \chi_n^J \rangle \approx V_{IJ}(\mathbf{R}_0)\langle \chi_m^I | \chi_n^J \rangle + \langle \chi_m^I | \mathbf{R} - \mathbf{R}_0 | \chi_n^J \rangle \frac{d}{d\mathbf{R}} V_{IJ}(\mathbf{R})|_{\mathbf{R}=\mathbf{R}_0}. \qquad (9)$$

The SPA resembles the Mulliken–Ruedenberg approximation used in electronic structure theory to approximately evaluate multicenter two-electron integrals.[69] It requires a single calculation of the potential energy for each matrix element, i.e. N_{TBF}^2 solutions of the electronic Schrödinger equation for AIMS at each time step. In practice, one can prescreen using the overlap matrix element in Eq. (7), and the number of ab initio calculations at each time step is much reduced, leading to nearly linear scaling with the number of TBFs.

The prescription given so far is essentially solution of the nuclear Schrödinger equation within a basis set of frozen Gaussians that follow classical trajectories. The coupled coherent states (CCS) method[48] of Shalashilin and Child follows this prescription, with the difference being that the force used in propagation of the phase space centers is averaged over the TBFs:

$$(\dot{P}_m^I)_{\text{CCS}} = -\left\langle \chi_m^I \left| \frac{\partial}{\partial R} V_{II} \right| \chi_m^I \right\rangle_R, \qquad (10)$$

which should be compared to the force used in FMS:

$$(\dot{P}_m^I)_{\text{FMS}} = -\left.\frac{\partial V_{II}}{\partial R}\right|_{R_m^I}. \qquad (11)$$

The force used in FMS is the zeroth order saddle-point evaluation of the CCS expression. Unfortunately, progressing beyond the zeroth order saddle-point approximation requires the calculation of second (and possibly higher) derivatives of the PESs at each time step. This is computationally costly, and we have not pursued it yet in the context of AIMS.

It is also possible to entirely abandon the classical equations of motion for the phase space centers within the framework given so far. In this case, one applies the time-dependent variational principle[70] (TDVP) to the phase space centers of the TBFs.[61,71,72] Worth and Burghardt have introduced an efficient set of equations for this purpose,[45,73] rooted in the multiconfiguration time-dependent Hartree (MCTDH) method.[74] They have called these the G-MCTDH and variational multiconfiguration Gaussian (vMCG) methods, according to whether the TBF widths are optimized or frozen, respectively. Variational solution of the equations of motion for the phase space centers incorporates further quantum mechanical effects in the evolution of the TBFs and may lead to faster convergence. However, the equations for the time evolution of the phase space centers of the TBFs become coupled to each other and also to the complex amplitudes, which can lead to numerical difficulties when integrating the equations of motion. Both the G-MCTDH and vMCG methods require the calculation of second derivatives of the PES at each time step, making a fully direct or AIMD approach costly, although it has been done for small molecules.[35,45]

It is worthwhile at this stage to summarize the relationship of these four approaches — the G-MCTDH, vMCG, CCS, and FMS methods. In fact, all four are closely related (as presented so far, *vide infra*) and correspond to Gaussian basis set expansions of the nuclear wavefunction. The G-MCTDH method starts with the Gaussian basis set ansatz and applies the TDVP to all basis set parameters. In the notation used in this article, the parameters which are varied so as to minimize the deviation from the time-dependent Schrödinger equation are $\mathbf{R}_m^I$, $\mathbf{P}_m^I$, c_m^I and α_m^I. The resulting integrals are evaluated either exactly or within the second-order saddle point approximation (which is equivalent to the local harmonic approximation[75] for the diagonal potential matrix elements). The other three methods can all be viewed as restrictions on the G-MCTDH ansatz, where some of the parameters are prescribed and the remainder are varied according to the TDVP. For example, the vMCG method fixes the TBF width parameter α_m^I and uses the TDVP to obtain equations of motion for the remaining parameters $\mathbf{R}_m^I$, $\mathbf{P}_m^I$, and c_m^I. The CCS and FMS methods further constrain the equations of motion for the phase space centers, and the complex amplitudes c_m^I are determined through the TDVP. In the CCS method, the phase space centers of the TBFs evolve by Hamilton's equations with a force averaged over each TBF as shown in Eq. (10). In applications of CCS so far,[76–78] the integrals have been evaluated exactly since the PESs were given in analytical form. Finally, in FMS, the phase space centers evolve via classical mechanics and the complex amplitudes are determined through application of the TDVP. In AIMS, a zeroth order saddle point approximation is further applied to facilitate evaluation of the required integrals.

2.3. Full multiple spawning dynamics — spawning

A key aspect of the FMS method which differentiates it from the G-MCTDH, vMCG and CCS methods is the adaptive nature of the nuclear basis set, i.e. spawning. There is no formal need for spawning — if one applies any of the G-MCTDH, vMCG, CCS, or FMS methods with a complete basis set (and with exact evaluation of all the required integrals), the resulting dynamics will be exact. However, this is impractical for all but the simplest problems. Furthermore, the number of TBFs required to describe the nuclear wavefunction is likely to change in time. One is often interested in cases where the wavefunction is initially well-described by a single multidimensional Gaussian, e.g. the lowest vibrational state of a molecule before photoexcitation. If the number of basis functions must be fixed over the duration of the simulation, one must then start with many more TBFs than needed in order to ensure that the dynamics can be modeled later in the simulation. This often leads to ill-conditioned equations that can be numerically difficult. Spawning avoids this by allowing the number of TBFs to expand during the simulation.

The key questions in adaptively increasing the basis set are when and where one should spawn. Although tunneling effects have been considered,[79] our primary focus (and that most relevant to the case of conical intersections) has been on the case of nonadiabatic transitions. In this case, the question of when to spawn is simplified. Specifically, the classical equations of motion for the TBF phase space centers will never generate TBFs on another electronic state. If the TBF basis set were complete, it would include TBFs on all electronic states and therefore there would be no need to expand the basis set to describe nonadiabatic transitions. The spawning idea proceeds backward from this notion. One imagines that the basis set is complete on all electronic states and then asks which basis functions would have been important in the near future. These are added to the basis set with vanishing complex amplitudes. One then rewinds the simulation, i.e. "back-propagates", in order to place the new "spawned" TBFs such that they will be in the "right place at the right time". This back-propagation applies only to the phase space centers, as it is only a means of ensuring that the appropriate TBFs are present to be populated if the nuclear Schrödinger equation so dictates. The simulation proceeds with the enlarged basis set, solving for both the evolution of the TBF phase space centers and also the complex amplitudes. It may happen that the spawned TBF is not populated during this propagation, i.e. its complex amplitude remains vanishingly small. This is not a problem, but simply indicates that the spawned TBF was not important after all.

In the case of nonadiabatic transitions, the question of when to spawn is relatively straightforward. The nonadiabatic coupling (NAC) vector can be calculated at the center of each TBF

$$\mathbf{d}_{IJ} = \left\langle \phi_I \left| \frac{\partial}{\partial R} \right| \phi_J \right\rangle_r, \qquad (12)$$

and the breakdown of the Born–Oppenheimer approximation is expected if the norm of this vector (or alternatively, its dot product with the nuclear velocity) gets large (where one of the electronic indices I or J corresponds to the basis function under consideration). We introduce a numerical threshold (determined by running preliminary dynamics simulations and choosing an acceptable value for each problem) and when the norm of the NAC vector exceeds this threshold, spawning should occur. In most applications of FMS so far, one introduces the concept of a nonadiabatic event defined as starting when the norm of the NAC vector exceeds the "spawning threshold" and ending when the norm of the NAC vector falls below this threshold. Then a child TBF is added to the simulation such that it has maximum overlap with the parent TBF when the NAC vector is largest. It is possible for a

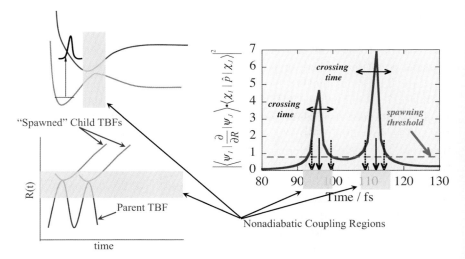

Fig. 1. Schematic depiction of the spawning procedure. Upper left shows a pair of model one-dimensional potential energy curves, where the avoided crossing region is shaded. Lower left shows the evolution of the center of position for the parent TBF, with the avoided crossing region again shaded. It also shows the evolution of the position center for two child TBFs that are spawned. The right panel shows the magnitude of the NAC vector as a function of time along the center of the parent TBF. When this exceeds the spawning threshold, a nonadiabatic event is occurring (denoted "crossing time" in the figure). Child TBFs are spawned during the crossing time, indicated by arrows pointing to the time axis. Three arrows are shown for each of two nonadiabatic events, and the number of child TBFs per event is a simulation parameter. The solid arrow denotes the child TBF which would be used if only a single child is spawned per event — this is placed at the maximum of the NAC during the event.

parent TBF to spawn multiple children during a nonadiabatic event, and the number of children per event is a simulation parameter. A sketch of the spawning procedure is provided in Fig. 1.

Recently, we have explored an alternative answer to the question of when one should spawn. The usual spawning method identifies a fixed number of times within each nonadiabatic event when TBFs should be spawned. This can be less than ideal because within a given simulation there may be both short and long nonadiabatic events. The former can be treated with few spawned child TBFs while the latter may require more. The "continuous spawning" method simply attempts to spawn a new TBF *every time step* during every nonadiabatic event.[80] This is only practical because child TBFs are rejected when they are linearly dependent with the existing TBFs. If a child TBF was just spawned, it is likely that an attempt to spawn again will generate a new child TBF which is almost identical and thus highly linearly dependent with the previous child. Thus, most spawning attempts do not lead to an increase in the number of TBFs in the continuous spawning method. Furthermore, the back-propagation step seems to be less important when continuous spawning is used. Continuous spawning has not yet been used in AIMS, but this is under investigation.

Once one has established *when* to spawn, the next question is *where* one should spawn. Specifically, what should be the phase space center of the child TBF? We demand that the child TBF have the same classical energy as its parent, in order to ensure conservation of the classical energy once the parent and child TBF are well separated. This is not strictly necessary (there is no restriction in quantum mechanics to forbid the inclusion of any basis functions), but rather is a restriction imposed to avoid amplifying errors due to basis set incompleteness. This constraint still leaves considerable freedom in where the child TBF can be placed. Two natural limits can be envisioned — placing the child TBF at the same position (momenta) as the parent TBF and adjusting the momenta (position) to satisfy energy conservation. We denote these as position-preserving and momentum-preserving spawns, respectively.

The position-preserving spawn is the choice most closely resembling current practice in surface hopping. In this case, Tully has suggested[38,81] adjusting the momentum along the direction of the NAC vector in order to satisfy energy conservation. Later, Herman showed that this choice can be justified from a semiclassical perspective.[82] In practice, the momentum of the child TBF is calculated by

$$\mathbf{P}^I_{new} = \mathbf{P}^J_{old} - D\hat{\mathbf{d}}^{IJ}, \quad (13)$$

where $\mathbf{P}^I_{new}$ is the momentum vector of the newly spawned child TBF and $\mathbf{P}^J_{old}$ is the momentum vector of the parent TBF. $\hat{\mathbf{d}}^{IJ}$ is a unit vector along the nonadiabatic coupling vector defined by

$$\hat{\mathbf{d}}^{IJ} = \frac{\mathbf{d}^{IJ}}{|\mathbf{d}^{IJ}|}, \quad (14)$$

and D is a scalar variable, the value of which is chosen to ensure energy conservation.

Note that, in some cases, it may happen that the surface to which a spawn should occur is not classically energetically accessible, i.e. there is no real and positive value of D which satisfies energy conservation. In the surface hopping method, such a failure is called a frustrated hop. The treatment of these frustrated hops has been a matter of some debate, but recently it has been argued[83,84] that they should be ignored (i.e. no transition should be made in these cases) in order to ensure detailed balance and correct equilibration.

In FMS, however, the detailed balance arguments do not hold. The spawning procedure provides the *possibility* of nonadiabatic transitions and it is only the integration of the nuclear Schrödinger equation through Eq. (5) that can turn this possibility into a reality by modulating the complex amplitudes of the child TBFs. It is the time-dependent Schrödinger equation, not the spawning event itself, that governs population transfer and thus maintains the detailed balance condition. Thus, there is no problem with "frustrated spawns" in FMS, although there are a number of different ways to implement the spawning algorithm when this happens. Specifically, one can (1) look for the closest time during the nonadiabatic event when the momentum adjustment is allowed, (2) accept the child TBF in spite of its energy being different from that of the parent, (3) adjust the position of the child TBF by a steepest descent procedure, or (4) adjust the momentum first along the NAC vector and then along other directions as needed. In many cases, we find that the resulting population transfer is not overly sensitive to the details here. However, one would prefer a more elegant approach, and this is the basis of the recently introduced "optimal spawning" procedure.[85]

The key to spawning optimally lies in pinpointing the ideal blend of position and momentum-preserving spawns. Heller and co-workers[86,87] noted the importance of hybrid jumps in position and momentum in their analysis of both radiative and non-radiative transitions. To our knowledge, a method acting on this intuition has thus far failed to appear. In part, this is certainly because the classical nature of the trajectories in surface hopping makes it difficult to see how the position can be adjusted. In AIMS, however, the basis functions have a phase space width and the TBFs will interact with each other even if they have different phase space centers. In order to define an optimal spawn, we appeal to the two criteria which have been implicit in our discussion so far: (1) the parent and child TBF should have the same classical energy and (2) the child TBF should be maximally coupled to the parent in order that the nonadiabatic event can be described with as few TBFs as possible. These two criteria can be encapsulated in a functional,

$$\lambda |E(\chi^I_{\text{parent}}) - E(\chi^J_{\text{child}})|^2 - |\langle \chi^I_{\text{parent}} | V_{IJ}(\mathbf{R}) | \chi^J_{\text{child}} \rangle_{\mathbf{R}}|, \qquad (15)$$

which should be minimized, where λ is formally a Lagrange multiplier to be optimized. Minimizing Eq. (15) is equivalent to jointly minimizing the energy difference and maximizing the coupling between parent and child basis functions (shown here in the diabatic electronic representation — the expression for the adiabatic representation is similar). We implement this minimization by a multiplier penalty approach, where the value of λ is fixed and $\mathbf{R}_{\text{child}}^{J}$ and $\mathbf{P}_{\text{child}}^{J}$ are varied to minimize Eq. (15). The value of λ is then steadily increased, each time repeating the optimization of $\mathbf{R}_{\text{child}}^{J}$ and $\mathbf{P}_{\text{child}}^{J}$ to minimize Eq. (15). The procedure ends when the classical energy difference falls below a small threshold. Each minimization cycle is performed with standard conjugate gradient techniques. A graphical depiction of the procedure for a two-dimensional model conical intersection problem is shown in Fig. 2.

With this procedure, the question of where to spawn is answered with no *ad hoc* assumptions. Preliminary results on small model systems suggest that convergence is more rapid with optimal spawning. It will be interesting to see the extent to which the optimal spawning procedure picks out the position-preserving spawn with momentum adjustment along the NAC vector, which corresponds most closely to the choice in surface hopping. It is easy to anticipate that this will often be the case when nonadiabatic transitions occur around strongly "peaked" intersections,[88–90] which have a pronounced funnel shape. In these cases, the wavepacket spends little time in the nonadiabatic coupling region and the transition is very well described as impulsive. However, when intersections have a more sloped topography and/or when population is funneled from a lower electronic state to an upper electronic state[91] (these are also the cases when one would observe more frustrated hops in a surface hopping method), the nonadiabatic events can be more prolonged and it is here that one expects the optimal spawning procedure to be an improvement. We find the ability to treat both of these limiting cases with a single prescription to be a compelling reason for optimal spawning.

2.4. *Full multiple spawning dynamics — initial conditions*

The last point which needs to be specified concerning the FMS method is the choice of initial conditions. We choose the phase space centers for the initial TBFs by sampling from the Wigner distribution of the desired initial wavefunction. Generating this Wigner distribution can itself be quite challenging, but is straightforward in the harmonic approximation (either

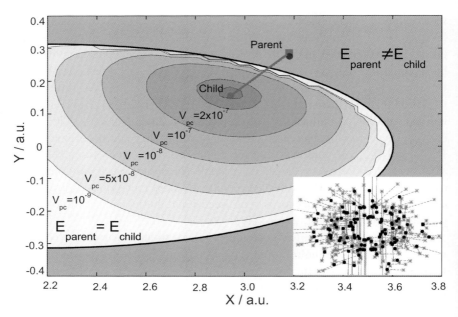

Fig. 2. Depiction of the optimal spawning algorithm for a model two-dimensional problem. The location of the parent TBF is shown as the gray square. Under the "standard" spawning algorithm (black dot), one adjusts the momentum of the spawned ("child") TBF to ensure conservation of classical energy and its position is the same as that of the parent. In optimal spawning, the child is placed at the phase space location that leads to the largest coupling matrix element between parent and child TBFs while simultaneously satisfying the classical energy conservation criteria. The graph shows the value of the maximum coupling matrix element as a function of the two coordinates (choosing the momenta which allow energy conservation and maximize the coupling). Coordinate locations where there is no possible choice of momenta satisfying energy conservation are shown in gray. Because the parent TBF lies in a gray region, this would be a "frustrated hop" in the surface hopping method. In the inset, we compare phase space locations for standard and optimal spawning. The x and y axes in this plot correspond to the X coordinate and the momentum along X. For standard spawning (black dots), only the momentum differs from the parent TBF. In contrast, both the momentum and position of the child TBF differ using optimal spawning (gray asterisks).

at zero temperature for the vibrational ground state or at any desired finite temperature). Once the phase space centers of the initial TBFs are identified, the initial complex amplitudes must be determined. This is done by least squares fitting:

$$c_m^I(0) = \sum_{n=1}^{N_I(0)} (S^{-1})_{mn}^{II} \langle \chi_n^I(t=0) | \Psi_{t=0}^{\text{target}} \rangle. \qquad (16)$$

Ideally, one starts with many TBFs which are chosen to reproduce the desired initial nuclear wavefunction as well as possible. In high-dimensional problems, this is often not warranted because the initial TBFs will rapidly separate and there will be little difference from averaging the results of many independent simulations, each with a single initial TBF (appropriately weighted). Indeed, the quantum coherence effects which FMS can most accurately reproduce are short-time coherences — after a child TBF is spawned, it propagates according to a different PES than its parent. Thus, the parent and child eventually separate and their interaction is minimal [this is most apparent in the zeroth-order SPA expression of Eq. (7), since the basis function overlap will decay as the basis functions separate]. These short-time coherence effects are critical — they are the entire reason for nonadiabatic transitions, which rely on the development of coherences that live long enough to be converted into finite population. However, long-time coherences are much more difficult to represent since the underlying classical dynamics of realistic multidimensional systems is often chaotic. This has been discussed in a related context by Walton and Manolopoulos.[92] It is likely that a large part of this effect is physically correct and corresponds to the decoherence phenomena which make it difficult to observe quantum interference in large systems.[93–95]

A schematic overview of the AIMS method is provided in Fig. 3. This shows the general flow of the program, highlighting the selection of initial conditions, the propagation of the TBFs and their associated complex amplitudes, and the spawning procedure which adaptively increases the size of the basis set.

2.5. *Ab initio multiple spawning — electronic structure considerations*

So far, we have discussed the FMS dynamics method which forms the core of AIMS. FMS was designed in order to be interfaced with *ab initio* electronic structure methods for nonadiabatic AIMD. Thus, the method can be interfaced with a large variety of methods for determining ground- and excited-state PESs. Indeed, we have carried out AIMS calculations using a variety of such methods, including wavefunction-based techniques like multireference configuration interaction (MRCI),[34] density-based methods like time-dependent density functional theory (TDDFT),[96] reparameterized semiempirical theories,[97,98] and hybrid quantum mechanics-molecular mechanics (QM/MM) approaches.[98,99] However, it is fair to say that the

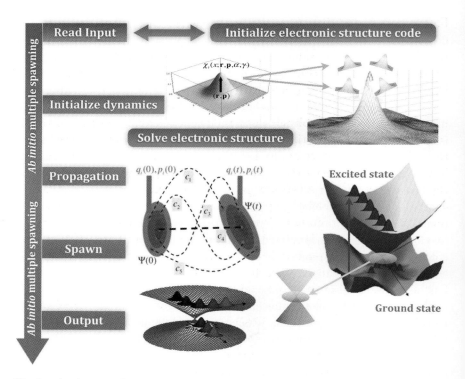

Fig. 3. A schematic depiction of *ab initio* multiple spawning dynamics. Trajectory basis functions are Monte Carlo sampled from a Wigner distribution representing the desired initial state. These TBFs are propagated using classical equations of motion for the phase space centers simultaneous with solution of the nuclear Schrödinger equation in the finite time-evolving basis set to obtain the complex amplitudes for each TBF. The spawning procedure introduces new TBFs when the nonadiabatic coupling is large and the code iterates between these propagation and spawning steps until the desired simulation time is reached.

solution of the electronic Schrödinger equation for excited states is far from routine and thus we comment here on the types of methods which are most suitable and the types of difficulties one may expect to encounter.

Ideally, the electronic structure method that is used should provide analytic gradients since these need to be computed at every time step. Likewise, analytic NAC vectors should be available, although this is less important since these only need to be computed if the energy gap between two states is small (e.g. <1 eV). Of course, it is possible to calculate gradients and NAC vectors numerically by finite difference and we have done this previously.[33,34,91] However, this is costly and not generally recommended.

Fortunately, the utility of analytic gradients is widely recognized and the technology is well developed.[100,101] Therefore, this is often not an obstacle, but there remain important cases where no implementation (e.g. high-order variants of coupled-cluster theory) or only a single implementation (e.g. MRSDCI[102]) of analytic gradients and NAC vectors exists.

The existence of analytic gradients and NACs is a practical matter, but more important from a physical perspective is the ability of the electronic structure method to describe at least qualitative aspects of the excited-state potential-energy surfaces and conical intersections. Here one must make a forceful distinction between methods which are appropriate for electronic spectroscopy and those which are appropriate for excited-state nonadiabatic dynamics. For most spectroscopic problems, it is only the description of the excited states in the Franck–Condon region which is relevant. However, if excited-state dynamics is of interest, a much wider range of the excited-state PES should be well described. This poses a serious challenge for single reference methods such as the most widely used coupled-cluster techniques.[103] This is especially important when the excited states are obtained by linear response. If the ground state is poorly described, there is little hope that a reasonable description of the excited state will ensue from such treatments. Thus, excitation energy equation-of-motion coupled-cluster (EE-EOM-CC) methods[104] are suspect when the ground-state coupled-cluster method has problems (which is often the case when bond rearrangement occurs). A further potential concern stems from the non-Hermitian nature of the eigenvalue problem which yields the excited-state energies. Near an intersection of two excited states, e.g. S_1/S_2, even small errors (for example, because the excitation level of the cluster operator is necessarily truncated) can lead to complex eigenvalues which cause severe difficulties for a dynamics method that expects the PESs to be real-valued.[105]

Furthermore, the excited-state energies are determined independently of the ground-state energy and thus a special problem arises for intersections with the ground state, e.g. S_0/S_1. Specifically, the topology of any intersections is not likely to be correct since the S_0 and S_1 energies are not obtained by diagonalization of the same Hamiltonian matrix. To see this, one can apply the same reasoning[106,107] that leads to the noncrossing rule and the need for two directions (and only two) that lift the degeneracy around a two-state conical intersection. The rationale is of course that there are two conditions to satisfy — equality of the diagonal elements of the Hamiltonian operator, $H_{11}(R) = H_{22}(R)$, and vanishing of the off-diagonal element, $H_{12}(R) = 0$. All three of these are independent functions and therefore R must have at least two independent degrees of freedom to satisfy the two constraints. Any method which solves for the ground-state energy first and then solves for the excited states separately without changing the ground-state energy may be considered (by whatever contrivance needed) to arise from a matrix eigenvalue problem of the form:

$$\begin{pmatrix} E_{S0} & 0 \\ 0 & \mathbf{A} \end{pmatrix} \mathbf{C}_i = E_i \mathbf{C}_i, \qquad (17)$$

where the indicated off-diagonal elements are restricted to always vanish. Thus, there is only one nontrivial condition for an intersection and therefore there is *only one* direction which breaks the degeneracy around the intersection. This means that the intersection is *not* conical, i.e. the dimensionality of the intersection is $N - 1$ and not $N - 2$, where N is the number of internal degrees of freedom. This argument is meant in the topological sense, i.e. it is not strictly a proof and yet it is certain that the slightest perturbation will bring any counterexamples into conformance.

On one level, the above reasoning makes a simple statement which may be considered as trivially profound or profoundly trivial[108]: no method which treats interacting states differently, i.e. independently, can possibly give rise to the correct topography around a conical intersection. The statement is trivial because a method which treats the region around conical intersections correctly should also treat the intersections themselves correctly. Since the two states are by construction degenerate at the intersection point, any method which treats them independently will tend to break the degeneracy.

None of the above considerations hold much sway when one is interested in the Franck–Condon region, where the S_0/S_1 gap is usually large (although when it is not, there can be surprises[109]). Thus, these comments are not incompatible with the statement that EE-EOM-CC is one of the most accurate and systematic methods for electronic spectroscopy.[110] They simply point out the difficulties which will be encountered when larger regions of the excited-state PES are explored.

It is also worth pointing out that the above considerations apply directly to linear response time-dependent density functional theory (LR-TDDFT).[111,112] One should therefore not be surprised when potential energies become complex near intersections between excited states or when the ground- and first-excited-state PESs are grossly distorted near intersections, as we have previously shown.[113] One way out of this dilemma is to ensure that the reference state is not of interest in the problem at hand. For example, one can build response equations which change the spin of the wavefunction or the number of electrons it describes. The spin-flip[114] and electron affinity EOM-CC (EA-EOM-CC)[115] approaches are examples of this strategy. Indeed, we have suggested[113] that a spin-flip variant of TDDFT might lead to improvements near intersections and this has been recently verified.[116]

The above discussion should make it clear that multireference methods are (at least at present) the most likely to describe the regions near intersections accurately. The MRCI method is one obvious choice and indeed the first AIMS simulations used this.[33,34,91,117] Recently, the Lischka group has carried out simulations using surface hopping and exploiting MRCI analytic gradients and NAC vectors.[31,118] Unfortunately, the expense associated with MRCI limits its application to rather small molecules, although recent developments may change this.[119,120]

Perhaps the simplest multireference method which could be used in AIMS is the complete active space self-consistent field (CASSCF) method.[121] The use of state-averaged variants[122] overcomes the root flipping problem which is otherwise often encountered when searching for excited states with CASSCF. The potential difficulty is that CASSCF does not effectively recover dynamic electron correlation and thus large errors (>1 eV) may be found for vertical excitation energies. More important for dynamics calculations than the absolute values of the vertical excitation energies is the *ordering* of the excited states. If this is grossly incorrect, it will be difficult to derive anything useful from dynamics simulations. A key point in this regard is that the choice of the active space in CASSCF calculations is somewhat arbitrary. Although in principle it is true that a larger active space is better, this is not always borne out by calculations on excited states.[96,123] The CASSCF method does include a varying degree of dynamic electron correlation as the active space is increased. This is clear because CASSCF becomes full configuration interaction (CI) (and hence exact within the chosen basis set) in the limit of an active space including all electrons and orbitals in the molecule. However, one is in practice always far from this limit, and the CASSCF method is well known to be a very inefficient way of including dynamic electron correlation effects. Thus, we take the pragmatic view that the active space should be chosen on energetic criteria. One useful way to determine an initial guess for the number of electrons and orbitals in the active space is to compute the vertical excitation energies using EE-EOM-CC. From analysis of the EE-EOM-CC coefficients for the low-lying electronic states, one can deduce a good guess at the required number of electrons and orbitals in the active space. Then, several calculations are carried out varying the number of electrons and orbitals around the EE-EOM-CC predicted values. From each of these, a CASPT2 calculation[124] (including dynamic electron correlation effects in the CASSCF through second-order multireference perturbation theory) is carried out. If the active spaces are all reasonable, the vertical excitation energies computed by CASPT2 will be nearly independent of the underlying CASSCF. From this procedure, one can collect a set of different active spaces which have the electronic states ordered correctly and also are apparently equally valid as judged by the resulting CASPT2 vertical excitation energy. Ideally, one continues the verification procedure by tracing out reaction paths on the excited state using both CASSCF and CASPT2 with the active spaces which remain under consideration. The CASSCF active space which gives the best agreement with CASPT2 is the one which is then used. The number of electronic states included in the state-averaging procedure can also be varied in this process, using the same general criteria that a reasonable CASSCF calculation will reproduce the features predicted by CASPT2. This type of procedure is much more effective at determining a robust active space than one based on chemical intuition, which often ignores the influence of dynamic electron correlation. When used in the context of dynamics, it is necessary to repeat this process several times. Essentially, one first determines which active spaces are reasonable at the Franck–Condon point as discussed above and then carries out a few dynamics simulations. From the observed dynamics, one can extract coordinates which are important and construct paths along which the CASSCF and CASPT2 potential-energy surfaces can be compared.

For small to medium sized molecules, it is now possible to supplant this procedure with dynamics and/or optimizations using CASPT2 directly.[125–127] Given the considerations above, it is critical in this context to use the "multistate" formulations of CASPT2,[128] where the final step is diagonalization of an effective Hamiltonian. In general, any multireference perturbation theory which ends in diagonalization should be appropriate.[129] Many applications of AIMS using MS-CASPT2 are underway and it will be interesting to see how well the "calibrated active space" approach has worked by direct testing.

3. Applications

Our primary focus in this article is methodological, but we also want to highlight one application of AIMS which is of considerable interest in a volume devoted to conical intersections. This concerns the dynamical effects of three-state intersections (see also Chapter 3 by Matsika in this volume) and the conclusions drawn here are likely to be quite general.

The existence of three-state intersections (3SIs, geometries where three adiabatic states of the same spin symmetry are simultaneously degenerate) has been known for some time,[130] but only recently have these been located in a number of "ordinary" molecules.[131–135] Using AIMS dynamics, we discovered a 3SI ($S_0/S_1/S_2$) in malonaldehyde,[123,136,137] a prototype for excited-state intramolecular proton transfer.[138] The electronic structure method used here was CASSCF, state averaging over the lowest three singlet states and using an active space with four electrons in four orbitals, i.e. SA-3-CAS(4/4). The electronic basis set used was the 6-31G* set. The total number of TBFs included in the AIMS simulations is nearly 5000 (distributed over 128 different initial conditions). The bright state in malonaldehyde corresponds to a $\pi \rightarrow \pi^*$ excitation and is the second excited state (S_2), lying above an optically dark $n \rightarrow \pi^*$ state (S_1). As shown schematically in Fig. 4, there are two easily accessible decay channels after excitation to S_2. The molecule can undergo hydrogen atom transfer (sometimes called proton transfer) while in a planar geometry, and in this case an S_2/S_1 intersection is encountered. Alternatively, it can twist around the C=C bond, which is more energetically favorable and leads to a 3SI. A surprising feature was that this 3SI was, as far as we could tell, the absolute minimum on the optically bright electronic excited state (see Fig. 5). Thus, population would be funneled towards the 3SI and one might wonder what the dynamical consequences would be. Going back to the noncrossing rule mentioned above, it is easy to see that the number of directions which lift the degeneracy around a 3SI will be exactly five. Furthermore, it is clear that in the neighborhood of a 3SI there will be numerous "normal" two-state conical intersections. Thus, even a wavepacket traveling directly towards a 3SI is highly likely to encounter two-state conical intersections on the way. The electronic population dynamics is shown in Fig. 6, and it can be seen that population appears to build up on S_1 before quenching to S_0, suggesting that the 3SI may not be reached directly. This is in line with the ideas mentioned above, i.e. that the 3SI is embedded in a sea of two-state intersections and it is exceedingly difficult for a wavepacket to reach the 3SI

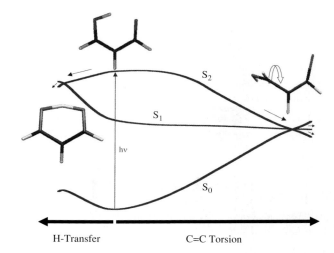

Fig. 4. Schematic description of the possible reaction paths in the excited state dynamics of malonaldehyde. Proceeding to the left, the molecule encounters a two-state intersection related to hydrogen atom transfer. Proceeding to the right, the molecule encounters a 3SI related to torsion about a C=C bond. The AIMS dynamics simulations predict that these two channels are not independent — the molecule can get close to the two-state conical intersection while still planar, but then proceed to twist and access the 3SI.

without first quenching through one of the surrounding two-state intersections. Furthermore, after quenching through a two state intersection, the wavepacket will be directed away from the intersection seam and thus very likely away from the 3SI. We can thus speculate that when 3SIs are involved in the dynamics, there will often be significant trapping of population on the intermediate state which may be longer lived than would be expected if population was funneled directly through the 3SI.

4. Conclusions

We have focused on the AIMS method for *ab initio* molecular dynamics around conical intersections in this article. The AIMS method combines FMS dynamics with *ab initio* electronic structure theory to treat excited-state dynamics including nonadiabatic transitions from first principles. Unlike mixed quantum-classical methods such as trajectory surface hopping,[38,39] FMS treats all degrees of freedom on the same footing and thus avoids many of the ambiguities which can otherwise arise. Although

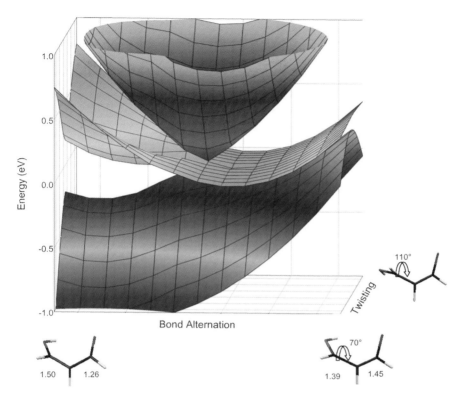

Fig. 5. Cut of the S_0, S_1, and S_2 potential energy surfaces around the 3SI in malonaldehyde. There are five directions which lift the degeneracy and only two are shown here.

AIMS was developed as an *ab initio* molecular dynamics method that can treat quantum effects associated with the breakdown of the Born–Oppenheimer approximation, it was also designed to exploit the ideas of semiclassical dynamics which provide an intuitive understanding of the physics behind chemical reactions and nonadiabatic effects.[139,140]

We presented one recent application of AIMS involving three-state conical intersections and suggested that the dynamics around three-state intersections may often look more like closely spaced two-state intersection dynamics. This does not in any way minimize the importance of three-state intersections — indeed, the only way one would ever expect to see $S_2 \to S_1$ decay followed within tens of femtoseconds by $S_1 \to S_0$ decay is if there were a 3SI nearby, since a 3SI necessarily induces the presence of many

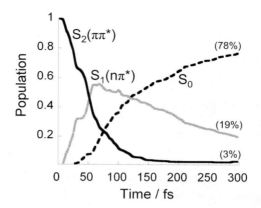

Fig. 6. Population dynamics in malonaldehyde after $S_0 \to S_2$ excitation, as predicted by AIMS. Population resides on S1 for a short time before proceeding to S_0, consistent with two-state S_2/S_1 intersections being involved in the dynamics before the $S_2/S_1/S_0$ 3SI is reached.

two-state intersections in its vicinity. It remains an open question as to what role geometric phase plays around 3SIs. We suggest that malonaldehyde may be a very interesting molecule for femtosecond experiments that can probe the electronic population dynamics. We also suggest that 3SIs may be quite common in molecules containing carbonyl groups conjugated to an unsaturated carbon backbone.

Acknowledgment

This work was supported in part under DOE Contract No. DE-AC02-7600515. We thank Joshua Coe, Benjamin Kaduk, and Benjamin Levine for their contributions to some of the work discussed herein.

References

1. W. Domcke, D.R. Yarkony and H. Köppel (Eds.) *Conical Intersections: Electronic Structure, Dynamics and Spectroscopy* (World Scientific, Singapore, 2004).
2. G.A. Worth and L.S. Cederbaum, *Annu. Rev. Phys. Chem.* **55**, 127 (2004).
3. D.R. Yarkony, *Rev. Mod. Phys.* **68**, 985 (1996).
4. D.R. Yarkony, *Acc. Chem. Res.* **31**, 511 (1998).
5. H. Gohlke and G. Klebe, *Angew. Chem. Internat. Ed.* **41**, 2644 (2002).
6. S.P. Edgcomb and K.P. Murphy, *Current Op. Biotechnol.* **11**, 62 (2000).

7. T. Lazaridis and M. Karplus, *Curr. Op. Struct. Biol.* **10**, 139 (2000).
8. M. Levitt and A. Warshel, *Nature* **253**, 694 (1975).
9. D.B. Gordon, S.A. Marshall and S.L. Mayo, *Curr. Op. Struct. Biol.* **9**, 509 (1999).
10. J. Mendes, R. Guerois and L. Serrano, *Curr. Op. Struct. Biol.* **12**, 441 (2002).
11. C.A. Rohl, C.E.M. Strauss, K.M.S. Misura and D. Baker, *Meth. Enz.* **383**, 66 (2004).
12. A.L. Lomize, I.D. Pogozheva and M.A. Lomize, *Protein Sci.* **15**, 1318 (2006).
13. T. Schaumann, W. Braun and K. Wutrich, *Biopolymers* **29**, 679 (1990).
14. A. Warshel, *Israel J. Chem.* **11**, 709 (1973).
15. J.W. Ponder and D.A. Case, *Adv. Chem. Phys.* **66**, 27 (2003).
16. A. Warshel, P.K. Sharma, M. Kato and W.W. Parson, *Biochim. Biophys. Acta* **1764**, 1647 (2006).
17. N. Gresh, G.A. Cisneros, T.A. Darden and J.-P. Piquemal, *J. Chem. Theory Comput.* **3**, 1960 (2007).
18. J.-P. Piquemal, G.A. Cisneros, P. Reinhardt, N. Gresh and T.A. Darden, *J. Chem. Phys.* **124**, 104101 (2006).
19. X. Chen and V.S. Batista, *J. Chem. Phys.* **125**, 124313 (2006).
20. J. Chen and T.J. Martínez, *Chem. Phys. Lett.* **438**, 315 (2007).
21. H. Yu and W.F. van Gunsteren, *Comp. Phys. Comm.* **172**, 69 (2005).
22. A.K. Rappe and W.A. Goddard, *J. Phys. Chem.* **95**, 3358 (1991).
23. A. Warshel and R.M. Weiss, *J. Am. Chem. Soc.* **102**, 6218 (1980).
24. R. Vuilleumier and D. Borgis, *Chem. Phys. Lett.* **284**, 71 (1998).
25. U.W. Schmitt and G.A. Voth, *J. Phys. Chem.* **102B**, 5547 (1998).
26. I.S. Ufimtsev, A.G. Kalinichev, T.J. Martinez and R.J. Kirkpatrick, *Phys. Chem. Chem. Phys.* **11**, 9420 (2009).
27. A.M.N. Niklasson, *Phys. Rev. Lett.* **100**, 123004 (2008).
28. R. Car and M. Parrinello, *Phys. Rev. Lett.* **55**, 2471 (1985).
29. M.E. Tuckerman, P.J. Ungar, T. van Rosevinge and M.L. Klein, *J. Phys. Chem.* **100**, 12878 (1996).
30. S.S. Iyengar, H.B. Schlegel, J.M. Millam, G.A. Voth, G.E. Scuseria and M.J. Frisch, *J. Chem. Phys.* **115**, 10291 (2001).
31. M. Barbatti, M. Ruckenbauer, J.J. Szymczak, A.J.A. Aquino and H. Lischka, *Phys. Chem. Chem. Phys.* **10**, 482 (2008).
32. E. Tapavicza, I. Tavernelli and U. Rothlisberger, *Phys. Rev. Lett.* **98**, 023001 (2007).
33. M. Ben-Nun, J. Quenneville and T.J. Martínez, *J. Phys. Chem. A* **104**, 5161 (2000).
34. M. Ben-Nun and T.J. Martínez, *Chem. Phys. Lett.* **298**, 57 (1998).
35. B. Lasorne, M.J. Bearpark, M.A. Robb and G.A. Worth, *Chem. Phys. Lett.* **432**, 604 (2006).
36. N.L. Doltsinis and D. Marx, *J. Theo. Comp. Chem.* **1**, 319 (2002).
37. S. Olsen, K. Lamothe and T.J. Martínez, *J. Am. Chem. Soc.* **132**, 1192 (2010).

38. J.C. Tully, *J. Chem. Phys.* **93**, 1061 (1990).
39. J.C. Tully and R.K. Preston, *J. Chem. Phys.* **55**, 562 (1971).
40. M. Ben-Nun and T.J. Martínez, *J. Chem. Phys.* **108**, 7244 (1998).
41. M. Ben-Nun and T.J. Martínez, *Adv. Chem. Phys.* **121**, 439 (2002).
42. T.J. Martínez, M. Ben-Nun and G. Ashkenazi, *J. Chem. Phys.* **104**, 2847 (1996).
43. T.J. Martínez, M. Ben-Nun and R.D. Levine, *J. Phys. Chem.* **100**, 7884 (1996).
44. B.G. Levine, J.D. Coe, A.M. Virshup and T.J. Martínez, *Chem. Phys.* **347**, 3 (2008).
45. G.A. Worth, M.A. Robb and I. Burghardt, *Faraday Disc.* **127**, 307 (2004).
46. Y. Wu and V.S. Batista, *J. Chem. Phys.* **124**, 224305 (2006).
47. Y.H. Wu, M.F. Herman and V.S. Batista, *J. Chem. Phys.* **122**, 114114 (2005).
48. D.V. Shalashilin and M.S. Child, *Chem. Phys.* **304**, 103 (2004).
49. E.J. Heller, *J. Chem. Phys.* **62**, 1544 (1975).
50. E.J. Heller, *J. Chem. Phys.* **64**, 63 (1976).
51. E.J. Heller, *J. Chem. Phys.* **75**, 2923 (1981).
52. T.J. Martínez, M. Ben-Nun and R.D. Levine, *J. Phys. Chem. A* **101**, 6389 (1997).
53. T.J. Martínez and R.D. Levine, *J. Chem. Soc. Faraday Trans.* **93**, 941 (1997).
54. R.C. Brown and E.J. Heller, *J. Chem. Phys.* **75**, 186 (1981).
55. D.J. Tannor and E.J. Heller, *J. Chem. Phys.* **77**, 202 (1982).
56. E.J. Heller, *Acc. Chem. Res.* **14**, 368 (1981).
57. G. Drolshagen and E.J. Heller, *J. Chem. Phys.* **82**, 226 (1985).
58. N.E. Henriksen and E.J. Heller, *J. Chem. Phys.* **91**, 4700 (1989).
59. R. Heather and H. Metiu, *Chem. Phys. Lett.* **118**, 558 (1985).
60. R. Heather and H. Metiu, *J. Chem. Phys.* **84**, 3250 (1986).
61. S.-I. Sawada, R. Heather, B. Jackson and H. Metiu, *J. Chem. Phys.* **83**, 3009 (1985).
62. S.-I. Sawada and H. Metiu, *J. Chem. Phys.* **84**, 227 (1986).
63. S.-I. Sawada and H. Metiu, *J. Chem. Phys.* **84**, 6293 (1986).
64. R.D. Coalson, *J. Chem. Phys.* **86**, 6823 (1987).
65. D. Dehareng, *Chem. Phys.* **120**, 261 (1988).
66. K.G. Kay, *J. Chem. Phys.* **91**, 170 (1989).
67. A.L. Thompson, C. Punwong and T.J. Martínez, *Chem. Phys.* **370**, 70 (2010).
68. T.J. Martínez and R.D. Levine, *J. Chem. Phys.* **105**, 6334 (1996).
69. K. Ruedenberg, *J. Chem. Phys.* **19**, 1433 (1951).
70. P. Kramer and M. Saraceno, *The Geometry of the Time-Dependent Variational Principle.* (Springer-Verlag, Berlin, 1981).
71. D. Hsu and D.F. Coker, *J. Chem. Phys.* **96**, 4266 (1992).
72. R.T. Skodje and D.G. Truhlar, *J. Chem. Phys.* **80**, 3123 (1984).
73. G.A. Worth and I. Burghardt, *Chem. Phys. Lett.* **368**, 502 (2003).
74. H.D. Meyer and G.A. Worth, *Theo. Chem. Acc.* **109**, 251 (2003).

75. E.J. Heller, R.L. Sundberg and D.J. Tannor, *J. Phys. Chem.* **86**, 1822 (1982).
76. D.V. Shalashilin and M.S. Child, *J. Chem. Phys.* **119**, 1961 (2003).
77. D.V. Shalashilin, M.S. Child and A. Kirrander, *Chem. Phys.* **347**, 257 (2008).
78. D.V. Shalashilin and M.S. Child, *J. Chem. Phys.* **121**, 3563 (2004).
79. M. Ben-Nun and T.J. Martínez, *J. Chem. Phys.* **112**, 6113 (2000).
80. M. Ben-Nun and T.J. Martínez, *Isr. J. Chem.* **47**, 75 (2007).
81. R.K. Preston and J.C. Tully, *J. Chem. Phys.* **54**, 4297 (1971).
82. M.F. Herman, *J. Chem. Phys.* **81**, 754 (1984).
83. P.V. Parandekar and J.C. Tully, *J. Chem. Phys.* **122**, 094102 (2005).
84. P.V. Parandekar and J.C. Tully, *J. Chem. Theory Comput.* **2**, 229 (2006).
85. S. Yang, J.D. Coe, B. Kaduk and T.J. Martínez, *J. Chem. Phys.* **130**, 134113 (2009).
86. E.J. Heller and D. Beck, *Chem. Phys. Lett.* **202**, 350 (1993).
87. E.J. Heller, B. Segev and A.V. Sergeev, *J. Phys. Chem. B* **106**, 8471 (2002).
88. B.G. Levine and T.J. Martínez, *Annu. Rev. Phys. Chem.* **58**, 613 (2007).
89. M. Ben-Nun, F. Molnar, K. Schulten and T.J. Martínez, *Proc. Natl. Acad. Sci. USA* **99**, 1769 (2002).
90. D.R. Yarkony, *J. Chem. Phys.* **114**, 2601 (2001).
91. T.J. Martínez, *Chem. Phys. Lett.* **272**, 139 (1997).
92. A.R. Walton and D.E. Manolopoulos, *Chem. Phys. Lett.* **244**, 448 (1995).
93. E.R. Bittner and P.J. Rossky, *J. Chem. Phys.* **103**, 8130 (1995).
94. E.R. Bittner and P.J. Rossky, *J. Chem. Phys.* **107**, 8611 (1997).
95. G.A. Fiete and E.J. Heller, *Phys. Rev. A* **68**, 022112 (2003).
96. B.G. Levine and T.J. Martínez, in *Quantum Dynamics and Conical Intersections*, edited by G.A. Worth and S.C. Allthorpe (CCP6, Daresbury, 2004).
97. A. Toniolo, A.L. Thompson and T.J. Martínez, *Chem. Phys.* **304**, 133 (2004).
98. A. Toniolo, S. Olsen, L. Manohar and T.J. Martínez, *Faraday Disc.* **127**, 149 (2004).
99. A.M. Virshup, C. Punwong, T.V. Pogorelov, B. Lindquist, C. Ko and T.J. Martínez, *J. Phys. Chem. B* **113**, 3280 (2009).
100. B.H. Lengsfield and D.R. Yarkony, *Adv. Chem. Phys.* **82**, 1 (1992).
101. Y. Yamaguchi, J.D. Goddard, Y. Osamura and H.F. Schaefer, *A New Dimension to Quantum Chemistry* (Oxford, Oxford, 1994).
102. H. Lischka, M. Dallos, P.G. Szalay, D.R. Yarkony and R. Shepard, *J. Chem. Phys.* **120**, 7322 (2004).
103. R.J. Bartlett and M. Musial, *Rev. Mod. Phys.* **79**, 291 (2007).
104. J.F. Stanton and R.J. Bartlett, *J. Chem. Phys.* **98**, 7029 (1993).
105. A. Kohn and A. Tajti, *J. Chem. Phys.* **127**, 044105 (2007).
106. E. Teller, *J. Phys. Chem.* **41**, 109 (1937).
107. J. von Neumann and E.P. Wigner, *Z. Physik* **30**, 467 (1929).
108. Adapted from "Chess is profoundly trivial and trivially profound...", George Steiner, source unknown.

109. J.F. Stanton, *J. Chem. Phys.* **115**, 10382 (2001).
110. H. Larsen, K. Hald, J. Olsen and P. Jorgensen, *J. Chem. Phys.* **115**, 3015 (2001).
111. E. Runge and E.K.U. Gross, *Phys. Rev. Lett.* **52**, 997 (1984).
112. M.E. Casida, in *Recent Advances in Density Functional Methods*, edited by D.P. Chong (World Scientific, Singapore, 1995).
113. B.G. Levine, C. Ko, J. Quenneville and T.J. Martínez, *Mol. Phys.* **104**, 1039 (2006).
114. A.I. Krylov, *Acc. Chem. Res.* **39**, 83 (2006).
115. A.I. Krylov, *Ann. Rev. Phys. Chem.* **59**, 433 (2008).
116. N. Minezawa and M.S. Gordon, *J. Phys. Chem. A* **113**, 12749 (2009).
117. J. Quenneville, M. Ben-Nun and T.J. Martínez, *J. Photochem. Photobio. A* **144**, 229 (2001).
118. G. Zechmann, M. Barbatti, H. Lischka, J. Pittner and V. Bonacic-Koutecky, *Chem. Phys. Lett.* **418**, 377 (2006).
119. T.S. Chwee, A.B. Szilva, R. Lindh and E.A. Carter, *J. Chem. Phys.* **128**, 224106 (2008).
120. T.S. Chwee and E.A. Carter, *J. Chem. Phys.* **132**, 074104 (2010).
121. B.O. Roos, *Adv. Chem. Phys.* **69**, 399 (1987).
122. H.J. Werner and W. Meyer, *J. Chem. Phys.* **74**, 5794 (1981).
123. J.D. Coe and T.J. Martínez, *J. Phys. Chem. A* **110**, 618 (2006).
124. B.O. Roos, *Acc. Chem. Res.* **32**, 137 (1999).
125. H. Tao, B.G. Levine and T.J. Martínez, *J. Phys. Chem. A* **113**, 13656 (2009).
126. B.G. Levine, J.D. Coe and T.J. Martínez, *J. Phys. Chem. B* **112**, 405 (2008).
127. J.D. Coe, B.G. Levine and T.J. Martínez, *J. Phys. Chem. A* **111**, 11302 (2007).
128. J. Finley, P.-A. Malmqvist, B.O. Roos and L. Serrano-Andres, *Chem. Phys. Lett.* **288**, 299 (1998).
129. H. Nakano, *J. Chem. Phys.* **99**, 7983 (1993).
130. J. Katriel and E.R. Davidson, *Chem. Phys. Lett.* **76**, 259 (1980).
131. S. Matsika and D.R. Yarkony, *J. Chem. Phys.* **117**, 6907 (2002).
132. S. Matsika and D.R. Yarkony, *J. Am. Chem. Soc.* **125**, 10672 (2003).
133. S. Matsika and D.R. Yarkony, *J. Am. Chem. Soc.* **125**, 12428 (2003).
134. K. Kistler and S. Matsika, *J. Chem. Phys.* **128**, 215102 (2008).
135. L. Blancafort and M.A. Robb, *J. Phys. Chem. A* **108**, 10609 (2004).
136. J.D. Coe, M.T. Ong, B.G. Levine and T.J. Martínez, *J. Phys. Chem. A* **112**, 12559 (2008).
137. J.D. Coe and T.J. Martínez, *J. Am. Chem. Soc.* **127**, 4560 (2005).
138. A.L. Sobolewski and W. Domcke, *J. Phys. Chem. A* **103**, 4494 (1999).
139. W.H. Miller, *Adv. Chem. Phys.* **25**, 69 (1974).
140. W.H. Miller, *J. Phys. Chem. A* **113**, 1405 (2009).

Chapter 10

Non-Born–Oppenheimer Molecular Dynamics for Conical Intersections, Avoided Crossings, and Weak Interactions

Ahren W. Jasper[*] and Donald G. Truhlar[†]

1. Introduction 376
2. Non-Born–Oppenheimer Molecular Dynamics 379
   2.1. Coupled potential energy surfaces 379
   2.2. Efficient integration of NBO trajectories 382
   2.3. Initial conditions for photochemistry 386
3. Fewest Switches with Time Uncertainty 388
4. Coherent Switches with Decay of Mixing 396
5. Summary of Recent Tests and Applications 401
6. Concluding Remarks 409
   Acknowledgments 410
   References 410

---

[*]Combustion Research Facility, Sandia National Laboratories, PO Box 969, Livermore, CA 94551-0969, USA.
[†]Department of Chemistry and Supercomputing Institute, University of Minnesota, 207 Pleasant Street S.E., Minneapolis, MN 55455-0431, USA.

1. Introduction

Processes involving nonradiative transitions between electronic states are ubiquitous in chemistry — from spin-forbidden reactions in combustion to light harvesting in solar cells — and they occur via a variety of elementary chemical mechanisms, such as intersystem crossing, internal conversion, and nonadiabatic electron transfer. The term "non-Born–Oppenheimer" (NBO) may be generally applied to these processes to emphasize the idea that the Born–Oppenheimer separation of the nuclear and electronic time scales breaks down and that potential energy surfaces other than the ground-electronic-state adiabatic potential energy surface play a role in the dynamics. A detailed understanding of NBO coupling of adiabatic electronic states and of the potential energy surfaces associated with them and the ability to predict the effect of this kind of coupling for real chemical systems remain significant challenges to current theories.

One may begin to understand NBO dynamics[1–6] in terms of features of the coupled potential energy surfaces, and in the past we have made the distinction between conical intersections (CIs) of adiabatic surfaces, avoided crossings (ACs) of adiabatic surfaces, and weak interactions (WIs)[7,8] of adiabatic electronic states.

The CIs are $(F-2)$-dimensional hyperseams of degenerate pairs of potential energy surfaces[9] where F is the number of internal nuclear degrees of freedom, which is $3N-6$ for general polyatomics, where N is the number of atoms. (Sometimes more than two surfaces intersect,[3,10] but this paragraph applies to the simplest case of two.) The surfaces form a double cone[4,11] in the two nondegenerate degrees of freedom, and the CI provides an ultrafast decay route from the higher-energy state in the coupled pair to the lower-energy one. The prominent role of conical intersections in promoting such radiationless decay routes was first emphasized by Teller[9]; and it was later used for mechanistic explanations of photochemical reactions.[12–16] Until recently, though, the organic photochemical literature usually associated these decay routes with avoided crossings and regions where potential surfaces approach closely but do not actually cross — such regions were called funnels or bifunnels, which are terms now usually applied to CIs.[17,18] However, the older arguments[9,19] that lead to a correct understanding of the dimensionality of avoided crossings also make it clear that the conical intersections are much more common than fully avoided crossings. Furthermore, since the crossings have a high dimensionality, the seam of crossings can extend over a wide range of geometries, and this can make the dynamics more complicated than that of a reaction dominated by a localized saddle point region or other localized topographical feature of a potential energy surfaces or set of potential energy surfaces.[21] The picturesque funnel language emphasizes the shape of the crossing in the two dimensions called the branching plane where the surfaces cross only at a point, but in many cases greater significance should be attached to the much larger number of dimensions in which the degeneracy is not broken. A picture describing how some coordinates break the degeneracy but others do not is an inverted cuspidal ridge rather than a funnel, or touching cuspidal ridges (an excited surface with ridge down and lower surface with ridge up, with the surfaces touching all along the ridges) rather than a bifunnel (a double cone, that is, a cone touching an inverted cone at a point).

The ACs are locations of nonzero minima in the energy gap as a function of local motion and are almost always associated with nearby CIs,[20,22,23] although those CIs may be energetically inaccessible. The most noteworthy WIs are characterized by wide regions of weak coupling between nearly parallel potential surfaces. Unlike CIs, there is no rule to prevent regions of coupling due to ACs and WIs from occurring in dimensionalities higher than $F - 2$. Although the edited volume in which this chapter appears focuses on CIs and their NBO dynamics, it is important to recognize that realistic potential energy surfaces featuring CIs contain chemically relevant nearby regions of ACs and may also contain regions of significant WIs. The methods presented in this chapter are general enough to treat all these cases.

The presence of a CI is often inferred when ultrafast decay is observed experimentally, and the CI is treated as a critical configuration connecting photoexcited reactants to quenched products when constructing mechanistic reaction coordinate diagrams of photochemistry. One can make a rough analogy to a transition state, but the analogy is at best imperfect and sometimes even deceptive because there are important differences between a CI and an adiabatic transition state as well as differences in the energetic accessibility of other critical regions of the potential energy surface in typical non-BO processes as compared to the kind of reaction where transition state theory is most useful.[21] Transition state dividing surfaces are of dimension $F-1$, and valid transition state dividing surfaces are such that all of the reactive flux must cross through them. Due to the reduced dimensionality of CIs, on the other hand, only a vanishingly small fraction of electronically nonadiabatic flux passes through a CI at the zero-gap intersection. Furthermore, quantitative studies of electronically nonadiabatic systems often require dynamical treatments that are more global than conventional transition state theories, and modeling multistate dynamics occurring via CIs is likely to require global dynamical methods as well. These considerations have motivated the development of trajectory-based methods for simulating NBO chemistry.

In NBO molecular dynamics, an ensemble of classical trajectories is used to model nuclear motions, electronic motion is treated quantum mechanically, and the nuclear and electronic subsystems are coupled according to semiclassical rules. Each trajectory in the ensemble may be thought of as representing a portion of a quantum mechanical wave packet, and taken together the evolution of the ensemble describes the flow of nuclear probability density over the coupled electronic surfaces. Alternatively, each trajectory in the ensemble may be thought of as a distinct chemical event, with its coordinates and momenta subject to the inherent indeterminacy of quantum mechanics.

NBO molecular dynamics is vulnerable to the same sources of error as conventional molecular dynamics, such as the errors associated with the neglect of tunneling through barriers, neglect of quantized vibrations and zero point energies, and neglect of coherences and resonances. NBO molecular dynamics is designed to incorporate one quantum mechanical effect into classical dynamics, namely that of the nonradiative electronic transitions. Accurate treatments of this quantum effect require consideration of tunneling and electronic coherence as well.

A variety of NBO molecular dynamics methods have been proposed. Here we discuss NBO molecular dynamics generally and focus our attention on two implementations: the fewest-switches with time uncertainty[24] (FSTU) surface hopping[25–27] method and the coherent switches with decay of mixing[28] (CSDM) method, a modification of the mean-field[29–31] formalism. The computational cost of these methods is close to that of conventional (i.e. electronically adiabatic) molecular dynamics, and the methods may be readily applied to study a wide variety of chemical processes in both small molecules and large ones.

The dynamics of each trajectory in an FSTU or CSDM ensemble is independent of the others, and transitions between electronic states are allowed anywhere that the electronic surfaces are coupled. Other classes of semiclassical NBO dynamics methods, such as those involving propagating coupled swarms of trajectories,[32–34] restricting hops to predetermined seams,[25,26,35] dressing classical trajectories with frozen Gaussians,[36–40] etc., are not considered in detail, nor are fully quantal calculations.[41–45]

The goal of this chapter is to describe in detail the latest implementations of the FSTU and CSDM methods, summarize the results of the tests used to validate and develop the methods, and describe several recent applications. Trajectory-based methods such as FSTU and CSDM are well suited for mechanistic interpretation, and a brief discussion of this application is also given.

2. Non-Born–Oppenheimer Molecular Dynamics

2.1. *Coupled potential energy surfaces*

In the NBO molecular dynamics simulations described here, an ensemble of independent classical trajectories for nuclear motion is propagated under the influence of a small number of coupled electronic states. The electronic energies (including nuclear repulsion) of each electronic state i provide a potential energy surface V_i for nuclear motion. When representing coupled electronic surfaces, one has a choice of electronic wave functions. The adiabatic electronic wave functions φ_i and energies V_i (where i labels the electronic states) are solutions of the electronic Schrödinger equation

$$H_0 \varphi_i = V_i \varphi_i, \qquad (1)$$

where H_0 contains the electronic kinetic energy and the Coulomb potential operators. When solving Eq. (1), the nuclear coordinates $\mathbf{Q}$ are treated parametrically, and $V_i(\mathbf{Q})$ are the adiabatic potential energy surfaces.

The nuclear kinetic energy operator is written as

$$T_n = -\frac{\hbar^2}{2M} \nabla_n^2, \qquad (2)$$

where ∇_n is a $3N$-dimensional gradient in the nuclear coordinates $\mathbf{Q}$, which are scaled to common reduced mass M (for example, M could be 1 amu). The total wave function of the system is written

$$\Psi = \sum_i \varphi_i(\mathbf{q}; \mathbf{Q}) \chi_i(\mathbf{Q}), \qquad (3)$$

where $\mathbf{q}$ is the collection of electronic coordinates, and χ_i is a wave function for nuclear motion.

If Eq. (3) is used to solve the full molecular Schrödinger equation with the Hamiltonian $H = T_n + H_0$, and if one neglects vibronic Coriolis coupling, one obtains a set of coupled equations for the nuclear motion[1,4,46]

$$\left(T_n + V_i - \frac{\hbar^2}{2M}G_{ii} - E\right)\chi_i = -\sum_{j\neq i}\left(\frac{\hbar^2}{2M}\mathbf{F}_{ij}\cdot\nabla_n + \frac{\hbar^2}{2M}G_{ij}\right)\chi_j, \quad (4)$$

where $G_{ii} = \langle\varphi_i|\nabla_n^2|\varphi_i\rangle$ are Born–Oppenheimer diagonal corrections (BODCs),[47–50] $\mathbf{F}_{ij} = \langle\varphi_i|\nabla_n|\varphi_j\rangle$ are nonadiabatic coupling vectors, $G_{ij} = \langle\varphi_i|\nabla_n^2|\varphi_j\rangle$ are 2nd-order nonadiabatic couplings, and Dirac brackets denote integration over the electronic variables. Although they are not necessarily negligible, G_{ii} and G_{ij} are often neglected, which is considered a semiclassical approximation, and this yields

$$(T_n + V_i - E)\chi_i = -\sum_{j\neq i}\frac{\hbar^2}{2M}\mathbf{F}_{ij}\cdot\nabla_n\chi_j. \quad (5)$$

Equation (5) is interpreted as coupling nuclear motion on the adiabatic surfaces V_i via the action of the nonadiabatic coupling vectors $\mathbf{F}_{ij}$.

Diabatic electronic wave functions may be generally defined as a linear combination of the adiabatic ones,[51,52]

$$\varphi_j^d = \sum_i d_{ij}\varphi_i, \quad (6)$$

that, unlike the adiabatic states, do not diagonalize H_0. Note that the d_{ij} are typically functions of $\mathbf{Q}$. The particular linear combination is often chosen such that the resulting diabatic potential energy surfaces

$$W_{ii} = \langle\varphi_i^d|E_0|\varphi_i^d\rangle \quad (7)$$

or diabatic states have some desirable property, such as smoothness. One may attempt to obtain to define a diabatic basis by minimizing the nonadiabatic coupling vectors; and the electronic basis where

$$\mathbf{F}_{ij}^d \equiv \langle\varphi_i^d|\nabla_n|\varphi_j^d\rangle = 0 \quad (8)$$

for all i and j is called the strictly diabatic basis. For real systems, no such strictly diabatic basis generally exists unless an infinite number of electronic states are considered.[53] The most useful diabatic states are those for which $\mathbf{F}_{ij}^d$ is small enough to neglect and where the infinities in $\mathbf{F}_{ij}$ associated with conical intersections have been transformed away. Such useful diabatic representations can be defined with manageable numbers of electronic states (even with only two). In the discussion that follows, we use "diabatic" both to refer to the artificial situation where Eq. (8) is satisfied and also to refer to the more general situation where $\mathbf{F}_{ij}^{\text{d}}$ is small and is neglected.

Some workers (including us at times) define the nonexistent set of diabatic states for which $\mathbf{F}_{ij}^{\text{d}}$ vanishes identically as "strictly diabatic" and define the states where $\mathbf{F}_{ij}^{\text{d}}$ is small or negligible as "quasidiabatic". Here, as just mentioned, we use a simpler notation, which is also in common use, of just calling all such states "diabatic". One should not think of "diabatic" as a synonym for "not adiabatic"; one could have states that are neither adiabatic nor diabatic. Such representations will be called "mixed".

Diabatic electronic wave functions are not eigenfunctions of H_0, and in general

$$W_{ij} = \langle \varphi_i^{\text{d}} | H_0 | \varphi_j^{\text{d}} \rangle \neq 0 \tag{9}$$

for $i \neq j$. If we write

$$\Psi = \sum_i \varphi_i^{\text{d}}(\mathbf{q}; \mathbf{Q}) \chi_i(\mathbf{Q}), \tag{10}$$

then the equation governing nuclear motion in the diabatic representation is

$$(T_n + W_{ii} - E)\chi_i = \sum_{j \neq i} W_{ij}\, \chi_j, \tag{11}$$

where the off-diagonal matrix elements of the electronic Hamiltonian W_{ij} couple the nuclear motion on the diabatic surfaces W_{ij}, and we have taken advantage of the assumed negligibility of $\mathbf{F}_{ij}^{\text{d}}$.

It is straightforward to employ a general electronic basis, where $\mathbf{F}_{ij}^{\text{d}}$ is not neglected and where $W_{ij} \neq 0$. This so-called mixed representation will not be explicitly considered, though the equations governing NBO dynamics in a mixed representation are straightforward extensions of the adiabatic and diabatic ones.

Adiabatic energies and couplings are readily calculated from the diabatic potential energy matrix elements W_{ij} and their gradients. The adiabatic energies V_i are the eigenvalues of the diabatic energy matrix $\mathbf{W}$, and the variables d_{ij} introduced already in Eq. (6) are the elements of a matrix whose columns are the eigenvectors of $\mathbf{W}$. The gradients of the adiabatic surfaces and the nonadiabatic couplings are

$$\nabla_n V_i = \sum_{j,k} d_{ij}^* d_{ik} \nabla_n W_{jk}, \tag{12}$$

$$\mathbf{F}_{ij} = \begin{cases} \dfrac{1}{V_j - V_i} \displaystyle\sum_{k,l} d_{ik}^* d_{jl} \nabla_n W_{kl} & (i \neq j) \\ 0 & (i = j). \end{cases} \quad (13)$$

On the other hand, if one knows the adiabatic energies and couplings, one may obtain diabatic energies and couplings, but due to the non-uniqueness of the diabatic representation, additional choices and approximations will be needed.[54,55] Procedures have also been developed for obtaining diabatic states without calculating the nonadiabatic coupling vectors.[56–60]

Spin-orbit coupling and other perturbative terms in the molecular Hamiltonian have not yet been considered. These terms may be readily treated using the dynamical methods to be described here, with one principal complexity being the need for a more complicated notation. When spin-orbit coupling is the dominant dynamical coupling and spin-free coupling is to be neglected, the adiabatic surfaces discussed above (which diagonalize the spin-free Hamiltonian H_0 and may be called *valence-adiabatic states*[61]) are often a convenient diabatic basis for the full Hamiltonian, e.g. a useful diabatic matrix for a spin-orbit-coupled two-state system might be

$$\begin{pmatrix} V_1 & U_{\text{SO}} \\ U_{\text{SO}} & V_2 \end{pmatrix}, \quad (14)$$

where U_{SO} is the spin orbit coupling. The eigenvalues of Eq. (14) are the adiabatic potential energy surfaces for the full Hamiltonian including spin. It is equally straightforward to include both spin-free and spin-orbit coupling, as in a recent application to the photodissociation of HBr.[62]

Throughout the rest of this chapter, it is assumed that global potential energy surfaces and their gradients and couplings are available or may be readily calculated for all the electronic states of interest in either the diabatic or the adiabatic representations.

2.2. *Efficient integration of NBO trajectories*

An NBO trajectory evolves independently from the other trajectories in the ensemble and according to classical equations of motion

$$\dot{\mathbf{P}} = -\nabla_n \bar{V}(\mathbf{Q}), \quad (15)$$

$$\dot{\mathbf{Q}} = \mathbf{P}/M, \quad (16)$$

where **P** is the vector of associated mass-scaled nuclear momenta, and the over-dot indicates time-differentiation. The time-dependence of **Q** defines a path through configuration space, and when $\bar{V}$ is the ground state adiabatic potential energy surface, $\mathbf{Q}(t)$ is a conventional classical trajectory. More general formulations of $\bar{V}$ are required to accurately model NBO nuclear-electronic coupling, as will be described in detail in Secs. 3 and 4 for the FSTU and CSDM methods.

The electronic state of the system at any time along an NBO trajectory may be represented as an electronic state density matrix with elements ρ_{ij} where the diagonal elements ρ_{ii} are the electronic populations of states i and the off-diagonal elements ρ_{ij} are coherences. The time evolution of the electronic density matrix elements ρ_{ij} is obtained by solving semiclassical equations along each NBO trajectory; this is sometimes called the classical path approximation. This approach is equivalent to solving for the quantum dynamics of the electronic subsystem in a time-dependent field, which in the present context is created by the nuclear motion. The electronic wave function may be expanded in the adiabatic basis

$$\Phi = \sum_i c_i \varphi_i, \tag{17}$$

where $c_i = a_i + i b_i$ are complex time-dependent expansion coefficients, and the electronic density matrix is defined by

$$\rho_{ij} = c_i^* c_j. \tag{18}$$

The evolution in time of Φ is obtained in this section by solving the time-dependent electronic Schrödinger equation

$$i\hbar \frac{\partial}{\partial t} \Phi = H_0 \Phi, \tag{19}$$

giving the classical path equation:

$$\dot{c}_i = -i\hbar^{-1} c_i V_i - \sum_j c_j \dot{\mathbf{Q}} \cdot \mathbf{F}_{ij}, \tag{20}$$

or, for the real and imaginary parts of c_i,

$$\dot{a}_i = \hbar^{-1} b_i V_i - \sum_j a_j \dot{\mathbf{Q}} \cdot \mathbf{F}_{ij}, \tag{21}$$

$$\dot{b}_i = -\hbar^{-1} a_i V_i - \sum_j b_j \dot{\mathbf{Q}} \cdot \mathbf{F}_{ij}, \tag{22}$$

where $\dot{\varphi}_i$ was evaluated using the "chain rule"[63]

$$\dot{\varphi}_i = \dot{\mathbf{Q}} \cdot \nabla_n \varphi_i, \qquad (23)$$

which is a semiclassical approximation. If a diabatic basis is used,

$$\Phi = \sum_i c_i^d \varphi_i^d, \qquad (24)$$

and

$$\dot{c}_i^d = -i\hbar^{-1} \sum_j c_j^d W_{ij}. \qquad (25)$$

or

$$\dot{a}_i^d = \hbar^{-1} \sum_j b_j^d W_{ij}, \qquad (26)$$

$$\dot{b}_i^d = -\hbar^{-1} \sum_j a_j^d W_{ij}. \qquad (27)$$

The time-dependence in Eqs. (20) and (25) contains an arbitrary phase factor that spins rapidly due to the action of V_i or W_{ii} on $\dot{c}_i$ or $\dot{c}_i^d$. This phase is readily analytically removed to simplify integration of the electronic variables by writing

$$c_i = \tilde{c}_i \exp(-i\theta_i), \qquad (28)$$

where

$$\theta_i = \int V_i dt, \qquad (29)$$

or

$$c_i^d = \tilde{c}_i^d \exp(-i\theta_i^d) \qquad (30)$$

and

$$\theta_i^d = \int W_{ii}^d dt. \qquad (31)$$

These substitutions give

$$\dot{\tilde{a}}_i = -\sum_{j \neq i} [\cos(\theta_j - \theta_i)\tilde{a}_j + \sin(\theta_j - \theta_i)\tilde{b}_j] \dot{\mathbf{Q}} \cdot \mathbf{F}_{ij}, \qquad (32)$$

$$\dot{\tilde{b}}_i = -\sum_{j \neq i} [\cos(\theta_j - \theta_i)\tilde{b}_j - \sin(\theta_j - \theta_i)\tilde{a}_j] \dot{\mathbf{Q}} \cdot \mathbf{F}_{ij}, \qquad (33)$$

and

$$\dot{\tilde{a}}_i^d = -\sum_{j \neq i} [\sin(\theta_j^d - \theta_i^d)\tilde{a}_j^d - \cos(\theta_j^d - \theta_i^d)\tilde{b}_j^d] W_{ij}, \qquad (34)$$

$$\dot{\tilde{b}}_i^d = -\sum_{j \neq i} [\sin(\theta_j^d - \theta_i^d)\tilde{b}_j^d + \cos(\theta_j^d - \theta_i^d)\tilde{a}_j^d] W_{ij}. \qquad (35)$$

Equation (20) neglects vibronic Coriolis coupling, which is discussed elsewhere.[3] In addition, it neglects electronic angular momentum.

It is straightforward to write equations for the time-dependence of the elements of the electronic density matrix by differentiating Eq. (18) and using Eqs. (20) or (25):

$$\dot{\rho}_{ij} = i\hbar^{-1}(\rho_{ii}V_{ii} - \rho_{jj}V_{jj}) + \sum_k \rho_{ik}\dot{\mathbf{Q}} \cdot \mathbf{F}_{ik} - \rho_{kj}\dot{\mathbf{Q}} \cdot \mathbf{F}_{kj}, \qquad (36)$$

$$\dot{\rho}_{ij}^d = -i\hbar^{-1}\sum_k \rho_{ik}^d W_{ik} - \rho_{kj}^d W_{kj}. \qquad (37)$$

The off-diagonal elements of ρ_{ii} are complex, and the real and imaginary parts must be integrated separately. The equations for the (real) electronic state populations may be further simplified

$$\dot{\rho}_{ii} = -2\sum_{j \neq i} \mathrm{Re}(\rho_{ij}\dot{\mathbf{Q}} \cdot \mathbf{F}_{ik}), \qquad (38)$$

$$\dot{\rho}_{ii}^d = 2\hbar^{-1}\sum_{j \neq i} \mathrm{Im}(\rho_{ij}^d W_{ik}). \qquad (39)$$

Quantum mechanical calculations without dynamical approximations and some NBO molecular dynamics methods are independent of the choice of electronic representation if no coupling terms are neglected. In general though, NBO simulations will be dependent on the choice of electronic representation, and both representations will be considered when the FSTU and CSDM methods are described in Secs. 3 and 4. Propagating an NBO trajectory for a system with N electronic states requires integrating the nuclear equations of motion [Eqs. (15) and (16)], as well as either the $2N$ real and imaginary parts of the adiabatic or diabatic electronic coefficients c_i [Eqs. (32) and (33) or (34) and (35)] and the N phases [Eqs. (29) or (31)] or the N^2 unique real and imaginary elements of the adiabatic or diabatic electronic density matrix ρ_{ij} [Eqs. (36) and (38) or (37) and (39)].

Although Eqs. (5) and (11) coupled to Eq. (20), (25), (36), or (37) are derived from the accurate Eqs. (4) and (19), the process of treating the nuclear equations of motion classically means that the quantum electronic subsystem is no longer explicitly coupled to a quantum mechanical environment. It is not correct to treat the electronic subsystem by eq 19 because it is not an isolated system; Eq. (19) is valid only for isolated systems. For subsystems coupled to a medium or environment, one must replace the time-dependent Schrödinger equation [Eq. (19)] by a nonunitary Liouville–von Neumann equation.[64–66] Here the system consists of the electronic degrees of freedom, and the medium consists of the nuclear degrees of freedom; the "nuclear degrees of freedom play the role of observers of the electronic degrees of freedom."[66] The effects of the medium may be broadly described as decoherence.[65–70] The effect of decoherence will be treated by a simple model[71] in Sec. 3 and by a more complete model[8,28,65] in Sec. 4.

2.3. Initial conditions for photochemistry

The ensemble of NBO trajectories is initiated with some distribution in coordinate and momentum space that is intended to simulate the width (or uncertainty) of a quantum mechanical wave packet or of a single-energy slice through a wave packet. The type of reaction and/or experimental situation being modeled determines the specific prescription for the selection of the initial conditions for each trajectory in the ensemble, and the techniques developed for single surface reaction dynamics[72–75] can be applied with minor modifications.

In one typical experimental situation, a chemical system is photoexcited from a well-characterized vibrational state of the ground electronic state to some excited target electronic state. A rigorous sampling scheme might involve calculating absorption cross sections[76,77] for the transitions of interest and sampling from the resulting distribution of quantized vibrational states of the excited electronic state or states. For systems with more than a few atoms the approximate methods used to calculate the ground state and excited state energy levels and the photoabsoprtion cross sections are likely to have significant uncertainty and/or computational cost.

A more efficient strategy for modeling this experimental situation and one that is likely suitable for NBO molecular dynamics of complex systems is as follows. One selects the initial nuclear coordinates and momenta from the ground-state wave function of interest using quasiclassical[73,74] initial conditions and then instantaneously promotes the trajectory to the target excited state. This scheme is equivalent to the Franck principle[78] (the semi-classical analog of the Franck–Condon principle); and it corresponds to exciting the sampled ground state wave function with "white light" that will generally result in an ensemble of trajectories with a relatively wide range of total energies.

An alternative to using quasiclassical initial conditions is to run a classical trajectory (often called molecular dynamics) on the ground-electronic state adiabatic surface and sample from that trajectory. This is done by several groups. One should note, however, that a purely classical trajectory does not retain the quantum distribution of zero point energy or thermal or state-specific vibrational excitation energy in the various vibrational modes, except in the nonquantal limit of vanishingly small vibrational motion. Therefore the quasiclassical initial conditions are preferred.

If the NBO dynamics are expected to be sensitive to energetic thresholds, it may be more appropriate to restrict the range of total energies. An alternative approach is to excite only that slice of the ground-state ensemble with energy gaps between the ground and target electronic states equal to the simulated photon energy within some tolerance. This scheme produces an arbitrarily narrow range of total energies, but it also limits the sampled configuration space.

When the initial conditions are selected from distributions associated with uncoupled regions of the potential energy surfaces, the electronic energies are independent of the choice of electronic representation and the initial electronic state may be assigned unambiguously. However, if an NBO simulation starts in a region where the initial electronic state is coupled to other electronic states, one has to choose both the initial electronic representation and the initial electronic state distribution. For example, it may be appropriate to compute initial distributions in the adiabatic representation. If the simulation is to be carried out in the diabatic representation, the initial adiabatic state i can be projected onto the diabatic states, with the initial diabatic state j selected with the weights d_{ij}^2 obtained from the adiabatic-to-diabatic transformation.

Although quasiclassical initial conditions are quite reasonable for modeling excited vibrational states, they are qualitatively incorrect for ground vibrational states.[79] Thus one reasonable strategy[80] for photodissociation is to use Wigner distributions[77] for vibrational modes with vibrational quantum number 0 and quasiclassical distributions for vibrational modes with quantum number greater than 0. Wigner distributions may also be more accurate than quasiclassical initial conditions for bimolecular collisions,[81,82] but they are only accurate for a short time,[83] and their higher quantum fidelity may be lost by the time the collision partners meet.

3. Fewest Switches with Time Uncertainty

The dynamics methods presented here may be applied in either an adiabatic or a diabatic representation. The results of accurate quantum mechanical calculations and *some* NBO molecular dynamics calculations are independent of the choice of electronic representation. In general, however, as already mentioned in Sec. 2.2, surface hopping and decay of mixing NBO molecular dynamics simulations carried out using the adiabatic representation will produce different results from those employing diabatic representations. When the equations governing the NBO molecular dynamics methods depend on the choice of electronic representation, two equations will be given with the equation numbers appended with "a" and "d" for the adiabatic and diabatic representations, respectively.

Trajectory surface hopping was first employed by Bjerre and Nikitin.[25] Shortly thereafter it was presented in more generality by Preston and Tully.[26] The generalization to allow hopping at any location was first turned into a general algorithm by Blais and Truhlar,[27] as discussed in the excellent review of Chapman.[84] Then Tully improved this procedure by introducing the fewest switches algorithm.[63] The method we will present below differs from the original fewest switches algorithm in three ways: (i) the introduction of time uncertainty,[24] leading to the FSTU method, (ii) the use of a grad V algorithm,[85] and (iii) the introduction of stochastic decay[71,86] (SD). The SD modification in the FSTU/SD method is similar to the method recently employed by Granucci and Persico.[70] These three enhancements to the method are explained in detail below.

In a surface hopping simulation, such as an FSTU simulation, trajectories are propagated under the influence of a single adiabatic or diabatic electronic surface which, for electronic state K, is given by

$$\bar{V} = V_K, \tag{40a}$$

$$\bar{V} = W_{KK}, \tag{40d}$$

but this propagation is interrupted by instantaneous surface switches, i.e. the state label K in Eq. (40), which denotes the currently occupied electronic state, changes at certain points along the trajectory. A change in K is called a surface hop, and at a hopping event the trajectory is instantaneously placed on a different potential energy surface. In general, the potential energy $\bar{V}$ will change discontinuously at a surface hop, and the kinetic energy is adjusted such that total energy and total nuclear angular momentum are conserved. (Electronic angular momentum is neglected.) The nuclear momenta $\mathbf{P}'$ after the hop from surface K to surface K' are given by

$$\mathbf{P}' = \mathbf{P} - \mathbf{P} \cdot \hat{\mathbf{h}}_{KK'}(1 - \sqrt{1 - \Delta_{KK'}/T_{KK'}})\hat{\mathbf{h}}_{KK'}, \quad (41)$$

where $\hat{\mathbf{h}}_{KK'}$ is a unit vector called the hopping vector,

$$\Delta_{KK'} = V_{K'} - V_K, \quad (42\text{a})$$

$$\Delta_{KK'} = W_{K'K'} - W_{KK}, \quad (42\text{d})$$

and

$$T_{KK'} = \frac{1}{2M}(\mathbf{P} \cdot \hat{\mathbf{h}}_{KK'})^2 \quad (43)$$

is the nuclear kinetic energy associated with $\hat{\mathbf{h}}_{KK'}$. The hopping vector determines the component of the nuclear momentum that is adjusted during a hop, and theoretical arguments[26,87] confirmed by numerical tests[42] show that a good choice is

$$\hat{\mathbf{h}}_{KK'} = \mathbf{F}_{KK'}/|\mathbf{F}_{KK'}|. \quad (44\text{a})$$

Because $\mathbf{F}_{KK'}$ is a vector of internal coordinates, the adjustment in Eq. (41) with the choice of Eq. (44a) conserves total angular momentum.

Using the nonadiabatic coupling vector as the hopping vector has been shown to provide accurate results for surface hopping calculations carried out in both the adiabatic and diabatic representations.[42] If the diabatic representation is used, $\mathbf{F}_{KK'}$ can be calculated directly from Eq. (13) for a two-state system. When more than two states are involved, $\mathbf{F}_{KK'}$ should not be used because the adiabatic and diabatic state labels do not generally correlate to a globally consistent pair of states. Instead, the hopping vector in the diabatic representation can be approximated as

$$\hat{\mathbf{h}}_{KK'} = \mathbf{F}^{\text{r}}_{KK'}/|\mathbf{F}^{\text{r}}_{KK'}|, \quad (44\text{d})$$

where $\mathbf{F}^{\text{r}}_{KK'}$ is the reduced nonadiabatic coupling for the submatrix

$$\mathbf{W}^{\text{r}} = \begin{pmatrix} W_{KK} & W_{KK'} \\ W_{K'K} & W_{K'K'} \end{pmatrix}, \quad (45)$$

i.e.,

$$\mathbf{F}^r_{KK'} = \frac{d^{r,*}_{KK}d^r_{K'K}\nabla_n W_{KK} + (d^{r,*}_{KK}d^r_{K'K'} + d^{r,*}_{KK'}d^r_{K'K})\nabla_n W_{KK'} + d^{r,*}_{KK'}d^r_{K'K'}\nabla_n W_{K'K'}}{W_+ - W_-},$$
(46)

and W_+ and W_- are the eigenvalues, and $d^r_{KK'}$ are the elements of matrices whose columns are the eigenvectors of $\mathbf{W}^r$ defined by Eq. (45).

Equation (41) cannot be solved if the radicand is negative, i.e., if the kinetic energy associated with the hopping vector is less than the required energy adjustment and the hop is an upward hop. (For downward hops, $\Delta_{KK'} < 0$ and Eq. (41) can always be solved.) When $\Delta_{KK'} > T_{KK'}$, the hop is declared "frustrated", and additional considerations are required, as discussed in detail below.

In early examples of trajectory surface hopping, hops were allowed only when a trajectory crossed a seam where W_{KK} crosses another diabatic surface,[25,26] but in later work[27,63] this was generalized so that stochastic hopping events may occur after each integration step Δt and anywhere along the trajectory where the currently occupied surface is coupled to one or more other surfaces. Tully provided an elegant and useful formulation[63] for the probability for hopping from the currently occupied electronic state K to some other state K'

$$P_{KK'}(t+\Delta t) = \max \left\{ \begin{array}{c} -\int_{t'=t}^{t+\Delta t} dt' b_{KK'}(t')/\rho_{KK}(t) \\ 0 \end{array} \right. , \qquad (47)$$

where

$$b_{KK'} = -2\mathrm{Re}(\rho_{KK'}\dot{\mathbf{Q}} \cdot \mathbf{F}_{KK'}), \qquad (48\mathrm{a})$$

$$b_{KK'} = 2\hbar^{-1}\mathrm{Im}(\rho^d_{KK'}W_{KK'}). \qquad (48\mathrm{d})$$

Equation (47) is the relative rate of change of the electronic population of state K due to coupling to the state K'. Hops away from state K are allowed only if ρ_{KK} is decreasing, and Eq. (47) is designed to maintain the populations of trajectories in each electronic state n_i according to ρ_{ii} with the fewest number of hops. (The self consistency of n_i and ρ_{ii} is generally *not* maintained, as discussed below.) Equation (47) is called the fewest switches (FS) hopping probability, and this scheme is also called molecular dynamics with quantum transitions (MDQT), which can be confusing because it is not the only scheme for molecular dynamics with quantum transitions.

The quantity $B_{KK'}(t) = \int_{t'=0}^{t} b_{KK'}(t')dt'$ can be integrated along with the nuclear and electronic variables, such that the hopping probability at time $t + \Delta t$ may be evaluated as

$$P_{KK'}(t + \Delta t) = \max \begin{cases} [B_{KK'}(t + \Delta t) - B_{KK'}(t)]/\rho_{KK}(t) \\ 0 \end{cases}. \quad (49)$$

Because surface hops are only allowed between time steps, and because hopping and nonhopping trajectories diverge from one another, the results of a surface hopping simulation must be converged with respect to the available hopping locations. Often, quite small step sizes are required when the electronic populations are changing rapidly, whereas larger step sizes (ultimately limited by the accuracy of the integration of the nuclear coordinates) may be used when propagating through uncoupled regions of potential surface. This situation benefits from variable-step-size integrators.

It may be difficult to converge the available hopping locations when using efficient adaptive-step-size integrators, as the integrator may step through regions where $\rho_{KK'}$ changes sign. Consider an example where a large step Δt is taken through a region where ρ_{KK} is locally quadratic and where $\rho_{KK}(t) = \rho_{KK}(t + \Delta t)$. The FS hopping probability for this step is 0, whereas if two steps of size $\Delta t/2$ are taken, the hopping probability will be finite for one of the steps. Many variable step size integrators can integrate quadratic functions exactly, and this example is of practical concern. A simple modification[42] provides a solution. Specifically, if the increasing

$$b_{KK'}^+ = \max(b_{KK'}, 0) \quad (50)$$

and decreasing

$$b_{KK'}^- = \min(b_{KK'}, 0) \quad (51)$$

parts of $b_{KK'}$ are integrated separately, the integrator is made to take small steps where $b_{KK'}$ changes sign and where $b_{KK'}^+$ and $b_{KK'}^-$ have discontinuous derivatives.

As mentioned above, the FS hopping probability attempts to populate the various electronic states with trajectories such that the fractions of trajectories in each electronic state $n_i \approx \rho_{ii}$ (with the accuracy limited by the finite number of trajectories that are sampled). This self consistency is maintained only when trajectories in the various electronic states do not diverge from one another, i.e. when the potential surfaces are degenerate. For real potential energy surfaces, trajectories in different electronic states diverge, and self consistency is not preserved, although it may be maintained in an ensemble averaged sense, i.e. $n_i \approx \langle \rho_{ii} \rangle$, where the brackets denote an average over the members of the ensemble of trajectories. When classically forbidden hops occur, only upward hops can be frustrated, and self consistency cannot be maintained.[88,89]

One may distinguish two sources of frustrated hops in trajectory simulations. First, the FS algorithm may be incomplete in some way that is causing it to predict finite hopping probabilities where hops should not be allowed. This argument is strengthened by studies showing that accurate results may sometimes be obtained when frustrated hops are simply ignored.[90] As pointed out in the original formulation[63] and further developed in later work,[28,65–71,91,92] one deficiency of the original FS method and other methods based on classical path electronic dynamics is that decoherence is not treated. (We will see below that including decoherence may reduce the number of frustrated hops by reducing unphysical amplitudes for unoccupied states in regions when such states are no longer strongly coupled.) Another possibility is that the FS method is correctly predicting energetically forbidden surface hops, but the hops are frustrated due to the limitations of classical mechanics. In this picture, a frustrated hop is a quantum mechanical attempt to tunnel into a classically forbidden region of an excited electronic state. Several improvements to the FS method based on both of the latter two considerations have been developed and are discussed in the remainder of this section.

One suggestion that was made for eliminating frustrated hops is to use modified velocities for the integration of the quantum amplitudes.[88,92] We do not employ this because comparison to accurate quantum dynamics shows[89] that it decreases the accuracy as compared to using the original unmodified velocities.

The first improvement to the FS method that we discuss is a simple modification designed to incorporate decoherence.[71] Prior to the first surface hop, the electronic variables are assumed to correctly evolve coherently along the trajectory according to the classical path equations. At a surface hop or an attempted surface hop, the system is imagined to split into two wave packets, one traveling on each of the surfaces involved in the surface hop. The system immediately begins to decohere with a first-order rate coefficient $\tau_{\rm SD}^{-1}$ obtained by considering the short time evolution of the overlap of two one-dimensional wave packets traveling in the different electronic states[93]

$$\tau_{\text{SD}}^{-1} = \frac{\pi}{2}\frac{\Delta f_{KK'}}{\bar{p}_{KK'}} + \sqrt{\left(\frac{\Delta p_{KK'}}{\hbar}\right)^2\frac{\Delta_{KK'}}{M} + \left(\frac{\pi\Delta f_{KK'}}{2\bar{p}_{KK'}}\right)^2}, \quad (52)$$

where

$$f_{KK'} = -\nabla_n(V_K - V_{K'}) \cdot \hat{\mathbf{h}}_{KK'} \quad (53)$$

is the difference in the forces of the two electronic states in the direction of the hopping vector,

$$\Delta p_{KK'} = (\mathbf{P} - \mathbf{P'}) \cdot \hat{\mathbf{h}}_{KK'} \quad (54)$$

is the difference in the nuclear momenta before and after the surface hop in the direction of the hopping vector, and

$$\bar{p}_{KK'} = \frac{1}{2}(\mathbf{P} + \mathbf{P'}) \cdot \hat{\mathbf{h}}_{KK'}. \quad (55)$$

If the decoherence event is initiated at a frustrated hop, $\mathbf{P'}$ cannot be calculated and is set to zero in Eqs. (54) and (55).

At each time step (of step size Δt) after the frustrated or successful hop, a stochastic decoherence (SD) probability is computed

$$P_{\text{SD}}(\Delta t) = \exp(-\Delta t/\tau_{\text{SD}}), \quad (56)$$

and P_{SD} is compared to a random number between 0 and 1. If the SD check is successful, the electronic state density matrix is reset to

$$\rho_{ij} = \begin{cases} 1 & \text{for } i,j = K, \\ 0 & \text{otherwise,} \end{cases} \quad (57)$$

where K is the currently occupied electronic state. After reinitialization, the electronic state populations evolve according to the coherent classical path equations. If a frustrated or successful hop occurs before decoherence is called for, τ_{SD} is updated and decoherence checks are continued.

The SD algorithm damps out coherence after some physically motivated time, which reduces the likelihood of the FS algorithm calling for surface hops in regions of weak coupling that are encountered between regions of strong coupling. This has the practical and intended physical effect of reducing frustrated hops in regions where the potential energy surfaces have different energies and/or shapes; decoherence is expected to be fast in such regions.

The next improvement to the FS method incorporates time-uncertainty (TU) hopping,[24] which simulates tunneling into classically forbidden regions of excited electronic states. Inspired by the time-energy version of the uncertainty principle, a frustrated hop occurring at some time t_f is allowed to hop at the nearest time t_h along the trajectory where a hop would be energetically allowed (if such a time exists) but only if

$$|t_f - t_h| \leq \hbar/2E_{\text{def}}, \qquad (58)$$

where

$$E_{\text{def}} = \Delta_{KK'}(t_f) - T_{KK'}(t_f) \qquad (59)$$

is the energy deficiency by which the attempted hop is frustrated. In this way, a TU hop may be thought of as allowing the trajectory to borrow an energy of E_{def} for some short time according to the uncertainty principle as it hops into the excited state. The FS method with TU hops was shown to significantly improve the accuracy of the surface hopping method for some systems, especially those with weakly coupled electronic surfaces.[24]

The FSTU method and the SD algorithm do not eliminate all frustrated hops. The remaining frustrated hops [i.e. those where a t_h satisfying Eq. (58) cannot be found] are attributed to the breakdown of the independent-trajectory approximation and are treated using the "grad V" prescription.[85] In the method, a frustrated trajectory instantaneously receives an impulse from the classically forbidden electronic state based on its gradient in the direction of the hopping vector. Specifically, at a frustrated hop that cannot be remedied by the TU method, the components of the nuclear momentum and force in the target electronic state in the direction of $\hat{\mathbf{h}}_{KK'}$ are calculated by

$$p_{K'} = \mathbf{P} \cdot \hat{\mathbf{h}}_{KK'}, \qquad (60)$$

$$f_{K'} = -\nabla_n V_{K'} \cdot \hat{\mathbf{h}}_{KK'}, \qquad (61a)$$

$$f_{K'} = -\nabla_n W_{K'K'} \cdot \hat{\mathbf{h}}_{KK'}. \qquad (61d)$$

If $p_{K'}$ and $f_{K'}$ have the same sign, the influence of the target electronic state is to accelerate the trajectory, and we choose to continue the trajectory in the currently occupied electronic state without making any adjustments to the nuclear momenta. If $p_{K'}$ and $f_{K'}$ have different signs, the target electronic state is thought to "reflect" the trajectory, and we choose to continue the trajectory in the currently occupied electronic state with the nuclear momentum reversed in the direction of $\hat{\mathbf{h}}_{KK'}$, i.e.

$$\mathbf{P}'' = \mathbf{P} - 2\,\mathbf{P} \cdot \hat{\mathbf{h}}_{KK'}\,\hat{\mathbf{h}}_{KK'}. \tag{62}$$

If the probability of an electronically inelastic event is very small because the probability of a hop is very small, e.g. 10^{-6}, it would typically require very extensive sampling to observe even one inelastic event and even more sampling to accumulate good statistics. For such cases special methods of rare-event sampling have been developed.[94]

It is interesting to examine the question of whether surface hopping methods can be improved by replacing the trajectories with wave packets. In principle the answer is yes, but so far no generally affordable method for doing so has been devised. The most widely employed method involving wave packets for photochemical calculations is the full multiple spawning (FMS) method.[36–39] The assumptions underlying this method have been examined in detail.[37] It was stated[37] that the basis set expansion method underlying FMS "is aimed only at describing quantum mechanical effects associated with electronic nonadiabaticity and *not* at correcting the underlying classical dynamics." One of many serious approximations in replacing an ensemble of trajectories with an ensemble of wave packets is that the wave packets must be coupled. In FMS, in order to keep the method practical, interference between the various initial wave packets that are required[95] to simulate the initial quantum state is neglected; this serious approximation is called the independent-first-generation approximation.[37] One of the features that makes trajectory calculations affordable for complex systems is that an ensemble of trajectories can be run independently of each other, without introducing approximations to accurate classical mechanics. In contrast, running wave packets independently is a serious approximation that is not overcome by spawning more packets or spawning them in a more physical way. Furthermore, FMS does not include decoherence in the treatment of electronic nonadiabaticity (it uses a unitary treatment of the electronic degrees of freedom, not a nonunitary one[65] as in the CSDM method discussed in the next section). Thus FMS is expected to have about the same accuracy as surface hopping without decoherence, which is consistent with our numerical tests.

Efforts to derive improved wave packet methods are underway in more than one group.[40,96] The reader is also referred to the multiconfiguration time-dependent Hartree method,[43–45] which is a variational time-dependent wave function expansion method designed to achieve converged quantum dynamics in an efficient way; it has had outstanding success for small enough systems and systems with particularly amenable Hamiltonians.

4. Coherent Switches with Decay of Mixing

One major deficiency of the FSTU method and of surface hopping methods in general is that the results of the NBO simulation may depend strongly on the choice of electronic representation, that is, adiabatic, diabatic, or mixed. Here we consider an alternative approach to NBO molecular dynamics based on a mean-field approximation. In its simplest form, the mean-field approximation under consideration here[29–31] is called the semiclassical Ehrenfest or SE approximation. It defines the semiclassical potential energy as a weighted average of the potential energy surfaces

$$\bar{V} \equiv \langle \Phi \mid H_0 \mid \Phi \rangle$$

$$= \sum_i \rho_{ii} V_{ii} \tag{63a}$$

$$= \sum_{i,j} \mathrm{Re}(\rho_{ij}^d) W_{ij}. \tag{63d}$$

Note that the gradient of the diabatic mean field energy is straightforward

$$\nabla \bar{V} = \sum_{i,j} \rho_{ij}^d \nabla W_{ij}, \tag{64d}$$

whereas the gradient of the adiabatic mean field energy is

$$\nabla \bar{V} = \sum_i \rho_{ii} \nabla V_i + 2 \sum_{i \neq j} \mathrm{Re}(\rho_{ij}) V_i \mathbf{F}_{ij}, \tag{64a}$$

with the second term on the right hand side arising semiclassically from the action of the nuclear gradient on ρ_{ij}.[97] More rigorous derivations of the equations governing mean-field motion in the adiabatic representation equivalent to Eq. (64a) have been given.[29] SE trajectories are independent of the choice of electronic representation.

In the SE model, trajectories propagating through regions of coupling are governed by an effective potential energy surface that is evolving as an appropriately weighted average of the coupled potential energy surfaces. Although this situation may be an accurate description of coupled-states semiclassical motion, a severe deficiency of the approach is that the trajectory remains in a coherent mixed state after the system leaves the region of coupling. This causes the SE method to predict molecular products to be in coherent superpositions of electronic states, which do not correspond to quantum mechanical or experimentally measured final states. Another less obvious but equally troubling consequence of fully coherent SE propagation is that the system does not "reset" electronically between regions of coupling, which may introduce errors into the dynamics.[91,92] Finally, SE trajectories are not able to explore some processes occurring with small probabilities, as the potential felt by an SE trajectory will be determined mainly by the potential energy surface associated with the higher-probability event.

The CSDM method[28] is a modification of the SE method designed to introduce decoherence outside regions of strong coupling, such that the predicted molecular products are formed in quantized final electronic states. As mentioned in Sec. 2.2, decoherence in the electronic equations of motion may be thought of as arising from the nuclear degrees of freedom acting as a bath, and the bath relaxes the electronic density matrix. An important feature of CSDM trajectories is that they behave similarly to SE trajectories in strong coupling regions, thus preserving much of the representation independence of the SE method.

The decay-of-mixing (DM) formalism collapses a coherent mixed state density matrix to a quantized pure state smoothly over time, and it includes both dephasing

$$\rho_{ij} \to 0 \ (i \neq j) \tag{65}$$

and demixing

$$\rho_{ii} \to \delta_{iK}, \tag{66}$$

where δ_{iK} is the Kronecker delta, and K labels the target decoherent state toward which the system is collapsing. The target decoherent state label K may change over time, as discussed below. When the electronic density matrix collapses to a quantized electronic state, the semiclassical potential energy surface [Eq. (63)] collapses to a pure one, thus providing realistic product internal energy distributions that may be compared with experimental and quantum mechanical ones.

Note that dephasing (as it is defined here as the damping of the off-diagonal elements of ρ_{ij}) is a physical effect, whereas demixing is a semiclassical choice. Dephasing and demixing are assumed to occur at the same rate τ_{iK}^{-1}, where

$$\tau_{iK} = \frac{\hbar}{|\Delta_{iK}|}\left(1 + \frac{E_0}{(\mathbf{P}\cdot\hat{\mathbf{s}}_i)^2/2M}\right), \tag{67}$$

where each state i other than K has its own decoherence time τ_{iK}, E_0 is a parameter typically chosen to be $0.1\,E_h = 2.72\,\text{eV}$, and $\hat{\mathbf{s}}_i$ is a unit vector called the decoherence vector. The decay time in Eq. (67) has a different functional form than the one used previously for the SD method [Eq. (52)] due to algorithmic requirements of the DM method. Equations (52) and (67) are expected to have similar magnitudes.[93] Alternatives to Eq. (67) have also been explored for CSDM calculations, and the results are not overly sensitive to the functional form.[98]

Decoherence and demixing are introduced into the NBO molecular dynamics by modifying the classical path electronic equations of motion,[28,67]

$$\dot{c}_i^{\text{DM}} = \dot{c}_i + \dot{c}_i^{\text{D}}, \tag{68}$$

where

$$\begin{aligned}\dot{c}_i^{\text{D}} &= \frac{1}{2}\frac{c_i}{\tau_{iK}} & i \neq K \\ &= \frac{1}{2}\frac{c_K}{\rho_{KK}}\sum_{j\neq K}\frac{\rho_{jj}}{\tau_{jK}} & i = K\end{aligned} \tag{69}$$

Equivalently, one may write the decoherence terms for the density matrix,

$$\dot{\rho}_{ij}^{\text{DM}} = \dot{\rho}_{ij} + \dot{\rho}_{ij}^{\text{D}}, \tag{70}$$

where

$$\begin{aligned}\dot{\rho}_{ii}^{\text{D}} &= -\frac{\rho_{ii}}{\tau_{iK}} & i \neq K \\ &= \sum_{j\neq K}\frac{\rho_{jj}}{\tau_{jK}} & i = K\end{aligned} \tag{71}$$

for the diagonal elements, and

$$\dot{\rho}_{ij}^{D} = -\frac{1}{2}\left(\frac{1}{\tau_{iK}} + \frac{1}{\tau_{jK}}\right)\rho_{ij} \qquad i,j \neq K$$

$$= \frac{1}{2}\left(\frac{1}{\rho_{KK}}\sum_{k \neq K}\frac{\rho_{kk}}{\tau_{kK}} - \frac{1}{\tau_{jK}}\right)\rho_{ij} \qquad i = K, j \neq K$$

$$= \frac{1}{2}\left(\frac{1}{\rho_{KK}}\sum_{k \neq K}\frac{\rho_{kk}}{\tau_{kK}} - \frac{1}{\tau_{iK}}\right)\rho_{ij} \qquad i \neq K, j = K \qquad (72)$$

for the off-diagonal elements. Equations (69), (71), and (72) can be derived by assuming first-order decay of the diagonal elements, and enforcing conservation of the electronic density and phase angle.[67]

As the system decoheres and demixes, the nuclear momenta are adjusted to conserve total energy

$$\dot{\mathbf{P}}^{\mathrm{DM}} = \dot{\mathbf{P}} + \dot{\mathbf{P}}^{\mathrm{D}}, \qquad (73)$$

where it is convenient to write the additional term as

$$\dot{\mathbf{P}}^{\mathrm{D}} = -\sum_{i \neq K}\frac{\dot{V}_{i}^{\mathrm{D}}}{(\mathbf{P} \cdot \hat{\mathbf{s}}_{i})/M}\hat{\mathbf{s}}_{i}. \qquad (74)$$

Equation (74) guarantees that decoherence is turned off as the momentum available in the decoherent direction $\hat{\mathbf{s}}_i$ goes to zero, with

$$\dot{V}_{i}^{\mathrm{D}} = \frac{\rho_{ii}}{\tau_{iK}}(V_K - V_i), \qquad (75\mathrm{a})$$

$$\dot{V}_{i}^{\mathrm{D}} = \frac{\rho_{ii}}{\tau_{iK}}W_{KK} - \left(\frac{\rho_{iK}}{\tau_{iK}} + \frac{\rho_{iK}}{\rho_{KK}}\sum_{j \neq K}\frac{\rho_{jj}}{\tau_{jK}}\right)W_{iK} - \frac{1}{2}\left(\frac{1}{\tau_{iK}} + \frac{1}{\tau_{jK}}\right)\rho_{ij}W_{ij}. \qquad (75\mathrm{d})$$

The decoherence vector determines the components of $\mathbf{P}$ into and out of which energy is exchanged as the system decoheres and demixes, and we choose

$$\mathbf{s}_i = (\mathbf{P} \cdot \hat{\mathbf{F}}_{iK}\mathbf{F}_{iK} + \mathbf{P}_{\mathrm{vib}}), \qquad (76\mathrm{a})$$

$$\mathbf{s}_i = (\mathbf{P} \cdot \hat{\mathbf{F}}_{iK}^{\mathrm{r}}\mathbf{F}_{iK}^{\mathrm{r}} + \mathbf{P}_{\mathrm{vib}}), \qquad (76\mathrm{d})$$

where ˆ denotes a unit vector (as it always does in this whole chapter), and $\mathbf{P}_{\text{vib}}$ is the vibrational momentum. In regions of strong coupling $\mathbf{s}_i \approx \mathbf{F}_{iK}$ or $\mathbf{F}_{iK}^{\text{r}}$, which is a physically reasonable choice, and when the coupling vanishes (where nonzero $\mathbf{F}_{iK}$ and $\mathbf{F}_{iK}^{\text{r}}$ are not defined) $\mathbf{s}_i \approx \mathbf{P}_{\text{vib}}$, which is a choice that conserves total angular momentum. The vibrational momentum can be calculated for polyatomics as[99]

$$\mathbf{P}_{\text{vib}}^{\alpha} = \mathbf{P}^{\alpha} - M\boldsymbol{\omega} \times \mathbf{Q}^{\alpha}, \tag{77}$$

where

$$\boldsymbol{\omega} = \mathbf{I}^{-1}\mathbf{J}, \tag{78}$$

$\mathbf{I}$ is the intertial tensor matrix, $\mathbf{J}$ is the total angular momentum vector, α labels atoms, and $\mathbf{P}_{\text{vib}}^{\alpha}$, $\mathbf{P}^{\alpha}$, $\boldsymbol{\omega}$, and $\mathbf{Q}^{\alpha}$ are three-dimensional vectors.

A quantum subsystem coupled to an environment does not actually decay to a pure state but rather to a classical, incoherent mixture of states,[100] each associated with a probability of occurring in an ensemble. To incorporate this into the present model, the decoherent state K is allowed to switch stochastically along a DM trajectory according to a fewest-switches criterion. In the coherent switches (CS) implementation of DM, equations similar to Eqs. (47) and (48) are used to switch K, with ρ_{ij} replaced by a locally coherent electronic density matrix ρ_{ij}^{CS}. The time evolution of ρ_{ij}^{CS} is fully coherent,

$$\dot{\rho}_{ij}^{\text{CS}} = \dot{\rho}_{ij}, \tag{79}$$

i.e. it does not include $\dot{\rho}_{ij}^{\text{D}}$, and ρ_{ij}^{CS} is made *locally* coherent by setting

$$\rho_{ij}^{\text{CS}} = \rho_{ij}^{\text{DM}} \tag{80}$$

when the trajectory experiences a local minimum in

$$D(t) = \sum_i |\mathbf{F}_{iK}|^2. \tag{81}$$

An ensemble of CSDM trajectories decays to a distribution of final electronic states, and this distribution is determined from the ensemble average, $\langle \rho_{ii}^{\text{CS}} \rangle$, obtained from the locally coherent solutions of the classical path equation.

In summary, the CSDM includes the quantum evolution of the electronic degrees of freedom as governed by a reduced density operator (density matrix), and it incorporates decoherence of the electronic degrees of freedom by the nuclear degrees of freedom. In strong interaction regions it is a mean-field method with the formal and practical advantage (such as representation independence) of the Ehrenfest method, but the decoherence mitigates the disadvantage of the mean-field approach. We note that the CSDM does not scale in a difficult way with system size, and it can easily be applied to large and complex systems.

5. Summary of Recent Tests and Applications

The FSTU method (with the SD algorithm and the grad V prescription for treating the remaining frustrated hops) and the CSDM method are the results of a long series of systematic studies of the NBO dynamics of triatomic and, more recently, polyatomic systems. The FSTU and CSDM methods are straightforward to implement, readily applicable to a wide variety of NBO molecular dynamics simulations with any number of atoms and any number of electronic states, and are available in the distributed computer code ANT.[101] Here we summarize the results of the validation studies that led to the improved methods, and we discuss recent applications. Before doing this we note that large systems may involve new features so there is no guarantee that methods found to be accurate for triatomic and tetraatomic cases are accurate in all cases, including large molecules; however, it is clear that methods that fail even for small molecules are not to be trusted for large molecules, and it would be hard to argue that they should ever be preferred. Anyway, as far as tests against accurate quantum dynamics for the same sets of potential energy surfaces and couplings, small-molecule tests are all we have at this point in time. Tests comparing NBO molecular dynamics with experimental results are tests against accurate quantum dynamics, but since the exact surfaces and couplings are not known and the extent of possible experimental error is often hard to estimate, such tests are not as straightforward to interpret as small-molecule tests where accurate quantum dynamics are available for given sets of surfaces and couplings.

The FSTU and CSDM NBO molecular dynamics methods, along with several variants and predecessors, were tested against accurate outgoing wave variational principle[41,102–104] quantum mechanical reactive scattering calculations on a series of two-state atom-diatom test cases. Full-dimensional test cases with prototypical AC,[105] WI,[89] and CI[7] interactions

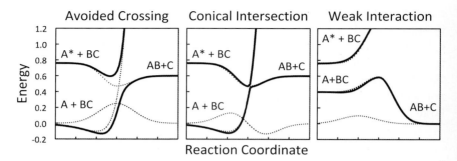

Fig. 1. Examples of the adiabatic (solid) and diabatic (dashed) potential energy surfaces along (left-to-right) the ground state reaction coordinate for the AC, WI, and CI families of test cases.

(as illustrated in reduced dimensionality in Fig. 1) were developed for this purpose, each of which describes a model reaction of the form

$$\begin{aligned} A^* + BC(v,j) &\to AB(E'_{\text{int}}) + C, \quad \text{reaction} \\ &\to A + BC(E''_{\text{int}}), \quad \text{quenching} \end{aligned} \quad \text{(R1)}$$

where the asterisk denotes electronic excitation, and the diatom is initially prepared in a quantized rovibrational state (v,j).

Six observables of the model reactions displayed in R1 were considered: the probability of reactive de-excitation (P_R), the probability of nonreactive de-excitation or quenching (P_Q), the total probability of a nonadiabatic event ($P_N = P_R + P_Q$), the reactive branching fraction $F_R = P_R/P_N$, and the average internal energies of the two diatomic fragments (E'_{int} and E''_{int}). Several test cases for each class of prototypical interactions were considered; they vary in the coupling strength of the model potential energy surfaces, the initial conditions, and/or the scattering conditions. By averaging over several test cases in each class, we obtain more robust and predictive error estimates. The results of these studies, which include errors for a total of six observables for each of 17 test cases, are summarized in Table 1.[8]

The 17 test cases in Table 1 include three cases of weak interaction (systems like $Br^* + HR \to Br + HR$ or $\to HBr + R$, where R is a radical; these cases are called WI cases or YHR cases), eight cases with accessible regions of avoided crossing but no accessible conical intersections (these are called AC cases or MXH cases), and five cases of accessible conical intersections (these are called CI cases or MCH cases). In each of these 17 cases we obtained accurate quantum dynamics results for a given realistic set of potential energy surfaces and couplings and compared these to

Table 1. Highly-averaged percentage errors for several NBO methods.

Method	Representation	AC	WI	CI	Overall
FS[a]	A	53	29	53	45
	D	42	289	43	125
	CC	53	29	40	41
SE	A/D/CC	74	c	66	c
FSTU[b]	A	43	25	56	41
	D	27	128	42	66
	CC	38	25	39	34
CSDM	A	20	18	42	27
	D	19	22	33	25
	CC	21	18	33	24

[a]Frustrated hops were ignored.
[b]Frustrated hops were treated using the TU and grad V prescriptions. The SD algorithm was not used because it had not been developed yet at the time that these calculations were carried out.
[c]The SE method fails for weakly coupled systems in that it does not produce all possible products; therefore average internal energies cannot be computed for the missing products, and an overall error cannot be computed.

the results of various semiclassical dynamics methods for the same potential energy surfaces and couplings and the same initial quantum states. The 17 cases differ from one another in the potential energy surfaces, the couplings, and/or the initial quantum state (for WI cases, there is one set of surfaces, and we ran the ground vibrational-rotational state of the reactants at two energies and one excited rotational state at one energy; for AC cases there are three different couplings surfaces — strong and broadly distributed, strong and localized, and weak and localized, and each was run for three initial rotational states; for CI cases there are five different sets of couplings). The accurate dynamics are independent of representation (adiabatic or diabatic), but the semiclassical results depend on the representation in which the dynamics are calculated; for each case and each representation we calculated the unsigned percentage error in each of the six physical observables mentioned in the previous paragraph by comparing the semiclassical results to the accurate quantum dynamics ones. Each column in the table shows mean unsigned percentage error averaged over the six observables in each of two representations for the cases in that column. The last column contains all 17 cases and so the mean unsigned percentage errors in the last column are averaged over $6 \times 2 \times 17 = 204$ absolute percentage errors.

The table shows that the results of the NBO molecular dynamics simulations are in general strongly dependent on the choice of electronic representation, and it shows that the adiabatic representation is usually more accurate than the diabatic one. The SE method is formally independent of the choice of representation, but it is less accurate than the other methods. Furthermore, it is unable to treat the small probability events occurring in the WI systems.[8]

The adiabatic representation is not always to be preferred, and the diabatic representation was found to be more accurate for some of the systems in the test set with ACs and CIs. A useful criterion for choosing between the adiabatic and diabatic representations is to prefer the representation where the diagonal surfaces are the least coupled to one another. One way to do this is to prefer the representation with the fewest number of attempted surface hops, and this representation is called the Calaveras County (CC) representation.[106] Results obtained using the CC representation are shown in Table 1. The CC is generally more accurate than using either the adiabatic or diabatic representations exclusively.

For larger systems than the ones considered here, it is likely that trajectories may sample some regions where the adiabatic representation is preferred and others where the diabatic representation is preferred in a single simulation. Invariance to the choice of electronic representation is therefore desirable, and it is encouraging that the CSDM method, which was designed with representation independence as a goal, is systematically less representation-dependent than the FSTU and other NBO molecular dynamics methods.

The overall accuracy of the best representations for each type of improved NBO method is generally good, and the CSDM method is the best method overall with an error of only $\sim$25%. Clearly the improvements made to the surface hopping approach and to the mean field approach have produced systematically improved methods of each type. Finally, we note that the improved methods work nearly equally well for the three types of interactions considered. Again, this robustness is important, as real systems are likely to feature more than one kind of interaction. Not only does the CSDM provide reasonably accurate final states, but — because of the explicit inclusion of decoherence with a physical time scale — it is expected to provide a realistic picture of the real-time process; the ability of semiclassical methods including decoherence to do this is expected to become more and more useful as shorter time scales[107,108] for studying the electron dynamics in molecules become accessible.

In another test, the accuracy of NBO MD methods for simulating deep quantum systems (i.e. systems with large electronic state energy gaps) was considered.[71] Typical energy gaps in the model AC and WI test cases are only a few tenths of an eV, whereas many real systems have much larger gaps. Quantum mechanical calculations of the photodissociation of the Na···FH van der Waals complex with a gap of ∼1.5 eV were carried out.[109] In the ground state, thermal excitation tends to break the weak van der Waals bond, producing the Na and HF products exclusively. Upon electronic excitation with visible light, however, the complex is promoted to a metastable complex called an exciplex. The exciplex is proposed to exhibit enhanced reactivity via the harpooning mechanism, where the change in the electronic structure results in a donation of partial charge from the Na atom to the F atom and promotes formation of the NaF + H products.

The FSTU and CSDM methods were shown to fairly accurately predict product branching and exciplex lifetimes for the photodissociation of the Na···FH system, as shown in Table 2, thus validating their use for deep quantum systems. The NBO classical and quantum dynamics simulations confirmed the enhanced reactivity of the harpooning mechanism, and NaF + H was predicted to be the dominant photodissociated bimolecular product.

In the course of this study, product branching in the NBO molecular dynamics simulations was found to be affected by a region of coupling where the excited state is classically energetically forbidden. An analysis of the NBO MD trajectories revealed that the results are sensitive to the treatment of decoherence. Figure 2 shows contour plots of the excited and ground electronic states, as well as hopping information for a subset of trajectories. The initial downward hops occur for a wide range of accessible geometries of the exciplex. More than three-fourths of the trajectories attempted to hop back into the exciplex after their first hop down, but many

Table 2. Product branching probabilities and half lives of the Na···FH exciplex.

Method	P_{Na+HF}	P_{NaF+H}	$t_{1/2}$, ps
Quantum	0.04	0.96	0.42
FSTU without SD	0.16	0.83	0.85
FSTU with SD	0.05	0.95	0.52
CSDM ($E_0 = 0.1\,E_h$)	0.29	0.71	0.76
CSDM ($E_0 = 0.001\,E_h$)	0.06	0.94	0.40

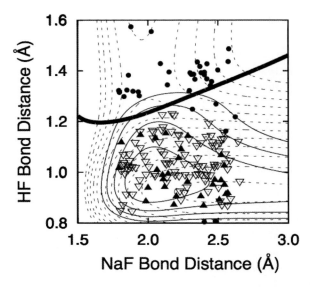

Fig. 2. Contour plots of the ground (dashed) and first excited (solid) potential energy surfaces for Na···FH. The initial hops down are shown as open triangles. Subsequent successful hops up are shown as solid triangles, and frustrated hops up are shown as black dots. The thick black line is the line of avoided crossings.

did so at geometries where the excited state is energetically forbidden. The majority of these frustrated hops occur near the line of avoided crossings, a region where the two electronic surfaces have very different shapes and where decoherence due to wave packet divergence may be expected to be significant. The use of the SD model for decoherence was found to reduce errors associated with frustrated hopping and to predict product branching and lifetimes in near quantitative agreement with the quantum mechanical results, as shown in Table 2. An analogous modification of the DM method resulting in faster decoherence in this critical region (obtained by decreasing the parameter E_0) was shown to give similarly improved results. This study highlighted the importance of accurate treatments of electronic decoherence in trajectory-based simulations of systems with coupled electronic states.

In another study,[62] the nonadiabatic photodissociation of HBr was modeled using several NBO trajectory methods. The calculated branching fractions for the H + Br($^2P_{3/2}$) and H + Br($^2P_{1/2}$) products were found to be in good agreement with experimental measurements[110] over a range of photon energies, as shown in Fig. 3.

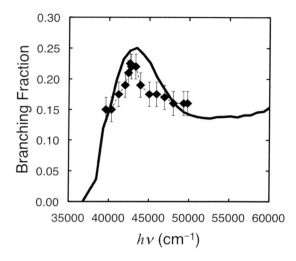

Fig. 3. HBr photodissociation branching fraction to form H+Br* as a function of photon energy ($h\nu$) obtained by the CSDM method (solid line) and experiment (diamonds).

Li et al. applied the CSDM method to several systems: the $D + H_2$ and $H + D_2$ reactions at collision energies up to 2 eV,[111] nonreactive and reactive charge transfer and reactive non-charge-transfer in $D^+ + H_2$ and $H^+ + D_2$ collisions,[111] and intersystem crossing in $O(^3P_{2,1,0}, {}^1D_2) + H_2$ reactive collisions yielding $OH(^2\Pi_{3/2,1/2}) + H(^2S)$.[112] For the first two reactions they employed a two-state electronic basis, for the next two a three-state electronic basis, and for the final two a four-state electronic basis (three triplet states and one singlet). For $D + H_2$ and $H + D_2$ they obtained very good agreement of reactive cross sections with accurate quantal dynamics over the whole energy range. For nonreactive and reactive charge transfer in $D^+ + H_2$ and $H^+ + D_2$, the CSDM cross sections provide overall trends in good agreement with accurate quantum dynamics, and for reactive non-charge-transfer the CSDM cross sections agree with accurate quantum dynamical ones over the whole energy range up to 2.5 eV, although in one case they are slightly lower. For $O(^3P_2) + H_2$, the cross sections to produce the $^2\Pi_{3/2}$ and $^2\Pi_{1/2}$ states are both in good agreement with accurate quantum dynamics over the whole range of collision energies, up to 28 kcal/mol, except that the cross section to produce the $^2\Pi_{3/2}$ state has a somewhat higher threshold.

The photodissociation of NH_3, which has been studied in detail experimentally,[113,114] was also recently modeled using NBO MD

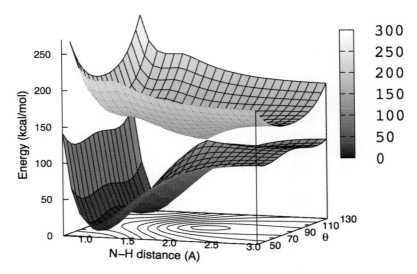

Fig. 4. A conical intersection between the ground and first excited states of NH_3 occurs at planar geometries and at an N–H distance of 2 Å.

simulations.[80,85] Analytic representations[115] of the coupled X and A states of NH_3 are shown in Fig. 4 as functions of one N–H distance and the umbrella angle θ. This system features a CI at extended N–H distances and planar geometries. Trends in the production of excited state amino radicals as a function of initial state preparation were computed and compared with experiment. The experimental results suggest an enhancement in the production of excited-state products when the antisymmetric stretch of NH_3 is excited, with the interpretation that excitation of the antisymmetric stretch causes the system to go around the CI and thus inhibits electronic state quenching. The NBO MD calculations predict that the production of excited state amino radicals depends on the total energy, and no state specificity is observed. The source of this discrepancy is unclear, although recent quantum mechanical wave packet results[116] are in fair agreement with the NBO trajectory results.

In addition to making semiquantitative predictions of product branching, lifetimes, and internal energy distributions, as discussed above, NBO molecular dynamics simulations are useful for studying chemical events in mechanistic detail, such as the role of conical intersections and avoided crossings in NBO dynamics. This analysis has been carried out for the FSTU and CSDM methods for the model CI and AC test cases,[7] Na···FH photodissociation,[71] and for the photodissociation of NH_3.[80,85]

From these studies, one can make some general comments about NBO trajectories and CIs. The capture efficiencies of a CI and an AC have been compared for similar potential energy surfaces, differing only in the interaction type. Trajectories with reasonable kinetic energies were found to be captured equally well by the CI and AC, i.e. the conical shape near the CI did not capture trajectories any more or less easily than an AC. The CI was shown to more efficiently move trajectories out of the interaction region than the AC, although the effect was small. Finally, it was noted that trajectories did not in general switch surfaces at zero-gap geometries. Instead, surface hops occurred over a range of energy gaps, geometries, and coupling strengths near and at the CI.

A similar analysis of the NH_3 trajectories was carried out to study the experimentally proposed mechanism of state specificity. The NBO MD trajectory results showed that the system is efficiently quenched via the seam of CIs when either the antisymmetric or symmetric stretches are excited. The distribution of energy gaps at surface hops was peaked at zero, but the average gap was 0.3 eV. The CI rapidly quenched photoexcited NH_3 nonreactively to form ground-electronic-state NH_3, which subsequently and much more slowly decayed to $NH_2 + H$. Only a small fraction of trajectories dissociated directly to the $NH_2 + H$ products. Furthermore, the number of direct trajectories avoiding the CI was not promoted by excitation into the antisymmetric stretch, in contrast to the experimentally proposed mechanism.

6. Concluding Remarks

Non-Born–Oppenheimer dynamics may be dominated by regions of conical intersections, by regions of avoided crossings, or by regions of weak interactions of electronic states. When a conical intersection seam or its neighborhood is dynamically accessible, the geometries in the neighborhood of the conical intersections seam will often provide an efficient route for excited state decay, as originally pointed out by Teller.[9] As mentioned in the introduction and as indicated by analyses of non-Born–Oppenheimer trajectories in Sec. 5, the seam of conical intersections — due to its dimensionality being two lower than the dimensionality of the full internal coordinate space — does not necessarily directly mediate electronic transitions; however, the conical intersection seam does anchor the loci of strong interaction of the potential energy surfaces. Because the conical intersection seam can be a very extended hypersurface, one must consider more than just the lowest-energy conical intersection, and because the conical intersection itself may be dynamically inaccessible, one must consider avoided crossings as well as conical intersections. Because conical intersections are surrounded by avoided crossing regions, it is important to consider the multidimensional character of the dynamics whenever a region of strong interaction of the electronic states is encountered; treatments based on treating the potentials along a trajectory path as one-dimensional avoided crossings ignore the fact that strong interaction regions in polyatomic systems have more complicated dynamics than the Landau–Zener behavior encountered in atom–atom collisions. The best zero-order model of the dynamics in a strong interaction region may be either diabatic or adiabatic. Furthermore, one must take into account the fact that decoherence may occur between successive visits to strong interaction regions. The semiclassical dynamics methods reviewed in this chapter take account of this decoherence; they have been validated in multidimensional studies for the treatment of photochemical dynamics in the vicinity of conical intersections and avoided crossing regions and also in weak interaction regions; they may be used for systems containing both predominantly diabatic and predominantly adiabatic regions of phase space; and rare-event sampling algorithms are available for treating processes with small transition probabilities. The CSDM method, in particular, is the culmination of a series of attempts to improve mean-field and surface hopping methods by combining the best features of both.

Acknowledgments

This work was supported in part by the National Science Foundation under grant No. CHE09-56776 and in part by the Division of Chemical Sciences, Geosciences, and Biosciences, Office of Basic Energy Sciences, U.S. Department of Energy under Contract No. DE-AC04-94-AL85000.

References

1. J.C. Tully, in *Modern Methods for Multidimensional Dynamics Computations in Chemistry*, edited by D.L. Thompson (World Scientific, Singapore, 1998), pp. 34–72.

2. E.E. Nikitin, *Annu. Rev. Phys. Chem.* **50**, 1 (1999).
3. A.W. Jasper, B.K. Kendrick, C.A. Mead and D.G. Truhlar, in *Modern Trends in Chemical Reaction Dynamics: Experiment and Theory, Part I*, edited by X. Yang and K. Liu (World Scientific, Singapore, 2004), pp. 329–391. [*Adv. Ser. Phys. Chem.* **14**, 329 (2004)].
4. G.A. Worth and L.S. Cederbaum, *Annu. Rev. Phys. Chem.* **55**, 127 (2004).
5. A.W. Jasper, S. Nangia, C. Zhu and D.G. Truhlar, *Acc. Chem. Res.* **39**, 101 (2006).
6. J.M. Bowman, *Science* **319**, 40 (2008).
7. A.W. Jasper and D.G. Truhlar, *J. Chem. Phys.* **122**, 044101 (2005).
8. A.W. Jasper, C. Zhu, S. Nangia and D.G. Truhlar, *Faraday Discuss. Chem. Soc.* **127**, 1 (2004).
9. E. Teller, *J. Phys. Chem.* **41**, 109 (1937).
10. K.A. Kistler and S. Matsika, *J. Chem. Phys.* **128**, 215102 (2008).
11. G.J. Atchity, S.S. Xantheas and K. Ruedenberg, *J. Chem. Phys.* **95**, 1862 (1991). D.R. Yarkony, *J. Phys. Chem. A* **105**, 985 (1996).
12. V. Bonacic-Koutecky, J. Koutecky and J. Michl, *Angew. Chem. Int. Ed.* **26**, 170 (1987).
13. M. Olivucci, I.N. Ragazos, F. Bernardi and M.A. Robb, *J. Am. Chem. Soc.* **115**, 371 (1993).
14. L. Seidner and W. Domcke, *Chem. Phys.* **186**, 27 (1994).
15. W. Fu, S. Lochbrunner, A.M. Müller, T. Schikarski, W.E. Schmid and S.A. Trushin, *Chem. Phys.* **232**, 16 (1998).
16. L. Serrano-Andrés, M. Merchán and A.C. Borin, *Proc. Nat. Acad. Sci. U. S. A.* **103**, 8691 (2004).
17. J. Michl, *Mol. Photochem.* **4**, 243 (1972).
18. H.E. Zimmerman, *Acc. Chem. Res.* **15**, 312 (1982).
19. E. Teller, *Israel J. Chem.* **7**, 227 (1969).
20. D.G. Truhlar and C.A. Mead, *Phys. Rev. A* **68**, 032501 (2003).
21. D.G. Truhlar, *Faraday Discuss. Chem. Soc.* **127**, 242 (2004).
22. T.C. Allison, G.C. Lynch, D.G. Truhlar and M.S. Gordon, *J. Phys. Chem.* **100**, 13575 (1996).
23. O. Tishchenko, D.G. Truhlar, A. Ceulemans and M.T. Nguyen, *J. Am. Chem. Soc.* **130**, 7000 (2008).
24. A.W. Jasper, S.N. Stechmann and D.G. Truhlar, *J. Chem. Phys.* **116**, 5424 (2002); **117**, 10427(E) (2002).
25. A. Bjerre and E.E. Nikitin, *Chem. Phys. Lett.* **1**, 179 (1967).
26. R.K. Preston and J.C. Tully, *J. Chem. Phys.* **54**, 4297 (1971).
27. N.C. Blais and D.G. Truhlar, *J. Chem. Phys.* **79**, 1334 (1983).
28. C. Zhu, S. Nangia, A.W. Jasper and D.G. Truhlar, *J. Chem. Phys.* **121**, 7658 (2004).
29. H.-D. Meyer and W.H. Miller, *J. Chem. Phys.* **70**, 3214 (1979).
30. A.D. Micha, *J. Chem. Phys.* **78**, 7138 (1983).
31. M. Amarouche, F.X. Gadea and J. Durup, *Chem. Phys.* **130**, 145 (1989).
32. A. Donoso and C.C. Martens, *Int. J. Quantum Chem.* **90**, 1348 (2002).

33. R. Kapral and G. Ciccotti, *J. Chem. Phys.* **110**, 8919 (1999).
34. M. Santer, U. Manthe and G. Stock, *J. Chem. Phys.* **114**, 2001 (2001).
35. H. Nakamura, *J. Phys. Chem. A* **110**, 10929 (2006).
36. T.J. Martinez, M. Ben-Nun and R.D. Levine, *J. Phys. Chem.* **100**, 7884 (1996).
37. M.D. Hack, A.M. Wensmann, D.G. Truhlar, M. Ben-Nun and T.J. Martinez, *J. Chem. Phys.* **115**, 1172 (2001).
38. M. Ben-Nun and T.J. Martinez, *Adv. Chem. Phys.* **121**, 439 (2002).
39. T.J. Martinez, *Acc. Chem. Res.* **39**, 119 (2006).
40. S. Yang, J.D. Coe, B. Kaduk and T.J. Martínez, *J. Chem. Phys.* **130**, 134113 (2009).
41. G.J. Tawa, S.L. Mielke, D.G. Truhlar and D.W. Schwenke, in *Advances in Molecular Vibrations and Collisional Dynamics*, edited by J.M. Bowman, (JAI, Greenwich, CT, 1994). pp. 45–116.
42. M.D. Hack, A.W. Jasper, Y.L. Volobuev, D.W. Schwenke and D.G. Truhlar, *J. Phys. Chem. A* **103**, 6309 (1999).
43. U. Manthe, H.-D. Meyer and L.S. Cederbaum, *J. Chem. Phys.* **97**, 3199 (1992).
44. H.-D. Meyer and G.A. Worth, *Theor. Chem. Acc.* **109**, 251 (2003).
45. S. Woittequand, C. Toubin, M. Monnerville, S. Briquez, B. Pouilly and H.-D. Meyer, *J. Chem. Phys.* **131**, 194303 (2009).
46. J.O. Hirschfelder and W.J. Meath, *Adv. Chem. Phys.* **12**, 3 (1967).
47. W. Kolos and L. Wolniewicz, *J. Chem. Phys.* **41**, 3663 (1964).
48. B.C. Garrett and D.G. Truhlar, *J. Chem. Phys.* **82**, 4543 (1985).
49. N.F. Zokov, O.L. Polyansky, C.R. LeSueur and J. Tennyson, *Chem. Phys. Lett.* **260**, 381 (1996).
50. D.W. Schwenke, *J. Phys. Chem. A* **105**, 2352 (2001).
51. T.F. O'Malley, *Adv. At. Mol. Phys.* **7**, 223 (1971).
52. H. Köppel, in *Conical Intersections: Electronic Structure, Dynamics and Spectroscopy*, edited by W. Domcke, D.R. Yarkony and H. Köppel (World Scientific, Singapore, 2004), pp. 175–204. [*Adv. Ser. Phys. Chem.* **15**, 175 (2004)].
53. C.A. Mead and D.G. Truhlar, *J. Chem. Phys.* **77**, 6090 (1982).
54. R. Abrol and A. Kuppermann, *J. Chem. Phys.* **116**, 1035 (2002).
55. B.N. Papas, M.S. Schuurman and D.R. Yarkony, *J. Chem. Phys.* **129**, 124104 (2008).
56. V.M. Garcia, M. Reguero, R. Caballol and J.P. Malrieu, *Chem. Phys. Lett.* **281**, 161 (1987).
57. G.J. Atchity and K. Ruedenberg, *Theor. Chem. Acc.* **97**, 47 (1997).
58. H. Nakamura and D.G. Truhlar, *J. Chem. Phys.* **115**, 10353 (2001).
59. H. Nakamura and D.G. Truhlar, *J. Chem. Phys.* **117**, 5576 (2002).
60. H. Nakamura and D.G. Truhlar, *J. Chem. Phys.* **118**, 6816 (2003).
61. R. Valero and D.G. Truhlar, *J. Phys. Chem. A* **111**, 8536 (2007).
62. R. Valero, D.G. Truhlar and A.W. Jasper, *J. Phys. Chem. A* **112**, 5756 (2008).
63. J.C. Tully, *J. Chem. Phys.* **93**, 1061 (1990).

64. V. May and O. Kühn, *Charge and Energy Transfer Dynamics in Molecular Systems* (Wiley-VCH, Berlin, 2000).
65. C. Zhu, A.W. Jasper and D.G. Truhlar, *J. Chem. Theory Comput.* **1**, 527 (2005).
66. D.G. Truhlar, in *Quantum Dynamics of Complex Molecular Systems*, edited by D.A. Micha and I. Burghardt (Springer, Berlin, 2007), pp. 227–243.
67. M.D. Hack and D.G. Truhlar, *J. Chem. Phys.* **114**, 9305 (2001).
68. B.J. Schwartz, E.R. Bittner, O.V. Prezhdo and P.J. Rossky, *J. Chem. Phys.* **104**, 5942 (1996).
69. O.V. Prezhdo and P.J. Rossky, *J. Chem. Phys.* **107**, 5863 (1997).
70. G. Granucci and M. Persico, *J. Chem. Phys.* **126**, 134114 (2007).
71. A.W. Jasper and D.G. Truhlar, *J. Chem. Phys.* **127**, 194306 (2007).
72. D.L. Bunker, *Methods Comp. Phys.* **10**, 287 (1971).
73. M. Karplus, R.N. Porter and R.D. Sharma, *J. Chem. Phys.* **43**, 3259 (1965).
74. D.G. Truhlar and J.T. Muckerman, in *Atom-Molecule Collision Theory: A Guide for the Experimentalist*, edited by R.D. Bernstein (Plenum, New York, 1979), pp. 505–566.
75. L.M. Raff and D.L. Thompson, in *Theory of Chemical Reaction Dynamics, Vol. 3*, edited by M. Baer (CRC Press, Boca Raton, 1985), pp. 1–121.
76. J.J. Sakurai, *Modern Quantum Mechanics* (Addison-Wesley, Reading, 1985), pp. 335–339.
77. R. Schinke, *Photodissociation Dynamics* (Cambridge University Press, Cambridge, 1993).
78. G. Herzberg, *Molecular Spectra and Molecular Structure. III. Electronic Spectra and Electronic Structure of Polyatomic Molecules* (van Nostrand Reinhold, Princeton, New York, 1966), pp. 149.
79. L. Pauling and E.B. Wilson, *Introduction to Quantum Mechanics: With Applications to Chemistry* (McGraw-Hill, New York, 1935), pp. 73–77.
80. D. Bonhommeau and D.G. Truhlar, *J. Chem. Phys.* **129**, 014302 (2008).
81. H.-W. Lee and M.O. Scully, *J. Chem. Phys.* **73**, 2238 (1980).
82. J.C. Gray and D.G. Truhlar, *J. Chem. Phys.* **76**, 5350 (1982).
83. E.J. Heller and R.C. Brown, *J. Chem. Phys.* **75**, 1048 (1981).
84. S. Chapman, *Adv. Chem. Phys.* **82**, 423 (1992).
85. A.W. Jasper and D.G. Truhlar, *Chem. Phys. Lett.* **369**, 60 (2003).
86. D. Bomhommeau, R. Valero, D.G. Truhlar and A.W. Jasper, *J. Chem. Phys.* **130**, 234303 (2009).
87. M.F. Herman, *J. Chem. Phys.* **81**, 754 (1984).
88. J.-Y. Fang and S. Hammes-Schiffer, *J. Chem. Phys.* **110**, 11166 (1999).
89. A.W. Jasper, M.D. Hack and D.G. Truhlar, *J. Chem. Phys.* **115**, 1804 (2001).
90. U. Müller and G. Stock, *J. Chem. Phys.* **107**, 6230 (1997).
91. M. Thachuk, M.Y. Ivanov and D.M. Wardlaw, *J. Chem. Phys.* **109**, 5747 (1998).
92. J.-Y. Fang and S. Hammes-Schiffer, *J. Phys. Chem. A* **103**, 9399 (1999).
93. A.W. Jasper and D.G. Truhlar, *J. Chem. Phys.* **123**, 064103 (2005).

94. S. Nangia, A.W. Jasper, T.F. Miller III and D.G. Truhlar, *J. Chem. Phys.* **120**, 3586 (2004).
95. R.T. Skodje and D.G. Truhlar, *J. Chem. Phys.* **80**, 3123 (1984).
96. G.A. Worth, M.A. Robb and B. Lasorne, *Mol. Phys.* **106**, 2077 (2008).
97. C. Zhu, A.W. Jasper and D.G. Truhlar, *J. Chem. Phys.* **120**, 5543 (2004).
98. S.C. Cheng, C. Zhu, K.K. Liang, S.H. Lin and D.G. Truhlar, *J. Chem. Phys.* **129**, 024112 (2008).
99. W.L. Hase, D.G. Buckowski and K.N. Swamy, *J. Phys. Chem.* **87**, 2754 (1983).
100. A. Bohm, *Quantum Mechanics: Foundations and Applications*, 3rd ed. (Springer-Verlag, New York, 1983), pp. 57–73.
101. Z.H. Li, A.W. Jasper, D.A. Bonhommeau, R. Valero and D.G. Truhlar, ANT–version 2009 (University of Minnesota, Minneapolis, 2009).
102. Y. Sun, D.J. Kouri, D.G. Truhlar and D.W. Schwenke, *Phys. Rev. A* **41**, 4857 (1990).
103. D.W. Schwenke, S.L. Mielke, G.J. Tawa, R.S. Friedman, P. Halvick and D.G. Truhlar, *Chem. Phys. Lett.* **203**, 565 (1973).
104. G.J. Tawa, S.L. Mielke, D.G. Truhlar and D.W. Schwenke, *J. Chem. Phys.* **100**, 5751 (1994).
105. Y.L. Volobuev, M.D. Hack, M.S. Topaler and D.G. Truhlar, *J. Chem. Phys.* **112**, 9716 (2002).
106. M.D. Hack and D.G. Truhlar, *J. Phys. Chem. A* **104**, 7917 (2000).
107. M. Wickenhauser, J. Burgdörfer, F. Krauz and M. Drescher, *Phys. Rev. Lett.* **94**, 023002 (2005).
108. F. Krausz and M. Ivanov, *Rev. Mod. Phys.* **81**, 163 (2009).
109. S. Garashchuk and V.A. Rassolov, *Chem. Phys. Lett.* **446**, 395 (2007).
110. P.M. Regan, S.R. Langford, A.J. Orr-Ewing and M.N.R. Ashfold, *J. Chem. Phys.* **110**, 281 (1999).
111. B. Li, T.-S. Chu and K.-L. Han, *J. Comput. Chem.* **31**, 362 (2010).
112. B. Li and K.-L. Han, *J. Phys. Chem. A* **113**, 10189 (2009).
113. A. Bach, J.M. Hutchison, R.J. Holiday and F.F. Crim, *J. Phys. Chem. A* **107**, 10490 (2003).
114. M.L. Hause, Y.H. Yoon and F.F. Crim, *J. Chem. Phys.* **125**, 174309 (2006).
115. Z.H. Li, R. Valero and D.G. Truhlar, *Theor. Chem. Acc.* **118**, 9 (2007).
116. W. Lai, S.Y. Lin, D. Xie and H. Guo, *J. Phys. Chem. A* **114**, 3121 (2010).

Chapter 11

Computational and Methodological Elements for Nonadiabatic Trajectory Dynamics Simulations of Molecules

Mario Barbatti[*], Ron Shepard[†] and Hans Lischka[‡]

1. Introduction 416
2. Nonadiabatic Dynamics Based on Classical Trajectories 417
   2.1. The time-dependent equations 417
   2.2. Implications of the local approximation 423
   2.3. The surface hopping approach 426
3. Initial Conditions 429
   3.1. Wigner and quasi-Wigner distributions 430
   3.2. Sampling with trajectory simulations 431
   3.3. Sampling in special points 431
   3.4. The excited-state distribution 432
4. Electronic Structure Methods 432

---

[*]Institute for Theoretical Chemistry, University of Vienna, Waehringerstrasse 17, A 1090 Vienna, Austria. Present address: Max-Planck-Institut für Kohlenforschung, Kaiser-Wilhelm-Platz 1, D-45470 Mülheim an der Ruhr, Germany. E-mail: barbatti@kofo.mpg.de

[†]Chemical Sciences and Engineering Division, Argonne National Laboratory, Argonne, IL 60439, USA. E-mail: shepard@tcg.anl.gov

[‡]Institute for Theoretical Chemistry, University of Vienna, Waehringerstrasse 17, A 1090 Vienna, Austria. E-mail: hans.lischka@univie.ac.at

5. Analytical Gradients 434
   5.1. MCSCF gradient 436
   5.2. State-averaged MCSCF gradient 441
   5.3. MRCI gradient 445
6. Nonadiabatic Coupling Terms 450
   6.1. NAC vectors in MCSCF and MRCI formalism 450
   6.2. Time-derivative NAC terms 454
7. Dynamics Close to Conical Intersections 455
8. Conclusions and Outlook 457
   Acknowledgments 458
   References 458

1. Introduction

Dynamics simulation is a powerful tool for investigation of nonadiabatic events in molecular processes. It allows the molecules to explore *by themselves* the configurational space, to find the main reaction paths and conical intersections, and to determine the different time scales at which the processes occur.

Nonadiabatic dynamics simulations are, however, computationally demanding. In particular, the complete quantum mechanical solution[1,2] given by wave packet propagation in full dimensionality reaches prohibitive levels even for relatively small molecules. For this reason, many different theoretical approaches based on different types of approximations[1,2] have been proposed such as independent trajectories,[3,4] Liouville dynamics,[5] trajectory-based Bohmian dynamics,[6] path integrals,[7,8] and multiple spawning.[9] In particular, nonadiabatic dynamics based on classical independent trajectories, the subject of this chapter, has a long history beginning with the origins of quantum mechanics with significant application to large molecules already by mid-1970s when Warshel investigated the excited-state relaxation of retinal using mean-field dynamics based on semi-empirical surfaces.[10]

The methodological background for nonadiabatic dynamics based on classical trajectories nowadays is not substantially different from that of the early 1990s when Tully proposed the Fewest Switches algorithm.[11] Although important developments in the field, such as the multiple spawning method,[9] the treatment of decoherence processes,[12–14] and the introduction of interactions with electric fields[15–17] have improved and opened new possibilities for simulations, most of advances have been achieved due to the development of computational capabilities and to methodological developments in quantum chemical methods. These developments include new multireference semi-empirical methods,[18,19] analytical gradients at MRPT2 level,[20] and analytical gradients and nonadiabatic coupling vectors at state-averaged MCSCF[21] and MRCI levels.[22–24] As a result, we are witnessing the dawn of many interesting applications in diverse fields of photochemistry and photophysics.[25–30]

In this chapter, we will take advantage of our experience developing methods,[23,24,38] implementing the programs COLUMBUS[39–41] and NEWTON-X,[42,43] and investigating a variety of photodynamical processes[31,44–47] to guide the reader into the main elements necessary to perform and to understand nonadiabatic trajectory dynamics simulations of molecules. These elements include the time-dependent mixed quantum-classical equations, the generation of initial conditions, and issues connected to conventional quantum chemistry, but of deep relevance for dynamics simulations, including electronic structure methods and the computation of energy gradients and nonadiabatic coupling terms. Finally, we discuss some general features revealed by nonadiabatic dynamics simulations of a series of molecules, which have helped to shape our understanding of how conical intersections contribute to nonadiabatic dynamics.

2. Nonadiabatic Dynamics Based on Classical Trajectories

2.1. *The time-dependent equations*

The basic problem in dynamics simulations of molecules is to solve the time-dependent Schrödinger equation (TDSE) for the complete molecular system

$$\left(i\hbar\frac{\partial}{\partial t} - H\right)\Psi(\mathbf{r},\mathbf{R},t) = 0, \qquad (1)$$

$$H = T_n + H_e, \qquad (2)$$

where the molecular wave function Ψ depends on time t, on the nuclear coordinates $\mathbf{R}$ and on the electronic coordinates $\mathbf{r}$. H is the total Hamiltonian consisting of the nuclear kinetic energy operator T_n and the electronic operator H_e which, by convention, includes the Coulombic nuclear–nuclear repulsion and electron–nuclear attractions.

The derivation of the time-dependent equations employed in nonadiabatic dynamics methods based on classical trajectories is usually presented by expanding the molecular wave function in a basis of time-dependent electronic wave functions.[3,11,14,48–51] In this chapter we approach this problem from a different point of view, by expanding the molecular wave function in a basis of time-dependent nuclear wave functions as it is usually done in wave packet propagation.[52] Although this is only a formal distinction that will lead to the same results as the alternative formulation, it has the advantage of allowing a more direct comparison with dynamics based on wave packets, of clarifying the approximations that are employed, and of suggesting possible ways to improve the method.

The nuclear motion can be evaluated by writing the molecular wave function as a Born–Oppenheimer expansion[52]

$$\Psi(\mathbf{r}, \mathbf{R}, t) = \sum_j \chi_j(\mathbf{R}, t)\, \Phi_j(\mathbf{r}; \mathbf{R}), \qquad (3)$$

$$\langle \Phi_k \mid \Phi_l \rangle_\mathbf{r} = \delta_{kl}, \qquad (4)$$

where χ_j and Φ_j are, respectively, the nuclear and electronic wave functions for electronic state j and the summation in Eq. (3) runs over all electronic states. χ_j is a function of the nuclear coordinates $\mathbf{R}$ and of time t, while Φ_j is a function of the electronic coordinates $\mathbf{r}$ and depends parametrically on $\mathbf{R}$.

Substituting Eq. (3) into Eq. (1), multiplying to the left by Φ_k^*, and integrating over the electronic coordinates leads to

$$i\hbar \frac{\partial \chi_k}{\partial t} + \frac{1}{2}\hbar^2 \nabla_M^2 \chi_k + \sum_j \left(-H_{kj} + i\hbar \mathbf{F}_{kj} \cdot \hat{\mathbf{v}} + \frac{1}{2}\hbar^2 G_{kj} \right) \chi_j = 0, \qquad (5)$$

where

$$\nabla_M^2 \equiv \sum_m^{N_{at}} \frac{\nabla_m^2}{M_m} = \frac{2}{\hbar^2} T_n, \qquad (6)$$

$$H_{kj}(\mathbf{R}) \equiv \langle \Phi_k \mid H_e \mid \Phi_j \rangle_\mathbf{r}, \qquad (7)$$

$$G_{kj}(\mathbf{R}) \equiv \langle \Phi_k \mid \nabla_M^2 \mid \Phi_j \rangle_\mathbf{r}. \qquad (8)$$

The index m in Eq. (6) runs over all N_{at} nuclei of the molecule. For atom m the velocity operator and the nonadiabatic coupling vector are defined, respectively, as

$$\hat{\mathbf{v}}_m \equiv -i\hbar \frac{\nabla_m}{M_m}, \tag{9}$$

$$\mathbf{F}_{kj}^m(\mathbf{R}) \equiv \langle \Phi_k | \nabla_m | \Phi_j \rangle_{\mathbf{r}}. \tag{10}$$

Until this point no approximation has been invoked, and different propagation methods will treat each of these terms in different ways. Note that to solve Eq. (5), the potential energy surfaces, nuclear wave functions, and coupling terms should be determined beforehand for the whole configuration space for a limited number of states j. This is what is usually done in wave packet propagation methods.[1,53] A series of approximations can be used to achieve a local version of this equation, which is dependent only on a single set of coordinates $\mathbf{R}^c$. In the Multiple Spawning approach for example,[9] this is done by the "saddle point" approximation, which treats the first and second terms in the brackets as

$$\langle \chi_k | O_{kj}(\mathbf{R}) | \chi_l \rangle \approx \langle \chi_k | \chi_l \rangle O_{kj}(\mathbf{R}^c) \equiv \langle \chi_k | \chi_l \rangle O_{kj}^c. \tag{11}$$

Equation (5) contains information on how the wave packet evolves on one potential energy surface as a function of time and also on how it is transferred to other states. For our purposes, this fact can be better explored if the wave function is written as[54,55]

$$\chi_k = A_k(\mathbf{R}, t) e^{i S_k(\mathbf{R}, t)/\hbar}, \tag{12}$$

where the amplitude A_k and the phase S_k are real functions. Substituting Eq. (12) into Eq. (5) and separating the real and imaginary terms, one obtains the following equations for the propagation of the probability density A_k^2 and the phase S_k

$$\frac{\partial S_k}{\partial t} + \frac{1}{2}(\nabla S_k)^2 = -H_{kk} + \frac{1}{2}\hbar^2 \left(\frac{1}{A_k} \nabla^2 A_k + G_{kk} \right) + \mathrm{Re}[\Omega_{kj}], \tag{13}$$

$$\frac{\partial A_k^2}{\partial t} + \nabla \cdot (A_k^2 \nabla S_k) + \frac{2}{\hbar} A_k^2 \,\mathrm{Im}[\Omega_{kj}] = 0, \tag{14}$$

where

$$\Omega_{kj} = \sum_{j \neq k} \left[-H_{kj} + i\hbar \mathbf{F}_{kj} \cdot \mathbf{v} + \hbar^2 \left(\frac{1}{A_j} \sum_m^{N_{at}} \mathbf{F}_{kj}^m \cdot \frac{\nabla_m A_j}{M_m} + \frac{1}{2} G_{kj} \right) \right]$$
$$\times \frac{A_j}{A_k} e^{i(S_j - S_k)/\hbar}. \tag{15}$$

In this equation, the velocity **v** is a vector composed by the nuclear velocity components

$$\mathbf{v}_m = \frac{\nabla_m S_j}{M_m}. \tag{16}$$

If the terms proportional to $\hbar^2$ and the nonadiabatic coupling term $\text{Re}[\Omega_{kj}]$ are neglected in Eq. (13), one obtains the Hamilton–Jacobi equation, where the phase S_k is the Hamilton's principal function.[56] In this limit, Eq. (13) describes a swarm of independent classical trajectories $\mathbf{R}^c$ evolving on the H_{kk} potential energy surface or, equivalently, driven by a force $-\nabla H_{kk}$ according to the Newton's law.[50] The diagonal terms proportional to $\hbar^2$ in Eq. (13) are responsible by the spatial correlation of the wave packet at a certain time and the last term is responsible for the nonadiabatic influence from other states on state k.

Equation (14) controls the probability density A_k^2 of state k. When the nonadiabatic coupling term $\text{Im}[\Omega_{kj}]$ is null, it reduces to the continuity equation, where the $A_k^2 \nabla S_k$ is the probability flux.[54] The term $\text{Im}[\Omega_{kj}]$ works as a sink term transferring the probability density between k and other states j.

For the trajectory-based dynamics methods, the limit when the terms proportional to $\hbar^2$ in Eqs. (13) and (15) can be neglected is especially interesting. In such case, Eq. (13) is reduced to

$$\frac{\partial S_k}{\partial t} + \frac{1}{2}(\nabla S_k)^2 = -H_{kk} + \text{Re}\left[\sum_{j\neq k} [-H_{kj} + i\hbar \mathbf{F}_{kj} \cdot \mathbf{v}]\frac{A_j}{A_k}e^{i(S_j - S_k)/\hbar}\right], \tag{17}$$

which represents a swarm of independent trajectories moving on the potential energy H_{kk} but subject to a nonadiabatic interactions given by the second term on the right hand side of the equation. In the regime of independent trajectories, the divergence of the probability flux is zero and Eq. (14) is reduced to

$$\frac{dA_k}{dt} + \frac{1}{\hbar}\text{Im}\left[\sum_{j\neq k}[-H_{kj} + i\hbar \mathbf{F}_{kj} \cdot \mathbf{v}]A_j e^{i(S_j - S_k)/\hbar}\right] = 0. \tag{18}$$

The interpretation of the wave packet propagation in terms of the Hamilton–Jacobi theory gives the basis for the trajectory-based methods. The time propagation is performed by integrating the Newton's equation for the swarm of trajectories, which is analogous to solving Eq. (17), while the population of each state is controlled by an equation analogous to Eq. (18).

To derive the actual equation controlling the population, we should turn to Eq. (5). First, the independent trajectory approximation is enforced by neglecting the terms proportional to $\hbar^2$. This approximation also motivates the trivial factorization of the nuclear wave function χ_k describing the wave packet around each trajectory $\mathbf{R}^c$ as

$$\chi_k(\mathbf{R}, \mathbf{R}^c, t) = c_k(t)\zeta_k(\mathbf{R} - \mathbf{R}^c), \qquad (19)$$

$$\langle \zeta_k | \zeta_k \rangle_\mathbf{R} = 1, \qquad (20)$$

where $\zeta_k(\mathbf{R} - \mathbf{R}^c)$ is a peaked function centered at each trajectory $\mathbf{R}^c$ of the swarm. Neglecting the terms proportional to $\hbar^2$, substituting Eq. (19) into Eq. (5), multiplying to the left by ζ_k^*, and integrating over the nuclear coordinates leads to

$$i\hbar \frac{dc_k}{dt} + \sum_j [-\langle \zeta_k | H_{ki} | \zeta_j \rangle_\mathbf{R} + i\hbar \langle \zeta_k | \mathbf{F}_{kj} \cdot \hat{\mathbf{v}} | \zeta_j \rangle_\mathbf{R}]c_j = 0. \qquad (21)$$

To obtain the final result, the following approximations are additionally employed. First, the velocity operator is substituted by the velocity function

$$\hat{\mathbf{v}}\zeta_k \approx \mathbf{v}(\mathbf{R})\zeta_k. \qquad (22)$$

Second, the saddle point approximation [Eq. (11)] is employed. Third, the overlap between the nuclear wave functions of different states is assumed to be

$$\langle \zeta_k | \zeta_l \rangle = 1, \qquad (23)$$

which is motivated by the fact that the amplitudes cannot be transferred between different $\mathbf{R}^c$ trajectories. With these approximations, whose implications are discussed in details in Sec. 2.2, Eq. (21) is reduced to

$$i\hbar \frac{dc_k}{dt} + \sum_j (-H^c_{kj} + i\hbar \mathbf{F}^c_{kj} \cdot \mathbf{v}^c) c_j = 0. \qquad (24)$$

Note that the nondiagonal terms in the parenthesis in this equation are the same nonadiabatic coupling terms, which appeared in Eqs. (17) and (18).

Equation (24) can be further simplified if either an adiabatic representation ($\{\Phi_k\} \,|\, H_{kj} = V_k \delta_{kj}$) or a diabatic representation ($\{\Phi_k^d\} \,|\, H_{kj} = W_{kj}, \mathbf{F}_{kj} = 0$) is adopted:

$$i\hbar \frac{\partial c_k}{\partial t} + \sum_j (-V_k^c \delta_{kj} + i\hbar \mathbf{F}_{kj}^c \cdot \mathbf{v}^c) c_j = 0 \quad \text{(adiabatic)}, \tag{25}$$

$$i\hbar \frac{\partial c_k}{\partial t} - \sum_j W_{kj}^c c_j = 0 \quad \text{(diabatic)}. \tag{26}$$

Although Eq. (17) establishes a clear conceptual justification for the classical propagation of independent trajectories, it is not apparent how this can be done in practical terms. This problem can be approached by invoking the Ehrenfest theorem,[54] which states that the center of the molecular wave packet moves like a classical set of particles, in which the motion of nucleus m is determined by Newton's law

$$\frac{d\langle \Psi \,|\, \hat{\mathbf{P}}_m \,|\, \Psi \rangle_{\mathbf{rR}}}{dt} + \langle \Psi \,|\, \nabla_m H_e \,|\, \Psi \rangle_{\mathbf{rR}} = 0, \tag{27}$$

where $\hat{\mathbf{P}}_m$ is the momentum operator for nucleus m. Using Eqs. (3), (19), (11) and (23), we can show that the force on m is

$$\mathscr{F}_m^c \equiv -\langle \Psi \,|\, \nabla_m H_e \,|\, \Psi \rangle_{\mathbf{rR}} \approx -\sum_{kj} c_k^* c_j \langle \Phi_k^c \,|\, \nabla_m H_e \,|\, \Phi_j^c \rangle_{\mathbf{r}}, \tag{28}$$

where the time-dependent coefficients c_i are also given by Eq. (24). In adiabatic and diabatic representations, the force is simplified to, respectively,

$$\mathscr{F}_m^c = -\sum_k |c_k|^2 \nabla_m V_k^c - \sum_{kj} c_j c_k^* (V_j^c - V_k^c) \mathbf{F}_{kj}^{c,m} \quad \text{(adiabatic)}, \tag{29}$$

$$\mathscr{F}_m^c = -\sum_{kj} c_j c_k^* \nabla_m W_{kj}^c \quad \text{(diabatic)}. \tag{30}$$

Using Eq. (28), the coordinates $\mathbf{R}^c$ can be determined from a classical trajectory given by the Newton's equations for each nucleus m

$$\frac{d^2 \mathbf{R}_m^c}{dt^2} - \frac{\mathscr{F}_m^c}{M_m} = 0. \tag{31}$$

Equations (24) and (31) define the mean-field or Ehrenfest method, which can also be derived via Hamilton–Jacobi theory, in a procedure analogous to that discussed above.[50,55] In this method, the evolution of each initial point in the phase space is determined by a single independent trajectory driven by an average potential energy surface involving many states like that defined in Eq. (28). The single trajectory feature of the Ehrenfest method represents a great computational advantage, which allowed the investigation of retinal dynamics in the middle 1970s.[10] But this feature is also the handicap of the method: if, for example, the wave packet splits half/half between two adiabatic states after leaving the region of strong nonadiabatic coupling, the single trajectory will also move on an average surface in the long run, which represents an unrealistic situation.

The reason for such wrong asymptotic behavior is discussed in Sec. 2.2. For now, we point out that the main alternative that has been proposed to overcome this problem is the surface hopping approach,[3] which propagates multiple trajectories from each initial point of the phase space. These trajectories are always driven by a single adiabatic potential energy surface, with the associated force

$$\mathscr{F}_m^c = -\nabla_m V_l^c, \qquad (32)$$

and the correct distribution among the states is achieved by allowing the trajectories to change the surface on which they are moving (the way this is done is discussed in Sec. 2.2). In the example above, with asymptotic half/half wave packet split, half of trajectories end up in one state and half in the other. A comparative review of the Ehrenfest and the surface hopping methods is given in Ref. 51.

In principle, the computational costs of the surface hopping approach are greater than that of the mean-field approach because of the multiple-trajectory feature of the first. In practical terms, however, the costs are just about the same. This happens because for a proper description of the time evolution, trajectories dynamics simulations with either method must be initiated from multiple initial points in the phase space. (The way this initial distribution is generated is discussed in details in Sec. 3.) Usually, only a single trajectory per initial point is carried out in surface hopping simulations and the state distribution information is approximately recovered by the multiple trajectories initiating at different points.

2.2. *Implications of the local approximation*

Equation (24) is a local version of the time-dependent Schrödinger equation for the nuclei. It describes the amplitude variations of a wave packet entirely localized at the coordinates $\mathbf{R}^c$. The terms in the parenthesis ($k \neq j$) determine the flow of population between different adiabatic states.

The approximations that led to Eq. (24), which usually are not explicit in the conventional derivation using time-dependent electronic wave functions, have different consequences for the method. In the local approximation, neglecting the first term proportional to $\hbar^2$ in Eq. (5) does not affect the transition probability because this term only contributes to the flux between the imaginary and real parts of c_k within the same state k. This term, however, is responsible for the phase information in different points of the space. The hypothesis that dynamics can be simulated by a swarm of independent trajectories rests on the validity of this approximation.[50]

The second term proportional to $\hbar^2$ in Eq. (5) affects the transition probability. However, its influence is expected to be minor not only because of the $\hbar^2$ factor, but also due to the usually small values of the G_{kj} coupling terms. In principle, this term could be reincorporated into the surface hopping approach by noting that the main contribution to the second derivative coupling term comes from the diagonal term G_{kk}, which can be approximated by[52]

$$G_{kk} \approx -\sum_j (\mathbf{F}_{kj}^c)^2. \tag{33}$$

The effect of the replacement of the velocity operator by the velocity function can be estimated by assuming that the nuclei can be described by a Gaussian wave packet[54] centered at the classical trajectory

$$\zeta_k(\mathbf{R}) = \prod_m^{N_{at}} \pi^{-3/4} d_m^{-3/2} \exp\left(i\mathbf{k}_m \cdot (\mathbf{R}_m - \mathbf{R}_m^c) - \frac{(R_m - R_m^c)^2}{2d_m^2}\right), \tag{34}$$

where $\mathbf{R}_m$ is the coordinate vector of nucleus m. In this case, the action of the velocity operator for atom m on ζ_k results in

$$\hat{\mathbf{v}}_m \zeta_k(\mathbf{R}) = \langle \mathbf{v}_m \rangle \zeta_k(\mathbf{R}) + \frac{i\hbar}{M_m d_m^2}(\mathbf{R}_m - \mathbf{R}_m^c)\zeta_k(\mathbf{R}). \tag{35}$$

If Eq. (19) is inserted into Eq. (5) and the action of the velocity operator is taken as given by Eq. (35), the relevant term becomes

$$i\hbar \mathbf{F}_{kj} \cdot \hat{\mathbf{v}} \zeta_j(\mathbf{R}) = i\hbar \mathbf{F}_{kj} \cdot \langle \mathbf{v} \rangle \zeta_j(\mathbf{R}) - \hbar^2 \sum_m^{N_{at}} \frac{1}{M_m d_m^2} \mathbf{F}_{kj}^m$$
$$\cdot (\mathbf{R}_m - \mathbf{R}_m^c) \zeta_j(\mathbf{R}). \tag{36}$$

The first term in the right side of Eq. (36) is the same as that present in Eq. (24). Therefore, the replacement of the operator by its expectation value implies neglecting a term proportional to $\hbar^2$. It has been noted in Ref. 12 that this approximation may not work well in low temperature regimes where zero-point motions become important.

The main implication of the saddle point approximation, Eq. (11), is that no global feature can be described within this approach, notably tunneling. Finally, the strong coherence between "classical wave packets" in different states caused by the independent trajectories hypothesis and that leads to the strong condition of Eq. (23) has important consequences that are often neglected in nonadiabatic trajectory dynamics. A quantum-mechanical system coupled to many degrees of freedom is expected to evolve from a pure state to a statistical mixture.[12,57] This phenomenon is governed by the evolution of the density matrix, whose non-diagonal terms quickly drop to zero. In trajectory-independent methods, decoherence cannot properly take place because the variation of the amplitudes is restricted to occur only between states at a single point $\mathbf{R}^c$. [See, for example, Eq. (18), which was derived under the hypothesis that the divergence of the spatial probability flux is zero.]

In the Ehrenfest method, the time evolution of the $c_k(t)$ coefficients determined by Eq. (24) along a single trajectory is fully coherent, which causes the wrong asymptotic behaviors discussed in Sec. 2.1. *Ad hoc* decoherence corrections that force a switch to a single state after leaving the region of strong nonadiabatic coupling have been proposed, showing largely improved results in comparison to the conventional Ehrenfest dynamics.[13,58,59]

In surface hopping methods, the problem of lack of decoherence is minimized first because trajectories are naturally forced to mixed asymptotic states and second because each trajectory (even starting at the same initial point) hops at different times due to the stochastic nature of the algorithm. However, already in the original formulation of the Fewest–Switches approach discussed in the next section, Tully recognized that this treatment of coherence might not be enough, which led him to propose the application of *ad hoc* damping terms to Eq. (24).[11] Indeed, it has been recently shown[14] that a long standing problem in surface hopping approaches, the divergence between occupation and population (see discussion in Sec. 2.3), is still caused by missing decoherence. It is also shown in Ref. 14 that if the "non-linear decay of mixing" model proposed in Ref. 59 in the context of Ehrenfest dynamics is applied to surface hopping dynamics, the occupation/

population problem is eliminated. These corrections are applied at every time step by transforming the solutions of Eq. (24) according to

$$c'_k = c_k \exp(-\Delta t/\tau_{kl}) \quad \forall k \neq l, \tag{37}$$

$$c'_l = c_l \left[\frac{1 - \sum_{k \neq l} |c'_k|}{|c_l|^2} \right]^{1/2}, \tag{38}$$

$$\tau_{kl} = \frac{\hbar}{|V_{kk} - V_{ll}|} \left(1 + \frac{\alpha}{E_{kin}}\right), \tag{39}$$

where l is the current state, E_{kin} is the nuclear kinetic energy, Δt is the integration interval, and α is an empirical parameter whose recommended value is $\alpha = 0.1$ hartree.

2.3. The surface hopping approach

In the surface hopping approach a swarm of independent trajectories determined by Eqs. (31) and (32) is run, each one moving always in a single state l. To guarantee the correct distribution of trajectories among the states, Eq. (24) is integrated simultaneously and c_k for each state is obtained. The transition probability between each two electronics states is evaluated and a stochastic algorithm decides whether the system remains on the same electronic surface or hops to another one. Therefore, if a swarm of N_T trajectories with the same initial condition is started, they will soon diverge due to the stochastic nature of the algorithm. A good surface hopping algorithm is expected to be self-consistent, which means that the fraction of trajectories in each state (occupation) at a certain time, which is given by

$$f_k(t) \equiv N_k(t)/N_T, \tag{40}$$

where $N_k(t)$ is the number of trajectories in state k in time t, should tend to the average value of $|c_k|^2$ over all trajectories (average population)

$$\bar{a}_{kk}(t) = \frac{1}{N_T} \sum_{n}^{N_T} |c_k^{(n)}(t)|^2. \tag{41}$$

There are several recipes for computing transition probabilities for the surface hopping approach.[3,11,34,60–64] Probably the most common is the fewest-switches method proposed by Tully.[11] In this algorithm, the number of hopping events within one time step Δt is minimized. Under this condition, the hopping probability between states l and k is

$$P_{l \to k} = \frac{\text{Population increment in } k \text{ due to flux from } l \text{ during } \Delta t}{\text{Population of } l}.$$

To compute this quantity, Eq. (24) is rewritten in terms of the density matrix elements $\rho_{kl} \equiv c_k c_l^*$:

$$i\hbar \frac{d\rho_{kl}}{dt} - \sum_j [(-H_{jl}^c + i\hbar \mathbf{F}_{jl}^c \cdot \mathbf{v}^c)\rho_{kj} - (-H_{kj}^c + i\hbar \mathbf{F}_{kj}^c \cdot \mathbf{v}^c)\rho_{jl}] = 0. \quad (42)$$

Assuming that the population flux from and to state l involves only state k, the population ρ_{ll} of state l is determined by

$$\begin{aligned} \frac{d\rho_{ll}}{dt} &= i\hbar^{-1}[(-H_{kl}^c + i\hbar \mathbf{F}_{kl}^c \cdot \mathbf{v}^c)\rho_{lk} - (-H_{lk}^c + i\hbar \mathbf{F}_{lk}^c \cdot \mathbf{v}^c)\rho_{kl}] \\ &= -2\hbar^{-1} H_{kl} \operatorname{Im}(\rho_{kl}) - 2\, \mathbf{F}_{kl}^c \cdot \mathbf{v}^c \operatorname{Re}(\rho_{kl}). \end{aligned} \quad (43)$$

The population increment in state k due to the flux from state l between t and $t + \Delta t$ is given by

$$\Delta \rho_{kk} = -\Delta \rho_{ll} = 2\Delta t(\hbar^{-1} \operatorname{Im}(\rho_{kl}) H_{kl} - \operatorname{Re}(\rho_{kl}) \mathbf{F}_{kl}^c \cdot \mathbf{v}^c). \quad (44)$$

If the counter-flux from k to l is neglected, the fewest-switches probability can be finally written as

$$P_{l \to k} = \max\left[0, \frac{2\Delta t}{\rho_{ll}} (\hbar^{-1} \operatorname{Im}(\rho_{kl}) H_{lk}^c - \operatorname{Re}(\rho_{kl}) \mathbf{F}_{kl}^c \cdot \mathbf{v}^c)\right]. \quad (45)$$

When adiabatic representation is employed, Eq. (45) simplifies to

$$P_{l \to k} = \max\left[0, \frac{-2\Delta t}{\rho_{ll}} \operatorname{Re}(\rho_{kl}) \mathbf{F}_{kl}^c \cdot \mathbf{v}^c\right] \text{ (adiabatic)}, \quad (46)$$

while in diabatic representation it simplifies to

$$P_{l \to k} = \max\left[0, \frac{2\Delta t}{\hbar \rho_{ll}} \operatorname{Im}(\rho_{kl}) W_{lk}^c\right] \text{ (diabatic)}. \quad (47)$$

In the evaluation of the hopping event, a transition from surface l to surface k takes place in time t if two conditions are simultaneously fulfilled:

(1) A uniformly selected random number r_t in the $[0, 1]$ interval is such that

$$\sum_{n=1}^{k-1} P_{l \to n}(t) < r_t \leq \sum_{n=1}^{k} P_{l \to n}(t). \qquad (48)$$

(2) The energy gap between the final and initial states satisfies[65]

$$V_k(\mathbf{R}^c(t)) - V_l(\mathbf{R}^c(t)) \leq \frac{\left(\sum_m^{N_{at}} \mathbf{v}_m^c \cdot \mathbf{F}_{kl}^{c,m}\right)^2}{2 \sum_m^{N_{at}} M_m^{-1} (\mathbf{F}_{kc}^{c,m})^2}. \qquad (49)$$

The second condition avoids the situation in which the total energy after hopping becomes larger than that before hopping. The situation where only Eq. (48) is satisfied is called a frustrated hopping (see Ref. 66 for a discussion about this topic). Equation (49) is derived under the condition that conservation of total energy after hopping is achieved by adding a quantity of linear momentum equivalent to $V_k - V_l$ to the direction of $\mathbf{F}_{kl}^c$. The correction along the direction of $\mathbf{F}_{kl}^c$ is motivated by the Pechukas force that occurs during the nonadiabatic transition, as one can see, for instance, in the second term on the right hand side of Eq. (29).[7,55] When $\mathbf{F}_{kl}^c$ is not explicitly computed, this correction can also be applied in the direction of $\mathbf{v}^c$. In this case, the second condition reads

$$V_k(\mathbf{R}^c(t)) - V_l(\mathbf{R}^c(t)) \leq E_{kin}(\mathbf{v}^c), \qquad (50)$$

where E_{kin} is the nuclear kinetic energy.

To avoid the cumbersome and computationally expensive procedures involved in the evaluation of $\mathbf{F}_{kj}^c$ for solving Eq. (24), some hopping algorithms just assume that the probability is unity if the energy gap between two states is smaller than some pre-defined energy threshold[34]; other algorithms take into account variations of wave function coefficients as a measurement of the nonadiabatic coupling,[62] or compute Landau–Zener transition probabilities.[64] Computational time can also be saved by neglecting the coupling terms that are expected to have minor contributions.[38] Although the usual formulation of surface hopping methods is derived to take into account exclusively transitions between states with the same multiplicity, a number of reformulations of the method to treat intersystem crossing processes have been reported.[67–69] Most of them are based on the computation of hopping probabilities based on Landau–Zener approach. Recently, intersystem crossing surface hopping based on a extension of the fewest-switches algorithm has been proposed in the context of interaction of molecules with metallic surfaces.[69]

As a summary, the methodology of surface hopping methods based on on-the-fly electronic structure calculations involves the following steps:

1. For a specific nuclear frame, solve the time-independent electronic Schrödinger equation (see Sec. 4) and obtain energies, energy gradients (Sec. 5), and nonadiabatic coupling terms (Sec. 6).
2. Still in the same nuclear frame, use the information from step 1 to integrate the *local* time-dependent Schrödinger equation for the nuclei [Eq. (24)]. This integration may be evaluated by standard algorithms to integrate first-order differential equations. Apply decoherence corrections [Eq. (37), (38), and (39)], evaluate the hopping probability [Eq. (45)], and determine the current state.
3. Determine the new nuclear arrangement by integrating the Newton's Equations in one time step [Eqs. (31) and (32)]. This integration is usually done with standard molecular dynamics algorithm such as the Velocity-Verlet.[70]
4. Go back to step 1 and repeat the procedure until the end of the trajectory.
5. Compute an ensemble of independent trajectories with different initial conditions (Sec. 3).

3. Initial Conditions

In order to integrate the Newton's equations for the nuclei, an ensemble of initial conditions needs to be prepared. These initial conditions should represent a classical phase space representation of the initially excited quantum wave packet. Usually, this problem is approached by building a phase space distribution in the electronic ground state and then projecting it onto the electronic excited states. The ground state distribution can be prepared either by a ground state trajectory simulation or from a probabilistic sampling. In principle, for an ergodic system, both approaches should produce similar results. Nevertheless, due to the classical nature of the trajectory simulations in the ground state and the quantum nature of typical distributions like that given by the Wigner function, the two sets may differ substantially.

3.1. *Wigner and quasi-Wigner distributions*

Assuming a quadratic approximation for the ground-state potential energy surface around the minimum, the $3N_{at} - 6$ internal coordinates can be described in terms of normal modes $\mathbf{Q}$ and the nuclear wave function can be approximated as that of a quantum harmonic oscillator. The classical phase space distribution can be approximated by a Wigner distribution[71]

$$P_W(Q^i, P^i) = (\pi\hbar)^{-1} \int d\eta \chi^0_{HO}(Q^i + \eta)^* \chi^0_{HO}(Q^i - \eta) e^{2i\eta P^i/\hbar}, \quad (51)$$

where χ^0_{HO} is the quantum harmonic oscillator wave function for the ground vibrational state and P^i is the momentum associated with the normal coordinate Q^i. The evaluation of this integral gives

$$P_W(Q^i, P^i) = (\pi\hbar)^{-1} \exp(-\mu^i \omega^i_{HO} Q^{i2}/\hbar) \exp(-P^{i2}/(\mu^i \omega^i_{HO} \hbar)), \quad (52)$$

where μ^i and ω^i_{HO} are, respectively, the reduced mass, the harmonic frequency and the equilibrium distance of normal mode i.

To sample coordinates and momentum, independent random values are assigned to Q^i and P^i and then the acceptance of the pair is evaluated according to the probability given by Eq. (52). This procedure will result in a Gaussian distribution in the (Q^i, P^i) space and because Q^i and P^i values are uncorrelated, the initial energy will be broadly distributed around the harmonic zero point value. This procedure should be repeated for each normal mode and then the normal coordinates and momentum are converted back to Cartesian coordinates.

If Eq. (51) is evaluated for a vibrationally excited level χ^n_{HO} instead for the ground vibrational state, the Wigner function can assume negative values and cannot be used as a distribution. To solve this problem, note that Eq. (52) can be written as

$$P_W(Q^i, P^i) = |\chi^0_{HO}(Q^i)|^2 |\xi^0_{HO}(P^i)|^2, \quad (53)$$

where ξ^0_{HO} is the harmonic oscillator wave function in the momentum representation. Even though Eq. (53) is valid only for the ground vibrational level, it motivates to write an analogous quasi-Wigner distribution for the excited vibrational states

$$P_{QW}(Q^i, P^i) = |\chi_{HO}^n(Q^i)|^2 |\xi_{HO}^n(P^i)|^2. \qquad (54)$$

Sampling of initial conditions based on Eq. (54) was used, for example, to create biased distributions towards specific normal modes in the applications of Ref. 46, where the goal was to investigate whether the activation of either in-plane or out-of-plane modes had different impact on internal conversion rates.

3.2. Sampling with trajectory simulations

An alternative to the probabilistic sampling is to perform a classical ground-state trajectory simulation and select points from it to initiate the excited state dynamics. For large systems where a normal mode analysis is not feasible, this can be the main procedure to generate initial conditions. Besides that, this procedure also allows inclusion of anharmonic effects in the initial conditions. One difficulty of this procedure is that it demands long simulation times to allow for an adequate sampling of the phase space. Additionally, the energy distribution among the degrees of freedom should be considered. Because of the classical nature of the trajectory propagation, there is no guarantee that one or more degrees of freedom will not have less than the zero point energy. For instance, in the simulations of oligophenylenes, it was found that ground-state trajectories thermalized at 300 K produced too cold torsional interring modes.[72]

3.3. Sampling in special points

Although our emphasis in this chapter is on dynamics simulations starting at the Franck–Condon region, two recent investigations are worth mentioning where the dynamics was initiated in other regions of the configuration space. In Ref. 33, dynamics simulations of thymine were initiated at a transition state along the S_2 state. The reason for that was to avoid the long initial dynamics of thymine moving from the Franck–Condon region to this transition state, a process that can take more than 2 ps.[73,74] In Ref. 36, dynamics simulations of pyrrole were started already at the conical intersection. In this case, the goal was to skip the whole excited state dynamics and to investigate the photochemistry induced by the molecule exiting a conical intersection.

3.4. The excited-state distribution

Having a ground-state distribution either via trajectory simulations or distribution sampling, the next step is to project this distribution onto the initial excited state. The simplest way of doing this is just to take the ground sate distribution as it is. This means to assume that the transition probability from the ground to the excited state is the same for every point in the distribution.

This approach can be refined by computing the transition probability for each point in the ground state distribution and evaluating whether it should be accepted as an initial condition or not according to this probability.[75] As for transition probability, one can take the oscillator strength, the Einstein coefficient B (Ref. 76), or some more advanced procedure that takes the shape of the electric field signal into account.[17] As a side product of these procedures, an approximation for the absorption spectrum can be directly obtained.[77]

Besides the computation of transition probabilities, the projection on the excited state can also take into account specific windows of transition energies to simulate laser excitations and transition into multiple excited states.

4. Electronic Structure Methods

Although many methods are available for electronic structure calculations, only few can be applied for nonadiabatic trajectories dynamics simulations. The first and most obvious condition to be satisfied by the method is that it should provide energies of electronically excited states. It should also be able to provide the energies at strongly distorted nuclear geometries and it should be fast enough to allow the computation of tens of thousands of single points composing the trajectory ensemble.

Besides energies, the method must provide energy gradients for excited states, preferentially computed by analytical procedures. This issue is discussed in more details in Sec. 5. The computation of nonadiabatic coupling terms or other quantities allowing the estimation of nonadiabatic transition probabilities is a final requirement to the method. This point is addressed in Sec. 6 (see also Sec. 2.3).

Nonadiabatic events usually occur in regions of the nuclear configuration space where the electronic wave function has strong multireference character. This means that in general single-reference methods fail to adequately describe the dynamics close to regions where nonadiabatic events occur, like conical intersections with the ground state and their neighborhoods. This limitation precludes the usage of methods based on single-reference coupled cluster theory (CCSD, CC2, EOM-CC, SAC-CI) and single-reference configuration interaction (CI) theory, although these methods can in principle still be used in investigations involving conical intersections between excited states.[78]

The computationally efficient time-dependent density functional theory (TD-DFT) method is inadequate to describe regions of multireference character as well. The problem arises not due to the single-reference character of the method, which can be compensated by inclusion of non-dynamic correlation in many exchange functionals, but mostly because of the linear response approximations in the time-dependent approach.[79] Another limitation of TD-DFT method is the poor description of charge-transfer states, an issue that has been addressed by development of new functionals.[80] DFT-based methods computing excited states by other ways than the time-dependent approach are also available. Examples are DFT/MRCI,[81] which has been shown to provide good description of potential energy surface even close to conical intersections, and the restricted open shell Kohn–Sham (ROKS) method[82] whose performance should be more limited. While ROKS has been used in nonadiabatic trajectory dynamics simulations for many different investigations,[82–85] the MRCI/DFT still awaits development of analytical gradients for being useful for dynamics.

Among semiempirical methods, similar inadequacies concerning the single-reference character are expected from methods like ZINDO/S and single reference CI expansions based on popular methods like PM3 and AM1. This has led to development of multireference approaches[18,19] opening the possibility of dynamics investigations of relatively large molecules.[37] An important issue is still to establish parametrizations adequate to describe large portions of conformational spaces of several classes of molecules.

Ab initio multireference (MR) methods based on the configuration interaction (CI) with singles and doubles or the coupled cluster (CC) ansatz are computationally demanding and dynamics investigations based on such methods are still restricted to relatively small molecules.[86] MRCI restricted to single excitations has demonstrated a good potential for dynamics simulations.[31] Recent development of analytical gradients for multireference perturbation theory (MRPT2) methods[20] including the popular complete active space (CASPT2) version is also a promising fact for dynamics simulations of medium sized molecules in the near future as soon as analytical nonadiabatic coupling terms are made available. In Ref. 87, efficient numerical procedures are proposed to obtain nonadiabatic couplings terms at CASPT2 level.

Currently, the most frequently used *ab initio* method applied in nonadiabatic trajectory dynamics simulations has been the multiconfigurational self-consistent field (MCSCF) method, in particular in its CASSCF version.[29,30,34,74,88,89] It allows dynamics simulations for few picosends of molecules of the size of important biological monomers like amino acids and nucleobases. The MCSCF method, however, has intrinsic problems, which have to be addressed in each simulation. First, ionic states are computed systematically too high, which leads to an overestimation of excitation energies into $^1\pi\pi^*$ states in the Franck–Condon region. The origin of the problem is related to the lack of σ/π correlation in usual CAS spaces.[90,91] Second, the missing dynamical correlation may lead to inadequate descriptions of excited-state barriers and minima,[92] which demands careful comparison between potential energy curves computed at MCSCF and correlated levels prior dynamics simulations. Third, orbital rotations between the virtual and doubly occupied spaces and the active space are often source of discontinuities in the potential energy surfaces during the simulations.[93] This last problem is much alleviated by more sophisticated construction of multiconfigurational spaces, like combinations of GVB-PP and CAS spaces.[36]

Whatever is the chosen electronic structure method, the size of the molecule will always represent a major limitation, which is especially dramatic for investigations in solvents, solid matrices and macromolecular environments. This limitation has been overcome by use of hybrid methods. Quantum-mechanical/molecular-mechanical (QM/MM) schemes[94] are starting to be used in nonadiabatic trajectory dynamics simulations[32,37,89] and have big potential for further investigations.

5. Analytical Gradients

In this section, we present an overview of some of the general features of MCSCF and MRCI analytic energy gradients. A more detailed and complete discussion is given in Refs. 95 and 96. Derivation of analytical gradients at the MRPT2 level is discussed in Ref. 20. The configuration state functions (CSF) expansion coefficients for these wave functions are variationally determined, and this allows the Hellmann–Feynman theorem, using second-quantized conventions for the Hamiltonian operator, to be exploited. The choice of orbitals for the MRCI expansion is somewhat arbitrary because of redundant orbital rotation variables associated with the MCSCF expansion space,[95,97,98] and in certain situations the resolution of these rotations must be accounted for when computing the MRCI energy gradient. One of the challenges is to develop and implement general efficient procedures that allow for all of these possibilities for both the MCSCF and MRCI wave functions.

A formal approach that meets this challenge of generality and efficiency is based on a sequence of successive geometry-dependent orbital transformations in which the effects of individual constraints of the orbitals may be considered individually. In the straightforward case, there would be four orbital basis sets.

$$\varphi^{[C]}(\mathbf{R}) = \chi(\mathbf{R})\,\mathbf{C}(0), \tag{55}$$

$$\varphi^{[S]}(\mathbf{R}) = \varphi^{[C]}(\mathbf{R})\,\mathbf{S}^{[C]}(\mathbf{R})^{-1/2}, \tag{56}$$

$$\varphi^{[K]}(\mathbf{R}) = \varphi^{[S]}(\mathbf{R})\exp(\mathbf{K}), \tag{57}$$

$$\varphi^{[Z]}(\mathbf{R}) = \varphi^{[K]}(\mathbf{R})\exp(\mathbf{Z}). \tag{58}$$

The basis $\chi(\mathbf{R})$ is the atom-centered basis; as the atom centers move within the molecular geometry, the associated basis functions move along with them. The $\mathbf{C}(0)$ matrix contains the fully optimized and resolved orbital coefficients at the reference geometry denoted, for convenience, $\mathbf{R} = 0$. The basis $\varphi^{[C]}(\mathbf{R})$ is a geometry-dependent basis that generally is orthonormal only at $\mathbf{R} = 0$. The symmetric positive-definite matrix $\mathbf{S}^{[C]}(\mathbf{R})$ is the orbital overlap matrix in the $\varphi^{[C]}(\mathbf{R})$ basis, and it is used to define the basis $\varphi^{[S]}(\mathbf{R})$, which is orthonormal at all $\mathbf{R}$. The basis $\varphi^{[K]}(\mathbf{R})$ is the energy-optimized orthonormal orbital basis defined in terms of the skew-symmetric matrix $\mathbf{K}$ whose nonzero elements correspond to the essential MCSCF orbital rotation parameters. Finally, $\varphi^{[Z]}(\mathbf{R})$ is the fully resolved orbital basis defined in terms of the skew-symmetric matrix $\mathbf{Z}$ whose nonzero elements correspond to the redundant MCSCF orbital rotation parameters. The two sets of orbital rotation parameters, essential and redundant, are disjoint in the sense that a nonzero K_{pq} element implies a zero Z_{pq} element, and a nonzero Z_{pq} element implies a zero K_{pq} element. It is this final orbital basis that is used to define the MRCI wave function at the reference geometry. In this formulation, the orbital bases $\varphi^{[K]}(\mathbf{R})$ and $\varphi^{[Z]}(\mathbf{R})$ are orthonormal because the transformation matrices $\exp(\mathbf{K})$ and $\exp(\mathbf{Z})$ are intrinsically orthogonal; thus no additional constraints need to be satisfied, and no additional optimization variables, particularly in the form of Lagrange multipliers, are introduced.

The general approach to computing a particular analytic energy gradient will be similar for all of the electronic structure methods discussed in this section. The energy as a function of $\mathbf{R}$ will be written in terms of the geometry-dependent one- and two-electron integrals in the most appropriate orbital basis. This will involve also the geometry-dependent density matrices, transition density matrices, and various other combinations of these quantities such as geometry-dependent Fock matrices. Expansion techniques will be used to determine the first-order dependence of the energy on the various geometry-dependent quantities at the reference geometry $\mathbf{R} = 0$. Finally, the transformation properties of these quantities, consistent with Eqs. (55)–(58), will be used as necessary to simplify the expressions, to isolate the geometry-dependent factors from the geometry-independent factors, and eventually to express the analytic energy gradient in the most computationally efficient form possible.

5.1. *MCSCF gradient*

We first summarize some of the important features of the MCSCF gradient for a single electronic state. A trial MCSCF wave function may be written in the $\varphi^{[S]}(\mathbf{R})$ orbital basis as

$$|\Phi^{\text{trial}}(\mathbf{R})\rangle = \exp(\hat{K}(\mathbf{R}))\exp(\hat{P}(\mathbf{R}))|ref(\mathbf{R});[S]\rangle, \qquad (59)$$

with

$$\hat{K}(\mathbf{R}) = \sum_{r,s} K(\mathbf{R})_{rs}\hat{E}_{rs} = \sum_{r>s} k(\mathbf{R})_{(rs)}\left(\hat{E}_{rs} - \hat{E}_{sr}\right), \qquad (60)$$

$$\hat{P}(\mathbf{R}) = |p(\mathbf{R})\rangle\langle ref(\mathbf{R})| - |ref(\mathbf{R})\rangle\langle p(\mathbf{R})|, \qquad (61)$$

$$|p(\mathbf{R})\rangle = \sum_{m}^{N_{csf}} p_m(\mathbf{R})|\tilde{m};[S]\rangle. \qquad (62)$$

The vector elements $k_{(rs)} = K_{rs} = -K_{sr}$ for $r > s$ are the unique essential orbital variation parameters. $|p(\mathbf{R})\rangle$ is some arbitrary wave function in the $N_{CSF} - 1$ dimensional subspace orthogonal to $|ref(\mathbf{R})\rangle$, which itself is an arbitrary, normalized, reference wave function. This reference wave function may be chosen to be either a ground or an excited electronic state; a ground state would correspond to the lowest Hamiltonian matrix eigenpair at $\mathbf{R} = 0$, whereas an excited state would correspond to a higher eigenpair. The operator $\hat{P}(\mathbf{R})$ may be written either in terms of orthonormal linear combinations of CSFs or the primitive CSF basis in which the computational procedures are most efficient. This trial wave function parameterization allows for arbitrary orbital variations and arbitrary CSF expansion coefficient variations relative to those of the reference function $|ref(\mathbf{R})\rangle$. It is convenient to take $|ref(\mathbf{R})\rangle$ to be the optimized MCSCF wave function at $\mathbf{R} = 0$, and at displaced geometries to be the wave function with the same normalized CSF expansion coefficients, $\|\mathbf{c}^{mc}(0)\|_2 = 1$, and represented in the corresponding orthonormal $\varphi^{[S]}(\mathbf{R})$ orbital basis. This reference wave function will be denoted $|mc(0);[S]\rangle$, and there is a corresponding reference PES associated with this reference wave function defined as

$$\begin{aligned}
E^{ref}(\mathbf{R}) &= \langle mc(0);[S] \,|\, \hat{H}^{[S]}(\mathbf{R}) \,|\, mc(0);[S]\rangle \\
&= \sum_{r,s} h_{rs}^{[S]}(\mathbf{R}) \langle mc(0);[S] \,|\, \hat{E}_{rs} \,|\, mc(0);[S]\rangle \\
&\quad + \frac{1}{2} \sum_{p,q,r,s} g_{pqrs}^{[S]}(\mathbf{R}) \langle mc(0);[S] \,|\, \hat{e}_{pqrs} \,|\, mc(0);[S]\rangle \\
&= Tr(\mathbf{h}^{[S]}(\mathbf{R})\mathbf{D}^{mc[S]}(0)) + \frac{1}{2} Tr(\mathbf{g}^{[S]}(\mathbf{R})\mathbf{d}^{mc[S]}(0)).
\end{aligned} \quad (63)$$

The last expression in particular shows that all of the geometry dependence of this reference energy surface derives from the geometry dependence of the one- and two-electron integrals that define the second-quantized Hamiltonian operator. The density matrices $\mathbf{D}^{mc[S]}(0)$ and $\mathbf{d}^{mc[S]}(0)$ are geometry independent because (1) they depend on the geometry-independent coupling coefficients, and (2) they depend on the fixed, reference-geometry, CSF expansion coefficients.

The trial energy expectation value may then be written

$$\begin{aligned}
E^{\text{trial}}(\mathbf{K},\mathbf{p};\mathbf{R}) &= \langle \Phi^{\text{trial}}(\mathbf{R}) \,|\, \hat{H} \,|\, \Phi^{\text{trial}}(\mathbf{R}) \rangle \\
&= \langle mc(0);[S] \,|\, \exp(-\hat{P}(\mathbf{R}))\exp(-\hat{K}(\mathbf{R}))\hat{H} \\
&\quad \times \exp(\hat{K}(\mathbf{R}))\exp(\hat{P}(\mathbf{R})) \,|\, mc(0);[S]\rangle.
\end{aligned} \quad (64)$$

The commutator expansion exposes the order-by-order dependence of the trial energy on the $K_{rs}(\mathbf{R})$ and $p_m(\mathbf{R})$ parameters:

$$E^{\text{trial}}(\mathbf{k}(\mathbf{R}), \mathbf{p}(\mathbf{R}); \mathbf{R}) = E^{ref}(\mathbf{R}) + (\mathbf{k}(\mathbf{R})^T \quad \mathbf{p}(\mathbf{R})^T) \begin{pmatrix} \mathbf{f}_{orb}(\mathbf{R}) \\ \mathbf{f}_{csf}(\mathbf{R}) \end{pmatrix}$$
$$+ \frac{1}{2}(\mathbf{k}(\mathbf{R})^T \quad \mathbf{p}(\mathbf{R})^T) \begin{pmatrix} \mathbf{G}_{orb,orb}(\mathbf{R}) & \mathbf{G}_{orb,csf}(\mathbf{R}) \\ \mathbf{G}_{csf,orb}(\mathbf{R}) & \mathbf{G}_{csf,csf}(\mathbf{R}) \end{pmatrix} \begin{pmatrix} \mathbf{k}(\mathbf{R}) \\ \mathbf{p}(\mathbf{R}) \end{pmatrix} + \cdots \quad (65)$$

Explicit expressions for the $\mathbf{f}(\mathbf{R})$ and $\mathbf{G}(\mathbf{R})$ subblocks have been given elsewhere (e.g. Ref. 95 and references therein). It is sometimes convenient to collect the parameters $\mathbf{k}(\mathbf{R})$ and $\mathbf{p}(\mathbf{R})$ together into a vector $\boldsymbol{\lambda}(\mathbf{R})$, which allows the trial energy to be written

$$E^{\text{trial}}(\boldsymbol{\lambda}(\mathbf{R}); \mathbf{R}) = E^{ref}(\mathbf{R}) + \boldsymbol{\lambda}(\mathbf{R}) \cdot \mathbf{f}(\mathbf{R}) + \frac{1}{2}\boldsymbol{\lambda}(\mathbf{R})^T \mathbf{G}(\mathbf{R}) \boldsymbol{\lambda}(\mathbf{R}) + \cdots \quad (66)$$

$\mathbf{f}(\mathbf{R})$ is the gradient of the energy with respect to wave function variations and consists of the $\mathbf{f}_{orb}(\mathbf{R})$ and $\mathbf{f}_{csf}(\mathbf{R})$ partitions, which correspond to orbital and CSF variations respectively. Similarly, the symmetric Hessian matrix $\mathbf{G}(\mathbf{R})$ consists of the four partitions indicated above. At any arbitrary $\mathbf{R}$, the MCSCF wave function parameters $\boldsymbol{\lambda}^{mc}(\mathbf{R})$ are those that satisfy the variational conditions

$$\left. \frac{\partial E^{\text{trial}}(\boldsymbol{\lambda}(\mathbf{R}); \mathbf{R})}{\partial \boldsymbol{\lambda}} \right|_{\boldsymbol{\lambda}^{mc}(\mathbf{R})} = \mathbf{0}. \quad (67)$$

This results in a coupled set of nonlinear equations

$$\mathbf{0} = \mathbf{f}^{mc}(\mathbf{R}) + \mathbf{G}^{mc}(\mathbf{R})\boldsymbol{\lambda}^{mc}(\mathbf{R}) + O(\boldsymbol{\lambda}^{mc}(\mathbf{R})^2)\cdots \quad (68)$$

that must be satisfied by the parameters $\boldsymbol{\lambda}^{mc}(\mathbf{R})$ at arbitrary $\mathbf{R}$. There is no closed-form solution to this equation. However, this equation is sufficient to determine the corresponding Taylor expansion of these geometry-dependent parameters relative to the reference geometry $\mathbf{R} = 0$ values. Differentiating Eq. (68) with respect to a displacement of a representative atomic center coordinate denoted x, evaluation at the reference geometry, and using the relation $\boldsymbol{\lambda}^{mc}(0) = \mathbf{0}$ gives

$$\boldsymbol{\lambda}^{mc}(0)^x = -\mathbf{G}^{mc}(0)^{-1}\mathbf{f}^{mc}(0)^x. \quad (69)$$

The superscript x denotes differentiation. This gives the first-order change in the MCSCF orbitals and CSF expansion coefficients at the reference geometry to the displacement along the coordinate direction labeled by x.

The MCSCF energy at arbitrary $\mathbf{R}$ is given by Eq. (59) with the specific $\mathbf{K}^{mc}(\mathbf{R})$ and $\mathbf{p}^{mc}(\mathbf{R})$ parameters determined from Eq. (68). Differentiation of this energy expression with respect to a geometry displacement and evaluation at $\mathbf{R} = 0$ gives an element of the MCSCF analytic energy gradient

$$E^{mc}(0)^x = E^{ref}(0)^x + \boldsymbol{\lambda}(0) \cdot \mathbf{f}(0)^x + \boldsymbol{\lambda}(0)^x \cdot \mathbf{f}(0) + O(\boldsymbol{\lambda}(0)^1). \tag{70}$$

Truncation follows from the relations $\mathbf{f}(0) = \mathbf{0}$ and $\boldsymbol{\lambda}(0) = \mathbf{0}$, and the result is an example of the Hellmann–Feynman theorem using second-quantized conventions for the definition of the Hamiltonian operator. That is, the first-order wave function does not contribute to the single-state MCSCF energy gradient. [As discussed later, the first-order wave function term in Eq. (70) will contribute for other energy expressions.] The MCSCF energy gradient may be written

$$\begin{aligned} E^{mc}(0)^x = E^{ref}(0)^x &= \langle mc(0); [S] \,|\, \hat{H}^{[S]}(0)^x \,|\, mc(0); [S] \rangle \\ &= \mathbf{c}^{mc}(0)^T \hat{\mathbf{H}}^{[S]}(0)^x \mathbf{c}^{mc}(0) \tag{71} \\ &= Tr(\mathbf{h}^{[S]}(0)^x \mathbf{D}^{mc[S]}(0)) + \frac{1}{2} Tr(\mathbf{g}^{[S]}(0)^x \mathbf{d}^{mc[S]}(0)). \end{aligned}$$

This last expression shows that the density matrices contain the displacement-independent factors of the energy gradient elements, and the derivative integrals contain all of the displacement-dependent factors. If the energy gradient were evaluated using this expression, the entire set of derivative terms $\mathbf{h}^{[S]}(0)^x$ and $\mathbf{g}^{[S]}(0)^x$ for all $3N_{at}$ possible displacement directions x would need to be computed (ignoring for the moment any simplifications due to the use of translational and rotational invariance and point group symmetry). This would require effort proportional to $3N_{at}$. A more efficient approach is to transform the gradient expression back to the original AO basis. Using the above sequence of orbital transformations, the gradient component may be written as

$$\begin{aligned} E^{mc}(0)^x = {}&Tr(\mathbf{h}^{[C]}(0)^x \mathbf{D}^{mc[C]}(0)) + \frac{1}{2} Tr(\mathbf{g}^{[C]}(0)^x \mathbf{d}^{mc[C]}(0)) \\ &- \frac{1}{2} Tr(\{\mathbf{h}^{[C]}(0); \mathbf{S}^{[C]}(0)^x\} \mathbf{D}^{mc[C]}(0)) \\ &- \frac{1}{4} Tr(\{\mathbf{g}^{[C]}(0); \mathbf{S}^{[C]}(0)^x\} \mathbf{d}^{mc[C]}(0)) \tag{72} \end{aligned}$$

$$= Tr(\mathbf{h}^{[C]}(0)^x \mathbf{D}^{mc[C]}(0)) + \frac{1}{2} Tr(\mathbf{g}^{[C]}(0)^x \mathbf{d}^{mc[C]}(0))$$
$$- Tr(\mathbf{S}^{[C]}(0)^x \mathbf{F}^{mc[C]}(0)) \tag{73}$$
$$= Tr(\mathbf{h}^{[\chi]}(0)^x \mathbf{D}^{mc[\chi]}(0)) + \frac{1}{2} Tr(\mathbf{g}^{[\chi]}(0)^x \mathbf{d}^{mc[\chi]}(0))$$
$$- Tr(\mathbf{S}^{[\chi]}(0)^x \mathbf{F}^{mc[\chi]}(0)). \tag{74}$$

The notation $\{\mathbf{A}; \mathbf{T}\}$ denotes a symmetrized one-index transformation of the array $\mathbf{A}$ with the transformation matrix $\mathbf{T}$. The derivative overlap terms, $\mathbf{S}^{[C]}(0)^x$ and $\mathbf{S}^{[\chi]}(0)^x$, appear here because the orbital basis $\varphi^{[C]}(\mathbf{R})$ is nonorthogonal at $\mathbf{R} \neq 0$. The trace relationships such as, for example, $\mathrm{Tr}((\mathbf{C}^T\mathbf{h}^{[\chi]}\mathbf{C})\mathbf{D}^{[C]}) = Tr(\mathbf{h}^{[\chi]}(\mathbf{C}\mathbf{D}^{[C]}\mathbf{C}^T))$ define implicitly the orbital basis transformation properties of the various density and Fock matrix arrays in these expressions.

This final expression is important because the two-electron Hamiltonian integrals are very sparse in the atom-centered AO basis. A particular two-electron repulsion integral depends on, at most, only four atom centers, or twelve Cartesian displacements, out of the $3N_{at}$ total possible displacements. Consequently, there are only about twelve times as many nonzero AO derivative integrals as undifferentiated AO integrals. By exploiting this sparseness, the trace operation may be computed in the AO basis with effort that is formally independent of N_{at}, i.e. $O(N_{at}^0) = O(1)$. This simplification affects the number of arithmetic operations required to evaluate the energy gradient and the total amount of memory and external storage space that is required for the computation. Thus, the analytic gradient procedure described above is both more efficient and more accurate than a finite difference approach, and it has similar advantages when these gradients are used to fit molecular potential energy surfaces, to optimize molecular geometries, or when they are used directly to compute classical dynamical trajectories. In Ref. 98, molecular geometry optimizations were performed for 20 molecules using MCSCF wave functions and with a variety of orbital basis sets and a wide range of CSF expansion spaces. The effort for the gradient evaluations for these molecules required between 8.0% and 84.4% of the total computational effort (including integral evaluation and wave function optimization), with a mean of 58.1%. This demonstrates that the computational procedure presented above is very efficient, that it is indeed independent of N_{at}, and that it may be applied to any molecule for which the MCSCF wave function optimization itself is practical.

5.2. State-averaged MCSCF gradient

It is often desirable to compute potential energy surfaces for several electronic states using a common set of molecular orbitals. This simplifies the computation of transition properties between electronic states (including nonadiabatic coupling elements), and because the individual energies correspond to eigenpairs of the same Hamiltonian matrix, the bracketing theorem allows the association of particular variational excited-state eigenpairs with the exact full-CI electronic state energies and wave functions. However, optimization of the orbitals for a specific electronic state can introduce bias in the computed energies in which one state is described preferentially to other states. This might, for example, artificially increase vertical excitation energies to higher states and artificially reduce excitation energies from lower states, or it might artificially introduce spurious Rydberg or ionic character from one state into the other states. This situation can be addressed by optimizing the orbitals to minimize an average energy of the states of interest

$$\bar{E}(\mathbf{R}) = \sum_{J}^{N_{av}} w_J E_J(\mathbf{R}). \tag{75}$$

N_{av} is the number of states included in the averaging procedure. The $E_J(\mathbf{R})$ energies are the Hamiltonian eigenvalues corresponding to the states of interest. The weight factors are positive constants, $w_J > 0$, independent of the displacement $\mathbf{R}$, with the normalization $\Sigma_J w_J = 1$. This complicates somewhat the specification of the wave function because the CSF expansion coefficients associated with each electronic state J are still optimized for the energy $E_J(\mathbf{R})$ of that specific state, but the orbitals, which are shared by all the states, are optimized to minimize $\bar{E}(\mathbf{R})$ rather than one of the individual states. We briefly summarize the impact of this state-averaging procedure on the computation of the energy gradients.

The trial wave function parameterization of Eqs. (59)–(62) is generalized as

$$\hat{P}(\mathbf{R}) = \sum_{J}^{N_{av}} \hat{P}^J(\mathbf{R}), \tag{76}$$

$$\hat{P}^J(\mathbf{R}) = |p^J(\mathbf{R})\rangle\langle mc^J(0)| - |mc^J(0)\rangle\langle p^J(\mathbf{R})|, \tag{77}$$

$$|p^J(\mathbf{R})\rangle = \sum_{m}^{N_{CSF}} p_m^J(\mathbf{R})|\tilde{m};[S]\rangle. \tag{78}$$

In analogy to the single-state case, the reference state $|mc^J(0)\rangle$ is defined with the $\varphi^{[S]}(\mathbf{R})$ orbitals and with the fixed CSF coefficients $\mathbf{c}^{J,mc}(0)$ corresponding to the Jth eigenpair at $\mathbf{R} = 0$. This generalization allows an averaged trial energy to be written

$$\bar{E}^{\text{trial}}(\mathbf{k}, \mathbf{p}^{1:N_{av}}; \mathbf{R}) = \sum_J^{N_{av}} w_J E_J^{\text{trial}}(\mathbf{k}, \mathbf{p}^{1:N_{av}}; \mathbf{R})$$

$$= \sum_J^{N_{av}} w_J \langle mc^J(0); [S] | \exp(-\hat{P}) \exp(-\hat{K}) \hat{H} \exp(\hat{K}) \exp(\hat{P}) | mc^J(0); [S] \rangle. \quad (79)$$

(The $\mathbf{R}$ dependence will be dropped for brevity in the notation in some of the expressions in this section.) The vector $\mathbf{p}^{1:N_{av},mc}(\mathbf{R})$ corresponds to the concatenation of the individual $\mathbf{p}^J(\mathbf{R})$ vectors for each of the states included in the state average. The commutator expansion reveals the order-by-order dependence of the trial energy on the wave function variation parameters:

$$\bar{E}^{\text{trial}}(\mathbf{k}, \mathbf{p}^{1:N_{av}}; \mathbf{R}) = \bar{E}^{ref}(\mathbf{R}) + (\mathbf{k}^T \quad \mathbf{p}^{1:N_{av},T}) \begin{pmatrix} \bar{\mathbf{f}}_{orb} \\ \mathbf{f}_{csf}^{1:N_{av}} \end{pmatrix}$$

$$+ \frac{1}{2} (\mathbf{k}^T \quad \mathbf{p}^{1:N_{av},T}) \begin{pmatrix} \bar{\mathbf{G}}_{orb,orb} & \mathbf{G}_{orb,csf}^{1:N_{av}} \\ \mathbf{G}_{csf,orb}^{1:N_{av}} & \mathbf{G}_{csf,csf}^{1:N_{av}} \end{pmatrix} \begin{pmatrix} \mathbf{k} \\ \mathbf{p}^{1:N_{av},T} \end{pmatrix} + \cdots, \quad (80)$$

with the state-averaged quantities

$$\bar{E}^{ref}(\mathbf{R}) = \sum_J^{N_{av}} w_J E_J^{ref}(\mathbf{R}), \quad (81)$$

$$\bar{\mathbf{f}}_{orb}(\mathbf{R}) = \sum_J^{N_{av}} w_J \mathbf{f}_{orb}^J(\mathbf{R}), \quad (82)$$

$$\bar{\mathbf{G}}_{orb,orb}(\mathbf{R}) = \sum_J^{N_{av}} w_J \bar{\mathbf{G}}_{orb,orb}^J, \quad (83)$$

defined in terms of their state-specific components. See Ref. 96 for further details of the formulation of these gradient and Hessian matrices. Imposing the stability condition on this trial energy results in the coupled nonlinear equation analogous to Eq. (68) that defines the optimal parameters $\mathbf{k}^{mc}(\mathbf{R})$ and $\mathbf{p}^{1:N_{av},mc}(\mathbf{R})$ at all $\mathbf{R}$. Differentiation of this expression with respect to an atom center displacement and evaluation at the reference geometry gives the first-order equation analogous to Eq. (69) for the orbital and CSF response in terms of these augmented gradient and Hessian matrices

$$\begin{pmatrix} \mathbf{k}(0)^x \\ \mathbf{p}^{1:N_{av}}(0)^x \end{pmatrix} = - \begin{pmatrix} \bar{\mathbf{G}}_{orb,orb}(0) & \mathbf{G}^{1:N_{av}}_{orb,csf}(0) \\ \mathbf{G}^{1:N_{av}}_{csf,orb}(0) & \mathbf{G}^{1:N_{av}}_{csf,csf}(0) \end{pmatrix}^{-1} \begin{pmatrix} \bar{\mathbf{f}}_{orb}(0)^x \\ \mathbf{f}^{1:N_{av}}_{csf}(0)^x \end{pmatrix}. \quad (84)$$

At the reference geometry $\bar{\mathbf{f}}_{orb}(0) = \mathbf{0}$, $\mathbf{f}^{1:N_{av}}_{csf}(0) = \mathbf{0}$, and $\mathbf{k}(0) = \mathbf{0}$. However, unlike the single-state situation, this does not mean that $\mathbf{f}^J_{orb}(0) = \mathbf{0}$ for any particular state J, only that the weighted summation Eq. (82) satisfies this condition. If all the states simultaneously are described well with the averaged orbitals, then the individual elements of the vectors $\mathbf{f}^J_{orb}(0)$ would be small in magnitude, whereas if there is strong competition among the states to describe the character of the orbitals, then the individual state gradient vectors might have large elements.

Substitution of the optimal parameters into the trial energy expression then results in $\bar{E}^{mc}(\mathbf{k}^{mc}, \mathbf{p}^{1:N_{av},mc}; \mathbf{R})$, the state-averaged energy at all $\mathbf{R}$. Differentiation with respect to a displacement and evaluation at $\mathbf{R} = 0$ gives, in principle, the state-averaged energy gradient $\bar{E}^{mc}(0)^x$ analogous to Eq. (71), in terms of the state averaged density matrices. However, it is not the gradient of the state-averaged $\bar{E}^{mc}(\mathbf{R})$ that is of interest; it is rather the gradients of the individual states that we seek. It is the energy gradients of the individual states that determine, for example, the classical trajectories on these PESs. This gradient is given by substituting the $\mathbf{k}(0)^x$ and $\mathbf{p}^J(0)^x$ for a particular state into the state-specific energy expression, Eq. (70). This gives

$$E^{mc}_J(0)^x = E^{ref}_J(0)^x + \mathbf{k}(0)^x \cdot \mathbf{f}^J_{orb}(0). \quad (85)$$

In contrast to Eq. (71), the generally nonzero state-specific orbital gradient terms $\mathbf{f}^J_{orb}(0)$ are seen to contribute to the energy gradient expression through the first-order change in the orbitals. The next step is to express this contribution to the gradient in a form that allows for efficient evaluation. To this end, Eq. (85) is written in the slightly modified form

$$E^{mc}_J(0)^x = E^{ref}_J(0)^x + (\mathbf{k}(0)^{x,T}, \mathbf{p}^{1:N_{av}}(0)^{x,T}) \begin{pmatrix} \mathbf{f}^J_{orb}(0) \\ \mathbf{0}^{1:N_{av}} \end{pmatrix} \quad (86)$$

and Eq. (84) is used to give[23,95,96,99,100] the following sequence of identities

$$E_J^{mc}(0)^x = E_J^{ref}(0)^x - (\bar{\mathbf{f}}_{orb}(0)^{x,T}, \mathbf{f}_{csf}^{1:N_{av}}(0)^{x,T})$$

$$\times \begin{pmatrix} \bar{\mathbf{G}}_{orb,orb}(0) & \mathbf{G}_{orb,csf}^{1:N_{av}}(0) \\ \mathbf{G}_{csf,orb}^{1:N_{av}}(0) & \mathbf{G}_{csf,csf}^{1:N_{av}}(0) \end{pmatrix}^{-1} \begin{pmatrix} \mathbf{f}_{orb}^{J}(0) \\ \mathbf{0}^{1:N_{av}} \end{pmatrix} \quad (87)$$

$$= E_J^{ref}(0)^x + (\bar{\mathbf{f}}_{orb}(0)^{x,T}, \mathbf{f}_{csf}^{1:N_{av}}(0)^{x,T}) \begin{pmatrix} \boldsymbol{\lambda}_{orb}^{J}(0) \\ \boldsymbol{\lambda}_{csf}^{1:N_{av},J}(0) \end{pmatrix}$$

$$= E_J^{ref}(0)^x + \bar{\mathbf{f}}_{orb}(0)^x \cdot \boldsymbol{\lambda}_{orb}^{J}(0) + \sum_{I}^{N_{av}} \mathbf{f}_{csf}^{I}(0)^x \cdot \boldsymbol{\lambda}_{csf}^{I,J}(0). \quad (88)$$

The vectors $\boldsymbol{\lambda}_{orb}^{J}(0)$ and $\boldsymbol{\lambda}_{csf}^{J}(0)$ in these expressions involve $\mathbf{G}(0)$ and $\mathbf{f}(0)$ arrays that consist only of undifferentiated quantities, namely integrals and density matrices at the reference geometry. These $\boldsymbol{\lambda}^{J}(0)$ vectors can be evaluated once and reused for all possible x, and the associated effort is thereby independent of N_{at}. As a practical matter, all N_{av} of the $\boldsymbol{\lambda}^{J}(0)$ vectors may be computed simultaneously during a single iterative procedure. As written, this expression for the energy gradient would require the computation of the $\bar{\mathbf{f}}_{orb}(0)^x$ and $\mathbf{f}_{csf}^{I}(0)^x$ vectors for all $3N_{at}$ possible displacements. We next examine the efficient computation of these last contributions to the energy gradient.

The $\boldsymbol{\lambda}_{orb}^{J}(0)$ term may be written

$$\bar{\mathbf{f}}_{orb}(0)^x \cdot \boldsymbol{\lambda}_{orb}^{J}(0) = -2Tr(\boldsymbol{\Lambda}_{orb}^{J} \bar{\mathbf{F}}(0)^x)$$
$$= -Tr(\mathbf{h}^{[S]}(0)^x \{\bar{\mathbf{D}}(0); \boldsymbol{\Lambda}_{orb}^{J}\})$$
$$- \frac{1}{2} Tr(\mathbf{g}^{[S]}(0)^x \{\bar{\mathbf{d}}(0); \boldsymbol{\Lambda}_{orb}^{J}\}) \quad (89)$$
$$= Tr(\mathbf{h}^{[S]}(0)^x \bar{\mathbf{D}}^{\Lambda_J}) + \frac{1}{2} Tr(\mathbf{g}^{[S]}(0)^x \bar{\mathbf{d}}^{\Lambda_J}),$$

with $\Lambda_{orb,pq}^{J} = -\Lambda_{orb,qp}^{J} = \boldsymbol{\lambda}_{orb}^{J}(0)_{(pq)}$. The $\boldsymbol{\lambda}_{csf}^{J}(0)$ term may be written

$$\sum_{I}^{N_{av}} \mathbf{f}_{csf}^{I}(0)^x \cdot \boldsymbol{\lambda}_{csf}^{I,J}(0) = 2 \sum_{m}^{N_{csf}} \sum_{I}^{N_{av}} \lambda_{csf}^{I,J}(0)_m \langle \tilde{m} | \hat{H}^{[S]}(0)^x | mc^{I}(0) \rangle$$
$$= Tr(\mathbf{h}^{[S]}(0)^x \bar{\mathbf{D}}^{J;\lambda}) + \frac{1}{2} Tr(\mathbf{g}^{[S]}(0)^x \bar{\mathbf{d}}^{J;\lambda}), \quad (90)$$

with

$$\bar{D}_{pq}^{J;\lambda} = \sum_{m}^{N_{csf}} \sum_{I}^{N_{av}} \lambda_{csf}^{I,J}(0)_m \langle \tilde{m} \,|\, \hat{E}_{pq} + \hat{E}_{qp} \,|\, mc^I(0) \rangle$$

$$\bar{d}_{pqrs}^{J;\lambda} = \frac{1}{2} \sum_{m}^{N_{csf}} \sum_{I}^{N_{av}} \lambda_{csf}^{I,J}(0)_m \langle \tilde{m} \,|\, \hat{e}_{pqrs} + \hat{e}_{qprs} + \hat{e}_{pqsr} + \hat{e}_{qpsr} \,|\, mc^I(0) \rangle. \tag{91}$$

These gradient contributions thereby assume the same general form as the $E_J^{ref}(0)^x$ expression with the effective density matrices denoted $\bar{\mathbf{D}}^{\Lambda_J}$ and $\bar{\mathbf{d}}^{\Lambda_J}$, which are constructed from one-index transformed averaged density matrices at $\mathbf{R} = 0$, and the averaged transition density matrices denoted $\bar{\mathbf{D}}^{J;\lambda}$ and $\bar{\mathbf{d}}^{J;\lambda}$. Combining these terms together allows the energy gradient for state J to be written as

$$\begin{aligned} E_J^{mc}(0)^x &= Tr(\mathbf{h}^{[S]}(0)^x(\mathbf{D}^J + \bar{\mathbf{D}}^{\Lambda_J} + \bar{\mathbf{D}}^{J;\lambda})) \\ &\quad + \frac{1}{2} Tr(\mathbf{g}^{[S]}(0)^x(\mathbf{d}^J + \bar{\mathbf{d}}^{\Lambda_J} + \bar{\mathbf{d}}^{J;\lambda})) \\ &= Tr(\mathbf{h}^{[S]}(0)^x \mathbf{D}^{J,\text{total}}) + \frac{1}{2} Tr(\mathbf{g}^{[S]}(0)^x \mathbf{d}^{J,\text{total}}). \end{aligned} \tag{92}$$

In this last form, it is clear that the analytic energy gradient can be written in the atom-centered AO basis using the same sequence of transformations as in Eqs. (71)–(74).

$$\begin{aligned} E_J^{mc}(0)^x &= Tr(\mathbf{h}^{[\chi]}(0)^x \mathbf{D}^{J,\text{total}[\chi]}(0)) + \frac{1}{2} Tr(\mathbf{g}^{[\chi]}(0)^x \mathbf{d}^{J,\text{total}[\chi]}(0)) \\ &\quad - Tr(\mathbf{S}^{[\chi]}(0)^x \mathbf{F}^{J,\text{total}[\chi]}(0)). \end{aligned} \tag{93}$$

As with the single-state wave function optimization case, this allows the energy gradient for each state J within the state-averaging procedure to be computed with effort that is formally independent of N_{at}. The additional effort corresponding to Eqs. (87)–(92) is comparable to that of a single iteration of the state-averaged MCSCF energy optimization procedure. Thus, we see that if the state-averaged wave functions and energy can be computed, then it is also practical to compute the energy gradients for the states of interest.

5.3. *MRCI gradient*

In most cases, the expansion space for a CI wave function consists of the union of the underlying MCSCF expansion space, along with all CSFs generated from certain excitations from these reference CSFs. For example, an MRCI-S expansion would include all single excitations from the MCSCF occupied orbitals into the secondary orbitals, an MRCI-SD expansion would include all single and double excitations, and so on. The CSF expansion coefficients are variationally optimized, which means that the eigenvalue equation

$$\hat{\mathbf{H}}^{[Z]}(\mathbf{R})\mathbf{c}^J(\mathbf{R}) = E_J^{ci}(\mathbf{R})\mathbf{c}^J(\mathbf{R}) \qquad (94)$$

is satisfied at all $\mathbf{R}$. Differentiating this expression with respect to an atomic center displacement and evaluation at $\mathbf{R} = 0$ results in

$$\begin{aligned}
E_J^{ci}(0)^x &= \mathbf{c}^J(0)^T \hat{\mathbf{H}}^{[Z]}(0)^x \mathbf{c}^J(0) \\
&= \langle ci^J(0);[Z] \,|\, \hat{H}^{[Z]}(0)^x \,|\, ci^J(0);[Z]\rangle \\
&= \sum_{m,n}^{N_{csf}} c_m^J(0) c_n^J(0) \langle \tilde{m};[Z] \,|\, \hat{H}^{[Z]}(0)^x \,|\, \tilde{n};[Z]\rangle \\
&= Tr(\mathbf{h}^{[Z]}(0)^x \mathbf{D}^{J,ci[Z]}(0)) + \frac{1}{2} Tr(\mathbf{g}^{[Z]}(0)^x \mathbf{d}^{J,ci[Z]}(0)),
\end{aligned} \qquad (95)$$

with the normalization $\|\mathbf{c}^J(0)\|_2 = \mathbf{1}$. As for the MCSCF energy gradient expression in Eq. (71), the first-order CI wave function response does not contribute to the energy gradient, and the Hellmann–Feynman theorem is seen to be satisfied for the CI energy gradient. In order to avoid the effort of constructing the derivative integrals in the $\varphi^{[Z]}(\mathbf{R})$ basis, we use the orbital transformation sequence in Eqs. (56)–(58) and the Hamiltonian commutator expansion to express the CI energy gradient in the $\varphi^{[S]}(\mathbf{R})$ basis.

$$\begin{aligned}
E_J^{ci}(0)^x &= \langle ci^J(0);[S] \,|\, \hat{H}^{[S]}(0)^x + [\hat{H}^{[S]}(0), K(0)^x] \\
&\quad + [\hat{H}^{[S]}(0), Z(0)^x] \,|\, ci^J(0);[S]\rangle \qquad (96) \\
&= \langle ci^J(0);[S] \,|\, \hat{H}^{[S]}(0)^x \,|\, ci^J(0);[S]\rangle + \mathbf{k}^{mc}(0)^x \mathbf{f}_{orb}^{J,ci}(0) \\
&\quad + \mathbf{z}^{mc}(0)^x \mathbf{f}_{orb}^{J,ci}(0). \qquad (97)
\end{aligned}$$

The first two terms are of the type previously considered in Eqs. (85)–(91) for the state-averaged MCSCF energy gradient, except that the CI density matrices and CI orbital rotation gradient vector elements are used rather than MCSCF density elements. Before discussing these terms further, we focus first on the $\mathbf{z}^{mc}(0)^x$ term.

The CI orbital rotation gradient vector is nonzero because the orbitals are optimized for either a single-state MCSCF wave function or for a state-averaged MCSCF energy rather than for the CI wave function. However, when the CSF expansion coefficients satisfy the eigenvalue equation of Eq. (94), then the $f_{orb,pq}^{J,ci}$ elements associated with CI-redundant orbital rotations are zero even without an orbital optimization being performed.[95] The most common wave function expansions discussed above, of the form MRCI-S, MRCI-SD, etc., do have redundant orbital rotations associated with the invariant orbital subspaces. These include, for example, the subspace of orbitals that are doubly occupied in all of the reference CSF, the subspace of orbitals that are unoccupied in all of the reference CSFs, and any subsets of the active MCSCF orbitals that correspond to full-CI subspaces. A more general discussion of invariant orbital subspaces is given in Ref. 95. The orbital rotations that are redundant for both the MCSCF and the CI wave functions do not contribute to the energy gradient through either the second or third terms in Eq. (97). Some MRCI expansions are chosen such that they have the exact same partitionings of essential and redundant rotation parameters as the associated MCSCF reference expansion; for these expansions, the third term in Eq. (94) does not contribute to the CI energy gradient, all nonzero contributions are through the first two terms only. Another situation that should be mentioned is a full-CI expansion in the full orbital basis; in this situation all orbital rotations are redundant in the CI wave function, $\mathbf{f}_{orb}^{J,ci}(0) = \mathbf{0}$, neither the second nor the third term in Eq. (97) contribute, and the CI energy gradient is computed entirely from just the first term in Eq. (97).

It is common to constrain the low-lying core orbitals to be doubly occupied in all of the CI expansion CSFs, or sometimes also some of the spectator valence orbitals are similarly constrained. Higher-lying virtual orbitals are sometimes constrained to be unoccupied in all CI expansion CSFs. These orbitals are called *frozen core* and *frozen virtual* orbitals respectively. This corresponds to a situation in which an orbital rotation would be redundant in the MCSCF expansion (a nonzero z_{pq}^{mc}) and essential in the CI expansion [a nonzero $f_{orb}^{J,ci}(0)_{pq}$]. Similarly, if an MCSCF active orbital with large occupation is treated as if it is doubly occupied when generating the CI expansion space, or if an MCSCF active orbital with small occupation is treated as if it is unoccupied when generating the CI expansion, then this can also result in redundant MCSCF rotations that are essential in the CI expansion. In these cases, the last term in Eq. (97) does contribute to the energy gradient expression.

There are several distinct resolutions (or canonicalizations) that have been implemented. We briefly summarize three of these: the natural orbital resolution and two different Fock matrix resolutions. Additional details of these orbital resolutions are given in Ref. 95. With natural orbital resolution, the orbitals within an invariant orbital subspace are transformed so that the one-particle density matrix $\mathbf{D}^{mc[Z]}(\mathbf{R})$ is diagonal. In the case of state-averaged orbital optimization, the diagonal condition would be imposed on the state-averaged one-particle density rather than the MCSCF density for a specific state. Differentiation with respect to a displacement and evaluation at the reference geometry results in a gradient contribution that may be written

$$\mathbf{z}^{mc}(0)^x \cdot \mathbf{f}_{orb}^{J,ci}(0) = \mathbf{p}^{mc}(0)^x \cdot \mathbf{f}_{csf}^{J,D}(0). \tag{98}$$

The effort for the computation of the effective gradient vector $\mathbf{f}_{csf}^{J,D}(0)$ is independent of N_{at}.

Natural orbital resolution is one of the typical resolutions imposed on the active orbitals in an MCSCF wave function. It cannot be imposed on the inactive orbitals (doubly occupied in all MCSCF expansion CSFs) because this subblock of the density matrix is invariant to orthogonal orbital transformations within this occupation-degenerate orbital subspace. Similarly, natural orbital resolution cannot be imposed on the MCSCF unoccupied orbitals or within invariant active orbital subspaces that happen to have occupation degeneracies. Consequently, other orbital resolutions must be imposed within these invariant orbital subspaces. One common resolution in these cases is diagonalization of the Fock matrix

$$Q_{pq}^{mc[Z]}(\mathbf{R}) = 2h_{pq}^{[Z]}(\mathbf{R}) + \sum_{r.s}(2g_{pqrs}^{[Z]}(\mathbf{R}) - g_{prqs}^{[Z]}(\mathbf{R}))D_{rs}^{mc[Z]}(\mathbf{R}). \tag{99}$$

Differentiation with respect to a displacement and evaluation at $\mathbf{R} = 0$ gives for these gradient contributions

$$\begin{aligned}\mathbf{z}^{mc}(0)^x \cdot \mathbf{f}_{orb}^{J,ci}(0) = &\ Tr(\mathbf{h}^{[S]}(0)^x \mathbf{D}^{J,Q}(0)) + \frac{1}{2}Tr(\mathbf{g}^{[S]}(0)^x \mathbf{d}^{J,Q}(0)) \\ &+ \mathbf{k}^{mc}(0)^x \cdot \mathbf{f}_{orb}^{J,Q}(0) + \mathbf{p}^{mc}(0)^x \cdot \mathbf{f}_{csf}^{J,Q}(0).\end{aligned} \tag{100}$$

The quantities $\mathbf{D}^{J,Q}(0)$, $\mathbf{d}^{J,Q}(0)$, $\mathbf{f}_{orb}^{J,Q}(0)$, and $\mathbf{f}_{csf}^{J,Q}(0)$ are all independent of the displacement coordinate x, they depend only on reference geometry integrals and density matrices, and the associated computation effort is independent of N_{at}. Resolution with the Fock matrix $\mathbf{Q}^{mc}(\mathbf{R})$

is applicable to MCSCF inactive, active, and unoccupied orbitals, but it cannot resolve individual orbitals that have degenerate $\mathbf{Q}^{mc}(\mathbf{R})$ matrix eigenvalues. In order to resolve orbitals in this case, diagonalization of the matrix $\mathbf{F}^{mc[Z]}(\mathbf{R})$ may be imposed. Differentiation with respect to a displacement and evaluation at $\mathbf{R} = 0$ gives for these gradient contributions

$$\mathbf{z}^{mc}(0)^x \cdot \mathbf{f}_{orb}^{J,ci}(0) = Tr(\mathbf{h}^{[S]}(0)^x \mathbf{D}^{J,F}(0)) + \frac{1}{2} Tr(\mathbf{g}^{[S]}(0)^x \mathbf{d}^{J,F}(0)) \\ + \mathbf{k}^{mc}(0)^x \cdot \mathbf{f}_{orb}^{J,F}(0) + \mathbf{p}^{mc}(0)^x \cdot \mathbf{f}_{csf}^{J,F}(0), \quad (101)$$

in terms of the displacement-independent quantities $\mathbf{D}^{J,F}(0)$, $\mathbf{d}^{J,F}(0)$, $\mathbf{f}_{orb}^{J,F}(0)$, and $\mathbf{f}_{csf}^{J,F}(0)$.

Combining Eqs. (98), (100), and (101) together allows the CI energy gradient in Eq. (97) to be written using state-averaged orbital and CSF gradients in the form

$$E_J^{ci}(0)^x = Tr(\mathbf{h}^{[S]}(0)^x \mathbf{D}^{J,\text{total}}) + \frac{1}{2} Tr(\mathbf{g}^{[S]}(0)^x \mathbf{d}^{J,\text{total}}) \\ + (\bar{\mathbf{f}}_{orb}^{mc}(0)^{x,T}, \mathbf{f}_{csf}^{1:N_{av},mc}(0)^{x,T}) \begin{pmatrix} \lambda_{orb}^{J,\text{total}}(0) \\ \lambda_{csf}^{1:N_{av},J,\text{total}}(0) \end{pmatrix}. \quad (102)$$

This is exactly the form that was considered in Eqs. (88)–(92), and it allows the CI energy gradient to be computed in the AO basis as in Eq. (93).

$$E_J^{ci}(0)^x = Tr(\mathbf{h}^{[\chi]}(0)^x \mathbf{D}^{J,\text{total}[\chi]}(0)) + \frac{1}{2} Tr(\mathbf{g}^{[\chi]}(0)^x \mathbf{d}^{J,\text{total}[\chi]}(0)) \\ - Tr(\mathbf{S}^{[\chi]}(0)^x \mathbf{F}^{J,\text{total}[\chi]}(0)). \quad (103)$$

The orbital resolutions discussed above may be combined in a quite flexible and general manner. Different invariant orbital subspaces may be resolved in different ways, and the corresponding effective operators and density matrices computed accordingly.

The total effort to construct the gradient vector is dominated, particularly for large CI expansions, by the effort to construct the CI density matrices $\mathbf{D}^{J,ci}(0)$ and $\mathbf{d}^{J,ci}(0)$. This requires roughly the same effort as that of a single Hamiltonian matrix-vector product operation during the iterative solution of the eigenvalue equation, Eq. (94), at the reference geometry. A consequence of this is that, unlike most other electronic structure methods, the CI energy gradient requires typically less effort than the computation of the CI wave function and energy itself. For the 20 molecules studied in Ref. 98, the effort for the MRCI gradient evaluations required between 2.5% and 52.2% of the total computational effort, with a mean of 9.9% over the whole set of molecules and basis sets. This demonstrates that the computational procedure presented above is very efficient, that it is independent of N_{at}, and that it may be applied to any molecule for which the CI wave function optimization itself is practical.

6. Nonadiabatic Coupling Terms

The nonadiabatic coupling vector $\mathbf{F}_{JI}(\mathbf{R})$ between two electronic states J and I at a molecular geometry $\mathbf{R}$ is given by Eq. (10), and a representative component of this vector will be denoted

$$f_{JI}(\mathbf{R})^x = \left\langle \Phi^J(\mathbf{R}) \left| \frac{\partial}{\partial x} \right| \Phi^I(\mathbf{R}) \right\rangle_\mathbf{r}, \qquad (104)$$

where x is one of the $3N_{at}$ coordinates of the atomic centers. In this section, we summarize the computation of this nonadiabatic coupling vector at the MCSCF and MRCI levels. This procedure is discussed in more detail in Ref. 24. The computation of the nonadiabatic coupling vector at the EOM-CC level is discussed in Ref. 78. The computation at the TD-DFT level is discussed in Refs. 101 and 102 and references therein. A numerical implementation especially adapted for dynamics simulations at the CASPT2 level is discussed in Ref. 87. The alternative use of time-derivative nonadiabatic coupling terms is discussed in Sec. 6.2.

6.1. *NAC vectors in MCSCF and MRCI formalism*

Expansion of the electronic wave function in a CSF basis in the fully optimized and resolved orbital basis $\varphi^{[Z]}(\mathbf{R})$ allows a wave function derivative to be written as

$$\begin{aligned}\frac{\partial}{\partial x}|\Phi^I(\mathbf{R})\rangle &= \frac{\partial}{\partial x}\sum_m^{N_{csf}} c_m^I(\mathbf{R})|\tilde{m}(\mathbf{R};[Z])\rangle \\ &= \sum_m^{N_{csf}} \left(\frac{\partial}{\partial x}c_m^I(\mathbf{R})\right)|\tilde{m}(\mathbf{R};[Z])\rangle + \sum_m^{N_{csf}} c_m^I(\mathbf{R})\frac{\partial}{\partial x}|\tilde{m}(\mathbf{R};[Z])\rangle.\end{aligned} \qquad (105)$$

The CSF expansion in Eq. (105) can be either an MCSCF expansion or a general MRCI expansion; with orbitals optimized for either single-state

MCSCF or state-averaged MCSCF energies, the following equations are the same in any case. The above separation in turn allows the nonadiabatic coupling element to be written as two contributions

$$f_{JI}(\mathbf{R})^x = f_{JI}^{ci}(\mathbf{R})^x + f_{JI}^{csf}(\mathbf{R})^x, \qquad (106)$$

with

$$\begin{aligned} f_{JI}^{ci}(\mathbf{R})^x &= \sum_{m,n}^{N_{csf}} c_m^J(\mathbf{R}) \left(\frac{\partial}{\partial x} c_n^I(\mathbf{R}) \right) \langle \tilde{m}(\mathbf{R};[Z]) \,|\, \tilde{n}(\mathbf{R};[Z]) \rangle \\ &= \mathbf{c}^J(\mathbf{R}) \cdot \mathbf{c}^I(\mathbf{R})^x, \end{aligned} \qquad (107)$$

$$f_{JI}^{csf}(\mathbf{R})^x = \sum_{m,n}^{N_{csf}} c_m^J(\mathbf{R}) c_n^I(\mathbf{R}) \langle \tilde{m}(\mathbf{R};[Z]) \,|\, \hat{X} \,|\, \tilde{n}(\mathbf{R};[Z]) \rangle \rangle, \qquad (108)$$

with

$$\begin{aligned} \hat{X}_{pq}^{[Z]}(\mathbf{R}) &= \sum_{p,q} X_{pq}^{[Z]}(\mathbf{R}) \hat{E}_{pq}, \\ X_{pq}^{[Z]}(\mathbf{R}) &= \int \varphi_p^{[Z]}(\boldsymbol{\tau};\mathbf{R}) \frac{\partial}{\partial x} \varphi_q^{[Z]}(\boldsymbol{\tau};\mathbf{R}) d\boldsymbol{\tau}. \end{aligned} \qquad (109)$$

These two contributions to the nonadiabatic coupling element will be examined separately. Differentiating Eq. (94) with respect to a coordinate and evaluation at $\mathbf{R} = 0$ gives an expression for the first-order response of the CSF expansion coefficients to a perturbation

$$(\hat{\mathbf{H}}^{[Z]}(0) - E_I^{ci}(0)\mathbf{1})\mathbf{c}^I(0)^x = -(\hat{\mathbf{H}}^{[Z]}(0)^x - E_I^{ci}(0)^x \mathbf{1})\mathbf{c}^I(0). \qquad (110)$$

Multiplication from the left by $\mathbf{c}^J(0)^T$ results in

$$\begin{aligned} f_{JI}^{ci}(0)^x &= (E_I^{ci}(0) - E_J^{ci}(0))^{-1} \mathbf{c}^J(0)^T \hat{\mathbf{H}}^{[Z]}(0)^x \mathbf{c}^I(0) \\ &= (E_I^{ci}(0) - E_J^{ci}(0))^{-1} \\ &\quad \times \left(Tr(\mathbf{h}^{[Z]}(0)^x \mathbf{D}^{JI,ci}(0)) + \frac{1}{2} Tr(\mathbf{g}^{[Z]}(0)^x \mathbf{d}^{JI,ci}(0)) \right) \\ &= (E_I^{ci}(0) - E_J^{ci}(0))^{-1} \Big(Tr(\mathbf{h}^{[S]}(0)^x \mathbf{D}^{JI,ci}(0)) \\ &\quad + \frac{1}{2} Tr(\mathbf{g}^{[S]}(0)^x \mathbf{d}^{JI,ci}(0)) + \mathbf{k}^{mc}(0)^x \cdot \mathbf{f}^{JI,ci}(0) \\ &\quad + \mathbf{z}^{mc}(0)^x \cdot \mathbf{f}^{JI,ci}(0) \Big). \end{aligned} \qquad (111) \\ (112)$$

Using the variational nature of $\mathbf{c}^J(\mathbf{R})$ in this manner avoids the explicit computation of $\mathbf{c}^I(0)^x$ in Eq. (107), which would be relatively expensive and also the effort would be proportional to N_{at}. Equation (111) is analogous to Eq. (95) but using the symmetrized transition density matrices $\mathbf{D}^{JI,ci}(0)$ and $\mathbf{d}^{JI,ci}(0)$ in place of the state-specific CI density matrices in the trace expressions and in the effective orbital gradient vectors $\mathbf{f}^{JI,ci}(0)$. As with the MCSCF and CI gradients discussed in the previous section, these density matrices are the displacement-independent factors whereas the derivative integrals are the displacement-dependent factors in this expression. The expression in Eq. (112) can be transformed to the AO basis using the same sequence of steps as in Eqs. (96)–(103) where the effective orbital gradient, Fock matrices, and other quantities are all written in terms of the symmetric transition density matrices rather than the state-specific density matrices in the various orbital basis sets. Before considering this transformation, the other contribution to the nonadiabatic coupling element is examined.

The $f_{JI}^{csf}(\mathbf{R})^x$ coupling of Eq. (108) may be written[24] in the $\varphi^{[Z]}(\mathbf{R})$ orbitals at $\mathbf{R} = 0$ as

$$f_{JI}^{csf}(0)^x = \sum_{p,q} X_{pq}^{[Z]}(0) \left\langle ci^J(0) \left| \hat{E}_{pq} \right| ci^I(0) \right\rangle. \tag{113}$$

The relation $0 = S_{pq}^{[Z]}(0)^x = X_{pq}^{[Z]}(0) + X_{qp}^{[Z]}(0)$ shows that the orbital matrix $\mathbf{X}^{[Z]}(0)$ is skew-symmetric. This allows the $f_{JI}^{csf}(0)^x$ coupling to be written in terms of the skew-symmetric CI one-particle density

$$\begin{aligned} f_{JI}^{csf}(0)^x &= \sum_{p,q} X_{pq}^{[Z]}(0) \frac{1}{2} \left\langle ci^J(0) \left| \hat{E}_{pq} - \hat{E}_{qp} \right| ci^I(0) \right\rangle \\ &= \sum_{p,q} X_{pq}^{[Z]}(0) D_{qp}^{(-)JI}(0) \\ &= Tr(\mathbf{X}^{[Z]}(0) \mathbf{D}^{(-)JI}(0)). \end{aligned} \tag{114}$$

Using Eqs. (56)–(58) and evaluation at $\mathbf{R} = 0$ gives

$$\begin{aligned} X_{pq}^{[Z]}(0) &= X_{pq}^{[C]}(0) - \frac{1}{2} S_{pq}^{[C]}(0)^x + K^{mc}(0)_{pq}^x + Z^{mc}(0)_{pq}^x \\ &= \frac{1}{2}(X_{pq}^{[C]}(0) - X_{qp}^{[C]}(0)) + K^{mc}(0)_{pq}^x + Z^{mc}(0)_{pq}^x. \end{aligned} \tag{115}$$

Because of the disjoint partitioning of the essential and redundant orbital rotation elements, only one of the last two terms in Eq. (115) can be nonzero for a given orbital index pair pq. Equation (114) can then be written as

$$f_{JI}^{csf}(0)^x = Tr(\mathbf{X}^{[C]}(0)\mathbf{D}^{(-)JI}(0)) + Tr(\mathbf{K}^{mc}(0)^x\mathbf{D}^{(-)JI}(0))$$
$$+ Tr(\mathbf{Z}^{mc}(0)^x\mathbf{D}^{(-)JI}(0)) \quad (116)$$
$$= Tr(\mathbf{X}^{[C]}(0)\mathbf{D}^{(-)JI}(0)) + \mathbf{k}^{mc}(0)^x \cdot \mathbf{f}_{orb}^{JI,D}(0) + \mathbf{z}^{mc}(0)^x \cdot \mathbf{f}_{orb}^{JI,D}(0),$$

with $f_{orb}^{JI,D}(0)_{(pq)} = 2D_{qp}^{(-)JI}(0)$. Upon comparing Eq. (116) with Eq. (112), it is clear that the common factors can be combined to give

$$f_{JI}(0)^x = f_{JI}^{ci}(0)^x + f_{JI}^{csf}(0)^x$$
$$= Tr(\mathbf{X}^{[C]}(0)\mathbf{D}^{(-)JI}(0))$$
$$+ (E_I^{ci}(0) - E_J^{ci}(0))^{-1}\left(Tr(\mathbf{h}^{[S]}(0)^x\mathbf{D}^{JI,ci[Z]}(0))\right.$$
$$\left. + \frac{1}{2}Tr(\mathbf{g}^{[S]}(0)^x\mathbf{d}^{JI,ci[Z]}(0))\right)$$
$$\quad (117)$$
$$+ (\mathbf{k}^{mc}(0)^x + \mathbf{z}^{mc}(0)^x) \cdot \left((E_I^{ci}(0) - E_J^{ci}(0))^{-1}\mathbf{f}^{JI,ci}(0) + \mathbf{f}^{JI,D}(0)\right)$$
$$= Tr(\mathbf{X}^{[C]}(0)\mathbf{D}^{(-)JI}(0))$$
$$+ Tr(\mathbf{h}^{[S]}(0)^x\mathbf{D}^{JI,eff}(0)) + \frac{1}{2}Tr(\mathbf{g}^{[S]}(0)^x\mathbf{d}^{JI,eff}(0))$$
$$+ \mathbf{k}^{mc}(0)^x \cdot \mathbf{f}_{orb}^{JI,eff}(0) + \mathbf{z}^{mc}(0)^x \cdot \mathbf{f}_{orb}^{JI,eff}(0),$$

where the effective density matrices and orbital gradient vector include the CI energy-difference factors as appropriate. The transformation steps of Eqs. (97)–(103) may be applied to transform this expression to the AO basis in which the derivative integral sparseness may be exploited.

$$f_{JI}(0)^x = Tr(\mathbf{h}^{[x]}(0)^x\mathbf{D}^{JI,total[x]}(0)) + \frac{1}{2}Tr(\mathbf{g}^{[x]}(0)^x\mathbf{d}^{JI,total[x]}(0))$$
$$- Tr(\mathbf{S}^{[x]}(0)^x\mathbf{F}^{JI,total[x]}(0)) + Tr(\mathbf{X}^{[x]}(0)^x\mathbf{D}^{(-)JI[x]}(0)). \quad (118)$$

Given the symmetric transition density matrices and effective orbital gradient vectors, the analytic energy gradient procedure may be applied, and the contributions from the first three terms in Eq. (118) may be computed in a straightforward manner. The last term in Eq. (118) is unique to the nonadiabatic coupling element. However, it involves only the skew-symmetric component of the one-particle transition CI density matrix, which requires an insignificant additional effort to compute along with the symmetric component that is used in the first terms. The above procedure may be applied to both MCSCF and general MRCI wave functions.

Timings for this procedure have been given previously. In Ref. 24, for example, it is seen that the effort required for all $3N_{at}$ components x of the nonadiabatic coupling vector for MRCI wave functions for H_2CO is almost exactly the same as that required for the computation of a single energy gradient vector, which, as shown in Sec. 5, is typically only a small fraction of the effort required for the energy and wave function optimization steps. Thus, for all practical purposes, if the wave functions and energies can be optimized for the states of interest, then the energy gradient vectors and the nonadiabatic coupling vectors can also be computed.

6.2. Time-derivative NAC terms

The integration of Eq. (24) does not demand explicit knowledge of the nonadiabatic coupling vector, but rather of its scalar product with the velocity vector. An alternative approach to computing nonadiabatic coupling terms during dynamics simulations, which takes advantage of this fact, is to rewrite the $\mathbf{F}_{kl} \cdot \mathbf{v}$ product in terms of time derivatives[61]:

$$\mathbf{F}_{kl} \cdot \mathbf{v} = \left\langle \Phi_k \left| \frac{\partial}{\partial t} \right| \Phi_l \right\rangle_{\mathbf{r}} \equiv \sigma_{kl}. \qquad (119)$$

The σ_{kl} coupling terms can be evaluated by numerical differences

$$\sigma_{kl}\left(t - \frac{3}{2}\Delta t\right) \approx \frac{1}{2\Delta t}[\langle \Phi_k(t - 2\Delta t) | \Phi_l(t - \Delta t) \rangle$$
$$- \langle \Phi_k(t - \Delta t) | \Phi_l(t - 2\Delta t) \rangle], \qquad (120)$$

$$\sigma_{kl}\left(t - \frac{1}{2}\Delta t\right) \approx \frac{1}{2\Delta t}[\langle \Phi_k(t - \Delta t) | \Phi_l(t) \rangle - \langle \Phi_k(t) | \Phi_l(t - \Delta t) \rangle], \qquad (121)$$

$$\sigma_{kl}(t) \approx \frac{1}{2}\left[3\sigma_{kl}\left(t - \frac{1}{2}\Delta t\right) - \sigma_{kl}\left(t - \frac{3}{2}\Delta t\right)\right]. \qquad (122)$$

Equations (120), (121), and (122) imply that the nonadiabatic coupling term can be approximately obtained by overlaps of electronic wave functions between subsequent time steps of the dynamics. This procedure was introduced by Hammes–Schiffer and Tully[61] in a context where nonadiabatic coupling vectors were not available at all. Recently, it has been employed in dynamics at TD-DFT,[103,104] TD-DFTB,[105] MCSCF,[38] and MRCI[38] levels. In particular, it is shown in Ref. 38 that the calculation of σ_{kl} can represent an effective reduction of computational costs in comparison to the computation of $\mathbf{F}_{kl}$. The use of wave function overlaps is also the basis for the approximated hopping algorithm discussed in Ref. 62 and for the local-diabatization method discussed in Ref. 18.

7. Dynamics Close to Conical Intersections

It is common practice to say that conical intersections act as funnels for radiationless transitions to other electronic states. Although true, this statement tends to oversimplify what really happens during the dynamical process of internal conversion. It neglects a series of questions such as how the molecule reaches the conical intersection, how long it takes to do so, which regions of the crossing seam are reached, how efficient for nonadiabatic transitions the intersections really are, and what happens after the nonadiabatic transition. Some of these questions are often addressed by conventional quantum chemical investigations, which are especially tailored to unveil the pathways connecting the Franck–Condon region where the molecule is initially photoexcited to the several conical intersections.[92,106] Dynamics simulations represent one step further in this analysis in the sense that they can give yields and time scales for activations of the pathways.

After the photoexcitation, the molecule can relax along different pathways. Even dynamics of very simple molecules, such as substituted ethylenes $H_2C=XH_2$ ($X = Si, N^+$), show complex behavior with splitting of trajectories between two different reaction pathways.[45,107] In the dynamics of a larger molecule like the protonated Schiff base $CH_2=(CH)_6=NH_2^+$, trajectories are distributed among several different pathways, reaching different conical intersections.[47] For cytosine, it has been observed that different reaction pathways can occur in completely distinct time scales.[88]

This competition between reaction pathways can be illustrated more clearly by restricting the discussion to a specific class of molecules, the small organic aromatic heterocycles like DNA/RNA nucleobases. Figure 1 shows schematically two sections of potential energy surface of the first singlet excited state (S_1), which are common to many of these molecules. After the excitation into the $^1\pi\pi^*$ optically active state (not necessarily the S_1 state), such heterocycles can usually relax within the $\pi\pi^*$ region of the S_1 surface as shown in path 1 of Fig. 1. This relaxation leads directly to a

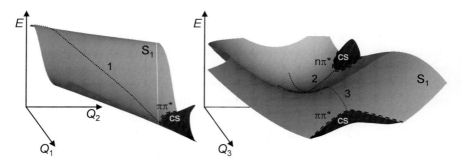

Fig. 1. Schematic representation of two sections of the S_1 potential energy surface of aromatic heterocycles. The S_1/S_0 seam of conical intersections is indicated by dashed lines. Reaction pathways discussed in the text are indicated by dotted lines. The sections of the surfaces labeled cs have closed shell character. Q_i represents generalized nuclear coordinates.

crossing with the ground state, whose closed shell (cs) configuration is destabilized along the same pathway. Based on the analysis of minimum energy paths, this direct path has been suggested to be the primary pathway for deactivation of pyrimidine nucleobases.[92] However, dynamics simulations have shown that the initial gradients are relatively unfavorable to activate this pathway[108]; instead, for several heterocycles like cytosine[88] and pyridone,[46] the initially excited molecule relaxes into the minimum of the S_1 state with $n\pi^*$ character (Fig. 1 right).

From the S_1 minimum, the molecule can usually reach a region of crossing seam between the $^1n\pi^*$ state and the ground state (path 2 in Fig. 1). But also from the S_1 minimum, the molecule can cross the barrier to the $\pi\pi^*$ region of the S_1 surface and then relax to the $\pi\pi^*/\text{cs}$ conical intersection (path 3). If the $n\pi^*/\pi\pi^*$ barriers and the $n\pi^*/\text{cs}$ conical intersection are both too high in energy, the molecule will remain trapped in the S_1 state until it fluoresces or undergoes a transition to a triplet state. The molecule can also remain trapped if a conical intersection is energetically accessible, but it is not efficient to promote nonadiabatic transitions. In this way, topographical details of the energy surfaces like gradients at the Franck–Condon region, height of barriers, or shape of conical intersections are main determining factors to rule the competition between the several deactivation pathways.

Since the seminal work of Atchity et al.,[109] it has been recognized that the shape of a conical intersection should play a relevant role in its efficiency for nonadiabatic transitions.[106,110] Dynamics simulations have shown that conical intersections like that between nπ^* and the closed shell state in Fig. 1 are especially inefficient for non-adiabatic transitions.[46] This happens because during the trajectory, the molecule reaches the region close to the crossing seam where a finite energy gap can still be observed.[111] The instantaneous transition probabilities in one time step [see Eq. (45)] are relatively small in such regions and the molecule should remain there for a while in order to deactivate. In the case of path 2, the shape of the surface is unfavorable for nonadiabatic transition because the molecule remains shortly close to the seam and then returns to the minimum. Pyridone is an extreme of such cases: although the conical intersection is energetically accessible, the molecule is still fluorescent.[46,112] An opposite situation occurs in paths 1 and 3 (Fig. 1). The crossing seam is close or coincident with the surface minimum, therefore when the molecule reaches this region, it remains trapped there until deactivation. This is what happens, for example, with adenine.[31]

Minimum energy path analysis revealed that the different photoproducts should be formed depending on the exit direction from a conical intersection,[106] which is confirmed by dynamics simulations showing strong correlation between products and initial velocity in the branching space.[36] This enforces the importance of a proper nonadiabatic treatment during the dynamics. For instance, simple surface hopping algorithms whose transition probability is determined by energy gaps tend to neglect the motion close to the seam. This can lead to a wrong distribution of exit directions and, therefore, also to wrong quantum yield of products.

8. Conclusions and Outlook

In this chapter, we discussed the main methodological elements necessary to perform nonadiabatic dynamics simulations for molecules based on classical trajectories. These elements included the derivation of the time-dependent equations, the generation of initial conditions, and the computation of electronic structure properties. These kinds of simulations have become an important tool in photochemical and photophysical investigations. They allow information to be obtained about several reaction pathways in competition during the excited-state relaxation, the time scales in which they are activated, and the efficiency of the conical intersections.

The main limitations of these methods are the high computational costs of the electronic structure methods used for computing excited states energies, energy gradients and nonadiabatic coupling terms. Progress has been made by the development of mutireference semiempirical methods,[18,19] derivation and implementation of analytical gradients and nonadiabatic coupling at *ab initio* levels,[14–16] and the usage of hybrid strategies to treat large molecular environments.[26,37,113] Although several attempts of working out nonadiabatic dynamics in the framework of density functional theory have been examined and proposed,[84,103,104,114] fully reliable DFT-based dynamics still depend on the development of functionals with proper treatment of charge transfer states,[80] more balanced account of static correlation,[115] and, for TD-DFT level, excited state calculations beyond linear response.[116] In spite of these limitations, nonadiabatic dynamics based on classical trajectories is becoming a routine tool in quantum chemical investigations, with a great potential to reveal details of the role of conical intersections.

Acknowledgments

This work has been supported by the Austrian Science Fund within the framework of the Special Research Program F41 (ViCom). RS acknowledges support by the Office of Basic Energy Sciences, Division of Chemical Sciences, Geosciences, and Biosciences, U.S. Department of Energy under contract number DE-AC02-06CH11357.

References

1. G.A. Worth, H.D. Meyer and L.S. Cederbaum, *Conical Intersections: Electronic Structure, Dynamics & Spectroscopy* (World Scientific Publishing Company, 2004).
2. G.A. Worth and L.S. Cederbaum, *Annu. Rev. Phys. Chem.* **55**, 127 (2004).
3. J.C. Tully and R.K. Preston, *J. Chem. Phys.* **55**, 562 (1971).
4. J.B. Delos, W.R. Thorson and S.K. Knudson, *Phys. Rev. A* **6**, 709 (1972).
5. M. Santer, U. Manthe and G. Stock, *J. Chem. Phys.* **114**, 2001 (2001).
6. I. Burghardt and G. Parlant, *J. Chem. Phys.* **120**, 3055 (2004).
7. P. Pechukas, *Phys. Rev.* **181**, 174 (1969).
8. J.S. Cao and G.A. Voth, *J. Chem. Phys.* **101**, 6168 (1994).
9. M. Ben-Nun, J. Quenneville, and T.J. Martínez, *J. Phys. Chem. A* **104**, 5161 (2000).
10. A. Warshel, *Nature* **260**, 679 (1976).
11. J.C. Tully, *J. Chem. Phys.* **93**, 1061 (1990).
12. O.V. Prezhdo and P.J. Rossky, *J. Chem. Phys.* **107**, 5863 (1997).

13. C. Zhu, A.W. Jasper and D.G. Truhlar, *J. Chem. Phys.* **120**, 5543 (2004).
14. G. Granucci and M. Persico, *J. Chem. Phys.* **126**, 134114 (2007).
15. K. Yagi and K. Takatsuka, *J. Chem. Phys.* **123**, 224103 (2005).
16. G.A. Jones, A. Acocella and F. Zerbetto, *J. Phys. Chem. A* **112**, 9650 (2008).
17. R. Mitric, J. Petersen and V. Bonacic-Koutecky, *Phys. Rev. A* **79**, 053416 (2009).
18. G. Granucci, M. Persico and A. Toniolo, *J. Chem. Phys.* **114**, 10608 (2001).
19. W. Weber and W. Thiel, *Theor. Chem. Acc.* **103**, 495 (2000).
20. P. Celani and H.J. Werner, *J. Chem. Phys.* **119**, 5044 (2003).
21. M.J. Bearpark, M.A. Robb and H.B. Schlegel, *Chem. Phys. Lett.* **223**, 269 (1994).
22. B.H. Lengsfield, P. Saxe and D.R. Yarkony, *J. Chem. Phys.* **81**, 4549 (1984).
23. R. Shepard, H. Lischka, P.G. Szalay, T. Kovar and M. Ernzerhof, *J. Chem. Phys.* **96**, 2085 (1992).
24. H. Lischka, M. Dallos, P.G. Szalay, D.R. Yarkony and R. Shepard, *J. Chem. Phys.* **120**, 7322 (2004).
25. A. Toniolo, G. Granucci and T.J. Martínez, *J. Phys. Chem. A* **107**, 3822 (2003).
26. G. Groenhof, M. Bouxin-Cademartory, B. Hess, S.P. deVisser, H.J.C. Berendsen, M. Olivucci, A.E. Mark and M.A. Robb, *J. Am. Chem. Soc.* **126**, 4228 (2004).
27. G. Stock and M. Thoss, *Conical Intersections: Electronic Structure, Dynamics & Spectroscopy* (World Scientific Publishing Company, 2004).
28. V. Bonacic-Koutecky and R. Mitric, *Chem. Rev.* **105**, 11 (2005).
29. L.M. Frutos, T. Andruniow, F. Santoro, N. Ferre and M. Olivucci, *Proc. Natl. Acad. Sci. U. S. A.* **104**, 7764 (2007).
30. O. Weingart, I. Schapiro and V. Buss, *J. Phys. Chem. B* **111**, 3782 (2007).
31. M. Barbatti and H. Lischka, *J. Am. Chem. Soc.* **130**, 6831 (2008).
32. C. Ciminelli, G. Granucci and M. Persico, *Chem. Phys.* **349**, 325 (2008).
33. D. Asturiol, B. Lasorne, M.A. Robb and L. Blancafort, *J. Phys. Chem. A* **113**, 10211 (2009).
34. S. Hayashi, E. Taikhorshid and K. Schulten, *Biophys. J.* **96**, 403 (2009).
35. Z. Lan, E. Fabiano and W. Thiel, *J. Phys. Chem. B* **113**, 3548 (2009).
36. B. Sellner, M. Barbatti and H. Lischka, *J. Chem. Phys.* **131**, 024312 (2009).
37. A.M. Virshup, C. Punwong, T.V. Pogorelov, B.A. Lindquist, C. Ko and T.J. Martinez, *J. Phys. Chem. B* **113**, 3280 (2009).
38. J. Pittner, H. Lischka and M. Barbatti, *Chem. Phys.* **356**, 147 (2009).
39. H. Lischka, R. Shepard, F.B. Brown and I. Shavitt, *Int. J. Quantum Chem.* **S.15**, 91 (1981).
40. H. Lischka, R. Shepard, R.M. Pitzer, I. Shavitt, M. Dallos, T. Müller, P.G. Szalay, M. Seth, G.S. Kedziora, S. Yabushita and Z.Y. Zhang, *Phys. Chem. Chem. Phys.* **3**, 664 (2001).
41. H. Lischka, R. Shepard, I. Shavitt, R.M. Pitzer, M. Dallos, T. Mueller, P.G. Szalay, F.B. Brown, R. Ahlrichs, H.J. Boehm, A.

Chang, D.C. Comeau, R. Gdanitz, H. Dachsel, C. Ehrhardt, M. Ernzerhof, P. Hoechtl, S. Irle, G. Kedziora, T. Kovar, V. Parasuk, M.J.M. Pepper, P. Scharf, H. Schiffer, M. Schindler, M. Schueler, M. Seth, E.A. Stahlberg, J.-G. Zhao, S. Yabushita, Z. Zhang, M. Barbatti, S. Matsika, M. Schuurmann, D.R. Yarkony, S.R. Brozell, E.V. Beck and J.-P. Blaudeau, *COLUMBUS, an ab initio electronic structure program, release 5.9.1*, www.univie.ac.at/columbus (2006).

42. M. Barbatti, G. Granucci, M. Persico, M. Ruckenbauer, M. Vazdar, M. Eckert-Maksic and H. Lischka, *J. Photochem. Photobiol.*, A **190**, 228 (2007).
43. M. Barbatti, G. Granucci, M. Ruckenbauer, J. Pittner, M. Persico and H. Lischka, *NEWTON-X: a package for Newtonian dynamics close to the crossing seam*, www.newtonx.org (2007).
44. M. Barbatti, G. Granucci, M. Persico, and H. Lischka, *Chem. Phys. Lett.* **401**, 276 (2005).
45. G. Zechmann, M. Barbatti, H. Lischka, J. Pittner and V. Bonačić-Koutecký, *Chem. Phys. Lett.* **418**, 377 (2006).
46. M. Barbatti, A.J.A. Aquino and H. Lischka, *Chem. Phys.* **349**, 278 (2008).
47. J.J. Szymczak, M. Barbatti and H. Lischka, *J. Phys. Chem. A* **113**, 11907 (2009).
48. M.H. Mittleman, *Phys. Rev.* **122**, 499 (1961).
49. H.-D. Meyer and W.H. Miller, *J. Chem. Phys.* **70**, 3214 (1979).
50. J.C. Tully, *Faraday Discuss.* **110**, 407 (1998).
51. M.D. Hack and D.G. Truhlar, *J. Phys. Chem. A* **104**, 7917 (2000).
52. M. Baer, *Beyond Born-Oppenheimer: Electronic Nonadiabatic Coupling Terms and Conical Intersections* (John Wiley & Sons, New Jersey, 2006).
53. X. Zhu and D.R. Yarkony, *J. Chem. Phys.* **132**, 104101 (2010).
54. J.J. Sakurai, *Modern Quantum Mechanics* (Addison-Wesley, Massachusetts, 1994).
55. D. Thompson, *Modern Methods for Multidimensional Dynamics Computations in Chemistry* (World Scientific Pub Co Inc, 1998).
56. H. Goldstein. Addison-Wesley, Massachusetts (1980).
57. S. Mukamel, *Principles of Nonlinear Optical Spectroscopy* (Oxford University Press, 1995).
58. C.Y. Zhu, S. Nangia, A.W. Jasper and D.G. Truhlar, *J. Chem. Phys.* **121**, 7658 (2004).
59. C. Zhu, A.W. Jasper and D.G. Truhlar, *JCTC* **1**, 527 (2005).
60. O. Weingart, A. Migani, M. Olivucci, M.A. Robb, V. Buss and P. Hunt, *J. Phys. Chem. A* **108**, 4685 (2004).
61. S. Hammes-Schiffer and J.C. Tully, *J. Chem. Phys.* **101**, 4657 (1994).
62. E. Fabiano, G. Groenhof and W. Thiel, *Chem. Phys.* **351**, 111 (2008).
63. N.C. Blais and D.G. Truhlar, *J. Chem. Phys.* **79**, 1334 (1983).
64. C. Lasser and T. Swart, *J. Chem. Phys.* **129**, 034302 (2008).
65. E. Fabiano, T.W. Keal and W. Thiel, *Chem. Phys.* **349**, 334 (2008).

66. A.W. Jasper, S.N. Stechmann, and D.G. Truhlar, *J. Chem. Phys.* **116**, 5424 (2002).
67. H. Tachikawa, K. Ohnishi, T. Hamabayashi and H. Yoshida, *J. Phys. Chem. A* **101**, 2229 (1997).
68. W. Hu, G. Lendvay, B. Maiti and G.C. Schatz, *J. Phys. Chem. A* **112**, 2093 (2008).
69. C. Carbogno, J. Behler, K. Reuter and A. Gross, *Phys. Rev. B* **81**, 035410 (2010).
70. W.C. Swope, H.C. Andersen, P.H. Berens and K.R. Wilson, *J. Chem. Phys.* **76**, 637 (1982).
71. R. Schinke, *Photodissociation Dynamics: Spectroscopy and Fragmentation of Small Polyatomic Molecules* (Cambridge University Press, Cambridge, 1995).
72. V. Lukes, R. Solc, M. Barbatti, M. Elstner, H. Lischka and H.-F. Kauffmann, *J. Chem. Phys.* **129**, 164905 (2008).
73. H.R. Hudock, B.G. Levine, A.L. Thompson, H. Satzger, D. Townsend, N. Gador, S. Ullrich, A. Stolow and T.J. Martinez, *J. Phys. Chem. A* **111**, 8500 (2007).
74. J.J. Szymczak, M. Barbatti, J.T. Soo Hoo, J.A. Adkins, T.L. Windus, D. Nachtigallova and H. Lischka, *J. Phys. Chem. A* **113**, 12686 (2009).
75. C. Ciminelli, G. Granucci and M. Persico, *Chem. Eur. J.* **10**, 2327 (2004).
76. R.C. Hilborn, *Am. J. Phys.* **50**, 982 (1982).
77. M. Barbatti, A.J.A. Aquino and H. Lischka, *Phys. Chem. Chem. Phys.* **12**, 4959 (2010).
78. A. Tajti and P.G. Szalay, *J. Chem. Phys.* **131**, 124104 (2009).
79. J.P. Perdew, A. Ruzsinszky, L.A. Constantin, J.W. Sun and G.I. Csonka, *JCTC* **5**, 902 (2009).
80. Y. Zhao and D.G. Truhlar, *J. Phys. Chem. A* **110**, 13126 (2006).
81. S. Grimme and M. Waletzke, *J. Chem. Phys.* **111**, 5645 (1999).
82. I. Frank, J. Hutter, D. Marx and M. Parrinello, *J. Chem. Phys.* **108**, 4060 (1998).
83. I. Frank and K. Damianos, *J. Chem. Phys.* **126**, 125105 (2007).
84. N.L. Doltsinis and D. Marx, *Phys. Rev. Lett.* **88**, 166402 (2002).
85. H. Nieber and N.L. Doltsinis, *Chem. Phys.* **347**, 405 (2008).
86. M. Vazdar, M. Eckert-Maksic, M. Barbatti and H. Lischka, *Mol. Phys.* **107**, 845 (2009).
87. H. Tao, B. G. Levine, and T. J. Martinez, *J. Phys. Chem. A* (2009).
88. H.R. Hudock and T. J. Martinez, *Chemphyschem* **9**, 2486 (2008).
89. G. Groenhof, L.V. Schafer, M. Boggio-Pasqua, M. Goette, H. Grubmuller and M.A. Robb, *J. Am. Chem. Soc.* **129**, 6812 (2007).
90. T. Müller, M. Dallos and H. Lischka, *J. Chem. Phys.* **110**, 7176 (1999).
91. C. Angeli, *J. Comput. Chem.* **30**, 1319 (2009).
92. M. Merchan, R. Gonzalez-Luque, T. Climent, L. Serrano-Andres, E. Rodriuguez, M. Reguero and D. Pelaez, *J. Phys. Chem. B* **110**, 26471 (2006).
93. P. Hurd, T. Cusati and M. Persico, *J. Comput. Phys.* **229**, 2109 (2010).

94. H. Lin and D.G. Truhlar, *Theor. Chem. Acc.* **117**, 185 (2007).
95. R. Shepard. in *Modern Electronic Structure Theory*, edited by D.R. Yarkony, (World Scientific, Singapore, 1995), p. 345.
96. H. Lischka, M. Dallos and R. Shepard, *Mol. Phys.* **100**, 1647 (2002).
97. R. Shepard. in *Ab Initio Methods in Quantum Chemistry II*, edited by K.P. Lawley, (Wiley, New York, 1987), pp. 63.
98. R. Shepard, G.S. Kedziora, H. Lischka, I. Shavitt, T. Müller, P.G. Szalay, M. Kállay and M. Seth, *Chem. Phys.* **349**, 37 (2008).
99. N.C. Handy and H.F. Schaefer, *J. Chem. Phys.* **81**, 5031 (1984).
100. J.E. Rice and R.D. Amos, *Chem. Phys. Lett.* **122**, 585 (1985).
101. C.P. Hu, H. Hirai and O. Sugino, *J. Chem. Phys.* **127** (2007).
102. R. Send and F. Furche, *J. Chem. Phys.* **132**, 044107 (2010).
103. E. Tapavicza, I. Tavernelli and U. Rothlisberger, *Phys. Rev. Lett.* **98**, 023001 (2007).
104. U. Werner, R. Mitric, T. Suzuki and V. Bonacic-Koutecký, *Chem. Phys.* **349**, 319 (2008).
105. R. Mitric, U. Werner, M. Wohlgemuth, G. Seifert and V. Bonacic-Koutecky, *J. Phys. Chem. A* **113**, 12700 (2009).
106. A. Migani and M. Olivucci. in *Conical Intersections: Electronic Structure, Dynamics & Spectroscopy*, edited by W. Domcke, D.R. Yarkony and H. Köppel, (World Scientific Publishing Company, Singapore, 2004.
107. M. Barbatti, M. Ruckenbauer, J.J. Szymczak, A.J.A. Aquino and H. Lischka, *Phys. Chem. Chem. Phys.* **10**, 482 (2008).
108. I. Antol, M. Vazdar, M. Barbatti and M. Eckert-Maksic, *Chem. Phys.* **349**, 308 (2008).
109. G.J. Atchity, S.S. Xantheas and K. Ruedenberg, *J. Chem. Phys.* **95**, 1862 (1991).
110. D.R. Yarkony, *J. Chem. Phys.* **114**, 2601 (2001).
111. M. Barbatti, M. Ruckenbauer and H. Lischka, *J. Chem. Phys.* **122**, 174307 (2005).
112. J.A. Frey, R. Leist, C. Tanner, H.M. Frey and S. Leutwyler, *J. Chem. Phys.* **125**, 114308 (2006).
113. P. Cattaneo, M. Persico and A. Tani, *Chem. Phys.* **246**, 315 (1999).
114. Y. Lei, S. Yuan, Y. Dou, Y. Wang and Z. Wen, *J. Phys. Chem. A* **112**, 8497 (2008).
115. N.C. Handy and A.J. Cohen, *Mol. Phys.* **99**, 403 (2001).
116. J.F. Dobson, M.J. Bunner and E.K.U. Gross, *Phys. Rev. Lett.* **79**, 1905 (1997).

Chapter 12

Nonadiabatic Trajectory Calculations with *Ab Initio* and Semiempirical Methods

Eduardo Fabiano[*], Zhenggang Lan[†],
You Lu[†] and Walter Thiel[†]

1. Introduction 464
2. Theory 467
   2.1. Propagation of quantum amplitudes 467
   2.2. Fewest switches algorithm 468
   2.3. Velocity adjustment 471
   2.4. Adiabatic representation 473
3. Implementation of Trajectory Surface Hopping 473
   3.1. Algorithm 473
   3.2. Initial sampling 477
4. Electronic Structure Methods 477
   4.1. *Ab initio* CASSCF and MRCI 478
   4.2. Semiempirical methods 479
   4.3. Hybrid QM/MM 483

---

[*]National Nanotechnology Laboratory (NNL), Istituto di Nanoscienze-CNR, Via per Arnesano 16, I-73100 Lecce, Italy.
[†]Max-Planck-Institut für Kohlenforschung, Kaiser-Wilhelm-Platz 1, D-45470 Mülheim an der Ruhr, Germany.

5. Applications 484
   5.1. Methaniminium cation 484
   5.2. 9H-adenine 486
6. Summary and Outlook 489
   Acknowledgments 491
   References 491

1. Introduction

The Born–Oppenheimer (BO) approximation[1] allows the separation of electronic and nuclear degrees of freedom and the definition of individual potential energy surfaces (PESs) for different electronic states. It provides a valid description of many chemical processes and is thus invoked in most quantum chemical calculations. However, there are also a number of phenomena that cannot be described within this framework. When two electronic states of the same multiplicity become degenerate, there will generally be a pronounced interstate coupling that leads to a strong electronic mixing and a breakdown of the BO approximation. In the region of such conical intersections, the strong coupling of the electronic and nuclear motion induces so-called nonadiabatic transitions between different electronic states which are at the heart of photochemistry, internal conversion, fluorescence quenching, and nonradiative energy dissipation processes.[2–9]

In recent years, many theoretical tools have been developed to study conical intersections and nonadiabatic phenomena. The location of minimum-energy conical intersections[4,10–16] and of conical intersection seams[17–20] can provide information on the topology of the relevant PESs and help to find geometrical configurations that are crucial for the dynamics. Excited-state reaction channels can be found by constructing minimum-energy reaction paths connecting the Franck–Condon region with different conical intersections, thus identifying favorable pathways and the associate energy barriers. A more complete characterization of nonadiabatic processes requires the direct simulation of nonadiabatic dynamics, which takes into account the full complexity of the relevant PESs and provides access not only to the available reaction channels, but also to reaction times and rates. Nonadiabatic molecular dynamics (MD) simulations are challenging because of the need for a self-consistent treatment of the electronic and nuclear degrees of freedom, beyond the BO approximation. In recent years, various methods have been developed to perform such simulations, ranging from fully-quantum to mixed quantum-classical treatments.[4]

At a fundamental level, the nonadiabatic dynamics at conical intersections can be studied using quantum-wavepacket methods.[4,7,21,22] In this approach, an analytical model of the relevant PESs is constructed on the basis of *ab initio* calculations, and the quantum nuclear wavepackets are propagated on the resulting coupled surfaces by direct solution of the time-dependent Schrödinger equation. This provides deep physical insight into the nonadiabatic dynamics since all quantum effects are included explicitly (for example, hydrogen-tunneling and geometric-phase effects), but the computational cost increases dramatically with the number of degrees of freedom. This problem is alleviated to some extent in the multi-configuration time-dependent Hartree (MCTDH) approach[8,23–26] and the multilayer MCTDH treatment[27,28] which have extended the applicability of quantum-wavepacket methods significantly.

Some of the practical limitations of the quantum-wavepacket approach can be overcome with the use of quantum dissipative theory,[4,7] in particular when employing the Redfield approach, which is based on a system-plus-bath model.[4,7] Here, only the relevant degrees of freedom are simulated explicitly, while the remaining ones are treated as an external bath. The interaction between the active and inactive modes is described by the system-bath coupling Hamiltonian, and the quantum evolution of the active degrees of freedom is governed by a reduced density operator, which can be obtained through a perturbation expansion in the case of weak system-bath coupling. Theoretical approaches based on quantum dissipative theory provide efficient tools to study the influence of external factors, such as vibrational relaxation, on the nonadiabatic dynamics, but they face severe limitations in terms of general applicability and the scaling of the computational effort.[29,30] Moreover, they normally employ rather simple models for the system-bath coupling Hamiltonian.

Nonadiabatic dynamics simulations of molecular systems can be performed using mixed quantum-classical approaches,[31] which comprise the mean-field Ehrenfest method,[31] surface-hopping methods,[31–69] quantum-classical Liouville descriptions,[31] and mapping procedures.[31] In the mixed quantum-classical methods, the degrees of freedom are again divided into two subsets, but in this case all of them are treated explicitly, i.e. no external bath is introduced. The relevant degrees of freedom (e.g. electrons, selected protons, or selected vibrational modes) are handled quantum mechanically, while the remaining ones are described by classical mechanics.

This allows for an efficient simulation of the nonadiabatic dynamics of large systems. However, quantum effects of nuclear motion are not captured for the degrees of freedom propagated by classical mechanics, and any of the associated quantum interferences are lost during the time propagation. These neglected quantum effects can be partially recovered for moderately sized systems, making use of semiclassical approaches like the Van-Vleck-Gutzwiller formulation[31,70,71] or the full multiple spawning method.[72–75] Comparisons between quantum methods, full multiple spawning, and trajectory surface hopping are available in the literature.[50,76]

In the mixed quantum-classical approaches, the quantum degrees of freedom are generally treated by the time-dependent Schröndinger equation. In a purely quantum system, the time evolution will then be reversible and will lead to a superposition of quantum states. This is no longer true in mixed quantum-classical systems where the quantum subsystem is coupled to a classical subsystem that plays the role of an observer in a measurement and will thus eventually cause the collapse of quantum wavefunction from a superposition state to a pure state.[77] As a consequence of such decoherence, the time evolution of the quantum subsystem is no longer unitary and reversible.[77] Efforts have been made to include decoherence effects in the treatment of mixed quantum-classical systems.[50,78–91] For instance, within mean-field approaches, the evolution of the quantum subsystem can be described by a Liouville–von Neumann equation including decoherence.[78–86] One of these approaches is the CSDM (coherent switches with decay of mixing) method which employs a phenomenological parameter to determine decoherence time constants.[78–83] Similar ideas to include decoherence corrections can also be implemented in surface hopping approaches.[50,87–91] Such decoherence corrections will not be discussed further here because they are described in detail in another chapter, with emphasis on the CSDM method and its application.[92]

In this chapter we will focus on the surface hopping approach[32–37] and in particular on the trajectory surface hopping (TSH) method of Tully,[38,39] where the nuclear degrees of freedom are propagated on independent classical trajectories and nonadiabatic effects are included by allowing hopping between different PESs. This yields a good description of the nonadiabatic dynamics at low computational cost. In particular, the use of well-defined classical trajectories in the TSH method makes it possible to employ direct dynamics,[93] i.e. an on-the-fly approach where the energies, gradients, and nonadiabatic couplings are calculated at each point along the trajectory, which obviates the need for precomputing and constructing the entire PESs.[42,55,64] The TSH method is nowadays one of the most popular tools for studying nonadiabatic phenomena in organic molecules.[38–69]

2. Theory

As in conventional MD calculations, the nuclear degrees of freedom are propagated in the TSH method on a single PES according to the classical equations of motion. The electronic degrees of freedom are propagated along the same trajectory according to the time-dependent Schrödinger equation. To maintain the self-consistency between the nuclear and the electronic time evolution, it is necessary to properly introduce nonadiabatic interactions. In the surface-hopping method this is achieved by allowing instantaneous hops from one PES to another, with the hopping probability at each time being controlled by a stochastic switching algorithm. To account for the random nature of the switching algorithm, a swarm of trajectories must be considered for each initial configuration of the dynamics, and the final results are obtained as an average over a large number of trajectories.

2.1. *Propagation of quantum amplitudes*

In the TSH method we assume that the nuclear motion is described by a classical trajectory $\mathbf{R}(t)$, which is determined by the solution of Newton's equations. The electronic Hamiltonian is

$$H_e(\mathbf{r}, \mathbf{R}) = -\frac{\hbar^2}{2} \sum_l \frac{1}{m_l} \nabla^2_{\mathbf{r}_l} + V_{rR}(\mathbf{r}, \mathbf{R}), \qquad (1)$$

where l labels the electronic degrees of freedom and V_{rR} is the potential including nuclear–electron and electron–electron interactions. The electronic Hamiltonian is time-dependent through $\mathbf{R}(t)$ and the electronic wave function $\Phi(\mathbf{r}, \mathbf{R}, t)$ is the solution of the time-dependent Schrödinger equation

$$i\hbar \frac{\partial \Phi(\mathbf{r}, \mathbf{R}, t)}{\partial t} = H_e \Phi(\mathbf{r}, \mathbf{R}, t). \qquad (2)$$

The electronic wave function can be expanded using a sum-over-states expression

$$\Phi(\mathbf{r}, \mathbf{R}, t) = \sum_i c_i(t) \phi_i(\mathbf{r}, \mathbf{R}), \qquad (3)$$

where $\{\phi_i\}$ is a set of known electronic functions and $c_i(t)$ denotes complex-valued expansion coefficients. Substitution of (3) into the time-dependent Schrödinger equation (2), multiplication on the left by $\phi_j^*(\mathbf{r},\mathbf{R})$ and integration over $\mathbf{r}$ gives

$$i\hbar \frac{dc_j(t)}{dt} = \sum_i c_i(t)[H_{ji} - i\hbar \dot{\mathbf{R}} \cdot \mathbf{d}_{ji}], \quad (4)$$

where $\dot{\mathbf{R}}$ is the vector of nuclear velocities and

$$H_{ji} \equiv \int d\mathbf{r} \phi_j^*(\mathbf{r},\mathbf{R}) \left[-\frac{\hbar}{2} \sum_l \frac{1}{m_l} \nabla_{\mathbf{r}_l}^2 + V_{rR}(\mathbf{r},\mathbf{R}) \right] \phi_i(\mathbf{r},\mathbf{R}), \quad (5)$$

$$\mathbf{d}_{ji} \equiv \int d\mathbf{r} \phi_j^*(\mathbf{r},\mathbf{R}) [\nabla_\mathbf{R} \phi_i(\mathbf{r},\mathbf{R})]. \quad (6)$$

In the derivation of the second term on the right hand side of Eq. (4) the chain rule

$$\frac{\partial}{\partial t} = \frac{\partial}{\partial \mathbf{R}} \frac{d\mathbf{R}}{dt} \quad (7)$$

was used to express the nonadiabatic coupling terms

$$F_{ij}(t) = \int d\mathbf{r} \phi_j^*(\mathbf{r},\mathbf{R}) \frac{\partial \phi_i(\mathbf{r},\mathbf{R})}{\partial t} \quad (8)$$

as the scalar product of the velocity vector $\dot{\mathbf{R}}$ and the nonadiabatic coupling elements $\mathbf{d}_{ji}$.

Equation (4) describes the time evolution of the quantum amplitude of each state ϕ_i along the trajectory $\mathbf{R}(t)$. This information is used to determine the hopping probability.

2.2. Fewest switches algorithm

A central point in the TSH method is the introduction of a hopping criterion to maintain the self-consistency between the classical and the quantum propagation. Although this criterion reflects the quantum nonadiabatic behavior of the system, it cannot be directly derived from the Schrödinger equation, because of the choice of treating the nuclear degrees of freedom classically. Thus, an *ad hoc* hopping criterion must be imposed, which satisfies the requirement that the classical occupation and the quantum population ($|c_i(t)|^2$) should be equal at any time. There is of course a large number of possible algorithms that satisfy this requirement (at least approximately), each one derivable under additional specific conditions.[50,65,92,94]

A popular choice is the *fewest switches algorithm* (FSA).[38] It is based on the hypothesis that different trajectories are totally independent and on the additional requirement that the quantum-classical self-consistency is maintained through the minimum possible number of hopping events. The latter condition is needed since a large number of hopping events would *de facto* lead to a mean-field description (similar to the Ehrenfest approach[31]). In that case the dynamics would in practice evolve on an average of the relevant PESs, while in the surface-hopping method the propagation occurs on a single physical PES at each time. The FSA is implemented assuming that the hopping events occur in an infinitesimal time. This causes state transitions to be discontinuous in a single trajectory. However, when the entire swarm of trajectories is considered a smooth transition is obtained, since hopping events will occur at different times for different trajectories.

In practice the FSA is not always able to guarantee the self-consistency between the quantum population and the classical occupation,[50,88,94–97] for two reasons. First, there are rejected hops (see Sec. 2.3), and second, despite the assumption of completely independent trajectories, in the derivation of the FSA all trajectories starting from the same initial conditions are considered to have the same quantum amplitudes when accessing conical intersections (see below). This simplifying approximation is not valid in general, but only in special cases (e.g. for one-dimensional systems with a single nonadiabatic coupling region or for systems with quasi-degenerate PESs, see Ref. 88 for a detailed discussion). Nevertheless, the FSA works very well in most situations and, owing to its simplicity, is probably the most widely used switching criterion in surface hopping studies.

There are other algorithms that attempt to improve upon the FSA in the description of the hopping event. An advanced example is FSTU/SD (fewest switches with time uncertainty and stochastic decay) with a special treatment of possible momentum changes at rejected hops.[87–91] This approach is described in detail in another chapter of this book.[92] A similar variant has been presented in Ref. 50. In the following, we only address the standard FSA.

In order to derive the FSA we consider a swarm of N trajectories. At each time the quantum-classical self-consistency is achieved if the fraction of trajectories classically evolving on the i-th PES is equal to the quantum population of state i, that is

$$\frac{N_i(t)}{N} = |c_i(t)|^2. \tag{9}$$

Suppose that at time $t + dt$ the classical occupation of state i is $N_i(t + dt)$ and that $N_i(t) > N_i(t + dt)$, then the minimum number of transitions needed for this change in occupation will be $N_i(t) - N_i(t + dt)$ hops from state i to any other state and zero hops from any state to i. The probability of a hopping out of the state i in the infinitesimal time interval dt is therefore

$$P_i(t)dt = \frac{N_i(t) - N_i(t + dt)}{N_i(t)}. \tag{10}$$

Using Eq. (9) we can rewrite Eq. (10) as

$$P_i(t)dt = \frac{|c_i(t)|^2 - |c_i(t + dt)|^2}{|c_i(t)|^2} = -\frac{\frac{d|c_i(t)|^2}{dt}dt}{|c_i(t)|^2} = -\frac{2\mathrm{Re}(c_i^* \frac{dc_i}{dt})}{|c_i(t)|^2}. \tag{11}$$

Use of Eq. (4) gives finally

$$P_i(t)dt = -\sum_j \frac{2\left[\hbar^{-1}\mathrm{Im}(c_i^* c_j H_{ij}) - \mathrm{Re}(c_i^* c_j \dot{\mathbf{R}} \cdot \mathbf{d}_{ji})\right]}{|c_i(t)|^2}. \tag{12}$$

If the time interval is not infinitesimal but finite (Δt), as for example in the practical implementation of MD simulations, the hopping probability is found by integrating the infinitesimal probability defined by Eq. (12) over this interval. The resulting expression for the hopping probability is

$$P_i = -\sum_j \frac{2\int_t^{t+\Delta t} dt \left[\hbar^{-1}\mathrm{Im}(c_i^* c_j H_{ij}) - \mathrm{Re}(c_i^* c_j \dot{\mathbf{R}} \cdot \mathbf{d}_{ji})\right]}{|c_i(t)|^2}, \tag{13}$$

where, in doing the integration, the denominator is considered constant, since $|c_i|^2$ cannot change during the time interval Δt (i.e. no hopping during Δt).

Equation (13) defines the probability that a hopping out of state i occurs after a time interval Δt. Because the probability of a transition from state i to any other state is the sum of all transition probabilities P_{ij} from state

i to any state j, the probability for a transition from i to j is

$$P_{ij} = -\frac{2\int_t^{t+\Delta t} dt \left[\hbar^{-1}\mathrm{Im}(c_i^* c_j H_{ij}) - \mathrm{Re}(c_i^* c_j \dot{\mathbf{R}} \cdot \mathbf{d}_{ji})\right]}{|c_i(t)|^2}. \tag{14}$$

The derivation of Eq. (14) assumes $N_i(t) > N_i(t+dt)$. For a correct definition of the transition probability it is thus necessary to discard negative (unphysical) values of P_{ij}. The hopping probability can then be defined as

$$g_{ij} = \max(P_{ij}, 0). \tag{15}$$

In order to determine whether a switch from state i to state k should occur, a uniform random number $0 < \xi < 1$ is selected at each time step and the hopping is performed if

$$\sum_{j=1}^{k} g_{ij} < \xi < \sum_{j=1}^{k+1} g_{ij}. \tag{16}$$

2.3. Velocity adjustment

When a hopping is performed, the total energy of the system is in general not conserved, since a discontinuous change occurs in the electronic energy. This unphysical behavior is due to the fact that the hopping event is imposed from outside, through an *ad hoc* hopping algorithm, and acts as an external perturbation on the system. Its effect must of course be removed since we aim at simulating a closed system for which the total energy must be conserved. To this end a velocity adjustment is usually performed after each hopping event. The velocity adjustment induces a change in the nuclear kinetic energy which is equal and opposite to the one produced by the hopping in the electronic energy, and thus the total energy is conserved. If the kinetic energy is too low to compensate for the energy variation caused by the electronic transition, the hopping is considered unphysical and is rejected. In this case the velocity of the system is reversed in analogy with elastic scattering.[39,98]

When the system jumps from state i to state j, the electronic energy changes from ϵ_i to ϵ_j. To compensate this energy change the velocity of the system is adjusted according to

$$\dot{\mathbf{R}}'_\beta = \dot{\mathbf{R}}_\beta - \gamma_{ij} \frac{\mathbf{w}_{ij}^\beta}{M_\beta}, \tag{17}$$

with M_β being the mass of nucleus β, γ_{ij} the scaling factor, and $\mathbf{w}$ a vector of dimension N_{atoms} along which the scaling is performed. The overall kinetic energy change is

$$\Delta T = \frac{1}{2}\sum_\beta M_\beta \left(\dot{\mathbf{R}}'_\beta\right)^2 - \frac{1}{2}\sum_\beta M_\beta \left(\dot{\mathbf{R}}_\beta\right)^2$$

$$= \frac{1}{2}\sum_\beta M_\beta \left[\gamma_{ij}^2 \frac{\mathbf{w}_{ij}^{\beta^2}}{M_\beta^2} - 2\gamma_{ij}\frac{\dot{\mathbf{R}}_\beta \cdot \mathbf{w}_{ij}^\beta}{M_\beta}\right]$$

$$= \gamma_{ij}^2 a_{ij} - \gamma_{ij} b_{ij}, \tag{18}$$

where in the last line we defined

$$a_{ij} \equiv \frac{1}{2}\sum_\beta \frac{\mathbf{w}_{ij}^{\beta^2}}{M_\beta}, \tag{19}$$

$$b_{ij} \equiv \sum_\beta \dot{\mathbf{R}}_\beta \cdot \mathbf{w}_{ij}^\beta. \tag{20}$$

Imposing the conservation of total energy we find

$$\gamma_{ij}^2 a_{ij} - \gamma_{ij} b_{ij} - (\epsilon_i - \epsilon_j) = 0. \tag{21}$$

If $b_{ij}^2 + 4a_{ij}(\epsilon_i - \epsilon_j) < 0$ there are no real solutions to Eq. (21), and the hopping cannot occur. In this case the nuclear velocities are reversed in the direction of $\mathbf{w}$ setting $\gamma_{ij} = b_{ij}/a_{ij}$. If $b_{ij}^2 + 4a_{ij}(\epsilon_i - \epsilon_j) \geq 0$, the scaling factor is

$$\gamma_{ij} = \frac{b_{ij} + \sqrt{b_{ij}^2 + 4a_{ij}(\epsilon_i - \epsilon_j)}}{2a_{ij}} \quad \text{if } b_{ij} < 0, \tag{22}$$

$$\gamma_{ij} = \frac{b_{ij} - \sqrt{b_{ij}^2 + 4a_{ij}(\epsilon_i - \epsilon_j)}}{2a_{ij}} \quad \text{if } b_{ij} \geq 0. \tag{23}$$

The $\mathbf{w}$ vector can be chosen to be any N_{atoms}-dimensional vector and many different choices are discussed in literature.[95,96,99–101] Usually, the velocity adjustment is performed in the direction of the nonadiabatic coupling elements vector $\mathbf{d_{ij}}$, since this choice is suggested by semiclassical analogies.[102–104] An alternative choice for the scaling direction is the gradient difference vector.[99–101]

2.4. Adiabatic representation

For the expansion of the total electronic wave function in Eq. (3), any set of known electronic functions can be used. In principle, the surface-hopping method can thus be implemented using either a diabatic or adiabatic representation. Most applications employ an adiabatic representation because this simplifies the formalism and facilitates the interpretation of the results by allowing comparisons with standard BO molecular dynamics.

In the adiabatic representation the electronic functions are eigenfunctions of the electronic Hamiltonian and the Hamiltonian matrix H_{ji}, see Eq. (5), is diagonal with non-zero elements equal to the eigenergies ϵ_i. That is,

$$H_{ji} = \epsilon_i \delta_{ji}. \tag{24}$$

Equation (4) is then simplified to

$$i\hbar \frac{dc_j(t)}{dt} = c_j(t)\epsilon_j - i\hbar \sum_i c_i(t)\dot{\mathbf{R}} \cdot \mathbf{d}_{ji}. \tag{25}$$

The time evolution of the quantum amplitude of each state is determined by a phase factor proportional to the eigenvalue of the electronic state and by the mixing between different adiabatic states (i.e. the transfer of adiabatic quantum population), which is driven by the nonadiabatic coupling term $\dot{\mathbf{R}} \cdot \mathbf{d}_{ji}$.

The expression of the hopping probability is also simplified in the adiabatic representation. Substitution of Eq. (24) into Eq. (14) gives

$$P_{ij} = -\frac{2 \int_t^{t+\Delta t} dt [\hbar^{-1}\mathrm{Im}(c_i^* c_j \epsilon_i \delta_{ji}) - \mathrm{Re}(c_i^* c_j \dot{\mathbf{R}} \cdot \mathbf{d}_{ji})]}{|c_i(t)|^2}. \tag{26}$$

The first term on the right-hand side vanishes if $i \neq j$, hence

$$P_{ij} = \frac{2 \int_t^{t+\Delta t} dt \mathrm{Re}(c_i^* c_j \dot{\mathbf{R}} \cdot \mathbf{d}_{ji})}{|c_i(t)|^2}. \tag{27}$$

3. Implementation of Trajectory Surface Hopping

3.1. Algorithm

The general algorithm for the implementation of the TSH molecular dynamics is as follows:

1. Initialization of velocity, gradients, and quantum amplitudes.

2. Time propagation of coordinates and velocities on the selected PES.
3. Computation of energies, gradients, and nonadiabatic coupling vectors of all relevant states at new position and velocity.
4. Time propagation of quantum amplitudes [Eq. (25)] and computation of hopping probabilities [Eq. (27)].
5. Random number generation and comparison with hopping probabilities according to Eq. (16).
6. If hopping is rejected: inversion of velocity (Sec. 2.3).
7. If hopping is performed: velocity adjustment (Sec. 2.3) and update of the active PES for molecular dynamics.
8. Back to point 2.

This scheme closely resembles the standard MD algorithm in the first steps (1–3), with the exception that the energy and gradient must be computed for several electronic states and the nonadiabatic couplings must be calculated in addition. The need for calculating the nonadiabatic couplings also restricts the possible choices for the nuclear propagator. In fact the nonadiabatic couplings depend not only on the coordinates of the system through the nonadiabatic coupling elements, but also directly on the velocity distribution. Therefore, the classical propagator must be able to yield both the coordinates and the velocity update at the same time. This means, for example, that the popular position-Verlet[105] and leap-frog[106] algorithms cannot be used for surface-hopping dynamics. A common choice for the nuclear propagator is the velocity-Verlet integration scheme.[107] In this algorithm the updates of position and velocity are given by

$$\mathbf{r}(t+\Delta t) = \mathbf{r}(t) + \mathbf{v}(t)\Delta t - \frac{\nabla E(t) \cdot \mathbf{M}^{-1}}{2}\Delta t^2, \tag{28}$$

$$\mathbf{v}(t+\Delta t) = \mathbf{v}(t) - \frac{[\nabla E(t) + \nabla E(t+\Delta t)] \cdot \mathbf{M}^{-1}}{2}\Delta t, \tag{29}$$

where $\mathbf{M}^{-1}$ is the vector of the reciprocal nuclear masses, i.e. $\mathbf{M}^{-1} = (M_1^{-1}, M_2^{-1}, \ldots, M_{N_{\text{atoms}}}^{-1})$. The updated coordinates and velocities are available simultaneously, and the nonadiabatic couplings can be readily calculated at each time step.

The calculation of the nonadiabatic coupling terms can be performed analytically, by expanding expression (6) in terms of the molecular orbital contributions. For a wave function in the form of a multireference configuration interaction expansion, the nonadiabatic coupling elements can be computed as[108,109]

$$\mathbf{d}_{ij} = \frac{\mathbf{h}_{ij}}{\epsilon_j - \epsilon_i} + \sum_{a,b}(\Gamma_{ij})_{ab}\left\langle \psi_a \left| \frac{\partial \psi_b}{\partial \mathbf{R}} \right.\right\rangle, \qquad (30)$$

where Γ_{ij} is the one-electron transition density matrix, ψ denotes molecular orbitals, and the interstate coupling vector is defined as

$$\mathbf{h}_{ij} = \mathbf{C}_i^\dagger \frac{\partial H_e}{\partial \mathbf{R}} \mathbf{C}_j, \qquad (31)$$

with $\mathbf{C}_i$ representing the vector of configuration interaction coefficients of state i. Alternatively, an approximate expression for the nonadiabatic coupling between two electronic states can be derived directly from the definition (8) which is written here as

$$F_{ij}\left(t + \frac{\Delta t}{2}\right) = \int d\mathbf{r} \phi_j^*\left(t + \frac{\Delta t}{2}\right)\left[\frac{\partial \phi_i(t)}{\partial t}\right]_{t+\frac{\Delta t}{2}}, \qquad (32)$$

using the short-hand notation $\phi_i(\mathbf{r}, \mathbf{R}(t)) = \phi_i(t)$. The two terms in the integral can be approximated as follows:

$$\phi_j^*\left(t + \frac{\Delta t}{2}\right) = \frac{\phi_j^*(t) + \phi_j^*(t + \Delta t)}{2}, \qquad (33)$$

$$\left[\frac{\partial \phi_i(t)}{\partial t}\right]_{t+\frac{\Delta t}{2}} = \frac{\phi_i(t + \Delta t) - \phi_i(t)}{\Delta t}. \qquad (34)$$

After substitution into Eq. (32) and use of the orthonormality of the wave functions, we find

$$F_{ij}\left(t + \frac{\Delta t}{2}\right) = \frac{\int d\mathbf{r} \phi_j^*(t)\phi_i(t + \Delta t) - \int d\mathbf{r} \phi_j^*(t + \Delta t)\phi_i(t)}{2\Delta t}. \qquad (35)$$

In this way the magnitude of the nonadiabatic coupling between two states can be calculated approximately at each time step with relatively small computational effort. We note however that the use of the analytical procedure for the computation of the nonadiabatic coupling term is advisable whenever it is affordable, because of its intrinsically superior accuracy and numerical stability.

The treatment of the nonadiabatic interactions is implemented in the second part of the TSH algorithm (steps 4–7). This involves the integration of Eq. (25) and the fewest switches algorithm (steps 5–7), which has already been discussed in Sec. 2.2. The integration of Eq. (25) is performed numerically and requires particular attention, since the quantum amplitudes are rapidly oscillating in time due to the presence of a phase factor proportional to the energy of each state. One possible solution involves a unitary propagator for the quantum amplitudes. For numerical convenience we use the new variables

$$\tilde{c}_i(t) = c_i(t) e^{-i \int_0^t d\tau \frac{\epsilon_0}{\hbar}} \tag{36}$$

that differ from the original quantum amplitudes only by a phase factor, which has no influence on any physical quantity. Substitution into Eq. (25) gives

$$i\hbar \frac{d\tilde{c}_j(t)}{dt} = \tilde{c}_j(t)(\epsilon_j - \epsilon_0) - i\hbar \sum_i \tilde{c}_i(t) \dot{\mathbf{R}} \cdot \mathbf{d}_{ji}, \tag{37}$$

which can be expressed in matrix form as

$$\frac{d}{dt}\underline{\tilde{\mathbf{c}}}(t) = i\underline{\mathbf{A}}(t)\underline{\tilde{\mathbf{c}}}(t) \tag{38}$$

with

$$A_{ji} = -\frac{1}{\hbar}(\epsilon_j - \epsilon_0)\delta_{ji} + i\dot{\mathbf{R}}(t) \cdot \mathbf{d}_{ji}(t). \tag{39}$$

Equation (38) has the formal solution

$$\underline{\tilde{\mathbf{c}}}(t) = \underline{\tilde{\mathbf{c}}}(t_0) e^{i \int_{t_0}^{t} d\tau \underline{\mathbf{A}}(\tau)}. \tag{40}$$

For small time intervals $\Delta t = (t - t_0)$, the integral in Eq. (40) can be approximated by $\underline{\mathbf{A}}((t+t_0)/2)\Delta t$. Thus we obtain the following formula for the time propagation of quantum amplitudes

$$\underline{\tilde{\mathbf{c}}}(t + \Delta t) = \underline{\tilde{\mathbf{c}}}(t) e^{i\underline{\mathbf{A}}(t+\Delta t/2)\Delta t}. \tag{41}$$

The matrix $e^{i\underline{\mathbf{A}}\Delta t}$ can be easily calculated using Silvester's formula, since the matrix $\underline{\mathbf{A}}$ is Hermitian and therefore $i\underline{\mathbf{A}}\Delta t$ is diagonalizable. Hence, we have

$$e^{i\underline{\mathbf{A}}\Delta t} = \underline{\mathbf{U}} \begin{pmatrix} e^{\omega_1} & 0 & \cdots & 0 \\ 0 & e^{\omega_2} & \cdots & 0 \\ \vdots & \vdots & \ddots & \vdots \\ 0 & 0 & \cdots & e^{\omega_n} \end{pmatrix} \underline{\mathbf{U}}^\dagger \tag{42}$$

with $\underline{\mathbf{U}}$ denoting the unitary matrix of eigenvectors and $\{\omega_i\}$ the corresponding eigenvalues.

3.2. Initial sampling

In analogy with standard BO dynamics, initial conditions (nuclear coordinates and momenta) for the dynamics must be sampled in TSH simulations to mimic the distribution function of the initial nuclear motion in the phase space. In addition, multiple trajectories must be run with the same initial conditions to exploit the stochastic nature of the switching algorithm. To simulate a photoinduced event, the starting configurations must be selected not only according to their ground-state distribution function but also according to the radiative coupling between the ground state and the relevant excited state.

The ground-state distribution function can be sampled either from a Wigner distribution[110, 111] (derived from normal mode analysis) or from a sufficiently long ground-state BO molecular dynamics run, according to the ergodic hypothesis.[112, 113] For each sampled configuration the probability of a radiative transition from state S_0 to S_k is[114]

$$P^{k0} = \frac{f^{k0}/\omega^{k0}}{\max(f^{k0}/\omega^{k0})} = \frac{|\mu_{k0}|^2}{\max(|\mu_{k0}|^2)}, \quad (43)$$

where μ_{k0} is the transition dipole moment, f^{k0} is the oscillator strength, and ω^{k0} is the energy difference between the two states. A stochastic procedure (similar to that outlined in Sec. 2.2) can thus be used to select or reject a given configuration. Note that the present approach goes beyond the Condon approximation because the f^{k0} and ω^{k0} values are calculated at every geometry.

4. Electronic Structure Methods

In the practical implementation of the TSH dynamics, it is extremely important to choose a suitable electronic structure method for calculating the energies, gradients, and nonadiabatic couplings. This constitutes in fact the main bottleneck of the calculations and at the same time largely determines the accuracy of the final results. The method should satisfy several requirements:

- it should yield accurate energies, gradients, and nonadiabatic couplings (interstate couplings) for the ground state and several different excited states;

- it should accurately describe multiconfigurational electronic states, as those encountered in the vicinity of conical intersections;
- it should give a balanced description of electronic states with different character (e.g. the ground state and $\pi\pi^*$ or $n\pi^*$ or charge-transfer excited states);
- it should be computationally efficient since a typical surface-hopping study requires several thousands of single-point calculations.

These requirements impose strong constraints on the choice of the electronic structure method to be used in surface-hopping studies. For example, single-reference post-Hartree–Fock methods are not eligible because they cannot properly describe multiconfigurational states. In a similar vein, the use of time-dependent density functional theory (TDDFT) is problematic because it is doubtful whether the currently available TDDFT approaches can treat conical intersections in an appropriate manner.[61,115,116] Furthermore, TDDFT yields an unbalanced description of excitations with different character.[116–119] We shall therefore focus on multiconfigurational wavefunction methods in the following.

4.1. *Ab initio CASSCF and MRCI*

At the *ab initio* level, the Complete Active Space Self-Consistent-Field (CASSCF) approach[120,121] is a popular choice to perform single-point calculations during TSH molecular dynamics. In CASSCF a limited number of orbitals is selected to define an active space, and the ansatz for the multiconfiguration wave function is the full configuration interaction (CI) expansion within the active space. Both the expansion coefficients and the orbitals are optimized in a multiconfigurational self-consistent-field procedure to yield the CASSCF wave function. This ensures the inclusion of static correlation effects arising from the quasi-degeneracy of the electronic states considered and yields a reasonably balanced description of such states with regard to static correlation effects. The CASSCF method is thus in principle appropriate for the study of conical intersections and for surface-hopping applications. However, in general, it does not properly describe dynamic correlation effects arising from the instantaneous interactions between the electrons. This can be relevant in TSH simulations where such effects may be quite different in the different electronic states considered.

The use of larger active spaces in CASSCF could cure these problems, but this straightforward solution is often not practical because of the steep scaling of the computational effort with the number of active orbitals and electrons. Instead, it may be feasible to further optimize the wave function by performing a multireference configuration interaction (MRCI) calculation using CASSCF reference orbitals.[43] In this case, a rather small MRCI expansion may be sufficient due to the high quality of the CASSCF reference orbitals so that the computational effort remains limited. An alternative to MRCI is provided by multireference perturbation theory.[121–124]

Generally speaking, MRCI calculations need not be based on a CASSCF reference, but can be performed using a large variety of suitably chosen orbitals and reference configurations. They provide a very flexible framework for capturing both static and dynamic correlation effects, and they can be made very accurate, albeit at very high computational costs.

Both CASSCF and MRCI methods have been employed in *ab initio* TSH excited-state dynamics studies. The technology for computing the required energies, gradients, and nonadiabatic couplings is available for both methods, and hence it is the balance between accuracy and computational effort that governs the choice between them. In current applications, *ab initio* TSH dynamics calculations are still restricted to small molecules when using accurate MRCI methods, while molecules of moderate size (up to about 20 atoms) can be handled at the CASSCF level.

4.2. *Semiempirical methods*

Given the high computational cost of *ab initio* TSH simulations (regardless whether based on CASSCF or MRCI), it is worthwhile to check the use of simpler alternatives for the electronic structure calculation. Semiempirical molecular orbital methods combined with a small MRCI treatment may provide such an alternative. Conceptually, dynamic electronic correlation is incorporated into these methods in an average manner by the use of suitably damped electron–electron interactions and by the parametrization against experiment, and therefore the MRCI calculation needs to account only for static near-degeneracy correlation effects which can be captured with small active spaces and a small number of interacting configurations. Such semiempirical MRCI computations are several orders of magnitudes faster than corresponding *ab initio* treatments, with substantial savings both at the SCF level (due to the use of a minimal basis and simple integral approximations) and at the MRCI level (since smaller expansions are sufficient). From a practical point of view, semiempirical MRCI methods are thus well suited for TSH dynamics of large molecules. Historically, the first

TSH study of electronically nonadiabatic dynamics was indeed performed at the semiempirical CI level.[125]

However, some care must be taken in the choice of the actual method applied, since the established NDDO-based semiempirical Hamiltonians have been developed and parameterized to reproduce only ground-state properties. The commonly used MNDO-type methods yield a poor description of electronically excited states and systematically underestimate the energy gaps between the occupied and unoccupied orbitals as well as the associated excitation energies. There are two strategies to correct this drawback. The first one is to employ a standard MNDO-type method and perform an *ad hoc* reparametrization to improve the description of excited states.[47–51] This approach is conceptually simple, but faces the difficulty of finding reliable reference data for the parametrization, and it also requires a cumbersome reparametrization for every specific system. The second option is to use an improved semiempirical Hamiltonian that is capable of describing both the ground state and the excited states well and in a balanced manner. This can be achieved by including orthogonalization corrections in the Fock matrix which account for Pauli exchange repulsion. They have the effect that the antibonding combination of two interacting orbitals is destabilized more than the bonding combination is stabilized, giving rise to an unsymmetrical splitting (as in the *ab initio* case), thus improving on MNDO-type methods with their inherently symmetrical splitting. Three variants of NDDO-based methods with orthogonalization corrections are available, namely OM1,[126] OM2,[127,128] and OM3,[129] which differ in the extent of the corrections applied and in the parametrization. The most elaborate one is OM2 which has been applied most widely. Systematic benchmarks on electronically excited states of 28 organic molecules (with a total of 167 valence excitations) have shown that the OM2 method combined with a standardized GUGA-MRCI treatment[130] reproduces the reference vertical excitation energies with a mean absolute deviation of 0.5 eV,[131] regardless of the electronic ($\pi\pi^*$ or $n\pi^*$) and spin (singlet or triplet) character of the states involved (see Fig. 1). By contrast, mean absolute deviations of about 1.5 eV are found for MNDO-type approaches without orthogonalization corrections.

Given the improvements over the established MNDO-type methods both with regard to the model and the numerical results, the OM2/MRCI approach appears to be a promising tool for TSH excited-state dynamics. The required energies, gradients, and nonadiabatic couplings are available,[132] and a corresponding implementation has been reported[64] along

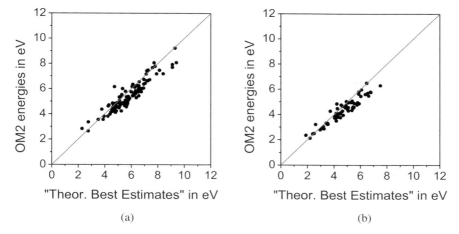

Fig. 1. Comparison of OM2 excitation energies with high-level *ab initio* results for (a) singlet and (b) triplet excitations.

with initial applications to DNA bases.[66–68] In the remainder of this section, we give a brief summary of the orthogonalization corrections that are included in OM2.

The standard semiempirical approaches solve a secular equation in canonical form (containing a unit overlap matrix), without ever performing an explicit transformation from the nonorthogonal set of atomic orbitals to an orthogonal set (consistent with the zero-differential overlap approximation). Performing this transformation in *ab initio* SCF treatments introduces orthogonalization corrections (for example, terms accounting for Pauli exchange repulsions) that are formally missing in MNDO-type methods where their effect on the total energy is mimicked by additional two-center terms in the core-core repulsions.[133] The basic idea of the OMx methods is to incorporate the dominant orthogonalization corrections explicitly in the one-electron part of the Fock matrix, using parametric formulas that are inspired by power series expansions in terms of overlap for the corresponding *ab initio* terms. For the sake of computational efficiency, the smaller corrections for the two-electron part of the Fock matrix are not treated explicitly, assuming that their influence can be captured in an average manner by a parametric scaling of the one-electron corrections. Corrections for the orthogonalization effects arising from the interaction of core and valence electrons are also included in the Fock matrix.

Within the NDDO approximation, the orthogonalization-corrected two-center one-electron Fock matrix elements $h_{\mu\lambda}$ can be expressed as

$$h_{\mu\lambda}^{\rm ORT} = \beta_{\mu\lambda} - \frac{1}{2}\sum_C \sum_{\nu\in C}(S_{\mu\nu}\beta_{\nu\lambda} + \beta_{\mu\nu}S_{\nu\lambda})$$
$$+ \frac{1}{8}\sum_C \sum_{\nu\in C} S_{\mu\nu}S_{\nu\lambda}(h_{\mu\mu} + h_{\lambda\lambda} - 2h_{\nu\nu}), \tag{44}$$

where C labels the nuclear centers; μ, λ, ν label the atomic orbitals, S is the off-diagonal part of the overlap matrix (i.e. $S_{\mu\mu} = 0 \ \forall \mu$), and β is the empirical resonance integral

$$\beta_{\mu\lambda} = \frac{1}{2}(\beta_\mu^A + \beta_\lambda^B)\sqrt{R_{\rm AB}}\, e^{-(\alpha_\mu^A + \alpha_\lambda^B)R_{\rm AB}^2}, \tag{45}$$

where A and B denote nuclear centers, α_μ^A and β_μ^A are parameters specific for the atomic orbitals, and $R_{\rm AB}$ is the distance between the two centers. Equation (44) can be rewritten in the more compact form

$$h_{\mu\lambda}^{\rm ORT} = \beta_{\mu\lambda} + \sum_C V_{\mu\lambda,C}^{\rm ORT} \tag{46}$$

using the pseudopotential-like term

$$V_{\mu\lambda,C}^{\rm ORT} \equiv -\frac{1}{2}\sum_{\nu\in C}(S_{\mu\nu}\beta_{\nu\lambda} + \beta_{\mu\nu}S_{\nu\lambda}) + \frac{1}{8}\sum_{\nu\in C} S_{\mu\nu}S_{\nu\lambda}(h_{\mu\mu} + h_{\lambda\lambda} - 2h_{\nu\nu}). \tag{47}$$

In the OM2 model this expression is scaled, introducing parameters to compensate for the lack of orthogonalization corrections in the two-electron part of the Fock matrix and other approximations in the formalism. Equation (47) is then rewritten as

$$V_{\mu\lambda,C}^{\rm ORT} \equiv -\frac{1}{2}G_1^{\rm AB}\sum_{\nu\in C}(S_{\mu\nu}\beta_{\nu\lambda} + \beta_{\mu\nu}S_{\nu\lambda})$$
$$+ \frac{1}{8}G_2^{\rm AB}\sum_{\nu\in C} S_{\mu\nu}S_{\nu\lambda}(h_{\mu\mu} + h_{\lambda\lambda} - 2h_{\nu\nu}), \tag{48}$$

where $G_1^{\rm AB}$ and $G_2^{\rm AB}$ are adjustable parameters. The resulting corrected Fock matrix is

$$F_{\mu\lambda}^{\rm ORT} = h_{\mu\lambda}^{\rm ORT} + L_{\mu\lambda} \tag{49}$$

$$= \beta_{\mu\lambda} - \frac{1}{2}G_1^{AB}\sum_C\sum_{\nu\in C}(S_{\mu\nu}\beta_{\nu\lambda} + \beta_{\mu\nu}S_{\nu\lambda})$$

$$+ \frac{1}{8}G_2^{AB}\sum_C\sum_{\nu\in C} S_{\mu\nu}S_{\nu\lambda}(h_{\mu\mu} + h_{\lambda\lambda} - 2h_{\nu\nu}) + L_{\mu\lambda} \quad (50)$$

with $L_{\mu\lambda}$ denoting the two-electron part of the Fock matrix.

An additional correction is provided to take into account the effects arising from the orthogonalization of core and valence orbitals. In OM2 and OM3, the form of these effective core potentials is derived by considering the corresponding orthogonalization terms in an *ab initio* SCF treatment with a minimal basis set. The corrected Fock matrix is given by

$$\tilde{F}_{\mu\lambda}^{ORT} = F_{\mu\lambda}^{ORT} + \sum_A V_{\mu\lambda,A}^{ECP} \quad (51)$$

with the effective core potential

$$V_{\mu\lambda,A}^{ECP} = -\sum_{\alpha\in A}(S_{\mu\alpha}G_{\alpha\lambda} + G_{\mu\alpha}S_{\alpha\lambda}) - \sum_{\alpha\in A}S_{\mu\alpha}F_{\alpha\alpha}S_{\alpha\lambda}, \quad (52)$$

where $G_{\mu\alpha} \equiv F_{\mu\alpha} - S_{\mu\alpha}F_{\alpha\alpha}$. In principle the $F_{\alpha\alpha}$ and $G_{\mu\alpha}$ terms appearing in Eq. (52) could be obtained from an all-electron SCF calculation. A detailed analysis[127,128] of these terms shows, however, that $F_{\alpha\alpha}$ is almost independent of the molecular environment and can thus be treated as an adjustable atomic parameter, while the $G_{\mu\alpha}$ terms can be approximated by the resonance-integral-like expression

$$G_{\mu\alpha} = \frac{1}{2}\left(\tilde{\beta}_\mu^A + \tilde{\beta}_\alpha^B\right)\sqrt{R_{AB}}\, e^{-(\tilde{\alpha}_\mu^A + \tilde{\alpha}_\alpha^B)R_{AB}^2}. \quad (53)$$

4.3. Hybrid QM/MM

Even with the use of semiempirical electronic structure methods, it is not possible to simulate the nonadiabatic dynamics of large molecules in the condensed phase, because of the overwhelming computational cost for the quantum-chemical calculations. Fortunately, photoinduced electronic nonadiabatic transitions are often localized in a small part of the system so that it is justified to apply a hybrid quantum mechanical/molecular mechanical (QM/MM) approach,[134,135] with a partitioning of the system into two subsystems. The smaller QM part contains the chromophore and is treated at an appropriate quantum level (be it *ab initio* or semiempirical), while the larger MM part comprises the remainder of the system and is represented by a molecular mechanics force field. The coupling between the QM and MM regions is described by the QM/MM interaction Hamiltonian that includes bonded terms as well as nonbonded van-der-Waals and electrostatic interactions between QM and MM atoms. The electrostatic terms cover the interaction between QM nuclei and MM atoms, which can be computed from classical electrostatics, and the mutual polarization of the QM and MM parts. There is a hierarchy of models for handling these electrostatic terms.[136] The most common approach is electronic embedding where the polarization of the QM region due to the environment is taken into account while the polarization of the MM atoms due to the QM part is neglected. Electronic embedding thus captures the polarization of the electronic wave function caused by the presence of the MM part and guarantees the state-selectivity of QM/MM interactions. This implies the need to perform the QM calculations in the external field produced by the MM part. In practical calculations this is done by incorporating the MM point charges in the one-electron QM Hamiltonian.

5. Applications

In this section we present some examples of TSH nonadiabatic dynamics simulations.

5.1. *Methaniminium cation*

The methaniminium cation ($CH_2NH_2^+$) is the smallest possible model system to investigate the nonadiabatic dynamics of protonated Schiff bases. Due to its small size it has also often been employed as a test system for TSH calculations at different levels of accuracy.[43, 62, 64, 137] Here we report a comparison of the results of TSH simulations performed on the isolated molecule using CASSCF and OM2/MRCI, as well as the results of simulations in aqueous solution.

Figure 2 shows the dynamics of the adiabatic population decay of $CH_2NH_2^+$ as obtained at the CASSCF level using the 6-31G* basis. After vertical excitation to the S_2 state, an ultrafast $S_2 \rightarrow S_1$ population transfer occurs with a time constant of 11 fs. This is accompanied by a slower $S_1 \rightarrow S_0$ internal conversion, with a decay time of 63 fs. Typical trajectories are depicted in Fig. 3 (a) and (b). Photoexcitation is followed by

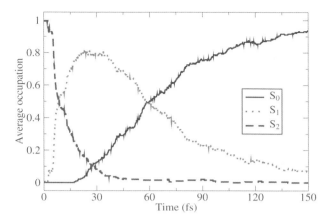

Fig. 2. Time-dependent average occupation of the low-lying adiabatic states of gas-phase $CH_2NH_2^+$ at the CASSCF level of theory.

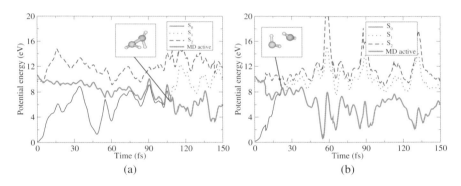

Fig. 3. Potential energies sampled by two typical trajectories during the nonadiabatic dynamics of gas-phase $CH_2NH_2^+$ (CASSCF level): (a) M-type trajectory: short C-N bond distance; (b) L-type trajectory: long C-N bond distance. The insets show the molecular geometries at the hopping points.

a sudden elongation of the C-N bond. The molecule reaches the S_2–S_1 conical intersection with a planar structure very fast. In this region the $S_2 \rightarrow S_1$ hopping occurs. Thereafter two possible decay paths are observed. For the first type of trajectories (hereafter denoted as M-type trajectories), the region of a S_1–S_0 conical intersection is reached through a torsional motion around the C-N axis eventually accompanied by minor distortions due to double bond twisting or pyramidalization of one or both end groups, without further C-N bond elongation. For the second type of trajectories (hereafter denoted as L-type trajectories), there is instead a further increase of the C-N bond distance, along with bipyramidalization, and the system accesses a different S_1–S_0 conical intersection. Both the M-type and L-type trajectories give almost equal contributions to the nonadiabatic dynamics of $CH_2NH_2^+$ in the gas phase.

Similar results are found at the OM2/MRCI level. In this case, the adiabatic population decay times are 15 and 83 fs for the S_2 and S_1 state, respectively, and the ratio between the M- and L-type trajectories is 7/3. The differences between the two approaches are thus small, suggesting that the OM2/MRCI approach can indeed provide a reliable description of the PESs and may be used to perform accurate dynamical simulations efficiently. In these gas-phase calculations of the nonadiabatic dynamics of $CH_2NH_2^+$, a single trajectory propagating to 150 fs with a time step of 0.05 fs (and a 200-times smaller time step for the integration of quantum amplitudes) typically takes about five hours at the CASSCF level of theory, compared with five minutes for OM2/MRCI (on a standard 64-bit single-CPU computer).

Solvation effects can be studied by performing TSH simulations of the $CH_2NH_2^+$ cation inside a water sphere at the QM/MM level, employing the OM2/MRCI method for the QM part (methaniminium cation) and the TIP3 model[138] to describe the surrounding MM water molecules. The computed decay times for the adiabatic populations are slightly shorter than that in the gas-phase, being 8 and 74 fs for the S_2 and S_1 state, respectively. Two types of trajectories are observed also in aqueous solution, but almost 80% of the trajectories are of M-type, since large L-type elongations of the C-N bond are rarely observed because of the steric repulsion with the surrounding water molecules.

5.2. 9H-adenine

9H-adenine (hereafter adenine) is a prototype for the study of the nonradiative decay of DNA bases. Here we present a summary of the results of TSH calculations performed at the OM2/MRCI level in the gas phase and in water[66] and a comparison with the available experimental results.

In the gas phase, at the OM2/MRCI level, the optically active state ($\pi\pi^*$ L_a state) in the Frank–Condon region is the second excited state (S_2), while the S_1 state is a dark $n\pi^*$ state. Upon photoexcitation only the S_2 state with $\pi\pi^*$ L_a character is populated. After 10–40 fs the molecule approaches the S_1–S_2 conical intersection and hops to the S_1 PES. Thereafter two possible

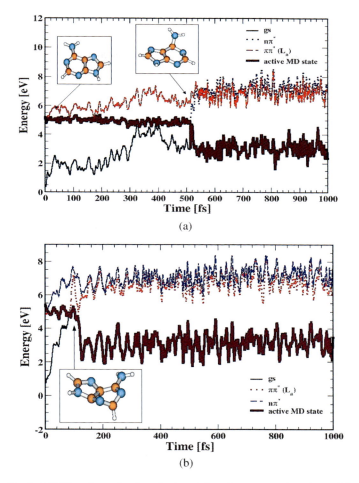

Fig. 4. Typical TSH trajectories for adenine: (a) trajectory passing the $n\pi^*/gs$ conical intersection; (b) trajectory passing the $\pi\pi^*/gs$ conical intersection. Molecular geometries at the hopping points are shown in the insets.

decay channels are observed. In about 90% of the trajectories (see Fig. 4(a) for a typical case) the S_1 state acquires $n\pi^*$ character and the system evolves towards a S_0–S_1 conical intersection of $n\pi^*/gs$ character. In the remaining trajectories (10%) the $\pi\pi^*$ character is retained after the $S_2 \to S_1$ hopping and the system quickly approaches a S_0–S_1 conical intersection of L_a/gs character (see Fig. 4(b) for a typical trajectory). The population decay dynamics can be analyzed using a bi-exponential fit. Decay times

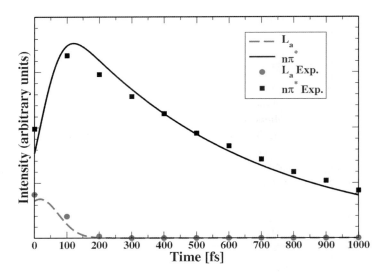

Fig. 5. Time-dependent relative populations of $\pi\pi^*$ and $n\pi^*$ excited states in adenine after photoexcitation. The calculated populations have been convoluted with a laser pulse (full width half-maximum 100 fs, centered at 0 fs) for direct comparison with the experimental data.[139]

of 15 and 560 fs are found for the L_a and $n\pi^*$ state, respectively. These results are compared in Fig. 5 with experimental results from time-resolved photoelectron spectroscopy.[139] Good agreement with experiment is found both for the decay times and for the intensity of the two decay channels.

When the effect of solvation in water is considered, by performing QM/MM simulations of an adenine molecule embedded into a sphere of MM TIP3[138] water molecules, rather different results are obtained. In the Franck–Condon region there is a strong electronic mixing between the S_1 and S_2 excited states, and after photoexcitation the two excited states are almost evenly populated. However, due to the strong nonadiabatic coupling, an ultrafast $S_2 \to S_1$ decay follows immediately, and after about 80 fs all trajectories evolve on the S_1 PES. Thereafter the nonadiabatic dynamics proceeds through two distinct channels. The vast majority of the trajectories (95%) evolves towards an S_0–S_1 conical intersection of $n\pi^*/gs$ character within 200–600 fs. A small fraction of trajectories (5%) moves, within 180 fs, towards a S_0–S_1 conical intersection of $\pi\pi^*/gs$ character. The overall population decay of the S_1 state occurs with a time constant of 410 fs, which is comparable with the published experimental decay times of 300–500 fs.[140–142]

6. Summary and Outlook

In this chapter, we have reviewed the methodology for on-the-fly surface-hopping simulations at the QM and QM/MM levels of theory. Two examples ($CH_2NH_2^+$ and adenine) have been discussed to illustrate the computational results that can be obtained for the nonadiabatic decay dynamics in the gas phase and the condensed phase. In these examples, both semiempirical OM2/GUGA-CI and *ab initio* CASSCF methods have been employed for the electronic structure calculations in the gas-phase surface-hopping simulations. For the condensed-phase dynamics in aqueous solution, OM2/GUGA-CI has been applied in a QM/MM framework using the TIP3P model of water. The simulations reveal the mechanisms of nonadiabatic decay and the role of the relevant conical intersections in the nonadiabatic dynamics of $CH_2NH_2^+$ and adenine. The results generally agree well with the available experimental data.

On-the-fly surface-hopping simulations are expected to become more popular because they provide detailed insight into the time evolution of a molecular system after photoexcitation, in particular into the nonadiabatic dynamics at conical intersections. Such simulations require a large number of trajectories to approach the proper statistical limit, and it is therefore essential to select an electronic structure method in such studies that offers a suitable compromise between accuracy and computational cost. High-level correlated *ab initio* methods are the best choice in terms of accuracy because they can in principle provide "the right answer for the right reason". In practice, the application of these methods is limited by the computational effort that rises steeply with system size, and the potentially very accurate *ab initio* MRCI methods can therefore still be used only for rather small molecules. The *ab initio* CASSCF approach remains practical for medium-size systems, but its accuracy is limited by the neglect of dynamic correlation and by the size of the active space that is affordable.

Given this situation, semiempirical methods such as OM2/GUGA-CI may offer an alternative for studying electronically excited states of larger molecules, because they may often generate reliable results (comparable to those from the high-level theories) at much lower computational costs. Such semiempirical investigations will start with a validation phase to ensure that the chosen method is able to reproduce key reference data from experiments or from high-level calculations (for example, vertical excitation energies, oscillator strengths, and the structure and energy of conical intersections). When using OM2/GUGA-CI, the optimum MRCI options will be determined in this initial phase (for example, the size of the active space and the number of reference configurations). After such validation, the OM2/GUGA-CI approach can be used with some confidence to treat the nonadiabatic dynamics of large molecules, using surface-hopping calculations at the QM level in the gas phase or at the QM/MM level in the condensed phase (solution or protein environment).

One should of course be aware of the limitations of semiempirical electronic structure methods in this context. As already noted, the established MNDO-type methods are not suitable for excited-state work without reparametrization. The more reliable OMx methods with explicit orthogonalization corrections have been parametrized only for the elements H, C, N, O and F up to now, so that they can be applied essentially only to organic molecules. All current semiempirical methods employ a minimal valence basis set and can thus not handle states with Rydberg or mixed valence/Rydberg character properly. Therefore, semiempirical surface hopping studies are restricted to low-lying valence excited states, and one needs to make sure that there are no dynamically important states with substantial Rydberg character in the region of interest. Finally, there is also the danger that the minimal basis set used in semiempirial methods may not be flexible enough to capture all the more intricate features of excited-state PESs (for example, very shallow minima that are dynamically important).

In view of these caveats, semiempirical TSH studies of excited-state dynamics should not be expected to give definitive answers. Their main purpose is an efficient initial exploration of the nonadiabatic dynamics of large molecules (either in the gas phase or in the condensed phase) to get an overview over possible reaction and decay paths and a qualitative feeling about the processes that may be important. This will generate insights and ideas that can be checked either experimentally or by higher-level calculations. On the computational side, we see such semiempirial TSH simulations as the first step to more comprehensive studies that will include higher-level calculations for improved accuracy. In this context, we adhere to a modular strategy in the development of the ChemShell software[143, 144] where all currently relevant electronic structure methods can be accessed in TSH simulations, either at the QM level (gas phase) or within a QM/MM framework (condensed phase) where the environment of the chromophore is described by a classical force field.

Although surface-hopping simulations provide a reasonable description of nonadiabatic dynamics, they do not include a number of quantum effects[31,78] that may play an important role during nonadiabatic decay processes. Therefore, it remains interesting and worthwhile to consider more rigorous semiclassical and quantum treatments to describe such phenomena.[31,78] In a similar vein, it is desirable to go beyond the results directly obtained from TSH simulations (concerning population transfer, reaction channels, and decay times) and to compute the actual observables measured experimentally, for example in the various forms of time-resolved ultrafast spectroscopy. Such a direct link between experiment and theory will improve our understanding of nonadiabatic dynamics at conical intersections.

Note Added in Proof

Since submission of this manuscript in November 2009, many papers with nonadiabatic trajectory calculations have been reported in the literature, including several OM2/GUGA-CI studies on excited-state dynamics.[145–148]

Acknowledgments

This work was supported by the Deutsche Forschungsgemeinschaft (SFB 663, project C4). We thank Thomas W. Keal for many helpful discussions and Mario R. Silva-Junior for providing the data shown in Fig. 1. Partial funding was provided by the ERC Starting Grant FP7 Project DEDOM, Grant Agreement No. 207441.

References

1. M. Born and R. Oppenheimer, *Ann. Phys.* **84**, 457 (1927).
2. J. Michl, *Top. Curr. Chem.* **46**, 1 (1974).
3. M. Klessinger and J. Michl, *Excited States and Photochemistry of Organic Molecules* (VCH Publishers, New York, 1995).
4. W. Domcke, D.R. Yarkony and H. Köppel (Eds.), *Conical Intersections: Electronic Structure, Dynamics and Spectroscopy* (World Scientific, Singapore, 2004).
5. M. Baer, *Beyond Born–Oppenheimer: Electronic Nonadiabatic Coupling Terms and Conical Intersections* (Wiley, New Jersey, 2006).
6. H. Köppel, W. Domcke and L.S. Cederbaum, *Adv. Chem. Phys.* **57**, 59 (1984).
7. W. Domcke and G. Stock, *Adv. Chem. Phys.* **100**, 1 (1997).
8. A.G. Worth and L.S. Cederbaum, *Ann. Rev. Phys. Chem.* **55**, 127 (2004).

9. M.K. Shukla and J. Leszczynski (Eds.), *Radition Induced Molecular Phenomena in Nucleic Acids* (Springer, 2008).
10. D.R. Yarkony, *Rev. Mod. Phys.* **68**, 985 (1996).
11. D.R. Yarkony, *Acc. Chem. Res.* **31**, 511 (1998).
12. D.R. Yarkony, *J. Phys. Chem. A* **105**, 6277 (2001).
13. M.J. Bearpark, M.A. Robb and H.B. Schlegel, *Chem. Phys. Lett.* **223**, 269 (1994).
14. F. Bernardi, M. Olivucci and M.A. Robb, *Chem. Soc. Rev.* **25**, 321 (1996).
15. C. Ciminelli, G. Granucci and M. Persico, *Chem. Eur. J.* **10**, 2327 (2004).
16. T.W. Keal, A. Koslowski and W. Thiel, *Theor. Chem. Acc.* **118**, 837 (2007).
17. D.R. Yarkony, *J. Chem. Phys.* **123**, 204101 (2005).
18. M.J. Paterson, M.J. Bearpark, M.A. Robb and L. Blancafort, *J. Chem. Phys.* **121**, 11562 (2004).
19. J.P. Paterson, M.A. Robb, L. Blancafort and A.D. DeBellis, *J. Phys. Chem. A* **109**, 7527 (2005).
20. I. Gomez, M. Reguero, M. Boggio-Pasqua and M.A. Robb, *J. Am. Chem. Soc.* **127**, 7119 (2005).
21. A. Kuppermann, in *Dynamics of Molecules and Chemical Reactions*, edited by R.E. Wyatt and J.Z. Zhang (Marcel Dekker, New York, 1996).
22. A. Kuppermann and R. Abrol, *Adv. Chem. Phys.* **124**, 283 (2002).
23. H.-D. Meyer, U. Manthe and L.S. Cederbaum, *Chem. Phys. Lett.* **165**, 73 (1990).
24. U. Manthe, H.-D. Meyer and L.S. Cederbaum, *J. Chem. Phys.* **97**, 3199 (1992).
25. M.H. Beck, A. Jäckle, G.A. Worth and H.-D. Meyer, *Phys. Rep.* **324**, 1 (2000).
26. G.A. Worth and M.A. Robb, *Adv. Chem. Phys.* **124**, 355 (2002).
27. H. Wang and M. Thoss, *J. Chem. Phys.* **119**, 1289 (2003).
28. H. Wang and M. Thoss, *J. Chem. Phys.* **124**, 034114 (2006).
29. U. Weiss, *Quantum Dissipative Systems*, 2nd edition (World Scientific, Singapore, 1999).
30. V. May and O. Kühn, *Charge and Energy Transfer Dynamics in Molecular Systems*, 2nd edition (Wiley-VCH, Weinheim, 2004).
31. G. Stock and M. Thoss, in *Conical Intersections: Electronic Structure, Dynamics and Spectroscopy*, edited by W. Domcke, D.R. Yarkony and H. Köppel (World Scientific, Singapore, 2004).
32. A. Bjerrre and E.E. Nikitin, *Chem. Phys. Lett.* **1**, 179 (1967).
33. J.C. Tully and R.K. Preston, *J. Chem. Phys.* **55**, 562 (1971).
34. R.K. Preston and J.C. Tully, *J. Chem. Phys.* **54**, 4297 (1971).
35. N.C. Blais and D.G. Truhlar, *J. Chem. Phys.* **79**, 1334 (1983).
36. F. Webster, P.J. Rossky and R.A. Friesner, *Comput. Phys. Commun.* **63**, 494 (1991).
37. S. Chapman, *Adv. Chem. Phys.* **82**, 423 (1992).
38. J.C. Tully, *J. Chem. Phys.* **93**, 1061 (1990).
39. S. Hammes-Schiffer and J.C. Tully, *J. Chem. Phys.* **101**, 4657 (1994).

40. M. Barbatti, M. Ruckenbauer and H. Lischka, *J. Chem. Phys.* **122**, 174307 (2005).
41. M. Barbatti, G. Granucci, M. Persico and H. Lischka, *Chem. Phys. Lett.* **401**, 276 (2005).
42. M. Barbatti, G. Granucci, M. Ruckenbauer, J. Pittner, M. Persico, H. Lischka, Newton-X: *A Package for Newtonian Dynamics Close to the Crossing Seam*, version 1.0, www.univie.ac.at/newtonx (2008).
43. M. Barbatti, G. Granucci, M. Persico, M. Ruckenbauer, M. Vazdar, M. Eckert-Maksić and H. Lischka, *J. Photochem. Photobiol. A* **190**, 228 (2007).
44. M. Barbatti, B. Sellner, A.J.A. Aquino and H. Lischka, in *Radition Induced Molecular Phenomena in Nucleic Acids*, edited by M.K. Shukla and J. Leszczynski (Springer, 2008).
45. R. Mitrić, V. Bonačić-Koutecký, J. Pittner and H. Lischka, *J. Chem. Phys.* **125**, 024303 (2006).
46. V. Bonačić-Koutecký and R. Mitrić, *Chem. Rev.* **105**, 11 (2005).
47. G. Granucci, M. Persico and A. Toniolo, *J. Chem. Phys.* **114**, 10608 (2001).
48. P. Cattaneo and M. Persico, *J. Am. Chem. Soc.* **123**, 7638 (2001).
49. G. Granucci and M. Persico, *Theor. Chem. Acc.* **117**, 1131 (2007).
50. G. Granucci and M. Persico, *J. Chem. Phys.* **126**, 134114 (2007).
51. M. Persico and G. Granucci, in *Continuum Solvation Models in Chemical Physics: From Theory to Applications*, edited by B. Mennucci and R. Cammi (Wiley, Chicester, 2007).
52. G. Groenhof, L.V. Schäfer, M. Boggio-Pasqua, M. Goette, H. Grubmüller and M.A. Robb, *J. Am. Chem. Soc.* **129**, 6812 (2007).
53. G. Groenhof, L.V. Schäfer, M. Boggio-Pasqua, H. Grubmüller and M.A. Robb, *J. Am. Chem. Soc.* **130**, 3250 (2008).
54. N.L. Doltsinis and D. Marx, *Phys. Rev. Lett.* **88**, 166402 (2002).
55. N.L. Doltsinis, in *Quantum Simulations of Complex Many-Body Systems: From Theory to Algorithms*, edited by J. Grotendorst, D. Marx and A. Muramatsu (NIC, Forschungzentrum Jülich, 2002).
56. H. Langer, N.L. Doltsinis and D. Marx, in *NIC Symposium 2004*, edited by D. Wolf, G. Münster and M. Kremer (NIC, Forschungzentrum Jülich, 2004).
57. N.L. Doltsinis, P.R.L. Markwick, H. Nieber and H. Langer, in *Radition Induced Molecular Phenomena in Nucleic Acids*, edited by M.K. Shukla and J. Leszczynski (Springer, 2008).
58. S. Grimm, C. Bräuchle and I. Frank, *ChemPhysChem* **6**, 1943 (2005).
59. I. Frank and K. Damianos, *J. Chem. Phys.* **126**, 125105 (2007).
60. I. Frank and K. Damianos, *Chem. Phys.* **343**, 347 (2008).
61. E. Tapavicza, I. Tavernelli, U. Röthlisberger, C. Filippi and M.E. Casida, *J. Chem. Phys.* **129**, 124108 (2008).
62. I. Tavernelli, E. Tapavicza and U. Röthlisberger, *J. Mol. Struct. (Theochem)* **914**, 22 (2009).

63. L.M. Frutos, T. Andruniów, F. Santoro, N. Ferré and M. Olivucci, *Proc. Nat. Acad. Sci. U.S.A.* **104**, 7764 (2007).
64. E. Fabiano, T.W. Keal and W. Thiel, *Chem. Phys.* **349**, 334 (2008).
65. E. Fabiano, G. Groenhof and W. Thiel, *Chem. Phys.* **351**, 111 (2008).
66. E. Fabiano and W. Thiel, *J. Phys. Chem. A* **112**, 6859 (2008).
67. Z. Lan, E. Fabiano and W. Thiel, *J. Phys. Chem. B* **113**, 3548 (2009).
68. Z. Lan, E. Fabiano and W. Thiel, *ChemPhysChem* **10**, 1225 (2009).
69. T.W. Keal, M. Wanko and W. Thiel, *Theor. Chem. Acc.* **123**, 145 (2009).
70. M. Thoss and H. Wang, *Ann. Rev. Phys. Chem.* **55**, 299 (2004).
71. G. Stock and M. Thoss, *Adv. Chem. Phys.* **131**, 243 (2005).
72. M. Ben-Nun and T.J. Martínez, *Adv. Chem. Phys.* **121**, 439 (2002).
73. A. Toniolo, B. Levine, A. Thompson, J. Quenneville, M. Ben-Nun, J. Owens, S. Olsen, L. Manohar and T.J. Martínez, in *Computational Methods in Organic Photochemistry*, edited by A. Kutateladze (Marcel-Dekker, New York, 2005).
74. B.G. Levine and T.J. Martínez, *Ann. Rev. Phys. Chem.* **58**, 613 (2007).
75. A.M. Virshup, C. Punwong, T.V. Pogorelov, B. Lindquist, C. Ko and T.J. Martínez, *J. Phys. Chem. B* **113**, 3280 (2009).
76. M.D. Hack, A.M. Wensmann, D.G. Truhlar, M. Ben-Nun and T.J. Martínez, *J. Chem. Phys.* **115**, 1172 (2001).
77. J. von Neumann, *Mathematical Foundations of Quantum Mechanics*, translated by R.T. Beyer (Princeton University Press, Princeton, 1983).
78. D.A. Micha and I. Burghardt (Eds.), *Quantum Dynamics of Complex Molecular Systems* (Springer Series in Chemical Physics, Vol. 83, Springer, Berlin, 2007).
79. C. Zhu, S. Nangia, A.W. Jasper and D.G. Truhlar, *J. Chem. Phys.* **121**, 7658 (2004).
80. A.W. Jasper, C. Zhu, S. Nangia and D.G. Truhlar, *Faraday Discuss.* **127**, 1 (2004).
81. C. Zhu, A.W. Jasper and D.G. Truhlar, *J. Chem. Theory Comput.* **1**, 527 (2005).
82. A.W. Jasper, S. Nangia, C. Zhu and D.G. Truhlar, *Acc. Chem. Res.* **39**, 101 (2006).
83. D.G. Truhlar, in *Quantum Dynamics of Complex Molecular Systems*, edited by D.A. Micha and I. Burghardt (Springer Series in Chemical Physics, Vol. 83, Springer, Berlin, 2007).
84. B.J. Schwartz, E.R. Bittner, O.V. Prezhdo and P.J. Rossky, *J. Chem. Phys.* **104**, 5942 (1996).
85. O.V. Prezhdo and P.J. Rossky, *J. Chem. Phys.* **107**, 5863 (1997).
86. M.J. Bedard-Hearn, R.E. Larsen and B.J. Schwartz, *J. Chem. Phys.* **123**, 234106 (2005).
87. A.W. Jasper, M.D. Hack and D.G. Truhlar, *J. Chem. Phys.* **115**, 1804 (2001).
88. A.W. Jasper, S.N. Stechmann and D.G. Truhlar, *J. Chem. Phys.* **116**, 5424 (2002).
89. A.W. Jasper and D.G. Truhlar, *Chem. Phys. Lett.* **369**, 60 (2003).
90. A.W. Jasper and D.G. Truhlar, *J. Chem. Phys.* **127**, 194306 (2007).

91. D. Bonhommeau, R. Valero, D.G. Truhlar and A.W. Jasper, *J. Chem. Phys.* **130**, 234303 (2009).
92. A.W. Jasper and D.G. Truhlar, in *Conical Intersections: Theory, Computation and Experiment*, edited by W. Domcke, D.R. Yarkony and H. Köppel (World Scientific, Singapore, 2011).
93. A. González-Lafont, T.N. Truong and D.G. Truhlar, *J. Phys. Chem.* **95**, 4618 (1991).
94. A. Bastida, C. Cruz, J. Zúñiga and A. Requena, *J. Chem. Phys.* **119**, 6489 (2003).
95. J.-Y. Fang and S. Hammes-Schiffer, *J. Phys. Chem. A* **103**, 9399 (1999).
96. U. Müller and G. Stock, *J. Chem. Phys.* **107**, 6230 (1997).
97. M. Thachuk, M.Y. Ivanov and D.M. Wardlaw, *J. Chem. Phys.* **109**, 5747 (1998).
98. W.H. Miller and T.F. George, *J. Chem. Phys.* **56**, 5637 (1972).
99. M.D. Hack and D.G. Truhlar, *J. Phys. Chem. A* **104**, 7917 (2000).
100. M.S. Topaler, T.C. Alison, D.W. Schwenke and D.G. Truhlar, *J. Chem. Phys.* **109**, 3321 (1998).
101. M.D. Hack, A.W. Jasper, Y.L. Volobuev, D.W. Schwenke and D.G. Truhlar, *J. Phys. Chem. A* **103**, 6309 (1999).
102. J.C. Tully, *Int. J. Quantum Chem. Symp.* **25**, 299 (1991).
103. M.F. Herman, *J. Chem. Phys.* **81**, 754 (1984).
104. D.F. Coker and L. Xiao, *J. Chem. Phys.* **102**, 496 (1995).
105. L. Verlet, *Phys. Rev.* **159**, 98 (1967).
106. M.A. Cuendet and W.F. van Gunsteren, *J. Chem. Phys.* **127**, 184102 (2007).
107. W.C. Swope, H.C. Andersen, P.H. Berens and K.R. Wilson, *J. Chem. Phys.* **76**, 637 (1982).
108. B.H. Lengsfield III, P. Saxe and D.R. Yarkony, *J. Chem. Phys.* **81**, 4549 (1984).
109. B.H. Lengsfield III and D.R. Yarkony, *Adv. Chem. Phys.* **82**, 1 (1992).
110. E. Wigner, *Phys. Rev.* **40**, 749 (1932).
111. H.-W. Lee, *Phys. Rep.* **259**, 147 (1995).
112. D.A. McQuarrie, *Statistical Mechanics* (Harper Collins, New York, 1976).
113. D. Chandler, *Introduction to Modern Statistical Mechanics* (Oxford University Press, New York, 1987).
114. D.J. Griffiths, *Introduction to Quantum Mechanics*, 2nd edition (Benjamin Cummings, San Francisco, 2004).
115. B.G. Levine, C. Ko, J. Quenneville and T.J. Martínez, *Mol. Phys.* **104**, 1039 (2006).
116. M. Wanko, M. Garavelli, F. Bernardi, T.A. Niehaus, T. Frauenheim and M. Elstner, *J. Chem. Phys.* **120**, 1674 (2004).
117. A. Dreuw, J.L. Weisman and M. Head-Gordon, *J. Chem. Phys.* **119**, 2943 (2003).
118. M.E. Casida, K.C. Casida and D.R. Salahub, *Int. J. Quantum Chem.* **70**, 933 (1998).
119. S. Grimme and M. Parac, *ChemPhysChem* **4**, 292 (2003).
120. B.O. Roos, *Adv. Chem. Phys.* **69**, 399 (1987).

121. K. Hirao (Ed.), *Recent Advances in Multireference Methods* (World Scientific, Singapore, 1999).
122. H. Nakano, T. Nakajima, T. Tsuneda and K. Hirao, in *Theory and Applications of Computational Chemistry: The First Forty Years*, edited by C.E. Dykstra, G. Frenking, K.S. Kim and G.E. Scuseria (Elsevier, Amsterdam, 2005).
123. H. Nakamura and D.G. Truhlar, *J. Chem. Phys.* **117**, 5576 (2002).
124. A. Devarajan, A. Gaenko, R. Lindh and P.-A. Malmqvist, *Int. J. Quantum Chem.* **109**, 1962 (2009).
125. D.G. Truhlar, J.W. Duff, N.C. Blais, J.C. Tully and B.C. Garrett, *J. Chem. Phys.* **77**, 764 (1982).
126. M. Kolb and W. Thiel, *J. Comput. Chem.* **14**, 775 (1993).
127. W. Weber, Ph.D. Thesis (Universität Zürich, Switzerland, 1996).
128. W. Weber and W. Thiel, *Theor. Chem. Acc.* **103**, 495 (2000).
129. M. Scholten, Ph.D. Thesis (Universität Düsseldorf, Germany, 2003).
130. A. Koslowski, M.E. Beck and W. Thiel, *J. Comput. Chem.* **24**, 714 (2003).
131. M.R. Silva-Junior and W. Thiel, *J. Chem. Theory Comput.* **6**, 1546 (2010).
132. S. Patchkovskii, A. Koslowski and W. Thiel, *Theor. Chem. Acc.* **114**, 84 (2005).
133. W. Thiel, *Adv. Chem. Phys.* **93**, 703 (1996).
134. H.M. Senn and W. Thiel, *Top. Curr. Chem.* **268**, 173 (2007).
135. H.M. Senn and W. Thiel, *Angew. Chem. Int. Ed.* **48**, 1198 (2009).
136. D. Bakowies and W. Thiel, *J. Phys. Chem.* **100**, 10580 (1996).
137. M. Barbatti, A.J.A. Aquino and H. Lischka, *Mol. Phys.* **104**, 1053 (2006).
138. W.L. Jorgensen, J. Chandrasekhar, J.D. Madura, R.W. Impey and M.L. Klein, *J. Chem. Phys.* **79**, 926 (1983).
139. S. Ullrich, T. Schultz, M.Z. Zgierski and A. Stolow, *J. Am. Chem. Soc.* **126**, 2262 (2004).
140. C.E. Crespo-Hernández, B. Cohen, P.M. Hare and B. Kohler, *Chem. Rev.* **104**, 1977 (2004).
141. T. Pancur, N.K. Schwalb, F. Renth and F. Temps, *Chem. Phys.* **313**, 199 (2005).
142. C.T. Middleton, K. de La Harpe, C. Su, Y.K. Law, C.E. Crespo-Hernández and B. Kohler, *Ann. Rev. Phys. Chem.* **60**, 217 (2009).
143. P. Sherwood, A.H. de Vries, M.F. Guest, G. Schreckenbach, C.R.A. Catlow, S.A. French, A.A. Sokol, S.T. Bromley, W. Thiel, A.J. Turner, S. Billeter, F. Terstegen, S. Thiel, J. Kendrick, S.C. Rogers, J. Casci, M. Watson, F. King, E. Karlsen, M. Sjøvoll, A. Fahmi, A. Schäfer and C. Lennartz, *J. Mol. Struct.* (Theochem) **632**, 1 (2003).
144. http://www.chemshell.org
145. Z. Lan, Y. Lu, E. Fabiano and W. Thiel, *ChemPhysChem* **12**, 1989 (2011).
146. Y. Lu, Z. Lan and W. Thiel, *Angew. Chem. Int. Ed.* **50**, 6864 (2011).
147. O. Weingart, Z. Lan, A. Koslowski and W. Thiel, *J. Phys. Chem. Lett.* **2**, 1506 (2011).
148. A. Kazaryan, Z. Lan, L.V. Schäfer, W. Thiel and M. Filatov, *J. Chem. Theory Comput.* **7**, 2189 (2011).

Chapter 13

Multistate Nonadiabatic Dynamics "on the Fly" in Complex Systems and Its Control by Laser Fields

Roland Mitrić[*], Jens Petersen[†] and
Vlasta Bonačić-Koutecký[†,‡]

1. Introduction 498
2. Nonadiabatic Dynamics "on the Fly" in the Framework of Time-Dependent Density Functional Theory (TDDFT) 503
   2.1. Tully's surface hopping in the framework of TDDFT 505
3. Simulation of Time-Resolved Photoelectron Spectra (TRPES) 512
   3.1. Wigner distribution approach for the simulation of time-resolved photoelectron spectra 513
   3.2. Stieltjes imaging approach for approximate description of photoionization continuum 514
4. Field-Induced Surface-Hopping Method (FISH) for Simulation and Control of Ultrafast Photodynamics 517

---

[*]Fachbereich Physik, Freie Universität Berlin, Arnimallee 14, D-14195 Berlin, Germany. E-mail: mitric@zedat.fu-berlin.de
[†]Institut für Chemie, Humboldt-Universität zu Berlin, Brook-Taylor-Straße 2, D-12489 Berlin, Germany.
[‡]E-mail: vbk@chemie.hu-berlin.de

5. Applications of the Nonadiabatic Dynamics "on the Fly" 524
   5.1. Ultrafast dynamics of photoswitching 524
   5.2. Ultrafast internal conversion through conical
        intersection in pyrazine 532
   5.3. Ultrafast photodynamics of adenine
        in microsolvated environment 540
6. Applications of the Field-Induced Surface-Hopping
   Method (FISH) for the Control of Ultrafast Dynamics
   by Shaped Laser Fields 547
   6.1. Optimal control of trans-cis isomerization
        in a molecular switch 547
   6.2. Optimal dynamic discrimination 551
7. Conclusions and Outlook 561
   Acknowledgments 562
   References 562

1. Introduction

Nonadiabatic processes involving the coupling between the nuclear and electronic motion occur typically in the regions of conical intersections and induce nonradiative transitions between electronic states that are responsible for fundamental photochemical processes such as internal conversion, isomerization, electron transfer or proton transfer.[1-4] In order to reveal the mechanisms of these processes, a joint effort in the development of experimental techniques as well as of accurate and efficient methods for the simulation of nonadiabatic dynamics and ultrafast spectroscopic observables in complex molecular systems has been undertaken in recent years.[4-11]

In particular, the essential contribution to the field of ultrafast science has been the development of techniques of femtosecond spectroscopy[4, 11-13] which allow for real-time investigation of electronic and nuclear dynamics during geometrical transformations also involving conical intersections. The common basis for all these techniques is the preparation of a coherent wavepacket by an ultrashort pump laser pulse and subsequent interrogation of its time evolution by monitoring processes such as fluorescence, resonant multiphoton ionization or photoelectron spectroscopy which are induced by a time-delayed probe pulse. This approach was pioneered by Zewail,[14-16] and one of its first applications was the observation of the nonadiabatic transition in the prototype system NaI exhibiting internal conversion between an ionic and a covalent electronic state.[15] In fact, such processes are ubiquitous in photochemistry and have been theoretically characterized in many classes of organic and bio-chromophores.[1] In particular, early theoretical work has revealed the presence of conical intersections involving biradicaloid species generated by partial breaking of double hetero bonds.[1,17] The general criteria for the occurence of conical intersections in biradicaloids have had a fundamental impact on establishing the importance of conical intersections and their characterization in different fields of photochemistry. This has stimulated extensive work on finding conical intersections in many molecular and cluster systems.[3,18–23] At the same time, the focus has shifted in recent years from the investigation of stationary electronic properties and locations of conical intersections between the potential energy surfaces towards the simulation and control of the ultrafast nonadiabatic dynamics involving also nonradiative transitions through conical intersections.[6,23–32]

The challenge for the theory in the field of ultrafast dynamics involves both the appropriate description of nonadiabatic processes due to the breakdown of the Born–Oppenheimer approximation in the vicinity of conical intersections or avoided crossings as well as the simulation of time-resolved spectroscopic signals for interesting systems taking into account all degrees of freedom. In this context, methods for the simulation of the nuclear dynamics based on classical trajectories are particularly convenient to use since they do not require the precalculation of global potential energy surfaces (PES) and can be carried out "on the fly". First-principles *ab initio* molecular dynamics (AIMD) "on the fly" has been pioneered by Car and Parrinello[33] for the ground state dynamics in the framework of the density functional method and plane-wave basis sets. The basic idea of the "on the fly" methods is to compute the forces acting on the nuclei from the electronic structure calculations only when they are needed during the propagation. This is in particular advantageous for systems which do not contain "chromophore-type" subunits and thus no separation in active and passive degrees of freedom is possible.

The conceptual framework for the development of semiclassical methods for the simulation of ultrafast spectroscopic observables is provided by the Wigner representation of quantum mechanics.[34,35] In this approach, the semiclassical limit of the Liouville–von Neumann equation for the time evolution of the vibronic density matrix has been formulated and developed for the simulation of ultrafast pump-probe spectroscopy using classical trajectories.[23,36–39] Our approach[23,37–39] is related to the Liouville space theory of nonlinear spectroscopy in the density matrix representation developed by Mukamel and his colleagues[40] and is characterized by the conceptual simplicity of classical mechanics and by the ability to approximately describe quantum phenomena such as optical transitions by means of averaging over the ensemble of classical trajectories. Moreover, the introduction of quantum corrections for the nuclear dynamics can be made in a systematic manner as recently proposed by Martens et al. in the framework of the "entangled trajectory method".[41,42] Quantum effects can be also incorporated into the nuclear dynamics in the framework of the multiple spawning method introduced by Martinez et al.[43] In general, trajectory-based methods require drastically less computational effort than full quantum mechanical calculations and provide physical insight in ultrafast processes. Additionally, they can be combined directly with quantum chemistry methods for the electronic structure calculations and allow to carry out multistate dynamics at different levels of accuracy.

In this context, one of the most efficient approaches applicable to a large variety of systems ranging from isolated molecules and clusters to complex nanostructures interacting with different environments is based on mixed quantum-classical dynamics in which the nonadiabatic transitions between electronic states occurring in the regions of strong coupling are simulated using the Tully's surface hopping (TSH) method.[44,45] This method is based on the propagation of classical trajectories in different electronic states which exhibit stochastic transitions between the states according to quantum mechanical hopping probabilities. The procedure for performing this has been designed in order to achieve consistency between the statistical fraction of trajectories in a specific state and the quantum mechanical state populations, using the minimal number of state switching events (fewest switching criterion). However, despite of its computational efficiency and conceptual simplicity the TSH approach has also a number of drawbacks such as for example the internal inconsistency which arises due to classically forbidden transitions and due to the divergence of independent trajectories, as analyzed in the literature.[46,47] The necessary ingredients to carry out TSH simulations are forces in ground and excited electronic states as well as nonadiabatic couplings. These can be calculated using the whole spectrum of methods such as *ab initio* "frozen ionic bond" approximation,[23] *ab initio* configuration interaction (CI),[24] restricted open-shell Kohn–Sham density functional theory (DFT),[28] linear response time-dependent density functional theory (TDDFT)[25–27,29,48–50] as well as semiempirical methods for the electronic structure[51–54] and can be employed in the framework of the TSH simulations. In addition, recently the applications have been extended to the mixed quantum mechanical–molecular mechanical (QM/MM) methods allowing to treat complex systems such as photoactive proteins[55–59] or chromophores interacting with the environment.[60]

An additional important aspect is the introduction of electric fields into molecular dynamics. This opens the perspective for controlling molecular processes by shaped laser pulses and allows for new applications in which the light is used as photonic catalyst in chemical reactions.[61,62] The idea to control the selectivity of product formation in a chemical reaction using ultrashort pulses employing either the proper choice of their phase or of time duration and the delay between the pump and the probe (or dump) step is based on exploitation of the coherence properties of laser radiation due to quantum mechanical interference effects. Early conceptual work by Tannor and Rice,[63,64] by Brumer and Shapiro[65–67] followed by variational optimization of electric fields[68,69] opened further application aspects.[69–74] Technological progress due to fs-pulse shapers allowed manipulation of ultrashort laser pulses.[75–79] Finally, closed-loop learning control (CLL) was introduced by Judson and Rabitz[80] and first realized experimentally by Gerber et al.,[81] Wöste et al. and others,[11] opening the possibility to apply optimal control to more complex systems. Since the potential energy surfaces (PES) of multidimensional systems are complicated and mostly not available, the idea was to combine a fs-laser system with a computer-controlled pulse shaper to produce specific fields acting on the system by initiating photochemical processes. After detection of the product, the learning algorithm is used to modify the field based on information obtained from the experiment and from the desired target. The shaped pulses are tested and improved iteratively until the optimal field for the chosen target is reached. Such a black-box procedure is extremely efficient but it does not provide information about the nature of the underlying processes which are responsible for the requested outcome.

Since tailored laser pulses have the ability to select pathways on those parts of the energy surfaces which optimally lead to the chosen target, their analysis should allow to determine the mechanism of the processes and at the end provide information about important parts of the PES. Therefore, developments of theoretical methods are needed which allow for the design of interpretable laser pulses for complex systems by establishing the connection between the underlying dynamics and the pulse shapes as well as between theoretically and experimentally optimized pulses. Until recently, the limitation was imposed by difficulties in precalculating multidimensional PES. To avoid this obstacle, *ab initio* adiabatic and in particular nonadiabatic MD "on the fly" without precalculation of the ground and the excited energy surfaces is particularly suitable provided that an accurate description of the electronic structure is feasible and practicable.[6] In addition, this approach offers the advantage that the MD "on the fly" can be applied to relatively complex systems and moreover it can be directly connected with different procedures for optimal control.[6,82,83] In the context with conical intersections the question can be raised whether it might be advantageous or not to choose pathways avoiding conical intersections by laser control in photochemical reactions. Moreover, as recently proposed by us, it is particularly convenient to introduce the field directly in the nonadiabatic dynamics "on the fly" which can be then optimized as desired.[84]

In this chapter we aim to present the development of theoretical methods for the simulation of nonadiabatic dynamics and its manipulation by laser fields in complex systems accounting for all degrees of freedom. Therefore, we will first describe nonadiabatic dynamics "on the fly" in the frame of time-dependent density functional theory (TDDFT) and its approximate tight-binding version (TDDFTB) in Sec. 2. Furthermore, we will briefly outline the procedure for the simulation of time-resolved photoelectron spectra based on the nonadiabatic dynamics "on the fly" in Sec. 3. Then, in Sec. 4 we will introduce the field-induced surface hopping method (FISH) which is based on the combination of quantum electronic state population dynamics with classical nuclear dynamics. For the propagation of classical trajectories in the frame of the FISH method, the whole spectrum of quantum chemical methods can be employed opening the possibility of broad applications for simulation of spectroscopic observables as well as to control dynamics employing shaped laser fields.

The applications of the above-mentioned theoretical approaches will be illustrated on a number of examples with the aim to show the scope and reliability the of methods with the focus on time-dependent density functional theory (TDDFT) nonadiabatic dynamics in Sec. 5 and field-induced surface hopping method (FISH) in Sec. 6. The choice of the presented examples serves to emphasize the importance of mutual interaction between theory and experiments. First, we use nonadiabatic dynamics to investigate ultrafast photoswitching in the prototype molecule benzylideneaniline for which the lifetimes of excited states were determined and the simulation of time-resolved photoelectron spectra was carried out, which allowed to identify time-resolved photoisomerization.[26] Furthermore, the simulation of time-resolved photoelectron spectra (TRPES) with Stieltjes imaging including the approximate description of the photoionization continuum is introduced, demonstrated for the ultrafast internal conversion in pyrazine and compared to the recent experimental TRPES measurements.[27,85] As an illustration of the applicability of our nonadiabatic dynamics in the framework of the time-dependent density functional tight-binding method (TDDFTB) to larger systems with biological relevance, we have chosen to present the ultrafast nonradiative relaxation of the microsolvated DNA base adenine.[86]

The scope of our FISH method[84] will be illustrated on two examples which are representative for different application areas. First we show that optimal pump-dump control can be used to efficiently drive the selective photoisomerization on the prototype Schiff base N-methylethaniminium through pathways avoiding the conical intersection[84] as an example for the broad application area in the photochemistry of biological molecules in different environments. Second, we wish to show that our FISH method can be used to reveal fundamental dynamical processes responsible for optimal dynamic discrimination (ODD)[87,88] between the two molecular species flavin mononucleotide (FMN) and riboflavin (RBF), which have almost identical spectroscopical features.[89,90] The selective identification of target molecules in the presence of a structurally and spectroscopically similar background using optimally shaped laser fields opens prospects for new applications in multiple areas of science and engineering. Finally, conclusions and outlook will be given in Sec. 7.

2. Nonadiabatic Dynamics "on the Fly" in the Framework of Time-Dependent Density Functional Theory (TDDFT)

The time-dependent density functional theory (TDDFT) represents an efficient generally applicable method for the treatment of the optical properties in complex systems whose performance and accuracy have been steadily improved.[91,92] Due to this fact, a significant effort has been recently invested to extend the applicability of TDDFT to the simulation of ultrafast nonadiabatic processes in complex molecular systems. In this context, a variety of approaches for performing nonadiabatic dynamics simulations "on the fly" have been developed and successfully applied in recent years.[25, 26, 28, 29, 48–50] Being aware of the drawbacks of the state-of-the-art density functionals concerning failure in the description of long range charge transfer transitions, dispersion interaction and multireference character, TDDFT is still one of the most practical means to address a large class of problems if the proper choice of the system is made. Moreover, recent developments of new hybrid functionals promise to substantially improve the description of long-range charge transfer transitions.[93–97] In connection with nonadiabatic processes, the ability of linear response TDDFT to describe conical intersections between excited states and the ground state has been critically examined in the literature.[49] The conclusion has been made that while the topology of the S_1-S_0 crossing region may be not exact, this does not substantially influence the relaxation pathways and photochemistry of the studied examples. Successful applications of TDDFT nonadiabatic dynamics steadily grow and have already significantly contributed towards understanding of the mechanisms of photochemical processes in complex systems[25–27, 49, 98] and have also been verified by comparison with experimental data.[27, 85]

Another attractive direction is the combination of the nonadiabatic dynamics with the approximate tight-binding density functional theory (DFTB)[99–102] which, due to its computational efficiency, is suitable for the treatment of large biological or supramolecular assemblies. In particular, the DFTB and its time-dependent version (TDDFTB) have been shown to provide a quite accurate description of both ground state[99–102] and excited state properties.[103, 104] Furthermore, recent implementation of analytic energy derivatives[105] and nonadiabatic couplings[86] in the framework of TDDFTB have made possible the extension to nonadiabatic dynamics "on the fly".[86] It should however be noted that while TDDFTB reproduces in many cases the accuracy of the TDDFT method it also shares all its deficiencies. The methodological developments sketched above should allow to investigate fundamental photochemical processes in complex molecular systems such as biomolecules interacting with e.g. solvents, surfaces, metallic nanostructures or protein environments.

In Sec. 2.1 we present our formulation of the nonadiabatic dynamics in the framework of TDDFT using localized Gaussian basis sets combined with Tully's surface hopping (TSH) method[44] in which the nonadiabatic couplings are calculated "on the fly". Furthermore, we also present the extension of the TDDFT nonadiabatic dynamics to the approximate TDDFTB method.

2.1. Tully's surface hopping in the framework of TDDFT

The simulation of nonadiabatic processes in the framework of the TSH procedure[44] relies on the propagation of ensembles of classical trajectories parallel with the solution of the time-dependent Schrödinger equation which determines the quantum-mechanical electronic state populations. For this purpose, along each classical trajectory an electronic wavefunction $|\Psi(\mathbf{r};\mathbf{R}(t))\rangle$ is defined and represented in terms of the adiabatic electronic states according to:

$$|\Psi(\mathbf{r};\mathbf{R}(t))\rangle = \sum_K C_K(t) |\Psi_K(\mathbf{r};\mathbf{R}(t))\rangle, \qquad (1)$$

where $|\Psi_K(\mathbf{r};\mathbf{R}(t))\rangle$ represents the wavefunction for the electronic state K while the $C_K(t)$ are the time-dependent expansion coefficients. The time evolution of the expansion coefficients for a given trajectory can be obtained by numerical solution of the time-dependent Schrödinger equation:

$$i\hbar \frac{dC_K(t)}{dt} = \sum_I C_I(t) \left(\langle \Psi_K | \hat{H}_{el} | \Psi_I \rangle - i\hbar \left\langle \Psi_K(\mathbf{r};\mathbf{R}(t)) \middle| \frac{\partial \Psi_I(\mathbf{r};\mathbf{R}(t))}{\partial t} \right\rangle \right). \qquad (2)$$

Choosing the adiabatic representation for the electronic states, the first term in brackets reduces to $E_K \cdot \delta_{KI}$ while the second term corresponds to the nonadiabatic coupling D_{KI} between the states I and K, which can be approximately calculated using the finite difference approximation for the time derivative[106]:

$$D_{KI}\left(\mathbf{R}\left(t+\frac{\Delta}{2}\right)\right) \approx \frac{1}{2\Delta}(\langle \Psi_K(\mathbf{r};\mathbf{R}(t)) | \Psi_I(\mathbf{r};\mathbf{R}(t+\Delta))\rangle$$
$$- \langle \Psi_K(\mathbf{r};\mathbf{R}(t+\Delta)) | \Psi_I(\mathbf{r};\mathbf{R}(t))\rangle), \qquad (3)$$

where Δ is the timestep used for the integration of the classical Newton's equations of motion. Since the nonadiabatic coupling D_{KI} is calculated only at the midpoint $t + \Delta/2$ between two nuclear timesteps [cf. Eq. (3)], the $D_{KI}(\tau)$ are obtained by linear interpolation in the interval $[t, t + \Delta/2]$ and extrapolation in the interval $[t + \Delta/2, t + \Delta]$.

The numerical solution of the Eq. (2), obtained e.g. using the fourth order Runge–Kutta procedure, provides the time-dependent electronic state coefficients $C_K(t)$ which can be used to define the hopping probabilities $P_{I \to K}$ that are needed for the electronic state switching procedure in the frame of the TSH approach. In order to increase the efficiency of the integration the rapidly oscillating part of the $C_K(\tau)$ can be eliminated by transforming these coefficients to the interaction representation. Therefore, only the slowly varying component of the $C_K(\tau)$ coefficients remains to be calculated, for which much larger timesteps can be used for the numerical integration, thus decreasing the computational demand. The hopping probabilities $P_{I \to K}$ can be either calculated after each nuclear dynamics time step Δ or, alternatively, after each small time step $\Delta\tau$ used for the integration of the electronic Schrödinger equation Eq. (2), as recently introduced by us.[26]

In the latter case, the hopping probability $P_{I \to K}$ is defined as:

$$P_{I \to K}(\tau) = -2 \frac{\Delta\tau [Re(C_K^*(\tau) C_I(\tau) D_{KI}(\tau))]}{C_I(\tau) C_I^*(\tau)}. \qquad (4)$$

It should be emphasized that our procedure is numerically more stable. The instabilities in other procedures can arise due to the fact that calculation of the hopping probabilities only in each nuclear time step Δ can lead to unphysical values greater than 1 if Δ is not sufficiently small. This occurs in particular in the region of strong coupling in which the change of the coefficients $C_K(t)$ can become very large within one nuclear time step. While the additional computational effort for determining the hopping probabilities at each time step during the integration [cf. Eq. (4)] is marginal, our approach allows to use much larger time steps for the nuclear dynamics, thus decreasing computational demand considerably, and therefore might be advantageous.

An alternative procedure for calculating the hopping probabilities can be also formulated based only on the occupations of the electronic states represented by diagonal density matrix elements $\rho_{II} = C_I^*(t) C_I(t)$ and $\rho_{KK} = C_K^*(t) C_K(t)$. The idea of this approach is to calculate separately the probability that the state I is depopulated and the state K is populated. The probability of depopulation of the state I is given by the rate of change of ρ_{II} and can be defined as

$$P_{I,\text{depopulation}} = \Theta(-\dot{\rho}_{II}) \frac{-\dot{\rho}_{II}}{\rho_{II}} \Delta. \qquad (5)$$

This probability is defined to be nonzero only if the population of the I th state is decreasing which is represented by the Θ function in Eq. (5). The rate of change of $\dot{\rho}_{II}$ can be calculated from the populations in successive nuclear time steps t and $t + \Delta$ using the finite difference approximation for the derivative. In the case that the population of the state I is decreasing, the probability to populate the particular state K can by defined as

$$P_{K,\text{populated}} = \frac{\Theta(\dot{\rho}_{KK})\dot{\rho}_{KK}}{\sum_L \Theta(\dot{\rho}_{LL})\dot{\rho}_{LL}}. \tag{6}$$

In this way, only those states can be populated whose population is growing between two successive nuclear time steps and the probability is defined by the rate of change of the population of the state K normalized to the total rate of change of the populations of all other states whose population is also growing [denominator in Eq. (6)]. Thus the total probability for hopping from state I to state K can be defined as:

$$\begin{aligned}P_{I\to K} &= P_{I,\text{depopulated}} P_{K,\text{populated}} \\ &= \Theta(-\dot{\rho}_{II})\Theta(\dot{\rho}_{KK})\frac{-\dot{\rho}_{II}}{\rho_{II}}\frac{\dot{\rho}_{KK}}{\sum_L \Theta(\dot{\rho}_{LL})\dot{\rho}_{LL}}\Delta.\end{aligned} \tag{7}$$

It should be pointed out that this equation is valid for any number of electronic states and requires only the calculation of the hopping probabilities at each nuclear time step since populations generally vary more slowly than the coherences which are employed in Eq. (4). This is particularly useful in the context of multistate dynamics during which the laser field is varying very fast as it will be shown in Sec. 6.2.

The necessary ingredients for carrying out TSH simulations are the forces (energy gradients) in the ground and excited electronic states as well as the nonadiabatic couplings $D_{KI}(\mathbf{R}(t+\frac{\Delta}{2}))$. While the calculation of excited state forces in the framework of TDDFT is already a standard procedure available in many commonly used quantum chemical program packages, the procedure for the calculation of nonadiabatic couplings in the framework of linear response TDDFT has been developed only recently using plane wave basis sets,[48–50] as well as using localized Gaussian basis sets.[25,26] Thus, in the following after introducing the representation of the electronic wave function within the Kohn–Sham linear response method, we briefly outline our approach for the calculation of the nonadiabatic couplings using localized Gaussian basis sets.

2.1.1. *Representation of the electronic wavefunction within the Kohn–Sham linear response method*

In order to calculate nonadiabatic couplings in the framework of the TDDFT method a representation of the wavefunction based on Kohn–Sham orbitals is required. Since in the linear response TDDFT method the time-dependent electron density contains only contributions from single excitations from the manifold of occupied KS orbitals to virtual KS orbitals, a natural ansatz for the excited state electronic wavefunction is the CIS-like expansion:

$$|\Psi_K(\mathbf{r};\mathbf{R}(t))\rangle = \sum_{i,a} c_{i,a}^K |\Phi_{i,a}^{CSF}(\mathbf{r};\mathbf{R}(t))\rangle, \qquad (8)$$

where $|\Phi_{i,a}^{CSF}(\mathbf{r};\mathbf{R}(t))\rangle$ represents a singlet spin adapted configuration state function (CSF) defined as:

$$|\Phi_{i,a}^{CSF}(\mathbf{r};\mathbf{R}(t))\rangle = \frac{1}{\sqrt{2}}(|\Phi_{i\alpha}^{a\beta}(\mathbf{r};\mathbf{R}(t))\rangle + |\Phi_{i\beta}^{a\alpha}(\mathbf{r};\mathbf{R}(t))\rangle), \qquad (9)$$

and $|\Phi_{i\alpha}^{a\beta}(\mathbf{r};\mathbf{R}(t))\rangle$ and $|\Phi_{i\beta}^{a\alpha}(\mathbf{r};\mathbf{R}(t))\rangle$ are Slater determinants in which one electron has been promoted from the occupied orbital i to the virtual orbital a with spin α or β, respectively. Notice, that while DFT can rigorously provide the wavefunction only for the lowest state of each symmetry, the approximate ansatz presented in Eq. (8) gives rise to the wavefunctions for an arbitrary number of excited states of any symmetry and represents a practicable way of defining an excited state wavefunction based on linear response TDDFT. This ansatz can be used to calculate the nonadiabatic coupling as described below, but more generally, it can provide the expectation values of any observable of interest. In particular, we will show in Sec. 3.2 how this approximate wavefunction can be used to calculate transition dipole moments between excited electronic states in the framework of TDDFT. In the context of nonadiabatic dynamics the accuracy of this representation of the wavefunction has been previously demonstrated in our work on pyrazine[25, 27] and benzylideneaniline.[26]

The expansion coefficients $c_{i,a}^K$ in Eq. (8) are determined on physical grounds by requiring that the wavefunction in Eq. (8) gives rise to the same density response as the one obtained by the linear response TDDFT procedure. The latter can be expressed as:

$$\rho(\mathbf{r},t) = \rho_0 + \sum_i (\phi_i^* \delta\phi_i + \phi_i \delta\phi_i^*), \qquad (10)$$

where ρ_0 represents the unperturbed ground state density, ϕ_i is an occupied Kohn–Sham orbital and $\delta\phi_i$ is the linear response of the KS orbital which can be further decomposed in terms of positive and negative frequency components as:

$$\delta\phi_i = \phi_i^+ e^{-i\omega_K t} + \phi_i^- e^{i\omega_K t}, \qquad (11)$$

where ω_K represents the transition frequency of the K'th excited state. Expanding the response orbitals ϕ_i^+ and ϕ_i^- in terms of virtual KS orbitals:

$$\phi_i^+ = \sum_a X_{ia}\phi_a \quad \text{and} \quad \phi_i^- = \sum_a Y_{ia}\phi_a, \qquad (12)$$

the time-dependent electron density response can be formulated as:

$$\rho(\mathbf{r},t) = \rho_0 + \sum_{ia}(X_{ia} + Y_{ia})\phi_i\phi_a^* e^{-i\omega_K t} + c.c., \qquad (13)$$

where $\mathbf{X}$ and $\mathbf{Y}$ represent the solution of the TDDFT eigenvalue problem.[107]

In the wavefunction picture, the time-dependent electron density response in Eq. (13) arises physically as a consequence of the coherent superposition of the ground and excited electronic state ($|\Psi_0\rangle$ and $|\Psi_K\rangle$) described by $|\Psi(t)\rangle = a|\Psi_0\rangle e^{-iE_0 t/\hbar} + b|\Psi_K\rangle e^{-iE_K t/\hbar}$. This wavefunction gives rise to the time-dependent electron density given by:

$$\rho(\mathbf{r},t) = |a|^2\langle\Psi_0|\hat{\rho}|\Psi_0\rangle + |b|^2\langle\Psi_K|\hat{\rho}|\Psi_K\rangle$$
$$+ a^*b\sum_{ia} c_{ia}^{K*}\phi_i(\mathbf{r})\phi_a^*(\mathbf{r})e^{-i\omega_K t} + c.c., \qquad (14)$$

where in the third term the Eq. (8) for $|\Psi_K\rangle$ has been used. From the expression (14) one can see that the density $\rho(\mathbf{r},t)$ consists of a time independent part which is just the sum of the ground and excited state densities and a time-dependent part ($e^{-i\omega_K t}$) which oscillates with the amplitudes determined by the CI coefficients c_{ia}^K. Assuming that in the linear response regime the ground state density is only slightly perturbed ($|a|^2 \approx 1 \gg |b|^2$), Eq. (14) can be further simplified to:

$$\rho(\mathbf{r},t) = \rho_0 + a^*b\sum_{ia} c_{ia}^{K*}\phi_i(\mathbf{r})\phi_a^*(\mathbf{r})e^{-i\omega_K t} + c.c.. \qquad (15)$$

By direct comparison of Eqs. (13) and (15) it can be seen that the coefficients $c_{i,a}^K$ can be taken to be proportional to $X_{ia}+Y_{ia}$ up to a normalization constant. Due to the fact that $\mathbf{X}$ and $\mathbf{Y}$ are solutions of a non-Hermitean eigenvalue problem they do not form an orthonormal set. Therefore, in order to obtain orthogonal electronic states, the non-Hermitean eigenvectors have to be transformed according to:

$$\mathbf{C} = (\mathbf{A} - \mathbf{B})^{-1/2}(\mathbf{X} + \mathbf{Y}), \tag{16}$$

where $\mathbf{A}$ and $\mathbf{B}$ are standard TDDFT matrices.[91] For non-hybrid functionals without exact exchange the coefficients $c_{i,a}^K$ giving rise to mutually orthogonal electronic states are given by:

$$c_{i,a}^K = (\epsilon_a - \epsilon_i)^{-1/2}(X_{ia} + Y_{ia}). \tag{17}$$

This allows to define the electronic wavefunction $|\Psi_K(\mathbf{r}; \mathbf{R}(t))\rangle$ which will be employed to calculate the nonadiabatic couplings.

2.1.2. *Nonadiabatic coupling in the framework of TDDFT and TDDFTB with localized basis sets*

The electronic structure of isolated molecular systems is most naturally described by using Gaussian type atomic orbitals (AO's) in contrast to plane waves, which represent the natural choice in extended periodic systems. In the latter case the nonadiabatic couplings in the framework of the TDDFT method have been formulated and implemented by Rothlisberger *et al.*[48,49] Recently, also a method for calculation of the nonadiabatic coupling vectors was introduced in this context by Rothlisberger *et al.*[50] Here we present the approach for the calculation of the nonadiabatic couplings using KS orbitals expanded in terms of localized Gaussian atomic basis sets. This formulation is particularly suitable since it can be combined with commonly used quantum chemical DFT codes.

In order to calculate the nonadiabatic couplings according to the discrete approximation given by Eq. (3) the overlap between two CI wavefunctions at times t and $t + \Delta$ along the nuclear trajectory $\mathbf{R}(t)$ is needed:

$$\langle \Psi_K(\mathbf{r}; \mathbf{R}(t)) | \Psi_I(\mathbf{r}; \mathbf{R}(t + \Delta)) \rangle$$
$$= \sum_{ia} \sum_{i'a'} c_{i,a}^{*K} c_{i',a'}^{I} \langle \Phi_{i,a}^{CSF}(\mathbf{r}; \mathbf{R}(t)) | \Phi_{i',a'}^{CSF}(\mathbf{r}; \mathbf{R}(t + \Delta)) \rangle. \tag{18}$$

The overlap of the CSF's in Eq. (18) can be expressed in terms of singly excited Slater determinants using Eq. (9), which can be further reduced to the overlap of spatial KS orbitals ϕ_i:

$$\langle \Phi_{i,a}^{CSF}(\mathbf{r}; \mathbf{R}(t)) | \Phi_{i',a'}^{CSF}(\mathbf{r}; \mathbf{R}(t+\Delta)) \rangle$$

$$= \begin{bmatrix} \langle \phi_1 | \phi_1' \rangle & \cdots & \langle \phi_1 | \underline{\phi_{i'}'} \rangle & \cdots & \langle \phi_1 | \phi_{n'}' \rangle \\ \vdots & & \vdots & & \vdots \\ \langle \underline{\phi_i} | \phi_1' \rangle & \cdots & \langle \underline{\phi_i} | \underline{\phi_{i'}'} | \rangle & \cdots & \langle \underline{\phi_i} | \phi_{n'}' \rangle \\ \vdots & & \vdots & & \vdots \\ \langle \phi_n | \phi_1' \rangle & \cdots & \langle \phi_n | \underline{\phi_{i'}'} \rangle & \cdots & \langle \phi_n | \phi_{n'}' \rangle \end{bmatrix}$$

$$\times \begin{bmatrix} \langle \phi_1 | \phi_1' \rangle & \cdots & \langle \phi_1 | \underline{\phi_{a'}'} \rangle & \cdots & \langle \phi_1 | \phi_{n'}' \rangle \\ \vdots & & \vdots & & \vdots \\ \langle \underline{\phi_a} | \phi_1' \rangle & \cdots & \langle \underline{\phi_a} | \underline{\phi_{a'}'} \rangle & \cdots & \langle \underline{\phi_a} | \phi_{n'}' \rangle \\ \vdots & & \vdots & & \vdots \\ \langle \phi_n | \phi_1' \rangle & \cdots & \langle \phi_n | \underline{\phi_{a'}'} \rangle & \cdots & \langle \phi_n | \phi_{n'}' \rangle \end{bmatrix}$$

$$+ \begin{bmatrix} \langle \phi_1 | \phi_1' \rangle & \cdots & \langle \phi_1 | \underline{\phi_{a'}'} \rangle & \cdots & \langle \phi_1 | \phi_{n'}' \rangle \\ \vdots & & \vdots & & \vdots \\ \langle \underline{\phi_i} | \phi_1' \rangle & \cdots & \langle \underline{\phi_i} | \underline{\phi_{a'}'} \rangle & \cdots & \langle \underline{\phi_i} | \phi_{n'}' \rangle \\ \vdots & & \vdots & & \vdots \\ \langle \phi_n | \phi_1' \rangle & \cdots & \langle \phi_n | \underline{\phi_{a'}'} \rangle & \cdots & \langle \phi_n | \phi_{n'}' \rangle \end{bmatrix}$$

$$\times \begin{bmatrix} \langle \phi_1 | \phi_1' \rangle & \cdots & \langle \phi_1 | \underline{\phi_{i'}'} \rangle & \cdots & \langle \phi_1 | \phi_{n'}' \rangle \\ \vdots & & \vdots & & \vdots \\ \langle \underline{\phi_a} | \phi_1' \rangle & \cdots & \langle \underline{\phi_a} | \underline{\phi_{i'}'} \rangle & \cdots & \langle \underline{\phi_a} | \phi_{n'}' \rangle \\ \vdots & & \vdots & & \vdots \\ \langle \phi_n | \phi_1' \rangle & \cdots & \langle \phi_n | \underline{\phi_{i'}'} \rangle & \cdots & \langle \phi_n | \phi_{n'}' \rangle \end{bmatrix}, \quad (19)$$

where $\langle \phi_i | \phi_{i'}' \rangle$ are the overlap integrals between two spatial KS orbitals $\phi_i(t)$ and $\phi_{i'}'(t+\Delta)$ at time steps t and $t+\Delta$. The underlined orbitals label the replacement of an occupied orbital by a virtual orbital such as $i \to a$ and $i' \to a'$ at t and $t+\Delta$, respectively. The spatial KS orbitals can be expressed in terms of atomic basis functions according to:

$$\langle \phi_i(t) | = \sum_{k=1}^{n} c_{ik}(t) \langle b_k(\mathbf{R}(t)) |, \quad (20)$$

$$|\phi'_j(t+\Delta)\rangle = \sum_{m=1}^{n} c'_{jm}(t+\Delta)|b'_m(\mathbf{R}(t+\Delta))\rangle, \qquad (21)$$

with the Gaussian basis functions $b_k(\mathbf{R}(t))$ and $b'_m(\mathbf{R}(t+\Delta))$ and the molecular orbital (MO) coefficients $c_{ik}(t)$ and $c'_{jm}(t+\Delta)$, respectively. Notice, that the two sets of functions $b_k(\mathbf{R}(t))$ and $b'_m(\mathbf{R}(t+\Delta))$ are centered at different positions $\mathbf{R}(t)$ and $\mathbf{R}(t+\Delta)$ and therefore do not form an orthonormal basis set. This leads to the final expression for the overlap integral of two spatial KS orbitals at times t and $t+\Delta$:

$$\langle \phi_i(t) | \phi'_{j'}(t+\Delta) \rangle = \sum_{k=1}^{n}\sum_{m=1}^{n} c_{ik}(t) c'_{jm}(t+\Delta) \langle b_k(\mathbf{R}(t)) | b'_m(\mathbf{R}(t+\Delta)) \rangle. \qquad (22)$$

Therefore, in order to calculate nonadiabatic couplings along each classical trajectory the overlap integrals between moving basis functions are calculated at successive nuclear time steps and the KS MO coefficients and linear response eigenvectors are utilized in order to transform the overlap integrals. In order to eliminate possible random phase variations of the nonadiabatic coupling, the phases of the CI-like wavefunction coefficients [cf. Eq. (8)] and of the Kohn–Sham orbital coefficients [cf. Eq. (20)] in each nuclear timestep are aligned to the phases of the previous one. While the calculation of the coupling matrix elements is formally analogous for both TDDFT and TDDFTB, the computational advantage of the TDDFTB method is that the integrals involved are not explicitly calculated but can, as usual in the DFTB procedure, be used in a tabulated form.[86]

3. Simulation of Time-Resolved Photoelectron Spectra (TRPES)

As mentioned in the Introduction, the time-resolved photoelectron spectroscopy represents a powerful approach for interrogation of nonadiabatic processes. The basic principle of the time-resolved spectroscopy involves the creation of a coherent superposition of the ground and excited electronic states of the studied system by an ultrashort laser pulse. This creates a wavepacket in the excited electronic states of the system whose time evolution is subsequently probed by the photoionization due to a time-delayed ultrashort probe pulse. The distribution of the kinetic energy of the photoelectrons created upon photoionization reflects therefore the composition of the electronic state which has been ionized. Since during the excited state dynamics the character of the electronic state can change, e.g. due to the passage through a conical intersection, the observables such as photoelectron kinetic energies (PKE) or photoelectron angular distribution offer a sensitive probe for the nonadiabatic transitions.[108]

In Sec. 3.1 we will shortly outline our approach for the simulation of TRPE spectra based on the Wigner distribution approach[6,37] which relies on the propagation of an ensemble of classical trajectories "on the fly". This approach provides a general tool for simulation of ultrafast processes and femtosecond signals in complex systems, involving both adiabatic and nonadiabatic dynamics.[6] Subsequently, in Sec. 3.2 we will introduce the approximate description of the ionization probability to the continuum states in the framework of the Stieltjes imaging procedure employing transition dipole moments between excited electronic states, which will be here formulated in the framework of TDDFT that is used for nonadiabatic dynamics "on the fly".[27]

3.1. *Wigner distribution approach for the simulation of time-resolved photoelectron spectra*

The TRPE spectra can be simulated in the framework of the Wigner distribution approach[6] by a modification of the equation for the zero kinetic energy pump-probe signals[6] which takes into account that a part of the probe-pulse energy E_{pr} changes into the kinetic energy of the electrons E

$$S(t_d, E) \sim \int\int d\mathbf{q}_0 d\mathbf{p}_0 \int_0^\infty d\tau_1 \exp\left\{-\frac{(\tau_1 - t_d)^2}{\sigma_{pu}^2 + \sigma_{pr}^2}\right\}$$

$$\times \exp\left\{-\frac{\sigma_{pr}^2}{\hbar^2}[E_{pr} - V_{21}(\mathbf{q}_1(\tau_1; \mathbf{q}_0, \mathbf{p}_0)) - E]^2\right\}$$

$$\times \exp\left\{-\frac{\sigma_{pu}^2}{\hbar^2}[E_{pu} - V_{10}(\mathbf{q}_0)]^2\right\} P_{00}(\mathbf{q}_0, \mathbf{p}_0). \quad (23)$$

In this expression σ_{pu} (σ_{pr}) and $E_{pu} = \hbar\omega_{pu}$ ($E_{pr} = \hbar\omega_{pr}$) are the pulse durations and excitation energies for the pump and probe step with time delay t_d. $V_{21}(\mathbf{q}_1(\tau_1; \mathbf{q}_0, \mathbf{p}_0))$ labels the time-dependent energy gap between the electronic state in which the dynamics takes place and

the electronic state that is used for probing, both obtained from the
ab initio MD "on the fly".[6] The initial coordinates and momenta $\mathbf{q}_0$ and
$\mathbf{p}_0$ needed for the dynamics simulation can be sampled from a canonical
Wigner distribution at the given temperature including all normal modes
according to:

$$P_{00}(\mathbf{q}_0, \mathbf{p}_0) = \prod_{i=1}^{N} \frac{\alpha_i}{\pi \hbar} \exp\left[-\frac{\alpha_i}{\hbar \omega_i}(p_{i0}^2 + \omega_i^2 q_{i0}^2)\right], \qquad (24)$$

where ω_i represents the frequency of the i'th normal mode and $\alpha_i = \tanh(\hbar\omega_i/2k_bT)$.[6] $V_{10}(\mathbf{q}_0)$ represents the excitation energies of the initial
ensemble. The signal is calculated by averaging over the whole initial
distribution $P_{00}(\mathbf{q}_0, \mathbf{p}_0)$ represented by the ensemble of trajectories. Notice,
that expression (23) is valid under the assumption of weak electric fields
due to the perturbation theory treatment.[6, 37] In addition we also assume
that the transition dipole moment for the ionization is independent of
the nuclear configuration (Condon approximation). In the next Sec. 3.2
we introduce the Stieltjes imaging approach which allows to take into
account the variation of the transition dipole moment along the nuclear
trajectories.

The simulation of the TRPES involves three steps: (i) The ensemble of
initial conditions is generated by sampling the Wigner distribution function
corresponding to the canonical ensemble at the given temperature. (ii) The
ensemble of trajectories is propagated using nonadiabatic MD "on the fly".
(iii) The TRPES is calculated by averaging over the ensemble of trajectories
employing the analytical expression (23) derived in the framework of the
Wigner distribution approach.[6]

3.2. *Stieltjes imaging approach for approximate description of photoionization continuum*

In order to describe accurately the photoionization process but avoiding
the extremely demanding solution of the scattering problem the Stieltjes
imaging (SI) procedure can be employed to reconstruct the photoionization
spectrum from the spectral moments. In principle, these can be obtained
either by diagonalization of the full Hamiltonian matrix or by using some
approximate approach which avoids the diagonalization as shown e.g. by
Gokhberg et al.[109] Since in linear response TDDFT the Hamiltonian matrix
is of a reasonable size this offers the opportunity to validate the quality of

SI spectra by comparison with the full spectral distribution obtained from the diagonalization of the TDDFT matrix. Notice that this diagonalization provides a set of neutral excited states above the ionization limit which we use as a discrete approximation for the true continuum states.

For the calculation of the photoionization probabilities the transition energies and transition dipole moments are needed which can be obtained from TDDFT calculations using the CI-like wavefunction ansatz in Eq. (8) for the excited electronic states of the neutral species. For this purpose, we introduce the transition dipole moments among excited states which are not available in the standard TDDFT procedure.[27]

The transition dipole matrix elements M_{IK} between two excited states I and K can be calculated according to:

$$\mathbf{M}_{IK} = \langle \Psi_I | \hat{\mu} | \Psi_K \rangle = \sum_{i,a} \sum_{j,b} c_{i,a}^{*I} c_{j,b}^{K} \langle \Phi_{i,a}^{CSF}(\mathbf{r}; \mathbf{R}(t)) | \hat{\mu} | \Phi_{j,b}^{CSF}(\mathbf{r}; \mathbf{R}(t)) \rangle, \quad (25)$$

where a and b indicate virtual and i and j occupied orbitals, respectively. The dipole matrix elements on the right hand side of Eq. (25) can be reduced to the transition dipoles between Kohn–Sham orbitals applying the standard Slater-Condon-Rules for matrix elements of one electron operators. Notice, that the calculation of the transition dipole moments between excited states within the linear response random phase approximation (RPA) has been developed by Yeager et al.,[110] which is related to our TDDFT based formalism. Using this procedure we obtain both the transition energies and oscillator strengths which represent a discrete approximation for the photoelectron spectrum. It should be emphasized that this description of the continuum states is extremely sensitive to the details of the quantum chemical description.

The discrete treatment of the photoionization continuum obtained in this way involving the transition energies and corresponding oscillator strengths is subsequently used in the SI procedure.[111–113] The basic quantity in the SI approach is the frequency-dependent polarizability α which contains both the information on the discrete part as well as on the continuum part of the ionization spectrum and can be represented by:

$$\alpha(z) = \sum_i^{\text{discrete}} \frac{f_i}{\epsilon_i^2 - z^2} + \int_{\epsilon_0}^{\infty} \frac{g(\epsilon) d\epsilon}{\epsilon^2 - z^2}. \quad (26)$$

In the above equation z is the complex frequency, $g(\epsilon)$ is the density of continuum states, ϵ_i are discrete transition energies, f_i are the oscillator strengths and ϵ_0 is the beginning of the ionization continuum. The frequency-dependent polarizability can be expanded into a series according to:

$$\alpha(z) = \sum_{k}^{\infty} S(-2k) z^{2k-2}, \qquad (27)$$

using the negative power spectral moments $S(-k)$:

$$S(-k) = \sum_{i=1}^{\text{discrete}} \frac{f_i}{(\epsilon_i)^k} + \int_{\epsilon_0}^{\infty} \frac{g(\epsilon) d\epsilon}{\epsilon^k}. \qquad (28)$$

It is important to notice that the series $\alpha(z)$ is divergent above ϵ_0 and thus cannot be calculated by direct summation. Therefore, an analytic continuation has to be introduced to calculate the $\alpha(z)$ in the continuum part of the spectrum.

In order to obtain approximate values of the spectral moments $S(-k)$ including discrete and continuum states, we use a similar relation between a power series expansion of the moments $S(-k)$ and the Stieltjes integral $\beta(z)$:

$$\beta(z) = \int_{\epsilon_0}^{\infty} \frac{\epsilon g(\epsilon) d\epsilon}{\epsilon - z} = \sum_{k=1}^{l} S(-k) z^{(k-1)} = \frac{P_{n-1}(z)}{Q_n(z)}. \qquad (29)$$

Furthermore, since this representation is also divergent in the continuum part of the spectrum on the real axis, we introduce the Padé approximation[113] on the right hand side of Eq. (29). The real roots of the polynomial $Q_n(z)$ are the pseudo ionization energies $\widetilde{\epsilon}_i$, and the corresponding oscillator strengths $\widetilde{f}_i$ are obtained from the residues of the Padé approximant for $z \to \widetilde{\epsilon}_i$:

$$\widetilde{f}_i = \frac{-1}{\widetilde{\epsilon}_i} \lim_{z \to \widetilde{\epsilon}_i} (\widetilde{\epsilon}_i - z) \frac{P_{n-1}(z)}{Q_n(z)} = \frac{-1}{\widetilde{\epsilon}_i} \frac{P_{n-1}(\widetilde{\epsilon}_i)}{Q_n'(\widetilde{\epsilon}_i)}, \qquad (30)$$

where $Q_n'(\widetilde{\epsilon}_i)$ is the derivative of the polynomial with respect to the transition energies.

The TRPE spectra are calculated by averaging over the ensemble of trajectories obtained using nonadiabatic dynamics "on the fly" employing the analytical expression (23). For this purpose, the ionization spectrum is approximated either by the discretized continuum (DC) represented by excited states of the neutral system above the ionization limit or by the Stieltjes imaging pseudo spectrum (SI) at selected timesteps in the nuclear dynamics.[27]

4. Field-Induced Surface-Hopping Method (FISH) for Simulation and Control of Ultrafast Photodynamics

The control of molecular processes by shaped laser fields opens a perspective for different applications. In particular, the closed-loop learning (CLL) scheme[80] has stimulated many experiments in which molecular fragmentation,[81,114] isomerization[115] or ionization[116] are controlled. The theoretical counterpart is the optimal control theory[69,71] which has contributed significantly to understand the mechanisms for the control of molecular fragmentation, ionization, and isotope selection. However, these theoretical achievements have so far been limited to low dimensional systems in which the explicit numerical solution of the time-dependent Schrödinger equation is feasible.

Recently, we have developed the Wigner distribution approach and successfully applied it to the simulation of time-resolved pump-probe spectra[6] as well as to the control of ground[83] and excited state[82] dynamics. However, due to the fact that the interaction with the laser field has been described using perturbation theory the method is limited only to processes in relatively weak fields. For this reason, the development of new theoretical methods for the simulation of laser-driven dynamics using moderately strong laser fields (below the multielectron ionization limit) is particularly desirable. Such fields open a very rich manifold of pathways for the control of ultrafast dynamics in complex systems.

Therefore, we present here our new semiclassical approach for the simulation and control of the laser-driven coupled electron-nuclear dynamics in complex molecular systems including all degrees of freedom. This stochastic "Field-Induced Surface Hopping" (FISH) method[84] is based on the combination of quantum electronic state population dynamics with classical nuclear dynamics carried out "on the fly" without precalculation of potential energy surfaces. The idea of the method is to propagate independent trajectories in the manifold of adiabatic electronic states and allow them to switch between the states under the influence of the laser field. The switching probabilities are calculated fully quantum mechanically. Thus, the purpose of this section is the presentation of a new generally applicable method for the treatment of the laser-driven photodynamics. The illustration of its scope will be presented in Sec. 6.

The description of the laser-driven multistate dynamics is based on the semiclassical limit of the quantum Liouville–von Neumann (LvN) equation

$$i\hbar\dot{\hat{\rho}} = [\hat{H}_0 - \vec{\mu} \cdot \vec{E}(t), \hat{\rho}] \tag{31}$$

for the density operator $\hat{\rho}$. $\hat{H}_0$ represents the field-free nuclear Hamiltonian for a molecular system with several electronic states in the Born–Oppenheimer approximation, and the interaction with the laser field $\vec{E}(t)$ is described using the dipole approximation. The semiclassical limit can be conveniently derived employing a phase space representation of quantum mechanics such as the Wigner representation.[34, 35] All quantum mechanical operators are thereby transformed to phase space functions of the coordinates $\mathbf{q}$ and momenta $\mathbf{p}$. In the lowest order, the commutators in the quantum LvN equation reduce to the classical Poisson brackets.[35] The equations of motion for the phase space representation of the diagonal ($\rho_{ii}(\mathbf{q}, \mathbf{p}, t)$) and off-diagonal ($\rho_{ij}(\mathbf{q}, \mathbf{p}, t)$) density matrix elements for an arbitrary number of states read:

$$\dot{\rho}_{ii} = \{H_i, \rho_{ii}\} - \frac{2}{\hbar} \sum_j \mathrm{Im}(\vec{\mu}_{ij} \cdot \vec{E}(t) \rho_{ji}), \tag{32}$$

$$\dot{\rho}_{ij} = -i\omega_{ij}\rho_{ij} + \frac{i}{\hbar} \vec{\mu}_{ij} \cdot \vec{E}(t)(\rho_{jj} - \rho_{ii})$$
$$+ \frac{i}{\hbar} \sum_{k \neq i,j} (\vec{\mu}_{ik} \cdot \vec{E}(t) \rho_{kj} - \vec{\mu}_{kj} \cdot \vec{E}(t) \rho_{ik}), \tag{33}$$

where the diagonal density matrix elements determine the quantum mechanical state populations and the off-diagonal elements describe the coherence. The curly braces denote the Poisson brackets, H_i are the Hamiltonian functions for the respective electronic state i. The quantity ω_{ij} is the energy gap between the electronic states i and j and $\vec{\mu}_{ij}$ denote the transition dipole moments. In the following, for clarity reasons we restrict the derivation to a system with two electronic states (denoted g and e) coupled by the laser field. In this case, the equations for the evolution of the diagonal (ρ_{ee} and ρ_{gg}) and non-diagonal (ρ_{ge}) density matrix elements are

$$\dot{\rho}_{gg} = \{H_g, \rho_{gg}\} - \frac{2}{\hbar} \mathrm{Im}(\vec{\mu}_{ge} \cdot \vec{E}(t) \rho_{eg}), \tag{34}$$

$$\dot{\rho}_{ge} = -i\omega_{ge}\rho_{ge} + \frac{i}{\hbar}\vec{\mu}_{ge}\cdot\vec{E}(t)(\rho_{ee}-\rho_{gg}), \tag{35}$$

$$\dot{\rho}_{ee} = \{H_e,\rho_{ee}\} - \frac{2}{\hbar}\mathrm{Im}(\vec{\mu}_{eg}\cdot\vec{E}(t)\rho_{ge}). \tag{36}$$

For calculating the population transfer between the states induced by the laser field, the coherence ρ_{ge} is needed. It can be obtained in analytic form by integrating Eq. (35):

$$\rho_{ge}(t) = \frac{i}{\hbar}\exp(i\omega_{eg}t)\int_0^t d\tau\,\exp(i\omega_{eg}\tau)\,\vec{\mu}_{ge}\cdot\vec{E}(\tau)(\rho_{ee}(\tau)-\rho_{gg}(\tau)). \tag{37}$$

Inserting this expression in Eq. (36) for the state e, the rate of change of the diagonal density matrix element $\dot{\rho}_{ee}$ which determines the excited state population becomes

$$\dot{\rho}_{ee} = \{H_e,\rho_{ee}\} - \frac{2}{\hbar^2}\mathrm{Re}\left\{\vec{\mu}_{ee}\cdot\vec{E}(t)\exp\left(i\omega_{eg}t\right)\int_0^t d\tau\,\exp\left(i\omega_{eg}\tau\right)\right.$$
$$\left.\times\,\vec{\mu}_{ge}\cdot\vec{E}(\tau)\left(\rho_{ee}(\tau)-\rho_{gg}(\tau)\right)\right\}. \tag{38}$$

The time evolution of the phase space function $\rho_{ee}(\mathbf{q},\mathbf{p},t)$ can now be separated into two physical contributions. The Poisson bracket $\{H_e,\rho_{ee}\}$ corresponds to the phase space density flow within the excited electronic state e while the second term in Eq. (38) describes the population transfer between the electronic states g and e.

In our FISH approach we represent the phase space functions $\rho_{ii}(\mathbf{q},\mathbf{p},t)$ by independent trajectories propagated in the ground and excited electronic states, respectively. Thus, if the finite number of trajectories N_{traj} is employed, $\rho_{ii}(\mathbf{q},\mathbf{p},t)$ can be represented by a swarm of time-dependent δ functions

$$\rho_{ii}(\mathbf{q},\mathbf{p},t) = \frac{1}{N_{\mathrm{traj}}}\sum_k \theta_i^k(t)\,\delta(\mathbf{q}-\mathbf{q}_k^i(t;\mathbf{q}_0,\mathbf{p}_0))\delta(\mathbf{p}-\mathbf{p}_k^i(t;\mathbf{q}_0,\mathbf{p}_0)), \tag{39}$$

where $(\mathbf{q}_k^i,\mathbf{p}_k^i)$ represents a trajectory propagated in the electronic state i and the parameter $\theta_i^k(t)$ has a value of one if the trajectory k resides in the state i, otherwise it has a value of zero.[6] The population transfer between the electronic states is achieved by a process in which the trajectories are allowed to switch between the states. This procedure is related to Tully's surface hopping method[44] which has been developed in order to describe field free nonadiabatic transitions in molecular systems. However, in our case the coupling between the states is induced by the applied laser field.

In the case of a two-state system the hopping probability $P_{g \to e}$ between the ground state g and excited state e can be calculated from the rate of change of the excited state population, normalized to the population of the ground state according to

$$P_{g\to e} = \Theta(\dot{\rho}_{ee})\frac{\dot{\rho}_{ee} - \{H_e, \rho_{ee}\}}{\rho_{gg}}\Delta t. \qquad (40)$$

The function $\Theta(\dot{\rho}_{ee})$ is defined to have a value of one for $\dot{\rho}_{ee} > 0$ and a value of zero for $\dot{\rho}_{ee} < 0$, guaranteeing that the hopping probability is nonzero only if the population of the excited state is increasing. Inserting the Eq. (36) for the rate of change of ρ_{ee}, the hopping probability for a two-state system can be expressed in analytic form as

$$P_{g\to e}(t+\Delta t) = -\frac{2\Delta t}{\hbar^2 \rho_{gg}} \operatorname{Re}\left(\vec{\mu}_{eg} \cdot \vec{E}(t)\exp(i\omega_{eg}t)\int_0^t d\tau \exp(i\omega_{eg}\tau)\vec{\mu}_{ge}\right.$$
$$\left. \times \vec{E}(\tau)(\rho_{ee}(\tau) - \rho_{gg}(\tau))\right), \qquad (41)$$

illustrating the direct dependence on the electric field $\vec{E}(t)$.

The simulation of the laser-induced dynamics in the framework of our FISH method using the above derived approach is performed in the following three steps:

(i) We generate initial conditions for an ensemble of trajectories by sampling e.g. the canonical Wigner distribution function [cf. Eq. (24)] or a long classical trajectory in the electronic ground state.

(ii) For each trajectory which is propagated in the framework of MD "on the fly", we calculate the density matrix elements ρ_{ij} by numerical integration. If the initial electronic state is a pure state as it is in our case, the set of equations (32)–(33) is equivalent to the time-dependent Schrödinger equation in the representation of adiabatic electronic states:

$$i\hbar \dot{c}_i(t) = E_i(\mathbf{R}(t))c_i(t) - \sum_j \vec{\mu}_{ij}(\mathbf{R}(t)) \cdot \vec{E}(t)c_j(t), \qquad (42)$$

where $c_i(t)$ are the expansion coefficients of the electronic wavefunction in the basis of adiabatic electronic states from which the density matrix elements can be calculated as $\rho_{ij} = c_i^* c_j$. It should be noticed that the adiabatic state energies as well as the transition dipole matrix elements are parametrically dependent on the nuclear trajectory $\mathbf{R}(t)$.

If the intrinsic nonadiabatic coupling of the electronic states also has to be taken into account, the equation (42) can be generalized to

$$i\hbar \dot{c}_i(t) = E_i(\mathbf{R}(t))c_i(t) - \sum_j (\vec{\mu}_{ij}(\mathbf{R}(t))) \cdot \vec{E}(t)$$
$$+ i\hbar \dot{\mathbf{R}}(t) \cdot \mathbf{d}_{ij}(\mathbf{R}(t)))c_j(t), \quad (43)$$

where $\mathbf{d}_{ij}(\mathbf{R}(t)) = \langle \Phi_i(\mathbf{R}(t)) | \nabla_R | \Phi_j(\mathbf{R}(t)) \rangle$ denotes the nonadiabatic coupling vector and $\dot{\mathbf{R}} \cdot \mathbf{d}_{ij}$ corresponds to the scalar coupling D_{ij} (cf. Sec. 2.1). Thus, after the duration of the applied field is over, field free multistate nonadiabatic dynamics can be further carried out. The Eqs. (42) or (43) are solved numerically using e.g. the fourth order Runge–Kutta procedure.

The nuclear trajectories $\mathbf{R}(t)$ are obtained by solution of Newton's equations of motion using the Verlet algorithm[117]:

$$M\ddot{\mathbf{R}}(t) = -\sum_i \theta_i(t) \nabla_R V_i(\mathbf{R}(t)). \quad (44)$$

In Eq. (44) $\theta_i(t)$ represents a parameter which has a value of one for the state in which the trajectory is propagated at the given time and zero for all other states and $V_i(\mathbf{R}(t))$ is the adiabatic potential energy of the electronic state i. The forces acting on the nuclei ($\nabla_R V_i(\mathbf{R}(t))$) are calculated "on the fly" when they are needed. In contrast to field free nonadiabatic dynamics, the energy of a molecular system is not conserved if an electric field is present, since energy exchange with the field occurs. Therefore, when exposed to a long intense laser pulse, molecules tend to accumulate energy and eventually get heated. For an isolated gas phase molecule, this can lead to fragmentation. However, if the molecule is interacting with an environment such as solution, the excess thermal energy can be dissipated. In order to approximately include the effect of the environment we use dissipative Langevin dynamics with the equation of motion

$$M\ddot{\mathbf{R}}(t) = -\sum_i \theta_i(t) \nabla_R V_i(\mathbf{R}(t)) - M\gamma \dot{\mathbf{R}}(t) + \mathbf{F}_{\text{rand}}(t) \quad (45)$$

instead of Eq. (44) for calculating the forces acting on the nuclei. Here, γ is an empirical friction coefficient and $\mathbf{F}_{\text{rand}}$ represents a random force. The solution of the Eqs. (44) or (45) provides continuous nuclear trajectories which reside in different electronic states according to the quantum mechanical occupation probabilities given by ρ_{ii}.

(iii) In order to determine in which electronic state the trajectory is propagated we calculate the hopping probabilities and decide if the trajectory is allowed to change the electronic state by using a random number generator. For a general number of states the hopping probability can be calculated according to Eq. (7). In contrast to the field-free nonadiabatic dynamics, the energy is not conserved during the interaction with the field, and thus after a hopping event no velocity rescaling is performed until the field has ceased.

Notice that while the trajectories jump between the electronic states at a given time, all density matrix elements are propagated continuously over the entire time according to Eqs. (32)–(33), or alternatively either (42) or (43) depending whether intrinsic nonadiabatic coupling between the states is treated or not. Although the individual trajectory is allowed to jump, the total number of trajectories in a given state representing ρ_{ii} is also a continuous function of time. The phase of the electronic wavefunction is preserved and our procedure gives rise to the full quantum mechanical coherent state population.

Therefore, our approach is able to mimic laser-induced processes such as coherent Rabi oscillations between the electronic states. In order to illustrate this, the FISH approach is compared to the full quantum mechanical treatment of laser-induced dynamics in a two state harmonic oscillator model system. The population dynamics shown in Fig. 1(a) clearly exhibits the coherent Rabi oscillations and is in perfect agreement with the populations obtained by full quantum dynamics presented in Fig. 1(b). It should be emphasized that the ability to describe the coherent electronic state dynamics is inherent to our approach and does not depend on the chosen model system.

In particular, the FISH procedure can be combined with the optimal control theory in order to steer molecular processes. For this purpose the electric field entering the Eq. (41) can be iteratively optimized using e.g. evolutionary algorithms[83, 119] as it will be illustrated later. Specifically, the field is parametrized either in the time or in the frequency domain, and the respective parameters as e.g. intensities, frequencies, pulse widths or chirp parameters can be iteratively optimized based on binary coding of the parameters and the usual selection, crossover and mutation operations.[119]

It should be pointed out that the theoretical optimization of laser fields within FISH dynamics, in particular if relatively long laser pulses are

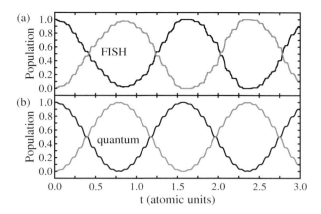

Fig. 1. Population dynamics in a two-electronic state harmonic model system. The ground state is given by $V_g(q) = 0.5\,q^2$ and the excited state by $V_e(q) = 0.5\,q^2 + 40$. The states are coupled by a resonant electric field with $E(t) = 4\,\sin(40\,t)$, the transition dipole moment is $\mu_{ge} = 1.0$. (a) Semiclassical populations of the ground (black) and excited (grey) states. (b) Quantum mechanical populations of the ground (black) and excited (grey) states. For quantum dynamics a grid-based numerical solution of the Schrödinger equation[118] was employed. Reprinted from Ref. 84. Copyright 2009, American Physical Society.

needed, is computationally quite demanding for complex systems since an ensemble of trajectories needs to be propagated many times using each field generated during the optimization procedure. The main part of the computational cost results from the necessity to calculate the energies and forces as well as the nonadiabatic couplings and transition dipole matrix elements between all electronic states. Efficient calculation of the needed quantities can e.g. be achieved by using either time-dependent density functional theory (TDDFT) or semiempirical quantum chemistry methods, provided they offer a sufficiently accurate description of the spectroscopic properties of the systems under study. The calculation of nonadiabatic couplings and transition dipole matrix elements in the frame of TDDFT is described in Secs. 2 and 3, respectively. For semiempirical methods, calculation of nonadiabatic couplings and transition dipole moments between excited states was recently introduced by Thiel et al.[53]

Our semiclassical FISH method is a valuable tool for the simulation and control of ultrafast laser-driven coupled electron-nuclear dynamics involving several electronically excited states in complex molecular systems. This approach combines classical MD simulations with the field-induced surface hopping for the electronic state population dynamics and can be used to simulate spectroscopic observables as well as to control the dynamics employing shaped laser fields. For the propagation of classical trajectories the whole spectrum of methods ranging from empirical force fields, semiempirical to *ab initio* quantum chemical methods can be employed. Moreover, the FISH method can be used to analyze and control laser-driven excited state dynamics in molecular systems in the gas phase as well as interacting with different environments such as solvent, bioenvironment, surfaces or metallic nanostructures. In particular, due to the density matrix formulation of the method, dissipative effects for nuclear and electronic motion can be taken into account. The FISH method allows not only to obtain optimized pulses but also to analyze their shapes on the basis of molecular dynamics "on the fly". In this way the comparison between theoretically optimized laser fields with those obtained from experiments, e.g. using the CLL procedure allows to assign the underlying processes to the specific forms of the pulses. In principle in this way the inversion problem can be addressed or important parts of the PES could be constructed. Altogether, the FISH method opens new avenues to perform the optimization of laser pulses for different exciting applications as it will be described in Sec. 6. Beyond the description of the electric dipole coupling, the FISH method represents a general approach which allows straightforward inclusion of any other kind of coupling between electronic states, e.g. magnetic dipole interactions in chiral systems or Förster type excitation energy transfer in molecular assemblies.

5. Applications of the Nonadiabatic Dynamics "on the Fly"

The choice of the applications presented here has the purpose to show the broad scope of our approach on the example of organic chromophores and biologically relevant chromophores interacting with the environment. The simulation of ultrafast observables such as TRPES allows to make direct comparison with experimental data and thus to reveal the dynamical processes involved in the excited state relaxation and their time scales. Moreover, the constantly developing methods for simulation of ultrafast processes challenge also the development of new experimental techniques allowing to address ultrafast phenomena at always increasing level of precision as it will be described below.

5.1. *Ultrafast dynamics of photoswitching*

The study of the ultrafast dynamics in benzylideneaniline (BAN) is of particular interest in the context of applications of molecular switches in molecular electronics and biosensing. Moreover, the example of the ultrafast nonadiabatic photoisomerization of BAN serves to illustrate the scope of our approach for the nonadiabatic dynamics based on the time-dependent density functional theory (TDDFT) described in Sec. 2.1. The aim of this study is twofold: First, we wish to establish the mechanism of the ultrafast photoinduced switching and to determine the nonradiative lifetime of the optically allowed S_1 state, gaining fundamental insight into selectivity and dynamics of molecular switches. Second, we present the simulation of TRPE spectra and identify the features that are characteristic for the excited state dynamics and for the nonadiabatic transitions. In addition, we investigate the influence of the manifold of cationic excited states on the general appearance of the TRPE spectra, since in the photoionization usually several cationic excited states can be reached. Our theoretical results serve also to motivate the experimental work on BAN in the frame of the TRPE spectroscopy.

5.1.1. *Computational*

For the description of the electronic structure we employ the non-hybrid gradient-corrected Perdew–Burke–Ernzerhof (PBE) exchange–correlation functional[120] combined with triple zeta valence plus polarization atomic basis sets (TZVP)[121] together with the Resolution-of-the-Identity (RI) approximation[122,123] for the calculation of energies and forces "on the fly" as implemented in Turbomole,[124] allowing to speed up the calculations considerably. The relatively large TZVP atomic basis set has been employed to decrease the influence of the basis set on the quality of the results.

In order to determine the accuracy of TDDFT for the description of the excited states of BAN we first calculated the stationary absorption spectra for the cis and trans isomer with RI-PBE and with the B3-LYP hybrid functional.[125,126] The transition energy for the S_1 state of the trans isomer calculated with RI-PBE is lower by approximately 0.4 eV than the one obtained with B3-LYP and lower by approximately 0.8 eV with respect to the available experimental results.[127] The character of the S_1 and S_2 transitions is identical for both functionals and no additional dark states appear below the dark S_2 state. In order to address the influence of the functional (B3-LYP vs. RI-PBE) we have performed excited state dynamics simulations using a small number of trajectories. The results have shown that the features of the excited state dynamics are not strongly dependent on the choice of the functional. Thus we employ the significantly computationally more efficient although less accurate non-hybrid PBE functional with the RI approximation for the nonadiabatic dynamics simulations.

Due to the well known problem of TDDFT in describing long range interactions, especially using non-hybrid functionals, we have checked the long range charge transfer character of the S_1 and S_2 states for the trans isomer. For this purpose we calculated the recently introduced quantity Λ,[128] which serves as an indicator for the long range charge transfer contribution to transitions. Λ is defined as:

$$\Lambda = \frac{\sum_{ia} c_{ia}^2 O_{ia}}{\sum_{ia} c_{ia}^2}, \qquad (46)$$

where O_{ia} represents the overlap of the moduli of an occupied and virtual orbital, $O_{ia} = \langle |\phi_i| | |\phi_a| \rangle$, while c_{ia} labels the contribution of the excitation $\phi_i \to \phi_a$ to a given transition. Λ is restricted to the range between 0 and 1, where low values indicate strong long range charge transfer character. Within the RI-PBE method both the S_1 and S_2 states of the trans isomer correspond to predominantly local transitions as evidenced by the Λ values of 0.67 and 0.47, respectively.[128] Thus, particularly in the S_1 state which is initially populated in the nonadiabatic dynamics, the charge transfer contribution is not dominant and therefore PBE offers an acceptable description.

For the simulation of the photodynamics of BAN using our TDDFT-based nonadiabatic dynamics approach elucidated in Sec. 2.1, an ensemble of 220 trajectories was excited to the S_1 state of the trans isomer. The initial coordinates and momenta were sampled from a long trajectory propagated at the temperature of 100 K in the electronic ground state. For the nuclear dynamics the classical Newton's equations of motion have been integrated using the velocity Verlet algorithm[117] with a timestep of 0.1 fs, and forces were calculated for the currently occupied state. The expansion coefficients $C_K(\tau)$ for the electronic wavefunction which are needed to calculate the hopping probabilities $P_{I \to K}(\tau)$ according to Eq. (4) were propagated using a fourth order Runge–Kutta procedure with a timestep of 10^{-5} fs. The time-dependent photoelectron spectrum (TRPES) has been simulated in the framework of the Wigner distribution approach (cf. Ref. 6) as described in Sec. 3.1.

Here we focus on the photoionization to the cationic ground electronic state which can be effectively reached both from the ground state as well as from the excited state of the neutral species. In order to assess the contribution of higher cationic states to the signal, we have also performed simulations for selected individual trajectories including additional eleven cationic excited states.

5.1.2. *Ultrafast photoswitching in benzylideneaniline (BAN)*

We present here the simulation of the photoinduced trans-cis isomerization in the prototype switchable Schiff base benzylideneaniline (BAN). In contrast to related photoswitchable molecules such as azobenzene and stilbene, the calculated equilibrium structure of the trans isomer of benzylideneaniline does not exhibit a planar geometry.[129,130] The equilibrium value for the dihedral angle of the phenyl ring bound to the nitrogen atom has the value of 42° while the phenyl ring bound to the carbon atom is almost coplanar with the C=N bond. These values are in good agreement with experimental gas phase electron diffraction measurements.[129] The cis isomer is 0.25 eV higher in energy and exhibits a structure in which the phenyl ring bound to the nitrogen atom is tilted by 76° with respect to the C=N bond plane while the other phenyl remains almost coplanar with a torsion angle of 14°. In order to verify the thermal stability of the cis isomer which is the product of the photoisomerization we have run a long constant temperature MD simulation at $T = 200$ K which did not exhibit any reverse isomerization. This shows that the cis isomer is stable well above the usual temperature of e.g. molecular beam experiments and therefore the photoisomerization should be easily experimentally detectable.

The lowest intense electronic transition of the trans isomer is centered around 3.2 eV as can be seen from the simulated thermally broadened absorption spectrum at $T = 100$ K shown in Fig. 2(a) (dashed line). The thermal ensemble obtained from MD simulations at $T = 100$ K shows that only the trans isomer is populated and the structure is relatively rigid [cf. inset in Fig. 2(a)]. The S_1 state has dominantly $\pi - \pi^*$ character. Notice, that the optical absorption of the cis isomer lies much higher in energy [cf. Fig. 2(b)] and therefore both isomers can be selectively excited which is a necessary requirement for reversible photoswitching. First, we present our results for the nonradiative lifetimes and then the simulation of the TRPE spectra.

For the calculation of lifetimes, the ensemble of 220 trajectories has been propagated starting in the S_1 state. The nonadiabatic dynamics has been performed in a manifold consisting of the ground electronic state and the two lowest electronically excited states. For the determination of the nonradiative lifetime the time-dependent electronic state populations have been calculated by monitoring the number of trajectories in each state during the dynamics simulation (cf. Fig. 3). After the initial $\pi - \pi^*$ excitation, the S_1 state's population decays approximately exponentially with a time

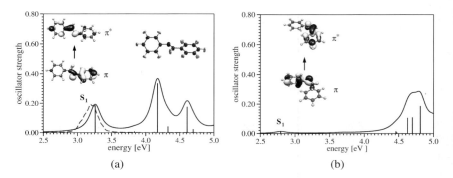

Fig. 2. Theoretical stationary absorption spectrum for (a) the trans isomer and (b) the cis isomer of benzylideneaniline obtained using the PBE functional with TZVP basis sets. The calculated discrete transition energies are convoluted with a Lorentzian function with a width of 0.2 eV. The character of the S_1 transition is indicated by the dominant KS-orbitals involved in the excitation. The dashed line in (a) denotes the theoretical thermally broadened stationary absorption spectrum for the S_1 state of trans benzylideneaniline obtained from the 300 initial conditions shown in the inset. The discrete absorption lines (not shown) have been convoluted with a Lorentzian function with a width of 0.04 eV. Reprinted with permission from Ref. 26, Copyright 2008, American Institute of Physics.

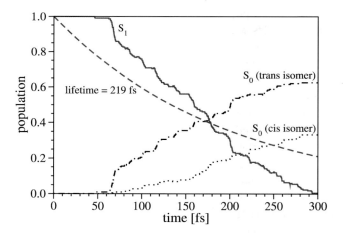

Fig. 3. Time-dependent population of the S_1 state (full line) and of the S_0 state of the cis (dotted line) and trans (dashed dotted line) isomer for benzylideneaniline after excitation of the trans isomer to the S_1 state. The lifetime of the S_1 state was determined by exponential fitting (dashed line) giving rise to the value of 219 fs. The structures are classified as cis, if the final C-N=C-C dihedral angle is in the range $[0.0°, 60°]$ and as trans, if it is in the range $[120.0°, 180.0°]$. Reprinted with permission from Ref. 26, Copyright 2008, American Institute of Physics.

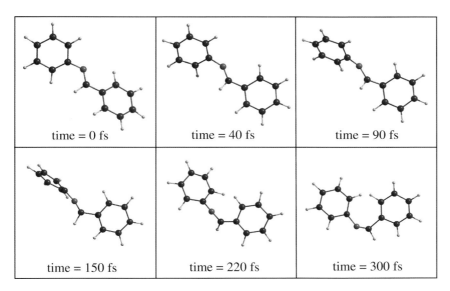

Fig. 4. Snapshots of the dynamics of benzylideneaniline for one trajectory exhibiting isomerization to the cis form. Reprinted with permission from Ref. 26, Copyright 2008, American Institute of Physics.

constant of 219 fs. The population transfer to the neutral ground state is completed within 300 fs while the second excited state S_2 is not populated at all. In order to determine the selectivity of the photoswitching we have decomposed the ground electronic state population into the contribution of cis and trans isomers. As can be seen from Fig. 3 the isomerization process is highly nonselective with a yield for the cis isomer of ∼33%. This low selectivity results from the high excess of kinetic energy gained during the propagation in the excited electronic state.

The mechanism of the photoisomerization is presented in Fig. 4 on the example of a typical trajectory isomerizing to the cis form. The snapshots of the dynamics show that the molecule reaches a semi-linear configuration with an almost linear C=N-C unit within the first 150 fs. This configuration is closely related to the transition state for the thermal isomerization[130] and thus can lead both to the cis (cf. Fig. 4) as well as to the trans isomer (not shown) after the nonadiabatic transition to the ground state. The complexity of the isomerization process is evident from Fig. 5 in which a selected trajectory exhibiting trans-cis isomerization has been projected onto the normal modes of the trans isomer. While in the first 70 fs only few normal modes are excited, during the subsequent dynamics an

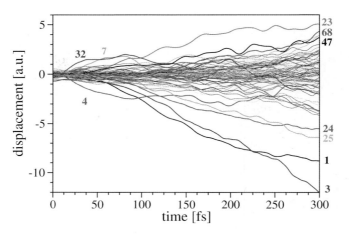

Fig. 5. Displacement of the normal coordinates of trans benzylideneaniline along one trajectory isomerizing to the cis form. The numbers indicate the corresponding normal modes with respect to increasing frequency. Reprinted with permission from Ref. 26, Copyright 2008, American Institute of Physics.

increasing number of normal coordinates is activated. This clearly demonstrates that the photoisomerization of benzylideneaniline cannot be accurately described using reduced models in which only few vibrational modes are explicitly taken into account.

In order to demonstrate the scope of the TRPE spectroscopy for the investigation of photoswitching processes we have simulated the TRPES of benzylideneaniline using our approach outlined in Sec. 3.1. For this purpose the energy gaps between the actual state in nonadiabatic dynamics of the neutral species and the cationic ground state have been calculated for the whole ensemble. The simulated TRPES shown in Fig. 6(a) initially exhibits coherent oscillations of the electron binding energy in the range between 4.75 and 6 eV. The period of the oscillations is approximately 20 fs and corresponds to the C=N stretching vibration, which is initially excited due to the electronic excitation into the antibonding π^* orbital (cf. Fig. 2). This origin of the oscillations is further confirmed by the analysis of the time-dependent internal coordinates corresponding to the C=N, C-C_{ph} and N-C_{ph} bond lengths shown in Fig. 7.

In addition to the oscillatory behavior, the electron binding energy also exhibits a systematic shift from 4.75 eV at $t = 0$ to 5.75 eV after ~100 fs. This shift is characteristic for the onset of the isomerization process during which the semi-linear transition state is reached. After 100 fs the signal at

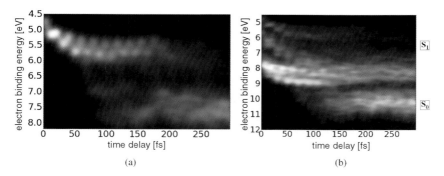

Fig. 6. Calculated time-resolved photoelectron spectrum (TRPES) of benzylideneaniline (a) for the cationic ground state averaged over 220 trajectories and (b) for the ground and eleven excited cationic states obtained from the nonadiabatic dynamics illustrating the population transfer from the S_1 to the S_0 state. Reprinted with permission from Ref. 26, Copyright 2008, American Institute of Physics.

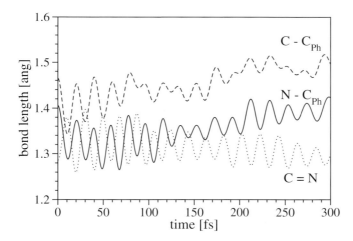

Fig. 7. Time-dependent bond length for the C=N (dotted line), C-C_{ph} (dashed line) and N-C_{ph} (full line) bonds of benzylideneaniline averaged over 220 trajectories. Reprinted with permission from Ref. 26, Copyright 2008, American Institute of Physics.

5.75 eV disappears and a new signal at the much higher electron binding energy of 7.5 eV starts to rise. This signal is characteristic for the ground state dynamics which takes place after the nonadiabatic transition. Notice, that in contrast to the excited state part of the TRPES the ground state signal does not show any vibrational structure due to the high excess of energy which is gained upon the nonadiabatic transition.

In order to investigate the influence of higher cationic states on the features of the time resolved photoelectron spectrum, we have also simulated TRPES including 11 excited cationic states which is presented in Fig. 6(b). The contribution of higher states which might correspond more realistically to the experimental situation smears out the vibrational structure and generally leads to the broadening of the signal. Nevertheless, the qualitative features of the nonadiabatic transition reflected in the shift of the electron binding energy remain preserved.

In summary, we have demonstrated on the example of the ultrafast photoswitching of BAN that TDDFT nonadiabatic dynamics is a suitable approach for simulation of time-resolved observables such as TRPES as well as for the determination of nonradiative lifetimes and mechanisms of the involved processes, provided that the description of the chosen system does not suffer from the known deficiencies of the TDDFT method. Our findings have motivated experimental work which is in progress. The predicted nonradiative lifetime of the S_1 state of BAN of $\sim$200 fs as well as the characteristic features of the TRPES are comparable with the experimental results for azobenzene[131] which is a structurally and electronically related molecule and has been already intensively investigated by others.[30, 47, 132–135]

5.2. Ultrafast internal conversion through conical intersection in pyrazine

Pyrazine is a prototype molecule for heterocyclic biochromophores and therefore represents a suitable system to study fundamental photochemical and photophysical processes. In particular, it has been intensively explored experimentally and theoretically as an example for ultrafast internal conversion through a conical intersection[25, 108, 136–138] and serves as a benchmark for experimental and theoretical work. Based on earlier quantum dynamical studies, it has been proposed to use the TRPES for the real-time observation of the $S_2 \rightarrow S_1$ internal conversion (IC) in pyrazine[136] since a systematic decrease of the PKE has been predicted to occur. However, experimental measurements[108, 137] revealed that the PKE distribution does not significantly change upon internal conversion. Our previous work on TRPES of pyrazine[25] based on nonadiabatic dynamics "on the fly" in the frame of TDDFT taking into account only the cationic ground state D_0 showed discrepancies from experimental TRPE spectra[108] in the time domain before the internal conversion occurs. Theoretical simulations using transition dipole moments for ionization to the D_0 and D_1 cationic states estimated from the experimental data and quantum dynamical simulations with reduced dimensionality have reproduced these experimental results.[139] Bearing in mind these findings as well as the improvement of experimental resolution we came to the conclusion that a general theoretical treatment accounting for an approximate description of the photoionization continuum is needed in order to address the growing interest in the interpretation of experimental TRPES results. In this section, we present our results on TRPES of pyrazine based on the Stieltjes imaging procedure described in Sec. 3.2 which represents a general tool for the investigation of ultrafast photoionization processes in complex systems and thus can be used to study their femtochemistry including all degrees of freedom.[27]

5.2.1. *Computational*

For the nonadiabatic dynamics we have used the hybrid B3LYP functional[125, 126] combined with triple zeta valence plus polarization atomic basis sets (TZVP)[121] as implemented in the Turbomole program.[124] As previously described, the B3LYP functional provides a reasonably accurate description of the lowest $\pi\pi^*$ and $n\pi^*$ states of pyrazine[25] as no long-range charge transfer or multireference character is involved in these transitions, and thus it can be used for the nonadiabatic dynamics simulation.

Investigation of the occupation of the vibrational states in pyrazine at 260 K revealed that even for the lowest normal mode exhibiting a vibrational temperature of 492 K the first excited vibrational state ($v = 1$) is only populated with 15% whereas for higher normal modes the occupation of $v = 1$ is less than 1%. Therefore, it can be assumed that at our simulation temperature only the vibrational ground state ($v = 0$) is occupied corresponding to the experimental conditions in a supersonic molecular beam.

In our nonadiabatic dynamics simulations the ground electronic state and the four lowest electronically excited states have been included. The Newton equations of motion have been integrated using the velocity Verlet algorithm[117] with a time step of 0.1 fs. For each dynamics time step the hopping probabilities were determined according to Eq. (4). The electronic state expansion coefficients $C_K(\tau)$ were evaluated by numerical integration of Eq. (2).

For the simulation of the TRPE spectra using both the Stieltjes imaging (SI) as well as the discretized continuum (DC) represented by neutral excited states above the ionization limit as described in Sec. 3, up to 200 transition energies and transition dipole moments [cf. Eq. (25)] have been calculated at selected timesteps, densely covering the energy range between the D_0 ionization threshold and the experimental probe pulse energy of 6.25 eV. In order to provide an adequate description of the electronic states above the ionization level the 6-311G**++ basis set containing diffuse functions has been employed. Subsequently, transition energies and transition dipole moments [cf. Eq. (25)] have been used for the SI procedure giving rise to photoionization energies and oscillator strengths according to Eq. (30). In order to obtain continuous spectra the discrete pseudo spectra at each time delay t_D have been folded in the energy domain with a Gaussian spectral profile with a width of 0.2 eV in accordance with the experiment, thus yielding the continuous Stieltjes pseudo spectrum $I(t_D, E)$. Employing such continuous photoelectron spectra the TRPES signal related to Eq. (23) has been calculated according to:

$$I_{TRPES}(t_D, E) = \int_0^\infty d\tau \exp\left[-\frac{1}{2} \cdot \frac{(\tau - t_D)^2}{\sigma^2}\right] I(\tau, E). \qquad (47)$$

To take into account the experimental time resolution we use the experimental full width at half-maximum value of 22 fs for the cross-correlation between the pump and the probe pulse which corresponds to $\sigma = 9.34$ fs.

5.2.2. *Simulation of TRPES for pyrazine and comparison with experimental spectra*

The comparison of the simulated TRPES based on the discretized continuum of neutral excited states above the ionization limit (DC) and the Stieltjes imaging approach (SI)[27] described in Sec. 3 with the experimental TRPES is presented in Fig. 8. This spectrum has been obtained using sub-20 fs deep UV pulses at 264 nm and 198 nm generated by four-wave mixing through filamentation by Suzuki et al. which allowed to investigate the ultrafast dynamics of polyatomic molecules in the framework of TRPE spectroscopy with unprecedented time resolution.[85]

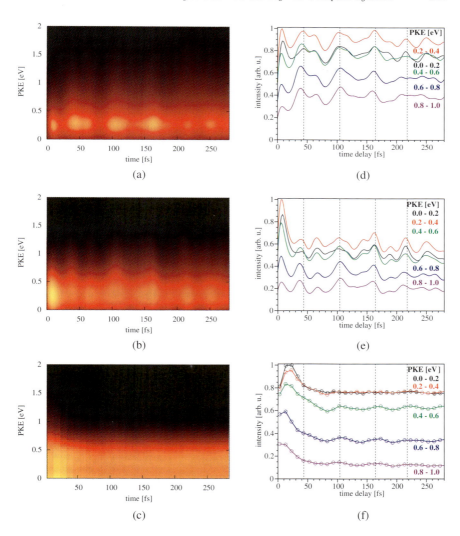

Fig. 8. Comparison of simulated TRPES for pyrazine (a) using discretized continuum represented by neutral excited states above the ionization limit (DC),[27] (b) using Stieltjes imaging (SI)[27] and (c) experimental spectrum (cf. Ref. 85). Time-dependent photoelectron signal intensities for different photoelectron kinetic energy (PKE) intervals obtained from (d) DC, (e) SI and (f) experiment. The signals (d)–(f) have been independently normalized with respect to the highest peak. Reprinted with permission from Ref. 27, Copyright 2010, American Institute of Physics.

Following the initial excitation to the S_2 excited state, the nonradiative transition to S_1 occurs with a time constant of $\sim$20 fs.[25] As shown in Fig. 8 the theoretical SI and the experimental spectrum exhibit a maximum within the first 20 fs which is less pronounced in the DC signal. After $\sim$50 fs the photoelectron signal remains in the PKE interval between 0 and 1 eV both in the DC [cf. Fig. 8(a)] and SI simulations [cf. Fig. 8(b)] as well as in the experiment [cf. Fig. 8(c)] in spite of the nonadiabatic $S_2 \rightarrow S_1$ transition. The reason for no significant shift of the PKE distribution is that the energy differences of the transitions from the S_2 state to the cationic excited state D_1 and from the S_1 state to the cationic ground state D_0, predominantly occurring in the ionization process, are very similar.

In order to perform a more detailed analysis of the simulated and experimental TRPE spectra we present in Fig. 8(d), (e) and (f) sections corresponding to selected PKE ranges. While both theoretical spectra are very similar after $\sim$50 fs, the SI spectrum exhibits a more pronounced maximum at $\sim$10–20 fs which is in agreement with the experiment. All three signals exhibit quantum beats in the range between 50 and 250 fs. However, the quantum beats are more pronounced in the theoretical spectra [cf. dashed lines in Fig. 8(d), (e) and (f)]. The average intensity is highest for the 0.0–0.2 eV and 0.2–0.6 eV intervals and decreases with increasing PKE above 0.6 eV in both the simulated and experimental spectra.

The influence of the approximate treatment of the photoionization on the appearance of the signal can be analyzed from the TRPE spectra sliced at selected timesteps (cf. Fig. 9) by comparing the results obtained with both presented approaches, DC (black line) and SI (blue line), results from our earlier work[25] which include only the cationic ground state D_0 (green dotted line) and experimental findings (red).[108] At time 4 fs both the experimental and the new theoretical signals based on the DC and SI procedures are located around 0.5 eV while the signal calculated taking into account only the cationic ground state D_0 is located at $\sim$2 eV. This clearly indicates that the ionization from the initially occupied S_2 state to D_0 (IP$_{\text{equ.}}$ = 4.02 eV, calculated at equilibrium geometry) occurs only weakly in the experiment as confirmed by the results obtained from the DC and SI approaches which show the dominance of the $S_2 \rightarrow D_1$ transition in this time period. As the time proceeds ($t > 24$ fs) a change in the ionization process takes place as can be seen from the experimental and theoretical results presented in Fig. 9. Within the period of 0 fs $< t <$ 24 fs the internal conversion from the S_2 to the S_1 state occurs (cf. Ref. 25) and subsequently for $t > 24$ fs ionization from S_1 to D_0 dominates. This is reflected

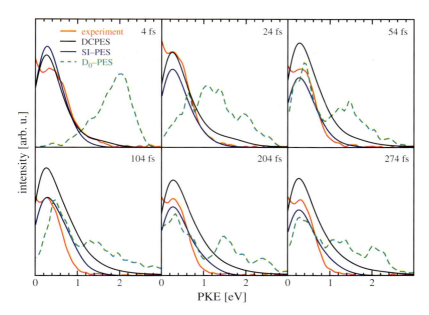

Fig. 9. Comparison of photoelectron spectra (PES) of pyrazine at selected timesteps obtained from DC (black), SI (blue line),[27] simulations only including the cationic ground state D_0 (green, dotted line)[25] and experiment (red).[85] Reprinted with permission from Ref. 27, Copyright 2010, American Institute of Physics.

in the similarity of all three theoretical and the experimental spectra for the time after 24 fs. The fact that several cationic states (D_1 at $t \sim 0$ fs and D_0 at $t > 24$ fs) are involved in the ionization process clearly demonstrates that the approximate description of the transition probabilities to ionization continua corresponding to D_0 and D_1 cationic states is necessary for the accurate description of the ionization process which is accounted for in the DC and SI procedures. Therefore both provide agreement with the experimental TRPE spectra in the whole measured time domain.

Further insight into the ionization process can be gained by analyzing a selected nonadiabatic trajectory which is presented in Fig. 10(a). As can be seen from the electron density difference during the dynamics the character of the electronic state changes from the $\pi\pi^*$ (S_2) to the $n\pi^*$ (S_1) state at $t = 20$ fs. The time-dependent energies of the four excited states of the single trajectory in Fig. 10(a) are well separated from the ground state within the simulation period which is in agreement with the absence of population transfer to the ground state within this time period. The character of the

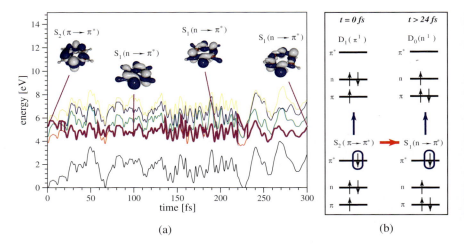

Fig. 10. (a) Energies of the four excited states and the ground state of pyrazine as a function of time obtained from the nonadiabatic dynamics along one selected trajectory. The thick violet line indicates the energy of the state in which the trajectory resides during the dynamics and the insets denote character of the dominant excitations. (b) Schematic analysis of the time-dependent ionization processes in pyrazine. Reprinted with permission from Ref. 27, Copyright 2010, American Institute of Physics.

states is also reflected in the dominant configurations shown in Fig. 10(b). As can be seen the dominant configuration at the beginning of the dynamics ($t = 0$ fs) corresponds to the S_2 state. In the one electron picture ionization of one electron out of the highest occupied molecular orbital leads to a configuration corresponding to the cationic D_1 state [cf. left hand side of Fig. 10(b)]. After 20 fs the main configuration in the dynamics has changed to the S_1 state which in the one electron picture ionizes to D_0 [cf. right hand side of Fig. 10(b)]. Notice, that the wavefunction is a superposition of many configurations and therefore an ionization process which is forbidden in the one-electron single determinant picture can become weakly allowed due to the minor contribution of other configurations.

The character of the excited state is also reflected in the time evolution of the diabatic populations, which can be estimated from the time-dependent oscillator strength averaged over the ensemble of trajectories presented in Fig. 11, since the $\pi\pi^*$ state exhibits a high oscillator strength compared to the $n\pi^*$ states. The oscillator strength decays within the first 50 fs corresponding to the depopulation of the diabatic $\pi\pi^*$ state and exhibits a recurrence at $\sim$75 fs.

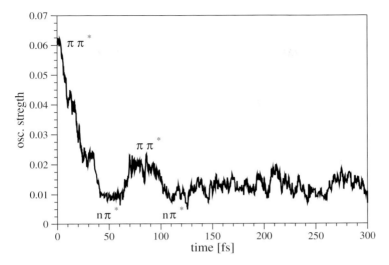

Fig. 11. Time-dependent oscillator strength for the transitions from the S_0 state to the excited state in which dynamics takes place in pyrazine averaged over the ensemble of trajectories. The population of the diabatic states ($\pi\pi^*/n\pi^*$) can be estimated, since the $\pi\pi^*$ state exhibits a high oscillator strength compared to the $n\pi^*$ states. Therefore, a high oscillator strength indicates population in the $\pi\pi^*$ state whereas low oscillator strength indicates population in the $n\pi^*$ state. Reprinted with permission from Ref. 27, Copyright 2010, American Institute of Physics.

In order to identify the normal modes responsible for the relaxation dynamics the time-dependent Cartesian coordinates of the ensemble of trajectories have been projected onto the normal modes of the equilibrium structure of pyrazine and the averaged values are presented in Fig. 12. As can be seen several modes are excited during the dynamics which in the Wilson notation are labeled by ν_1, ν_2, ν_{6a}, ν_{8a}, ν_{9a} and ν_{10a}. The dominant mode is the totally symmetric ν_{6a} mode which involves mainly the in-plane motion of the carbon and nitrogen atoms with a period of ∼60 fs. This mode is mainly responsible for the quantum beats present in the experimental and theoretically simulated TRPE spectra (cf. Fig. 8).

In summary, we have presented a theoretical approach for the simulation of TRPE spectra combining the nonadiabatic dynamics "on the fly" with approximate description of the electronic continuum which is used in the Stieltjes imaging approach. Our approach has been implemented in the framework of TDDFT and has been applied to study the ultrafast internal conversion (IC) between the S_2 and S_1 states in pyrazine. The IC takes place on a timescale of 20 fs leading to a change in the transition

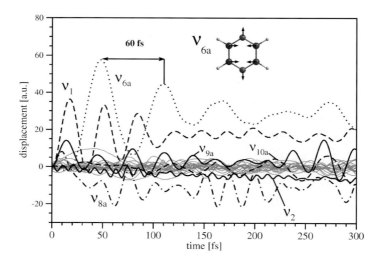

Fig. 12. Time-dependent normal mode displacements obtained by projection onto the equilibrium ground state normal coordinates of pyrazine averaged over 60 trajectories. The displacement of the dominant normal mode ν_{6a} exhibits a periodicity of ~60 fs. Reprinted with permission from Ref. 27, Copyright 2010, American Institute of Physics.

probabilities so that the initial ionization from the S_2 to the D_1 state turns over continuously to the ionization from the S_1 to the D_0 state. Since both ionization channels exhibit similar ionization energies the maximum of the simulated photoelectron distribution remains in an almost constant energy range which is in agreement with experimental findings. The agreement of simulated TRPE spectra obtained from both approaches DC and SI validates the use of SI in particular if the spectral moments can be efficiently calculated without full diagonalization of the Hamiltonian matrix. In summary, our results demonstrate that the approximate description of electronic continuum together with Stieltjes imaging provides quantitative agreement with the experimentally measured TRPE spectrum. Thus our approach represents a viable tool for the investigation and interpretation of the time-dependent photoionization processes in complex systems.

5.3. *Ultrafast photodynamics of adenine in microsolvated environment*

The ultrafast dynamics of DNA bases has been intensively studied in recent years in order to reveal molecular features which are responsible for their intrinsic photostability.[140] In the case of adenine, experimental studies have shown that isolated adenine in the gas phase returns nonradiatively to the ground state within about one picosecond after photoexcitation of the strongly absorbing $\pi - \pi^*$ electronic state.[141–143] In order to identify the mechanism of the nonradiative relaxation several theoretical studies have been performed with the aim to assign the conical intersections which dominate the relaxation process.[144–147] Recently, mixed quantum-classical dynamics simulations both using high level *ab initio* multireference CI[32] as well as semiempirical CI[52] have been performed giving a dynamical picture of the relaxation process. According to these studies, the relaxation proceeds in a two step mechanism. First, the initially excited state (S_3[32] or S_2[52]) relaxes rapidly to the lowest excited state (S_1) with a time constant of about 20 fs. The second slower step corresponds to the transition from the S_1 state to the electronic ground state with a time constant of about 500 fs.

A fundamental issue in the photophysics of nucleobases is the role played by water. The experimental study of adenine in solution[148] using femtosecond transient absorption spectroscopy has revealed that the lifetime of the S_1 state of adenine in water of 180 fs is about 50% shorter than in acetonitrile (440 fs) and is in general much shorter than in the gas phase (1.2 ps).[143] This shows that the realistic description of nucleobase dynamics requires the explicit inclusion of solvent effects which is a challenging task from the theoretical point of view. We wish to show here that the TDDFTB nonadiabatic dynamics represents a general and highly efficient method which can be used to simulate the nonadiabatic dynamics of biochromophores solvated by a large number of water molecules which are not accessible to high level *ab initio* efforts. Due to its accuracy which is comparable to the full TDDFT method, TDDFTB nonadiabatic dynamics can be used to investigate nonadiabatic processes in a whole variety of complex systems such as solvated biochromophores, photoreceptors or nanostructures which are of interest for materials science applications.

5.3.1. *Computational*

The system studied here consists of the adenine molecule solvated by 26 water molecules. The initial structure has been prepared in several steps starting with the equilibration of a water box consisting of 561 water molecules with 30 Å × 30 Å × 30 Å dimensions using constant temperature molecular dynamics at $T = 300$ K and the TIP-3P force field.[149] After equilibration, the adenine molecule has been inserted in the center of the water

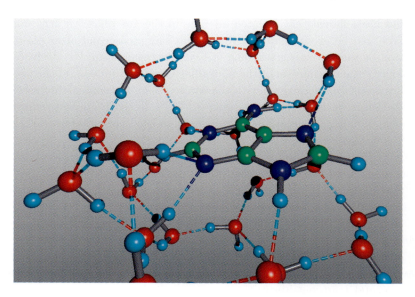

Fig. 13. The DFT/B3LYP optimized structure of microsolvated adenine. Reprinted with permission from Ref. 86, Copyright 2009, American Chemical Society.

box and further equilibrated by using the AMBER force field[150] for adenine. Subsequently, the first solvation shell has been isolated and optimized in the framework of DFT using the B3LYP functional[125] combined with the triple zeta valence plus polarization basis set (TZVP)[123] as well using the DFTB method. In order to check the accuracy of the TDDFTB method, the absorption spectrum for the DFT optimized structure shown in Fig. 13 has been calculated both using TDDFT with the same functional and basis set as above as well as using TDDFTB. The comparison of the absorption spectra in Fig. 14 demonstrates a qualitative agreement and gave us confidence to carry out the nonadiabatic dynamics simulations in the framework of TDDFTB. The first intense absorption line which corresponds to the $\pi - \pi^*$ transition in adenine is located at 250 nm in TDDFT [cf. arrow in Fig. 14(a)] and at 260 nm in TDDFTB [cf. arrow in Fig. 14(b)]. While these line positions are very similar in both methods, the $n - \pi^*$ transition is located at 240 nm within TDDFT and very close to the $\pi - \pi^*$ transition, at 268 nm, within TDDFTB. In fact, the relative position of the $n - \pi^*$ and $\pi - \pi^*$ transition within TDDFTB is analogous to the finding obtained by the *ab initio* multireference perturbation configuration interaction method (CIPSI) combined with the PCM-IEF solvation model.[151] However, notice

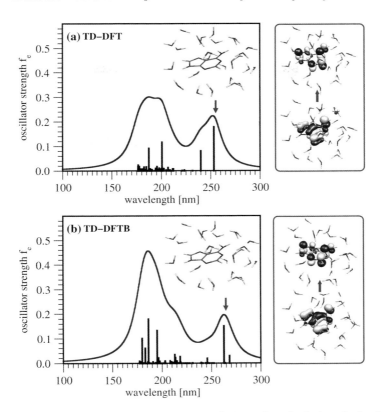

Fig. 14. Comparison of the absorption spectra of microsolvated adenine obtained using (a) full TDDFT with the hybrid B3LYP functional and (b) TDDFTB method. The insert shows the B3LYP/TZVP optimized structure used for spectrum calculation. The right panels show the character of the main excitation contributing to the first intense transition marked in the spectra by arrows. Reprinted with permission from Ref. 86, Copyright 2009, American Chemical Society.

that the proper description of excited states of chromophores interacting with water molecules is a particularly difficult issue in the framework of TDDFT and TDDFTB due to the possibility of long range charge transfer between water and the chromophore. In the case of adenine the long range charge transfer does not play a significant role at least in the equilibrium geometry. In order to study the photodynamics 100 initial conditions have been sampled from a 10 ps classical trajectory at $T = 300$ K. The trajectories have been propagated using our TDDFTB nonadiabatic dynamics starting in the third excited state S_3. Totally, seven excited states and the ground electronic state have been included in the simulation.

5.3.2. Influence of solvation on the nonradiative relaxation of adenine

In order to illustrate the applicability of our TDDFTB nonadiabatic dynamics for the investigation of complex systems we present here the simulation of the nonadiabatic relaxation of microsolvated adenine. From the experimental work in water solution it is known that adenine in water assumes two tautomeric forms termed 9H-adenine and 7H-adenine.[148] However, since 9H-adenine is dominant (~78%) we limit ourselves here only to study the photodynamics of that form. The optimized structure of microsolvated adenine presented in Fig. 13 shows that all nitrogen atoms of adenine as well as the two hydrogen atoms of the amino group are saturated by hydrogen bonds.

The time-dependent excited state populations obtained from the ensemble of 100 trajectories are shown in Fig. 15. The initially populated S_3 state is depopulated with a time constant of 16 fs and the population is transiently transferred to the lower lying S_2 and S_1 states as well as to a lesser extent to several other energetically close-lying states (S_4–S_7). Notice that no direct population transfer from the initially occupied S_3 state to the ground electronic state occurs. The S_0 state begins to be continuously populated from the S_2 and S_1 states starting at ~20 fs

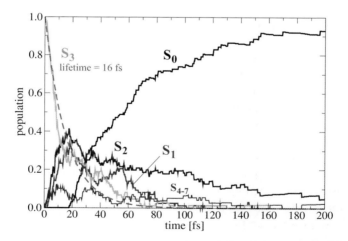

Fig. 15. Population of the ground and excited electronic states during the nonadiabatic dynamics simulation for microsolvated adenine. Reprinted with permission from Ref. 86, Copyright 2009, American Chemical Society.

with the full population transfer occurring on the time scale of 200 fs. It should be emphasized that the nature of the nonadiabatic dynamics of the microsolvated adenine is similar to that of isolated adenine simulated by using both the high level *ab initio* CI method[32] and the semiempirical CI method.[52] However, the transition to the electronic ground state in microsolvated adenine is significantly faster than in the gas phase (200 fs versus $\sim$550 fs[32,52]). Notice that previously theoretically calculated lifetimes of adenine in the gasphase are significantly shorter than the experimental ones.[143] However, the experimental values are strongly wavelength dependent and lie in the range between 1.2 ps and 9 ps (cf. Ref. 143). For the purpose of comparison we have also calculated the lifetimes of gasphase adenine in the frame of TDDFTB. The transition to the ground state is again a two step process where the S_2 to S_1 transition occurs with the time constant of 120 fs and S_1 to S_0 exhibits a time constant of 11 ps. However, the S_1 to S_0 transition is strongly wavelength dependent which means that it is highly sensitive to quantitative features of the potential energy surfaces which are difficult to reproduce by the available methods. The general trend of the shortening of the lifetime in water is in agreement with the experiments on adenine in solution.[148]

Thus the nonradiative relaxation of microsolvated adenine occurs in a two step process in which first the initially excited $\pi - \pi^*$ state (S_3) is depopulated on the time scale of 16 fs and subsequently the ground state S_0 is populated with a time constant of 200 fs. Notice that the populations of S_1 and S_2 states grow parallel (cf. Fig. 15) and both of them are depopulated as the population of the ground state grows.

Further insight in the relaxation process can be gained by examining the character of the excited electronic states and their relative energies along a selected trajectory as presented in Fig. 16. As can be seen, in the initial stage of the dynamics several excited states are very close to the initially excited S_3 state. After the initial excitation to the $\pi - \pi^*$ electronic state, within the first 10 fs the character of the electronic state changes to $n - \pi^*$ (cf. inserts at 0.25 fs and 7.25 fs in Fig. 16). This proximity of electronic states with different character leads to the coupling which induces several state switchings before the system reaches the S_1 electronic state after $\sim$75 fs. Subsequently, within $\sim$25 fs the crossing with the ground electronic state is reached and the trajectory continues to propagate in the ground electronic state. The higher density of electronic states in microsolvated adenine compared to the gas phase increases the number of pathways which

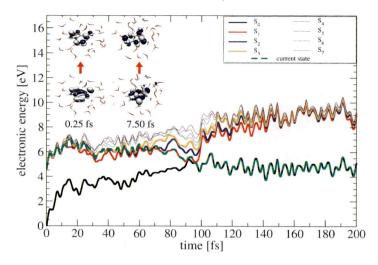

Fig. 16. The electronic state energy along a selected nonadiabatic trajectory in microsolvated adenine. The actual state in which the trajectory resides is labeled by the green dashed line. The insert on the left hand side shows the dominant electronic configurations at 0.25 ($\pi - \pi^*$) and 7.25 fs ($n - \pi^*$). Reprinted with permission from Ref. 86, Copyright 2009, American Chemical Society.

can lead to the crossing with the ground electronic state and thus causes faster nonradiative relaxation.

In summary, we have presented the combination of the tight-binding time-dependent density functional theory (TDDFTB) with Tully's surface hopping method on the example of the nonradiative relaxation of microsolvated adenine taking into account the first solvation shell. Our simulations have revealed that the nonradiative transition to the ground electronic state proceeds according to a two step mechanism involving the ultrafast relaxation of the initially excited $\pi - \pi^*$ state with a lifetime of 16 fs and subsequent transition to the ground state within 200 fs. Overall, the dynamics of microsolvated adenine is thus faster than the one of gas phase adenine.

Our results demonstrate that the TDDFTB nonadiabatic dynamics represents a useful qualitative tool for the investigation of photodynamics in complex systems which are beyond the reach of *ab initio* methods. This opens the possibility to investigate photoinduced dynamics in systems such as for example biochromophores interacting with the protein environments or solvent, light harvesting systems, biosensors, photonic nanoarchitectures, polymers etc. The knowledge about the mechanisms for nonradiative relaxation in these complex systems is mandatory in order to tune their properties for future applications.

6. Applications of the Field-Induced Surface-Hopping Method (FISH) for the Control of Ultrafast Dynamics by Shaped Laser Fields

The scope of the FISH method will be illustrated on two prototype examples for different application areas using shaped laser pulses for control of photoisomerization products and for optimal dynamic discrimination of similar molecules. In Sec. 6.1 we present the optimal pump-dump control of the trans-cis isomerization of a prototype Schiff base molecular switch with the aim to achieve selective isomerization. In Sec. 6.2 we theoretically investigate the optimal dynamic discrimination by shaped laser pulses between the very similar biomolecules flavin mononucleotide and riboflavin with identical absorption and emission spectra. Our FISH simulations utilize experimentally optimized laser fields and show that the fluorescence depletion ratio between two molecules can be manipulated with such fields, thus achieving discrimination between them. Moreover, these results validate for the first time the experimental optimal control technique applied on complex systems.

6.1. *Optimal control of trans-cis isomerization in a molecular switch*

Switchable molecules are used by Nature as photoreceptors in the vision process and can be also employed as building blocks for molecular electronic devices. Thus the control and the mechanism of selective photoswitching by tailored laser fields is an interesting general issue which can contribute to design of more effective photoswitches with desired functionality. The first application of our FISH method to the control of laser driven dynamics will be illustrated on the example of the optimal pump-dump photoisomerization of N-Methylethaniminium (N-MEI) with the chemical composition $[CH_3NH = CHCH_3]^+$. This molecule is representative for a broad class of Schiff base photoswitches and serves to demonstrate that our FISH method can be used to design laser pulses which induce selective trans-cis isomerization. Moreover, we show that selective photoswitching is realized by selecting isomerization pathways which avoid the region of conical intersections.[84]

6.1.1. *Computational*

In order to control the trans-cis isomerization in N-MEI we use our FISH method described in Sec. 4. For this purpose we employ the semiempirical AM1 configuration interaction (CI) method[152] for the description of the electronic states and for propagation of trajectories in the framework of MD "on the fly". Inclusion of all degrees of freedom in molecular dynamics is of conceptual importance even in the cases that few degrees of freedom might appear to dominate the dynamics. The AM1 CI method reproduces reasonably accurately both the spectroscopic properties as well as the shape of the potential energy surfaces of small Schiff base molecules.[1] Therefore we use this procedure which is computationally practicable for optimization of laser fields.

The simulation of the laser-induced dynamics is performed by (i) generating initial conditions for an ensemble of 30 trajectories sampled from a 10 K canonical Wigner distribution [Eq. (24)], (ii) for each trajectory propagated in the framework of AM1 CI MD "on the fly" the density matrix elements $\rho_{ij} = c_i^* c_j$ are calculated by numerical integration of the time-dependent Schrödinger equation (42) and accounting for the hopping to the ground state for small energy gaps between ground and excited state after the duration of the applied field is over. The nuclear trajectories are obtained by solution of Newton's equations of motion (44), (iii) finally, in order to determine in which electronic state the trajectory is propagated, the hopping probabilities are calculated according to Eq. (40) and the decision is made whether the trajectory is allowed to change the electronic state by using a random number generator.

We have optimized the analytically parametrized laser field using Eq. (41) with a genetic algorithm described briefly in Sec. 4 in order to selectively populate the cis isomer. The laser field entering Eq. (41) is parametrized in the time domain as a Gaussian pulse train according to

$$E(t) = \sum_n E_n \exp(-\eta_n (t - t_n)^2) \sin((\omega_n + \gamma_n (t - t_n))t),$$

where E_n represents the field amplitude, η_n describes the temporal width, t_n is the center of the Gaussian, ω_n is the pulse frequency, and γ_n is the linear chirp parameter. The optimization was achieved by minimizing the target functional $J[t_f] = (180 - |\phi(t_f)|) + 500\, E_{kin}(t_f)$ accounting both for a maximal torsion angle ϕ and a minimal kinetic energy of the molecule, preventing thermal back isomerization to the trans isomer, thus populating the cis isomer. For further details cf. Ref. 84. The convergence has been tested by applying the optimal pulse to an ensemble of 100 trajectories. The shape of our optimized pulse as well as field-induced population changes of the excited and ground states will be discussed below.

6.1.2. Selective trans-cis isomerization of N-MEI

The N-MEI has two isomers in the ground electronic state. Its global minimum structure is the trans isomer while the energy of the cis isomer is 0.13 eV higher. The optimal pulse inducing trans-cis isomerization is shown in the upper part of Fig. 17 and consists of two parts which are nearly overlapping. The maximum intensity of the optimal pulse is

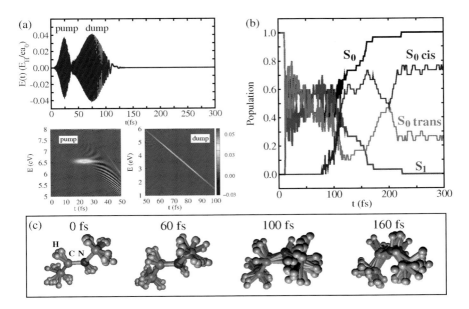

Fig. 17. (a) (Upper): Optimal pump-dump pulse driving the trans-cis isomerization of N-MEI, (lower): Wigner-Ville transform of the optimal pump (left) and dump (right) pulse showing the temporal distribution of the pulse energies. The intensity is represented by a greyscale bar. The Wigner–Ville transform is defined as $W(t,\omega) = 2\mathrm{Re}\int_0^\infty d\tau\, e^{-i\omega\tau} E^*(t+\tau/2) E(t-\tau/2)$. (b) Time-dependent populations of the S_0 and S_1 electronic states. For the ground state, also the populations of the trans and the cis isomers are shown. The Rabi oscillations are present during the first 100 fs. (c) Snapshots of the laser-induced dynamics. Reprinted from Ref. 84. Copyright 2009, American Physical Society.

$1.7 \times 10^{14}\,\mathrm{Wcm}^{-2}$ which is in the regime of strong but not ultrastrong fields. The time-energy structure of the optimal pulse obtained by the Wigner–Ville transformation [cf. bottom part of Fig. 17(a)] shows that the pump subpulse has constant energy centered around 6.6 eV while the dump pulse is linearly down chirped. The energy of the dump pulse varies from 6 eV to less than 2 eV in the time interval between 50 and 100 fs [cf. Fig. 17(a)]. Such large bandwidth has been recently realized by white light continuum pulse shaping.[153] The laser induced population dynamics presented in Fig. 17(b) shows that the pump pulse depopulates the ground state after ∼20 fs. During the subsequent 100 fs the populations of the ground and excited states exhibit Rabi oscillations around the average value of 50%. Within this period, the dump pulse successively depopulates the excited electronic state, before the energy gap closes and the conical intersection is reached, thus steering the dynamics towards the cis isomer. Since during the excited state dynamics the energy gap between the excited and the ground electronic state becomes smaller as the system performs the rotation around the C=N bond [cf. Fig. 17(c)] the energy of the dump pulse decreases with time (down-chirp) in order to satisfy the resonance condition. The selectivity of the isomerization process is reflected in the time-dependent population of the cis and trans isomers [cf. Fig. 17(b)] giving rise to the final occupation of the cis isomer of ∼75%. The occupation of the cis isomer is not 100% due to competing pathways through the conical intersection which start to dominate after the pulse terminates. This is the reason why the population of the cis isomer changes after the laser pulse has been switched off. Notice, that excitation with an unshaped pump pulse and subsequent field free isomerization through the conical intersection between the first excited singlet state and the ground state leads to the cis isomer only with a yield of ∼30%. The reason for this is that the excess of energy gathered after passing through the conical intersection induces a hot ground state and the return to the trans isomer prevails. This is an interesting prototype example showing that it might be difficult to achieve the desired photochemical products by passing through a conical intersection. However, the new pathways reachable by shaped laser fields can easily suppress the passage through the conical intersection and can eventually lead to selective isomerization.

The snapshots of the laser-controlled dynamics shown in Fig. 17(c) illustrate that the cis isomer is reached within 160 fs and together with the time evolution of the electronic state population [cf. Fig. 17(b)] confirm that the pathway through the conical intersection is suppressed. Thus, the optimal pump dump control can be used to efficiently drive the selective photoisomerization of molecular switches.

6.2. Optimal dynamic discrimination

We wish to reveal the processes responsible for the discrimination of the two very similar molecules flavin mononucleotide (FMN) and riboflavin (RBF) using optimally shaped laser fields[90] which has been recently demonstrated experimentally. The general concept of the optimal dynamic discrimination (ODD) has been recently proposed by Rabitz and Wolf et al.[87,88] The idea of the ODD relies on a theoretical analysis which has shown that quantum systems differing even infinitesimally may be distinguished by means of their dynamics when a suitably shaped ultrafast control field is applied. In the case of the two similar flavins (differing only by replacement of H by $PO(OH)_2$) the controlled depletion of the fluorescence signal has been used as a discriminating observable.[89] The schematic representation of the discrimination process is presented in Fig. 18. In general,

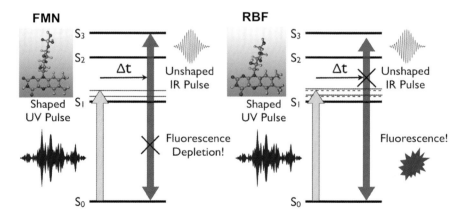

Fig. 18. Schematic illustration of the discrimination of FMN and RBF by fluorescence depletion. Excitation with a shaped UV laser pulse leads to transition from S_0 to S_1 state, as indicated by the light grey arrow. After a time-delay Δt during which dynamical processes take place, an unshaped IR pulse is applied. In the case of FMN (left part of the Figure), this leads to transitions to higher excited states where irreversible processes such as ionization can occur (dark arrow), consequently the fluorescence gets depleted (crossed dark arrow). For RBF (right part of the Figure), excitation to higher states is less favorable (crossed dark arrow), and fluorescence will remain stronger than in FMN (dark arrow). With differently shaped UV pulses, also the reverse situation is possible.

a shaped UV pulse excites both molecules to the S_1 state and induces ultrafast dynamics which can follow slightly different pathways in both molecules. After a specified time delay Δt a second unshaped IR pulse excites the molecule further to higher excited states and can induce dissipative processes such as ionization which lead to irreversible depopulation of the S_1 state, and thus to depletion of the fluorescence signal in one of the species (cf. left part of Fig. 18) and not in the other one (cf. right part of Fig. 18). Since for both molecules depletion can be minimized and maximized independently the total fluorescence yield obtained can be used to quantitatively determine the amounts of both species.[89] Although in this study only flavins have been considered, the results should be broadly applicable to control systems whose static spectra show essentially indistinguishable features. In particular, this should allow in the future the selective identification of target molecules in the presence of structurally and spectroscopically similar background. This is an important issue in multiple areas of science and engineering. Our FISH method offers a unique opportunity not only to perform multistate dynamics "on the fly" and to optimize the laser pulses but also to apply directly the experimentally optimized pulses and thus to reveal the processes which enable discrimination of similar chromophores.

6.2.1. *Computational*

In order to achieve discrimination between FMN and RBF in the framework of our FISH method we use experimentally optimized pulses obtained in the framework of ODD[87,88] which maximize and minimize the fluorescence depletion ratio of both molecules, respectively.

For the multistate dynamics in the ground and the nine lowest excited singlet states (S_0–S_9) under the influence of the shaped laser fields, we describe the electronic structure employing the semiempirical PM3 CI method.[154] We use an active space of 11 occupied and 6 virtual orbitals, taking into account all single excitations out of 12 reference configurations which have been identified as the leading configurations in the electronic states under consideration. In this way, the spectroscopic properties of the two flavin molecules are reasonably accurately reproduced as evidenced by the comparison with TDDFT results obtained using the hybrid B3LYP functional. They are also in agreement with experimental findings as shown in Fig. 19. We calculate the needed energies, forces, nonadiabatic couplings, and transition dipole moments "on the fly" regarding all degrees

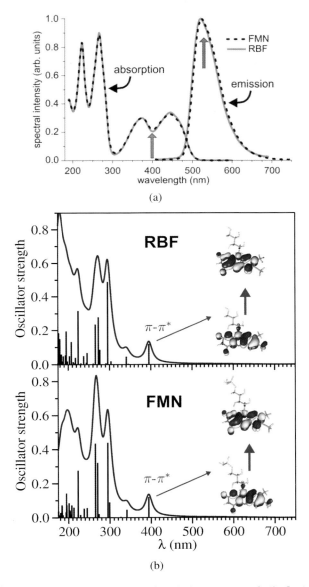

Fig. 19. (a) Experimental absorption and emission spectra of riboflavin (RBF) and flavin mononucleotide (FMN). Reprinted with permission from Ref. 89. Copyright 2009, American Physical Society. The arrows at 400 nm and 530 nm indicate the wavelengths of the shaped UV control pulses and the collected fluorescence, respectively. (b) Theoretical absorption spectra of RBF and FMN obtained using the semiempirical PM3 CI method. The leading HOMO-LUMO excitation for the lowest energy electronic transition, which has $\pi - \pi^*$ character, is depicted.

of freedom. The nonadiabatic couplings and transition dipole moments are needed between all states and are obtained using the method developed by Thiel et al.[53]

For the simulation of shaped field-induced surface-hopping dynamics, the following three steps are needed: First, initial conditions (30 coordinates and momenta) are generated by sampling a 10 ps long ground state trajectory at 300 K which was obtained by using the semiempirical PM3 method[154] for both molecules. Second, the density matrix elements formulated in Eqs. (32) and (33) were calculated along the trajectories. For this purpose, numerical integration of the time-dependent Schrödinger equation (43) accounting for nonadiabatic couplings has been carried out. This ensures that after the duration of the applied field is over, multistate nonadiabatic dynamics can be further performed. The nuclear dynamics has been carried out using the Langevin equation of motion [Eq. (45)], which is integrated employing a modified version of the velocity Verlet algorithm.[155] For the atomic friction, an empirical coefficient of $\gamma = 91.0\,\mathrm{ps}^{-1}$ for water environment is used. In this way, dissipative effects on the nuclear motion are approximately accounted for and the comparison of theoretical results with the experiment which was carried out in water can be approximately made. Moreover, in the case of long pulses, the dissipative effects play a role due to the excess of energy gained during the dynamics. Third, in order to determine in which state the trajectories are propagated, the hopping probabilities are calculated using Eq. (7) on the basis of the solution of the time-dependent Schrödinger equation (43).

The laser fields obtained from experiment[89] are used in the simulations according to Eq. (43). For the shaped UV pulses, we use 50 experimental spectral phases and amplitudes in order to reconstruct the field in the time domain according to

$$E(t) = \sum_n A_n e^{i(\omega_n t + \phi_n)}, \qquad (48)$$

where A_n represents the spectral amplitude, ω_n is the frequency, and ϕ_n represents the spectral phase. The wavelengths corresponding to the frequencies $\left(\lambda = \frac{2\pi c}{\omega}\right)$ lie in the range between 394.6 nm and 405.6 nm. The reconstructed pulses have a duration in the time domain of ~5 ps and a maximum amplitude of $\sim 6 \cdot 10^{11}\,\mathrm{W\,cm^{-2}}$. The unshaped infrared probe pulse with a wavelength of 800 nm has a maximum amplitude of $\sim 3 \cdot 10^{12}\,\mathrm{W\,cm^{-2}}$ and a Gaussian envelope with a full width at half maximum of 100 fs.

Since the fluorescence depletion relies on irreversible processes such as ionization, these effects must be introduced approximately in the Schrödinger equation (43) for the electronic states. This is possible to model by adding an imaginary component $i\Gamma$ to the energy of the highest excited state S_9 which lies close to the experimentally determined ionization limit in water.[156] The parameter Γ can be chosen in order to achieve the irreversible population decay from S_9. Subsequently, the time-dependent coefficients along the trajectories have to be recalculated and the hopping from the S_9 state to the ionized state has to be considered. By averaging over the ensemble of trajectories the ionized populations P_{ioniz} are obtained. This allows for the description of the fluorescence depletion after the duration of the pulses is over which corresponds to the population of the excited states decreased by ionization. Specifically, this is achieved in the experiment by taking the difference between the two fluorescence intensities corresponding to the shaped UV pulse alone and to both the shaped UV and the IR probe pulse, normalized to the fluorescence intensity that corresponds to the shaped UV pulse alone. In order to calculate this depletion signal from our populations, we use the ionized population P_{ioniz} to determine the fluorescence depletion D as

$$D = \frac{P_{\text{ioniz}}(UV + IR) - P_{\text{ioniz}}(UV)}{1 - P_{\text{ioniz}}(UV)}.$$

6.2.2. *Discrimination of RBF and FMN by shaped laser pulses*

The RBF and FMN molecules represent particularly challenging systems for the optical discrimination since they have nearly identical absorption and fluorescence spectra as can be seen from Fig. 19(a). The theoretically calculated absorption spectra of RBF and FMN presented in Fig. 19(b) exhibit very strong similarities as well. The electronic spectroscopy of flavins is primarily associated with their common chromophore $\pi - \pi^*$ type transitions localized on the isoalloxazine ring and is influenced indirectly and only very slightly by the chemical moieties (H versus $PO(OH)_2$) on the side chains. In order to discriminate both species experimentally, the laser control employing a shaped ultraviolet component at 400 nm followed by an unshaped infrared 100 fs component at 800 nm with a time delay of 500 fs is used. The excitation with an unshaped UV component leads to indistinguishable fluorescence depletion signals of 26% for RBF and FMN as shown in Fig. 20(a). The optimal UV pulses allowing for discrimination have been obtained experimentally by closed-loop optimization of the fluorescence depletion ratio of FMN over RBF and vice versa. Specifically, the maximization of the FMN over RBF ratio yielded a shaped UV/IR pulse pair (termed pulse 1 in the following) that leads to distinguishable fluorescence depletions values of 12.6% for RBF and 16.4% for FMN [cf. blue and gold histograms on the left side of Fig. 20(a)]. Oppositely, minimization of the FMN over RBF ratio has yielded a second pulse pair (termed pulse 2 in the following) that achieves approximately the same level of discrimination but reverses the ordering of the depletion signals [cf. gold and blue histograms on the right side of Fig. 20(a)]. The temporal shapes of the experimental pulses achieving the minimization and maximization of the depletion ratio are shown in Fig. 20(b). The pulses have been reconstructed by extracting phases and amplitudes of the spectral components from the experimental FROG trace. The time-wavelength structure of both pulses obtained by Wigner–Ville transform is shown as well in the Fig. 20(c).

The population dynamics induced by both pulses in RBF and FMN is presented in Figs. 21 and 22. Both pulses 1 and 2 lead to a smaller population of the higher excited states (S_{2-9}) in RBF before the IR component has been applied. It should be pointed out, that if no irreversible processes such as ionization from the higher excited states are taken into account both pulses invoke similar population dynamics and lead to the return of the population from higher excited states to the S_1 state after the pulses have ceased. Therefore, in order to describe fluorescence depletion the irreversible population of the higher excited states has to be introduced. We achieve this by adding an imaginary component to the energy of highest S_9 excited state which lies above the ionization limit as described in Sec. 6.2.1.

Fig. 20. (a) Absolute experimental RBF and FMN depletion signals for optimized UV pulse shapes at time delay for the IR pulse of 500 fs. Reprinted with permission from Ref. 89. Copyright 2009, American Physical Society. Left part: Depletion for the optimal pulse for maximizing the ratio D(FMN)/D(RBF), right part: Depletion for the optimal pulse for minimizing the ratio D(FMN)/D(RBF). Absolute depletions induced by the transform-limited pulse for both RBF (black) and FMN (grey) are statistically equivalent at 26%. Optimal pulses pull apart the RBF (blue) and FMN (gold) distributions to achieve discrimination between the two molecules. (b) Temporal structure of the two shaped pulses for maximizing (left part) and minimizing (right part) the depletion ratio D(FMN)/D(RBF). (c) Wigner–Ville transforms of the two shaped pulses, left: maximization, right: minimization of D(FMN)/D(RBF). Red corresponds to positive, blue to negative values.

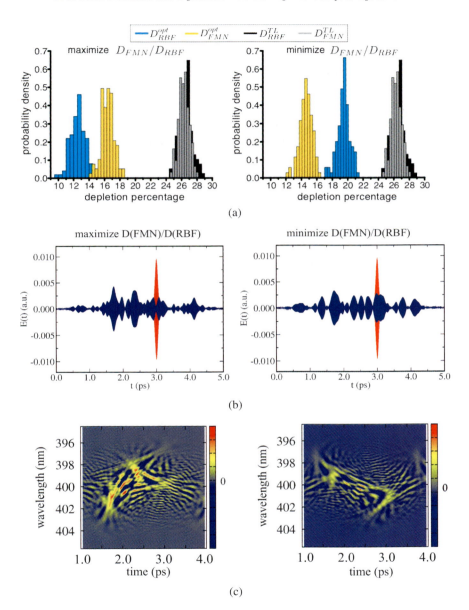

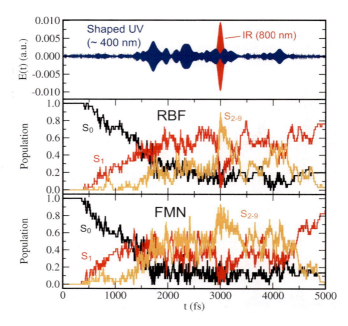

Fig. 21. Upper panel: Temporal structure of the shaped pulse 1 for maximization of D(FMN)/D(RBF) (blue) and of the unshaped IR probe pulse (red). Middle panel: Time-dependent populations of the electronic states S_0 (black), S_1 (red), and $S_2 - S_9$ (orange) in RBF driven by the pulses shown in the upper panel. Lower panel: Time-dependent populations of the electronic states S_0 (black), S_1 (red), and $S_2 - S_9$ (orange) in FMN driven by the pulses shown in the upper panel.

The value of the imaginary component has been calibrated for both molecules so that with an unshaped UV pulse both molecules exhibit identical depletion ratios. Subsequently, we have used these values to calculate the ionized populations presented in Fig. 23 using the shaped pulses 1 and 2. The application of both pulses leads to the ionization which sets in RBF after 3 ps and is systematically lower for pulse 1 than for pulse 2. The situation is opposite for FMN where the pulse 1 causes stronger ionization than the pulse 2. This means that for pulse 1 for maximizing the ratio D(FMN)/D(RBF) the fluorescence depletion is stronger for FMN whereas for pulse 2 for minimizing that ratio the fluorescence depletion is stronger in RBF. The experimental and theoretical fluorescence depletion values are shown in Fig. 24. As can be seen, the calculated depletion ratios of FMN and RBF are reversed by different pulses in agreement with the experimental depletion ratios. Notice, that our method nicely validates

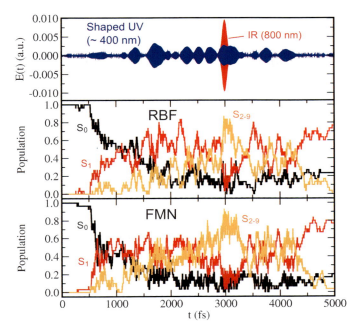

Fig. 22. Upper panel: Temporal structure of the shaped pulse 2 for minimization of D(FMN)/D(RBF) (blue) and of the unshaped IR probe pulse (red). Middle panel: Time-dependent populations of the electronic states S_0 (black), S_1 (red), and $S_2 - S_9$ (orange) in RBF driven by the pulses shown in the upper panel. Lower panel: Time-dependent populations of the electronic states S_0 (black), S_1 (red), and $S_2 - S_9$ (orange) in FMN driven by the pulses shown in the upper panel.

the general trend of the depletion ratios induced by pulse 1 and pulse 2.[90] However, quantitative agreement should not be expected due to the approximate consideration of ionization which has been introduced only at a model level. Furthermore, the measurements have been performed in water solution and the theoretical approach accounts for the solvent effects only at the level of Langevin dynamics.

In order to reveal the physical background responsible for successful discrimination we have calculated the averaged transition dipole moments to higher excited states (cf. Fig. 25). The pulse 1 induces dynamical pathways which exhibit systematically higher transition dipole moments for FMN than for RBF which explains the higher depletion of the fluorescence in FMN. In contrast, for the pulse 2 at later times after 3 ps this behavior is reversed leading to higher transition dipole moments for RBF. Thus we can conclude that successful optical discrimination relies on driving the system

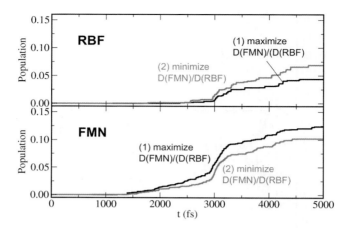

Fig. 23. Upper panel: Ionized populations of RBF for the dynamics driven by pulse 1 (black) and pulse 2 (grey). Lower panel: Ionized populations of FMN for the dynamics driven by pulse 1 (black) and pulse 2 (grey).

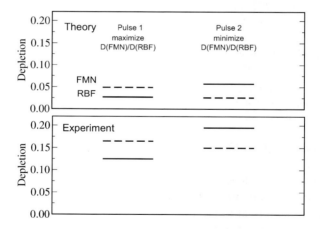

Fig. 24. Comparison of theoretical (upper panel) and experimental (lower panel) depletion ratios for FMN (dashed line) and RBF (full line) achieved by pulse 1 (left part) and pulse 2 (right part), respectively.

to these regions of the potential energy surfaces where the one or the other molecule exhibits systematically higher transition dipole moments and thus can be more effectively excited.[90] This might represent a general feature that can be exploited for the discrimination between similar molecules.

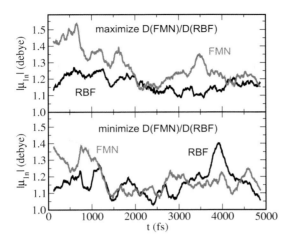

Fig. 25. Upper panel: Time-dependent averaged transition dipole moments from S_1 to the higher excited states $S_2 - S_9$ for the dynamics driven by pulse 1 for FMN (grey) and RBF (black). Lower panel: Time-dependent averaged transition dipole moments from S_1 to the higher excited states $S_2 - S_9$ for the dynamics driven by pulse 2 for FMN (grey) and RBF (black).

In summary, our FISH method accounting for the full complexity of the dynamical behavior of the FMN and RBF molecules in their environment, using ODD experimental pulses, has validated the conceptual scope of discrimination between almost identical molecules by shaped laser fields.

7. Conclusions and Outlook

We have presented a general theoretical approach for the simulation and control of ultrafast processes in complex molecular systems. Our methodological developments are based on the combination of quantum chemical nonadiabatic dynamics "on the fly" with the Wigner distribution approach for simulation and control of laser-induced ultrafast processes. Specifically, we have developed an approach for the nonadiabatic dynamics in the framework of TDDFT and TDDFTB using localized basis sets, which is generally applicable to a large class of molecules and clusters.

Furthermore, the "field-induced surface hopping" (FISH) method is introduced, allowing us to include laser fields directly into the nonadiabatic molecular dynamics simulations and thus to realistically model their influence on ultrafast processes. In particular, this approach has been combined with genetic algorithms to design shaped laser pulses which can drive a variety of processes.

The applications of our approaches have been illustrated on selected examples which serve to demonstrate their general scope as well as the ability to accurately simulate experimental ultrafast observables and to assign them to underlying dynamical processes. In particular, an approach for the simulation of TRPES has been developed representing a powerful tool to identify nonadiabatic processes involving conical intersections. For the first time, we have demonstrated that experimentally optimized laser fields can be directly used in the framework of the FISH method to reveal dynamical processes behind the optimal control. In addition, the FISH method combined with the optimal control theory allows us to predict forms of laser fields capable to steer molecular dynamics in complex systems such as large molecules and nanosystems in different environments. Altogether, our approaches based on the classical molecular dynamics and accounting for electronic transitions induced by both nonadiabatic effects as well as by light open new avenues for studying the femtochemistry of attractive molecular and nano-systems which were not accessible earlier due to their complexity.

Acknowledgments

We are grateful to Ute Werner and Matthias Wohlgemuth who have contributed to parts of the nonadiabatic dynamics applications presented here. We also acknowledge the contribution of our experimental and theoretical partners Prof. T. Suzuki, Prof. G. Seifert and Prof. H. Lischka. We extend our thanks to Prof. J.-P. Wolf and Prof. H. Rabitz for stimulating cooperation and for providing us with experimental results on the discrimination of flavin molecules. Prof. W. Thiel we thank for providing us with the nonadiabatic MNDO code. Finally, we would also like to acknowledge the financial support from the Deutsche Forschungsgemeinschaft in the frame of SFB 450, the Emmy Noether Program, MI-1236 (R.M.), as well as the Fonds der Chemischen Industrie (J.P.).

References

1. J. Michl and V. Bonačić-Koutecký, *Electronic Aspects of Organic Photochemistry* (John Wiley & Sons Inc., New York, 1990).

2. W. Domcke, D.R. Yarkony and H. Köppel, (Eds.), *Conical Intersections. Electronic Structure, Dynamics and Spectroscopy*, (World Scientific, Singapore, 2004).
3. M.A. Robb, M. Garavelli, M. Olivucci and F. Bernardi, *Rev. Comp. Chem.* **15**, 87 (2000).
4. A.H. Zewail, *J. Phys. Chem. A* **104**, 5660 (2000).
5. M. Ben-Nun and T.J. Martinez, *Adv. Chem. Phys.* **121**, 439 (2002).
6. V. Bonačić-Koutecký and R. Mitrić, *Chem. Rev.* **105**, 11 (2005).
7. A. Stolow, *Annu. Rev. Phys. Chem.* **54**, 89 (2003).
8. A. Stolow, A.E. Bragg and D.M. Neumark, *Chem. Rev.* **104**, 1719 (2004).
9. A. Stolow and J. Underwood, *Adv. Chem. Phys.* **139**, 497 (2008).
10. T. Suzuki, *Annu. Rev. Phys. Chem.* **57**, 555 (2006).
11. L. Wöste and O. Kühn, (Eds.), *Analysis and Control of Ultrafast Photoinduced Reactions* (Springer Series in Chemical Physics 87, 2007).
12. J. Manz and L. Wöste, (Eds.), *Femtosecond Chemistry Vol. 1 and 2* (VCH Verlagsgesellschaft mbH, Weinheim, Germany, 1995).
13. I.V. Hertel and W. Radloff, *Rep. Prog. Phys.* **69**, 1897 (2006).
14. M. Dantus, R.M. Bowman, M. Gruebele and A.H. Zewail, *J. Chem. Phys.* **91**, 7437 (1989).
15. A. Mokhtari, P. Cong, J.L. Herek and A.H. Zewail, *Nature* **348**, 225 (1990).
16. A.H. Zewail, *Faraday Discuss. Chem. Soc.* **91**, 207 (1991).
17. V. Bonačić-Koutecký, J. Koutecký and J. Michl, *Angew. Chem. Int. Ed. Engl.* **26**, 170 (1987).
18. F. Bernardi, M. Olivucci and M.A. Robb, *Chem. Soc. Rev.* **25**, 321 (1996).
19. M. Ben-Nun and T.J. Martinez, *Chem. Phys.* **259**, 237 (2000).
20. H. Lischka, M. Dallos and R. Shepard, *Mol. Phys.* **100**, 1647 (2002).
21. H. Lischka, M. Dallos, P.G. Szalay, D.R. Yarkony and R. Shepard, *J. Chem. Phys.* **120**, 7322 (2004).
22. D.R. Yarkony, *J. Phys. Chem. A* **105**, 6277 (2001).
23. M. Hartmann, J. Pittner and V. Bonačić-Koutecký, *J. Chem. Phys.* **114**, 2123 (2001).
24. R. Mitrić, V. Bonačić-Koutecký, J. Pittner and H. Lischka, *J. Chem. Phys.* **125**, 024303 (2006).
25. U. Werner, R. Mitrić, T. Suzuki and V. Bonačić-Koutecký, *Chem. Phys.* **349**, 319 (2008).
26. R. Mitrić, U. Werner and V. Bonačić-Koutecký, *J. Chem. Phys.* **129**, 164118 (2008).
27. U. Werner, R. Mitrić and V. Bonačić-Koutecký, *J. Chem. Phys.* **132**, 174301 (2010).
28. N.L. Doltsinis and D. Marx, *Phys. Rev. Lett.* **88**, 166402 (2002).
29. C.F. Craig, W.R. Duncan and O.V. Prezhdo, *Phys. Rev. Lett.* **95**, 163001 (2005).
30. C. Ciminelli, G. Granucci and M. Persico, *Chem Eur. J.* **10**, 2327 (2004).
31. M. Barbatti, G. Granucci, M. Persico, M. Ruckenbauer, M. Vazdar, M. Eckert-Maksić and H. Lischka , *J. Photochem. Photobiol. A.: Chem.* **190**, 228 (2007).

32. M. Barbatti and H. Lischka, *J. Am. Chem. Soc.* **130**, 6831 (2008).
33. R. Car and M. Parrinello, *Phys. Rev. Lett.* **55**, 2471 (1985).
34. E. Wigner, *Phys. Rev.* **40**, 749 (1932).
35. M. Hillery, R.F. O'Connel, M.O. Scully and E.P. Wigner, *Phys. Rep.* **106**, 121 (1984).
36. Z. Li, J.-Y. Fang and C.C. Martens, *J. Chem. Phys.* **104**, 6919 (1996).
37. M. Hartmann, J. Pittner, V. Bonačić-Koutecký, A. Heidenreich and J. Jortner, *J. Chem. Phys.* **108**, 3096 (1998).
38. M. Hartmann, J. Pittner, V. Bonačić-Koutecký, A. Heidenreich and J. Jortner, *J. Phys. Chem. A* **102**, 4069 (1998).
39. M. Hartmann, J. Pittner and V. Bonačić-Koutecký, *J. Chem. Phys.* **114**, 2106 (2001).
40. S. Mukamel, *Principles of Nonlinear Optical Spectroscopy* (Oxford University Press, 1995).
41. A. Donoso and C.C. Martens, *Phys. Rev. Lett.* **87**, 223202 (2001).
42. A. Donoso, Y. Zheng and C. Martens, *J. Chem. Phys.* **119**, 5010 (2003).
43. M. Ben-Nun, J. Quenneville and T.J. Martínez, *J. Phys. Chem. A* **104**, 5161 (2000).
44. J.C. Tully, *J. Chem. Phys.* **93**, 1061 (1990).
45. J.C. Tully, *Faraday Discuss.* **110**, 407 (1998).
46. J-Y. Fang and S. Hammes-Schiffer, *J. Chem. Phys.* **110**, 11166 (1999).
47. G. Granucci and M. Persico, *J. Chem. Phys.* **126**, 134114 (2007).
48. E. Tapavicza, I. Tavernelli and U. Rothlisberger, *Phys. Rev. Lett.* **98**(2), 023001 (2007).
49. E. Tapavicza, I. Tavernelli, U. Rothlisberger, C. Filippi and M.E. Casida, *J. Chem. Phys.* **129**, 124108 (2008).
50. I. Tavernelli, E. Tapavicza and U. Rothlisberger, *J. Chem. Phys.* **130**, 124107 (2009).
51. G. Granucci, M. Persico and A. Toniolo, *J. Chem. Phys.* **114**, 10608 (2001).
52. E. Fabiano and W. Thiel, *J. Phys. Chem. A* **112**, 6859 (2008).
53. A. Koslowski, M.E. Beck and W. Thiel, *J. Comput. Chem.* **24**, 714 (2003); S. Patchkovskii, A. Koslowski and W. Thiel, *Theor. Chem. Acc.* **114**, 84 (2005).
54. Z. Lan, E. Fabiano and W. Thiel, *J. Phys. Chem. B* **113**, 3548 (2009).
55. A. Toniolo, S. Olsen, L. Manohar and T.J. Martinez, *Farad. Discuss.* **127**, 149 (2004).
56. A.M. Virshup, C. Punwong, T.V. Pogorelov, B.A. Lindquist, C. Ko and T.J. Martinez, *J. Phys. Chem. B* **113**, 3280 (2009).
57. L.V. Schäfer, G. Groenhof, M. Boggio-Pasqua, M.A. Robb and H. Grubmüller, *PLoS Comp. Biol.* **4**, e1000034 (2008).
58. G. Groenhof, L.V. Schäfer, M. Boggio-Pasqua, H. Grubmüller and M.A. Robb, *J. Am. Chem. Soc.* **130**, 3250 (2008).
59. M. Boggio-Pasqua, M.A. Robb and G. Groenhof, *J. Am. Chem. Soc.* **131**, 13580 (2009).
60. C. Ciminelli, G. Granucci and M. Persico, *Chem. Phys.* **349**, 325 (2008).
61. T. Brixner and G. Gerber, *ChemPhysChem* **4**, 418 (2003).

62. M. Dantus and V.V. Lozovoy, *Chem. Rev.* **104**, 1813 (2004).
63. D.J. Tannor and S.A. Rice, *J. Chem. Phys.* **83**, 5013 (1985).
64. D.J. Tannor and S.A. Rice, *Adv. Chem. Phys.* **70**, 441 (1988).
65. P. Brumer and M. Shapiro, *Faraday Discuss. Chem. Soc.* **82**, 177 (1986).
66. M. Shapiro and P. Brumer, *J. Chem. Phys.* **84**, 4103 (1986).
67. M. Shapiro and P. Brumer, *Int. Rev. Phys. Chem.* **13**, 187 (1994).
68. D.J. Tannor and S.A. Rice, *J. Chem. Phys.* **85**, 5805 (1986).
69. A.P. Peirce, M.A. Dahleh and H. Rabitz, *Phys. Rev. A* **37**, 4950 (1988).
70. S. Shi and H. Rabitz, *Chem. Phys.* **139**, 185 (1989).
71. R. Kosloff, S.A. Rice, P. Gaspard, S. Tersigni and D.J. Tannor, *Chem. Phys.* **139**, 201 (1989).
72. S.H. Tersigni, P. Gaspard and S.A. Rice, *J. Chem. Phys.* **93**, 1670 (1990).
73. H. Rabitz and S. Shi, *Adv. Mol. Vibr. Collision Dyn.* **1A**, 187 (1991).
74. W.S. Warren, H. Rabitz and M. Dahleh, *Science* **259**, 1581 (1993).
75. T. Baumert, T. Brixner, V. Seyfried, M. Strehle and G. Gerber, *Appl. Phys. B* **65**, 779 (1997).
76. D. Yelin, D. Meshulach and Y. Silberberg, *Opt. Lett.* **22**, 1793 (1997).
77. A. Efimov, M.D. Moores, N.M. Beach, J.L. Krause and D.H. Reitze, *Opt. Lett.* **23**, 1915 (1998).
78. E. Zeek, K. Maginnis, S. Backus, U. Russek, M.M. Murnane, G. Mourou, H.C. Kapteyn and G. Vdovin, *Opt. Lett.* **24**, 493 (1999).
79. E. Zeek, R. Bartels, M.M. Murnane, H.C. Kapteyn, S. Backus and G. Vdovin, *Opt. Lett.* **25**, 587 (2000).
80. R.S. Judson and H. Rabitz, *Phys. Rev. Lett.* **62**, 1500 (1992).
81. A. Assion, T. Baumert, M. Bergt, T. Brixner, B. Kiefer, V. Seyfried, M. Strehle and G. Gerber, *Science* **282**, 919 (1998).
82. R. Mitrić, M. Hartmann, J. Pittner and V. Bonačić-Koutecký, *J. Phys. Chem. A* **106**, 10477 (2002).
83. R. Mitrić and V. Bonačić-Koutecký, *Phys. Rev. A* **76**, 031405(R) (2007).
84. R. Mitrić, J. Petersen and V. Bonačić-Koutecký, *Phys. Rev. A* **79**, 053416 (2009).
85. Y.I. Suzuki, T. Fuji, T. Horio, T. Suzuki, *J. Chem. Phys.* **132**, 174302 (2010).
86. R. Mitrić, U. Werner, M. Wohlgemuth, G. Seifert and V. Bonačić-Koutecký, *J. Phys. Chem. A* **113**, 12700 (2009).
87. B. Li, G. Turinici, V. Ramakrishna and H. Rabitz, *J. Phys. Chem. B* **106**, 8125 (2002).
88. B. Li, H. Rabitz and J.-P. Wolf, *J. Chem. Phys.* **122**, 154103 (2005).
89. M. Roth, L. Guyon, J. Roslund, V. Boutou, F. Courvoisier, J.-P. Wolf and H. Rabitz, *Phys. Rev. Lett.* **102**, 253001 (2009).
90. J. Petersen, R. Mitrić, V. Bonačić-Koutecký, J.-P. Wolf, J. Roslund and H. Rabitz *Phys. Rev. Lett.* **105**, 073003 (2010).
91. A. Dreuw and M. Head-Gordon, *Chem. Rev.* **105**, 4009 (2005).
92. S. Kümmel and L. Kronik, *Rev. Mod. Phys.* **80**, 3 (2008).
93. T. Leininger, H. Stoll, H.-J. Werner and A. Savin, *Chem. Phys. Lett.* **275**, 151 (1997).

94. H. Iikura, T. Tsuneda, T. Yanai and K.J. Hirao, *Chem. Phys.* **115**, 3540 (2001).
95. T. Yanai, D.P. Tew and N. C Handy, *Chem. Phys. Lett.* **393**, 51 (2004).
96. R. Baer and D. Neuhauser, *Phys. Rev. Lett.* **94**, 043002 (2005).
97. T. Stein, L. Kronik and R. Baer, *J. Am. Chem Soc.* **131**, 2818 (2009).
98. O.V. Prezhdo, *Acc. Chem. Res.* **42**, 2005 (2009).
99. D. Porezag, T. Frauenheim, T. Kohler, G. Seifert and R. Kaschner, *Phys. Rev. B* **51**, 12947 (1995).
100. G. Seifert, D. Porezag and T. Frauenheim, *Int. J. Quant. Chem.* **58**, 185 (1996).
101. M. Elstner, D. Porezag, G. Jungnickel, J. Elsner, M. Haugk, T. Frauenheim, S. Suhai and G. Seifert, *Phys. Rev. B* **58**, 7260 (1998).
102. T. Frauenheim, G. Seifert, M. Elstner, T. Niehaus, C. Köhler, M. Amkreutz, M. Sternberg, Z. Hajnal, A. Di Carlo and S. Suhai, *J. Phys.: Condens. Matter* **14**, 3015 (2002).
103. J. Fabian, L.A. Diaz, G. Seifert and T. Niehaus, *J. Mol. Struc. THEOCHEM* **594**, 41 (2002).
104. T. Niehaus, S. Suhai, F. Della Salla, P. Lugli, M. Elstner and G. Seifert, *Phys. Rev. B* **63**, 5108 (2001).
105. D. Heringer, T. Niehaus, M. Wanko and T. Frauenheim, *J. Comput. Chem.* **28**, 2589 (2007).
106. S. Hammes-Schiffer and J.C. Tully, *J. Chem. Phys.* **101**(6), 4657 (1994).
107. M.E. Casida, In *Recent Advances in Density Functional Methods*, D.P. Chong, Eds.; World Scientific, Singapore, 1995; page 155.
108. T. Horio, T. Fuji, Y.-I. Suzuki and T. Suzuki, *J. Am. Chem Soc.* **113**, 10392 (2009).
109. K. Gokhberg, V. Vysotskiy, L.S. Cederbaum, L. Storchi, F. Tarantelli and V. Averbukh, *J. Chem. Phys.* **130**, 064104 (2009).
110. D.L. Yeager, M.A.C. Nascimento and V. McKoy, *Phys. Rev. A* **11**, 1168 (1975).
111. P.W. Langhoff and C.T. Corcoran, *J. Chem. Phys.* **61**, 146 (1974).
112. T.N. Rescigno, C.F. Bender, B.V. McKoy and P.W. Langhoff, *J. Chem. Phys* **68**, 970 (1978).
113. I. Cacelli, V. Carravetta, A. Rizzo and R. Moccia, *Phys. Rep.* **205**, 283 (1991).
114. C. Daniel, J. Full, L. Gonzalez, C. Lupulescu, J. Manz, A. Merli, S. Vajda and L. Wöste, *Science* **299**, 536 (2003).
115. G. Vogt, G. Krampert, P. Niklaus, P. Nuernberger and G. Gerber, *Phys. Rev. Lett.* **94**, 068305 (2005).
116. B. Schäfer-Bung, R. Mitrić, V. Bonačić-Koutecký, A. Bartelt, C. Lupulescu, A. Lindinger, S. Vajda, S.M. Weber and L. Wöste, *J. Phys. Chem. A* **108**, 4175 (2004).
117. L. Verlet, *Phys. Rev.* **159**, 98 (1967).
118. R. Kosloff, *J. Phys. Chem.* **92**, 2087 (1988).
119. D.E. Goldberg, *Genetic Algorithms in Search, Optimization and Machine Learning;* Addison-Wesley, 1989.

120. J.P. Perdew, K. Burke and M. Ernzerhof, *Phys. Rev. Lett.* **77**, 3865 (1996).
121. A. Schäfer, C. Huber and R.R. Ahlrichs, *J. Chem. Phys.* **100**, 5829 (1994).
122. B.I. Dunlap, J.W.D. Connolly and J.R. Sabin, *J. Chem. Phys.* **71**, 3396 (1979).
123. K. Eichkorn, O. Treutler, H. Öhm, M. Häser and R. Ahlrichs, *Chem. Phys. Lett.* **240**, 283 (1995).
124. R. Ahlrichs, M. Bär, M. Häser, H. Horn and M. Kölmel, *Chem. Phys. Lett.* **162**, 165 (1989).
125. A.D. Becke, *J. Chem. Phys.* **98**, 5648 (1993).
126. A.D. Becke, *J. Chem. Phys.* **107**, 8554 (1997).
127. T. Bally, E. Haselbach, S. Lanyiova, F. Marschner and M. Rossi, *Helv. Chim. Acta* **59**, 486 (1976).
128. M.J.G. Peach, P. Benfield, T. Helgaker and D.J. Tozer, *J. Chem. Phys.* **128**, 044118 (2008).
129. M. Traetteberg, I. Hilmo, R.J. Abraham and S. Ljunggren, *J. Mol. Struc.* **48**, 395 (1978).
130. A.V. Gaenko, A. Devarajan, L. Gagliardi, R. Lindh and G. Orlandi, *Theo. Chem. Acc.* **118**, 271 (2007).
131. T. Schultz, J. Quenneville, B.G. Levine, A. Toniolo, T.J. Martinez, S. Lochbrunner, M. Schmitt, J.P. Schaffer, M.Z. Zgierski and A. Stolow, *J. Am. Chem. Soc.* **125**, 8098 (2003).
132. H. Rau and E. Lüddecke, *J. Am. Chem. Soc.* **104**, 1616 (1982).
133. N. Tamai and H. Miyasaka, *Chem. Rev.* **100**, 1875 (2000).
134. A. Toniolo, C. Ciminelli, M. Persico and T.J. Martinez, *J. Chem. Phys.* **123**, 234208 (2005).
135. I. Conti, M. Garavelli and G. Orlandi, *J. Am. Chem. Soc.* **130**, 5216 (2008).
136. M. Seel and W. Domcke, *J. Chem. Phys.* **95**, 7806 (1991).
137. V. Stert, P. Farmanara and W. Radloff, *J. Chem. Phys.* **112**, 4460 (2000).
138. P. Puzari, B. Sarkar and A. Satrajit, *J. Chem. Phys.* **125**, 194316 (2006).
139. S. Hahn and G. Stock, *Phys. Chem. Chem. Phys.* **3**, 2331 (2001).
140. A.L. Sobolewski, W. Domcke, C. Ledonder-Lardeux and C. Jouvet, *Phys. Chem. Chem. Phys.* **4**, 1093 (2002).
141. H. Kang, K.T. Lee, B. Jung, Y.J. Ko and S.K. Kim, *J. Am. Chem. Soc.* **124**, 12958 (2002).
142. S. Ullrich, T. Schultz, M.Z. Zgierski and A. Stolow, *J. Am. Chem. Soc.* **126**, 2262 (2004).
143. C.Z. Bisgaard, H. Satzger, S. Ullrich and A. Stolow, *ChemPhysChem* **10**, 101 (2009).
144. S. Perun, A.L. Sobolewski and W. Domcke, *J. Am. Chem. Soc.* **127**, 6257 (2005).
145. S. Perun, A.L. Sobolewski and W. Domcke, *Chem. Phys.* **313**, 107 (2005).
146. L. Blancafort, B. Cohen, P.M. Hare, B. Kohler and M.A. Robb, *J. Phys. Chem. A* **109**, 4431 (2005).
147. L. Serrano-Andres, M. Merchan and A. Borin, *Proc. Natl. Acad. Sci.* **103**, 8691 (2006).

148. B. Cohen, P.M. Hare and B. Kohler, *J. Am. Chem. Soc.* **125**, 13594 (2003).
149. W.L. Jorgensen, J. Chandrasekhar, J.D. Madura, R.W. Impey and M.L. Klein, *J. Chem. Phys.* **79**, 926 (1983).
150. W.D. Cornell, P. Cieplak, C.I. Bayly, I.R. Gould, K.M. Merz, D.M. Ferguson, D.C. Spellmeyer, T. Fox, J.W. Caldwell and P.A. Kollman, *J. Am. Chem. Soc.* **117**, 5179 (1995).
151. B. Mennucci, A. Toniolo and J. Tomasi, *J. Am. Chem. Soc.* **125**, 13594 (2003).
152. M.J.S. Dewar, E.G. Zoebisch, E.F. Healy and J.J.P. Stewart, *J. Am. Chem. Soc.* **107**, 3902 (1985).
153. B.E. Schmidt, W. Unrau, A. Mirabal, L. Shaohui, M. Krenz, L. Wöste and T. Siebert, *Opt. Express* **16**, 18911 (2008).
154. J.J.P. Stewart, *J. Comput. Chem.* **10**, 209 (1989); A. Koslowski, M.E. Beck and W. Thiel, *J. Comput. Chem.* **24**, 714 (2003).
155. W.F. van Gunsteren and H.J.C. Berendsen, *Molec. Phys.* **45**, 637 (1982).
156. N. Getoff, S. Solar and D.B. McCormick, *Science* **201**, 616 (1978).

Chapter 14

Laser Control of Ultrafast Dynamics at Conical Intersections

Yukiyoshi Ohtsuki* and Wolfgang Domcke[†]

1. Introduction 569
2. Quantum Control Theory 572
   2.1. Time-dependent Schrödinger equation 572
   2.2. Minimal model analysis 574
   2.3. Optimal control theory (OCT) 579
3. Case Studies 585
   3.1. Wave packet shaping utilizing CI-induced
        nonadiabatic transitions 585
   3.2. Geometric phase effects on coherent control
        of CI-induced dynamics 590
4. Conclusions 596
   Acknowledgments 596
   References 596

1. Introduction

Laser control of molecular dynamics actively exploits the coherence of laser pulses in addition to their high intensity and well-defined power

---

*Department of Chemistry, Graduate School of Science, Tohoku University, Sendai 980-8578, Japan.
[†]Department of Chemistry, Technical University of Munich, D-85747 Garching, Germany.

spectrum.[1-4] These properties of laser pulses are encoded in a molecular wave function to create an objective state using constructive quantum interference, while undesirable populations are minimized by using destructive interference. The same principles of manipulating quantum interference are employed in various research areas ranging from control of chemical reactions to quantum information processing.

Ultrafast dynamics at conical intersections (CIs)[5] has been reported for various photochemical reactions, including the photochemical ring opening of 1,3-cyclohexadiene,[6,7] photodissociation of ammonia,[8,9] hydrogen transfer in the excited ammonia dimer,[10] and H-atom elimination in the excited states of phenol and pyrrole.[11-13] Ultrafast radiationless decay through CIs has been reported for the S_1 state of azulene[14-16] and the S_2 state of adenine,[17] as well as for other heteroaromatic molecules and their biological analogues.[18,19] Evidence of vibrational coherence, even after the internal conversion through CIs, has been observed in some time-resolved experiments.[20-23] Sorgues et al.[20] performed a femtosecond pump-probe experiment on tetrakis(dimethylamino)ethylene and observed coherent oscillations associated with the umbrella mode (coupling mode) of the amino groups. In the photoisomerization of retinal in rhodopsin (the first step in vision), time-dependent oscillations in the absorption band of the photoproducts were observed.[21-23] These experiments indicate that vibrational coherence can be created by and/or survive nonadiabatic transitions at CIs.[24,25] This implies in turn that laser pulses can act cooperatively with CIs by utilizing them as "wave-packet cannons," thereby making CIs unique from the viewpoint of coherent control. As CIs often connect different conformers with high quantum yield, they could also be regarded as molecular optical switches for triggering peptide folding, electronic devices, etc.

When controlling CI-induced dynamics, it is necessary to simultaneously manipulate different kinds of wave packet dynamics to achieve physical objectives. These wave packet dynamics include the propagation of a wave packet from the optically active region to the CI region and that from the CI region to the target state. In the CI region, a portion of the wave packet is transferred to another potential-energy surface (PES) while being distorted during the nonadiabatic transition. The wave-packet components that remain on the original PES may receive a π-phase shift, which is called the geometric (Berry) phase.[26-30] Through these processes, the local topography of the CI can significantly affect the subsequent dynamics. Between the initial and the objective state, the wave packet obviously encounters many undesirable states. In addition, almost all the dynamics occurs in a dark region that is not directly optically accessible. Thus, it is challenging to realize a physical objective with a high degree of accuracy by adjusting the laser pulse shape as control knobs.

If we restrict ourselves to varying the amplitude of each frequency component of a laser pulse, the number of control knobs is limited. However, if we actively use the relative phases among the frequency components of a laser pulse, the number of knobs increases exponentially. At least in principle, this huge amount of information could be encoded in a wave packet and its subsequent dynamics programmed to enable it to reach an objective state by utilizing quantum interferences. This is the basic strategy of laser control of molecular dynamics. Our primary concern is to find a way to implement a huge number of control knobs in real molecular systems.

Closed-loop experiments are useful for this purpose, as they overcome the problem of laboratory implementation by using a molecule as a simulator to evaluate the "quality" of a control pulse.[31–35] Such experiments have successfully been applied to both isolated systems and to molecules in condensed phases.[36–40] As schematically illustrated in Fig. 1, the success of these experiments depends largely on the feedback control they employ, which is implemented by so-called learning algorithms that require little information about the molecular Hamiltonian.[31] In other words, they cannot provide direct information about the molecular Hamiltonian, control mechanisms, etc. It is natural to ask what we can learn from these feedback experiments.

From a theoretical/computational viewpoint, one essential step of laser control is to design a pulse shape to achieve a specified objective. Pulse design based on physical considerations is usually limited because of the high complexity of molecules (although these ideas still play an important role in understanding the underlying control mechanisms). More general pulse design schemes are required that are based on control theory such as

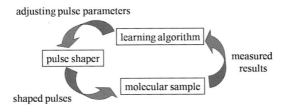

Fig. 1. Schematic illustration of closed-loop (feedback) experiments.

optimal control theory.[41,42] Once the molecular Hamiltonian has been specified, these control theories can design laser pulses by maximizing (or minimizing) quantitative criteria. Because computational cost often imposes limitations on simulations, it is essential to develop efficient methods for solving the pulse-design equations.[43–51] However, such feedforward control schemes do not always result in a high degree of control in real systems because of the lack of complete information about PESs especially for polyatomic molecules.

These theoretical/computational and experimental situations illustrate the primary reason why both numerical analyses (simulations) with model systems and closed-loop experiments complement each other when clarifying control mechanisms. By performing simulations and experiments, it is expected to be possible to derive a set of basic rules for laser control for manipulating quantum phenomena such as ultrafast dynamics at CIs and for obtaining information about PESs. The primary purpose of this chapter is to introduce and explain simulation techniques based on quantum control theory by presenting case studies.

2. Quantum Control Theory

2.1. Time-dependent Schrödinger equation

We assume semiclassical interaction between molecules and laser fields, in which the laser fields are treated classically and the molecules are treated quantum-mechanically. As the laser pulses used in ultrafast experiments always have longer wavelengths than any characteristic length associated with molecular systems, we can safely neglect the spatial dependence of the laser field; this is known as the dipole approximation. By using the gauge transformation of the potentials of the laser field, $\mathbf{E}(t)$, we derive the following total Hamiltonian

$$H_T^t = H_T + V^t = H_e + T_n - \boldsymbol{\mu} \cdot \mathbf{E}(t), \tag{1}$$

where $V^t = -\boldsymbol{\mu} \cdot \mathbf{E}(t)$ describes the "length-form" interaction with $\boldsymbol{\mu}$ being the electric dipole moment operator. In Eq. (1), H_e denotes the electronic Hamiltonian and T_n is the nuclear kinetic energy operator. The whole system obeys the time-dependent Schrödinger equation,

$$i\hbar \frac{\partial}{\partial t} |\Psi_T(t)\rangle = H_T^t |\Psi_T(t)\rangle, \tag{2}$$

which includes all degrees of freedom (i.e. electronic, vibrational, and rotational).

Concerning the nuclear degrees of freedom, we disregard the center of mass and rotational motion in this chapter. The rotational degrees of freedom describe the alignment/orientation of a molecule with respect a space-fixed frame (i.e. the polarization vector of a laser pulse). A change in the molecular orientation alters the torque imposed by the laser field, which leads to rotational transitions through the orientational angles appearing in the optical interaction, $V^t = -\boldsymbol{\mu} \cdot \mathbf{E}(t)$. However, if we are interested in the intramolecular dynamics induced by a laser pulse, it is often possible to neglect the rotational motion to a good approximation, provided there appear no highly excited rotational states in the dynamics.

We introduce basis functions to solve Eq. (2). We usually assume the Born–Oppenheimer approximation, which involves adiabatic separation of the electronic and nuclear degrees of freedom. The adiabatic basis, $\{|\Phi_i(\mathbf{R})\rangle\}$, which describes the electronic eigenstate, satisfies the eigenvalue problem for a given nuclear configuration, $|\mathbf{R}\rangle$:

$$H_e |\Phi_i\, \mathbf{R}\rangle = V_i(\mathbf{R}) |\Phi_i\, \mathbf{R}\rangle, \tag{3}$$

with $|\Phi_i\, \mathbf{R}\rangle = |\Phi_i(\mathbf{R})\rangle |\mathbf{R}\rangle$, and the $\mathbf{R}$-dependent eigenvalue, $V_i(\mathbf{R})$, which is called the ith adiabatic PES. Employing the closure relation

$$1 = \sum_i \int |\Phi_i\, \mathbf{R}\rangle d\mathbf{R} \langle \Phi_i\, \mathbf{R}|, \tag{4}$$

we expand the total wave function, $|\Psi_T(t)\rangle$,

$$|\Psi_T(t)\rangle = \sum_i \int |\Phi_i\, \mathbf{R}\rangle d\mathbf{R} \langle \Phi_i\, \mathbf{R}|\Psi_T(t)\rangle = \sum_i \int d\mathbf{R} |\Phi_i\, \mathbf{R}\rangle \psi_i(\mathbf{R}, t). \tag{5}$$

The wave packets, $\{\psi_i(\mathbf{R}, t) \equiv \langle \mathbf{R}|\psi_i(t)\rangle\}$, describe the time evolution of the nuclear dynamics on the adiabatic PESs, $\{V_i(\mathbf{R})\}$, and they interact with each other through nonadiabatic interactions and/or optical transitions. They obey the equations of motion

$$\begin{aligned} i\hbar \frac{\partial}{\partial t} |\psi_i(t)\rangle &= [T_n + V_i(\mathbf{R}) - \boldsymbol{\mu}_{ii}(\mathbf{R}) \cdot \mathbf{E}(t)] |\psi_i(t)\rangle \\ &\quad + \sum_{j(\neq i)} [V_{ij}(\mathbf{R}) - \boldsymbol{\mu}_{ij}(\mathbf{R}) \cdot \mathbf{E}(t)] |\psi_j(t)\rangle, \end{aligned} \tag{6}$$

where $V_i(\mathbf{R})$ may include the correction term, $\langle \Phi_i(\mathbf{R})|T_n\Phi_i(\mathbf{R})\rangle$. In Eq. (6), the nonadiabatic interaction is specified by $V_{ij}(\mathbf{R})$, and the transition (permanent) dipole moment operator is defined by

$$\boldsymbol{\mu}_{ij}(\mathbf{R}) = \langle \Phi_i(\mathbf{R})|\boldsymbol{\mu}|\Phi_j(\mathbf{R})\rangle. \tag{7}$$

The adiabatic PES, $V_i(\mathbf{R})$, determines the vibrational states in the ith electronic state, $\{|\chi_{i\mathbf{v}}\rangle\}$, that satisfies the eigenvalue problem

$$[T_n + V_i(\mathbf{R})]|\chi_{i\mathbf{v}}\rangle \equiv H_i|\chi_{i\mathbf{v}}\rangle = \hbar\omega_{i\mathbf{v}}|\chi_{i\mathbf{v}}\rangle, \tag{8}$$

whereby the adiabatic eigenstate with the eigenvalue $\hbar\omega_{i\mathbf{v}}$ is expressed as

$$|\Phi_{i\mathbf{v}}\rangle = \int d\mathbf{R}|\Phi_i\,\mathbf{R}\rangle\langle\mathbf{R}|\chi_{i\mathbf{v}}\rangle. \tag{9}$$

2.2. *Minimal model analysis*

Before discussing the formal treatment of quantum control problems, we introduce typical procedures by adopting a minimal model that consists of two electronic states. The equations of motion are expressed by

$$i\hbar\frac{\partial}{\partial t}\begin{bmatrix}|\psi_a(t)\rangle \\ |\psi_b(t)\rangle\end{bmatrix} = \begin{bmatrix}H_a & -\mu_{ab}E(t)+V_{ab} \\ -\mu_{ba}E(t)+V_{ba} & H_b\end{bmatrix}\begin{bmatrix}|\psi_a(t)\rangle \\ |\psi_b(t)\rangle\end{bmatrix}, \tag{10}$$

with the initial condition $[|\psi_a(t_0)\rangle\,|\psi_b(t_0)\rangle] = [|\psi_a^0\rangle\,|\psi_b^0\rangle]$. The subscripts, a and b, indicate the ground and excited electronic states, respectively. As we are interested in the time-dependent laser field, $E(t)$, we account for optical interactions through the transition dipole moment operator, μ_{ab} (μ_{ba}), and neglect the permanent dipole moment operators, μ_{aa} and μ_{bb}. The nonadiabatic interaction is expressed by $V_{ab} = V_{ba}^\dagger$.

2.2.1. *First-order optical transitions*

If we assume that the molecule is initially in the lowest vibrational state in the ground electronic state, $|\chi_{a\mathbf{0}}\rangle$, the first-order solution with respect to the optical interaction is given by

$$|\psi_b^{(1)}(t)\rangle = \frac{i}{\hbar}\int_{t_0}^{t} d\tau\, e^{-iH_b(t-\tau)/\hbar}\mu_{ba}E(\tau)e^{-iH_a(\tau-t_0)/\hbar}|\chi_{a\mathbf{0}}\rangle, \tag{11}$$

where we have neglected the nonadiabatic transitions. The wave packet, $|\psi_b^{(1)}(t)\rangle$, is expressed as a superposition of the components excited at various instants, $\tau \in [t_0, t]$, which contain amplitude and phase information of the electric field. In other words, the time evolution of the wave packet can be controlled by adjusting the shape of the laser field. To discuss the control procedures in detail, we focus on the time after the laser pulse. Then Eq. (11) is reduced to (except for an unimportant phase factor, $e^{i\omega_{a\mathbf{0}}t_0}$)

$$|\psi_b^{(1)}(t)\rangle = \frac{i}{\hbar} e^{-iH_b t/\hbar} \tilde{E}(H_b/\hbar - \omega_{a\mathbf{0}})\mu_{ba}|\chi_{a\mathbf{0}}\rangle, \quad (12)$$

where

$$\tilde{E}(H_b/\hbar - \omega_{a\mathbf{0}}) = \int_{-\infty}^{\infty} d\tau \, e^{i(H_b - \hbar\omega_{a\mathbf{0}})\tau/\hbar} E(\tau). \quad (13)$$

Here, we consider the extreme case in which the Fourier components of the laser field are virtually independent of the transition frequencies, $\omega_{b\mathbf{v},a\mathbf{0}} = \omega_{b\mathbf{v}} - \omega_{a\mathbf{0}}$, for all vibronic states $\{|\chi_{b\mathbf{v}}\rangle\}$ that have non-zero transition matrix elements with the initial state. Because $\tilde{E}(\omega_{b\mathbf{v},a\mathbf{0}}) \simeq \tilde{E}_0$, Eq. (12) is reduced to

$$|\psi_b^{(1)}(t)\rangle = \frac{i}{\hbar} \tilde{E}_0 \, e^{-iH_b t/\hbar} \mu_{ba}|\chi_{a\mathbf{0}}\rangle, \quad (14)$$

which is called a Franck–Condon wave packet. If we further ignore the coordinate dependence of the transition moment function (the Condon approximation), the laser pulse creates a replica of the initial state on the electronic excited state that evolves in time according to H_b.

The control achievement is evaluated by the expectation value of a target operator, A. An operator, A, which commutes (does not commute) with H_b, is often called a stationary (nonstationary) target. A typical example of a stationary target is the population of a specified vibrational state in the b-state,

$$P_{b\mathbf{v}} = |\langle\chi_{b\mathbf{v}}|\psi_b^{(1)}(t)\rangle|^2 = \frac{1}{\hbar^2}|\tilde{E}(\omega_{b\mathbf{v},a\mathbf{0}})|^2|\mu_{b\mathbf{v},a\mathbf{0}}|^2, \quad (15)$$

where $\mu_{b\mathbf{v},a\mathbf{0}} = \langle\chi_{b\mathbf{v}}|\mu_{ba}|\chi_{a\mathbf{0}}\rangle$. The degree of control is determined by the power spectrum of the laser pulse.

We consider the case where the laser pulse is expressed as a sum of two Gaussian subpulses with a temporal width of $2\sqrt{\ln 2}\,\tau_G$ and an amplitude of E_G^0. Assuming that its central frequency, ω_0, is close to the transition frequency, we extract resonance components to obtain

$$P_{b\mathbf{v}}(\tau_d) = \frac{\pi}{\hbar^2}(E_G^0 \tau_G)^2 |\mu_{b\mathbf{v},a\mathbf{0}}|^2 e^{-\tau_G^2(\omega_{b\mathbf{v},a\mathbf{0}}-\omega_0)^2}[1 + \cos(\omega_{b\mathbf{v},a\mathbf{0}}\tau_d)], \quad (16)$$

where τ_d represents the delay of the second pulse with respect to the first one. This is a real-time version of Young's double-slit experiment, which is called a quantum interferometer.[52–55] The interference in Eq. (16) originates from the fact that the "which-path" information is not disturbed because we do not know which subpulse actually creates the state $|\chi_{b\mathbf{v}}\rangle$.[54,55]

In a more general situation, we often aim to maximize the control achievement for a nonstationary target at a specified time t

$$J = \frac{\langle \psi_b^{(1)}(t)|A|\psi_b^{(1)}(t)\rangle}{\langle \psi_b^{(1)}(t)|\psi_b^{(1)}(t)\rangle}. \tag{17}$$

If the operator, A, has a finite maximum eigenvalue (the eigenstate of which is denoted by $|A_0\rangle$), an obvious solution to the maximization of J is to realize the eigenstate $|A_0\rangle$. This is sometimes called a wave packet shaping problem, which is one of the fundamental targets in quantum control. For a nonstationary target, the phases of the laser pulse play an essential role in controlling the wave packet. As an illustrative example of the phase control, we choose the Franck–Condon wave packet as a target; i.e. $|A_0\rangle = \mu_{ba}|\chi_{a\mathbf{0}}\rangle$. (Here, we do not consider the trivial solution of a delta-function pulse.) As the denominator of Eq. (17) is independent of the pulse phases, we focus on the numerator, which can be expressed as

$$J = \langle \psi_b^{(1)}(t)|A|\psi_b^{(1)}(t)\rangle = \frac{1}{\hbar^2}|\sum_{\mathbf{v}}|\tilde{E}(\omega_{b\mathbf{v},a\mathbf{0}})||\mu_{b\mathbf{v},a\mathbf{0}}|e^{-i\omega_{b\mathbf{v}}t - i\delta_{b\mathbf{v},a\mathbf{0}}}|^2, \tag{18}$$

where $\delta_{b\mathbf{v},a\mathbf{0}}$ is the phase associated with the transition frequency, $\omega_{b\mathbf{v},a\mathbf{0}}$, and includes the pulse phase. To maximize J in Eq. (18), all the phases should be identical to each other; i.e. $\omega_{b\mathbf{v}}t + \delta_{b\mathbf{v},a\mathbf{0}} = \delta_0 \bmod (2\pi)$. Considering the difference between two adjacent phases, we have

$$\Delta\omega_{b\mathbf{v}}t + \Delta\delta_{b\mathbf{v},a\mathbf{0}} = 0 \bmod (2\pi), \tag{19}$$

where $\Delta\omega_{b\mathbf{v}} = \omega_{b\mathbf{v}} - \omega_{b\mathbf{v}-1}$ and $\Delta\delta_{b\mathbf{v},a\mathbf{0}} = \delta_{b\mathbf{v},a\mathbf{0}} - \delta_{b\mathbf{v}-1,a\mathbf{0}}$, the latter of which can be adjusted by varying the phases of the frequency components in a laser pulse. If we consider the bound states of an anharmonic oscillator, $\Delta\omega_{b\mathbf{v}}$ decreases with the vibrational quantum number. To satisfy the condition expressed by Eq. (19), $\Delta\delta_{b\mathbf{v},a\mathbf{0}}$ should increase to cancel the reduction in $\Delta\omega_{b\mathbf{v}}$. If we introduce such phase modulations into the control pulse, it becomes a negatively chirped pulse; i.e. the higher vibrational states are excited before the lower states (this has been confirmed experimentally; e.g. by Wilson's group[56]).

2.2.2. First-order nonadiabatic transitions

To examine the effects of nonadiabatic transitions on laser control, we consider a special case in which at the initial time (t_1) and the final time (t_f), the wave packets are spatially localized far from the nonadiabatic transition region. Assuming the initial condition, $|\psi_b(t_1)\rangle = \sum_{\mathbf{v}} |\chi_{b\mathbf{v}}\rangle C_{b\mathbf{v}}(t_1)$, we calculate the first-order solution with respect to the nonadiabatic interaction, whereby the probability amplitude $\langle \chi_{a\mathbf{u}}|\psi_a^{(1)}(t_f)\rangle$ is given by

$$\langle \chi_{a\mathbf{u}}|\psi_a^{(1)}(t_f)\rangle = -\frac{i}{\hbar} \sum_{\mathbf{v}} D(\omega_{b\mathbf{v},a\mathbf{u}}; t_f - t_1)$$
$$\times \langle \chi_{a\mathbf{u}}|V_{ab}(\mathbf{R})|\chi_{b\mathbf{v}}\rangle C_{b\mathbf{v}}(t_1) e^{-i\omega_{b\mathbf{v}}(t_f-t_1)}. \quad (20)$$

Equation (20) consists of three parts. The first term, $D(\omega_{b\mathbf{v},a\mathbf{u}}; t_f - t_1)$, describes the free propagation that includes all the interference from portions of the amplitude transferred to the a-state at one instant and those transferred at another instant. From the definition, its magnitude is expressed as

$$|D(\omega_{b\mathbf{v},a\mathbf{u}}; t_f - t_1)| = \left| \int_0^{t_f-t_1} d\tau e^{i\omega_{b\mathbf{v},a\mathbf{u}}\tau} \right| = \left| \frac{\sin[\omega_{b\mathbf{v},a\mathbf{u}}(t_f - t_1)/2]}{\omega_{b\mathbf{v},a\mathbf{u}}/2} \right|. \quad (21)$$

This has a nonzero value only when the energy difference, $\omega_{b\mathbf{v},a\mathbf{u}}$, satisfies the relation $|\omega_{b\mathbf{v}} - \omega_{a\mathbf{u}}| \leq \pi/(t_f - t_1)$. As t_f increases, only the initial vibronic states with eigenvalues close to that of $|\chi_{a\mathbf{u}}\rangle$ can contribute to the amplitude in Eq. (20); that is, this first term is associated with energy conservation.

For the second term in Eq. (20), $\langle \chi_{a\mathbf{u}}|V_{ab}(\mathbf{R})|\chi_{b\mathbf{v}}\rangle$, we can consider two limiting cases. If we ignore the nuclear-coordinate dependence of $V_{ab}(\mathbf{R})$, such that $V_{ab}(\mathbf{R}) = V_{ab}^0$, we have

$$\langle \chi_{a\mathbf{u}}|V_{ab}(\mathbf{R})|\chi_{b\mathbf{v}}\rangle = V_{ab}^0 \langle \chi_{a\mathbf{u}}|\chi_{b\mathbf{v}}\rangle, \quad (22)$$

which is called the Condon approximation and there appears the Franck-Condon overlap integral, $\langle \chi_{a\mathbf{u}}|\chi_{b\mathbf{v}}\rangle$. Around a CI, on the other hand, $V_{ab}(\mathbf{R})$ typically depends on the nuclear coordinates considerably because of the strong vibronic interaction. For the purpose of illustration, let us consider another limiting case, in which the nonadiabatic transition occurs at the point $\mathbf{R}_0$. This fictitious situation is modeled by $V_{ab}(\mathbf{R}) = V_{ab}^0 \delta(\mathbf{R} - \mathbf{R}_0)$ and we have

$$\langle \chi_{a\mathbf{u}}|V_{ab}|\chi_{b\mathbf{v}}\rangle = \langle \chi_{a\mathbf{u}}|\mathbf{R}_0\rangle V_{ab}^0 \langle \mathbf{R}_0|\chi_{b\mathbf{v}}\rangle. \quad (23)$$

Equation (23) means that the initial wave packet components, $\{|\chi_{b\mathbf{v}}\rangle\}$, that satisfy the energy conservation [Eq. (21)] are launched into the a-state at $\mathbf{R}_0$; i.e., V_{ab} acts as a "wave-packet cannon." The narrow active region of $V_{ab}(\mathbf{R})$ leads to a narrower spatial distribution, but to a wider momentum distribution of the wave packet after the nonadiabatic transitions.

The remaining (third) term in Eq. (20) indicates that both the initial amplitude and the phase of $C_{b\mathbf{v}}(t_1)$ play an essential role in controlling the probability amplitude $\langle \chi_{a\mathbf{u}}|\psi_a^{(1)}(t_f)\rangle$. As any physical target in the a-state is expressed in terms of these amplitudes, the shape of the initial wave packet, $|\psi_b(t_1)\rangle = \sum_{\mathbf{v}} |\chi_{b\mathbf{v}}\rangle C_{b\mathbf{v}}(t_1)$, which can be designed using laser pulses, is important to improve the control achievement.

2.2.3. *Pre-target state*

From the analysis in Sec. 2.2.2, it would be natural to think about the "optimum" pre-target state, $|\psi_b(t_1)\rangle = \sum_{\mathbf{v}} |\chi_{b\mathbf{v}}\rangle C_{b\mathbf{v}}(t_1)$, that realizes the highest control achievement.[57-60] In this subsection, we consider this problem in a general manner without using a perturbative treatment for nonadiabatic transitions. Let $U_0(t_f, t_1)$ be an operator that describes the time evolution of a molecular system from t_1 to t_f, i.e. $|\psi_a(t_f)\rangle = U_0(t_f, t_1)|\psi_b(t_1)\rangle$. The control achievement is expressed as

$$J = \langle \psi_a(t_f)|A|\psi_a(t_f)\rangle = \langle \psi_b(t_1)|U_0^\dagger(t_f, t_1) A\, U_0(t_f, t_1)|\psi_b(t_1)\rangle. \qquad (24)$$

If the optical interactions during this period can be neglected, the pre-target state evolves in time according to the field-free molecular Hamiltonian. We apply calculus of variations to Eq. (24) to determine the shape of the "optimum" pre-target state. Introducing the Lagrange multiplier, ξ, which represents the normalization condition of the pre-target state, we have the unconstrained functional to be maximized,

$$\bar{J} = \langle \psi_b(t_1)|P\, A(t_f, t_1) P|\psi_b(t_1)\rangle - \xi[\langle \psi_b(t_1)|P|\psi_b(t_1)\rangle - 1], \qquad (25)$$

where $A(t_f, t_1) \equiv U_0^\dagger(t_f, t_1) A\, U_0(t_f, t_1)$ is a Hermitian operator. The projector, P, is introduced to specify the optically accessible states as the pre-target state is required to be created by laser pulses. From Eq. (25), we have the eigenvalue equation

$$P\, A(t_f, t_1) P|\psi_b(t_1)\rangle = \xi\, P|\psi_b(t_1)\rangle. \qquad (26)$$

The pre-target state with the maximum eigenvalue leads to the maximum expectation value of J in Eq. (24).

The idea of introducing a pre-target state is general and is not restricted to quantum control via nonadiabatic transitions. The pre-target state can provide the information on the path to the target state within an optically dark region, which is useful for understanding the control mechanisms. However, this does not imply that introducing a pre-target state is always justified. In addition to the numerical difficulties associated with solving Eq. (26), introducing a pre-target state imposes a strong constraint on a control problem. For example, we explicitly assume that the optical and nonadiabatic interactions are temporally separated, which could considerably affect the control mechanisms. Despite these deficiencies, several studies have fruitfully utilized the concept of the pre-target state.

In laser control, Gross et al.[57] pioneered the idea of the pre-target state to determine a bound target state in the electronic ground state using the Rayleigh-Ritz principle. We applied this approach to an unbound wave packet in the excited electronic state of NaI[58] and to the adsorption mode of NO/Pt under the influence of a specified external field.[59] This approach was also employed to determine a wave packet with a long lifetime by eliminating radiationless transitions in the $S_1 \leftarrow S_2$ internal conversion via a CI in pyrazine.[60] This is an extension of previous studies on electronically localized eigenstates of vibronically coupled PESs of pyrazine, which are related to the concept of quasi-periodic orbits and have lifetimes longer than several nanoseconds in a two-dimensional model.[61,62]

2.3. Optimal control theory (OCT)

Optimal control theory (OCT) provides general and flexible pulse-design algorithms to achieve a specified physical objective.[41–51] In an optimal control simulation, we calculate the molecular dynamics while designing the control pulse according to OCT-based algorithms, which provide clues for clarifying control mechanisms, etc. Here, we introduce the formal procedure and provide some numerical techniques that are useful for optimal control simulations.

2.3.1. Optimal control simulation

An optimal control pulse is designed to maximize an objective functional that quantitatively evaluates the control achievement by the calculus of variations. In this subsection, we explain the procedure by presenting a case study. We express the equations of motion of a molecular system as

$$i\hbar \frac{\partial}{\partial t}|\psi(t)\rangle = [H_0 - \mu E(t)]|\psi(t)\rangle = H^t|\psi(t)\rangle, \quad (27)$$

where $|\psi(t)\rangle$ are sets of nuclear wave packets, H_0 is a field-free Hamiltonian that includes nonadiabatic interactions and μ is the (transition) dipole moment operator. The initial state is given by $|\psi(t_0)\rangle = |\psi_0\rangle$. In a typical example,[45] the control objective is specified in terms of two non-negative Hermitian operators, A and $B(t)$, so that the objective functional can be expressed as

$$J = \langle \psi(t_f)|A|\psi(t_f)\rangle + \int_{t_0}^{t_f} dt \langle \psi(t)|B(t)|\psi(t)\rangle - \int_{t_0}^{t_f} dt \frac{1}{\hbar \lambda(t)}[E(t)]^2. \quad (28)$$

This objective functional consists of three terms: the first term specifies the objective state at a specified final time, t_f, the second term represents a constraint on the intermediate states, and the third term is a penalty due to pulse fluence with a positive function $\lambda(t)$ that weighs the physical significance of the penalty.

By introducing the Lagrange multiplier $|\xi(t)\rangle$ that represents the constraint of the equation of motion in Eq. (27), we obtain the unconstrained objective functional

$$\bar{J} = J - 2\mathrm{Re} \int_{t_0}^{t_f} dt \langle \xi(t)| \left(\frac{\partial}{\partial t} + \frac{i}{\hbar} H^t \right) |\psi(t)\rangle, \quad (29)$$

where $\mathrm{Re}\langle \cdots \rangle$ is the real part of $\langle \cdots \rangle$. The first-order variation is expressed as

$$\delta \bar{J} = -\int_{t_0}^{t_f} dt \frac{2}{\hbar \lambda(t)} \{E(t) + \lambda(t) \mathrm{Im} \langle \xi(t)|\mu|\psi(t)\rangle\} \delta E(t)$$

$$+ 2\mathrm{Re}\{\langle \psi(t_f)|A|\delta \psi(t_f)\rangle - \langle \xi(t_f)|\delta \psi(t_f)\rangle\} + 2\mathrm{Re} \int_{t_0}^{t_f} dt$$

$$\times \left\{ \left\langle \frac{\partial}{\partial t} \xi(t)|\delta \psi(t) \right\rangle - \frac{i}{\hbar} \langle \xi(t)|H^t|\delta \psi(t)\rangle + \langle \psi(t)|B(t)|\delta \psi(t)\rangle \right\}, \quad (30)$$

whereby pulse-design equations are derived. [For simplicity, we have neglected the derivatives with respect to $|\xi(t)\rangle$ and $\langle \xi(t)|$ as they just lead to Eq. (27).] The optimal pulse is expressed as

$$E(t) = -\lambda(t) \mathrm{Im} \langle \xi(t)|\mu|\psi(t)\rangle, \quad (31)$$

where $\text{Im}\langle\cdots\rangle$ is the imaginary part of $\langle\cdots\rangle$. The equation of motion for the Lagrange multiplier associated with the constraint, Eq. (27), is given by

$$i\hbar\frac{\partial}{\partial t}|\xi(t)\rangle = [H_0 - \mu E(t)]|\xi(t)\rangle - i\hbar B(t)|\psi(t)\rangle, \tag{32}$$

with a final condition, $|\xi(t_f)\rangle = A|\psi(t_f)\rangle$. If we simultaneously solve the pulse-design equations given by Eqs. (27), (31) and (32), we can perform optimal control simulations: i.e. we determine the optimal laser pulse together with the time evolution of the molecular system for the optimal pulse.

In Sec. 2.2.3, we introduced a control scheme that utilizes a pre-target state. Here, we discuss the scheme from the viewpoint of OCT. In the pre-target scheme, optical excitation processes are temporally separated from the ensuing wave-packet propagation with the field-free molecular Hamiltonian. These two processes are connected by a pre-target state, $|\psi(t_1)\rangle = |\psi_1\rangle$, at a specified time, t_1. To realize such a control in Eq. (28), we need to choose the intermediate target operator as $B(t) = |\psi_1\rangle\delta(t-t_1)\langle\psi_1|$. To prevent the optimal pulse from appearing after t_1, we need to replace $\lambda(t)$ with $\lambda(t)\theta(t_1-t)$, where $\theta(t_1-t) = 1$ for $t_1 > t$ and $\theta(t_1-t) = 0$ for $t_1 < t$. These conditions apparently impose strong restrictions on the control mechanisms, although this may not always be the case.

In optimal control simulations, we have to solve the coupled nonlinear differential equations of Eqs. (27), (31) and (32) (i.e. the pulse-design equations), which generally requires iteration methods. Because the simulations are computationally intensive, it is frequently essential to develop efficient solution algorithms to perform the simulations.[43-51] If A and $B(t)$ are non-negative operators, there exist monotonically convergent algorithms,[45,47-49] whereby the above pulse-design equations at the kth iteration step can be summarized as follows:

$$i\hbar\frac{\partial}{\partial t}|\xi^{(k)}(t)\rangle = [H_0 - \mu\bar{E}^{(k)}(t)]|\xi^{(k)}(t)\rangle - i\hbar B(t)|\psi^{(k-1)}(t)\rangle \tag{33}$$

with the final condition $|\xi^{(k)}(t_f)\rangle = A|\psi^{(k)}(t_f)\rangle$, and

$$i\hbar\frac{\partial}{\partial t}|\psi^{(k)}(t)\rangle = [H_0 - \mu E^{(k)}(t)]|\psi^{(k)}(t)\rangle, \tag{34}$$

with the initial condition $|\psi^{(k)}(t_0)\rangle = |\psi_0\rangle$. The electric fields at the kth step are given by

$$\bar{E}^{(k)}(t) = -\lambda(t)\text{Im}\langle\xi^{(k)}(t)|\mu|\psi^{(k-1)}(t)\rangle, \tag{35}$$

and

$$E^{(k)}(t) = -\lambda(t)\text{Im}\langle \xi^{(k)}(t)|\mu|\psi^{(k)}(t)\rangle. \tag{36}$$

In Eq. (35), the wave packet is replaced by the wave packet obtained in the previous step, so that Eq. (33) has a closed form with respect to the Lagrange multiplier. Similarly, Eq. (34) is decoupled from the Lagrange multiplier and has a closed form with respect to the wave packet. The nonlinear terms with respect to the state vectors in Eqs. (33) and (34) are often removed by employing the linearization approximation, in which the state vectors are expanded in terms of those obtained in the previous time step (e.g. $|\psi^{(k)}(t)\rangle = |\psi^{(k)}(t-\Delta t)\rangle + |d\psi^{(k)}(t-\Delta t)/dt\rangle \Delta t + O(\Delta t)^2$).

When we consider spatially delocalized dynamics such as (half) collisions, the wave packet often spreads beyond the grid region. As further propagation causes fictitious reflection from the grid edge, we usually introduce a certain cutoff function and divide the wave packet into two components, $|\psi(t)\rangle = |\psi_P(t)\rangle + |\psi_{1-P}(t)\rangle$.[63,64] Here, $|\psi_P(t)\rangle$ ($|\psi_{1-P}(t)\rangle$) is the component inside (outside) the grid region. When the time interval is divided into N steps such that the nth time step is specified by $t_n = t_0 + n\Delta t$ with $\Delta t = (t_f - t_0)/N$ ($n = 0, 1, 2, \ldots, N$), the algorithm can be summarized as

$$U(t_n, t_{n-1})|\psi_P(t_{n-1})\rangle = |\psi_P(t_n)\rangle + |\psi_{1-P}(t_n)\rangle, \tag{37}$$

where $U(t_n, t_{n-1})$ is the time-evolution operator associated with Eq. (27).

Due to the nature of the optimal control problem, the pulse-design equations have the form of an inverse problem. However, there is no general method to determine the final condition for the Lagrange multiplier in Eq. (32) for spatially delocalized dynamics. Therefore, we have to modify the optimal control simulation. To illustrate the modification for the Lagrange multiplier, we consider the simple case in which the control objective is specified solely by A at the final time. If we restrict ourselves to the target operator A that is expressed as a spatially delocalized quasi-projector,[65] we can numerically integrate the Lagrange multiplier using the following algorithm

$$|\xi_P(t_n)\rangle = U(t_n, t_{n+1})[|\xi_P(t_{n+1})\rangle + A|\psi_{1-P}(t_{n+1})\rangle]. \tag{38}$$

Because Eq. (32) has the same form as Eq. (27) when $B(t) = 0$, they have the same time-evolution operator. If the optically active region lies within the grid region, the optimal pulse is expressed as

$$E(t_n) = -\lambda(t_n)\text{Im}\langle \xi_P(t_n)|\mu|\psi_P(t_n)\rangle. \tag{39}$$

As the pulse-design equations have a closed form with respect to the components within the grid region, we can solve them using a monotonically convergent algorithm.[65,66]

Because of the high computational cost, optimal control simulations typically employ one- or two-dimensional models. To extend it to treat higher-dimensional models, the multiconfiguration time-dependent Hartree (MCTDH) method has been proposed.[67–69] The overall procedure is the same as that explained above, except that the wave function, $|\psi(t)\rangle$, and its associated Lagrange multiplier, $|\xi(t)\rangle$, have MCTDH forms. That is, their vibrational wave functions are expressed as product sums of single-particle wave functions, each of which belongs to a single vibrational coordinate. Although the MCTDH equations of motion contain nonlinear terms with respect to time-dependent expansion coefficients, monotonically convergent algorithms are sometimes used to approximately solve the coupled pulse-design equations.[69]

So far, we have restricted ourselves to the case where the time evolution of an isolated molecular system is described by the Schrödinger equation. However, a molecular system may not be isolated or we may be interested in specific degrees of freedom rather than all the degrees of freedom of a molecule. In these cases, the whole system can be divided into a relevant system and a reservoir.[70] The relevant system is described by a reduced density matrix and it satisfies the quantum Liouville equation that includes relaxation operators with/without memory. These relaxation operators represent decoherence processes caused by the entanglement between the relevant system and the reservoir. Optimal pulses are derived using the calculus of variations under the constraint of the quantum Liouville equation.[44,46–48] From a mathematical viewpoint, the objective functional in the density matrix formalism can have a linear form with respect to the state vector (density operator), while that in the wave function formalism has a bilinear form as given by Eq. (28). It is worth noting that for arbitrary target operators, A and $B(t)$, there are monotonically convergent algorithms for solving the optimal control problems that are expressed by linear functionals.

2.3.2. *Other (approximate) control methods*

Roughly speaking, other (approximate) control methods can be classified into three categories. Control methods belonging to the first category use the formal theory of optimal control and the equations of motion with certain approximations are assumed. Examples include the above-mentioned MCTDH factorization of a molecular wave function.[69] Some of them use (semi)-classical descriptions of molecular motion,[71–73] which can reduce the computational effort because they disregard the nonlocal nature of quantum dynamics.

The second category of control methods assumes weak laser fields, which allows the pulse-design equations to be approximated up to a certain order with respect to the optical interaction. The first-order approximation provides an analytical expression for the optimal pulse.[74] The second-order approximation leads to an eigenvalue problem, in which the eigenvector with the maximum eigenvalue corresponds to the optimal pulse.[75–77] In both these cases, optimal pulses are obtained without performing iterative calculations.

The third category of methods consists of local control methods that are characterized by instantaneous feedback.[42,78–81] We explain the method by using the equation of motion expressed by Eq. (27) and by assuming that the control achievement is evaluated by a time-independent target operator, A,

$$J = \langle \psi(t_f)|A|\psi(t_f)\rangle = \int_{t_0}^{t_f} dt \frac{d}{dt} \langle \psi(t)|A|\psi(t)\rangle + \langle \psi(t_0)|A|\psi(t_0)\rangle. \quad (40)$$

The substitution of Eq. (27) into the integrand in Eq. (40) yields

$$\frac{d}{dt}\langle \psi(t)|A|\psi(t)\rangle = \frac{i}{\hbar}\langle \psi(t)|\{[H_0, A] - [\mu, A]E(t)\}|\psi(t)\rangle. \quad (41)$$

If the target operator commutes with the molecular Hamiltonian (a stationary target) and the control pulse is expressed as

$$E(t) = -\lambda(t)\mathrm{Im}\langle \psi(t)|A\mu|\psi(t)\rangle, \quad (42)$$

the derivative, $d\langle \psi(t)|A|\psi(t)\rangle/dt$, will always be positive.[42] We need to choose an appropriate positive function, $\lambda(t)$, to make $\langle \psi(t_f)|A|\psi(t_f)\rangle$ as large as possible. The restriction due to the stationary target can be removed by solving the problem inversely because of the time reversibility of the Schrödinger equation. That is, we can calculate the control pulse by backward time propagation in which the nonstationary target state, $|A_0\rangle$, and the "true" initial state, $|\psi_0\rangle$, are regarded as formal initial and target states, respectively. Let $U(t_f, t_0)$ be the time evolution operator that includes the designed local control pulse. Because $|\langle \psi_0|U(t_0, t_f)|A_0\rangle|^2 =$

$|\langle A_0|U(t_f,t_0)|\psi_0\rangle|^2$, the control achievement is guaranteed to have the same value as that calculated by the formal solution obtained from the backward time propagation.[80]

3. Case Studies

Two examples of optimal control simulations are presented.[66,82] CIs are characterized by efficient nonadiabatic transitions and geometric (Berry) phase effects. These effects are inseparably related with each other; however, it is convenient to classify previous studies into two categories according to their primary interest. In Sec. 3.1, we focus on shaping bound wave packets under the influence of CI-induced nonadiabatic transitions.[82] In Sec. 3.2, geometric phase effects on the branching ratio of photodissociation products are examined.[66] In these case studies, we discuss how to actively use CIs as wave packet cannons and/or π-phase shifters (geometric phase effects).

In addition to these case studies, there have been several optimal control simulations,[83–88] which mainly focus on the efficient nonadiabatic transitions caused by CIs and which include the cis-trans photoisomerization via a CI of the Na-H_2 collision complex[83,85] or cyclohexadiene.[86] These studies aimed to create a localized wave packet near the CI to maximize the nonadiabatic transition probability and the selectivity of the products, although the target packet is chosen based on physical intuition rather than by the calculus of variation. In another study, a set of pump and dump pulses are obtained as the optimal pulse to improve the product yield.[87] Ndong et al.[88] reported the implementation of a (classical) NOT logical gate in a *cis-trans* photoisomerization model, in which the two isomers were regarded as the two states of a bit.

3.1. *Wave packet shaping utilizing CI-induced nonadiabatic transitions*

The first example considered here is the optimal control of ultrafast cis-trans photoisomerization of retinal in rhodopsin. Femtosecond time-resolved experiments reveal that the *trans* product is formed within ~ 200 fs with a high quantum yield[21–23]; this has been qualitatively explained in quantum chemical studies by the involvement of a CI.[89–93] The experiments also found that a wave packet oscillation accompanies the photoisomerization.[22,23] This wave packet oscillation survives up to 2 ps in rhodopsin, but not in solution, suggesting that rhodopsin may provide partial dynamic guidance in a decoherence-free environment. To consistently explain these experimental observations, Hahn and Stock proposed a two-dimensional, two-electronic-state CI model that consists of the "effective" reaction (ϕ) and coupling (x) coordinates, which correspond largely to the torsional angle and the stretching coordinate of the polyene chain, respectively.[24,25] In the diabatic representation, the (field-free) molecular Hamiltonian is expressed as

$$H_0 = -\frac{\hbar^2}{2I}\frac{\partial^2}{\partial \varphi^2} - \frac{\hbar^2}{2m}\frac{\partial^2}{\partial x^2} + \begin{bmatrix} W_{11} & W_{12} \\ W_{21} & W_{22} \end{bmatrix}, \quad (43)$$

where I (m) is the reduced moment of inertia (the reduced mass) and $\{W_{ij}; i,j = 1,2\}$ are diabatic potentials. By diagonalizing the diabatic-potential matrix in Eq. (43), we obtain the adiabatic potentials, which are shown in Fig. 2. The CIs appear at $(\phi, x) = (0.52\pi, 0)$ and $(1.48\pi, 0)$. According to this model, the experimentally measured coherent signal is attributed to wave packet oscillation along the reaction coordinate on the diabatic potential.

In Ref. 82, an optimal control simulation was performed with the aim of determining how to efficiently manipulate a wave packet after nonadiabatic transitions via CIs. Wave packet localization along the reaction coordinate, $\phi \in [0.9\pi, 1.1\pi]$, in the ground electronic state of the *trans* isomer was

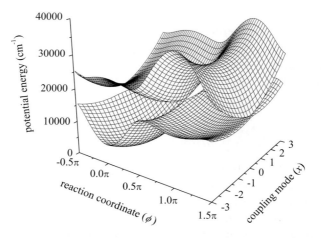

Fig. 2. Two dimensional, two-electronic-state model PESs in the adiabatic representation, in which the CIs appear at $(\phi, x) = (0.52\pi, 0)$ and $(1.48\pi, 0)$.[82] The parameters are taken from Ref. 24.

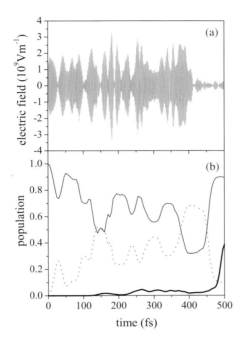

Fig. 3. (a) Optimal control pulse. (b) Time evolution of the target expectation value (bold solid line), and that of populations of the adiabatically electronic ground (thin solid line) and excited (dotted line) states.[82]

chosen as a prototype control target, while no restriction was imposed on the coupling mode. The control time was set to $t_f = 500$ fs, which is longer than the photoisomerization time (~ 200 fs). Figure 3 shows the optimal pulse and the time-dependent target population. The control pulse consists of several pulse trains. The pulse trains before ~ 350 fs adjust the wave packet shape through multiple electronic transitions within the Franck–Condon region of the *cis* isomer. The shaped wave packet is launched through CIs to the target state at ~ 400 fs. After ~ 400 fs, the laser field is virtually zero, indicating that the optimal pulse cooperatively works well with the CI as a wave-packet cannon. At the control time, 39% of the population is transferred in the target region. This successful control achievement is mainly due to the wave packet energy not being explicitly specified by the target. In fact, if we choose one of the bound states of the *trans* isomer as a target, the control pulse appears after the nonadiabatic transitions to adjust the wave packet energy.[82]

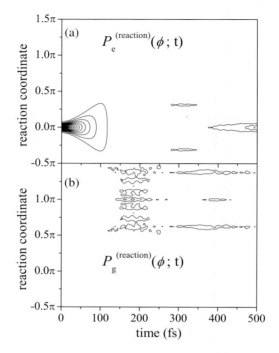

Fig. 4. Time evolution of the reaction-coordinate wave packets of the Franck–Condon (uncontrolled) wave packet: (a) $P_e^{(\text{reaction})}(\phi;t)$ and (b) $P_g^{(\text{reaction})}(\phi;t)$.[82]

To discuss the control mechanisms in detail, it is convenient to introduce reaction-coordinate wave packets, $P_{e,g}^{(\text{reaction})}(\varphi;t)$, and coupling-mode wave packets, $P_{e,g}^{(\text{coupling})}(x;t)$, which are obtained by tracing out the coupling mode and the reaction coordinate, respectively. Here, the suffixes e and g denote the adiabatic excited and ground electronic states, respectively. Figure 4 shows the time evolution of the Franck–Condon (uncontrolled) wave packet along the reaction coordinate, which rapidly loses its initial localized distribution because of the anharmonicity of the PES. In contrast, as shown in Fig. 5, the controlled reaction-coordinate wave packets are characterized by localized distributions both on the excited and ground PESs. Breathing oscillations in the ground electronic state are induced and gradually enhanced through multiple electronic transitions before ~ 350 fs. The larger amplitude of the breathing oscillation leads to a narrower distribution at one of the turning points, which is used to create a localized wave packet in the excited electronic state around 350–400 fs. There is also a considerable

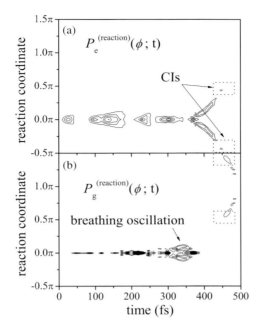

Fig. 5. Time evolution of the reaction-coordinate wave packets of the controlled wave packet: (a) $P_e^{(\text{reaction})}(\phi;t)$ and (b) $P_g^{(\text{reaction})}(\phi;t)$.[82]

difference in the behaviors of the controlled and uncontrolled coupling-mode wave packets. Specifically, the control pulse suppresses the coupling mode excitation to minimize dephasing,[94] which can be detrimental for quantum control. (Here, the term "dephasing" refers to a wave packet spread over an undesirable region.) As indicated by the dotted squares in Fig. 5, the CIs transfer almost all the wave packet components in the excited PES [Fig. 5(a)] to the ground PES while maintaining the spatially localized distribution along the reaction coordinate [Fig. 5(b)], which shows one of the characteristics of the CI as a wave-packet cannon. Because the CIs minimize the dephasing associated with the nonadiabatic transitions, the vibrational coherence implemented by the control pulse can survive to yield a high degree of achievement.

The efficient use of a CI as a wave packet cannon naturally leads to the question about the effects of the local topography around the CI. As changing the local topography will lead to different active areas associated with the CI, it may result in the wave packet being launched in a different way. For comparison, an optimal control simulation was performed using a

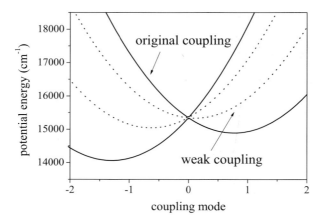

Fig. 6. Adiabatic potential cuts along the coupling mode at $\phi = 0.52\pi$. The solid and dotted lines represent those in the original- and weak-coupling cases, respectively.[82]

smaller coupling parameter. Figure 6 shows the potential cuts around the CIs; the original-coupling and weak-coupling cases are indicated by solid and dotted lines, respectively. One of the potential cuts has a double-well structure in the original coupling case, whereas in the weak-coupling case, its shape varies smoothly as a function of the reaction coordinate, resulting in weaker dephasing. In fact, about 70% of the population reaches the target region at the control time, which is about two times higher than the original value. Wave packet calculations for reduced coupling indicate that we can achieve similar control mechanisms as those for the original-coupling case. However, the optimal pulse does not remove coupling-mode excitation, suggesting that the weaker dephasing imposes a weaker restriction on the preparation of the wave packet for the CI cannon.

3.2. *Geometric phase effects on coherent control of CI-induced dynamics*

In addition to efficient nonadiabatic transitions, the geometric (Berry) phase effect is another consequence associated with CI-induced dynamics.[26–30] When the nuclear wave packet cycles around the CI, the electronic wave function changes sign, as it adiabatically depends on the nuclear motion. For the total molecular wave function to be single-valued, the nuclear wave packet must acquire the opposite phase of the

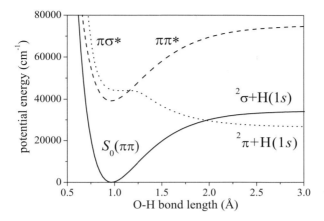

Fig. 7. Diabatic potential cuts along the OH stretching coordinate for planar phenol ($\theta = 0$).

electronic wave function. This is a typical example of a geometric phase. Here, we consider its effects on coherent control of chemical reactions by examining the case study of photodissociation of phenol.[66,95,96] Phenol is a prototype molecule for investigating the photochemistry of aromatic biomolecules.[95–103]

A two-dimensional, three-electronic-state model is taken from Ref. 96 to describe the photodissociation of phenol. It includes the OH-stretching coordinate (r) as a reaction coordinate and the CCOH dihedral angle (θ) as a coupling mode. Figure 7 shows the potential cuts of the diagonal diabatic components along the reaction coordinate at $\theta = 0$. The electronic states involved are the $\pi\pi^*$, $\pi\sigma^*$, and $S_0(\pi\pi)$ states. The model is characterized by two CIs, resulting in two dissociation channels that lead to the $^2\pi$ and $^2\sigma$ phenoxyl radicals and H(1s). This reduced-dimensionality model is consistent with the experimental observations of Tseng et al.,[13] who measured the translational energy distribution of hydrogen atoms released from photoexcited phenol and found that the photoexcitation energy does not spread significantly over the intramolecular vibrational modes. In addition, within this model, the Franck-Condon wave packet is predicted to have insufficient energy to exceed the barrier to reach the CI.[96] As the initial vibrational energy can compensate for this shortfall, we expect that the photodissociation probability could be enhanced by using vibrational-mediated passive control. Such effects have recently been observed by Crim's group.[104]

The aim of the simulation is to achieve coherent control of the branching ratio of photodissociation products through electronic transitions, in which the laser pulse is designed to maximize either the $^2\sigma + \text{H}(1s)$ dissociation probability or the $^2\pi + \text{H}(1s)$ dissociation probability. When enhancement of the $^2\sigma + \text{H}(1s)$ dissociation channel is chosen as the objective, the quasi-projector that specifies this target is expressed as

$$A = \int dr d\theta |[^2\sigma + \text{H}(1s)]r\theta\rangle a(r)\langle [^2\sigma + \text{H}(1s)]r\theta|, \qquad (44)$$

where

$$a(r) = 1 - \frac{1}{1 + \exp[\beta(r - r_d)]} \qquad (45)$$

with $\beta = 20.0\,\text{Å}^{-1}$ and $r_d = 3.0\,\text{Å}$. The latter value is obtained from the assumption that phenol is dissociated when the OH bond is longer than $3.0\,\text{Å}$. When the control achievement is evaluated by this operator

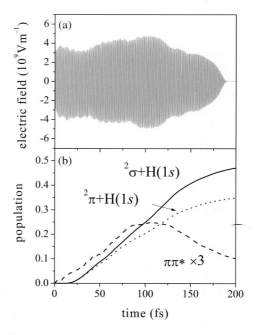

Fig. 8. (a) Optimal pulse that enhances the dissociation probability of the $^2\sigma + \text{H}(1s)$ channel. (b) Time evolution of the dissociation probability of the $^2\sigma + \text{H}(1s)$ channel (solid line) and that of the $^2\pi + \text{H}(1s)$ (dotted line), and that of the population in the diabatic $\pi\pi^*$ state (dashed line).[66]

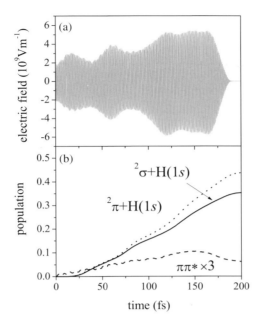

Fig. 9. (a) Optimal pulse that enhances the dissociation probability of the $^2\pi + \mathrm{H}(1s)$ channel. (b) Time evolution of the dissociation probability of the $^2\sigma + \mathrm{H}(1s)$ channel (solid line) and that of the $^2\pi + \mathrm{H}(1s)$ (dotted line), and that of the population in the diabatic $\pi\pi^*$ state (dashed line).[66]

at time $t_f = 200$ fs, we have the optimal pulse in Fig. 8(a) that leads to 47.1% [34.9%] of the population to the $^2\sigma + \mathrm{H}(1s)[^2\pi + \mathrm{H}(1s)]$ dissociation channel. The optimal pulse has a simple structure with a central frequency of 44380 cm^{-1}.

In a similar manner, we obtain the optimal pulse that enhances the $^2\pi + \mathrm{H}(1s)$ dissociation, which is shown in Fig. 9(a). Its central frequency is 43980 cm^{-1}, which is ∼500 cm^{-1} lower than that in Fig. 8(a). From Fig. 8(b), the total dissociation probability is 79.1%, which is similar to that in Fig. 8(b) (82.0%); however, 43.8% [35.3%] of the population dissociates to the $^2\pi + \mathrm{H}(1s)$ [$^2\sigma + \mathrm{H}(1s)$] channel. The optimal pulse changes the branching ratio [$^2\pi + \mathrm{H}(1s)$]/[$^2\sigma + \mathrm{H}(1s)$] from 0.81 (Fig. 8) to 1.35 (Fig. 9). As the calculated optimal pulses in Figs. 8(a) and 9(a) have simple structures, control is expected to be mainly achieved by actively utilizing the CIs as wave packet cannons/π-phase shifters (geometric phase effects).

The population in the diabatic $^1\pi\pi^*$ state in Fig. 9(b) is reduced to one-third of that in Fig. 8(b), although the total dissociation probabilities

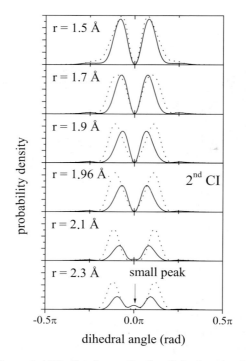

Fig. 10. Cuts of the probability density on the S_1 adiabatic potential along the CCOH dihedral angle θ for various OH bond length r at $t = 120$ fs. The solid [dotted] lines represent those created by the optimal pulse that enhances the $^2\sigma + \text{H}(1s)$ [$^2\pi + \text{H}(1s)$] dissociation. In each figure, the probability densities are plotted in the range of $[0, 8 \times 10^{-6}]$.[66]

are similar in both cases. That is, in Fig. 9(b), a portion of the excited population is created in the optically dark $^1\pi\sigma^*$ state because of vibronic coupling with the optically active $^1\pi\pi^*$ state (known as the intensity-borrowing effect). The central frequency of the optimal pulse is chosen either to prevent vibronic coupling (Fig. 8) or to efficiently cause it (Fig. 9). The control mechanisms are also supported by the cuts of the wave packets on the first-excited adiabatic PES (S_1) along the CCOH dihedral angle (θ) in Fig. 10, in which the solid (dotted) lines show the cuts associated with Fig. 8 (Fig. 9). The wave packet associated with Fig. 8 (solid lines) is an even function of θ immediately after the optical excitation to $^1\pi\pi^*$ state as the transition moment function has a symmetric structure with respect

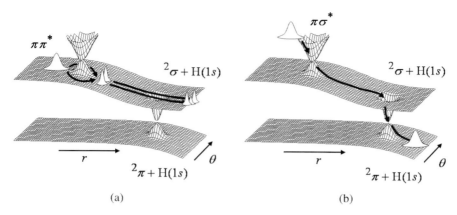

Fig. 11. Schematic illustrations of the control mechanisms associated with (a) Fig. 8 and (b) Fig. 9.

to θ. When the wave packet encounters the first CI at $r = 1.16$ Å, it is bifurcated into two components with opposite phases because of geometric phase effects. They introduce a node at $\theta = 0$ into the wave packet, which reduces the probability density around the second CI at $r = 1.96$ Å. In turn, this suppresses the nonadiabatic transitions around the second CI and enhances the dissociation probability to the $^2\sigma + \mathrm{H}(1s)$ channel. After the second CI, the geometric phase effects are canceled in the remaining wave packet components on the S_1 PES, resulting in the small peak in the probability density around $\theta = 0$. On the other hand, the wave packet associated with Fig. 9 (dotted lines) is an odd function of θ immediately after the optical excitation to the $^1\pi\sigma^*$ state, as it is created by vibronic coupling, which is expressed by a linear function of θ around $\theta = 0$. The wave packet is nonadiabatically transferred onto the S_1 PES through the CI while being slightly distorted from its original structure. Because of wave packet broadening due to the anharmonicity of the S_1 PES, there exists a certain population within the active region of the second CI, which enhances the dissociation in the $^2\pi + \mathrm{H}(1s)$ channel. Based on these results, the control mechanisms are schematically illustrated in Fig. 11(a) and (b), which correspond to those associated with Figs. 8 and 9, respectively. A slight change in the population distribution around the second CI can lead to efficient control of the branching ratio of the dissociation products.

4. Conclusions

Laser control of ultrafast dynamics at CIs has been discussed. In Sec. 2, we first explained the basic idea of coherent control by adopting a two-electronic-state model within the first-order perturbative approximation with respect to optical and non-adiabatic interactions. We then introduced more general control procedures based on a pre-target scheme, optimal control theory and local control theory. Numerical techniques frequently used in optimal control simulations were also examined. In Sec. 3, we discussed control mechanisms by presenting case studies of two optimal control simulations, i.e. a wave packet shaping in the photoisomerization of retinal in rhodopsin and the control of branching ratio of photodissociation channels of phenol. These simulations show that optimally designed laser pulses cooperate well with CIs to yield a high degree of achievement by actively using CIs as wave packet cannons and/or π-phase shifters (geometric phase effects).

Acknowledgments

We wish to thank Ms. M. Abe, Dr. Z. Lan and Prof. Emeritus Y. Fujimura for stimulating discussions on numerical applications. This work was in part supported by a Grant-in-Aid for Scientific Research (B) (20350001).

References

1. S.A. Rice and M. Zhao, *Optical Control of Molecular Dynamics* (Wiley, New York, 2000).
2. M. Shapiro and P. Brumer, *Principles of the Quantum Control of Molecular Processes* (Wiley, Hoboken, 2003).
3. D.J. Tannor, *Introduction to Quantum Mechanics: A Time-Dependent Perspective* (University Science Books, Sausalito, 2007).
4. W.S. Warren, H. Rabitz and M. Dahleh, *Science* **259**, 1581 (1993).
5. *Conical Intersections*, edited by W. Domcke, D.R. Yarkony and H. Köppel, (World Scientific, Singapore, 2004).
6. W. Fuss, T. Schikarski, W.E. Schmid, S. Trushin and K.L. Kompa, *Chem. Phys. Lett.* **262**, 675 (1996).
7. S. Trushin, W. Fuss, T. Schikarski, W.E. Schmid and K.L. Kompa, *J. Chem. Phys.* **106**, 9386 (1997).
8. D.H. Mordaunt, M.N.R. Ashfold and R.N. Dixon, *J. Chem. Phys.* **104**, 6460 (1996).

9. M.L. Hause, Y.H. Yoon and F.F. Crim, *J. Chem. Phys.* **125**, 174309 (2006).
10. P. Farmanara, W. Radloff, V. Stert, H.-H. Ritze and I.V. Hertel, *J. Chem. Phys.* **111**, 633 (1999).
11. S. Ishiuchi, K. Daikoku, M. Saeki, M. Sakai, K. Hashimoto and M. Fujii, *J. Chem. Phys.* **117**, 7077 (2002).
12. J. Wei, A. Kuczmann, J. Riedel, F. Renth and F. Temps, *Phys. Chem. Chem. Phys.* **5**, 315 (2003).
13. C.-M. Tseng, Y.T. Lee and C.-K. Ni, *J. Chem. Phys.* **121**, 2459 (2004).
14. A.J. Wurzer, T. Wilhelm, J. Piel and E. Riedle, *Chem. Phys. Lett.* **299**, 296 (1999).
15. D. Zhong, E.W.-G. Diau, T.M. Bernhardt, S.D. Feyter, J.D. Roberts and A.H. Zewail, *Chem. Phys. Lett.* **298**, 129 (1998).
16. E.W.-G. Diau, S.D. Feyter and A.H. Zewail, *J. Chem. Phys.* **110**, 9785 (1999).
17. H. Kang, B. Jung and S.K. Kim, *J. Chem. Phys.* **118**, 6717 (2003).
18. M.N.R. Ashfold, B. Cronin, A.L. Devine, R.N. Dixon and M.G.D. Nix, *Science* **312**, 1637 (2006).
19. C.E. Crespo-Hernández, B. Cohen, P.M. Hare and B. Kohler, *Chem. Rev.* **104**, 1977 (2004).
20. S. Sorgues, J.M. Mestdagh, J.P. Visticot and B. Soep, *Phys. Rev. Lett.* **91**, 103001 (2003).
21. R.W. Schoenlein, L.A. Peteanu, R.A. Mathies and C.V. Shank, *Science* **254**, 412 (1991).
22. L.A. Peteanu, R.W. Schoenlein, Q. Wang, R.A. Mathies and C.V. Shank, *Proc. Natl. Acad. Sci. USA*, **90**, 11762 (1993).
23. Q. Wang, R.W. Schoenlein, L.A. Peteanu, R.A. Mathies and C.V. Shank, *Science* **266**, 422 (1994).
24. S. Hahn and G. Stock, *J. Phys. Chem. B*, **104**, 1146 (2000).
25. S. Hahn and G. Stock, *Chem. Phys.* **259**, 297 (2000).
26. G. Herzberg and H.C. Longuet-Higgins, *Discuss. Faraday Soc.* **35**, 77 (1963).
27. H.C. Longuet-Higgins, *Proc. Roy. Soc. London Ser. A* **344**, 147 (1975).
28. M.V. Berry, *Proc. Roy. Soc. London Ser. A* **392**, 45 (1984).
29. C.A. Mead, *Rev. Mod. Phys.* **64**, 51 (1992).
30. J.C. Juanes-Marcos, S.C. Althorpe and E. Wrede, *Science* **309**, 1227 (2005).
31. R.S. Judson and H. Rabitz, *Phys. Rev. Lett.* **68**, 1500 (1992).
32. A. Assion, T. Baumert, M. Bergt, T. Brixner, B. Kiefer, V. Seyfried, M. Strehle and G. Gerber, *Science* **282**, 919 (1998).
33. H. Rabitz, R. de Vivie-Riedle, M. Motzkus and K. Kompa, *Science* **288**, 824 (2000).
34. R.J. Levis, G.M. Menkir and H. Rabitz, *Science* **292**, 709 (2001).
35. T. Brixner, G. Krampert, T. Pfeifer, R. Selle, G. Gerber, M. Wollenhaupt, O. Graefe, C. Horn, D. Liese and T. Baumert, *Phys. Rev. Lett.* **92**, 208301 (2004).

36. T. Hornung, R. Meier, D. Zeidler, K.-L. Kompa, D. Proch and M. Motzkus, *Appl. Phys. B* **71**, 277 (2000).
37. C.J. Bardeen, V.V. Yakovlev, K.R. Wilson, S.D. Carpenter, P.M. Weber and W.S. Warren, *Chem. Phys. Lett.* **280**, 151 (1997).
38. T.C. Weinacht, J.L. White and P.H. Bucksbaum, *J. Phys. Chem. A* **103**, 10166 (1999).
39. J.L. Herek, W. Wohlleben, R.J. Cogdell, D. Zeidler and M. Motzkus, *Nature*, **417**, 533 (2002).
40. T. Brixner, N.H. Damrauer, B. Kiefer and G. Gerber, *J. Chem. Phys.* **118**, 3692 (2003).
41. A.P. Peirce, M.A. Dahler and H. Rabitz, *Phys. Rev. A* **37**, 4950 (1988).
42. R. Kosloff, S.A. Rice, P. Gaspard, S. Tersigni and D.J. Tannor, *Chem. Phys.* **139**, 201 (1989).
43. W. Zhu, J. Botina and H. Rabitz, *J. Chem. Phys.* **108**, 1953 (1998).
44. Y. Ohtsuki, W. Zhu, H. Rabitz, *J. Chem. Phys.* **110**, 9825 (1999).
45. Y. Ohtsuki, K. Nakagami, Y. Fujimura, W. Zhu and H. Rabitz, *J. Chem. Phys.* **114**, 8867 (2001).
46. Y. Ohtsuki, *J. Chem. Phys.* **119**, 661 (2003).
47. Y. Ohtsuki, G. Turinici and H. Rabitz, *J. Chem. Phys.* **120**, 5509 (2004).
48. Y. Ohtsuki, Y. Teranishi, P. Saalfrank, G. Turinici and H. Rabitz, *Phys. Rev. A* **75**, 033407 (2007).
49. Y. Ohtsuki and K. Nakagami, *Phys. Rev. A* **77**, 033414 (2008).
50. J. Somlói, V.A. Kazakov and D.J. Tannor, *Chem. Phys.* **172**, 85 (1993).
51. A. Bartana, R. Kosloff and D.J. Tannor, *J. Chem. Phys.* **106**, 1435 (1997).
52. H. Metiu and V. Engel, *J. Opt. Soc. Am. B* **7**, 1709 (1990).
53. N.F. Scherer, A.J. Ruggiero, M. Du and G.R. Fleming, *J. Chem. Phys.* **93**, 856 (1990).
54. K. Ohmori, Y. Sato, E.E. Nikitin and S.A. Rice, *Phys. Rev. Lett.* **91**, 243003 (2003).
55. K. Ohmori, *Ann. Rev. Phys. Chem.* **60**, 511 (2009) and references therein.
56. B. Kohler, V.V. Yakovlev, J. Che, J.L. Krause, M. Messina, K.R. Wilson, N. Schwenter, R.M. Whitnell and Y. Yan, *Phys. Rev. Lett.* **74**, 3360 (1995).
57. P. Gross, A.K. Gupta, D.B. Bairagi and M.K. Mishra, *J. Chem. Phys.* **104**, 7045 (1996).
58. K. Hoki, Y. Ohtsuki and Y. Fujimura, *J. Phys. Chem.* **103**, 6301 (1999).
59. K. Nakagami, Y. Ohtsuki and Y. Fujimura, *Chem. Phys. Lett.* **360**, 91 (2002).
60. P.S. Christopher, M. Shapiro and P. Brumer, *J. Chem. Phys.* **123**, 064313 (2005).
61. M. Sukharev and T. Seideman, *Phys. Rev. Lett.* **93**, 093004 (2004).
62. M. Sukharev and T. Seideman, *Phys. Rev. A* **71**, 012509 (2005).
63. R. Heather and H. Metiu, *J. Chem. Phys.* **86**, 5009 (1987).
64. R. Kosloff, *J. Phys. Chem.* **92**, 2087 (1988).
65. K. Nakagami, Y. Ohtsuki and Y. Fujimura, *J. Chem. Phys.* **117**, 6429 (2002).

66. M. Abe, Y. Ohtsuki, Y. Fujimura, Z. Lan and W. Domcke, *J. Chem. Phys.* **124**, 224316 (2006).
67. H.-D. Meyer, U. Manthe and L.S. Cederbaum, *Chem. Phys. Lett.* **165**, 73 (1990).
68. U. Manthe, H.-D. Meyer and L.S. Cederbaum, *J. Chem. Phys.* **97**, 3199 (1992).
69. L. Wang, H.-D. Meyer and V. May, *J. Chem. Phys.* **125**, 014102 (2006).
70. K. Blum, *Density Matrix Theory and Applications* (Plenum Press, New York, 1981).
71. C.D. Schwieters and H. Rabitz, *Phys. Rev. A* **48**, 2549 (1993).
72. J. Botina, H. Rabitz and N. Rahman, *J. Chem. Phys.* **102**, 226 (1995).
73. A. Kondorskiy, G. Mil'nikov and H. Nakamura, *Phys. Rev. A* **72**, 041401 (2005).
74. V. Dubov and H. Rabitz, *J. Chem. Phys.* **103**, 8412 (1995).
75. D.J. Tannor and S.A. Rice, *J. Chem. Phys.* **83**, 5013 (1985).
76. Y. Yan, R.E. Gillian, R.M. Whitnell, K.R. Wilson and S. Mukamel, *J. Phys. Chem.* **97**, 2320 (1993).
77. J.L. Krause, R.M. Whitnell, K.R. Wilson, Y. Yan and S. Mukamel, *J. Chem. Phys.* **99**, 6562 (1993).
78. R. Kosloff, A.D. Hammerich and D.J. Tannor, *Phys. Rev. Lett.* **69**, 2172 (1992).
79. A. Bartana, R. Kosloff and D.J. Tannor, *J. Chem. Phys.* **99**, 196 (1993).
80. Y. Ohtsuki, H. Kono and Y. Fujimura, *J. Chem. Phys.* **109**, 9318 (1998).
81. M. Sugawara, *J. Chem. Phys.* **118**, 6784 (2003).
82. M. Abe, Y. Ohtsuki, Y. Fujimura and W. Domcke, *J. Chem. Phys.* **123**, 144508 (2005).
83. R. de Vivie-Riedle, K. Sundermann and M. Motzkus, *Faraday Discuss.* **113**, 303 (1999).
84. Y. Ohtsuki, K. Ohara, M. Abe, K. Nakagami and Y. Fujimura, *Chem. Phys. Lett.* **369**, 525 (2003).
85. D. Geppert, A. Hofmann and R. de Vivie-Riedle, *J. Chem. Phys.* **119**, 5901 (2003).
86. D. Geppert, L. Seyfarth and R. de Vivie-Riedle, *Appl. Phys. B* **79**, 987 (2004).
87. F. Grossmann, L. Feng, G. Schmidt, T. Kunert and R. Schmidt, *Europhys. Lett.* **60**, 201 (2002).
88. M. Ndong, L. Bomble, D. Sugny, Y. Justum and M. Desouter-Lecomte, *Phys. Rev. A* **76**, 043424 (2007).
89. R.M. Weiss and A. Warshel, *J. Am. Chem. Soc.* **101**, 6131 (1979).
90. M. Garavelli, P. Celani, F. Bernardi, M.A. Robb and M. Olivucci, *J. Am. Chem. Soc.* **119**, 6891 (1997).
91. M. Garavelli, T. Vreven, P. Celani, F. Bernardi, M.A. Robb and M. Olivucci, *J. Am. Chem. Soc.* **120**, 1285 (1998).
92. F. Molnar, M. Ben-Num, T.J. Martinez and K. Schulten, *J. Mol. Struct. (Theochem)*, **506**, 169 (2000).
93. M. Nonella, *J. Phys. Chem. B* **104**, 11379 (2000).

94. M. Abe, Y. Ohtsuki, Y. Fujimura and W. Domcke, in *Ultrafast Phenomena XIV*, edited by T. Kobayashi, T. Okada, T. Kobayashi, K.A. Nelson and S. De Silvestri (Springer, Berlin, 2005), p. 613.
95. A.L. Sobolewski, W. Domcke, C. Dedonder-Lardeux and C. Jouvet, *Phys. Chem. Chem. Phys.* **4**, 1093 (2002).
96. Z. Lan, W. Domcke, V. Vallet, A.L. Sobolewski and S. Mahapatra, *J. Chem. Phys.* **122**, 224315 (2005).
97. D.R. Yarkony, *J. Phys. Chem. A* **105**, 6277 (2001).
98. J. Quenneville and T.J. Martinez, *J. Phys. Chem. A* **107**, 829 (2003).
99. W. Domcke and A.L. Sobolewski, *Science* **302**, 1693 (2003).
100. M.J. Paterson, M.A. Robb, L. Blancafort and A.A. De Bellis, *J. Am. Chem. Soc.* **126**, 2912 (2004).
101. V. Vallet, Z. Lan, S. Mahapatra, A.L. Sobolewski and W. Domcke, *J. Chem. Phys.* **123**, 144307 (2005).
102. A.L. Sobolewski, W. Domcke and C. Hättig, *Proc. Natl. Acad. Sci. USA* **102**, 17903 (2005).
103. A.L. Sobolewski and W. Domcke, *Chem. Phys. Chem*, **7**, 561 (2006).
104. M.L. Hause, Y.H. Yoon, A.S. Case and F.F. Crim, *J. Chem. Phys.* **128**, 104307 (2008).

Part III

Experimental Detection of Dynamics at Conical Intersections

Chapter 15

Exploring Nuclear Motion Through Conical Intersections in the UV Photodissociation of Azoles, Phenols and Related Systems

Thomas A. A. Oliver, Graeme A. King, Alan G. Sage and
Michael N. R. Ashfold*

1.	Introduction	604
2.	Experimental Methods	604
	2.1. Photofragment translational spectroscopy	604
	2.2. Ion imaging methods	606
	2.3. H Rydberg atom photofragment translational spectroscopy	608
3.	Results	609
	3.1. Azoles	609
	3.2. Phenols	615
	3.3. Larger analogues	623
4.	Future Prospects	627
	References	629

School of Chemistry, University of Bristol, Bristol BS8 1TS, UK.
*mike.ashfold@bris.ac.uk

1. Introduction

There has been much recent interest in the role of $\pi\sigma^*$ and $n\sigma^*$ excited states (i.e. states that are formed by $\sigma^* \leftarrow \pi$ and $\sigma^* \leftarrow n$ electron excitations) in promoting the non-radiative decay of azoles (e.g. pyrrole, imidazole, *etc.*), phenols, and larger heteroaromatic molecules built from such units (e.g. nucleobases like adenine, guanine, *etc.*, and aromatic amino acids like histidine, tryptophan and tyrosine).[1] The strong UV absorptions associated with these chromophores are generally attributable to $\pi^* \leftarrow \pi$ transitions, but these molecules also possess $\pi\sigma^*$ (and, in some cases, $n\sigma^*$) excited states. For brevity, we will focus attention on $\pi\sigma^*$ excited states, unless the discussion necessitates otherwise. $\sigma^* \leftarrow \pi$ transitions typically have much smaller absorption cross-sections than $\pi^* \leftarrow \pi$ transitions, but $\pi\sigma^*$ states can be populated — either by direct photo-excitation (as in imidazole and pyrrole), or by radiationless transfer following initial excitation to an 'optically bright' $\pi\pi^*$ state (as in phenols, indoles, *etc.*). Sobolewski, Domcke and colleagues[2,3] showed that the diabatic potential energy surfaces (PESs) associated with such $\pi\sigma^*$ states will be repulsive with respect to X–H (X=N, O, *etc.*) bond extension. Population in a $\pi\sigma^*$ excited state could thus decay by X–H bond fission and/or by radiationless transfer to a lower electronic state (typically the ground state) via a conical intersection (CI) between their respective diabatic PESs.

This chapter surveys recent (mainly experimental) studies of the photofragmentation dynamics of various azoles and phenols and heavier analogues based on such units, highlighting ways in which CIs promote dissociation, and/or influence the dissociation dynamics. We start by presenting an overview of some of the variants of photofragment translational spectroscopy (PTS) by which such experimental data can be obtained, before proceeding to more detailed descriptions of the experimental findings for the chosen molecules. The chapter concludes with a brief prospective section.

2. Experimental Methods

2.1. *Photofragment translational spectroscopy*

An 'ideal' PTS experiment might involve the following: photo-excitation of the parent molecule of interest, in a known quantum state, at a well-defined instant in time, with linearly polarized radiation (with associated polarization vector $\varepsilon_{\mathbf{phot}}$), and measurement of the nascent recoil velocities (speeds *and* directions) of all of the resulting fragments, with internal quantum state resolution. Almost inevitably, some compromises are necessary in order to realise such studies, but modern PTS experiments can (in favourable cases) get quite close to this ideal. Wilson and colleagues[4] pioneered the so-called 'universal detection' PTS experiment, which has since been refined and exploited by many other groups.[5] This employs pulsed laser photolysis of a selected precursor molecule in a skimmed molecular beam issuing from a rotatable source into a chamber maintained at high vacuum. Those fragments that 'fly' a known distance into a known (small) solid angle under collision-free conditions are ionized by electron bombardment prior to entering a mass spectrometer and detection. Fragment speeds and thus kinetic energies (E_{kin}) are deduced from the measured times-of-flight (TOFs), with the laser pulse defining the zero of time. Angular distributions are determined by observing how the measured fragment flux varies as $\varepsilon_{\mathbf{phot}}$ is rotated, and characterised in terms of an angular anisotropy parameter, β. Photo-excitation preferentially selects molecules that are aligned so that their transition moment ($\boldsymbol{\mu}$) is parallel to $\varepsilon_{\mathbf{phot}}$. Direct dissociation occurs on a timescale that is much faster than a classical rotational period. The resulting fragments will recoil along the axis of the breaking bond in the photo-excited molecule, i.e. their recoil anisotropy will reflect the original $\boldsymbol{\mu}\cdot\varepsilon_{\mathbf{phot}}$ interaction. The angular distribution of the recoiling fragments is given by

$$I(\theta) = (1 + \beta P_2(\cos\theta))/4\pi,$$

where θ is the angle between $\varepsilon_{\mathbf{phot}}$ and the TOF axis, and $P_2(x) = (3x^2 - 1)/2$ is the second order Legendre polynomial. β takes limiting values of $+2$ in the case of prompt dissociation following excitation via a parallel transition (i.e. $\boldsymbol{\mu}$ lies along the breaking bond) and -1 in the case of a perpendicular transition (i.e. $\boldsymbol{\mu}$ is orthogonal to the breaking bond). Less anisotropic fragment recoil distributions (i.e. with $\boldsymbol{\beta}$ closer to 0) are observed in the case of direct dissociations brought about by exciting a transition with $\boldsymbol{\mu}$ at an intermediate angle to the bond of interest, and in the case of predissociations — i.e. where the excited state lifetime is comparable to, or longer than, the rotational period of the parent molecule. Isotropic recoil velocity distributions are one signature of a dissociation occurring over an extended timescale, e.g. from highly vibrationally excited ground state molecules formed by radiationless transfer from an excited electronic state.

Blank et al.[6] applied these traditional PTS methods to the case of pyrrole photolysis at $\lambda_{\text{phot}} = 248$ nm — a wavelength at which N–H bond fission was shown to be the main decay process. Let us assume that the energy of the photolysis photon (E_{phot}) is known, and that it is possible to 'cool' the parent pyrrole molecules through use of a skimmed molecular beam so that they are all in their zero-point ($v = 0$) vibrational level and in a narrow distribution of (low) J rotational states [i.e. the internal energy, $E_{\text{int}}(\text{pyrrole}) \sim 0$]. Under these circumstances, determining E_{kin} for the H atoms (or the pyrrolyl co-fragments) provides, via momentum conservation arguments, a measure of the total kinetic energy release (TKER). The population distribution over the *internal* energy states of the radical fragment can thus be determined via the energy conservation equation (1):

$$E_{\text{phot}} + E_{\text{int}}(\text{pyrrole}) = D_0(\text{H–pyrrolyl}) + \text{TKER} + E_{\text{int}(\text{pyrrolyl})}, \quad (1)$$

where $D_0(\text{H–pyrrolyl})$ is the N–H bond strength in pyrrole. This type of PTS experiment is very general, since a wide range of atomic and molecular fragments can be detected by mass spectrometry but, as Fig. 1(a) shows, it offers limited resolution and sensitivity and will rarely be quantum state specific (except, for example, when monitoring simple atomic products like H or D). Nonetheless, the data in Fig. 1(a) already offer one key result: the TKER distribution of the H + pyrrolyl products from photolysis of pyrrole at 248 nm peaks far from zero, implying that dissociation occurs on a repulsive, excited state PES and not by unimolecular decay of hot ground state molecules.

2.2. *Ion imaging methods*

Ion imaging techniques[7,8] provide another means of studying molecular photodissociation processes. Again, the molecule of interest is typically delivered in the form of a jet-cooled, skimmed molecular beam directed towards the centre of the imaging detector and photolysed with the output of a pulsed laser. The resulting fragments will recoil from the interaction region with a range of velocities — determined by the fragmentation dynamics. The first novel feature of the ion imaging technique is that the fragment of interest is ionized, at source, and (in many cases) with quantum-state selectivity, by resonance enhanced multiphoton ionization (REMPI) methods. In the ideal limit, ionization will cause minimal perturbation of the nascent velocity distribution of the photofragment of interest.

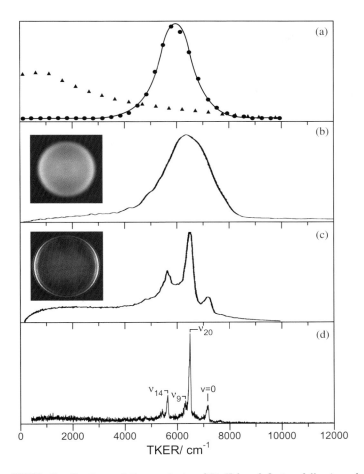

Fig. 1. TKER distributions of the products of N–H bond fission following photolysis of pyrrole at: (a) $\lambda_{\text{phot}} = 248$ nm derived from a traditional PTS study, with the circles and triangles representing deduced distributions from dissociation on the $^1\pi\sigma^*$ and S_0 PESs, respectively; (b) $\lambda_{\text{phot}} = 243.1$ nm, obtained using velocity map imaging methods; (c) $\lambda_{\text{phot}} = 250$ nm, obtained by slice imaging methods — with $\varepsilon_{\text{phot}}$ aligned vertically, in the plane of the page, in both cases; and (d) $\lambda_{\text{phot}} = 250$ nm, obtained by HRAPTS methods. The vibrational quantum numbers of the pyrrolyl products associated with the prominent features in (d) are labelled. (Adapted from Refs. 6, 9, 11 and 15.)

The expanding ion cloud is then accelerated through a sequence of electric fields so as to impact on a time and position sensitive detector. The resulting image of ion intensity versus position is a two-dimensional (2-D) depiction of the full 3-D velocity distribution of the state-selected fragment ion of interest which, in a well-designed experiment, can be recovered by image reconstruction methods.

Temps and co-workers[9,10] imaged the H atom fragments resulting from photolysis of pyrrole at various UV wavelengths including λ_{phot} = 243.1 nm. As Fig. 1(b) shows, image analysis returned a TKER distribution for the H + pyrrolyl fragments that was very similar to that determined in the earlier PTS study (at λ_{phot} = 248 nm). Imaging methods offer some significant advantages. Ionization at source encourages high collection efficiencies; all ionized fragments with the mass to charge (m/z) ratio of interest contribute to the measured image. Second, the entire photofragment velocity (i.e. speed and angular) distribution is carried in a single image. The image shown in Fig. 1(b) suggests that the recoil velocity distribution of the H atoms from pyrrole photolysis at λ_{phot} = 243.1 nm is relatively isotropic, but Wei et al.[9] deduced some preference for recoil along axes perpendicular to ε_{phot} — i.e. for dissociation characterised by a negative recoil anisotropy parameter, $\beta = -0.37 \pm 0.05$. Imaging methods also offer the advantage of fairly widespread applicability: all atoms and many small radical fragments are amenable to detection by REMPI methods. As Fig. 1(b) hints, the achievable kinetic energy resolution remains a challenge for imaging methods. Resolution may be constrained by a number of factors including the following: the spread of internal (or transverse translational) energies in the parent molecular beam; space charge effects (arising from the Coulombic repulsions between ions in the interaction region); the precision with which the point of ion impact on the position sensitive detector can be determined; and the fidelity with which the measured 2-D image can be transformed into the 3-D velocity distribution of interest. Several strategies for minimising such limitations have now been demonstrated.[8] The slice image shown in Fig. 1(c), and the TKER spectrum derived from analysis of this image, illustrate the current state of the art with regard to resolution achieved using imaging methods to study the UV photolysis of pyrrole.[11]

2.3. *H Rydberg atom photofragment translational spectroscopy*

In the specific case of H atom photofragments, such as those arising from X–H bond fission in the case of azoles, phenols, *etc.*, the technique of H Rydberg atom (HRA) PTS[12,13] is without peer in terms of kinetic energy resolution. This technique shares elements in common with both universal detection PTS and ion imaging methods. The hydrogen containing molecule of interest (seeded in an inert carrier gas like He or Ar) is introduced into the interaction region in the form of a pulsed, skimmed molecular beam directed along an axis orthogonal to the detection axis, where it is subjected to pulsed laser photolysis. As in ion imaging, the nascent H ($n = 1$) atoms are immediately 'tagged' at source (i.e. prior to their escaping from the interaction region). Tagging in this case, however, involves two colour double resonant excitation via the $n = 2$ state to a Rydberg state with high principal quantum number ($n \sim 80$). These Rydberg atoms are still neutral species, and thus immune from the effects of space charge blurring. The Rydberg electron has only a small binding energy and can easily be field ionized and the H atom fragments detected (as protons) after travelling a known distance d to a detector. The collection efficiency is intrinsically low since, in order to have any chance of being detected, the nascent H atom photofragment must recoil into the small solid angle subtended by the detector. Furthermore, to determine the fragment recoil anisotropy it is necessary to make at least three separate TOF measurements (at $\theta = 0$, 54.7 and 90°).[14] But, as Fig. 1(d) shows, the TKER spectrum of the H + pyrrolyl fragments from photolysis of pyrrole (at $\lambda_{\text{phot}} = 250\,\text{nm}$) measured using the HRAPTS technique shows much greater detail: features attributable to the formation of pyrrolyl radicals in specific vibrational modes are readily apparent.[15] Furthermore, the sharpness of these peaks indicates that the pyrrolyl fragments are formed with a narrow spread of rotational energies. The remainder of this chapter highlights many of the dynamical insights that follow given such detailed information about the product energy disposal following UV photo-excitation of a range of azoles, phenols and heavier analogues.

3. Results

3.1. *Azoles*

Ab initio calculations of selected portions of the PESs of the ground and first few excited states of pyrrole (an illustrative azole) have played a very important role in rekindling interest in the photochemical importance of $^1\pi\sigma^*$ states.[2,3] This section surveys recent findings involving simple azoles like pyrrole, imidazole and pyrazole, the structures of which are shown in Fig. 2. The photochemistry exhibited by the various azoles shows many features in common, but also some important differences.

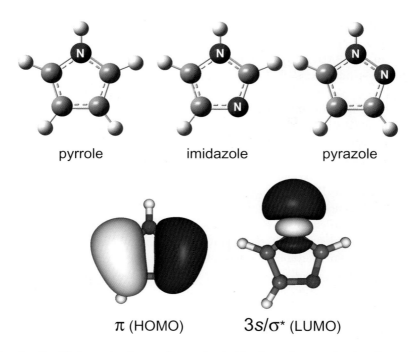

Fig. 2. Equilibrium structures of pyrrole, imidazole and pyrazole, together with illustrations of the π HOMO and $3s/\sigma^*$ LUMO of imidazole.

We start by considering imidazole, which is in many regards the simplest system. The first excited singlet state of this molecule is a $^1\pi\sigma^*$ state (of $^1A''$ symmetry) formed by electron promotion from the highest occupied molecular orbital (HOMO) — a π orbital (see Fig. 2) — to the lowest unoccupied molecular orbital (LUMO). The latter orbital possesses substantial Rydberg ($3s$) character in the vertical Franck–Condon (vFC) region but acquires progressively more σ^* anti-bonding valence and, eventually, H($1s$) character upon extending the N–H bond (R_{N-H}). As with ammonia,[13,16] the $^1\pi\sigma^*$ PES is repulsive with respect to R_{N-H}, and displays a CI with the ground state (henceforth labelled S_0) PES in this coordinate at planar geometries as depicted in Fig. 3(b). Photoexcitation to the $^1\pi\sigma^*$ state (at $\lambda_{\text{phot}} \leq 240$ nm) results in prompt dissociation, and formation of ground ($\tilde{X}$) state imidazolyl radical (plus H atom) products, with an anisotropic recoil velocity distribution wherein the H atoms recoil preferentially in the plane perpendicular to $\varepsilon_{\textbf{phot}}$. As Fig. 4(a) shows, most of the photon energy in excess of that required to dissociate the N–H bond [determined as

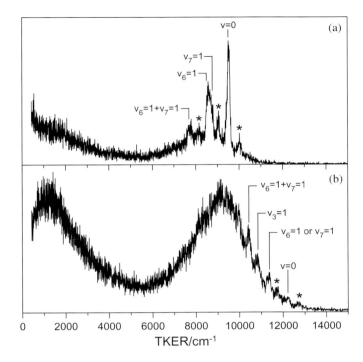

Fig. 4. TKER distributions of the products of N–H bond fission following photolysis of imidazole at λ_{phot} =(a) 234 nm and (b) 220 nm, obtained by HRAPTS. The vibrational quantum numbers associated with the dominant imidazolyl product peaks are indicated; peaks marked with an asterisk arise from photolysis of imidazole molecules carrying one quantum of the N–H out-of-plane wagging mode, ν_{24}.

pathways following excitation to the $^1\pi\pi^*(^1A')$ state. One involves transfer to the $^1\pi\sigma^*(^1A'')$ PES, via one or more CIs accessed by appropriate distortions of the heavy atom ring, and subsequent fragmentation on the $^1\pi\sigma^*$ PES yielding fast H + imidazolyl($\tilde{X}$) co-fragments — reminiscent of behaviour seen at longer λ_{phot} [Fig. 4(a)] where the $^1\pi\sigma^*$ PES is populated directly. The second involves radiationless transfer to the S_0 PES — e.g. by successive $^1\pi\pi^*(^1A') \leadsto {^1\pi\sigma^*}(^1A'') \leadsto S_0(^1A')$ couplings, mediated by CIs between the relevant PESs, and subsequent unimolecular decay, or via alternative CIs between the $\pi\pi^*(^1A')$ and S_0 PESs, which could facilitate ring opening (i.e. C–N bond fission) and/or H atom loss.[22]

Pyrazole, like imidazole, is a five-membered heterocycle containing two nitrogen atoms. Upon extending $R_{\text{N–H}}$, its ground state PES correlates diabatically with radical products in the second ($^2B_2$) excited electronic state. The predicted energy separation between the excited states arising from the $\sigma^* \leftarrow \pi_{HOMO}$ and $\sigma^* \leftarrow \pi_{HOMO-1}$ promotions in the vFC region is $\sim 0.6\,\text{eV}$, but the energy separation between the ground ($^2A_2$) and first excited ($^2B_1$) states of the pyrazolyl radical to which these $^1\pi\sigma^*$ states correlate is just $0.033\,\text{eV}$.[23, 24] Both $^1\pi\sigma^*$ PESs exhibit CIs with the diabatic S_0 PES at extended R_{N-H} — as shown in Fig. 3(c). Relative to pyrrole and imidazole, the photophysics of pyrazole is further complicated by the fact that, in the vFC region, both $^1\pi\sigma^*$ excited states lie above the minimum of the first $^1\pi\pi^*$ state. The long wavelength onset of UV absorption by pyrazole is $\sim 225\,\text{nm}$. The H atom products formed when exciting at $\lambda_{phot} > 214\,\text{nm}$ exhibit a broad velocity distribution peaking at low TKER — characteristic of excitation to (and indirect decay from and/or multiphoton excitation via) the $^1\pi\pi^*$ state.[23] Fast H atoms, displaying anisotropic recoil velocity distributions characteristic of excitation to, and prompt decay on, a $^1\pi\sigma^*$ PES are also observed when pyrazole is excited at shorter wavelengths, however. The partner pyrazolyl radicals are formed in selected vibrational levels of both the $^2A_2$ and $^2B_1$ electronic states with relative probabilities that likely reflect the coupling between the two $^1\pi\sigma^*$ states at extended R_{N-H}.

As Fig. 3 showed, the $^1\pi\sigma^*$ PES(s) of the various azoles all form CIs with the S_0 PES at extended R_{N-H}. However, the influence of these CIs on the fragmentation dynamics is not that obvious. In the case of the direct $\sigma^* \leftarrow \pi$ excitations, the ground and excited molecules, and the eventual radical products, all have planar equilibrium geometries (the nitrogen atoms remain sp^2 hybridised). Thus there is little reason for the dissociating molecules to diverge from the minimum energy path on the $^1\pi\sigma^*$ PES (i.e. to follow the diabatic path) and evolve through the CI to ground state radical products. The minimum energy geometries of the $^1\pi\pi^*$ states are also planar, but radiationless transfer from a $^1\pi\pi^*$ state to a dissociative $^1\pi\sigma^*$ PES will, again, involve flux traversing through a CI.[20, 22] Symmetry conservation requires that any transfer to a $^1\pi\sigma^*$ PES must necessarily involve an out-of-plane (a'') coupling mode. On average, therefore, molecules that evolve on the $^1\pi\sigma^*$ PES after radiationless transfer from the $^1\pi\pi^*$ state are more likely to sample the CI at large R_{N-H} with non-planar geometries, and to follow the adiabatic path towards electronically excited ($^2B_2$) state radical products. Such products have not been recognised experimentally — quite possibly because the dissociating molecules have insufficient energy in the R_{N-H} coordinate to reach the excited asymptote and thus turn back towards the CI and transfer to the S_0 PES, as envisioned in the original theoretical comprising a set of resolved peaks at TKER $\sim 6000\,\mathrm{cm}^{-1}$ and a slower component centred at TKER $\sim 2000\,\mathrm{cm}^{-1}$ [see Fig. 6(b)]. Both recoil velocity components are isotropic. Broadly similar spectra are observed upon reducing λ_{phot} (i.e. exciting higher vibrational levels of the $^1\pi\pi^*$ state) — though the vibronic structure of the phenoxyl products within the higher TKER feature ceases to be resolvable. This is illustrated in Fig. 6(c), which shows the TKER spectrum derived from H atom TOF measurements at $\lambda_{\mathrm{phot}} = 242.0\,\mathrm{nm}$. This wavelength lies just within the short wavelength regime, and careful inspection of this spectrum reveals additional, weak resolved features at high TKER. These become increasingly evident as λ_{phot} is reduced further [Fig. 6(d)], before eventually coalescing into a broad unresolved feature centred at TKER $\sim 12000\,\mathrm{cm}^{-1}$. The recoil velocity distribution of these fast H atoms is anisotropic ($\beta \sim -0.5$), implying that O–H bond fission occurs on a timescale that is fast compared with the parent molecular rotational period.[31] Recent femtosecond pump-probe studies of phenol–h_6 photolysis at $\lambda_{\mathrm{phot}} = 200\,\mathrm{nm}$ return a time constant of $103 \pm 30\,\mathrm{fs}$ for the high TKER feature,[32] reinforcing the conclusion that dissociation following excitation in the short wavelength regime can occur on an ultrafast timescale.

Analysis of numerous TKER spectra recorded in the range $275.113 \geq \lambda_{\mathrm{phot}} \geq 206\,\mathrm{nm}$ led to the conclusion that, in no case, does dissociation result in H + phenoxyl($\tilde{X}$,v=0) fragments.[31] Only by assuming that the fastest peak in Fig. 6(b) is associated with formation of phenoxyl fragments carrying one quantum in vibrational mode ν_{16a}[a] is it possible to arrive at an internally consistent value of the O–H bond strength: D_0(H-phenoxyl–h_5) = $30015 \pm 40\,\mathrm{cm}^{-1}$. ν_{16a} is an out-of-plane (a'') ring torsion mode. The pattern of product peaks evident in TKER spectra obtained within the short λ_{phot} regime [e.g. Fig. 6(d)] is similarly offset. In this latter case, the fastest feature is attributable to formation of phenoxyl radicals carrying one quantum of the ring puckering vibration, ν_{16b}.

Analogous studies involving phenol-d_6 reveal no fast D atom products when exciting at the $^1\pi\pi^* \leftarrow S_0$ origin though, as with phenol-h_6, D atom loss is observed once $\lambda_{\mathrm{phot}} < 248\,\mathrm{nm}$.[31] The partner (phenoxyl-$d_5(\tilde{X})$) fragments are formed in a broad spread of vibrational levels, the identities of which are more clearly revealed in related photolysis studies of phenol-d_5.[33] All involve a quantum of ν_{16b} and one or more quanta in

---

[a]For phenolic systems we use Wilson mode labelling notation, as it allows a convenient way of highlighting the mapping between the parent and radical modes of vibration.

ν_{18b} (the in-plane C–O wag) and/or ν_{19a} (a combination of in-plane ring breathing and C–O stretching motions). Studies of phenol-d_5 photolysis at much shorter wavelengths ($\lambda_{\text{phot}} = 193$ nm) have revealed formation of electronically excited phenoxyl-$d_5(\tilde{B})$ products at low TKER. The calculated CASPT2(12/11)/aug-cc-pVTZ PESs (fig. 6(a)) suggest that this asymptote is accessed by photoexciting to the second $^1\pi\pi^*$ state and subsequent coupling to the second $^1\pi\sigma^*$ PES via the relevant $^1\pi\pi^*/^1\pi\sigma^*$ CI. Electronically excited phenoxyl-$d_5(\tilde{A})$ products have also been reported by Crim and coworkers following vibrationally mediated photodissociation of phenol-d_5.[34] This study involved initial excitation of the ground state O–H stretch fundamental prior to UV excitation to total energies above that of the first $^1\pi\pi^*/^1\pi\sigma^*$ CI. These experimental results lend support to earlier theoretical predictions[35] that the introduction of O–H stretching excitation should encourage dissociating molecules to follow the adiabatic path at the $^1\pi\sigma^*/S_0$ CI (to H + phenoxyl-$d_5(\tilde{A})$ products).

The $^1\pi\pi^*_{(v=0)}$ levels of phenol–h_6 and phenol-d_1 (C_6H_5OD) exhibit markedly different fluorescence lifetimes ($\tau_f \sim 2$ ns and ~ 16 ns, respectively).[36–38] Pino et al.[39] noted a clear correlation between τ_f and the calculated $^1\pi\pi^*/^1\pi\sigma^*$ energy gaps for a range of substituted phenols and their complexes with NH_3 and explained the measured variations in τ_f, and the differing O–H/D bond dissociation probabilities, by assuming that phenol molecules excited to low-lying vibrational levels within the $^1\pi\pi^*$ state dissociate by H atom tunnelling through the barrier under the $^1\pi\pi^*/^1\pi\sigma^*$ CI.

Symmetry conservation provides a rationale for the foregoing observations. The S_0 and $^1\pi\pi^*$ states of phenol both have A' electronic symmetry in C_s, whilst the $^1\pi\sigma^*$ state has A'' symmetry. Earlier MRCI calculations suggested that O–H torsion, $\tau_{\text{O-H}}$ (of a'' vibrational symmetry), has the largest non-adiabatic matrix coupling element at both the $S_0/^1\pi\sigma^*$ and $^1\pi\pi^*/^1\pi\sigma^*$ CIs,[40] but did attempt to address the specific vibrational energy disposal observed in the phenoxyl products. Our more recent analysis[41] recognises the need to consider phenol in the non-rigid molecular symmetry group G_4 (isomorphous with C_{2v}), rather than C_s. In this representation, $\tau_{\text{O-H}}$ and ν_{16a} are distinguished by their respective symmetries (b_1) and (a_2). Only the latter has the appropriate symmetry to couple the $^1\pi\pi^*(^1B_2)$ and $^1\pi\sigma^*(^1B_1)$ states and result in excitation of odd quanta of ν_{16a}. The measured energy disposals thus provide further support for the $^1\pi\pi^* \leadsto {}^1\pi\sigma^*$ route to H + phenoxyl($\tilde{X}^2B_1$) products, despite the necessity of tunnelling through the substantial energy barrier under the $^1\pi\pi^*/^1\pi\sigma^*$ CI.

Figure 7 serves to highlight the similarities between the internal energy (E_{int}) spectra of the radical products obtained following excitation of a number of different phenols at their respective $^1\pi\pi^*$–S_0 origins. The E_{int} spectrum measured in the case of phenol-h_6 [Fig. 7(a), $\lambda_{\text{phot}} = 275.113\,\text{nm}$] can be assigned by reference to the calculated (DFT/B3LYP/6-311+G**) anharmonic wavenumbers for the various product normal modes. The peaks demonstrate selective formation of phenoxyl-h_5 products with $v_{16a} = 1, 3$ and 5, built on zero, one and two quanta of ν_{18b} (the C–O in plane wagging mode); all of these levels have a_2 (in G_4) overall vibrational symmetry. Activity in ν_{18b} is likely the result of modest changes in the C–C–O angle on O–H bond extension. A range of behaviours are observed when exciting to $^1\pi\pi^*$ state levels with v > 0. In some cases, the mode promoted in excitation behaves as a 'spectator' to the O–H bond fission and maps through into the radical product; in other cases we see clear evidence of mode mixing, with parent nuclear motion introduced in the $^1\pi\pi^*$–S_0 excitation appearing in other vibrational modes of the product, or released as product translation.

Studies of substituted phenols provide a route to deeper understanding of the dynamics involved in coupling between the various PESs of phenol. TKER spectra obtained following long wavelength photolysis of phenol-d_5 show many similarities with those from phenol-h_6. Again, excitation at the $^1\pi\pi^*$–S_0 origin yields a well-resolved set of peaks at TKER $\sim 6000\,\text{cm}^{-1}$. Close examination of the populated product modes reveal some differences, however, for example, the reduced extent of the progression in ν_{16a} [Fig. 7(b)].[33]

Introducing an electron withdrawing halogen (Y) atom at the para (p-) position in phenol leads to a reduction in the yield of O–H bond fission products across the series F > Cl > Br.[42] Fast H atoms consistent with eventual dissociation on the $^1\pi\sigma^*$ PES were observed when exciting at the $^1\pi\pi^*$–S_0 origins of p-fluorophenol and (weakly) p-chlorophenol, but not from p-bromophenol. The E_{int} spectrum of the p-fluorophenoxyl fragments [Fig. 7(c)] shows selective population of levels with v_{16a} = odd, and in modes ν_{18b} and ν_{9b} (an asymmetric C–F and C–O wagging motion) — both of which are understandable on dFC grounds. Fast H atom products were detected from all three of these p-halophenols, however, when exciting within their respective short wavelength regimes ($\lambda_{\text{phot}} < 238\,\text{nm}$).[42] The Y dependent differences in behaviour at long excitation wavelengths can be understood by recognising the possible alternative C–Y bond fission pathways — which ion imaging experiments show to be significant fragmentation channels in both p-bromophenol and p-iodophenol.[43]

Introducing a methyl group to the phenol ring can be expected to increase the overall vibrational state density and thus the likelihood of intramolecular vibrational redistribution.[44,45] Yet, as Fig. 7(d) shows for

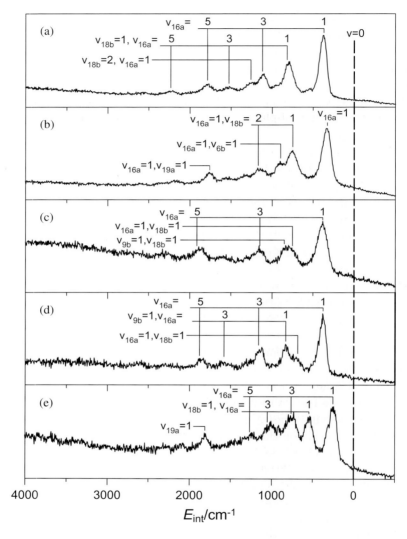

Fig. 7. Internal energy (E_{int}) spectra of the radical products resulting from O–H fission following excitation at the $^1\pi\pi^* \leftarrow S_0$ origins in: (a) phenol-h_6 ($\lambda_{phot} = 275.113$ nm), (b) phenol-d_5 ($\lambda_{phot} = 273.815$ nm), (c) p-fluorophenol ($\lambda_{phot} = 284.768$ nm), (d) p-methylphenol ($\lambda_{phot} = 283.023$ nm), and (e) 4-hydroxyindole ($\lambda_{phot} = 284.893$ nm). Vibrational quantum numbers associated with the various product peaks are indicated by the superposed combs.

the case of p-methylphenol, the vibrational energy disposal in the p-methylphenoxyl fragments formed by exciting p-methylphenol at the origin of its $^1\pi\pi^* - S_0$ transition ($\lambda_{\text{phot}} = 283.023\,\text{nm}$) is very reminiscent of that in the p-fluorophenoxyl($\tilde{X}$) products from p-fluorophenol. Again, this fragmentation is now considered to proceed via initial $\pi^* \leftarrow \pi$ excitation, and subsequent tunnelling through the energy barrier under the $^1\pi\pi^*/^1\pi\sigma^*$ CI (facilitated by out-of-plane mode ν_{16a}) leading to eventual O–H bond fission.

Our final example of a phenolic system is unusual in that the obvious chromophore is an indole. Huang et al.[46] suggested that the short fluorescence lifetime of 4-hydroxyindole in its $^1\pi\pi^*$ state ($\tau_f = 0.18 \pm 0.1\,\text{ns}$) was due to efficient keto-enol isomerisation on the excited state PES. As Fig. 7(e) shows, however, excitation at the $^1\pi\pi^*$–S_0 origin yields translationally excited H atom photoproducts. The vibrational energy disposal in the radical co-fragment is very reminiscent of that found for phenol-h_6, with additional activity in product mode ν_{19a} (as seen in other p-substituted phenols), suggesting that, again, dissociation involves tunnelling through the energy barrier under the $^1\pi\pi^*/^1\pi\sigma^*$ CI and eventual dissociation on the $^1\pi\sigma^*$ PES.[47] 4-hydroxyindole presents a further potential complication, since the N–H bond on the indole ring might also be expected to dissociate upon UV photoexcitation (see Sec. 3.3). Given the evident similarities between the various E_{int} spectra shown in Fig. 7, and the good agreement between experiment and the results of companion CASPT2(10/9)/aug-cc-pVTZ PES calculations (both with regard to the product mode frequencies and the parent dissociation energy), we can be confident that the observed products arise as a result of O–H bond fission. As shown in Sec. 3.3, the threshold energy for the competing N–H bond fission channel is likely to fall at shorter UV excitation wavelengths.

Replacing the oxygen atom in phenol by a sulphur atom changes the heteroatom lone pairs from $2p$ to $3p$ orbitals, with the consequence that the HOMO of thiophenol is an admixture of the sulphur $3p_x$ lone pair and the $2p\pi$ benzene system with much of the amplitude localised on the sulfur. The decreased conjugation (relative to that in phenol) reduces the energy separation between the ground ($\tilde{X}^2B_1$) and first excited ($\tilde{A}^2B_2$) states of the thiophenoxyl radical (to $\sim 3000\,\text{cm}^{-1}$, cf $\sim 8100\,\text{cm}^{-1}$ in phenoxyl). Excitation of thiophenol at $\lambda_{\text{phot}} = 285.8\,\text{nm}$, within the longest wavelength absorption band associated with its $^1\pi\pi^* - S_0$ transition, yields H atom photoproducts with a structured, bimodal and isotropic TKER distribution — as shown in Fig. 8.[44,48] The bimodality reflects branching into

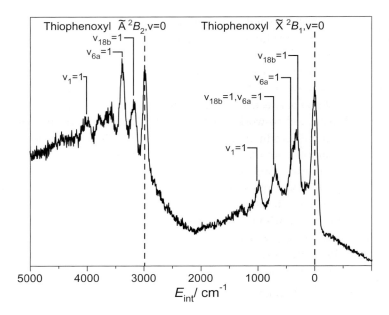

Fig. 8. Internal energy (E_{int}) spectrum of the thiophenoxyl radical products resulting from photolysis of thiophenol at $\lambda_{phot} = 285.8$ nm. Vibrational quantum numbers associated with the various product peaks are indicated by the superposed combs.

both $\tilde{X}$ and $\tilde{A}$ states of the thiophenoxyl product. Similar TKER spectra are observed at shorter λ_{phot}, but the H atom products now show clear recoil anisotropy ($\beta \sim -0.8$). This evolving behaviour has been explained within the framework of *ab initio* PESs for the ground and first few excited states of thiophenol.[49] Dissociation following excitation at the $\pi^* \leftarrow n/\pi$ origin is deduced to involve predissociation, by tunnelling to the dissociative $^1\pi\sigma^*$ PES, and subsequent branching at the $^1\pi\sigma^*/S_0$ CI at extended R_{S-H} whereas, at shorter λ_{phot}, the $^1\pi\sigma^*$ state is assumed to be populated directly.

Inspection of Fig. 8 reveals that the thiophenoxyl ($\tilde{X}$) products are formed in the $v = 0$ level and in excited vibrational levels involving ν_{18b}, ν_{6a} (both of which motions were active also in phenoxyl radicals resulting from long wavelength photolysis of phenol) and ν_1 (a ring breathing motion); all have a' vibrational symmetry. Recent photodetachment experiments[50] provide the first definitive determination of the energy splitting between the $v = 0$ levels of the $\tilde{X}$ and $\tilde{A}$ states of the thiophenoxyl radical. This has impacted on, and allowed re-evaluation of, our previous assignment[48] of the vibrational structure associated with thiophenoxyl ($\tilde{A}$) radicals resulting from UV photolysis of thiophenol. Specifically, the re-analysis implies population of similar vibrational levels in both the $\tilde{A}$ and $\tilde{X}$ state radicals. Such energy disposal can be understood if the out-of-plane torsion τ_{S-H} (which disappears upon S–H bond fission) is the nuclear motion controlling the branching between H+ thiophenoxyl ($\tilde{A}$ and $\tilde{X}$) products at the $^1\pi\sigma^*/S_0$ CI. Kim and co-workers have shown that the electronic branching in the thiophenoxyl products can be tuned by specific substitutions at the p-position.[51] Substituents that, in the S_0 state, cause the S–H bond to move out of the ring plane are found to favour $\tilde{A}$ state radical products upon photolysis — as expected, if such substitutions encourage the dissociating flux to approach the $^1\pi\sigma^*/S_0$ CI at non-planar geometries and thus favour the adiabatic path to the excited product asymptote. Our recent full wavelength studies of a range of *para*-substituted thiophenols[52] show this branching to be both substituent *and* excitation wavelength dependent.

3.3. *Larger analogues*

Azoles and phenols illustrate two generic classes of behaviour. In the former case, as noted in Sec. 3.1, the $^1\pi\sigma^*$ state can be populated directly by photon absorption. Dissociation occurs on the $^1\pi\sigma^*$ PES, the topology of which — for these molecules — favours the retention of planar geometry. The dissociating molecules thus funnel straight through from the upper cone of the $^1\pi\sigma^*/S_0$ CI, yielding ground state radical products (or, in the case of pyrazole, radicals in their ground and low lying first excited states). In the phenols, however, absorption is primarily to optically bright $^1\pi\pi^*$ states; passage through a CI is required if these molecules are to dissociate on a $^1\pi\sigma^*$ PES. At long wavelengths, the consensus view is now that H atom loss from the photo-excited $^1\pi\pi^*$ molecules involves tunnelling through the energy barrier associated with the $^1\pi\pi^*/^1\pi\sigma^*$ CI. In contrast to the azoles, therefore, the energy disposal in the products from photodissociation of phenols will be sensitive not just to any optically induced changes in equilibrium geometry, but also to the nuclear motions that promote coupling to the $^1\pi\sigma^*$ PES (e.g. out-of-plane skeletal modes like ν_{16a}, as implicated in the HRAPTS studies of a range of phenols).

The available data suggests that these generic behaviours extrapolate to larger analogues of both the azoles and phenols. Indole, for example, can be viewed as a pyrrole unit fused to a benzene ring at the 2 and 3 positions, and fast H atoms attributable to N–H bond dissociation on a $^1\pi\sigma^*$ PES have been identified following UV excitation of this molecule at $\lambda < 263$ nm.[53] Fast CH_3 radicals have also been reported following 248 nm photolysis of N-methylindole, and attributed to $N-CH_3$ bond fission on the $^1\pi\sigma^*$ PES.[26] The nucleobase adenine can be pictured as a variant of indole in which the CH groups at the 1, 3 and 7 positions have been replaced by N atoms. It can exist in no fewer than 12 tautomers, of which the N_9-H tautomer (shown in Fig. 9) is the most stable. TKER spectra derived from H atom TOF data

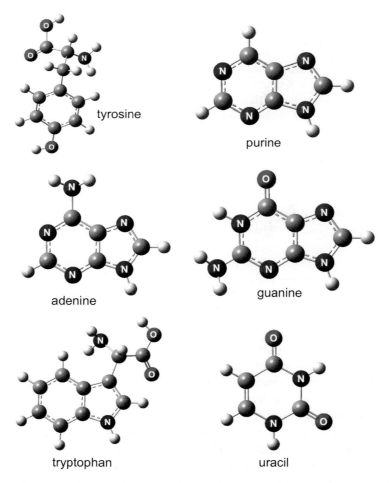

Fig. 9. Structural formulae of tyrosine (phenol), purine (imidazole), adenine, guanine, tryptophan (indole), and uracil, that serve to illustrate the base chromophore from which each derives (indicated in brackets).

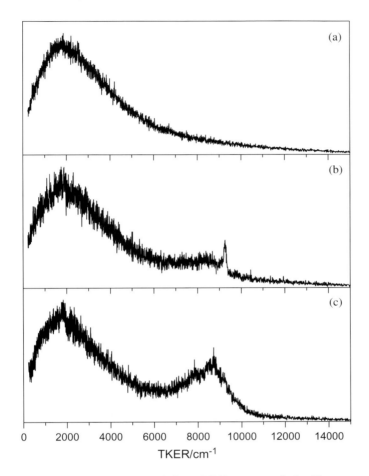

Fig. 10. TKER distributions derived from TOF spectra of the H atom products resulting from photolysis of adenine at λ_{phot} = (a) 250 nm, (b) 228 nm, and (c) 224 nm.

recorded following photolysis of adenine at $\lambda_{\text{phot}} > 233$ nm display isotropic profiles that peak at low TKER [$\sim 1800 \text{ cm}^{-1}$, as shown in Fig. 10(a)] and are generally insensitive to the choice of photolysis wavelength. TKER spectra obtained at $\lambda_{\text{phot}} \leq 233$ nm show additional fast structure, as illustrated by the λ_{phot} = 228 nm and 224 nm data displayed in Figs. 10(b) and 10(c).[54] Analogy with recent studies of 4-methoxyindole suggests that these H atoms arise as a result of direct (vibronically enhanced) excitation to, and dissociation on, the $^1\pi\sigma^*$ PES.[47] This is of course just one possible source of H atom photoproducts; dissociation of the amino N_{10}–H bond is another. The available *ab initio* data[55–58] suggests that the latter bond fission on the $^1\pi\sigma^*$ PES is hampered by a barrier in the $R_{\text{N10–H}}$ coordinate, with the result that the energetic threshold for this channel following excitation in the vFC region lies above that for N_9–H bond fission. Thus we might anticipate a role for N–H bond fission from within the NH_2 group of adenine at shorter λ_{phot}, but recognise that any H atoms arising via this channel could well be buried within the dominant, broad, unresolved feature at low TKER. Two sets of time-resolved pump-probe studies of adenine — monitoring, respectively, photoelectrons[59,60] and ionized fragments[61] — have appeared that imply a role for the $^1\pi\sigma^*$ excited state at significantly lower excitation energies. One factor guiding the former discussion was an evident difference in the time-resolved photoelectron spectra of adenine and 9-methyladenine, which was rationalised by assuming that methylation blocks the site of the $^1\pi\sigma^*$ state of interest.[59] As shown above, however, in the context of pyrrole and N-methylpyrrole, and indole and N-methylindole, such an assumption may well be questionable. Indeed very recent time-resolved photoelectron spectroscopy studies of adenine following excitation at several different UV wavelengths imply that the long wavelength threshold for populating the $^1\pi\sigma^*$ PES in adenine lies at $\lambda_{\text{phot}} < 238$ nm.[62]

H atom loss has also been reported following UV photolysis of uracil (at several wavelengths in the range $230 < \lambda_{\text{phot}} < 270$ nm)[63] and thymine (at $235 < \lambda_{\text{phot}} < 268$ nm).[64] Both yield isotropic recoil velocity distributions centred at low TKER and consistent with initial excitation to, and subsequent non-radiative decay from, comparatively long-lived $^1n\pi^*/^1\pi\pi^*$ excited states. The interpretations offered for the respective distributions differ in some regards, but neither distribution shows any evidence of structured signal at high TKER such as might be expected from dissociation on a repulsive $^1\pi\sigma^*$ PES. *Ab initio* calculations for excited states of the biologically relevant 9H-keto-amino tautomer of guanine (another of the nucleobases) have identified CIs linking the PESs of the optically 'bright' $^1\pi\pi^*$ states to those of both the ground state and the N_9–H bond centred $^1\pi\sigma^*$ state,[65] however. It remains to be seen if uracil and/or thymine yield fast H atoms, characteristic of dissociation on a $^1\pi\sigma^*$ PES, if excited at shorter wavelengths.

The photolysis of anisole ($C_6H_5OCH_3$) has been investigated by multimass imaging methods at 248 nm and at 193 nm, and translationally excited CH_3 radicals — consistent with O–CH_3 bond fission on a $^1\pi\sigma^*$ PES — identified at both wavelengths.[26] Again, the analogy with phenol is striking and it is tempting to suggest that such O–C bond fissions will extend to heavier analogues also.

4. Future Prospects

The central role of $^1\pi\sigma^*$ (and $^1n\sigma^*$) excited states in promoting bond fission is becoming ever more widely recognised — not just in azoles and phenols, but also in many other families of molecule.[1] The azoles exhibit a CI between their $^1\pi\sigma^*$ and ground state PESs at planar geometries and extended R_{N-H}. Long wavelength excitation of pyrrole, imidazole, etc populates the $^1\pi\sigma^*$ state directly, at planar (or very near planar) geometries. The dissociating molecules follow the diabatic $^1\pi\sigma^*$ PES through the CI, yielding ground state radical products. In these cases, the CI appears to have little effect on the eventual product energy disposal. Shorter wavelength excitation of imidazole and pyrrole populates one or more $^1\pi\pi^*$ states. These can predissociate by coupling via CIs with the $^1\pi\sigma^*$ PES and/or the S_0 PES. Once again, much of the flux evolving on the $^1\pi\sigma^*$ PES appears to pass straight through the CI at extended R_{N-H} but, in this case, the CI may also play an increased role in funnelling population from the $^1\pi\sigma^*$ to the S_0 PES, as suggested in some of the early theoretical discussions.[2,3] As discussed previously, however, the increase in the relative yield from dissociations on the S_0 PES suggested by the experimental data may also be due to decay via other CIs that link the $^1\pi\pi^*$ and S_0 PESs at non-planar geometries.[22] The photochemistry of phenols is largely determined by CIs linking the $^1\pi\pi^*$ and $^1\pi\sigma^*$ PESs. Excitation at long wavelengths leads to predissociation, by H atom tunnelling through the energy barrier under the $^1\pi\pi^*/^1\pi\sigma^*$ CI. At shorter excitation wavelengths, the $^1\pi\sigma^*$ state could be accessed directly, or by radiationless transfer from photo-excited $^1\pi\pi^*$ levels. In both cases, the demands of symmetry conservation have an obvious effect on the product vibrational energy disposal (in the case of phenoxyl products from the various phenols) and/or vibronic energy disposal (in the case of thiophenoxyl radicals from thiophenol).

As noted in Sec. 3.3, azole and phenol units are key components in many larger biomolecules like purine, guanine, and tyrosine (Fig. 9) and one future challenge is to explore the efficiency of X–H bond fission in these larger systems when excited at UV wavelengths appropriate for accessing $^1\pi\sigma^*$ PESs. One significant experimental impediment to studying these systems in the gas phase is their involatility. Simply heating the sample, as used successfully in the UV photolysis studies of adenine,[54] will more typically cause decomposition (by pyrolysis); pulsed laser desorption currently looks the most promising route to producing densities of intact molecules sufficient for subsequent photofragmentation studies.[66] The increased density of states will likely provide further challenges: the increased electronic state density will likely increase the number of CIs, and thus the number of relaxation pathways competing with excited ($^1\pi\sigma^*$) state bond fission, while the increased vibrational state density reduces the probability of achieving vibrational state resolution in the resulting products. Nonetheless, there is accumulating evidence that $^1\pi\sigma^*$ state induced dissociation remains important in larger biomolecules. For example, efficient coupling to a $^1\pi\sigma^*$ state and subsequent N–H bond fission has also been advanced to explain the very fast decay ($\tau \sim 380$ fs) of protonated tryptophan following $\pi^* \leftarrow \pi$ excitation at 266 nm.[67]

The effect of solvation on $^1\pi\sigma^*$ excited states is now also starting to be explored — by experiments involving molecular clusters (generally involving azoles or phenols), and by *ab initio* theory. Photolysis studies of size-selected pyrrole clusters (at $\lambda = 243$ and 193 nm) reveal that, relative to the bare monomer (recall Fig. 1), the distribution of slow H atom photofragments increases at the expense of the fast component.[68] Calculations of pyrrole complexed with an Ar atom, or with another pyrrole moiety, suggest that the 'solvent' adduct tends to destabilise the $^1\pi\sigma^*$ state, raising the energy of its PES (relative to that of the $^1\pi\pi^*$ state) and thereby 'closing' the N–H dissociation coordinate.[68] A similar apparent 'blocking' of the N–H bond fission channel was observed when photolysing clusters formed during expansion of pyrrole in a beam of Xe.[11] Pump-probe experimental studies of $(NH_3)_n$[69] and phenol-$(NH_3)_n$[70,71] clusters, and companion theory, provide contradictory hints regarding $^1\pi\sigma^*$ state induced photochemistry, however. In the case of phenol-$(NH_3)_3$ clusters, for example, the calculated minimum energy structure has the hydroxyl H atom forming a hydrogen-bond with a terminal N atom in a 'wire' of three NH_3 molecules, that are themselves linked by hydrogen-bonds. Transient IR absorption measurements following UV photo-excitation of phenol-$(NH_3)_3$ clusters at $\lambda \sim 281$ nm (i.e. on a resonance close above the $\pi^* \leftarrow \pi$ origin of the phenol chromophore) reveal the presence of $NH_4(NH_3)_2$ products, the formation of which is explained by 'excited state H transfer' — i.e. photo-induced H atom (or coupled electron + proton) transfer — from the phenol to the adjacent NH_3. Theory in this case suggests that the $^1\pi\sigma^*$ state of phenol is stabilized (relative to the $^1\pi\pi^*$ state) by complexation with the $(NH_3)_3$ unit, and that the

$^1\pi\pi^*/^1\pi\sigma^*$ CI thus promotes O–H bond fission in the cluster at longer excitation wavelengths than in bare phenol.[71] Experimental studies of the effect of complexing H_2O molecules to protonated tryptophan molecules suggest a different picture, however.[72] The long wavelength action spectrum for forming fragment ions from the unsolvated protonated tryptophan cation is broad and unstructured, but addition of just two H_2O molecules causes the action spectrum to 'sharpen up', to the extent that it shows clear vibronic structure — an observation that has been explained by assuming that the presence of two water molecules hydrogen-bonded to the amino group raises the relative energy of the σ^* orbital, to the extent that the dominant absorption is now to the (longer lived) $^1\pi\pi^*$ state of the cation.

More detailed studies of $^1\pi\sigma^*$ state induced photochemistry, in a broader range of molecular clusters, and with a sufficient variety of solvent species, will be needed before we have a clear and predictive understanding of the effects of solvation on the energies and properties of these states. The extent to which conclusions drawn from studies involving selected well-defined clusters and cluster geometries transfer to the condensed phase is a yet larger question; indeed, extrapolating the rapidly growing appreciation of $^1\pi\sigma^*$ state induced photochemistry obtained from gas phase studies into the condensed phase remains a major challenge. Real-time pump-probe measurements of the formation of an increasing range of condensed phase photo-products are becoming feasible, however, under conditions where unintended multiphoton absorption effects can (hopefully) be avoided or, at least, recognised and distinguished. Recent ultrafast pump-probe studies of the UV photodissociation of *para*-methylthiophenol in ethanol solution that demonstrate prompt S–H bond fission represent a first step in this direction.[73] We anticipate rapid future growth of such studies given the range of applications that could benefit from such improved understanding — including photo-damage and/or the photostability of biomolecules and nucleobases,[74] photo-acidity,[75] even the possibility of 'photo'synthesising selected target chemicals — and confidently predict that azoles and phenols will feature largely in many such studies.

References

1. M.N.R. Ashfold, G.A. King, D. Murdock, M.G.D. Nix, T.A.A. Oliver and A.G. Sage, *Phys. Chem. Chem. Phys.* **12**, 1218 (2010).
2. A.L. Sobolewski and W. Domcke, *Chem. Phys.* **259**, 181 (2000).

3. A.L. Sobolewski, W. Domcke, C. Dedonder-Lardeux and C. Jouvet, *Phys. Chem. Chem. Phys.* **4**, 1093 (2002).
4. G.E. Busch, R.T. Mahoney, R.I. Morse and K.R. Wilson, *J. Chem. Phys.* **51**, 449 (1969).
5. See, for example, A.M. Wodtke and Y.T. Lee, in "Molecular Photodissociation Dynamics", edited by M.N.R. Ashfold and J.E. Baggott (Royal Society of Chemistry, London, UK, 1989), p 31.
6. D.A. Blank, S.W. North and Y.T. Lee, *Chem. Phys.* **187**, 35 (1994).
7. D.W. Chandler and P.L. Houston, *J. Chem. Phys.* **87**, 1445 (1986).
8. M.N.R. Ashfold, N.H. Nahler, A.J. Orr-Ewing, O.P.J. Vieuxmaire, R.L. Toomes, T.N. Kitsopoulos, I. Anton-Garcia, D. Chestakov, S.-M. Wu and D.H. Parker, *Phys. Chem. Chem. Phys.* **8**, 26 (2006) and references therein.
9. J. Wei, A. Kuczmann, J. Riedel, F. Renth and F. Temps, *Phys. Chem. Chem. Phys.* **5**, 315 (2003).
10. J. Wei, J. Riedel, A. Kuczmann and F. Temps, *Faraday Discuss.* **127**, 267 (2004).
11. L. Rubio-Lago, D. Zaouris, Y. Sakellariou, D. Sofitikis, T.N. Kitsopoulos, F. Wang, X. Yang, B. Cronin, A.L. Devine, G.A. King, M.G.D. Nix, M.N.R. Ashfold and S.S. Xantheas, *J. Chem. Phys.* **127**, 064306 (2007).
12. L. Schnieder, W. Meier, K.H. Welge, M.N.R. Ashfold and C.M. Western, *J. Chem. Phys.* **92**, 7027 (1990).
13. M.N.R. Ashfold, D.H. Mordaunt and S.H.S. Wilson, in "Advances in Photochemistry", edited by D.C. Neckers, D.H. Volman and G. von Bunau, **21**, 217 (1996).
14. P.A. Cook, S.R. Langford, M.N.R. Ashfold and R.N. Dixon, *J. Chem. Phys.* **113**, 994 (2000).
15. B. Cronin, M.G.D. Nix, R.H. Qadiri and M.N.R. Ashfold, *Phys. Chem. Chem. Phys.* **6**, 5031 (2004).
16. R. Polák, I. Paidarová, V. Spirko and P.J. Kuntz, *Int. J. Quant. Chem.* **57**, 429 (1996).
17. A.L. Devine, B. Cronin, M.G.D. Nix and M.N.R. Ashfold, *J. Chem. Phys.* **125**, 184302 (2006).
18. B.J. Roos, P.-Å. Malmqvist, V. Molina, L. Serrano-Andrés and M. Mérchan, *J. Chem. Phys.* **116**, 7528 (2002).
19. H. Lippert, H.-H. Ritze, I.V. Hertel and W. Radloff, *Chem. Phys. Chem.* **5**, 1423 (2004).
20. M. Barbatti, M. Vazdar, A.J.A. Aquino, M. Eckart-Maksić and H. Lischka, *J. Chem. Phys.* **125**, 164323 (2006).
21. M. Vazdar, M. Eckart-Maksić, M. Barbatti and H. Lischka, *Mol. Phys.* **107**, 845 (2009).
22. M. Barbatti, H. Lischka, S. Salzmann and C.M. Marian, *J. Chem. Phys.* **130**, 034305 (2009).
23. G.A. King, T.A.A. Oliver, M.G.D. Nix and M.N.R. Ashfold, *J. Chem. Phys.* **132**, 064305 (2010).
24. T. Ichino, A.J. Gianola, W.C. Lineberger and J.F. Stanton, *J. Chem. Phys.* **125**, 084312 (2006).
25. A.G. Sage, M.G.D. Nix and M.N.R. Ashfold, *Chem. Phys.* **347**, 300 (2008).

26. C.-M. Tseng, Y.T. Lee and C.-K. Ni, *J. Phys. Chem. A* **113**, 3881 (2009).
27. G. Piani, L. Rubio-Lago, M.A. Collier, T.N. Kitsopoulos and M. Becucci, *J. Phys. Chem. A* **113**, 14554 (2009).
28. A.L. Sobolewski and W. Domcke, *J. Phys. Chem. A* **105**, 9275 (2001).
29. Z.G. Lan, W. Domcke, V. Vallet, A.L. Sobolewski and S. Mahapatra, *J. Chem. Phys.* **122**, 224315 (2005).
30. C.-M. Tseng, Y.T. Lee and C.-K. Ni, *J. Chem. Phys.* **121**, 2459 (2004).
31. M.G.D. Nix, A.L. Devine, B. Cronin, R.N. Dixon and M.N.R. Ashfold, *J. Chem. Phys.* **125**, 133318 (2006).
32. A. Iqbal, L.-J. Pegg and V.G. Stavros, *J. Phys. Chem. A* **112**, 9531 (2008).
33. G.A. King, T.A.A. Oliver, M.G.D. Nix and M.N.R. Ashfold, *J. Phys. Chem. A* **113**, 7984 (2009).
34. M.L. Hause, Y.H. Yoon, A.S. Case and F.F. Crim, *J. Phys. Chem.* **128**, 104307 (2008).
35. Z. Lan, W. Domcke, V. Vallet, A.L. Sobolewski and S. Mahapatra, *J. Chem. Phys.* **122**, 224315 (2005).
36. R.J. Lipert and S.D. Colson, *J. Phys. Chem.* **93**, 135 (1989).
37. R.J. Lipert, G. Bermudez and S.D. Colson, *J. Phys. Chem.* **92**, 3801 (1988).
38. A. Sur and P.M. Johnson, *J. Chem. Phys.* **84**, 1206 (1986).
39. G.A. Pino, A.N. Olfdani, E. Marceca, M. Fujii, S.-I. Ishiuchi, M. Miyazaki, M. Broquier, C. Dedonder and C. Jouvet, *J. Chem. Phys.* **133**, 124313 (2010).
40. O.P.J. Vieuxmaire, Z. Lan, A.L. Sobolewski and W. Domcke, *J. Chem. Phys.* **129**, 224307 (2008).
41. R.N. Dixon, T.A.A. Oliver, G.A. King and M.N.R. Ashfold, *J. Chem. Phys.* **134**, 194303 (2011).
42. A.L. Devine, M.G.D. Nix, B. Cronin and M.N.R. Ashfold, *Phys. Chem. Chem. Phys.* **9**, 3749 (2007).
43. A.G. Sage, Ph.D. thesis, University of Bristol, 2010.
44. M.N.R. Ashfold, A.L. Devine, R.N. Dixon, G.A. King, M.G.D. Nix and T.A.A. Oliver, *Proc. Nat. Acad. Sci.* **105**, 12701 (2008).
45. G.A. King, A.L. Devine, M.G.D. Nix, D.E. Kelly and M.N.R. Ashfold, *Phys. Chem. Chem. Phys.* **10**, 6417 (2008).
46. Y. Huang and M. Sulkes, *Chem. Phys. Lett.* **254**, 242 (1996).
47. T.A.A. Oliver, G.A. King and M.N.R. Ashfold, *Phys. Chem. Chem. Phys.* **13**, 14646 (2011).
48. A.L. Devine, M.G.D. Nix, R.N. Dixon and M.N.R. Ashfold, *J. Phys. Chem. A* **112**, 9563 (2008).
49. I.S. Lim, J.S. Lim, Y.S. Lee and S.K. Kim, *J. Chem. Phys.* **126**, 034306 (2007).
50. J.B. Lim, T.I. Yacovitch, C. Hock and D.M. Neumark, *Phys. Chem. Chem. Phys.* DOI: 10.1039/C1CP22211B (2011).
51. J.S. Lim, Y.S. Lee and S.K. Kim, *Angew. Chem., Int. Ed.* **47**, 1853 (2008).
52. T.A.A. Oliver, G.A. King, D.P. Tew, R.N. Dixon and M.N.R. Ashfold, *in preparation*.
53. M.G.D. Nix, A.L. Devine, B. Cronin and M.N.R. Ashfold, *Phys. Chem. Chem. Phys.* **8**, 2610 (2004).

54. M.G.D. Nix, A.L. Devine, B. Cronin and M.N.R. Ashfold, *J. Chem. Phys.* **126**, 124312 (2007).
55. A.L. Sobolewski and W. Domcke, *Chem. Phys. Lett.* **315**, 293 (1999).
56. S. Perun, A.L. Sobolewski and W. Domcke, *Chem. Phys.* **313**, 107 (2005).
57. W.C. Chung, Z. Lan, Y. Ohsuki, N. Shimakura, W. Domcke and Y. Fujimura, *Phys. Chem. Chem. Phys.* **9**, 2075 (2007).
58. I. Conti, M. Garavelli and G. Orlandi, *J. Am. Chem. Soc.* **131**, 16108 (2009).
59. H. Satzger, D. Townsend, M.Z. Zgierski, S. Patchkovskii, S. Ullrich and A. Stolow, *Proc. Nat. Acad. Sci.* **103**, 10196 (2006).
60. C.Z. Bisgaard, H. Satzger, S. Ulrich and A. Stolow, *Chem. Phys. Chem.* **10**, 101 (2009) and references therein.
61. K.L. Wells, G.M. Roberts and V.G. Stavros, *Chem. Phys. Lett.* **446**, 20 (2007).
62. N.L. Evans and S. Ullrich, *J. Phys. Chem. A* **114**, 11225 (2010).
63. M. Schneider, R. Maksimenka, F.J. Buback, T. Kitsopoulos, L.Rubio-Lago and I. Fischer, *Phys. Chem. Chem. Phys.* **8**, 3017 (2006).
64. M. Schnieder, C. Schon, I. Fischer, L. Rubio-Lago and T. Kitsopoulos, *Phys. Chem. Chem. Phys.* **9**, 6021 (2007).
65. S. Yamazaki, W. Domcke and A.L. Sobolewski, *J. Phys. Chem. A* **112**, 11965 (2008).
66. A.M. Rijs, M.S. de Vries, J.S. Hannam, D.A. Leigh, M. Fanti, F. Zerbetto, W.J. Buma. *Angew. Chem., Int. Edn.* **47**, 3174 (2008).
67. H. Kang, C. Jouvet, C. Dedonder-Lardeux, S. Martrenchard, G. Grégoire, C. Desfrançois, J.-P. Schermann, M. Barat and J.A. Fayeton, *Phys. Chem. Chem. Phys.* **7**, 394 (2005).
68. V. Poterya, V. Profant, M. Fárnik, P. Slaviéek and U. Buck, *J. Chem. Phys.* **127**, 064307 (2007).
69. S. Nonose, T. Taguchi, F. Chen, S. Iwata and K. Fuke, *J. Phys. Chem. A* **106**, 5242 (2002).
70. K. Daigoku, S. Ichiuchi, M. Sakai, M. Fujii and K. Hashimoto, *J. Chem. Phys.* **119**, 5149 (2003).
71. S.-I. Ishiuchi, M. Sakai, K. Daigoku, K. Hashimoto and M. Fujii, *J. Chem. Phys.* **127**, 234304 (2007) and references therein.
72. S.R. Mercier, O.V. Boyarkin, A. Kamariotis, M. Guglielmi, I. Tavernelli, M. Cascella, U. Rothlisberger and T.R. Rizzo, *J. Amer. Chem. Soc.* **128**, 16938 (2006).
73. T.A.A. Oliver, Y. Zhang, M.N.R. Ashfold and S.E. Bradforth, *Faraday Discuss.* **150**, 439 (2011).
74. C.T. Middleton, K. de La Harpe, C. Su, Y.K. Law, C.E. Crespo-Hernandez and B. Kohler, *Ann. Rev. Phys. Chem.* **60**, 217 (2009) and references therein.
75. N. Agmon, *J. Phys. Chem. A* **109**, 13 (2005).

Chapter 16

Interrogation of Nonadiabatic Molecular Dynamics via Time-Resolved Photoelectron Spectroscopy

Michael S. Schuurman and Albert Stolow

1. Introduction 634
2. Wavepacket Dynamics 636
   2.1. Born–Oppenheimer wavepacket dynamics 636
   2.2. Nonadiabatic wave-packet dynamics 641
   2.3. Selection of experimental observables 644
3. Principles of Time-Resolved Photoelectron
   Spectroscopy 645
   3.1. Photoelectron spectroscopy as a probe 645
   3.2. Description of electronic and nuclear
        degrees of freedom 649
4. Applications of Time-Resolved Photoelectron
   Spectroscopy 657
   4.1. Experimental demonstration of TRPES 657
   4.2. Interrogation of molecular dynamics
        employing TRPES 660
5. Concluding Remarks 663
   Acknowledgments 664
   References 664

Steacie Institute for Molecular Sciences, National Research Council, Canada.

1. Introduction

The time evolution of photoexcited molecular systems invariably involves the complicated coupling of nuclear and electronic dynamics, resulting in a myriad of potential radiationless processes such as internal conversion, isomerization and proton transfer reactions.[1-8] The unambiguous elucidation of the photodynamics of polyatomic molecular systems will require a set of theoretical and experimental tools robust enough to disentangle these coupled dynamical processes.

The primary reason that dynamics in the excited states of molecules are, in general, qualitatively different from dynamics in the ground state, and correspondingly more complicated to study, is due to the frequent breakdown of the Born–Oppenheimer approximation (BOA) in these regimes. The BOA states, in part, that the electronic energy depends only parametrically on the nuclear coordinates.[9] The justification for this approximation lies in the different mass scales between the relatively heavy nuclei versus the lightweight electrons. In this picture, the electronic distribution relaxes instantaneously upon displacement of the nuclei, thereby defining the potential energy surfaces on which the nuclei move. The BOA, in the context of nuclear motion on a given electronic state, is also a statement about how energetically isolated a particular electronic state is from the manifold of all electronic states. A breakdown of the BOA occurs when the parametric dependence on the nuclei becomes a functional one, which will occur when electronic states come into close energetic proximity. In addition, and complimentarily, this situation is associated with a concomitant increase in the derivative coupling between the now energetically proximal electronic states. One understands the difference between this parametric and functional dependence intuitively given that when the BOA is valid, small nuclear perturbations generate small perturbations in the electronic energy, whereas in a region near an electronic degeneracy, small nuclear perturbations may result in qualitative changes in the electronic character of the nearby states.

It is because excited electronic states of polyatomic molecules generally occur in manifolds, as opposed to energetically well separated potential energy surfaces, that the breakdown of the BOA is a standard characteristic of excited state dynamics. A particularly relevant breakdown of the BOA occurs when two (or more) electronic states are degenerate at a given nuclear configuration. If the degeneracy of the electronic states is lifted linearly in the vicinity of this point via a specific set of nuclear displacements, it is termed a conical intersection. For an intersection of two electronic states these nuclear displacements, in general, correspond to motion along exactly two coordinates which span the branching[10] or g-h space.[11] These directions are readily determined from *ab initio* gradient information, where the g direction is defined as the energy difference gradient between the two states, while the h direction is collinear with the nonadiabatic coupling gradient. Similarly, if nuclear motion is to be treated in terms of normal mode vibrations, generally only a subset of vibrational modes will be identified with nonadiabatic transitions and are termed tuning or coupling modes.[8,12] Analogous to the internal coordinate treatment, here the former identify those vibrational modes which affect the energy difference between the states of interest, while the latter is associated with of modulation the nonadiabatic coupling. Thus, passages through conical intersections that result in transitions from one electronic state to another are effectively described by only a small subset of vibrational modes/internal coordinates.

Conical intersections are extremely efficient at facilitating nonradiative transitions between electronic states, and when found in the Franck–Condon region, can lead to transitions between electronic states on femtosecond timescales following the absorption of the excitation photon. These theoretical concepts have proven useful in the characterization and simulation of nonadiabatic dynamics in molecular systems, but direct spectroscopic observation of passages through conical intersections, and nonadiabatic transitions in general, in polyatomic systems with many vibrational degrees of freedom remains a more significant challenge.

One experimental approach that has been developed to interrogate complex excited state dynamics is gas-phase time-resolved photoelectron spectroscopy (TRPES).[13–25] This pump-probe spectroscopic technique first prepares a non-stationary state on the excited state manifold, then employs a time-delayed ionizing probe laser pulse to determine photoelectron spectra as a function of time. Assuming the excited state potential energy surface differs from that of the ground state, the time-dependent evolution of the prepared wavepacket will necessarily involve nuclear motion, a process which may be observed in the evolution of the Franck–Condon spectra that define the vibrational structure of the bands in the photoelectron spectrum. If, however, the wavepacket undergoes a nonadiabatic transition to a different electronic state, one could expect a dramatic change in the dominant ionization channel, which would be observed as a change in the kinetic energy spectrum of the emitted photoelectron. We note that the processes we call *chemistry* involve the rearrangement of the valence (as opposed to inner shell) electrons, and therefore we expect that photoelectron spectroscopies (i.e. UV as opposed to X-ray) which are directly sensitive to the valence electrons will confer the greatest insights. Additionally, the photoelectron angular distribution, an information-rich observable that can be employed to help elucidate the electronic character of the ionized state, would also be expected to display a time dependence that is sensitive to electronic states on which the wavepacket is propagating.[26]

Thus, TRPES techniques have in principle the ability to detect both changes in the motion of the nuclei, as well as the evolution of the valence electronic structure of the time-dependent wavepacket. Some details of the method, and potential complications, will be discussed in further detail below. In this discussion, our language shall assume that the excited state dynamics occur on a manifold of neutral electronic states, which upon ionization yield a singly charged cationic species. While this is the simplest and most straightforward implementation of the technique, TRPES is of course not limited to these scenarios. In the following, we present in Sec. 2 the wavepacket formalism necessary for a conceptual understanding of the motivations and results of TRPES, while in Sec. 3 we discuss limiting pedagogical cases of molecular systems that may be studied using this technique. In Sec. 4 we present representative experimental applications, and we close with a prognosis for the future of the approach and the increasingly important interplay between theory and experiment.

2. Wavepacket Dynamics

2.1. *Born–Oppenheimer wavepacket dynamics*

In order to discuss the physical processes involved in the preparation of a molecular excited electronic state and the subsequent evolution of the nuclei on the corresponding potential energy surfaces, the most conceptually straightforward approach is to invoke the notion of time-dependent vibronic wave packets. A pump-probe experiment of the type considered here involves three distinct steps: the pump, in which ground state population is transferred to an excited state, the evolution of the excited state population, and finally, the probe, in which the excited state wavepacket is projected onto a manifold of final states.

Under this ansatz, the evolving non-stationary state is represented as a wavepacket, which is defined as a coherent superposition of time-independent eigenstates:

$$|\psi(t)\rangle = \sum_\alpha C_\alpha(t)|\Psi_a\rangle, \qquad (1)$$

where

$$C_\alpha(t) = A_\alpha e^{-iE_\alpha t/\hbar}. \qquad (2)$$

As Eq. (2) shows, the coefficients $C_\alpha(t)$ are complex valued and time dependent. The initial phase and amplitude of the wavepacket component state α prepared by the pump laser pulse is contained in the complex-valued coefficient A_α. Since the exact eigenstate $|\Psi_\alpha\rangle$ is time-independent, the evolution of the wavepacket is wholly determined by the changes in the coefficients $C_\alpha(t)$.

During the probe process, the wavepacket is lifted up onto the manifold of final states, where the transition probability to a given final state $|\Psi_f\rangle$ is given by:

$$\begin{aligned}
I_f(t;\omega) &= |\langle\Psi_f|E(\omega)\cdot\mu|\psi(t)\rangle|^2 \\
&= \left|\sum_\alpha A_\alpha\langle\Psi_f|E(\omega)\cdot\mu|\Psi_\alpha\rangle e^{-iE_\alpha t/\hbar}\right|^2 \\
&= \left|\sum_\alpha B_\alpha(\omega)e^{-iE_\alpha t/\hbar}\right|^2. \qquad (3)
\end{aligned}$$

If one expands the norm shown in Eq. (3), employing standard trigonometric identities, the resulting series can be expressed as a double sum:

$$I_f(t;\omega) = \sum_{\alpha,\beta}|B_\alpha||B_\beta|\cos([E_\alpha - E_\beta]t/\hbar + \Phi(\alpha,\beta)), \qquad (4)$$

where $\Phi(\alpha,\beta)$ contains the phase difference between the $B_{\alpha,\beta}$ coefficients, each of which contain the initial phase of the component states, α and β, of the wavepacket and the phase of the transition dipole matrix elements connecting each state to the final state $|\Psi_f\rangle$. Figure 1 illustrates both the initial preparation of the excited state wavepacket, and the subsequent projection of this superposition of states onto a manifold of final states via a probe photon.

Since the time dependence in the observed transition intensity shown in Eq. (4) is multiplied by the energy difference $E_\alpha - E_\beta$, the total signal,

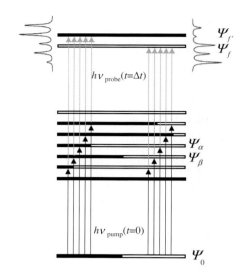

Fig. 1. Schematic for wavepacket creation and subsequent simultaneous projection onto a manifold of final states. Following excitation from the initial state ψ_0, the excited state wavepacket is composed of states (denoted ψ_α, ψ_β, etc.) with varying degrees of "black" and "white" character (for example, bond stretch and angle bend vibrational modes). The choice of final state, either "white" (ψ_f) or "black" ($\psi_{f'}$), will differentially project out different components of the wavepacket based on the degree of overlap between the two states.

shown to be a coherent sum of all interferences between pairs of states in the wavepacket, will display modulations at frequencies corresponding to all the level spacings present in the neutral state wavepacket. Therefore, the Fourier transform of the time-domain signal will yield a power spectrum in which energy level spacings within the prepared neutral state may be observed. However, the amplitude with which these frequencies appear in the transformed spectrum depends on the modulation depth of the signal, i.e. the amplitude is given by the product $|B_\alpha||B_\beta|$. From Eq. (3), one observes that the magnitude of $|B_\alpha|$ depends on the magnitude of the A_α, but more importantly, on the magnitude of the transition dipole matrix element between the neutral state Ψ_α with the final state Ψ_f. If the norm of these terms of is approximately equal, then the interference between the two transitions will be alternatively totally destructive or constructive, yielding a strong signal in the power spectrum at this frequency, with intensity proportional to these amplitudes. Conversely, if one of the amplitudes is small due to poor overlap with the final state (e.g. a forbidden transition), the interference will be minimal and only a weak signal at the frequency corresponding to this level spacing would be expected.

Since, for the case of photoionization detection, it is likely that many final states will be accessible given the photon energy of the probe laser pulse, the probe step is actually the projection of the wavepacket onto numerous final states, each resulting in a distinct set of interferences. It is precisely this simultaneous projection onto multiple states in the cationic manifold that enables the characterization of the excited state wavepacket. If the characters of the final states in the projection are known, the wavepacket may be successfully probed based on the differential signal with each level. For example, Fig. 1 shows coherent two photon transitions in which the final state is either of "black" or "white" character. Each of these states exhibit favorable overlap with neutral states that are primarily of the same character, as evinced by the spectral profiles shown in the diagram. When coherent two-photon transitions to the final state of white character are considered, the resultant power spectrum will display strong signal corresponding to the level spacings between the predominantly white states of the neutral, while the converse will be true for the primarily black final state. The level spacings corresponding to eigenstates of mixed character will arise with varying intensities depending on the relative (and absolute) magnitudes of the overlap with the chosen final state. In this way, the final states onto which the wavepacket is projected are effectively *filters* through which components of the wavepacket may be viewed. Through a reasonable selection of final vibronic states onto which the projection is performed, both the electronic and nuclear character of the evolving excited state wavepacket may be successfully interrogated as a function of time.

While the exact non-Born–Oppenheimer eigenstates employed in Eq. (1) are of use for in a formal discussion of wavepacket dynamics, they are in practice seldom accessible. Instead, these exact states are expanded in a basis-set of known functions. These expansions, in addition to being necessary for a useful theoretical description and simulation of wavepacket dynamics, are also often of pedagogical utility, as the essential physics of the dynamical processes can be made more transparent.

To this end, we will begin by partitioning the total Hamiltonian of the system into a zeroth-order component and a perturbation, such that:

$$(H_0 + V)\Psi_\alpha = E_\alpha \Psi_\alpha. \tag{5}$$

The exact eigenstates, and thus the wavepacket defined in Eq. (1), may then be expressed in a perturbation theory expansion, where the zeroth-order functions are those that diagonalize H_0. Employing the standard results of perturbation theory, one may obtain expressions for the energy of the system, such that:

$$H_0|\phi_\alpha^{(0)}\rangle = E_\alpha^{(0)}|\phi_\alpha^{(0)}\rangle, \qquad (6)$$

$$E_\alpha = E_\alpha^{(0)} + E_\alpha^{\text{corr}} = E_\alpha^{(0)} + \langle\phi_\alpha^{(0)}|V|\phi_\alpha^{(0)}\rangle + \sum_{\beta\neq\alpha}\frac{|\langle\phi_\beta^{(0)}|V|\phi_\alpha^{(0)}\rangle|^2}{E_\beta^{(0)} - E_\alpha^{(0)}} + \cdots \qquad (7)$$

$$\Psi_\alpha = \phi_\alpha^{(0)} + c_\alpha^{(1)}\phi_\alpha^{(1)} + c_\alpha^{(2)}\phi_\alpha^{(2)} + \cdots \qquad (8)$$

Note that the second-order contribution to the perturbation energy is determined via coupling of zeroth-order states through the perturbation. Inserting these expressions into Eq. (1), one obtains

$$|\psi(t)\rangle = \sum_\alpha A_\alpha e^{-i\left(E_\alpha^{(0)} + E_\alpha^{\text{corr}}\right)t/\hbar}\phi_\alpha^{(0)} + \sum_\alpha C_\alpha(t)\sum_{n=1} c_\alpha^{(n)}\phi_\alpha^{(n)}. \qquad (9)$$

If the perturbation, V, is small, then the first term in Eq. (9) will initially predominate and the short time behavior of the wavepacket will be well described by the zeroth-order states. However, it is precisely the coupling between zeroth-order states shown in Eq. (7), coupling that arises due to the terms excluded in the zeroth-order Hamiltonian, which give rise to time-evolution of the wavepacket in the zeroth-order basis.

To give a pedagogical example, a common choice for H_0 in molecular vibration problems is the harmonic oscillator Hamiltonian, which would result in the functions $\phi_\alpha^{(0)}$ being the normal modes of the system and the zeroth-order energies given by $E_i^{(0)} = \hbar\omega_i(n_i + \frac{1}{2})$. However, the potential energy surfaces of molecular systems are to varying degrees anharmonic, which would be manifested in V containing cubic, quartic and higher-order potential terms. Additionally, many molecular systems, particularly those in excited electronic states, will display nonadiabatic transitions. The Hamiltonian for these systems will necessarily involve coupling between different electronic states, so that V may include the effect of coupling to other electronic states. The list of additional non-negligible potential couplings (i.e. spin-orbit, rotation-vibration, etc.) is virtually endless. Thus, the coupling between the zeroth-order states that arises due to the excluded terms in H_0 ensures that the states will evolve in time such that any initially prepared zeroth-order state will generally dephase into a complex superposition.

2.2. Nonadiabatic wave-packet dynamics

As discussed above, the initial preparation of an excited state wavepacket via a pump laser pulse creates a coherent superposition of states on the excited state manifold. Under the Born–Oppenheimer approximation,[9,27,28] the electronic energy is presumed to display only a parametric dependence (indicated by a semi-colon below) on the nuclear coordinates, $\mathbf{R}$, so that the electronic wavefunctions are given by:

$$H = T_e + V_e(\mathbf{r};\mathbf{R}) + T_N = H_e + T_N, \quad (10)$$

$$H_e \psi_e^i(\mathbf{r};\mathbf{R}) = E_e^i(\mathbf{R}) \psi_e^i(\mathbf{r};\mathbf{R}), \quad (11)$$

where T_e and T_N are the electronic and nuclear kinetic energy operators, respectively, and V_e is the coulombic potential terms of the electrons and nuclei. If the total wavefunction is expanded in a zeroth-order basis in which the electronic component is described in terms of adiabatic electronic states, one obtains

$$\Psi_\alpha(\mathbf{r}, \mathbf{R}) = \sum_i \chi_\alpha^i(\mathbf{R}) \psi_e^i(\mathbf{r};\mathbf{R}). \quad (12)$$

Equation (12) is known as the Born–Huang expansion,[29] and the nuclear functions, $\chi_\alpha^i(\mathbf{R})$, may be viewed as the expansion coefficients. While Eq. (12) is formally exact, since the entire set of adiabatic electronic states form a complete basis, in practice the summation is truncated to include only the states of physical relevance to the molecular system under consideration.

In keeping with the discussion from the previous section, the time evolution of a wavepacket expanded in a basis of zeroth-order states is due to couplings between the basis functions, which in this case correspond to the Born–Oppenheimer states. One may construct a set of coupled equations for the expansion coefficients, $\chi_\alpha^i(\mathbf{R})$, by inserting the wavefunction expansion in Eq. (12) into the Schrödinger equation for the total system and left multiplying by $\psi_e^{j*}(\mathbf{r};\mathbf{R})$,

$$\sum_i \psi_e^{j*}(\mathbf{r};\mathbf{R}) H_e \psi_e^i(\mathbf{r};\mathbf{R}) \chi_\alpha^i(\mathbf{R}) + \sum_i \psi_e^{j*}(\mathbf{r};\mathbf{R}) T_N \psi_e^i(\mathbf{r};\mathbf{R}) \chi_\alpha^i(\mathbf{R})$$
$$= E_\alpha \sum_i \psi_e^{j*}(\mathbf{r};\mathbf{R}) \psi_e^i(\mathbf{r};\mathbf{R}) \chi_\alpha^i(\mathbf{R}), \quad (13)$$

followed by integration over the electronic coordinates:

$$E_e^j(\mathbf{R}) \chi_\alpha^j(\mathbf{R}) + \sum_i \langle \psi_e^{j*}(\mathbf{r};\mathbf{R}) | T_N | \psi_e^i(\mathbf{r};\mathbf{R}) \rangle \chi_\alpha^i(\mathbf{R}) = E_\alpha \chi_\alpha^j(\mathbf{R}). \quad (14)$$

Since the nuclear kinetic energy is given by $T_N = -\frac{1}{2\mu}\nabla \cdot \nabla$, where μ is a reduced mass whose precise definition is coordinate system dependent, it is straightforward to obtain expressions for the matrix element in Eq. (14) given by:

$$\langle \psi_e^{j*}(\mathbf{r};\mathbf{R})|T_N|\psi_e^i(\mathbf{r};\mathbf{R})\rangle = -\frac{1}{2\mu}(\delta_{ji}\nabla^2 + 2\mathbf{F}^{ji}\cdot\nabla + G_{ji})$$
$$= T_N \delta_{ji} - \Lambda_{ji}. \qquad (15)$$

Finally, inserting Eq. (15) into Eq. (14), and converting to matrix notation yields

$$(\mathbf{T}_N + \mathbf{E}_e(\mathbf{R}) - \mathbf{\Lambda} - E_\alpha \mathbf{I})\chi(\mathbf{R}) = \mathbf{0}, \qquad (16)$$

where $\mathbf{T}_N$ and $\mathbf{E}_e(\mathbf{R})$ are diagonal matrices, with the elements of the latter given by the adiabatic electronic energies. Most germane to this discussion is the derivative coupling term, $\mathbf{\Lambda}$, which couples the adiabatic electronic states via the nuclear kinetic energy operator. The matrix elements that define of this term are given by

$$\begin{aligned}\mathbf{F}^{ji} &= \langle \psi_e^j(\mathbf{r},\mathbf{R})|\nabla\psi_e^i(\mathbf{r},\mathbf{R})\rangle,\\ G^{ji} &= \langle \psi_e^j(\mathbf{r},\mathbf{R})|\nabla^2\psi_e^i(\mathbf{r},\mathbf{R})\rangle.\end{aligned} \qquad (17)$$

$\mathbf{F}^{ji}$ and G^{ji} are the first- and second-derivative couplings, respectively, where $\mathbf{F}^{ji}$ is a vector quantity whose length is determined by the number of nuclear degrees of freedom and G^{ji} is a scalar. Since the scalar couplings can be expressed as

$$\mathbf{G} = \nabla \cdot \mathbf{F} + \mathbf{F} \cdot \mathbf{F}, \qquad (18)$$

the total nonadiabatic coupling term is fully determined by the $\mathbf{F}$ matrices, which will subsequently be denoted simply as the derivative coupling. The magnitude of $\mathbf{F}^{ji}$ can be considerable even at significant distances from a point of degeneracy.[30] If the gradient operator is applied to Eq. (11) for states i and j, one can show that $\mathbf{F}^{ji}$ is given by:

$$\mathbf{F}^{ji} = \frac{\langle \psi_e^j(\mathbf{r},\mathbf{R})|\nabla H_e|\psi_e^i(\mathbf{r},\mathbf{R})\rangle}{E_e^i - E_e^j}. \qquad (19)$$

As implied by the energy difference denominator, this term becomes particularly large as the energy difference between states i and j becomes small, and will be singular when the two electronic states are degenerate. If the degeneracy is lifted linearly (i.e. not due to the Renner–Teller effect),[31,32] the crossing between states i and j may be termed a conical intersection.

The computation of accurate electronic wavefunctions has been one of the primary foci of *ab initio* quantum chemistry over the past few decades. In particular, advances in the computation of analytic energy gradients and derivative couplings now allow for the efficient *ab initio* determination of the derivative coupling terms in Eq. (19), thus enabling the quantitative description of nonadiabatic processes in general, and points of conical intersection in particular.

In order to determine the electronic wavefunctions defined in Eq. (11), the adiabatic electronic states are expanded in a configuration state function (CSF) basis such that

$$\psi_e^i(\mathbf{r};\mathbf{R}) = \sum_{k=1}^{N^{CSF}} c_k^i(\mathbf{R})\phi_k(\mathbf{r};\mathbf{R}), \qquad (20)$$

where $\phi_k(\mathbf{r};\mathbf{R})$ are spin-adapted linear combinations of Slater determinants and $c_k^i(\mathbf{R})$ are the configuration interaction (CI) coefficients.[30,33,34] While the electronic structure aspects of the following discussion will assume CI type wavefunctions and will thus employ this nomenclature, it should be noted that equation-of-motion coupled cluster (EOM-CC) methods[35,36] are also being successfully employed, with increasing frequency, to study the influence of nonadiabatic processes on molecular dynamics.[37–39]

To illustrate the concepts introduced thus far, we close this section by considering the time evolution of a hypothetical wavepacket on a manifold of excited electronic states which display regions of strong nonadiabatic coupling. At time $t = 0$, following the creation of the wavepacket on an excited electronic state, the initial wavepacket is well represented by a single zeroth-order, adiabatic electronic state. The wavepacket will then evolve in time due to the motion of the nuclei, corresponding to the forces on the potential energy surface. At some later time, dependent on the topography of the potential energy surface on which the wavepacket is evolving, the wavepacket will approach a region in which the derivative coupling between the current adiabatic state and another state becomes large. This scenario is generally associated with a decrease in the energy difference between the two states. As the coupling increases, the total wavefunction can no longer be associated with a single zeroth-order state, but is rather a linear combination of electronic states due to the mixing engendered by the (now large) derivative coupling terms. If the coupling is indeed strong, there is the possibility that the wavepacket will undergo a qualitative change in character, which can be uniquely identified as a transition to a different adiabatic electronic state. That is to say, as the wavepacket vacates the coupling region, it is now well represented by a different zeroth-order electronic state. The ability to describe and predict the nature of these changes in wavepacket character, as well as the time scales on which they occur, continues to be a growing area of modern research in theoretical/computational chemical physics. Employing spectroscopic methods to access this information, particularly in general cases for "large" molecules, presents a hosts of challenges.

2.3. *Selection of experimental observables*

To this point, we have not discussed the determination of any particular observables. Specifically, the nature of the process by which the wavepacket is projected onto a set of final states, i.e. the computation of an expectation value. The determination of an appropriate experimental observable is an important point not just for experimentalists, but for theorists as well. The simulation of molecular dynamics for chemically relevant molecular systems requires that a number of significant approximations be made at a number of levels. These include the level of rigour in the description of the electronic structure of the excited states, the number of nuclear degrees of freedom that will be explicitly considered, the degree to which nonadiabatic coupling is included, etc. Often these approximations are relatively trivial and can be easily justified and/or convincingly validated. However, they can just as often be motivated by computational efficacy or other concerns not related to the physics of the problem. As a result, multiple mechanistic pictures, or narratives, may be constructed for the same molecular system, depending on the approximations employed. The availability of experimentally determined observables, therefore, provides an invaluable constraint on the results of computational simulations. However, it should be noted that spectroscopically determined observables are of maximum utility when compared to independently computed observable quantities using theoretical methods, as opposed to direct comparisons to mechanistic narratives which may be consistent with multiple spectroscopic measurements.

The choice of observable employed to study these processes is a degree of freedom available to the investigator. The ultimate selection of an observable must be motivated by the questions that one wishes to answer. In the case of interest here, the observable of choice should be sensitive to *both* the reorganization of electronic charge and the motion of the nuclei: the degrees of freedom relevant to the dynamics of the nonadiabatic process under investigation. Furthermore, the observable should be non-vanishing for the degrees of freedom being studied on the time scales of the experiment. If one were to employ electronic absorption spectroscopy to monitor the evolution of a vibronic wavepacket, possible changes in the relevant selection rules would have to be known and anticipated. For example, if the wavepacket were to undergo intersystem crossing, evolving from singlet to triplet character, the selection rules governing the allowed transitions would change drastically, leading to a drop off in signal (i.e. formation of the so-called "dark state"). In the absence of a measurable response, little could be said regarding the behavior of the wavepacket subsequent to the intersystem crossing.

In the section that follows, it is argued that the photoelectron spectrum is an observable of particular utility for studying nonadiabatic wavepacket dynamics on excited molecular electronic states. When measured on femtosecond timescales, time-resolved photoelectron spectroscopy offers insights into molecular dynamics with a level of detail that was once exclusively the domain of theoretical simulation.

3. Principles of Time-Resolved Photoelectron Spectroscopy

3.1. *Photoelectron spectroscopy as a probe*

The following discussion will argue that the energy resolved, angle integrated photoelectron spectrum is a simple but particularly powerful spectroscopic observable to employ in the study of excited state wavepacket dynamics. While even more information may be obtained from energy-angle-resolved spectra,[40] discussion of these techniques is beyond the scope of the present chapter and the interested reader is referred elsewhere for a more detailed treatment.[24,26,40–44] In addition to more conceptual concerns, any experimental probe of excited state dynamics should ideally have a number of practical characteristics. If the technique is to be applied to problems of broad chemical interest, it must be general enough to accommodate the countless complexities that define molecular systems, but sensitive enough to distinguish the often subtle differences in the response of a time-evolving vibronic wavepacket. It is argued here that photoelectron spectroscopy is precisely such a multiplexed spectroscopic approach.

3.1.1. Motivation

Firstly, concerning the generality of the approach, there are no "dark states" in photoelectron spectroscopy, as ionization is practically always an allowed process. Treating the free electron as a spectator (an over-simplification), the intensity of a particular ionizing transition depends on the degree of overlap between the neutral and cationic state wavefunctions. This intensity may be described as a product of the overlap between the electronic wavefunctions and a Franck–Condon factor describing overlap of the vibrational component of the total wavefunction. The extent to which these factors can be considered independent enables the construction of experiments that are designed to disentangle nuclear and electronic degrees of freedom.

Secondly, the ability to unambiguously assign an observed spectrum to the molecular species that produces the signal (via the charge-to-mass ratio) is the signature of a powerful spectroscopic approach. The determination of a photoelectron spectrum involves not only the production of a photoelectron, the kinetic energy of which gives rise to the spectrum, but also a cationic molecular species that can be identified via mass spectrometry. If these spectra are measured in coincidence, the parent neutral source of the photoelectron spectrum can be uniquely identified. These photoelectron-photoion coincidence (PEPICO) experiments have proven successful in identifying the ion parent of a photoelectron spectrum in femtosecond resolved experiments.[45–47] Given the propensity of excited cationic states to fragment, the ability to differentiate parent ion signal from fragment ion signal is a significant advantage.

Furthermore, the presence multiphoton ionization processes, which can often obscure the nature of initially prepared wavepacket, as well as the identity of the ionized species, can be readily identified. Since the photoelectron spectrum is recorded at each time step, the kinetic energy of the emitted electron provides a decisive fingerprint for not only the final cationic channel the molecular species is ionized into, but also the total number of photons involved in the ionizing photon absorption process. Electrons generated via multiphoton process will be shifted to higher kinetic energies.

Many of these attractive features of a photoelectron spectroscopy probe of molecular dynamics are a result of employing a highly differential detection approach. Rather than measuring integrated quantities, such as ion yields, each measurement retains the energy resolution of the detected species. While the complications that arise in this approach may result in reduced signal levels, electron detection is nevertheless highly sensitive and amenable to data-intensive measurement of many photoelectron spectra.

3.1.2. Description of photoionization

The ionization of a polyatomic molecule during the probe step of a TRPES experiment involves the absorption of a photon in order to produce a cationic molecular species and a free electron (here treated as a spectator), where the kinetic energy of the latter is analyzed to construct the photoelectron spectrum. Assuming that the wavepacket at any time t corresponds largely to a particular zeroth-order state Ψ_0, then the matrix element that describes this projection at a given time on the cationic continuum state such that

$$I_f(t;\omega) \propto \left|\langle \Psi_f^{\text{cat}} | \mu(\mathbf{r}) | \Psi_0 \rangle\right|^2. \quad (21)$$

It is common to represent the final cationic state as a spin-adapted, anti-symmetrized product of an $N-1$ electronic wavefunction with a single free electron wavefunction:

$$\Psi_f^{\text{cat}} = \frac{\left(1 - \sum_{i=1}^{N-1} \hat{P}_{iN}\right)}{\sqrt{2N}} \left(\Psi_\alpha^{\text{cat}}(r_1,\ldots,r_{N-1})\phi_\beta(r_N)\right.$$
$$\left. - \Psi_\beta^{\text{cat}}(r_1,\ldots,r_{N-1})\phi_\alpha(r_N)\right), \quad (22)$$

where the subscripts α, β on Ψ^{cat} and ϕ indicate the spin of electrons r_{N-1} and r_N, respectively, and where the operator $\hat{P}_{iN}$ interchanges electrons i and N. Given this ansatz for the cationic wavefunction, and since the electrons are indistinguishable and the wavefunctions anti-symmetric, one may determine the photoelectron matrix element in Eq. (21) to be:

$$\left|\langle \Psi_f^{\text{cat}} | \mu(r) | \Psi_0 \rangle\right|$$
$$= \sqrt{2N} \left[\begin{array}{l} \langle \phi_\beta(r_N) | \mu(r_N) | \Psi_\alpha^{\text{cat}}(r_1,\ldots,r_{N-1})\Psi_0 \rangle \\ - \sum_{i=1}^{N-1} \langle \hat{P}_{iN} \Psi_\alpha^{\text{cat}}(r_1,\ldots,r_{N-1})\phi_\beta(r_N) | \mu(r_N) | \Psi_0 \rangle \end{array} \right].$$
$$(23)$$

The second term on the right-hand side of Eq. (23) will be non-zero when the free-electron function is not orthogonal to the neutral state wavefunction Ψ_0. In the case of photoionization, the overlap between the relatively localized neutral electronic state and the free electron will necessarily vanish in the asymptotic limit, an observation which is generally employed to justify a strong orthogonality approximation[48,49] and the corresponding neglect of this term.

The first term on the right-hand side of Eq. (23) contains the so-called Dyson orbital, defined as:

$$\phi_{fi}^D(r) = \sqrt{N} \int \Psi_f(r_1, \ldots, r_{N-1}) \Psi_i(r_1, \ldots, r_N) dr_1 \ldots dr_{N-1}, \quad (24)$$

which when inserted into Eq. (23), in concert with strong orthogonality approximation, yields

$$\langle \Psi_f^{\text{cat}} | \mu(r) | \Psi_0 \rangle = \sqrt{2} \langle \phi_\beta(r_N) | \mu(r_N) | \phi_{f0}^D(r_N) \rangle. \quad (25)$$

Note that the definition in Eq. (24) requires the identification of an initial neutral electronic state and a final cationic state, yielding a distinct Dyson orbital for each ionization channel. The quantitative calculation of these photoelectron matrix elements, as well as the analogous recombination matrix elements that simply interchange the definition of initial and final states, is an active area of research.[42,50-60] To rigorously address the scattering problems exemplified by the above expression often represents a significant computational challenge, particularly for molecules of chemical and/or biological interest.

The more modest goal of this chapter is to employ the above expressions in order to identify and anticipate the preferred ionization continuum for a given molecular system. In particular, the current discussion is primarily interested with the *relative* rates of ionization into different cationic electronic states from neutral state vibronic wavepackets. Thus, it is the relative norms of the Dyson orbitals that yield qualitatively accurate insights into the preferred ionization channels for wavepackets on a given electronic state to specific states of the cation.

While the measured photoionization signal will be dependent on the degree of overlap between electronic states of the neutral species and the energetically accessible electronic states of the cation, it will also be sensitive to changes in the vibrational character of the time-evolving wavepacket. As the wavepacket moves upon a single electronic potential energy surface, the nuclear motion that characterizes the kinetic energy of the wavepacket will influence the Franck–Condon spectra that determine the vibrational structure of the observed photoelectron bands. In summary, the observed photoionization signal will be sensitive not only to the electronic character of the evolving wavepacket, via the overlap of the electronic wavefunction of the neutral species with the accessible ionization continua, but also to nuclear motion through the evolution of the Franck–Condon spectrum. As discussed above, it is this ability to experimentally monitor precisely these characteristics of the evolving wavepacket that makes photoelectron spectroscopy particularly well suited for the study of nonadiabatic dynamics in polyatomic molecules. The utilization of these sensitivities to elucidate dynamical processes, as well as the interpretation of the resultant spectra, will be discussed in more detail below.

3.2. Description of electronic and nuclear degrees of freedom

In the preceding sections, it has been shown that the degree to which particular final states in a photoelectron spectroscopy experiment view the character of a time-evolving wavepacket is defined by the projection of the wavepacket onto the set of accessible states in the molecular ionization continuum. In practice, the observed signal from a time-resolved photoelectron spectroscopy experiment can be interpreted in terms of contributions from electronic and vibrational scaling factors. Specifically, the ability to ionize a neutral excited state of a molecule on which a nuclear wavepacket is evolving depends on the final and initial electronic states as well as the vibrational character of the wavepacket and the energetically accessible vibronic levels in the cationic manifold.

Turning first to the electronic scale factor, this term is determined by the degree of overlap between the neutral electronic state and the cationic core to which the molecule is ionized. This overlap is quantified via the computation of Dyson orbitals and photoionization cross sections, as discussed above; the tendency of particular neutral electronic states to ionize to specific electronic states of the cation can be readily understood in terms of simple Koopmans' correlations.[61,62]

The Koopmans' approximation states simply that the first ionization energy of a molecule is equal to negative the orbital energy of the ionized electron. This approximation assumes that there is no orbital relaxation in the cationic manifold, and thus the same set of orbitals that define the initial state wavefunction are employed in the description of the final cationic state. What the approximation implies is that photoionization occurs most efficiently when it is a one-photon, one-electron process. This simple picture of photoionization is a valuable pedagogical tool for understanding why molecules will preferentially ionize into different cationic continuum channels. For example, it is possible that the removal of an electron from the highest occupied molecular orbital (HOMO) of a neutral molecular species will produce an ionic core with an electronic configuration differing from the corresponding neutral only in the electron hole created left in the HOMO. These two states would be termed Koopmans' correlated and the ionization probability would be predicted to be high. In contrast, a neutral electronic configuration in which both HOMO electrons are promoted to the lowest unoccupied orbital (LUMO) would be predicted to be poorly correlated to the same cationic state, as removal of a single from the frontier orbital would produce an excited cationic electronic configuration. Following the seminal work of Domcke,[61] this interplay, and at times competition, between the electronic and vibrational correlation factors will be explored below employing a series of test cases that illustrate a wide range of behaviors.

3.2.1. *Complementary Koopmans' correlations (Type I)*

For time-resolved experiments involving nonadiabatic transitions between multiple excited states of the neutral species, the evolution of the electronic dynamics can be readily monitored if the different electronic states preferentially ionize to different states of the corresponding cation. Under these favorable circumstances, the different excited states under investigation can be said to display complementary ionization correlations. In this situation, the kinetic energy of the departing electron will generally be uniquely assigned to a particular cationic manifold, assuming the energy differences between the electronic states of the cation accessible by the probe photon energy are large enough to be distinguished. Figure 2 presents a schematic of an experiment of this type, in which the time-scale of nonadiabatic passage from one electronic state to another is uniquely correlated with a switch in dominant ionization channel, resulting in a sudden change in the kinetic energy spectrum of the photoelectrons.

Paradigmatic examples of such molecular systems are the linear polyenes.[63] The ten carbon chain decatetraene (DT), in particular, was the subject of TRPES studies.[64] The lowest-lying singlet excited states of

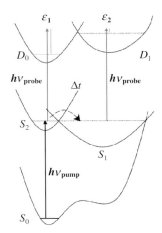

Fig. 2. A prototypical TRPES experiment in which a wavepacket is initially created at time $t = 0$ on state S_2, which preferentially ionizes to state D_0. The wavepacket evolves on this excited state for a time Δt, at which point it undergoes a nonadiabatic transition to state S_1. However, this electronic state preferentially ionizes to cationic state D_1, resulting in a change in the threshold ionization energy, from ε_1 to ε_2.

DT are identified as a bright S_2 state, accessed via a $\pi\pi^*$ transition, and an optically dark S_1 state, which can be excited directly in the Franck-Condon region only via multiphoton processes. As Table 1 shows,[65,66] the electronic structure of these two states is significantly different, with the S_2 state arising from a HOMO–LUMO transition, while the S_1 state involves double excitations out of the closed shell ground state. As such, each state correlates predominantly to a different electronic state in the cationic manifold. The first ionization threshold for the S_2 state corresponds to removal of the π^* electron, yielding the ground state of the cation, D_0. Conversely, ionization of the S_1 will preferentially form the first excited state of the cation, D_1.

If the Type I Koopmans' correlation picture provides a reasonable context for understanding TRPES measurements, once would expect a sudden switch in ionization channels as the vibronic wave packet undergoes a nonadiabatic transition from S_2 to S_1. Given that the excitation energy from D_0 to D_1 is approximately 1.5 eV, these transitions would correspond to a sudden decrease in the kinetic energy of the ejected photoelectrons of approximately the same magnitude and thus a shift of the photoelectron spectrum to lower energies. As Fig. 3 shows, this is indeed what is observed.

Table 1. Predominant molecular orbital configurations for the low-lying neutral and cationic electronic states of all-trans 2,4,6,8-decatetraene. The rightmost column indicates the Koopmans' correlated cationic state for each neutral configuration, assuming a single-photon, single-electron ionization event.

Electronic state	Energy (eV)	Molecular orbital occupancy					Weight[a]	Correlated state
		$1b_g$	$2a_u$	$2b_g$	$3a_u$	$3b_g$		
Neutral								
S_0, $1\ ^1A_g$	0.0	2	2	2	0	0	89%	D_0
S_1, $2\ ^1A_g$	3.6	2	2	0	2	0	35%	D_1, D_2
		2	1	2	1	0	19%	D_1, D_2
		2	2	1	0	1	19%	D_0
S_2, $1\ ^1B_u$	4.3	2	2	1	1	0	95%	D_0
Cation								
D_0, $1\ ^2B_g$	7.3	2	2	1	0	0	96%	
D_1, $1\ ^2A_u$	8.7	2	1	2	0	0	53%	
		2	2	0	1	0	41%	
D_2, $2\ ^2A_u$	9.6	2	1	2	0	0	38%	
		2	2	0	1	0	43%	
D_3, $2\ ^2B_g$	9.7	1	2	2	0	0	56%	
		2	2	0	0	1	20%	

[a]Configuration weights computed employing semi-empirical QCFF/PI + CISD level of theory (Ref. 55).

The decrease in signal due to ionization of S_2 can be fit to exponential decay functions in order to determine a lifetime of the S_2 state of 402 ± 65 fs. The nonadiabatic transitions that give rise to the sub-picosecond excited state lifetimes, like those observed for DT, are very often facilitated by passage through seams of conical intersection.

3.2.2. *Corresponding Koopmans' correlations (Type II)*

While decatetraene is a limiting case for Type I (complementary) ionization correlations, many molecular systems will involve excited state dynamics where multiple electronic states will correlate to the same state of the cation. In these cases, the evolution of the vibronic wavepacket may not be readily disentangled on the basis of the kinetic energy of the ionized electron alone. When multiple electronic states of the neutral preferentially ionize to the same cationic state, they may be said to display corresponding ionization correlations (Type II).

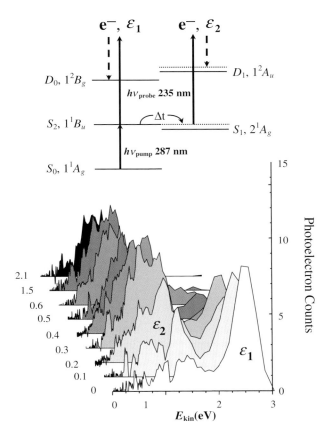

Fig. 3. (Top) Energy level scheme for TRPES of all-trans decatetratene, an example of a Type I molecule. The pump laser prepares the optically bright state S_2. Due to ultrafast internal conversion, this state converts to the lower lying state S_1 with $\sim 0.7\,\text{eV}$ of vibrational energy. The expected complementary Type I Koopmans' correlations are shown: $S_2 \to D_0 + \text{e}^-(\varepsilon_1)$ and $S_1 \to D_1 + \text{e}^-(\varepsilon_2)$. (Bottom) TRPES spectra of DT pumped at 287 nm and probed at 235 nm. There is a rapid shift ($\sim 400\,\text{fs}$) from ε_1, an energetic band at 2.5 eV due to photoionization of S_2 into the D_0 cation ground electronic state, to ε_2, a broad, structured band at lower energies due to photoionization of vibrationally hot S_1 into the D_1 cation first excited electronic state. The structure in the low energy band reflects the vibrational dynamics in S_1.

An example of such Type II systems can be found in the spectroscopic studies on the naphthalene and phenanthrene,[67] two rigid, fused ring molecular systems. As elucidated in Table 2, the leading electronic configurations of the pertinent excited states S_1 and S_2 have reasonable

Table 2. Predominant molecular orbital configurations for the low-lying neutral and cationic electronic states of phenanthrene. The rightmost column indicates the Koopmans' correlated cationic state for each neutral configuration, assuming a single-photon, single-electron ionization event.

Electronic state	Energy (eV)	Molecular orbital occupancy					Weight[a]	Correlated state
		a_2	a_2	b_1	a_2	b_1		
Neutral								
S_0, $1\ ^1A_1$	0.0	2	2	2	0	0	100%	D_0
S_1, $2\ ^1A_1$	3.6	2	2	1	0	1	50%	D_0
		2	1	2	1	0	50%	D_1
S_2, $1\ ^1B_2$	4.4	2	2	1	1	0	75%	D_0
		2	1	2	0	1	25%	D_1
Cation								
D_0, $1\ ^2B_1$	7.9	2	2	1	0	0	100%	
D_1, $1\ ^2A_2$	8.4	2	1	2	0	0	100%	
D_2, $2\ ^2A_2$	9.3	1	2	2	0	0	100%	

[a]Configuration weights computed employing semi-empirical QCFF/PI + CISD level of theory (Ref. 58).

overlap with the energetically accessible D_0 and D_1. Thus, one may expect to observe four different spectral bands in the photoelectron spectrum corresponding to ionization to the lowest two cationic states from both singlet excited states.

As Fig. 4 shows, there is a large degree of overlap between the observed ionization bands. However, the rigidity of the fused ring structures ensures that geometric changes between the ground, excited and cationic states are relatively small, thereby limiting the extent of vibrational progressions and ensuring that the $\Delta\nu = 0$ peak will dominate the photoelectron spectrum. Thus, the band origin of the $D_0 \leftarrow S_2$ ionizing transition, and its subsequent decay as a function of time, is clearly visible amongst the overlapping bands.

These same features of the potential energy surfaces aid in the identification of the rise in signal due to the ionization of S_1 state population. The adiabatic energy difference between the S_1 and S_2 states is approximately 0.75 eV. Thus, in order for population to transfer nonadiabatically from S_2 to S_1, the most favorable overlap of states will result in the creation of a wavepacket on S_1 with approximately 0.75 eV of vibrational energy. Since this vibrational energy should be largely conserved upon ionization to D_0, the S_1 ionization band should be shifted by the above amount relative to the S_2 ionization band, even though both neutral states ionize to the

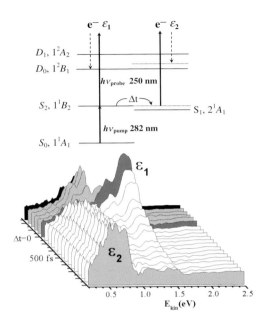

Fig. 4. (Top) Energy level scheme for TRPES of phenanthrene, an example of a Type II molecule. The pump laser prepares the optically bright state S_2. Due to ultrafast internal conversion, this state converts to the lower lying state S_1 with $\sim 0.74\,\text{eV}$ of vibrational energy. The expected corresponding Type II Koopmans' correlations are shown: $S_2 \rightarrow D_0 + e^-(\varepsilon_1)$ and $S_1 \rightarrow D_0 + e^-(\varepsilon_2)$. (Bottom) TRPES spectra of phenanthrene for a pump wavelength of 282 nm and a probe wavelength of 250 nm. The disappearance of band ε_1 at $\sim 1.5\,\text{eV}$ and growth of band at ε_2 at $\sim 0.5\,\text{eV}$ is a direct measure of the S_2–S_1 internal conversion time (520 fs). Despite the unfavorable Type II correlation, the rigidity of this molecule allows for direct observation of the internal conversion via vibrational propensities alone.

same cationic state. This is what is observed and an exponential fit of the decay of the S_2 state yields an excited state lifetime of $522 \pm 16\,\text{fs}$.

In this case it was possible to take advantage of the specifics of the molecular system in order to effectively analyze the energy domain photoelectron spectra to observe the nonadiabatic transfer of electronic population and determine excited state lifetimes. However, it is straightforward to envision situations in which multiple excited states ionize to the same manifold of cationic states, or overlapping complementary Koopmans' correlated ionization channels. In these cases, an analysis of the photoelectron angular distributions may offer further information regarding the evolving electronic character of excited state vibronic wavepacket.

3.2.3. Franck–Condon Spectra

While the above sections have addressed the ability of photoelectron spectroscopy to monitor changes in the electronic character of excited state wavepackets, the evolution of the nuclear component of the wavepacket on a particular potential energy surface can also be tracked via changes in the vibrational structure of the corresponding photoelectron band. Clearly, it is the overlap of the vibrational component of the wavepacket with the energetically accessible vibrational levels of the cation that determines the appearance of this spectrum.

If the vibrational component of the excited state wavepacket is expanded in a basis of eigenfunctions of the harmonic oscillator Hamiltonian, than the nuclear motion is at zeroth-order is described by motion along a set of normal mode coordinates. The total vibrational wavefunction will then be a direct product of harmonic oscillator functions and the Franck–Condon factor is given by

$$d_{m,n} = \left\langle \prod_{i=1}^{N^{\text{vib}}} \chi_{m_i}^{\text{cation}}(Q_i) \bigg| \prod_{j=1}^{N^{\text{vib}}} \chi_{n_j}^{\text{neutral}}(Q_j) \right\rangle, \qquad (26)$$

where the harmonic oscillator eigenfunctions are defined by

$$\chi_i^j(Q_j) = \frac{1}{2^i i!} \left(\frac{\omega_j}{\pi}\right)^{\frac{1}{4}} H_i(\sqrt{\omega_j} Q_j) e^{-\omega_j Q_j^2/2}, \qquad (27)$$

and ω_j are the harmonic frequencies associated with normal mode Q_j, and $H_i(x)$ are the Hermite polynomials of degree i. The efficient evaluation of these integrals are discussed elsewhere, including this current volume.[68–70] From the definitions above, these intensity factors are seen to be functions of the nuclear coordinates only. As the wavepacket evolves on an electronic state, its motion governed by the local energy gradients on potential energy surfaces, there will be a complementary reorganization of the nuclear coordinates which the Franck–Condon overlap integrals will be sensitive to. The schematic given in Fig. 1 can be readily applied to illustrate the simultaneous projection of the neutral excited state wavepacket onto a manifold of cationic vibrational energy levels. For example, one may associate the "white" and "black" states shown in the figure with bond stretch and angle bend vibrational energy levels, respectively, of the neutral and cation. If, at some initial time, the vibrational component of the wavepacket wavefunction is well described by a normal mode that is predominantly of bond stretch character, then observed Franck–Condon spectrum will be largely due to those cationic vibronic energy levels that project out bond stretch modes, i.e. have large Franck–Condon overlap with bond stretching normal modes. As the wavepacket evolves and vibrational energy is redistributed to other modes via IVR (intramolecular vibrational energy redistribution) processes, one may observe that at some later time the vibrational wavefunction resembles an angle bend normal mode. The Franck–Condon spectrum would reflect this evolution, as the observed signal could now be assigned largely to the angle bend vibrational levels in the cation manifold.

In summary, the intensity of the time-dependent energy-resolved angle-integrated photoelectron spectrum can be approximated as the product of two scaling factors: the electronic transition dipole matrix element connecting the neutral excited state to the cationic continuum state, and the vibrational Franck–Condon factor. The latter is the basis for the projection of the vibrational component of the wavepacket onto the vibronic manifold of the cation, thereby giving rise to the vibrational structure of a given ionization band. The ability to project both the electronic and nuclear components of the wavepacket enables, in favorable cases, the disentanglement of these degrees of freedom, thereby simplifying the analysis of information rich spectral measurements.

4. Applications of Time-Resolved Photoelectron Spectroscopy

4.1. *Experimental demonstration of TRPES*

We have argued that TRPES can be a powerful method for studying excited state nonadiabatic processes in polyatomic molecules. By choosing observables obtained from photoionization, we expect to extract information about both charge and vibrational energy flow, and their coupling, during these ultrafast dynamical processes. Ultimately, however, the theoretically motivated ideas discussed above must be corroborated by experiment. In the following, we present experimental demonstrations, using femtosecond laser pulses in a pump-probe scheme with energy-resolved photoelectron detection, of the two limiting cases of Koopmans' correlations for TRPES.

An illustration of Type I Koopmans' correlations is ultrafast internal conversion in the linear polyene *all-trans* 2,4,6,8 decatetraene (DT). The first optically allowed transition in DT is (S_2) $1^1B_u \leftarrow (S_0)$ 1^1A_g. As discussed above, the S_2 state is a singly excited configuration. In contrast, the lowest excited state is the dipole forbidden $(S_1)^1 A_g$ state which arises from configuration interaction between singly and doubly excited A_g configurations. Nonadiabatic vibronic coupling, promoted by b_u symmetry vibrational motions, leads to ultrafast internal conversion from S_2 to S_1. As discussed above, the S_2 excited state electronically correlates with the (D_0) $1\ ^2 B_g$ ground electronic state of the cation whereas the S_1 state correlates predominately with the (D_1) $1\ ^2 A_u$ first excited state of the cation. Therefore, DT is an example of a Type I Koopmans' correlation.[63]

In Fig. 3 (top) we show the DT energy level scheme. A femtosecond pump pulse at 287 nm (4.32 eV) prepared the excited S_2 state in its vibrationless electronic origin. It then evolves into a vibrationally hot (~ 0.7 eV) S_1 electronic state via internal conversion. The rapidly evolving electronic states are observed by projecting the wavepacket onto several cation electronic states using a UV probe photon of sufficient energy (here, 235 nm, 5.27 eV). As discussed above, the nonadiabatic coupling evolves the electronic character of the wavepacket, switching the photoionization electronic channel. This, in turn, leads to large shifts in the time-resolved photoelectron spectrum. Indeed, the experimental photoelectron kinetic energy spectra in Fig. 3 (bottom) show a rapid shift from a narrow high energy band ($\varepsilon_1 = 2.5$ eV) to a broad, structured low energy band (ε_2). This is the direct observation of the evolving electronic character induced by nonadiabatic coupling: the 2.5 eV band is due to ionization of S_2 into the D_0 ion state; the broad, low energy band arises from photoionization of S_1 into the D_1 ion state. Integration of the two photoelectron bands directly reveals the S_2 to S_1 internal conversion time scale of 386 ± 65 fs. Furthermore, the vibrational structure within each photoelectron band yields information about the vibrational dynamics which promote and tune the electronic population transfer.

We now consider Type II Koopmans' correlations where such a favorable separation of electronic from vibrational dynamics might not be expected. Examples of Type II systems include the polyaromatic hydrocarbons, a specific example being S_2–S_1 internal conversion in phenanthrene (PH). In the case of PH, Table 2 shows that both the S_2 and the S_1 states correlate similarly with the electronic ground state as well as the first excited state of the cation.[67]

In PH, the nonadiabatic coupling of the bright electronic state (S_2) with the dense manifold of zeroth order S_1 vibronic levels leads to the non-radiative "decay" (dephasing) of the zeroth order S_2 state. The energy gap between these two excited states is large and the density of S_1 vibronic levels is extremely large compared to the reciprocal electronic energy spacing. In this case, the dark state forms an apparently smooth quasicontinuum, the so-called statistical limit of the radiationless transition problem. The statistical limit is phenomenologically characterized by Lorentzian absorption bands where the apparent homogeneous width Γ is related to the apparent 'lifetime' of the bright state, $\tau \sim \hbar/\Gamma$.

We turn now to the femtosecond pump-probe photoelectron probing of excited state dynamics in PH. As illustrated in Fig. 4 (top), we excited PH from the $S_0\ ^1A_1$ ground state to the origin of the $S_2\ ^1B_2$ state with a 282 nm (4.37 eV) fs pump pulse. The excited molecules are then ionized after a time delay Δt using a 250 nm (4.96 eV) probe photon. The $S_2\ ^1B_2$ state rapidly internally converts to the lower lying $S_1\ ^1A_1$ state at 3.63 eV, transforming electronic into vibrational energy. In PH, both the $S_2\ ^1B_2$ and $S_1\ ^1A_1$ states correlate with the $D_0\ ^2B_1$ ion ground state, producing the ε_1 and ε_2 photoelectron bands. In Fig. 4 (bottom) we show time-resolved photoelectron spectra for PH, revealing a rapidly decaying but energetically narrow peak at $\varepsilon_1 \sim 1.5$ eV, due to photoionization of the vibrationless $S_2\ ^1B_2$ state into the ionic ground state $D_0\ ^2B_1$, with a decay time constant of 522 ± 16 fs. The broad band, centered at about ~ 0.7 eV is due to ionization of vibrationally hot molecules in the S_1 state, formed by the $S_2 - S_1$ internal conversion. At times $t > 1500$ fs or so (i.e. after internal conversion), the photoelectron spectrum is exclusively due to S_1 ionization, the S_1 state itself being long lived on the time scale of the experiment. Despite the fact that Type II molecules present an unfavorable case for disentangling electronic from vibrational dynamics, we can still see in PH a dramatic shift in the photoelectron spectrum as a function of time. This is due to the fact that PH is a rigid molecule and the S_2, S_1 and D_0 states all have similar geometries. The photoionization probabilities are therefore expected to be dominated by small Δv transitions. Hence, the 0.74 eV of vibrational energy in the populated S_1 state should be roughly conserved upon ionization into the D_0 ionic state. In PH, the small geometry change favors conservation of vibrational energy upon ionization, permitting the observation of the excited state electronic population dynamics. More generally, however, significant geometry changes will lead to overlapping photoelectron bands, complicating the disentangling of vibrational from electronic dynamics.

4.2. Interrogation of molecular dynamics employing TRPES

4.2.1. Excited state intramolecular proton transfer

Proton transfer is one of the most important chemical processes. Of interest here, excited state intramolecular proton transfer (ESIPT) processes are of particular importance since they allow for a detailed comparison of experiment with theory. o-hydroxybenzaldehyde (OHBA) is the smallest aromatic molecule displaying ESIPT. We used TRPES to study ESIPT in OHBA, the monodeuterated ODBA and the analogous two-ring system hydroxyacetonaphthone (HAN) as a function of pump laser wavelength, tuning over the entire enol $S_1(\pi\pi^*)$ absorption.[67]

In Fig. 5 (left), we show the energetics of OHBA. Excitation with a tunable pump laser $h\nu_{pump}$ forms the enol tautomer in the $S_1(\pi\pi^*)$ state. ESIPT leads to ultrafast population transfer from the S_1 enol to the S_1 keto tautomer. On a longer time scale, the S_1 keto population decays via internal conversion back to the ground state. Both the enol and keto excited state populations are probed by photoionization with photon energies $h\nu_{probe}$, producing the two photoelectron bands ε_1 and ε_2. In Fig. 5 (right), we present TRPES spectra of OHBA at an excitation wavelength of 326 nm, showing the two photoelectron bands ε_1 and ε_2. Band ε_1 is due to photoionization of the initially populated S_1 enol tautomer, band ε_2 to the photoionization of the S_1 keto tautomer. Both bands were observed across the whole absorption range (286–346 nm) of the S_1 state.

The decay of band ε_1 corresponds to the decay of the S_1 enol population and contains information about the proton transfer dynamics, with an estimated upper limit of 50 fs for the lifetime of the S_1 enol tautomer. Proton transfer reactions often proceed via tunnelling of the proton through a barrier and deuteration of the transferred proton should then significantly prolong the lifetime of the S_1 enol tautomer. In experiments with ODBA, we did not observe an isotope effect — i.e. the ESIPT reaction was again finished within 50 fs, showing that the barrier in the OH stretch coordinate must be very small or non-existent. TRPES also reveals the dynamics on the "dark" state, the S_1 keto state. The picosecond decay of band ε_2 corresponds to the loss of the S_1 keto population due to internal conversion to the ground state. The wavelength-dependent S_1 keto internal conversion rates in OHBA and ODBA likewise revealed no significant isotope effect. The measured internal conversion rates are very fast (1.6–6 ps over the range 286–346 nm) considering the large energy gap of 3.2 eV between

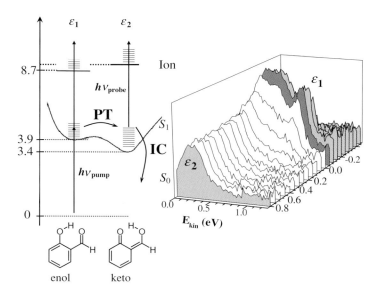

Fig. 5. (Left) Energetics for excited state proton transfer in OHBA, showing the enol and keto forms. Excitation with a pump laser E_{pump} forms the enol tautomer in the $S_1(\pi\pi^*)$ state. ESIPT leads to ultrafast population transfer from the S_1 enol to the S_1 keto tautomer. On a longer time scale, the keto S_1 population decays via internal conversion to the keto ground state. Both the enol and keto excited state populations are probed via ionization with a probe laser E_{probe}, producing the two photoelectron bands ε_1 and ε_2. (Right) TRPES spectra of OHBA at an excitation wavelength of 326 nm and a probe laser wavelength of 207 nm. Two photoelectron bands were observed: ε_1 due to ionization of the S_1 enol; ε_2 due to ionization of the S_1 keto. Band ε_1 indicates a sub-50 fs timescale for the proton transfer. Band ε_2 displayed a wavelength-dependent lifetime in the picosecond range corresponding to the energy dependent internal conversion rate of the dark S_1 keto state formed by the proton transfer.

the ground and excited state. These results suggest that interactions with a nearby $n\pi^*$ state may play an important role in the keto internal conversion. This example illustrates how TRPES can be used to study the dynamics of biologically relevant processes such as proton transfer, revealing details of both the proton transfer step and the subsequent dynamics in the "dark" state formed after the proton transfer.

4.2.2. Photostability of DNA bases

The UV photostability of the DNA bases is determined by the competition between a number of ultrafast excited-state electronic relaxation

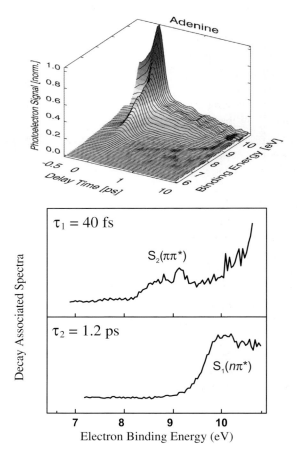

Fig. 6. Femtosecond TRPES spectra of adenine (top) recorded at 267 nm pump and 200 nm probe. The time axis is plotted on a logarithmic scale. The decay-associated spectra of the 40 fs and 1.2 ps channels, extracted from a biexponential global 2D fit, are shown (bottom). The 40 fs time constant spectrum corresponds to $\pi\pi^* \to D_0\ (\pi^{-1})$ photoionization. The 1.2 ps time constant spectrum corresponds to $n\pi^* \to D_1\ (n^{-1})$ photoionization. This confirms that the optically bright but short-lived $\pi\pi^*$ state decays via internal conversion to the dark but longer lived $n\pi^*$ state.

processes, some of which may be destructive to the molecule. In order to protect against these potentially harmful processes, nature designed ultrafast protective mechanisms that convert dangerous electronic energy to less dangerous vibrational energy. The purine base adenine is a heterocycle, having a strong $\pi\pi^*$ UV absorption band and, due to the lone electron pairs on the heteroatoms, has an additional low lying $n\pi^*$ state. In the gas phase, the 9H tautomer of isolated adenine is the lowest energy and most abundant form. The relative importance of the electronic relaxation channels in adenine has been a matter of some debate.[71,72] A TRPES study of the excited-state decay dynamics of adenine at 267 nm excitation, shown in Fig. 6, required a biexponential fit using two time constants: a fast component decaying in 40 fs followed by a slower component with a 1.2 ps lifetime.[73] A broad photoelectron band in the 7.5–10.8 eV range decays quickly and transforms into a second band spanning the 8.5–10.8 eV range. This second band grows smoothly between 8.5–9.6 eV and then decays slowly. Beyond 6 ps, no remaining photoelectron signal was observed. The two lowest electronic states of the adenine cation are the D_0 (π^{-1}) cation ground state and the D_1 (n^{-1}) cation excited state. The expected Koopmans' correlations are: $\pi\pi^* \to D_0$ (π^{-1}) and $n\pi^* \to D_1$ (n^{-1}). The spectrum of the 40 fs component corresponds to the $\pi\pi^* \to D_0$ (π^{-1}) ionizing transition whereas that of the 1.1 ps component to the $n\pi^* \to D_1$ (n^{-1}) ionizing transition. These TRPES results confirm that, in adenine, the short-lived state formed upon UV photoexcitation is the $\pi\pi^*$ state, which, upon internal conversion, forms the longer lived $n\pi^*$ state, itself decaying back to the ground state. This ultrafast radiationless decay to the ground state represents therefore a rapid conversion of "dangerous" electronic energy to "less dangerous" vibrational energy, since the former may lead to photochemistry whereas the latter may be rapidly quenched by dissipation to the surrounding water solvent.

5. Concluding Remarks

The ability of time-resolved photoelectron spectroscopy to interrogate vibronic wavepackets on nonadiabatically coupled molecular potential energy surfaces presents an excellent opportunity to take advantage of the deep and growing relationship between experiment and theoretical simulation. The nature and quality of the information available from a well designed TRPES experiment can be directly comparable to the highly detailed predictions of *ab initio* computation. Through the exploitation of the tendency for different excited molecular electronic states to preferentially ionize into different continuum channels, the dependence of observed Franck–Condon spectra on the motion of the nuclei, in the determination of molecular frame photoelectron angular distributions, and with time resolution on the order of femtoseconds, this spectroscopic technique has the potential to achieve deep insights into the nature of excited state polyatomic molecular dynamics. In fact, insights that were, until recently, exclusively the domain of theoretical simulation may now be gleaned from experimental results.

The nature of this detailed interplay between theory and experiment has farther reaching implications than simply the employment of wavepacket simulations to aid in the explanation of experimental results or in the validation of a given level of theory. Rather, we believe that the correlated development of experimental and *ab initio* molecular dynamics methods will play a key role in developing simplified "rules" that govern the behavior of excited polyatomic molecules. It might be said that conical intersections play a role in excited state dynamics analogous to that of the transition state in ground state dynamics. For ground state reaction dynamics, the Polanyi rules relate the topographical features of the transition state to the outcomes and energy disposal of the reaction.[74] For example, ground state A + BC reactions with "late" transition state barriers are move favorably traversed by vibrationally excited BC reactants, as opposed to "early" barriers that are more favorably crossed by high velocity collisions. By analogy with the Polanyi rules, we might expect that specific vibrational dynamics at conical intersections will be as important to the dynamics as are the topographical feature of the conical intersection itself. We look forward to the experimental–theoretical developments of many research groups which will give birth to an unprecedentedly detailed picture of dynamical processes in polyatomic molecules.

Acknowledgments

We thank our co-workers and collaborators Prof. Todd J. Martinez, M. Z. Zgierski, S. Patchkovskii, A. Boguslavskiy, O. Schalk, P. Hockett, O. Clarkin and G. Wu for numerous insightful discussions.

References

1. M. Bixon and J. Jortner, *J. Chem. Phys.* **48**, 715 (1968).
2. J. Jortner, S.A. Rice and R.M. Hochstrasser, *Adv. Photochem.* **7**, 149 (1969).
3. S.R. Henry and W. Siebrand, in *Organic Molecular Photophysics*, edited by J.B. Birks (John Wiley & Sons, Inc., London, 1973), Vol. 1, pp. 152.

4. K.F. Freed, in *Radiationless Processes in Molecules and Condensed Phases*, edited by F.K. Fong (Spring-Verlag, Berlin, 1976), pp. 23.
5. G. Stock and W. Domcke, *Adv. Chem. Phys.* **100**, 1 (1997).
6. G.A. Worth and L.S. Cederbaum, *Annu. Rev. Phys. Chem.* **55**, 127 (2004).
7. M. Klessinger and J. Michl, *Excited States and Photochemistry of Organic Molecules*. (VCH, New York, 1994).
8. H. Köppel, W. Domcke and L.S. Cederbaum, *Adv. Chem. Phys.* **57**, 59 (1984).
9. M. Born and J.R. Oppenheimer, *Ann. Phys. (Leipzig)* **84**, 457 (1927).
10. G.J. Atchity, S.S. Xantheas and K. Ruedenberg, *J. Chem. Phys.* **95**, 1862 (1991).
11. D.R. Yarkony, *Acc. Chem. Res.* **31**, 511 (1998).
12. H. Köppel, W. Domcke and L.S. Cederbaum, in *Conical Intersections*, edited by W. Domcke, D.R. Yarkony and H. Koppel (World Scientific, New Jersey, 2004), Vol. 15, pp. 323.
13. A. Stolow, A. Bragg and D.M. Neumark, *Chem. Rev.* **104**, 1719 (2004).
14. I. Fischer, M. Vrakking, D. Villeneuve and A. Stolow, in *Femtosecond Chemistry*, edited by M. Chergui (World Scientific, Singapore, 1996).
15. A. Stolow, *Philos. Trans. R. Soc. Lodon, Ser. A*. **356**, 345 (1998).
16. C.C. Hayden and A. Stolow, in *Advanced Physical Chemistry*, edited by C.-Y. Ng (World Scientific, Singapore, 2000), Vol. 10.
17. W. Radloff, in *Advanced Physical Chemistry*, edited by C.-Y. Ng (World Scientific, Singapore, 2000), Vol. 10.
18. K. Takatsuka, Y. Arasaki, K. Wang and V. McKoy, *Faraday Discuss* **115**, 1 (2000).
19. D.M. Neumark, *Annu. Rev. Phys. Chem.* **52**, 255 (2001).
20. A. Stolow, *Annu. Rev. Phys. Chem.* **54**, 89 (2003).
21. A. Stolow, *Inter. Rev. Phys. Chem.* **22**, 377 (2003).
22. T. Suzuki, T. Seideman and M. Stener, *J. Chem. Phys.* **120**, 1172 (2004).
23. V. Wollenhaupt, M. Engel and T. Baumert, *Annu. Rev. Phys. Chem.* **56**, 25 (2005).
24. A. Stolow and J.G. Underwood, *Adv. Chem. Phys.* **139**, 497 (2008).
25. T. Suzuki, *Annu. Rev. Phys. Chem.* **57**, 555 (2006).
26. T. Seideman, *Annu. Rev. Phys. Chem.* **53**, 41 (2002).
27. L.S. Cederbaum, in *Conical Intersections: Electronic Structure, Dynamics and Spectroscopy*, edited by W. Domcke, D.R. Yarkony and H. Köppel (World Scientific, Singapore, 2004), Vol. 15, pp. 3.
28. T. Pacher, L.S. Cederbaum and H. Köppel, *Adv. Chem. Phys.* **84**, 293 (1993).
29. M. Born and K. Huang, *Dynamical Theory of Crystal Lattices*. (Oxford University Press, 1954).
30. D.R. Yarkony, in *Conical Intersections: Electronic Structure, Dynamics and Spectroscopy*, edited by W. Domcke, D.R. Yarkony and H. Köppel (World Scientific, Singapore, 2004), Vol. 15, pp. 41.
31. R. Renner, *Z. Phys.* **92**, 172 (1934).
32. C. Jungen and A.J. Merer, in *Modern Spectroscopy: Modern Research*, edited by K.N. Rao (Academic, New York, 1976), Vol. II.

33. H. Lischka, M. Dallos, P. Szalay, D.R. Yarkony and R. Shepard, *J. Chem. Phys.* **120**, 7322 (2004).
34. M. Dallos, H. Lischka, P. Szalay, R. Shepard and D.R. Yarkony, *J. Chem. Phys.* **120**, 7330 (2004).
35. J.F. Stanton and R.J. Bartlett, *J. Chem. Phys.* **98**, 7029 (1993).
36. A.I. Krylov, *Annu. Rev. Phys. Chem.* **59**, 433 (2008).
37. J.F. Stanton, *J. Chem. Phys.* **115**, 10382 (2001).
38. T. Ichino, A.J. Gianola, W.C. Lineberger and J.F. Stanton, *J. Chem. Phys.* **125**, 084312 (2006).
39. J.F. Stanton, *J. Chem. Phys.* **126**, 134309 (2007).
40. K.L. Reid, *Annu. Rev. Phys. Chem.* **54**, 397 (2003).
41. K. Wang and V. McKoy, *Annu. Rev. Phys. Chem.* **46**, 275 (1995).
42. Y. Arasaki, K. Takatsuka, K. Wang and V. McKoy, *J. Chem. Phys.* **112**, 8871 (2000).
43. R.R. Lucchese, *J. Electron Spectrosc. Relat. Phenom.* **141**, 201 (2004).
44. P. Hockett, M. Staniforth, K.L. Reid and D. Townsend, *Phys. Rev. Lett.* **102**, 253002 (2009).
45. V. Stert, W. Radloff, T. Freudenberg, F. Noack, I.V. Hertel, C. Jouvet, C. Dedonder-Lardeux and D. Solgadi, *Europhys. Lett.* **40**, 515 (1997).
46. V. Stert, W. Radloff, C.P. Schulz and I.V. Hertel, *Eur. Phys.* **5**, 97 (1999).
47. I.V. Hertel and W. Radloff, *Rep. Prog. Phys.* **69**, 1897 (2006).
48. B.T. Pickup, *Chem. Phys.* **19**, 193 (1977).
49. Y. Öhrn and G. Born, *Adv. Quantum Chem.* **13**, 1 (1981).
50. P. Langhoff, *Chem. Phys. Lett.* **22**, 60 (1973).
51. R.K. Nesbet, *Phys. Rev. A* **14**, 1065 (1976).
52. P. Langhoff, in *Electron-Molecule and Photon-Molecule Collisions*, edited by T. Rescigno, V. McKoy and B. Schneider (Plenum, New York, 1979), pp. 183.
53. K. Gokhberg, V. Vysotskiy, L.S. Cederbaum, L. Storchi, F. Tarantelli and V. Averbukh, *J. Chem. Phys.* **130**, 064104 (2009).
54. R.R. Lucchese, K. Takatsuka and V. McKoy, *Phys. Rep.* **131**, 147 (1986).
55. Y. Arasaki, K. Takatsuka, K. Wang and V. McKoy, *J. Chem. Phys.* **119**, 7913 (2003).
56. C.M. Oana and A.I. Krylov, *J. Chem. Phys.* **127**, 234106 (2007).
57. C.M. Oana and A.I. Krylov, *J. Chem. Phys.* **131**, 124114 (2009).
58. S. Patchkovskii, Z. Zhao, T. Brabec and D.M. Villenueve, *Phys. Rev. Lett.* **97**, 123003 (2006).
59. C. Jin, A.-T. Le, S.-F. Zhao, R.R. Lucchese and C.D. Lin, *Phys. Rev. A* **81**, 033421 (2010).
60. Y. Arasaki, K. Takatsuka, K. Wang and V. McKoy, *J. Chem. Phys.* **132**, 124307 (2010).
61. M. Seel and W. Domcke, *J. Chem. Phys.* **95**, 7806 (1991).
62. M. Seel and W. Domcke, *Chem. Phys.* **151**, 59 (1991).
63. V. Blanchet, M.Z. Zgierski and A. Stolow, *J. Chem. Phys.* **114**, 1194 (2001).
64. V. Blanchet, M.Z. Zgierski, T. Seideman and A. Stolow, *Nature* **401**, 52 (1999).
65. A. Warshel and M. Karplus, *Chem. Phys. Lett.* **17**, 7 (1972).

66. F. Zerbetto, M.Z. Zgierski, F. Negri and G. Orlandi, *J. Chem. Phys.* **89**, 3681 (1988).
67. M. Schmitt, S. Lochbrunner, J.P. Shaffer, J.J. Larsen, M.Z. Zgierski and A. Stolow, *J. Chem. Phys.* **114**, 1206 (2001).
68. E.V. Doktorov, I.A. Malkin and V.I. Man'ko, *J. Mol. Spec.* **64**, 302 (1977).
69. D. Gruner and P. Brumer, *Chem. Phys. Lett.* **138**, 310 (1987).
70. A. Hazra and M. Nooijen, *Int. J. Quant. Chem.* **95**, 643 (2003).
71. L. Serrano-Andrés, M. Merchán and A.C. Borin, *Proc. Natl. Acad. Sci. USA* **103**, 8691 (2006).
72. M. Barbatti and H. Lischka, *J. Am. Chem. Soc.* **130**, 6831 (2008).
73. C.Z. Bisgaard, H. Satzger, S. Ullrich and A. Stolow, *ChemPhysChem* **10**, 101 (2009).
74. J.C. Polanyi, *Acc. Chem. Res.* **5**, 161 (1972).

Chapter 17

Pump-Probe Spectroscopy of Ultrafast Vibronic Dynamics in Organic Chromophores

Nina K. Schwalb, Ron Siewertsen, Falk Renth
and Friedrich Temps

1. Introduction 669
2. Femtosecond Time-Resolved Pump-Probe Schemes
   for Organic Chromophores 672
3. Radiationless Electronic Deactivation
   in Free DNA Bases 675
4. Electronic Deactivation in H-Bridged Base Pairs,
   π-Stacked Dimers and Oligonucleotides 681
5. Conformer-Specific Photochemistry of Furylfulgides 688
6. Ultrafast Photoisomerization of Azobenzenes 694
7. Conclusions 700
   Acknowledgments 701
   References 702

1. Introduction

The ubiquitous recognition within the past decade that conical intersections (CIs) among the potential-energy (PE) hypersurfaces of molecules

Institut für Physikalische Chemie, Christian-Albrechts-Universität zu Kiel, Olshausenstr. 40, D-24098 Kiel, Germany.

in excited electronic states or between the PE hypersurfaces of an excited state (ES) and the electronic ground state (GS) are the rule rather than the exception has stimulated immense interest in detailed studies of the vibronic dynamics of organic chromophores following photoexcitation by ultraviolet or visible (UV/VIS) light. The topography of CIs determines the distinctive nonadiabatic transition pathways that the electronically excited, highly energized molecules take on return to their initial state by ultrafast internal conversion or en route to new minimum energy conformations and chemically different reaction products on the global PE hypersurfaces.[1-5] Crucial photochemical outcomes, like fluorescence quantum yields, fractions of electronically deactivated molecules, and yields of photoproducts, are thereby closely linked and governed by the detailed molecular motions in the vicinity of the CIs that the molecules visit. The picture of CIs as "photochemical funnels"[1] nicely summarizes this new paradigm of organic photochemistry in its many facets.

Unlike the situation for the (normally) stationary vibrational states of molecules, there exists no single standard spectroscopy to experimentally "detect" a CI on a PE surface. Its presence and influence may only be deduced from the effect it exerts on the observable photophysics and photochemistry. State-of-the-art experiments that explore the nonadiabatic dynamics of photo-excited molecules involving CIs in considerable detail are discussed in the chapters of part III of this book. Fundamentally, because electronic transitions mediated through CIs are usually ultrafast, i.e. take place in femtoseconds or picoseconds, femtosecond time-resolved pump-probe spectroscopies in various forms are needed for monitoring the electronic relaxation processes and chemical transformations of the photoexcited molecules in real time.

This chapter illuminates recent discoveries by ultrafast pump-probe spectroscopy in two areas of photochemistry that arguably surged in the last decade more than virtually any other: The photostability of DNA and its building blocks and the dynamics of photochromic molecular switches. The first field is of vital interest regarding the integrity of the genomic information, the second has gained prevalent attention for the enormous application potential of miniature light-driven molecular-size switches and actuators.

UV radiation is a major cause for DNA damage, which can lead to genomic instabilities, cellular malfunctions, premature cell death, and, in the worst case, carcinogenesis and malignant cell growth.[6-9] The nucleobases adenine (Ade or A), cytosine (Cyt or C), guanine (Gua or G) and thymine (Thy or T), the letters of the genetic code written by the DNA, have strong UV absorption bands at wavelengths below $\lambda \lesssim 300$ nm. The small amount of UV-B (280–320 nm) light transmitted through the Earth's atmosphere thus affects all living organisms exposed to the sun. Humans are highly vulnerable in particular to skin cancer, which is increasing in many countries at alarming rates,[7] but UV light-induced mutagenesis is indeed generally thought to be one of the two main factors determining the rate of de novo spontaneous base mutations in the genomes of prokaryotic and eukaryotic organisms.[10]

Evolution has developed ingenious active DNA repair mechanisms,[9,11] but the involved complex and energetically costly enzymatic machinery of a living cell would be dramatically overburdened, if it alone had to account for the genomic integrity. It is of vital importance that the natural DNA bases themselves exhibit remarkable intrinsic photostabilities due to ultrafast radiationless electronic deactivation pathways, by which the energy of an absorbed UV photon is dissipated before chemical reactions in the UV-excited state can lead to profound molecular damage.[12] Despite intense work in many laboratories worldwide, however, the complex electronic structures and electronic dynamics in the macromolecular DNA remain highly controversial.[13–17] The mutual interactions among the many chromophores of a DNA strand by, e.g. dipole-dipole coupling, π-stacking, and via the hydrogen bonds between the complementary bases of the double helix strongly affect the electronic dynamics through formation of exciton, exciplex, and partial charge-transfer states in as yet barely understood ways.[16,17] The elucidation of the underlying elementary photophysical and photochemical mechanisms in the DNA and its building blocks constitutes one of the most important present challenges in photobiophysical chemistry.

Light-driven molecular switches have inspired extensive research in recent years because of their potential application as, e.g. tiny molecular machines,[18–22] for high-density optical data storage,[20,22–24] or in super-resolution imaging below the Abbe diffraction limit.[25–28] The switching is triggered by excitation of the molecules to an excited electronic state using light at suitable wavelengths and typically occurs as a result of a photochemical reaction such as E/Z isomerization, electrocyclic ring closure or opening, valence isomerization, or proton transfer. The photo-induced interconversion between two isomeric forms leads to reversible changes of the absorption spectra, i.e. photochromism.[29,30] Compounds with favorable switching properties include stilbenes, hemi-stilbenes and, in particulur, azobenzenes[20,31] (E/Z isomerization), fulgides[32] and diarylethenes[33] (cyclization of hexatriene backbone), chromenes[34] (valence isomerization) and spiropyranes[35] (ring opening to merocyanine), norbornadiene (valence isomerization to quadricylcane),[36] and numerous molecules showing photo-induced H-atom transfer.[37,38] Resolved absorption spectra, optical bistability, and full photoreversibility are highly desirable molecular properties for applications, but there are also strict demands regarding thermal stability and low photochemical fatigue. The latter often rests on the ultrafast nature of the ES transformations to rule out competing degrading processes.[39] A rational design of optimized functional molecules requires detailed understanding of the underlying ultrafast molecular processes, including the effects of chemical modifications of the chromophores and the molecular environment.

Since the molecules of interest typically exist in their functional forms in solution in water or in an organic solvent, this chapter focuses on research by time-resolved UV/VIS pump-probe fluorescence and absorption spectroscopy as the two most versatile methods for probing nonadiabatic dynamics in large organic chromophores in liquids. In combination, they allow one to monitor the vibronic transformations in the molecules from the initially photo-excited state(s) all the way to the thermally equilibrated final products.

2. Femtosecond Time-Resolved Pump-Probe Schemes for Organic Chromophores

To survey briefly the main experimental techniques available to elucidate nonadiabatic molecular dynamics, Merkt pushed high-resolution photoelectron spectroscopy to a level that comes close to a "spectroscopy of conical intersections", but is problematic with very short-lived excited states.[40] For dissociative processes via ultrafast vibrationally mode-specific nonadiabatic transitions, photofragment imaging[41,42] and, especially, high-resolution H-Rydberg atom translational energy spectroscopy[43] monitor product vibrational state distributions that echo the dynamics in the photoexcited molecules. Where vibronic symmetry selection rules limit the set of allowed product states, the latter method allows researchers to identify the vibrational coupling modes that span a CI along a reaction coordinate,[44] a huge advantage when it comes to rationalizing the observed dynamics on the basis of theoretical models. Femtosecond-resolved mass spectrometry reveals the lifetimes of electronically excited molecules in the gas phase in real time.[45, 46] An illustrative example is the electronic relaxation of hexafluorobenzene from its first $\pi\pi^*$ excited state,[47] where a low-frequency out-of-plane vibration in the populated $\pi\sigma^*$ state is reflected by an oscillating time signal. Moreover, Stolow applied time-resolved molecular beam photoelectron spectroscopy as a nearly ideal means for following ES populations and mapping ensuing nonradiative electronic transformations.[48] Eventually, femtosecond photoelectron imaging spectroscopy is rapidly gaining in popularity as an experimentally very straightforward modern alternative.[49]

The above methods are, however, applicable mostly only to molecules of small to moderate size. When it comes to the photodynamics in large organic chromophores and biomolecules, femtosecond UV/VIS absorption[50–55] and fluorescence[56–60] spectroscopy of the samples in solution invariably become the widely preferred techniques. Both experiments rest on the pump-probe principle as illustrated schematically in Fig. 1: an ultrashort pump laser pulse photo-excites the sample molecules in a thin (< 1 mm) flow cell. The resulting molecular response is probed by a second laser pulse that is temporally delayed with respect to the pump by a well-defined optical path difference. The evolution of the molecular system

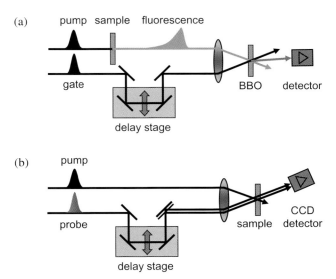

Fig. 1. Basic schemes of (a) a femtosecond fluorescence up-conversion experiment and (b) a pump-probe transient absorption experiment.

following excitation is monitored by taking measurements as function of the pump-probe delay time. The temporal resolution of the experiments thus depends in principle only on the durations of the pump and probe pulses.

The required femtosecond laser pulses are conveniently supplied by a regeneratively amplified Ti:Sa femtosecond laser at $\lambda \approx 800$ nm with ≈ 100 fs pulse duration at 1 kHz repetition rate. Wavelength tunability of the pump and probe pulses for selective excitation and monitoring of different chromophores is achieved using nonlinear optical processes, the simplest being second and third harmonic generation (SHG and THG) of the Ti:Sa laser fundamental. Broadly tunable pump pulses in the VIS and NIR range are readily available from noncollinear optical parametric amplifiers (NOPAs),[61–63] which have the added advantage of a significant shortening of the pulses compared to the Ti:Sa SHG/THG pulses and improving the experimental time resolution. Tunable UV pulses are derived by frequency doubling the NOPA signal or by sum frequency generation (SFG) of the NOPA and the Ti:Sa fundamental or SHG pulses.

Fluorescence up-conversion (Fig. 1a) is the standard approach for femtosecond time-resolved emission spectroscopy.[56,57,59] The temporal resolution is achieved by spatially and temporally overlapping the fluorescence with an intense gate beam, usually the Ti:Sa laser fundamental, in a nonlinear optical crystal that enables SFG. Since the up-converted light intensity is proportional to the fluorescence at the given temporal delay between the pump and gate, detection of the SFG intensity as function of the pump-gate delay allows one to record the fluorescence-time profile. In general, only a limited range of fluorescence wavelengths can be up-converted simultaneously due to the type I or type II phase matching restrictions for the SFG process. The up-converted fluorescence is detected using single-photon counting technology at a selected wavelength through a double monochromator. Time-sliced fluorescence spectra may be derived from the time profiles at discrete wavelengths using a spectral reconstruction technique.[64–66] Broadband up-conversion variants have been developed to cover larger windows of the VIS spectrum,[57,67] but are less common in the UV. Kerr-gating provides a favorable alternative for two-dimensional, simultaneously time- and wavelength-resolved fluorescence detection.[60,68–70]

Broadband transient absorption spectroscopy (Fig. 1b) is ideally suited for the study of organic chromophores which show significant photo-induced changes in their spectra. Spectrally broad probe pulses allowing one to monitor the complete spectro-temporal evolution of a system are generated by super-continuum generation in optically transparent nonlinear media such as sapphire and CaF_2.[54,55,71,72] Whereas fluorescence reflects directly the population in the optically bright ES, the two-dimensional wavelength-time maps normally arise from miscellaneous contributions, namely ground-state bleaching (GSB), excited-state absorption (ESA), stimulated emission (SE), and hot ground state absorption (HGSA) or product absorption (PA) signals, which may superimpose each other more or less. Additional artifacts require careful data analyses, especially for eliminating the cross-phase modulation (XPM) signal from the simultaneous interaction of the pump and probe fields with the sample.[50–52,73] In principle, however, broadband absorption spectroscopy is capable of detecting transient populations of nonfluorescent optically dark states and has the invaluable advantage of allowing one to map the dynamics from the initially ES all the way to the equilibrated GS products. Eventually, given a characteristic vibrational marker mode that is changed by the photo-induced molecular transformation, broadband IR[74] and Raman probes[75] are very powerful additional, but experimentally more demanding tools.

Being ultrafast and intramolecular, the nonadiabatic vibronic dynamics of organic molecules in solution are, in principle, similar to the gas phase. But even in the absence of specific solute–solvent interactions (e.g. H-bond formation), the solvent as polarizable dielectric continuum affects the energies of the excited PE hypersurfaces and may even alter their energetic order. Strong effects arise in the case of excited states with large dipole moments, especially in the case of charge-transfer (CT). Time-resolved pump-probe spectroscopy often reveals such effects.

3. Radiationless Electronic Deactivation in Free DNA Bases

Ultrafast nonradiative electronic deactivation processes in the free DNA bases (Scheme 1) have been surmised for some time from the observed extremely low fluorescence quantum yields ($\leq 10^{-4}$).[76] Time-resolved measurements which became possible only in the last 10 years finally established the ES lifetimes of the bases quantitatively.[12]

Ade has arguably been studied in most detail. It thus defines a benchmark for the analysis of the dynamics in DNA sub-structures like

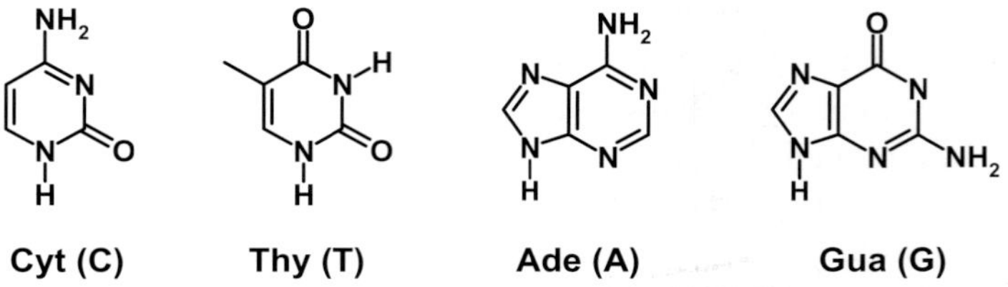

Cyt (C)   **Thy (T)**   **Ade (A)**   **Gua (G)**

Scheme 1.

base pairs and π-stacked dimers or longer oligonucleotides. In the gas phase, Ade exhibits a few resolved vibronic states near its $\pi\pi^*$ origin with lifetimes up to some tens of picoseconds.[77–84] A broad unstructured absorption continuum reflects the onset of ultrafast internal conversion right above.[77,85–87] The dynamics established by transient absorption[88–91] and fluorescence[58,91–94] measurements on Ade and Ado (≡ adenosine) in aqueous solution is illustrated by experimental data[58] in Figs. 2(a) and (b). As all nucleobases, Ade can exist in several tautomeric forms. The naturally occuring "canonical" form is 9H-Ade, but the minor 7H-tautomer coexists in water in equilibrium at room temperature and contributes to the experimental data. The sub-picosecond first time constant of $\tau \approx 0.3$ ps describing the fast initial decay of the measured biphasic time profiles reflects the ES lifetime of the 9H-Ade tautomer, because the same value fits the experimental data for Ado, where the ribose moiety blocks the tautomerization. The almost 30-fold longer second time constant of $\tau = 8.4$ ps that is missing in the Ado case gives the 7H-Ade lifetime.

Figure 2(c) sketches the minimum energy pathways (MEPs) for electronically excited 9H-Ade in accordance with high-level ab initio calculations.[95–109] The theoretical studies have recently been reviewed.[110] The applied pump laser pulses in the strong first UV absorption band around $\lambda = 260$ nm projects the molecules to the second $\pi\pi^*$ ES (L_a), which carries the bulk of the oscillator strength. L_a is dielectrically stabilized by solvation in water compared to the gas phase owing to its high dipole moment so that it becomes almost degenerate with the first $\pi\pi^*$ ES (L_b).[101,106,111,112] The prepared wavepacket departs from the Franck–Condon (FC) region in only some ≈ 100 fs, at the limit of experimental time resolution.[91,93] The main relaxation pathway thereafter takes the wavepacket along a steep PE gradient on the $\pi\pi^*(L_a)$ hypersurface to a direct CI with the GS involving an out-of-plane puckering deformation of the C^2–H unit.[98,99,101,102] Hinting at the partially localized characters of

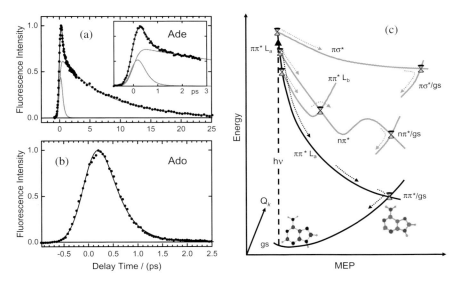

Fig. 2. (a, b) Experimental fluorescence-time profiles for Ade and Ado in water following excitation at $\lambda = 269$ nm.[58] The solid lines give the best fits to the data points using a single exponential (Ado) or two exponentials (9H-Ade and 7H-Ade, gray lines) convoluted with the instrument response function (IRF). (c) Electronic deactivation routes for 9H-Ade. Absorption of a UV photon prepares the molecules primarily in the $\pi\pi^*(L_a)$ state. The main relaxation channel from there leads to a direct $\pi\pi^*(L_a)$/gs CI along a barrierless MEP. Minor secondary pathways involving the nearby $\pi\pi^*(L_b)$ and $n\pi^*$ states and, at higher excitation, the $\pi\sigma^*$ state are sketched in gray.

the unpaired π^* and π electrons at the critical configuration, this unique route has been named the "biradical channel".[104] Although the electronic characters are not easily distinguishable in the region, where the $\pi\pi^*(L_a)$, (L_b) and $n\pi^*$ states cross,[98, 100, 103, 104, 107, 108] consensus has emerged that this biradical CI is reached on a barrier-less pathway, to which the measured $\tau \approx 0.3$ ps ES lifetime is overwhelmingly assigned.

The $\pi\pi^*(L_b)$ and $n\pi^*$ states in 9H-Ade can provide only secondary relaxation pathways (Fig. 2(c), gray lines), which may take the molecules to the GS through less effective other CIs, but may play a larger role in the gas phase. The L_b state is optically accessed from the GS in the red wing of the first absorption band.[111] Participation of a predissociative $\pi\sigma^*$ state involving the N^9–H bond[95–98, 105] or an out-of-plane motion of the NH_2 group[108] appear to require higher initial excitation levels. The $\pi\sigma^*$ state has, however, been claimed to be responsible for a relaxation channel in the gas phase, where multiphoton ionization experiments,[87] photoelectron spectra,[113–115] and H-atom photofragment translational energy measurements[116–118] disclosed bi-exponential decay behaviors with electronic lifetimes of < 0.3 ps. Differences in the dynamics in the gas phase and in solution are expected by the solvent shifts on the excited states.[106, 112]

A highly striking feature of Ade is that only minor modifications of its chromophore strongly alter the electronic dynamics. As mentioned, 7H-Ade for example takes almost thirty times longer ($\tau \approx 8.4$ ps) to return to its GS than 9H-Ade, although it differs only by the position of an H atom. Its $\pi\pi^*(L_a)$ state is supposed to have a plateau-like minimum in the region near the crossings with the $\pi\pi^*(L_b)$ and $n\pi^*$ states,[101, 119] which trap the ES population and let the molecule return to the GS only over a low PE barrier. 2-Aminopurine (2AP) is a highly fluorescent structural isomer of Ade ($\equiv$ 6-aminopurine) that is widely used as a fluorescence marker in DNA studies.[120–125] It has an ES lifetime of ≈ 2 ns and a fluorescence quantum yield of ≈ 0.66.[120, 121, 126, 127] The replacement of the light H atom at C^2 by the heavy NH_2 group raises the biradical CI above the minimum on the $\pi\pi^*(L_a)$ PE hypersurface,[102, 104] preventing an ultrafast return to the GS. The arguably most eye-catching Ade derivative is, however, N^6, N^6-dimethyladenine (DMAde), which emits so-called dual fluorescence.[128–130]

DMAde exists in solution at room temperature virtually only in its 9H form because of the steric demands by the dimethylamino group. Its strong $\pi\pi^*$ UV absorption is very similar to that of Ade. But while the fluorescence of Ade is in the near UV and shows only a small, normal Stokes shift,[93] DMAde exhibits two widely separated emission bands (Fig. 3a, b). The first peaks in the near UV at $\lambda_{fl} \approx 330$ nm, resembles the emission from Ade, and is therefore attributed to the local excited (LE) state associated with the purine chromophore. The main fluorescence band of DMAde is, however, in the visible spectrum, where it has its maximum between $\lambda_{fl} \approx 450$–550 nm depending on the solvent. The visible emission strongly predominates especially in nonpolar aprotic solvents, but the red-shift is most pronounced in more polar solvents. Because it resembles the prototypical case of dimethylaminobenzonitrile (DMABN),[131] the dual fluorescence of DMAde has been rationalized by a transformation from the LE state to an intramolecular charge transfer (ICT) state involving the dimethylamino group and the purine ring as electron donor and acceptor, respectively.[128–130] The question is whether the LE $\rightarrow$ ICT transition in DMAde can be better described as

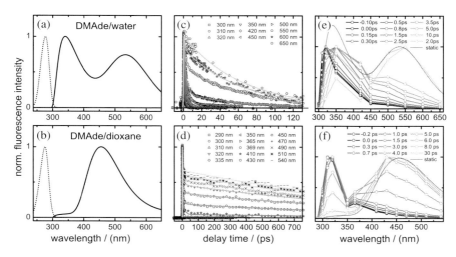

Fig. 3. (a, b) Static absorption (dotted) and fluorescence (solid lines) spectra, (c, d) fluorescence-time profiles, and (e, f) time-sliced fluorescence spectra of DMAde in water and in dioxane, respectively, after excitation of the molecules at $\lambda_{\text{pump}} = 258$ nm.[59,132]

a unimolecular reaction by crossing a transition state at the maximum of a PE barrier on a double-minimum hypersurface, or whether a nonadiabatic transition through a CI provides the more appropriate picture, with a MEP that may even have no PE barrier at all.

Schwalb and Temps investigated the electronic relaxation dynamics of DMAde in water and in dioxane using femtosecond fluorescence up-conversion spectroscopy following UV excitation at wavelengths from close to the electronic origin ($\lambda_{\text{pump}} = 294$ nm) to vibronic excess energies of $\approx 5400\,\text{cm}^{-1}$ above ($\lambda_{\text{pump}} = 258$ nm).[59,132] The measured fluorescence-time profiles (Fig. 3c, d), which they recorded in a wide wavelength range, from $290\,\text{nm} \leq \lambda_{\text{fl}} \leq 650\,\text{nm}$, turned out to reflect a surprisingly complex sequence of distinctive molecular transformations taking place on at least five well-defined time scales. Sub-100 fs and 500 fs lifetimes were found to predominate at the shortest UV and blue emission wavelengths in water, 1.5 and 3.0 ps components at intermediate wavelengths, and a 62 ps value in the red region of the spectrum. In dioxane, these lifetimes changed to < 270 and 600 fs at UV, 1.5 and 11 ps in a wide range of intermediate, and the high value of 1.4 ns at the longest wavelengths. The ensuing sequential relaxation processes are underscored by the time-dependent fluorescence spectra (Fig. 3(e, f)). Within experimental errors, the respective time constants are

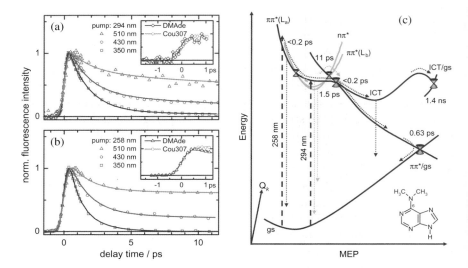

Fig. 4. (a, b) Comparison of the experimental fluorescence-time profiles for DMAde in dioxane following excitation at $\lambda = 294$ and 258 nm at three selected emission wavelengths.[59] The insets show the rise times of the red emission. (c) Proposed electronic deactivation routes (time constants apply to dioxane).

surprisingly independent of the excitation wavelength (Fig. 4(a,b)), even following excitation just above the electronic origin ($\lambda_{pump} = 294$ nm). Moreover, the strongly red-shifted fluorescence was observed without a time delay. Those findings suggest relaxation via practically barrier-less pathways through a hierarchy of CIs. A tentative four-state model for the electronic dynamics in DMAde is sketched in Fig. 4(c). It assumes similar $\pi\pi^*(L_a)$, $\pi\pi^*(L_b)$, and $n\pi^*$ states as in Ade plus an added ICT state that is responsible for the long-wavelength fluorescence and has no counterpart in Ade.

Closely related to Ade is the other purine base, Gua, but experimental[87–89, 92, 93, 133, 134] and theoretical[110, 135–140] studies on this base are not as numerous and are still conflicting. Gua exhibits two somewhat separated absorption bands, which stem from the $\pi\pi^*(L_a)$ and $\pi\pi^*(L_b)$ transitions. Keto-hydroxy and amino-imino tautomers of Gua have been investigated by UV/UV and IR/UV double resonance spectroscopy.[141–144] Moreover, there have been numerous pump-probe investigations of the pyrimidine bases, which show pronounced bi-exponential ES decay profiles in aqueous solution and in the gas phase.[87–90, 92, 93, 145–154] Only Cyt has a molecular beam spectrum with vibronically resolved bands near the $\pi\pi^*$ origin transition.[81] Initial time-resolved experiments[88,89,92,93] pointed at straightforward pico- to sub-picosecond return of the spectroscopically prepared $\pi\pi^*$ ES to the GS, but newer pump-probe absorption measurements, which monitored the refilling of the GS population at shorter UV wavelength than the pump,[148] and by UV pump/IR probe experiments[155,156] revealed the striking existence of a parallel second and significantly slower deactivation pathway on the tens of ps time scale. Significant fractions of population apparently follow that detour.[17,148] Theory pictures a bifurcation of the wavepacket that starts in the FC region of the spectroscopic $\pi\pi^*$ state. A fraction of the population evolves through a direct $\pi\pi^*$/GS CI. The second route takes the molecules to the optically dark $n\pi^*$ state(s).[110,148,157–165] The subsequent deactivation from there to the GS is affected in subtle ways by the molecular substituents and by the solvent environment, hinting at energetic shifts of the $n\pi^*$ state and a PE barrier along the MEP. The pyrimidine bases also receive attention due to the availability of triplet states[148,155,166–168] and their controversial role in the formation of cyclobutane pyrimidine dimers (CPDs), the most prevalent DNA photolesions.[17,169–174]

4. Electronic Deactivation in H-Bridged Base Pairs, π-Stacked Dimers and Oligonucleotides

The shape and function of DNA beyond the primary base sequence is determined by electrostatic stabilization by counterions and the solvent, dipole-dipole and π-interactions among the stacked coplanar aromatic rings of the bases, the H-bridges between the complementary single strands of the double helix, and, eventually, interactions with surrounding proteins. The individual bases linked in a DNA chain in a long sequence therefore cannot be considered in isolation. Pump-probe spectroscopy revealed strong cooperative effects between the neighboring bases in DNA that affect and alter the photobiophysical dynamics in dramatic, previously unforeseen, and still highly controversial manners.[13–17]

Base pairing and base stacking are the two key structural motifs of the DNA double helix. Deciphering the ensuing complex photo-induced dynamics in DNA thus seems impossible without understanding the photophysics in H-bridged base pairs and in stacked nucleobase assemblies. This section gives a status report.

H-bonded nucleobase complexes have been intensely studied in the gas phase by mass spectrometry[175] and by molecular beam UV multiphoton ionization[81,176] and IR/UV double resonance[177–181] spectroscopies. Pioneering work was carried out by the de Vries and Kleinermanns groups, who developed sophisticated pulsed molecular beam sources using gentle laser vaporization schemes to avoid excessive fragmentation of the fragile bases and their nucleosides in conventional thermal vaporization sources.[81,176] The structures of various isomers were distinguished by comparison of the observed vibrational frequencies with quantum chemical predictions. To much surprise, however, the measured spectra lack fingerprints of the naturally occuring "canonical" Watson–Crick (WC) base pairs and all belong to non-WC isomers.[179–181] In the case of A·T, the missing record was rationalized by quantum chemical calculations, which showed that its WC form is not the energetically most favorable structure in the gas phase.[179] But although there are numerous other isomers in reach within the first ≈ 80 kJ/mol of energy, the WC form is the most stable structure of G·C.[180,181] Only recently has a broad unstructured electronic spectrum been tentatively attributed to the G·C WC complex.[181] Abo-Riziq et al., who reported the spectrum, hypothesized that the observed broad bands might be evidence for a much shorter ES lifetime of the G·C WC complex than for the other isomers. This suggestion was supported by theoretical results by Sobolewski et al.[182,183] who computed an ultrafast internal conversion pathway for the electronically excited G·C WC pair by a coupled G-to-C electron-proton transfer mechanism.

The spectra of Abo-Riziq et al.[181] and calculations by Sobolewski et al.[182,183] triggered a run for direct time-domain experimental data on the dynamics of H-bridged nucleobase complexes. Because of the presence of several isomeric complexes in molecular beams, pump-probe measurements by femtosecond mass spectrometry were primarily conducted on nucleobase mimics. Dimeric 2-aminopyridine was the first example found to have a much shorter ES lifetime ($\tau = 65$ ps) than either the monomer ($\tau = 1.5$ ns) or some three- and four-membered larger aggregates,[184,185] presumably by a similar mechanism as in G·C. Homodimers of A and T, complexes with water, as well as A·T base pairs were explored as well.[186–188]

H-bridged base pairs are also produced from the monomers in aprotic solvents like chloroform ($CHCl_3$), which strongly favors the formation of H-bonded dimers, while stacked aggregates are disfavored.[189,190] Following this strategy, Schwalb and Temps reported femtosecond fluorescence up-conversion experiments that revealed the dynamics of the optically excited electronic states of the dimers of G and C by direct means in the time domain.[191,192] Accidentally, CHCl$_3$, which does not lend itself to H-bonding, has a dielectric constant ($\epsilon = 4.9$) similar to that inside the DNA double helix ($\epsilon \approx 3$–5).[193] The H-bonded complexes were prepared in CHCl$_3$ by employing *tert*-butyldimethylsilyl (TBDMS) derivatives of the nucleosides.[194] The products were characterized by FTIR spectroscopy together with quantum chemical vibrational frequency calculations. With an association constant of $K_{G \cdot C} \approx 3.4 \times 10^4 \, M^{-1}$ compared to $K_{C \cdot C} = 43$ and $K_{G \cdot G} = 1000 \, M^{-1}$ ($T = 298$ K), the expected G·C WC pair turned out as the by far most stable H-bonded complex of G and C in CHCl$_3$. The time-resolved experiment[191] recorded the fluorescence decay profiles at $\lambda_\text{fl} = 350$ nm of the molecules and complexes after excitation at $\lambda_\text{pump} = 283$ nm (Fig. 5(a)). As can be seen, the data demonstrate a drastic acceleration of the fluorescence decay for a G-C mixture compared to C or G alone. Under the conditions used, C by itself is present in CHCl$_3$ practically only as monomer, whereas G shows some self-association. Both have fluorescence decay profiles that require two resp. three exponentials for fitting, with time constants between 0.4–1.2 ps, 4–10 ps, and

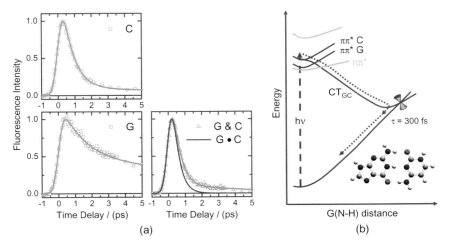

Fig. 5. (a) Fluorescence decay profiles of C and G (left) and an equimolar mixture of C and G (right) in CHCl$_3$ at $c_0 = 0.1$ mM for each nucleoside after excitation at $\lambda_\text{pump} = 283$ nm. The black solid line gives the decay profile for the G·C base pair after correction for the residual monomers.[191] (b) Sketch of the PE profiles of the electronic ground state and the lowest excited states of the G·C WC pair as function of the G imino bond distance.[182,183]

several 100 ps.[191, 192] In contrast, the high degree of association of G·C ($\beta_{\text{G·C}} = 0.59$ at the applied concentrations of $c_0 = 0.1$ mM of C and G) results in an almost single-exponential time profile for the G-C mixture. After correction of the raw data for the residual monomers, the measurement gave a value for the fluorescence lifetime of the G·C WC complex of $\tau_{\text{G·C}} = 300$ fs.

The ultrashort ES lifetime of the G·C WC complex was subsequently confirmed by measurements at other pump wavelengths from close to the electronic origin ($\lambda_{\text{pump}} = 296$ nm) up to considerably higher vibronic excitation levels ($\lambda_{\text{pump}} = 262$ nm) and at other nucleoside concentrations.[192] The additional experiments also suggested substantial changes in the lifetimes for H-bridged self-complexes of G.[192] It is pointed out here that the self-association of G is probably more complex than initially thought. A closer look at FTIR and NMR data indicates that more than one G·G association motif may be encountered and that an association of G with traces of H_2O in the solvent may play a role as well. Further work on the structures and the dynamics of self- and water-complexes of G and C appears to be needed. The complexity is illustrated by the case of C in n-hexane, where large self-complexes (e.g. trimers, tetramers, extended tapes) are formed, which exhibit substantially different vibronic dynamics.[195]

Unambiguously, however, the results for the G·C WC pair[191, 192] demonstrate the operation of an optically dark, nonfluorescent gate state that is rapidly accessed from the optically prepared bright fluorescing ES by an ultrafast nonadiabatic vibronic transition. This radiationless process appears to be active only in the base pair, where it initiates the electronic relaxation, but not in monomeric G or C. The time-resolved fluorescence experiment alone does not reveal the character of the gate state. It is possible, on the one hand, that the excited wavepacket is trapped in some energetically accessible dark state, one that could even be rather long-lived, before it returns to the GS. On the other hand, the coupled electron-proton transfer mechanism proposed by Sobolewski and Domcke[182, 183] provides a highly attractive and consistent picture that allows one to rationalize the broad gas phase spectra[181] as well as the time-dependent solution phase measurements[191, 192] on G·C. As sketched in Fig. 5(b), it is supposed that there is a G-to-C CT state close to the $\pi\pi^*$ optically ES of G·C that is populated without PE barrier. The G-to-C electron transfer is followed by a proton shift along the central N−H···N hydrogen bond, which lowers the ES. The ensuing biradical state leads to a CI with the GS, through which the original WC structure is recovered. Accidentally, the experimentally found ultrashort time scale (≈ 300 fs) has also been seen in Car-Parrinello[196] and in QM/MM dynamics calculations.[197] The reader is referred to Chapter 4 of this book for more details on this coupled electron-proton transfer deactivation mechanism in G·C and in numerous other systems by their discoverers.

A central question arising with these results pertains to role of the coupled electron-proton transfer mechanism to the electronic relaxation in DNA molecules. Most of the work in the literature focused on pure $d(A)_n \cdot d(T)_n$ or $d(G)_n \cdot d(C)_n$ and alternating $d(AT)_n \cdot d(AT)_n$ DNA oligomers,[13–15, 198] where the discovery of comparatively slow GS recoveries from the UV-excited states attributed to the stacking of the bases dominates the ongoing debate. But all DNA studies commonly also show sub-picosecond contributions to the electronic relaxation; these are usually attributed to "unstacked" bases. It has rarely been appreciated that those fast processes could also originate from H-bridging. A recent report of a sizable H/D isotope effect on the electronic deactivation rate in an alternating $d(GC)_9 \cdot d(CG)_9$ duplex[199] thus came as a surprise. The strong acceleration of the fluorescence decay in the alternating $d(AG)_9A \cdot d(TC)_9T$ duplex, which is almost as fast as in the mononucleotides, points at some specific role of G·C pairs in DNA as well.[16] Eventually, it has been noted that a 5-bromouracil tracer separated from a G·C hotspot by a few A·T pairs experiences photo-damage by charge transfer from G in the same strand or in the complementary strand with almost the same efficiency.[200]

Base stacking clearly exerts an overarching influence on the photodynamics in DNA. It is assumed that UV excitation of stacked nucleobases leads to partially delocalized dipole-dipole coupled (Frenkel) exciton states. The initial excitation may therefore extend over two or several adjacent bases.[201–206] As suggested by the slow GS recoveries in DNA[17] and supported by first theoretical results,[206–208] these states may subsequently transform to excimer- or exciplex-like states, which should essentially appear optically dark and evidently return to the GS much more slowly than the free, isolated bases. Simple dinucleotides, where excimer/exciplex states have been surmised from strongly red-shifted emission bands in static fluorescence spectra,[209] constitute ideal model systems for elucidating the vibronic dynamics in stacked nucleobase assemblies. Pump-probe spectroscopy can reveal the lifetimes of the supposed excimer/exciplex states of the dimer and show how the second chromophore in close proximity

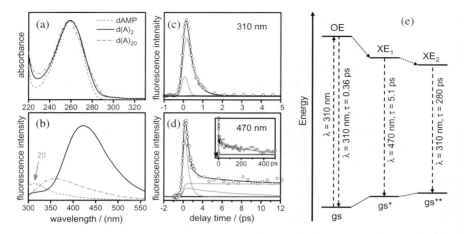

Fig. 6. (a) UV absorption spectra of dAMP, d(A)$_2$, and d(A)$_{20}$. (b) Static fluorescence spectra of dAMP, d(A)$_2$, and d(A)$_{20}$. (c, d) Measured temporal fluorescence decay profiles of d(A)$_2$ at $\lambda_\text{fl} = 310$ and 470 nm (open circles: data points, solid black line: fitted decay curve, thinner gray lines: contributing exponentials). (e) Sketch of the sequential relaxation from the optically excited (OE) state to lower-lying excimer excited (XE) states.

modulates the known CIs and nonadiabatic vibronic deactivation routes of the monomers.

Static absorption and fluorescence spectra and time-resolved fluorescence measurements by Stuhldreier et al.[210] shed light on the electronic dynamics in the adenine dinucleotide d(A)$_2$. As indicated by its circular dichroism (CD),[211] the dinucleotide preferentially adopts a B-DNA like configuration, albeit with much structural flexibility. The UV spectrum of d(A)$_2$ plotted in Fig. 6(a) on close inspection shows the slight blue-shift of the absorption maximum near 260 nm compared to the monophosphate dAMP that is typical for an excitonic state with dipole-dipole coupled transition moments.[201–203] Additionally, the dinucleotide exhibits a continuing weak absorption to the red of the main absorption band. The static fluorescence spectrum given in Fig. 6(b) is dominated by a strongly red-shifted emission band peaking at $\lambda_\text{fl} \approx 430$ nm with much higher intensity than the monomer-like emission around $\lambda_\text{fl} = 320$ nm.

Time-resolved measurements made within a wide range of emission wavelengths ($\lambda_\text{fl} = 290$–470 nm) reflect the underlying dynamics with striking clarity. As can be seen from the data at two representative wavelengths displayed in Fig. 6(c) and (d), the fluorescence decay profile at

$\lambda_{fl} = 310$ nm closely resembles that of dAMP. A global fit gave lifetimes of $\tau_1 = 0.10$ and $\tau_2 = 0.36$ ps. In contrast, the observed multi-exponential decay curve at $\lambda_{fl} = 470$ nm reveals substantially slower dynamics. From a global fit, the long-lived states have lifetimes of $\tau_3 = 5$ and $\tau_4 = 280$ ps. These results, which are in nice agreement with pump-probe absorption data,[212] suggest the stepwise relaxation of the optically excited (OE) $\pi\pi^*$ state(s) of $d(A)_2$ to energetically more favorable excimer-like states (XE_1, XE_2) sketched in Fig. 6(e). These processes evidently take place besides the ultrafast direct electronic deactivation from more monomer-like (unstacked) $\pi\pi^*$ ES via the biradical CI of the Ade. The two excimer states may differ in the relative orientations of the bases. CASSCF/CASPT2 calculations in the author's laboratory suggest that there are four $\pi\pi^*$-like states with substantially mixed electronic characters in the dinucleotide in the vicinity of the optically excited FC state(s), of which at least two have lower energy than the FC state(s) and appear to be accessible without a PE barrier.[213] The four states arise from the $\pi\pi^*(L_b)$ and $\pi\pi^*(L_a)$ states of the two monomers. The lower two are more closely and more strongly bound than the GS, i.e. have a smaller equilibrium distance between the two monomer units and a higher bond energy and may therefore fluoresce only at significantly red-shifted wavelengths. These are main criteria for excimer states. Once the population has been trapped in the PE minimum of the excimer, the biradical CI that is responsible for the ultrafast electronic deactivation in Ade can be reached only via a sizable PE barrier.[207,208] It is this substantial energetic stabilization of the XE_2 state that explains its long (280 ps) ES lifetime.

The photodynamics in longer DNA oligo- and polynucleotides have been explored by time-resolved fluorescence[14,16,214–219] and absorption[13,15,91,198,199,212,219–222] measurements. The review of Middleton et al.[17] may be consulted for deeper insight beyond the scope of this book. As has been recognized, the photo-excited DNA molecules return to their GS in part only after several 100 ps,[13,15,17,212,221] two to three orders of magnitude more slowly than the free nucleobases. To complicate matters further, the lifetimes of the optically bright states of DNA exhibit a pronounced base sequence dependence.[16] Additionally, as the base sequence induces higher-order structure, various secondary structural effects on the electronic dynamics were revealed.[16] Not surprisingly, the diversity of these effects results in very complex ES properties and dynamics. For instance, base stacking interactions seem to be of major importance especially in pure $d(A) \cdot d(T)$ or $d(AT) \cdot d(TA)$ strands,[13–15,91,198,219] quadruplex formation and G-stacking mechanisms seem to affect the dynamics in d(G)·d(C) repeats,[16,223,224] whereas inter-strand coupling through G·C H-bonds may accelerate electronic relaxation in alternating d(AG)·d(TC)[16] and d(GC)·d(CG)[199] runs.

In principle, the photophysical and photochemical dynamics in DNA should be derived from the active primary mechanisms seen in the basic structural motifs: monomeric nucleobase chromophores, stacked assemblies such as di- and trinucleotides, and H-bridged base pairs. Conical intersections likely govern site-specific electronic deactivation channels of localized excited states in DNA as they do in the elementary building blocks, but the cooperative coupling effects modulate the ES PE surfaces extensively. The well-stacked coplanar aromatic ring systems of the bases in an extended DNA strand experience π-orbital overlap as well as dipolar couplings not only within one strand, but also between the two strands of the double helix. The delocalized nature of the resulting photo-excited states in DNA is as yet far from being understood. How delocalized excitations in specific base sequences are trapped depends, among other factors, on the oxidation potentials of the different bases. The reader is referred to the extensive literature on charge transport in DNA for insights.[225,226] Last but not least, a DNA molecule is not static but highly flexible. Sizable structural fluctuations will break the otherwise ensuing cooperative mechanisms. Interestingly, UV/IR pump-probe studies of photo-induced thymine CPD formation[171,172] show that these harmful photolesions can be formed via the singlet channel on the ≈ 1 ps time scale, but the two reacting bases must have the right pre-orientation with respect to each other for the cyclization to occur. Since this is not the case in the regular B-DNA configuration, the small CPD quantum yield is likely due to the small fraction of adjacent pyrimidine bases with the required conformation.

5. Conformer-Specific Photochemistry of Furylfulgides

Fulgides and their derivatives are important photochromic switches, because they are thermally irreversible and can be switched back and forth purely photochemically by light of two different colors.[29,32,227] This makes them ideal candidates for numerous applications, including light-driven molecular machines,[20] optical memories,[23,24,228,229] electron/energy transfer modulators,[230,231] nonlinear optical materials,[232] and photo-switchable fluorescence imaging beacons.[25-28]

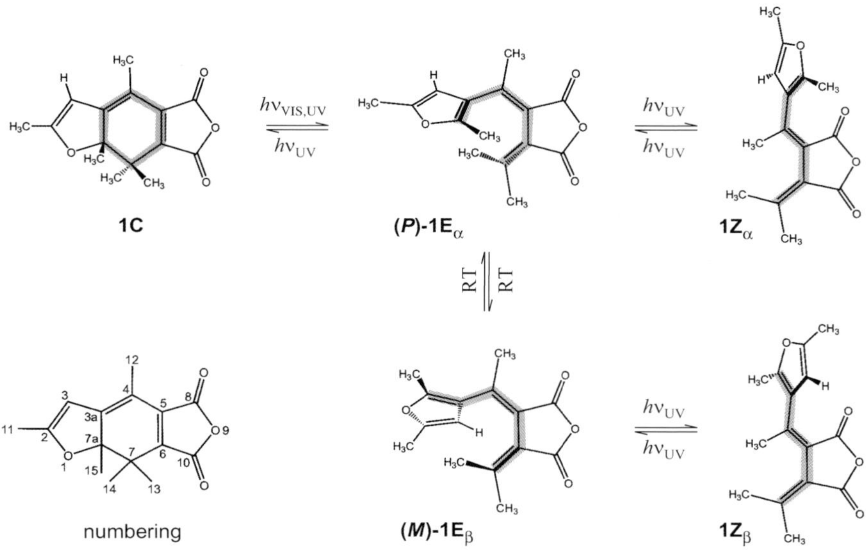

Scheme 2.

Scheme 2 shows the reaction cycle for the prototypical furylfulgide **FF1** (2-[1-(2,5-dimethyl-3-furyl)-ethylidene]-3-isopropylidene succinic anhydride) that is highlighted in this section. The photochromism of **FF1** rests on the UV-initiated conrotatory electrocyclic $E \to C$ ring closure of the open **FF1E** to the cyclic **FF1C** isomer, and the reverse $C \to E$ conrotatory ring-opening that can be achieved by irradiation of **FF1C** with visible light. The competing $E \to Z$ isomerization is a usually unwanted side reaction that also occurs upon irradiation with UV light[32, 233] and can have a strong impact on the functionality in practical applications with many switching cycles. Because of its roughly comparable $E \to C$ and $E \to Z$ quantum yields of 20 resp. 12%,[234] furylfulgide **FF1** offers a unique opportunity for studying the molecular mechanisms of the $E \to C/E \to Z$ branching. This is not only of practical, but also of considerable fundamental interest, for example regarding the role of CIs and the location of the branching point on the ES PE surface[45, 235–238] and the ability to influence the branching by coherent control techniques.[239, 240]

As can be seen from Scheme 2, **FF1E** exists as two conformers ($\mathbf{E}_\alpha$ and $\mathbf{E}_\beta$). Both are helically chiral. Thermal equilibrium by enantiotopomerization between the two conformers is established at room temperature via rotation about the C^4-C^5 single bond.[32, 241] This is indicated in

Scheme 2 by (P)-**1E**$_\alpha$ and (M)-**1E**$_\beta$ (the other enantiomeric forms and the chirality of **1C** have been left out for clarity). Only the **FF1E**$_\alpha$ conformer can react to **FF1C**, but both (E)-rotamers can form their respective (Z)-variants. This poses the question, whether the observed branching is due to conformer-specific photoreactions of **FF1E**$_\alpha$ and **FF1E**$_\beta$, or whether the two reaction pathways start from a common origin in the **FF1E**$_\alpha$ ES.

Exploiting the unique photophysical and conformational properties of furylfulgide **FF1**, Siewertsen et al. investigated the role of conformer-specific photochemistry and the ultrafast dynamics of the competing reactions in this photochrome by femtosecond broadband absorption and fluorescence up-conversion spectroscopy.[54] The complex transient absorption changes following excitation of **FF1E** in n-hexane solution at $\lambda_{\text{pump}} = 335$ nm are displayed in Fig. 7(a). They are caused by several superimposed wavelength-dependent and temporally varying contributions: The negative transient absorption at probe wavelengths of $\lambda_{\text{probe}} \leq 365$ nm arises mainly from the GSB and possibly SE of the (E)-isomer with overlapping positive contributions from ESA at short delay times. At wavelengths of $\lambda_{\text{probe}} \geq 365$ nm, two very broad and overlapping ESA bands with maxima at $\lambda_{\text{probe}} \approx 390$ and ≈ 500 nm are observed. Their instrument-limited rise times and subsequent ultrafast decays indicate extraordinarily short electronic deactivation times, for a molecule the size of **FF1**. The ESA decay leaves a broad positive HGSA band from vibrationally excited (E)-, (Z)- and (C)-molecules after transformation to their electronic ground states. As is visible at $\lambda_{\text{probe}} \leq 450$ nm, the dynamics are accompanied by superimposed weak damped oscillations. Most importantly, however, the signature of the (C)-product that dominates at $\lambda_{\text{probe}} \geq 450$ nm is fully established within $\Delta t < 0.25$ ps, confirming the ultrafast electronic deactivation time scale deduced from the ESA decay. HGSA at shorter wavelengths ($\lambda_{\text{probe}} \approx 360$ nm) is mainly due to the (Z)-product and occurs slightly later. The subsequent minor and slower spectro-temporal changes reflect vibrational cooling in the electronic ground state, leading to rising absorptions by thermalized products (PA). The permanent absorption changes established in the transient spectrum taken at $\Delta t = 20$ ps (Fig. 7b) can be described well by a weighted sum of the steady-state UV/VIS absorption spectra of all three isomers with a $(C) : (Z)$ product ratio of $\approx 2 : 1$ in excellent agreement with the published isomerization quantum yields.[234]

The spectro-temporal evolution map demonstrates impressively how the $E \rightarrow C$ and the $E \rightarrow Z$ isomerization reactions of furylfulgide **FF1** occur in

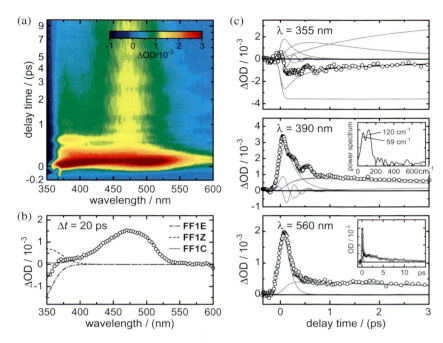

Fig. 7. Transient absorption changes ΔOD after excitation of **FF1E** at $\lambda_{\text{pump}} = 335$ nm.[54] (a) Two-dimensional map for probe wavelengths between 350 nm $\leq \lambda_{\text{pump}} \leq$ 600 nm and delay times up to $\Delta t = 10$ ps. (b) Deconvolution of the transient spectrum at $\Delta t = 20$ ps into the contributions from the (E), (Z), and (C) isomers. (c) Time profiles at $\lambda_{\text{probe}} = 355$, 390, and 560 nm and their non-linear least-squares fits (solid black lines). The contributions from GSB, ESA/SE, HGSA, PA, and the damped oscillations are shown by thin gray lines. The inset for $\lambda_{\text{probe}} = 390$ nm shows the Fourier power spectrum of the oscillating parts.

parallel and on sub-picosecond time scales. Quantitative insight is provided by a global least-squares fit to the data. A consecutive kinetic scheme,

$$A \xrightarrow{+h\nu} A^* \xrightarrow{\tau_1} B \xrightarrow{\tau_2} C,$$

is suggested, where A^* denotes molecules in the ES, B vibrationally hot species after deactivation to the GS, and C the **FF1C**, **FF1Z**, and **FF1E** molecules after their eventual vibrational cooling. Time constant τ_1 represents the decay of ESA/SE and rise of HGSA, time constant τ_2 likewise describes the HGSA decay and PA rise.

Representative time profiles at three probe wavelengths, $\lambda_{\text{probe}} = 355$, 390, and 560 nm, and the fits to the data are shown in Fig. 7(c). The main results can be summarized as follows: (1) The data at some wavelengths (e.g. $\lambda_{\text{probe}} = 355$ nm) are highly complex due to contributions to the transients from all three isomers (cf. Fig. 7b). To disentangle the spectrally overlapping components and produce deconvoluted data, the amplitude ratios of the PA contributions and the initial ($\Delta t = 0$) GSB were therefore fixed using the stationary spectra and known quantum yields. (2) It turned out that the spectral features reflecting the formation of **FF1C** and **FF1Z** could not be described by single values for τ_1 and τ_2. This clearly indicates different dynamics to those products. (3) In particular, the profiles at $\lambda_{\text{probe}} \leq 450$ nm were described with an ESA decay time of $\tau_1 = 0.25$ ps and a second decay time $\tau_2 = 4.0$ ps. The latter is obviously related to vibrational cooling of the (Z)-product. Time constant τ_1 thus applies to the $E \to Z$ isomerization reaction. (4) The two superimposed oscillations reflect a more complex than a simple one-dimensional motion of the ES wavepacket that is damped with τ_1. The fitted frequencies of $\nu = 64$ and $114\,\text{cm}^{-1}$ agree well with respective values from a Fourier power spectrum (see inset in Fig. 7(c) for $\lambda_{\text{probe}} = 390$ nm). (5) At $\lambda_{\text{probe}} \geq 450$ nm, the fits yielded a significantly shorter ESA decay value of $\tau_1' = 0.10$ ps compared to the above τ_1 and also a longer a vibrational cooling time of $\tau_2' = 7.5$ ps. The simultaneous HPA rise from the (C)-product with τ_1' indicates that the $E \to C$ isomerization happens on that time scale. (6) Final evidence supporting the conclusions stems from a fluorescence-time profile of **FF1E** recorded at $\lambda_{\text{fl}} = 525$ nm following 387 nm excitation, which gave a lifetime for the optically bright ES of **FF1E** of $\tau \approx 0.08$ ps, barely less than τ_1'. (7) There is no evidence for an additional slow (16 ps) indirect $E \to C$ isomerization pathway proposed in a previous study of **FF1E**.[242]

All results together within reasonable experimental errors ($\leq 20\%$) clearly indicate that the photo-excited (E)-isomer of furylfulgide **FF1** shows parallel $E \to C$ ring closure in only ≈ 0.1 ps (τ_1') and $E \to Z$ isomerization in ≈ 0.25 ps (τ_1) with a 2 : 1 product ratio. There are two feasible explanations for the distinct time constants. First, the observed ultrafast dynamics might be explained by a branching of reaction pathways with a common origin in the FC region of **FF1E**$_\alpha$ (**FF1E**$_\beta$ does not seem to be a sensible starting structure for the $E \to C$ photoisomerization, cf. Scheme 2). In that case, the distinctive ESA bands must arise from different regions of the ES PE hypersurface en route to two different CIs to the products to give two different time constants, since ESA from the FC region should kinetically decay with a single first-order time constant. A similar case of branching has been supposed for, e.g. the competing ultrafast *cis-trans* isomerizations and ring closures of cyclohepta-1,3-diene and cyclo-octa-1,3-diene.[237] Alternatively and likely, given the observed kinetics, the two products are due (mostly) to conformer-specific photoisomerizations of the **FF1E**$_\alpha$ and **FF1E**$_\beta$ rotamers, i.e. the two time constants τ_1 and τ'_1 belong to two entirely separate ES reaction pathways, one leading from photo-excited **FF1E**$_\alpha$ to the (C)-isomer, the other leading from **FF1E**$_\beta$ to the (Z)-isomer. The latter scenario requires a comparable stability with a thermal equilibrium between the conformers in the electronic ground state. This is supported by ^{1}H and ^{13}C NMR data in agreement with literature.[54, 241, 243] Additionally, density functional theory (DFT) calculations at the B3LYP/def2-TZVP level showed **FF1E**$_{\beta,\text{eq}}$ to be less stable by $\Delta E = 0.03$ eV than **FF1E**$_{\alpha,\text{eq}}$, which should thus be the dominant conformer ($\approx 70\%$, assuming equal entropies). As in the similar cases of, e.g. *cis*-1-(2-naphthyl)-2-phenylethene or previtamin D,[244, 245] the conformer-specific photochemistry relies on the so-called NEER principle (nonequilibration of electronically excited rotamers).[246] Because this appears reasonable in view of the fast reaction times, the pump-probe results are consistent with conformer-specific photoreactions of **FF1E**. That conclusion, however, does not rule out a parallel $E \to Z$ isomerization reaction of the **FF1E**$_\alpha$ conformer in addition to the $E \to C$ ring closure; such a competition is indeed suggested by dynamics calculations.[247]

The hexatriene/cyclohexadiene (HT/CHD) system[235, 248–251] has been considered as model for the $E \to C$ ring closure of fulgides or fulgimides.[54, 238, 252–259] The $E \to Z$ photoisomerizations of ethylene[260] and polyenes[261, 262] likewise are prototypes for the fulgide $E \to Z$ transformation. In the HT/CHD example, it is assumed that the excitation leads to the FC region of the optically bright $1B_2$ state, from where the ES wavepacket first crosses to the optically dark $2A_1$ state via a conical intersection CI$_1$, before it reaches a second intersection CI$_2$ with the ground state ($1A_1$).[235, 248–250, 261, 262] The extended and branched π-electron system of the furylfulgide is significantly more complex, however, casting some doubt on the applicability of the HT/CHD model.

Better insight into the photoreaction pathways of the fulgides was gained by time-dependent density functional (TDDFT) calculations at the B3LYP/def2-SVP level and second-order approximate coupled-cluster (CC2/def2-SVP) calculations.[54] A geometry optimization found a direct, barrier-less pathway from the vertically excited **FF1E**$^*_{\alpha,\text{FC}}$ structure to a S_1/S_0 CI at a C^7–C^{7a} distance of $r \approx 2.3$ Å, much shorter than the initial 3.6 Å and about halfway to the final 1.6 Å in the **FF1C** product.

The observed damped oscillations suggest that the motion of the excited wavepacket along the PE gradient also involves low-frequency torsional degrees of freedom. The ES of the (C)-isomer has a shallow PE minimum, but the above CI is lower in energy than the **FF1C*** FC state and may therefore also be involved in the photo-induced $C \to E$ ring opening.[54] Those results were fully confirmed by a following CASPT2//CASSCF ab initio study,[263] which also showed that the S_1 state of the fulgide is zwitterionic and does not correspond to the optically ES in the HT/CHD case.

The vibronic dynamics of furylfulgide **FF1** may, however, be even more complex. An inspection of its π-electron system reveals that the molecules may in principle have three potential reaction sites, each one connected with a specific CI that connects the excited PE surface to a respective product or mediates electronic deactivation back to the GS: (a) The central HT system entertains the electrocyclic $E \to C$ ring closure between C^7 and C^{7a}, (b) the ethylenic double bond between C^4 and C^{3a} offers the $E \to Z$ isomerization pathway via a roughly 90° twisted configuration, and (c) the isopropylenic double bond between C^6 and C^7 may provide an additional $E \to Z$ isomerization route for this group via its respective 90° twisted configuration (cf. Scheme 2). Since electronic deactivation to the GS can compete through all three CIs, and the isopropylidene E/Z isomerization (c) always leads to deactivation because its product is indistinguishable from the reactant, one can qualitatively understand the relatively poor photoreaction quantum yields of furylfulgide **FF1**. Dynamics calculations elucidating the routes that the photo-excited molecules follow are on the way.[247]

Last but not least, the insight into conformer-specific photochemistry gained from the pump-probe experiments opens new horizons for chemical syntheses of advanced fulgides with improved photoswitching properties. Sterically demanding substituents suppress the β-conformers and enforce ring closure reactions.[264] Another route is opened by the stabilizing electron donor and acceptor characters of the furyl and anhydride moieties, respectively, on the zwitterionic S_1 state.[263]

6. Ultrafast Photoisomerization of Azobenzenes

The $E \to Z$ and $Z \to E$ photoisomerization reactions of azobenzenes (ABs) form the basis for numerous applications of photochromic switches relying on large changes of molecular shape, size, or dipole moment and on the low photochemical fatigue of ABs.[18, 20, 265–267]

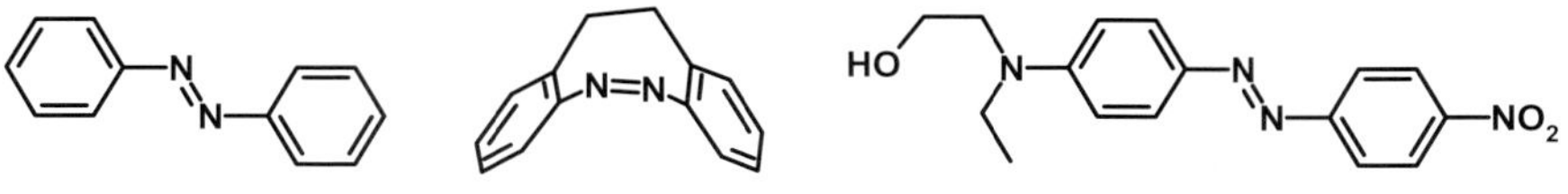

AB **brAB** **DR1**

Scheme 3.

Scheme 3 shows the structures of some prototypical compounds as examples. The respective phototransformations are affected by the topologies of the molecules, as in the heavily conformationally restricted bridged **brAB** or other cyclic ABs, by additional substitutents as in electron donor-acceptor substituted AB dyes like Disperse Red 1 (**DR1**), by the surrounding solvent, or by steric constraints and external forces, when the AB chromophore is embedded in supra- or macromolecular architectures, in cyclic peptides and DNA strands, or attached to surfaces.[268–280] For a rational design of optimized functional AB devices, it is mandatory to have detailed knowledge of the ensuing molecular dynamics under these different circumstances.

Pump-probe studies of the parent compound AB have shown that the isomerization quantum yield and mechanism are closely related.[267, 281–292] For (E)-AB in solution, the quantum yields for visible $S_1(n\pi^*)$ excitation are higher than for $S_2(\pi\pi^*)$ excitation in the UV, but they become wavelength-independent, when the free rotation about the central NN-bond is hindered.[293–295] This was long interpreted in favor of a dual isomerization mechanism with in-plane "inversion" in the S_1 state and "rotation" in the S_2 state,[293, 294, 296] but the previously accepted view has been contradicted in recent years. Time-resolved studies with S_1 excitation of (E)-AB yielded characteristic times of ≈ 0.3 ps assigned to the initial departure from the FC region on the S_1 PE hypersurface and ≈ 3 ps for the actual isomerization.[282, 285–287] Direct evidence for the rotatory pathway was provided by time-resolved fluorescence anisotropy measurements in n-hexane as solvent.[287] For (Z)-AB, an even faster reaction time (≈ 1 ps) was found.[286, 297] The ultrafast molecular dynamics was rationalized by a barrier-less rotational pathway involving a S_1/S_0 CI,[289–292, 298–302] but an additional concerted inversion channel involving the in-plane symmetric CNN-bending mode was proposed as well.

After S_2 excitation, AB first undergoes ultrafast $S_2 \rightarrow S_1$ conversion and then also isomerizes via the S_1/S_0 CI.[281–283, 290, 303–306] The lower isomerization quantum yields for S_2 compared to S_1 excitation are attributed to differences between the initial geometries in the S_1 state in both cases.[290, 305, 306]

Significant increases of the isomerization time up to an order of magnitude were measured for azobenzophanes[307] and an azobenzene capped with a crown ether,[308] where rotational freedom should be restricted. For the latter, a large-amplitude rotation of the phenyl rings was ruled out by comparison with a chemically similar open derivative.[308] This is in line with calculations, where the constraints in these hindered ABs were shown to only impede the large-scale rotation, but not the torsion about the CNNC dihedral angle along a more involved isomerization coordinate.[288, 309] Restriction of the opening of the NNC angle rather than rotation was suggested to lead to higher quantum yield optical molecular switches.[288, 290, 309]

The bridged molecule **brAB** (Scheme 3) is an interesting new model system, where the short ethylenic bridge in *ortho*-position of the phenyl rings restricts the conformational freedom much more than in previously studied hindered ABs. Highly unusual, **brAB** shows a greater thermodynamic stability of its (Z)-isomer due to the large ring strain for the (E)-isomer.[310] Both structures are displayed in Fig. 8(a). Moreover, unlike normal AB, where the $n\pi^*$ absorptions of both isomers virtually coincide, the (Z)- and (E)-isomers have well-separated $n\pi^*$ absorption bands. As shown in Fig. 8(b), the respective absorption maxima are at $\lambda = 404$ nm and at $\lambda = 490$ nm. Eventually, the photoisomerization quantum yields, $\Phi_{E \to Z} = 0.50$ and $\Phi_{Z \to E} = 0.72$, are remarkably higher than for AB, where $\Phi_{E \to Z} = 0.24$ and $\Phi_{Z \to E} = 0.53$.[294] Thus, the heavily constrained molecule has much enhanced photochromic properties, while it still shows low photochemical fatigue.[310] Its structure appears ideal for the development of a molecular tweezer.

Motivated by the drastic changes of its static properties with respect to AB, Siewertsen *et al.*[311] studied the $Z \to E$ and $E \to Z$ isomerization dynamics of $n\pi^*$-excited **brAB** in *n*-hexane as solvent by femtosecond pump-probe spectroscopy and by quantum chemical calculations. The observed transient absorption changes following excitation of (Z)-**brAB** at $\lambda_{\text{pump}} = 387$ nm and of (E)-**brAB** at $\lambda_{\text{pump}} = 490$ nm are compared in Figs. 8(c), (d) and 8(e), (f).

The spectro-temporal maps (Fig. 8(c) and (e)) both reveal negative transient absorption contributions from GSB, which mirror the static absorption spectra of the respective isomers and occur at probe wavelengths

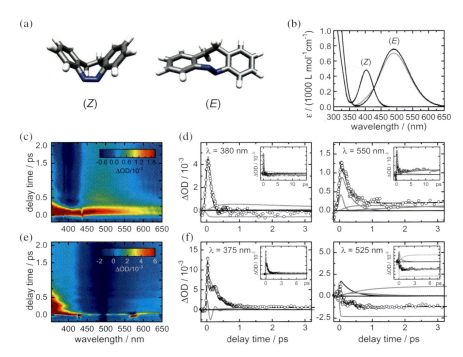

Fig. 8. (a) DFT structures of the (Z)- and (E)-isomers of **brAB**. (b) UV/VIS absorption spectra of the (Z)- and (E)-isomers of **brAB** (black lines). The gray line shows the spectrum in the photostationary state at $\lambda = 385$ nm.[310] (c, d) Transient absorption changes (ΔOD) following excitation of (Z)-**brAB** at $\lambda_{pump} = 387$ nm.[311] (e, f) Transient absorption changes (ΔOD) following excitation of (E)-**brAB** at $\lambda_{pump} = 490$ nm. Panels (c) and (e) show two-dimensional plots of ΔOD for probe wavelengths from 355 to 650 nm and delays between −0.2 and 2.1 ps, panels (d) and (f) display the extracted transient absorption time profiles at selected probe wavelengths (open circles) and the global non-linear least squares fits to the data (black lines; gray lines indicate the different components).

between 360–440 nm in the (Z)-case and as a very broad GSB band betweeen 425–600 nm in the (E)-case. ESA leads to positive transient absorptions over the entire probe wavelength range for both isomers at early delay times ($\Delta t \lesssim 0.2$) ps. Its amplitude is largest for $\lambda_{probe} < 400$ nm and decays within only a few 100 fs in both cases. The transient absorption following excitation of (Z)-**brAB** shows a positive absorption band from 450 nm $\leq \lambda_{probe} \leq$ 600 nm caused by the HGSA of the (E)-product. The corresponding band after (E)-**brAB** excitation is centered around $\lambda_{probe} \approx 400$ nm. The rise of those bands matches the ESA decays in

Table 1. Parameters derived from a global fit to the absorption-time profiles for the (Z)- and (E)-isomers of **brAB**.[a,b]

Reaction	λ_{pump}/nm	τ_1/fs	τ_2/fs	τ_3/ps	ν_{osc}/cm^{-1}	τ_d/ps
$Z \to E$	387	70(1)	270(30)	5(1)	—	—
$E \to Z$	490	< 50	320(100)	5[c]	113(12)	0.13(7)

[a] The amplitudes of GSB and PA were fixed using the known steady-state UV/VIS spectra and quantum yields.
[b] 2σ standard deviations in parentheses.
[c] Fixed.

both cases, proving the sub-picosecond time scale of the $Z \to E$ and $E \to Z$ isomerizations. Slower spectral evolution towards the final spectra at $\Delta t = 20$ ps reflects vibrational cooling.

The parameters[311] extracted by global fitting to the absorption-time profiles at selected wavelengths (Fig. 8d and f) are compiled in Table 1. In both cases, an ultrafast and only partially resolved sub-100 fs decay component (τ_1) and a sub-picosecond second decay component ($\tau_2 \approx 300$ fs) were found.

Since time-resolved fluorescence up-conversion measurements with excitation at $\lambda_{pump} = 387$ nm (Z) and 500 nm (E) gave decay constants $\tau_{fl} < 150$ fs within the available time resolution, τ_1 is attributed to the initial departure of the FC region, and τ_2 to the subsequent transition to the GS through the presumed S_1/S_0 CI. This assignment implies that the $Z \to E$ and $E \to Z$ photoisomerizations happen within ≈ 300 fs. Thus, the photoreaction of the (Z)-isomer of **brAB** occurs at a similar rate as in (Z)-AB, but the photoisomerization of the (E)-isomer is accelerated compared to (E)-AB six to eight times. This contrasts with the substantial slowing of the photoisomerization by a factor of ≈ 6 relative to (E)-AB in the rotation-restricted crown ether-capped AB.[308]

Taking into account theoretical studies on azobenzophanes and a capped AB, which show that a "rotatory" isomerization pathway is possible in restricted AB derivatives,[288,309] the observed isomerization times suggest a torsional type of mechanism for both **brAB** isomers despite the severe constraints. CASPT2//CASSCF calculations for the first ES gave vertical excitation energies of 2.90 eV (427 nm) for the (Z)-isomer and 2.43 eV (511 nm) for the (E)-isomer, in excellent agreement with the measured UV/VIS absorption maxima at 404 and 490 nm. The optimized ES reaction pathways lead to a S_1/S_0 CI at a twisted structure with a dihedral CNNC angle slightly larger than 90°. That CI can be reached directly from the $n\pi^*$-excited (Z)- and likewise from the $n\pi^*$-excited (E)-molecules along steep downhill PE gradients. The torsion of the central NN-bond is accompanied by simultaneous structural changes of other coordinates. The very efficient photochemical funnel provided by that CI along with the forces due to the steep PE gradients explains the high quantum yields and short isomerization times.[311]

Unlike for unsubstituted AB and the cyclic ABs, the dynamics of push-pull ABs have as yet been much less well established, despite several time-resolved studies.[312–315] **DR1** (Scheme 3) is of particlar interest in this context because it can be easily covalently linked to methacrylate polymers by copolymerization in miniemulsions, where it maintains its photoswitching properties.[316] The electron donor-acceptor substitution induces a strong CT character in the $\pi\pi^*$ ES that leads to a large red-shift of its intense absorption band such that it overlaps with the weaker $n\pi^*$ transition.[313,315,317,318] In this situation, the correct energetic order of the $n\pi^*$ and $\pi\pi^*$ excited states has long been uncertain, but newer work supports the $S_1(n\pi^*) < S_2(\pi\pi^*)$ sequence,[315,317,318] unless the solvent forms strong H-bonds with the solute. In any case, both states are very close and radiationless interconversion between them should be feasible and fast.

Bahrenburg et al.[319] recently initiated a comprehensive study of the ultrafast photo-induced dynamics of **DR1**. Typical transient absorption and fluorescence data for **DR1** in solution in 2-fluorotoluene (2FT) at room temperature after excitation at $\lambda_{\text{pump}} = 475\,\text{nm}$ illustrating her work are displayed in Fig. 9. The experimental spectro-temporal evolution map (Fig. 9a) showcases a pronounced negative transient absorption band due to the GSB around $\lambda_{\text{probe}} = 475\,\text{nm}$ and positive transient absorptions due to ESA elsewhere. The rise of the ESA band is instrument-limited only for probe wavelengths around $\lambda_{\text{probe}} = 560\,\text{nm}$, but slightly delayed at other wavelengths. This is clearly evident from the transient spectra at the shortest delay times in Fig. 9(b) and strongly suggests a step-wise isomerization and electronic deactivation. ESA decay, HGSA rise, and a partial refilling of the GSB happen in ≈1 ps and reflect the electronic deactivation. The slower HGSA decay is due to vibrational cooling in the GS.

The measured fluorescence-time profiles (Fig. 9c) also decay bi-exponentially within ≈1 ps. The rapidly decreasing amplitude of the faster component with increasing emission wavelengths suggests that the fluorescence stems from two different excited states. The energetically higher one shows the faster decay and probably populates the lower one by an ultrafast internal conversion through a conical intersection CI_1. Conical intersection CI_2 afterwards mediates the return to the GS and formation of the $E \rightarrow Z$ isomerization product.

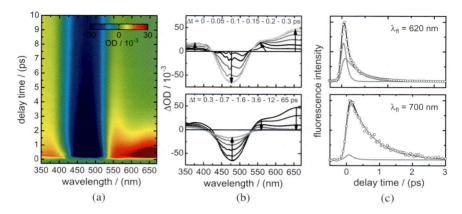

Fig. 9. Time-resolved transient absorption and fluorescence data following excitation of **DR1** in 2FT at $\lambda_{\text{pump}} = 475$ nm.[319] (a) Spectro-temporal evolution map, (b) transient spectra at selected times (the arrows indicate the temporal evolution), (c) fluorescence decay profiles (circles) at $\lambda_{\text{fl}} = 620$ and 700 nm and global non-linear least squares fits to the data (solid black lines, participating exponentials in gray).

Global fits to the transient absorption and fluorescence time profiles with a consecutive kinetic model yielded two time constants, $\tau_1 = 0.10$ ps and $\tau_2 = 1.0$ ps related to the ES dynamics (each with $\pm 15\%$ error limits) and a third time constant $\tau_3 = 7.4$ ps describing the vibrational cooling. The first two time constants and the virtually identical ES dynamics observed in absorption and in emission strongly support a sequential isomerization/ electronic relaxation mechanism. It is concluded that the excitation pulse projects the **DR1** molecule to the FC region of the optically bright $\pi\pi^*$ state (S_2). Internal conversion through CI_1 then takes the excited wavepacket to the energetically lower $n\pi^*$ state (S_1) in only ≈ 100 fs (τ_1). This is followed by isomerization and electronic deactivation through CI_2 to the GS in 1 ps (τ_2). This scenario, which differs from an earlier proposal,[313,314] resembles that of a previous study at lower time resolution,[315] is in accordance with the apparent energetic ordering of the excited states deduced from static fluorescence spectra,[317,318] and is supported by theoretical calculations on model compounds.[320,321]

7. Conclusions

The selected examples highlighted in this chapter reveal the power of UV/VIS pump-probe absorption and fluorescence spectroscopy for elucidating ultrafast photo-induced transformations of large organic chromophores and biomolecules in solution. The examples are of outstanding interest on the one hand for the stability of the genomic information of life on Earth (DNA building blocks) and on the other hand because of their huge application potential (molecular switches).

The selection by nature of Ade, Cyt, Gua, and Thy for its genetic code has been thought of as being a consequence of the evolutionary pressure caused by the high intensity of UV radiation at the origin of life. Pump-probe measurements established the ultrashort excited-state lifetimes of the free nucleobases. Distinctive nonadiabatic electronic deactivation pathways have been identified in the purines and in the pyrimidines. The ensuing photobiophysical dynamics in a DNA oligo- or polynucleotide are, however, dramatically altered. Base pairing and base stacking are the two key structural motifs of the DNA double helix, and pump-probe spectroscopy allowed us to decipher the ensuing mechanisms in isolated H-bridged base pairs and in stacked dinucleotides as relevant DNA building blocks.

The studies of photochromic molecules contribute towards the rational design of improved light-driven molecular switches by understanding the details of the underlying ultrafast photodynamics and the factors affecting it. The results for the prototypical furylfulgide highlight the role of conformer-specific photochemistry and are of fundamental interest for the understanding of competing ultrafast reactions in polyatomic molecules, with implications for applications and coherent control of chemical reactivity. They indicate that favorable shifts of conformer equilibria, e.g. by sterically demanding substituents, may lead to optimized compounds with higher quantum yields. For the push-pull azobenzene **DR1**, a sequential isomerization and electronic relaxation is suggested. As demonstrated for a bridged azobenzene, steric effects may accelerate the molecular dynamics and increase the switching efficiency contrary to chemical intuition. Restricting molecular degrees of freedom may provide improved switches, and ultrafast pump-probe spectroscopy helps to find the relevant ones.

Acknowledgments

The work described in this chapter has been supported by the Deutsche Forschungsgemeinschaft through individual research grants and the Sonderforschungsbereich 677 "Function by Switching" and by the Fonds der Chemischen Industrie. The authors would like to express their sincere thanks to all former and present femto-group members, who made substantial contributions over the years, especially Drs. Thomas Pancur, Harald Studzinski, and Magdalena Foca, who set up much of the equipment in the Kiel Ultrafast Spectroscopy Laboratory. Most of the work described in the text has been carried out in the frame of the Ph.D. projects of N.K.S. and R.S. Mayra Stuhldreier, Katharina Röttger, Julia Bahrenburg, and Carmen Schüler contributed some newer pieces, which we credited in the text. Over the years, furthermore, we have enjoyed many stimulating discussions with friends and colleagues. We would especially like to acknowledge our fruitful collaborations with B. Hartke, R. Herges, J. Mattay, and F. Sönnichsen, as well as our partners in ongoing other research projects.

References

1. M. Klessinger and J. Michl, *Excited States and Photochemistry of Organic Molecules* (VCH, Weinheim, 1995).
2. W. Domcke, D.R. Yarkony and H. Köppel (eds.), *Conical Intersections: Electronic Structure, Dynamics and Spectroscopy* (World Scientific, Singapore, 2004).
3. A.L. Sobolewski, C. Woywod and W. Domcke, *J. Chem. Phys.* **98**, 5627 (1993).
4. A.L. Sobolewski and W. Domcke, *Chem. Phys.* **259**, 181 (2000).
5. A.L. Sobolewski, W. Domcke, C. Dedonder-Lardeux and C. Jouvet, *Phys. Chem. Chem. Phys.* **4**, 1093 (2002).
6. J.-S. Taylor, *Pure Appl. Chem.* **67**, 183 (1995).
7. B.K. Armstrong and A. Kricker, *J. Photochem. Photobiol. B* **63**, 8 (2001).
8. Y. Matsumura and H.N. Ananthaswamy, *Front. Biosci.* **7**, d765 (2002).
9. R.P. Sinha and D.-P. Häder, *Photochem. Photobiol. Sci.* **1**, 225 (2002).
10. S. Ossowski, K. Schneeberger, J.I. Lucas-Lledó, N. Warthmann, R.M. Clark, R.G. Shaw, D. Weigel and M. Lynch, *Science* **327**, 92 (2010).
11. E.C. Friedberg, G.C. Walker, W. Siede, R.D. Wood, R.A. Schultz and T. Ellenberger, *DNA Repair and Mutagenesis*, 2nd edn. (ASM Press, Washington, 2006).
12. C.E. Crespo-Hernández, B. Cohen, P.M. Hare and B. Kohler, *Chem. Rev.* **104**, 1977 (2004).
13. C.E. Crespo-Hernández, B. Cohen and B. Kohler, *Nature* **436**, 1141 (2005).
14. D. Markovitsi, F. Talbot, T. Gustavsson, D. Onidas, E. Lazzarotto and S. Marguet, *Nature* **441**, E7 (2006).
15. C.E. Crespo-Hernández, B. Cohen and B. Kohler, *Nature* **441**, E8 (2006).
16. N.K. Schwalb and F. Temps, *Science* **322**, 243 (2008).
17. C.T. Middleton, K. de La Harpe, C. Su, Y.K. Law, C.E. Crespo-Hernández and B. Kohler, *Annu. Rev. Phys. Chem.* **60**, 217 (2009).

18. I. Willner and S. Rubin, *Angew. Chem. Int. Ed.* **35**, 367 (1996).
19. I. Willner, *Acc. Chem. Res.* **30**, 347 (1997).
20. B.L. Feringa (ed.), *Molecular Switches* (Wiley-VCH, Weinheim, 2001).
21. V. Balzani, A. Credi, B. Ferrer, S. Silvi and M. Venturi, *Top. Curr. Chem.* **262**, 1 (2005).
22. V. Balzani, A. Credi and M. Venturi, *Molecular Devices and Machines: Concepts and Perspectives for the Nanoworld*, 2nd edn. (Wiley-VCH, Weinheim, 2008).
23. Y.C. Liang, A.S. Dvornikov and P.M. Rentzepis, *Proc. Natl. Acad. Sci. U.S.A.* **100**, 8109 (2003).
24. A.S. Dvornikov, E.P. Walker and P.M. Rentzepis, *J. Phys. Chem. A* **113**, 13633 (2009).
25. S.W. Hell, *Science* **316**, 1153 (2007).
26. S.W. Hell, *Nature Methods* **6**, 24 (2009).
27. M. Sauer, *Proc. Natl. Acad. Sci. U.S.A.* **102**, 9433 (2005).
28. M. Heilemann, P. Dedecker, J. Hofkens and M. Sauer, *Laser Photonics Rev.* **3**, 180 (2009).
29. H. Bouas-Laurent and H. Dürr, *Pure Appl. Chem.* **73**, 639 (2001).
30. H. Dürr and H. Bouas-Laurent (Eds.), *Photochromism: Molecules and Systems* (Elsevier, Amsterdam, 2003).
31. M. Müri, K.C. Schuermann, L. De Cola and M. Mayor, *Eur. J. Org. Chem.* 2562 (2009).
32. Y. Yokoyama, *Chem. Rev.* **100**, 1717 (2000).
33. M. Irie, *Chem. Rev.* **100**, 1685 (2000).
34. B. Van Gemert, in *Organic Photochromic and Thermochromic Compounds*, Eds. J. Crano and R. Guglielmetti (Plenum, New York, 1998).
35. G. Berkovic, V. Krongauz and V. Weiss, *Chem. Rev.* **100**, 1741 (2000).
36. R. Herges and T. Winkler, *Eur. J. Org. Chem.* 4419 (2001).
37. M. Wiechmann, H. Port, F. Laermer, W. Frey and T. Elsaesser, *Chem. Phys. Lett.* **165**, 238 (1990).
38. L. Lapinski, M.J. Nowak, J. Nowacki, M.F. Rode and A.L. Sobolewski, *ChemPhysChem* **10**, 2290 (2009).
39. N. Tamai and H. Miyasaka, *Chem. Rev.* **100**, 1875 (2000).
40. F. Merkt, S. Willitsch and U. Hollenstein, High-resolution photoelectron spectroscopy, in *Handbook of High-Resolution Spectroscopy*, Eds. M. Quack and F. Merkt (Wiley, 2010)
41. J. Wei, J. Riedel, A. Kuczmann, F. Renth and F. Temps, *Faraday Discuss.* **127**, 267 (2004).
42. D. Townsend, W. Li, S.K. Lee, R.L. Gross and A.G. Suits, *J. Phys. Chem. A* **109**, 8661 (2005).
43. M.N.R. Ashfold, G.A. King, D. Murdock, M.G.D. Nix, T.A.A. Oliver and A.G. Sage, *Phys. Chem. Chem. Phys.* **12**, 1218 (2010).
44. B. Cronin, A.L. Devine, M.G.D. Nix and M.N.R. Ashfold, *Phys. Chem. Chem. Phys.* **8**, 3440 (2006).
45. W. Fuß, S. Lochbrunner, A.M. Müller, T. Schikarski, W.E. Schmid and S.A. Trushin, *Chem. Phys.* **232**, 161 (1998).

46. W. Fuß, W.E. Schmid and S.A. Trushin, *J. Chem. Phys.* **112**, 8347 (2000).
47. H. Studzinski and F. Temps, *J. Chem. Phys.* **128**, 164314 (2008).
48. A. Stolow, *Annu. Rev. Phys. Chem.* **54**, 89 (2003).
49. T. Suzuki, *Annu. Rev. Phys. Chem.* **57**, 555 (2006).
50. S.A. Kovalenko, A.L. Dobryakov, J. Ruthmann and N.P. Ernsting, *Phys. Rev. A* **59**, 2369 (1999).
51. A.L. Dobryakov, S.A. Kovalenko and N.P. Ernsting, *J. Chem. Phys.* **119**, 988 (2003).
52. A.L. Dobryakov, S.A. Kovalenko and N.P. Ernsting, *J. Chem. Phys.* **123**, 044502 (2005).
53. G. Cerullo, C. Manzoni, L. Lüer and D. Polli, *Photochem. Photobiol. Sci.* **6**, 135 (2007).
54. R. Siewertsen, F. Renth and F. Temps, *Phys. Chem. Chem. Phys.* **11**, 5952 (2009).
55. U. Megerle, I. Pugliesi, C. Schriever, C.F. Sailer and E. Riedle, *Appl. Phys. B* **96**, 215 (2009).
56. J. Shah, *IEEE J. Quantum Electron.* **24**, 276 (1988).
57. L. Zhao, J.L. Pérez Lustres, V. Farztdinov and N.P. Ernsting, *Phys. Chem. Chem. Phys.* **7**, 1716 (2005).
58. T. Pancur, N.K. Schwalb, F. Renth and F. Temps, *Chem. Phys.* **313**, 199 (2005).
59. N.K. Schwalb and F. Temps, *J. Phys. Chem. A* **113**, 13113 (2009).
60. B. Schmidt, S. Laimgruber, W. Zinth and P. Gilch, *Appl. Phys. B* **76**, 809 (2003).
61. T. Wilhelm, J. Piel and E. Riedle, *Opt. Lett.* **22**, 1494 (1997).
62. G. Cerullo, M. Nisoli and S. de Silvestri, *Appl. Phys. Lett.* **71**, 3616 (1997).
63. G. Cerullo and S. De Sivestri, *Rev. Sci. Instrum.* **74**, 1 (2003).
64. M. Maroncelli, J.-M. MacInnis and G.R. Fleming, *Science* **243**, 1674 (1989).
65. R.S. Fee and M. Maroncelli, *Chem. Phys.* **183**, 235 (1994).
66. N. Eilers-König, T. Kühne, D. Schwarzer, P. Vöhringer and J. Schroeder, *Chem. Phys. Lett.* **253**, 69 (1996).
67. A. Cannizzo, O. Bräm, G. Zgrablic, A. Tortschanoff, A. Ajdarzadeh Oskouei, F. van Mourik and M. Chergui, *Opt. Lett.* **32**, 3555 (2007).
68. R. Nakamura and Y. Kanematsu, *Rev. Sci. Instrum.* **75**, 636 (2004).
69. C. Ma, W.M. Kwok, W.S. Chan, P. Zuo, J.T.W. Kan, P.H. Toy and D.L. Phillips, *J. Am. Chem. Soc.* **127**, 1463 (2005).
70. M. Sajadi, A.L. Dobryakov, E. Garbin, N.P. Ernsting and S.A. Kovalenko, *Chem. Phys. Lett.* **489**, 44 (2010).
71. R.R. Alfano, *The Supercontinuum Laser Source* (Springer, New York, 2005).
72. P.J.M. Johnson, V.I. Prokhorenko and R.J.D. Miller, *Opt. Express* **17**, 21488 (2009).
73. A.L. Dobryakov, J.L. Pérez Lustres, S.A. Kovalenko and N.P. Ernsting, *Chem. Phys.* **347**, 127 (2008).
74. E.T. Nibbering, H. Fidder and E. Pines, *Annu. Rev. Phys. Chem.* **56**, 337 (2005).

75. P. Kukura, D.W. McCamant and R.A. Mathies, *Annu. Rev. Phys. Chem.* **58**, 461 (2007).
76. P.R. Callis, *Annu. Rev. Phys. Chem.* **34**, 329 (1983).
77. N.J. Kim, G. Jeong, Y.S. Kim, J. Sung, S.K. Kim and Y.D. Park, *J. Chem. Phys.* **113**, 10051 (2000).
78. D.C. Lührs, J. Viallon and I. Fischer, *Phys. Chem. Chem. Phys.* **3**, 1827 (2001).
79. C. Plützer, E. Nir, M.S. de Vries and K. Kleinermanns, *Phys. Chem. Chem. Phys.* **3**, 5466 (2001).
80. E. Nir, K. Kleinermanns, L. Grace and M.S. de Vries, *J. Phys. Chem. A* **105**, 5106 (2001).
81. E. Nir, C. Plützer, K. Kleinermanns and M. de Vries, *Eur. Phys. J. D* **20**, 317 (2002).
82. C. Plützer and K. Kleinermanns, *Phys. Chem. Chem. Phys.* **4**, 4877 (2002).
83. N.J. Kim, H. Kang, Y.D. Park and S.K. Kim, *Phys. Chem. Chem. Phys.* **6**, 2802 (2004).
84. Y. Lee, M. Schmidt, K. Kleinermanns and B. Kim, *J. Phys. Chem. A* **110**, 11819 (2006).
85. H. Kang, K.T. Lee, B. Jung, Y.J. Ko and S.K. Kim, *J. Am. Chem. Soc.* **124**, 12958 (2002).
86. H. Kang, B. Jung and S.K. Kim, *J. Chem. Phys.* **118**, 6717 (2003).
87. C. Canuel, M. Mons, F. Piuzzi, B. Tardivel, I. Dimicoli and M. Elhanine, *J. Chem. Phys.* **122**, 074316 (2005).
88. J.-M. Pecourt, J. Peon and B. Kohler, *J. Am. Chem. Soc.* **122**, 9348 (2000).
89. J.-M. Pecourt, J. Peon and B. Kohler, *J. Am. Chem. Soc.* **123**, 10370 (2001).
90. B. Cohen, P.M. Hare and B. Kohler, *J. Am. Chem. Soc.* **125**, 13594 (2003).
91. W.-M. Kwok, C. Ma and D.L. Phillips, *J. Am. Chem. Soc.* **128**, 11894 (2006).
92. J. Peon and A.H. Zewail, *Chem. Phys. Lett.* **348**, 255 (2001).
93. D. Onidas, D. Markovitsi, S. Marguet and T. Gustavsson, *J. Phys. Chem. B* **106**, 11367 (2002).
94. T. Gustavsson, A. Sharonov, D. Onidas and D. Markovitsi, *Chem. Phys. Lett.* **356**, 49 (2002).
95. A.L. Sobolewski and W. Domcke, *Eur. Phys. J. D* **20**, 369 (2002).
96. S. Perun, A.L. Sobolewski and W. Domcke, *Chem. Phys.* **313**, 107 (2005).
97. S. Perun, A.L. Sobolewski and W. Domcke, *J. Am. Chem. Soc.* **127**, 6257 (2005).
98. C.M. Marian, *J. Chem. Phys.* **122**, 104314 (2005).
99. S.B. Nielsen and T.I. Solling, *ChemPhysChem* **6**, 1276 (2005).
100. H. Chen and S. Li, *J. Phys. Chem. A* **109**, 8443 (2005).
101. L. Serrano-Andrés, M. Merchán and A.C. Borin, *Chem. Eur. J.* **12**, 6559 (2006).
102. L. Serrano-Andrés, M. Merchán and A.C. Borin, *Proc. Natl. Acad. Sci. U.S.A.* **103**, 8691 (2006).

103. L. Blancafort, *J. Am. Chem. Soc.* **128**, 210 (2006).
104. M.Z. Zgierski, S. Patchkovskii and E.C. Lim, *Can. J. Chem.* **85**, 124 (2007).
105. W.C. Chung, Z. Lan, Y. Ohtsuki, N. Shimakura, W. Domcke and Y. Fujimura, *Phys. Chem. Chem. Phys.* **9**, 2075 (2007).
106. S. Yamazaki and S. Kato, *J. Am. Chem. Soc.* **129**, 2901 (2007).
107. I. Conti, M. Garavelli and G. Orlandi, *J. Am. Chem. Soc.* **131**, 16108 (2009).
108. M. Barbatti and H. Lischka, *J. Am. Chem. Soc.* **130**, 6831 (2008).
109. Y. Lei, S. Yuan, Y. Dou, Y. Wang and Z. Wen, *J. Phys. Chem. A* **112**, 8497 (2008).
110. L. Serrano-Andrés and M. Merchán, *J. Photochem. Photobiol. C* **10**, 21 (2009).
111. B. Cohen, C.E. Crespo-Hernández and B. Kohler, *Faraday Discuss.* **127**, 137 (2004).
112. B. Mennucci, A. Toniolo and J. Tomasi, *J. Phys. Chem. A* **105**, 4749 (2001).
113. S. Ullrich, T. Schulz, M.Z. Zgierski and A. Stolow, *J. Am. Chem. Soc.* **126**, 2262 (2004).
114. H. Satzger, D. Townsend, M.Z. Zgierski, S. Patchkovskii, S. Ullrich and A. Stolow, *Proc. Natl. Acad. Sci. U.S.A.* **103**, 10196 (2006).
115. C.Z. Bisgaard, H. Satzger, S. Ullrich and A. Stolow, *ChemPhysChem* **10**, 101 (2009).
116. M.G.D. Nix, A.L. Devine, B. Cronin and M.N.R. Ashfold, *J. Chem. Phys.* **126**, 124312 (2007).
117. K. Wells, G. Roberts and V. Stavros, *Chem. Phys. Lett.* **446**, 20 (2007).
118. K.L. Wells, D.J. Hadden, M.G.D. Nix and V.G. Stavros, *J. Phys. Chem. Lett.* **1**, 993 (2010).
119. C.M. Marian, M. Kleinschmidt and J. Tatchen, *Chem. Phys.* **347**, 346 (2008).
120. D.C. Ward, E. Reich and L. Stryer, *J. Biol. Chem.* **244**, 1228 (1969).
121. O.J.G. Somsen, van A. Hoek and van H. Amerongen, *Chem. Phys. Lett.* **402**, 61 (2005).
122. J.M. Jean and K.B. Hall, *Biochemistry* **43**, 10277 (2004).
123. J.M. Jean and B.P. Krueger, *J. Phys. Chem. B* **110**, 2899 (2006).
124. R.K. Neely and A.C. Jones, *J. Am. Chem. Soc.* **128**, 15952 (2006).
125. E.Y.M. Bonnist and A.C. Jones, *ChemPhysChem* **9**, 1121 (2008).
126. S.K. Pal, J. Peon and A.H. Zewail, *Chem. Phys. Lett.* **363**, 57 (2002).
127. S.K. Pal, L. Zhao, T. Xia and A.H. Zewail, *Proc. Nat. Acad. Sci. USA* **100**, 13746 (2003).
128. B. Albinsson, *J. Am. Chem. Soc.* **119**, 6369 (1997).
129. J. Andreásson, A. Holmén and B. Albinsson, *J. Phys. Chem. B* **103**, 9782 (1999).
130. A.B.J. Parusel, W. Rettig and K. Rotkiewicz, *J. Phys. Chem. A* **106**, 2293 (2002).
131. Z.R. Grabowski, K. Rotkiewicz and W. Rettig, *Chem. Rev.* **103**, 3899 (2003).

132. N.K. Schwalb and F. Temps, Phys. Chem. Chem. Phys. **8**, 5229 (2006).
133. V. Karunakaran, K. Kleinermanns, R. Improta and S.A. Kovalenko, J. Am. Chem. Soc. **131**, 5839 (2009).
134. F.-A. Miannay, T. Gustavsson, A. Banyasz and D. Markovitsi, J. Phys. Chem. A **114**, 3256 (2010).
135. H. Chen and S. Lu, J. Chem. Phys. **124**, 154315 (2006).
136. C.M. Marian, J. Phys. Chem. A **111**, 1545 (2007).
137. S. Yamazaki and W. Domcke, J. Phys. Chem. A **112**, 7090 (2008).
138. S. Yamazaki, W. Domcke and A. L. Sobolewski, J. Phys. Chem. A **112**, 11965 (2008).
139. L. Serrano-Andrés, M. Merchán and A.C. Borin, J. Am. Chem. Soc. **130**, 2473 (2008).
140. Z. Lan, E. Fabiano and W. Thiel, ChemPhysChem **10**, 1225 (2009).
141. E. Nir, C. Janzen, P. Imhof, K. Kleinermanns and M.S. de Vries, J. Chem. Phys. **115**, 4604 (2001).
142. E. Nir, I. Hünig, K. Kleinermanns and M.S. de Vries, ChemPhysChem **5**, 131 (2004).
143. M. Mons, F. Piuzzi, I. Dimicioli, L. Gorb and J. Leszczynski, J. Phys. Chem. A **110**, 10921 (2006).
144. J. Zhou, O. Kostko, C. Nicolas, X.T.L. Belau, M.S. de Vries and M. Ahmed, J. Phys. Chem. A **113**, 4829 (2009).
145. R.J. Malone, A.M. Miller and B. Kohler, Photochem. Photobiol. **77**, 158 (2003).
146. L. Blancafort, B. Cohen, P.M. Hare, B. Kohler and M.A. Robb, J. Phys. Chem. A **109**, 4431 (2005).
147. P.M. Hare, C.E. Crespo-Hernández and B. Kohler, J. Phys. Chem. B **110**, 18641 (2006).
148. P.M. Hare, C.E. Crespo-Hernández and B. Kohler, Proc. Natl. Acad. Sci. U.S.A. **104**, 435 (2007).
149. T. Gustavsson, A. Sharonov and D. Markovitsi, Chem. Phys. Lett. **351**, 195 (2002).
150. A. Sharonov, T. Gustavsson, V. Carre, E. Renault and D. Markovitsi, Chem. Phys. Lett. **380**, 173 (2003).
151. T. Gustavsson, N. Sarkar, E. Lazzarotto, D. Markovitsi and R. Improta, Chem. Phys. Lett. **429**, 551 (2006).
152. T. Gustavsson, Á. Bányász, E. Lazzarotto, D. Markovitsi, G. Scalmani, M.J. Frisch, V. Barone and R. Improta, J. Am. Chem. Soc. **128**, 607 (2006).
153. T. Gustavsson, N. Sarkar, Á. Bányász, D. Markovitsi and R. Improta, Photochem. Photobiol. **83**, 595 (2007).
154. T. Gustavsson, A. Bányász, N. Sarkar, D. Markovitsi and R. Improta, Chem. Phys. **350**, 186 (2008).
155. P.M. Hare, C.T. Middleton, K.I. Mertel, J.M. Herbert and B. Kohler, Chem. Phys. **347**, 383 (2008).
156. S. Quinn, G.W. Doorley, G.W. Watson, A.J. Cowan and M.W. George, Chem. Comm. 1182 (2007).

157. L. Blancafort and M.A. Robb, *J. Phys. Chem. A* **108**, 10609 (2004).
158. M. Merchán, R. González-Luque, T. Climent, L. Serrano-Andrés, E. Rodríguez, M. Reguero and D. Peláez, *J. Phys. Chem. B* **110**, 26471 (2006).
159. S. Matsika, *J. Phys. Chem. A* **109**, 7358 (2005).
160. K.A. Kistler and S. Matsika, *J. Phys. Chem. A* **111**, 2650 (2007).
161. K.A. Kistler and S. Matsika, *J. Phys. Chem. A* **111**, 8708 (2007).
162. K.A. Kistler and S. Matsika, *J. Chem. Phys.* **128**, 215102 (2008).
163. H.R. Hudock and T.J. Martinez, *ChemPhysChem* **9**, 2486 (2008).
164. Z. Lan, E. Fabiano and W. Thiel, *J. Phys. Chem. B* **113**, 3548 (2009).
165. D. Asturiol, B. Lasorne, M.A. Robb and L. Blancafort, *J. Phys. Chem. A* **113**, 10211 (2009).
166. C.M. Marian, F. Schneider, M. Kleinschmidt and J. Tatchen, *Eur. Phys. J. D* **20**, 357 (2002).
167. S. Matsika, *J. Phys. Chem. A* **108**, 7584 (2004).
168. M. Etinski, T. Fleig and C.M. Marian, *J. Phys. Chem. A* **113**, 11809 (2009).
169. S. Marguet and D. Markovitsi, *J. Am. Chem. Soc.* **127**, 5780 (2005).
170. W.-M. Kwok, C. Ma and D.L. Phillips, *J. Am. Chem. Soc.* **130**, 5131 (2008).
171. W.J. Schreier, T.E. Schrader, F.O. Koller, P. Gilch, C.E. Crespo-Hernández, V.N. Swaminathan, T. Carell, W. Zinth and B. Kohler, *Science* **315**, 625 (2007).
172. W.J. Schreier, J. Kubon, N. Regner, K. Haiser, T.E. Schrader, W. Zinth, P. Clivio and P. Gilch, *J. Am. Chem. Soc.* **131**, 5038 (2009).
173. L. Blancafort and A. Migani, *J. Am. Chem. Soc.* **129**, 14550 (2007).
174. M. Boggio-Pasqua, G. Groenhof, L.V. Schaefer, H. Grubmüller and M.A. Robb, *J. Am. Chem. Soc.* **129**, 10996 (2007).
175. M. Dey, F. Moritz, J. Grotemeyer and E.W. Schlag, *J. Am. Chem. Soc.* **116**, 9211 (1994).
176. E. Nir, K. Kleinermanns and M.S. de Vries, *Nature* **408**, 949 (2000).
177. C. Plützer, I. Hünig and K. Kleinermanns, *Phys. Chem. Chem. Phys.* **5**, 1158 (2003).
178. E. Nir, C. Janzen, P. Imhof, K. Kleinermanns and M.S. de Vries, *Phys. Chem. Chem. Phys.* **4**, 740 (2002).
179. C. Plützer, I. Hünig, K. Kleinermanns, E. Nir and M.S. de Vries, *ChemPhysChem* **4**, 838 (2003).
180. E. Nir, C. Janzen, P. Imhof, K. Kleinermanns and M.S. de Vries, *Phys. Chem. Chem. Phys.* **4**, 732 (2002).
181. A. Abo-Riziq, L. Grace, E. Nir, M. Kabeláč, P. Hobza and M. de Vries, *Proc. Natl. Acad. Sci. U.S.A.* **102**, 20 (2005).
182. A.L. Sobolewski and W. Domcke, *Phys. Chem. Chem. Phys.* **6**, 2763 (2004).
183. A.L. Sobolewski, W. Domcke and C. Hättig, *Proc. Natl. Acad. Sci. U.S.A.* **102**, 17903 (2005).
184. T. Schultz, E. Samoylova, W. Radloff, I.V. Hertel, A.L. Sobolewski and W. Domcke, *Science* **306**, 1765 (2004).
185. E. Samoylova, W. Radloff, H.-H. Ritze and T. Schultz, *J. Phys. Chem. A* **113**, 8195 (2009).

186. E. Samoylova, H. Lippert, S. Ullrich, I.V. Hertel, W. Radloff and T. Schulz, *J. Am. Chem. Soc.* **127**, 1782 (2005).
187. H.-H. Ritze, H. Lippert, E. Samoylova, V.R. Smith, I. Hertel, W. Radloff and T. Schulz, *J. Chem. Phys.* **122**, 224320 (2005).
188. A. Samoylova, T. Schultz, I.V. Hertel and W. Radloff, *Chem. Phys.* **347**, 376 (2008).
189. R.M. Hamlin, R.C. Lord and A. Rich, *Science* **148**, 1734 (1965).
190. Y. Kyogoku, R.C. Lord and A. Rich, *Science* **154**, 518 (1966).
191. N.K. Schwalb and F. Temps, *J. Am. Chem. Soc.* **129**, 9272 (2007).
192. N.K. Schwalb and F. Temps, *J. Phys. Chem. B* **113**, 16365 (2009).
193. K. Siriwong, A.A. Voityuk, M.D. Newton and N. Rösch, *J. Phys. Chem. B* **107**, 2595 (2003).
194. P. Carmona, M. Molina, A. Lasagabaster, R. Escobar and A.B. Altabef, *J. Phys. Chem.* **97**, 9519 (1993).
195. N.K. Schwalb and F. Temps, *J. Photochem. Photobiol. A* **208**, 164 (2009).
196. P.R.L. Markwick and N.L. Doltsinis, *J. Chem. Phys.* **126**, 175102 (2007).
197. G. Groenhof, L.V. Schäfer, M. Boggio-Pasqua, M. Goette, H. Grubmüller and M.A. Robb, *J. Am. Chem. Soc.* **129**, 6812 (2007).
198. I. Buchvarov, Q. Wang, M. Raytchev, A. Trifonov and T. Fiebig, *Proc. Natl. Acad. Sci. U.S.A.* **104**, 4794 (2007).
199. K. de La Harpe, C.E. Crespo-Hernández and B. Kohler, *J. Am. Chem. Soc.* **131**, 17557 (2009).
200. T. Watanabe, R. Tashiro and H. Sugiyama, *J. Am. Chem. Soc.* **129**, 8163 (2007).
201. B. Bouvier, T. Gustavsson, D. Markovitsi and P. Millié, *Chem. Phys.* **275**, 75 (2002).
202. B. Bouvier, J.-P. Dognon, R. Lavery, D. Markovitsi, P. Millié, D. Onidas and K. Zakrzewska, *J. Phys. Chem. B* **107**, 13512 (2003).
203. E. Emanuele, D. Markovitsi, P. Millié and K. Zakrewska, *ChemPhysChem* **6**, 1387 (2005).
204. E. Emanuele, K. Zakrzewska, D. Markovitsi, R. Lavery and P. Millié, *J. Phys. Chem. B* **109**, 16109 (2005).
205. F. Santoro, V. Barone and R. Improta, *ChemPhysChem* **9**, 2531 (2008).
206. A.W. Lange and J.M. Herbert, *J. Am. Chem. Soc.* **1331**, 3913 (2009).
207. G. Olaso-González, M. Merchán and L. Serrano-Andrés, *J. Am. Chem. Soc.* **131**, 4368 (2009).
208. C.R. Kozak, K.A. Kistler, Z. Lu and S. Matsika, *J. Phys. Chem. B* **114**, 1674 (2010).
209. J. Eisinger, M. Gueron, R.G. Shulman and T. Yamane, *Proc. Natl. Acad. Sci. U.S.A.* **55**, 1015 (1966).
210. M.C. Stuhldreier, C. Schüler, J. Kleber and F. Temps, in: *Ultrafast Phenomena XVII*, pp. 553–555, M. Chergui, D. Jonas, E. Riedle, R.W. Schoenlein, A. Taylor (Eds.), (Oxford University Press, 2011).
211. C.S.M. Olsthoorn, L.J. Bostelaar, J.F.M. De Rooij, J.H. van Boom and C. Altona, *Eur. J. Biochem.* **115**, 309 (1981).

212. T. Takaya, C. Su, K. de La Harpe, C.E. Crespo-Hernández and B. Kohler, *Proc. Natl. Acad. Sci. U.S.A.* **105**, 10285 (2008).
213. N. Öksüz, B. Hartke and F. Temps, unpublished results.
214. D. Markovitsi, A. Sharonov, D. Onidas and T. Gustavsson, *ChemPhysChem* **4**, 303 (2003).
215. D. Markovitsi, D. Onidas, T. Gustavsson, F. Talbot and E. Lazzarotto, *J. Am. Chem. Soc.* **127**, 17130 (2005).
216. D. Onidas, T. Gustavsson, E. Lazzarotto and D. Markovitsi, *J. Phys. Chem. B* **111**, 9644 (2007).
217. D. Onidas, T. Gustavsson, E. Lazzarotto and D. Markovitsi, *Phys. Chem. Chem. Phys.* **9**, 5143 (2007).
218. F.-A. Miannay, A. Bányász, T. Gustavsson and D. Markovitsi, *J. Am. Chem. Soc.* **129**, 14574 (2007).
219. W.-M. Kwok, C. Ma and D.L. Phillips, *J. Phys. Chem. B* **113**, 11527 (2009).
220. C.E. Crespo-Hernández and B. Kohler, *J. Phys. Chem. B* **108**, 11182 (2004).
221. C.E. Crespo-Hernández, K. de La Harpe and B. Kohler, *J. Am. Chem. Soc.* **130**, 10844 (2008).
222. K. de La Harpe, C.E. Crespo-Hernández and B. Kohler, *ChemPhysChem* **10**, 1421 (2010).
223. D. Markovitsi, T. Gustavsson and A. Sharanov, *Photochem. Photobiol.* **79**, 526 (2004).
224. F.-A. Miannay, A. Bányász, T. Gustavsson and D. Markovitsi, *J. Phys. Chem. B* **113**, 11760 (2009).
225. C. Prunkl, S. Berndl, C. Wanninger-Weiß, J. Barbaric and H.-A. Wagenknecht, *Phys. Chem. Chem. Phys.* **12**, 32 (2010).
226. J.C. Genereux and J.K. Barton, *Chem. Rev.* **110**, 1642 (2010).
227. H.G. Heller and S. Oliver, *J. Chem. Soc. Perkin Trans. 1* 197 (1981).
228. T. Kardinahl and H. Franke, *Appl. Phys. A* **61**, 23 (1995).
229. S.Z. Janicki and G.B. Schuster, *J. Am. Chem. Soc.* **117**, 8524 (1995).
230. F.M. Raymo and M. Tomasulo, *Chem. Soc. Rev.* **34**, 327 (2005).
231. J. Cusido, E. Deniz and F.M. Raymo, *Eur. J. Org. Chem.* 2031 (2009).
232. P. Seal and S. Chakrabarti, *J. Phys. Chem. A* **114**, 673 (2010).
233. M.A. Wolak, C.J. Thomas, N.B. Gillespie, R.R. Birge and W.J. Lees, *J. Org. Chem.* **68**, 319 (2003).
234. E. Uhlmann and G. Gauglitz, *J. Photochem. Photobiol. A* **98**, 45 (1996).
235. P. Celani, F. Bernardi, M.A. Robb and M. Olivucci, *J. Phys. Chem.* **100**, 19364 (1996).
236. C. Dugave and L. Demange, *Chem. Rev.* **103**, 2475 (2003).
237. W. Fuß, S. Panja, W.E. Schmidt and S.A. Trushin, *Mol. Phys.* **105**, 1133 (2006).
238. F. Renth, M. Foca, A. Petter and F. Temps, *Chem. Phys. Lett.* **428**, 62 (2006).
239. H. Tamura, S. Nanbu, T. Ishida and H. Nakamura, *J. Chem. Phys.* **125**, 034307 (2006).

240. O. Kühn and L. Wöste (Eds.), *Analysis and Control of Ultrafast Photoinduced Reactions* (Springer, New York, 2007).
241. Y. Yokoyama, K. Ogawa, T. Iwai, K. Shimazaki, Y. Kajihira, T. Goto, Y. Yokoyama and Y. Kurita, *Bull. Chem. Soc. Jpn.* **69**, 1605 (1996).
242. M. Handschuh, M. Seibold, H. Port and H.C. Wolf, *J. Phys. Chem. A* **101**, 502 (1997).
243. Y. Yokoyama, T. Iwali, Y. Yokoyama and Y. Kurita, *Chem. Lett.* 225 (1994).
244. J. Saltiel, N. Tarkalanov and D.F. Sears Jr., *J. Am. Chem. Soc.* **117**, 5586 (1995).
245. J. Saltiel, L. Cires and A.M. Turek, in *CRC Handbook of Organic Photochemistry and Photobiology*, Eds. W. Horspool and F. Lenzi (CRC Press, Boca Raton, 2004) pp. 27/1–27/22.
246. H.J.C. Jacobs and E. Havinga, *Adv. Photochem.* **11**, 305 (1979).
247. J.B. Schönborn and B. Hartke, to be published.
248. A. Hofmann and R. de Vivie-Riedle, *J. Chem. Phys.* **112**, 5054 (2000).
249. M. Garavelli, C.S. Page, P. Celani, M. Olivucci, W.E. Schmid, S.A. Trushin and W. Fuß, *J. Phys. Chem. A* **105**, 4458 (2001).
250. A. Nenov, P. Kölle, M.A. Robb and R. de Vivie-Riedle, *J. Org. Chem.* **75**, 123 (2010).
251. K. Kosma, S.A. Trushin, W. Fuß, and W.E. Schmid, *Phys. Chem. Chem. Phys.* **11**, 172 (2008).
252. J. Voll, T. Kerscher, D. Geppert and R. De Vivie-Riedle, *J. Photochem. Photobiol. A* **190**, 352 (2007).
253. Y. Ishibashi, M. Murakami, H. Miyasaka, S. Kobatake, M. Irie and Y. Yokoyama, *J. Phys. Chem. C* **111**, 2730 (2007).
254. B. Heinz, S. Malkmus, S. Laimgruber, S. Dietrich, C. Schulz, K. Rück-Braun, M. Braun, W. Zinth and P. Gilch, *J. Am. Chem. Soc.* **129**, 8577 (2007).
255. F.O. Koller, W.J. Schreier, T.E. Schrader, S. Malkmus, C. Schulz, S. Dietrich, K. Rück-Braun and M. Braun, *J. Phys. Chem. A* **112**, 210 (2008).
256. T. Cordes, T.T. Herzog, S. Malkmus, S. Draxler, T. Brust, J. DiGirolamo, W.J. Lees and M. Braun, *Photochem. Photobiol. Sci.* **8**, 528 (2009).
257. T. Brust, S. Draxler, A. Popp, X. Chen, W.J. Lees, W. Zinth and M. Braun, *Chem. Phys. Lett.* **477**, 298 (2009).
258. T. Brust, S. Malkmus, S. Draxler, S.A. Ahmed, K. Rück-Braun, W. Zinth and M. Braun, *J. Photochem. Photobiol. A* **207**, 209 (2009).
259. S. Draxler, T. Brust, S. Malkmus, J.A. DiGirolamo, W.J. Lees, W. Zinth and M. Braun, *Phys. Chem. Chem. Phys.* **11**, 5019 (2009).
260. M. Barbatti, J. Paier and H. Lischka, *J. Chem. Phys.* **121**, 11614 (2004).
261. F. Bernardi and M. Olivucci, *Chem. Soc. Rev.* **25**, 321 (1996).
262. W. Fuß, Y. Haas and S. Zilberg, *Chem. Phys.* **256**, 273 (2000).
263. G. Tomasello, M.J. Bearpark, M.A. Robb, G. Orlandi and M. Garavelli, *Angew. Chem. Int. Ed.* **49**, 2913 (2010).
264. F. Strübe, R. Siewertsen, F.D. Sönnichsen, F. Renth, F. Temps and J. Mattay, *Eur. J. Org. Chem.* 1947 (2011).

265. Z.F. Liu, K. Hashimoto and A. Fujishima, *Nature* **347**, 658 (1990).
266. T. Ikeda and O. Tsutsumi, *Science* **268**, 1873 (1995).
267. H. Rau, in *Photochromism: Molecules and Systems*, Eds. H. Dürr and H. Bouas-Laurent (Elsevier, Amsterdam, 2003)
268. L. Ulysse, J. Cubillos and J. Chmielewski, *J. Am. Chem. Soc.* **117**, 8466 (1995).
269. O. Pieroni, A. Fissi, N. Angelini and F. Lenci, *Acc. Chem. Res.* **34**, 9 (2001).
270. S. Spörlein, H. Carstens, H. Satzger, C. Renner, R. Behrendt, L. Moroder, P. Tavan, W. Zinth and J. Wachtveitl, *Proc. Natl. Acad. Sci. U.S.A.* **99**, 7998 (2002).
271. A. Archut, F. Vögtle, L.D. Cola, G.C. Azzellini, V. Balzani, P.S. Ramanujam and R.H. Berg, *Chem. Eur. J.* **4**, 699 (1998).
272. F. Vögtle, M. Gorka, R. Hesse, P. Ceroni, M. Maestri and V. Balzani, *Photochem. Photobiol. Sci.* **1**, 45 (2002).
273. G.S. Kumar and D.C. Neckers, *Chem. Rev.* **89**, 1915 (1989).
274. A. Natansohn and P. Rochon, *Chem. Rev.* **102**, 4139 (2002).
275. T. Hugel, N.B. Holland, A. Cattani, L. Moroder, M. Seitz and H.E. Gaub, *Science* **296**, 1103 (2002).
276. N.B. Holland, T. Hugel, G. Neuert, A. Cattani-Scholz, C. Renner, D. Oesterhelt, L. Moroder, M. Seitz and H.E. Gaub, *Macromolecules* **36**, 2015 (2003).
277. H. Asanuma, X. Liang, H. Nishioka, D. Matsunaga, M. Liu and M. Komiyama, *Nature Protocols* **2**, 203 (2007).
278. H. Kang, H. Liu, J.A. Phillips, Z. Cao, Y. Kim, Y. Chen, Z. Yang, J. Li, and W. Tan, *Nano Lett.* **9**, 2690 (2009).
279. U. Jung, B. Baisch, D. Kaminski, K. Krug, A. Elsen, T. Weineisen, D. Raffa, J. Stettner, C. Bornholdt, R. Herges and O. Magnussen, *J. Electroanal. Chem.* **619–620**, 152 (2008).
280. B. Baisch, D. Raffa, U. Jung, O.M. Magnussen, C. Nicolas, J. Lacour, J. Kubitschke and R. Herges, *J. Am. Chem. Soc.* **131**, 442 (2009).
281. I.K. Lednev, T.-Q. Ye, R.E. Hester and J.N. Moore, *J. Phys. Chem.* **100**, 13338 (1996).
282. I.K. Lednev, T.-Q. Ye, P. Matousek, M. Towrie, P. Foggi, F.V.R. Neuwahl, S. Umapathy, R.E. Hester and J.N. Moore, *Chem. Phys. Lett.* **290**, 68 (1998).
283. T. Fujino, S.Y. Arzhantsev and T. Tahara, *Bull. Chem. Soc. Jpn.* **75**, 1031 (2002).
284. T. Fujino, S.Y. Arzhantsev and T. Tahara, *J. Phys. Chem. A* **105**, 8123 (2001).
285. Y.-C. Lu, C.-W. Chang and E.W.-G. Diau, *J. Chin. Chem. Soc.* **49**, 693 (2002).
286. H. Satzger, S. Spörlein, C. Root, J. Wachtveitl, W. Zinth and P. Gilch, *Chem. Phys. Lett.* **372**, 216 (2003).
287. C.-W. Chang, Y.-C. Lu, T.-T. Wang and E.W.-G. Diau, *J. Am. Chem. Soc.* **126**, 10109 (2004).

288. C. Nonnenberg, H. Gaub and I. Frank, *ChemPhysChem* **7**, 1455 (2006).
289. G. Granucci and M. Persico, *Theor. Chem. Acc.* **117**, 1131 (2007).
290. I. Conti, M. Garavelli and G. Orlandi, *J. Am. Chem. Soc.* **130**, 5216 (2008).
291. S. Yuan, Y. Dou, W. Wu, Y. Hu and J. Zhao, *J. Phys. Chem. A* **112**, 13326 (2008).
292. Y. Ootani, K. Satoh, A. Nakayama, T. Noro and T. Taketsugu, *J. Chem. Phys.* **131**, 194306 (2009).
293. H. Rau and E. Lüddecke, *J. Am. Chem. Soc.* **104**, 1616 (1982).
294. H. Rau, *J. Photochem.* **26**, 221 (1984).
295. P. Bortolus and S. Monti, *J. Phys. Chem.* **91**, 5046 (1987).
296. S. Monti, G. Orlandi and P. Palmieri, *Chem. Phys.* **71**, 87 (1982).
297. T. Nägele, R. Hoche, W. Zinth and J. Wachtveitl, *Chem. Phys. Lett.* **272**, 489 (1997).
298. T. Ishikawa, T. Noro and T. Shoda, *J. Chem. Phys.* **115**, 7503 (2001).
299. A. Cembran, F. Bernardi, M. Garavelli, L. Gagliardi and G. Orlandi, *J. Am. Chem. Soc.* **126**, 3234 (2004).
300. C. Ciminelli, G. Granucci and M. Persico, *Chem. Eur. J.* **10**, 2327 (2004).
301. E.W.-G. Diau, *J. Phys. Chem. A* **108**, 950 (2004).
302. P. Sauer and R.E. Allen, *Chem. Phys. Lett.* **450**, 192 (2008).
303. T. Schultz, J. Quenneville, B. Levine, A. Toniolo, T.J. Martinez, S. Lochbrunner, M. Schmitt, J.P. Shaffer, M.Z. Zgierski and A. Stolow, *J. Am. Chem. Soc.* **125**, 8098 (2003).
304. H. Satzger, C. Root and M. Braun, *J. Phys. Chem. A* **108**, 6265 (2004).
305. S. Yuan, W.F. Wu, Y.S. Dou and J.S. Zhao, *Chin. Chem. Lett.* **19**, 1379 (2008).
306. L.X. Wang, W.L. Xu, C.H. Yi and X.G. Wang, *J. Molec. Graphics Modeling* **27**, 792 (2009).
307. Y.-C. Lu, E.W.-G. Diau and H. Rau, *J. Phys. Chem. A* **109**, 2090 (2005).
308. T. Pancur, F. Renth, F. Temps, B. Harbaum, A. Krüger, R. Herges and C. Näther, *Phys. Chem. Chem. Phys.* **7**, 1985 (2005).
309. C. Ciminelli, G. Granucci and M. Persico, *J. Chem. Phys.* **123**, 174317 (2005).
310. R. Siewertsen, H. Neumann, B. Buchheim-Stehn, R. Herges, C. Näther, F. Renth and F. Temps, *J. Am. Chem. Soc.* **131**, 15594 (2009).
311. R. Siewertsen, J.B. Schönborn, B. Hartke, F. Renth and F. Temps, *Phys. Chem. Chem. Phys.* **13**, 1054 (2011).
312. M. Hagiri, N. Ichinose, C.L. Zhao, H. Horiuchi, H. Hiratsuka and T. Nakayama, *Chem. Phys. Lett.* **391**, 297 (2004).
313. B. Schmidt, C. Sobotta, S. Malkmus, S. Laimgruber, M. Braun, W. Zinth and P. Gilch, *J. Phys. Chem. A* **108**, 4399 (2004).
314. F. Koller, C. Sobotta, T. Schrader, T. Cordes, W. Schreier, A. Sieg and P. Gilch, *Chem. Phys. Lett.* **341**, 258 (2007).
315. M. Poprawa-Smoluch, J. Baggerman, H. Zhang, H.P.A. Maas, L.D. Cola and A.M. Brouwer, *J. Phys. Chem. A* **110**, 11926 (2006).
316. J.A. Delaire and K. Nakatani, *Chem. Rev.* **100**, 1817 (2000).

317. C. Toro, A. Thibert, L. De Boni, A.E. Masunov and F.E. Hernandez, *J. Phys. Chem. B* **112**, 929 (2008).
318. L. De Boni, C. Toro, A.E. Masunov and F.E. Hernández, *J. Phys. Chem. A* **112**, 3886 (2008).
319. J. Bahrenburg, F. Renth and F. Temps, to be published.
320. C.R. Crecca and A.E. Roitberg, *J. Phys. Chem. A* **110**, 8188 (2006).
321. L.X. Wang and X.G. Wang, *J. Mol. Struc. Theochem.* **847**, 1 (2007).

Chapter 18

Femtosecond Pump-Probe Polarization Spectroscopy of Vibronic Dynamics at Conical Intersections and Funnels

William K. Peters, Eric R. Smith and David M. Jonas

1. Introduction 716
   1.1. The polarization anisotropy 717
   1.2. Tuning and coupling coordinates 721
2. Theory 723
   2.1. Model system and Hamilitonian 723
   2.2. Reduced density matrix and Redfield equations 725
   2.3. Effect of electronic relaxation on the anisotropy 727
3. Experiment 732
   3.1. A D_{4h} silicon naphthalocyanine (SiNc) 732
   3.2. A D_{2h} free-base naphthalocyanine (H_2Nc) 735
4. Discussion 740
5. Conclusion 742
   References 743

Department of Chemistry and Biochemistry, University of Colorado, Boulder, CO 80309-0215, USA.

1. Introduction

This chapter reviews use of the polarization dependence of femtosecond pump-probe spectroscopy to probe the coupled electronic and vibrational dynamics at conical intersections and funnels.[1-6] Typically, femtosecond pump-probe spectroscopy creates and probes wavepackets in Franck–Condon active vibrations which modulate the electronic absorption and emission frequencies; this requires pulses that are both abrupt with respect to the vibrational period and frequency selective with respect to the electronic spectrum.[7] In contrast, the experiments discussed here create and probe electronic wavepackets which are a coherent superposition of the vibrational-electronic states involved in a conical intersection.[4] Such electronic wavepackets have much in common with the electronic wavepackets observed in atoms,[8] but are modified by the vibrational-electronic coupling. These interactions provide signatures of a conical funnel which can be detected in what might otherwise be considered separate vibrational and electronic contributions to the pump-probe signal.[2,3,5] Because two electronic states with differently oriented transition moments are involved, vibrational-electronic coupling produces signatures in the pump-probe polarization anisotropy.

In contrast to Franck–Condon vibrational wavepacket signatures, these polarization signatures survive under spectrally non-selective impulsive excitation, so electronic motions and coupled vibrations can be preferentially detected. This approach has provided experimental determination of vibrational symmetry (from the polarization anisotropy of vibrational quantum beats), measurement of electronic-vibrational coupling (from the amplitude of vibrational quantum beats), and independent characterization of the electronic motion at the conical intersection. The experiments described here probe a symmetry required Jahn–Teller conical intersection[9,10] in a four-fold symmetric (D_{4h}) silicon napthalocyanine and a pseudo Jahn–Teller effect[11] that can be modeled as a conical funnel where the point of degeneracy is either absent or energetically inaccessible in a lower symmetry (D_{2h}) free-base naphthalocyanine. The naphthalocyanine chromophores (see Fig. 1) have weak electronic-vibrational coupling, which allows vibronic dynamics slow enough to measure with femtosecond spectroscopy. Interestingly, the timescale for electronic motion at a conical intersection or funnel can be faster than the coupled vibrational motion, and can be dictated by the strength of the vibrational-electronic coupling and the width of the vibrational wavepacket. This role of vibrational wavepacket

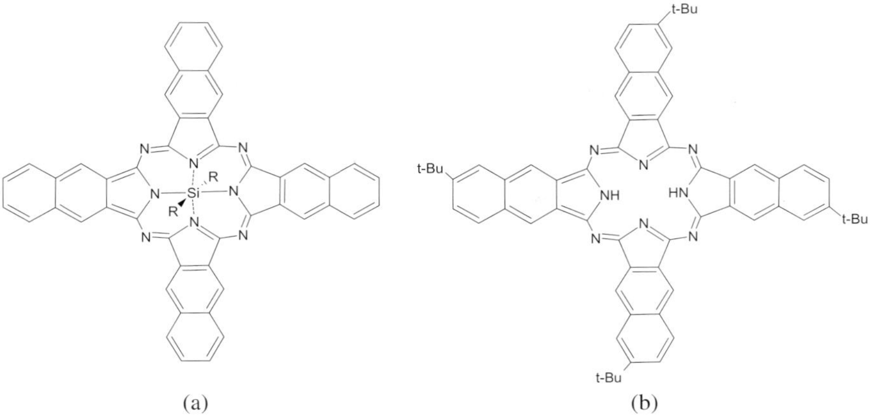

Fig. 1. Structures for (a) silicon 2,3-naphthalocyanine bis(trihexylsilyloxide) (SiNc) and (b) 2,11,20,29-tetra-*tert*-butyl-2,3-naphthalocyanine (H$_2$Nc). R = -O-Si-((CH$_2$)$_5$CH$_3$)$_3$. t-Bu = -C(CH$_3$)$_3$. Each *tert*-butyl group is likely attached with equal probability to either available naphthalene β position on its given "arm" of the molecule. Lines between the central pyrrole nitrogen atoms (bound to silicon in SiNc) define the x–y axes (with x along NH bonds in H$_2$Nc) in the molecular frame.

width is inherently missing in the semiclassical Landau–Zener approach[12,13] to curve crossing.

1.1. *The polarization anisotropy*

Time-resolved spectroscopy with polarized light can reveal the dynamics of molecular rotation.[14–16] The principle of the measurement is that a weak, linearly polarized, pump pulse, resonant with a single vibronic transition, excites an aligned $\cos^2(\theta)$ angular distribution of molecules, where θ is the angle between the vibronic transition dipole moment and the optical electric field of the pulse.[15,16] A probe pulse resonant with the vibronic transition excited by the pump will stimulate emission (Excited State Emission — ESE) from a $\cos^2(\theta)$ distribution of excited molecules and not be absorbed by the $\cos^2(\theta)$ distribution of molecules missing from the ground state (Ground State Bleach — GSB). Initially, stimulated emission and reduced absorption contribute equally to increasing probe transmission. If changes in transmission are measured with probe pulses polarized parallel and perpendicular to the pump, the parallel signal is initially 3× greater than the perpendicular signal.

Experimentally, rotational alignment is quantified using the anisotropy,

$$r \equiv (S_\| - S_\perp)/(S_\| + 2S_\perp), \tag{1.1}$$

where $S_\|$ is the signal for parallel pulses and $S_\perp$ is the signal for perpendicular pulses.[15] The division by $S_\| + 2S_\perp \equiv 3S_{iso}$ removes isotropic dynamics, such as decay of excited state population. For a single dipolar vibronic transition, a 3:1 signal ratio leads to $r = 2/5$. If a signal arises from several sources with non-zero isotropic strength, the total signal has the average anisotropy[17]

$$\langle r \rangle = \sum_i r^i S^i_{iso} \Big/ \sum_i S^i_{iso}. \tag{1.2}$$

For a change in transition dipole angle of θ',

$$r(\theta') = (1/5)[3\cos^2(\theta') - 1]. \tag{1.3}$$

After rotation and collisions destroy the alignment created by the pump, the distribution of θ' becomes isotropic in three dimensions, so that the ensemble average $\langle \cos^2(\theta') \rangle = 1/3$ yields $r = 0$. Equation (1.3) can also be applied to changes in transition dipole within the molecular frame; for example, a change in state with $\theta' = \pi/2$ yields $r = -1/5$. Different distributions of θ' can produce the same anisotropy; for example, the distribution $p = 1/2$ for $\theta' = 0$ and $p = 1/2$ for $\theta' = \pi/2$ has the same anisotropy ($r = 1/10$) as the two-dimensionally isotropic distribution of θ' with $\langle \cos^2(\theta') \rangle = 1/2$ arising from randomized rotation in a plane.

For a single transition between nondegenerate vibronic states, the transition dipole direction is fixed in the molecular frame and has the same direction regardless of the transition probability. This is not the case when more states are involved. Consider a hydrogen atom coherently excited from the $1s$ state to the $2p$ level by a pulse linearly polarized along Z; the excited state will always be a $2p_z$ state. This should enhance the initial emission anisotropy, yielding $r_{ESE} = 1$. Similarly, a doubly degenerate excited state (or a dimer with perpendicular chromophores) should lead to greater alignment than a single dipole; for this case, calculations by the groups of Knox[18] and Hochstrasser[19] predicted $r_{ESE} = 7/10$ for an initially isotropic sample in 1993. These emission anisotropies are sensitive to coherent motion of the excited state superposition (which can change the direction of the emission dipole), transfer of population between the degenerate excited states, and loss of coherence between them.

Conflicting reports about pump-probe anisotropies greater than 0.4 in experiments involving coherent excitation of several transitions caused confusion about the above theory for some time.[17,20,21] Then, in 2001, Albrecht Ferro and Jonas showed that the emission anisotropy is not the same as the pump-probe anisotropy when coherent excitation of more than one transition is involved.[1] The emission result does not apply to the pump-probe anisotropy because, relative to a single nondegenerate transition, the anisotropy of the transmission increase from depopulation of the ground state (GSB) is not more aligned, but less.[1,2] For example, in the hydrogen atom, a depletion of the $1s$ state population causes equal transmission increases for pulses polarized along the laboratory X, Y, and Z axes, so that the anisotropy of the GSB contribution to the signal is $r_{GSB} = 0$. Furthermore, the strength of the emission and bleach signals are no longer equal; in hydrogen, the transitions to $2p_x$, $2p_y$, and $2p_z$ are all bleached while only one emits (yielding a transmission increase equal to one of the three bleached transitions). The result is that the initial pump-probe polarization anisotropy should be reduced to $r = 1/4$. Similarly, for a doubly degenerate transition, both orthogonal transitions to the excited state are bleached, so that $r_{GSB} = 1/10$.[1] In the absence of molecular rotation, the ground state bleach anisotropy is not time-dependent.

Absorption transitions starting from the initially excited states and terminating on highly excited states (Excited State Absorption: ESA) can also contribute to pump-probe signals (they cause a transmission decrease) and the time-dependence of the pump-probe anisotropy. Indeed, such transitions must be included to recover the known experimental results for isolated chromophores as a limiting case of the theory for coherent excitation of coupled chromophores.[1,2] When excited state absorption transitions to states with two excited chromophores are included, the initial anisotropy for an aggregate is also 2/5, consistent with most experiments.[1,17,22]

Clearly, changes in electronic state character caused by nonadiabatic transitions between electronic states are reflected in the anisotropy whenever there is a change in transition dipole direction. In the vicinity of a conical intersection, the adiabatic electronic states have a strong variation in electronic character as a function of the vibrational coordinates. As a result, adiabatic changes in electronic state character during vibrational motion can also alter the polarization anisotropy.[1] However, these effects occur against a backdrop of anisotropic "quantum beats" in coherent experiments. Before discussing these, we remark that, if a vibrational wavepacket on one adiabatic electronic state reaches the vicinity of a conical intersection, the nonadiabatic coupling will turn it into a coherent superposition state with amplitude on both adiabatic surfaces. This coherent superposition state will have "quantum beat" dynamics similar to those discussed here.

These quantum beat phenomena can be illuminated by first order time-dependent perturbation theory. In an initially isotropic sample, the pump pulse excites molecules from state $|g\rangle$ to the coherent superposition state

$$|\psi(t)\rangle = |g\rangle + \sum_e (i/\hbar) e^{-i\omega_{eg} t} [\vec{\mu}_{eg} \cdot \hat{\mathcal{E}}(\omega_{eg})] |e\rangle, \quad (1.4)$$

where g is the ground state, the states e are excited, $\vec{\mu}_{eg} = \langle e|\hat{\mu}|g\rangle$ is the vibronic transition dipole moment, $\hat{\mathcal{E}}(\omega)$ is the inverse Fourier transform of the time domain electric field, and $\omega_{eg} = (E_e - E_g)/\hbar$ is the Bohr frequency.[23] The expectation value of the dipole moment in the molecular frame is

$$\langle \vec{\mu}(t) \rangle = \langle \psi(t) | \hat{\mu} | \psi(t) \rangle$$
$$= \sum_e ((i/\hbar) e^{-i\omega_{eg} t} [\vec{\mu}_{eg} \cdot \hat{\mathcal{E}}(\omega_{eg})] \vec{\mu}_{ge} + c.c.). \quad (1.5)$$

If the basis is chosen so the transition dipoles are real-valued, and the pulses have real-valued transforms (e.g. cosinusoidal fields), Eq. (1.5) simplifies so that, in the case of two states with equal magnitude transition dipoles along x and y in the molecular frame,

$$\langle \vec{\mu}(t) \rangle = (2\mu^2/\hbar)[(\hat{\mathcal{E}} \cdot \hat{x}) \sin(\omega_{xg} t) \hat{x} + (\hat{\mathcal{E}} \cdot \hat{y}) \sin(\omega_{yg} t) \hat{y}].$$

The dipole has a $\pi/2$ phase lag behind the excitation field, and has an initial direction given by the projection of the field onto the xy plane, but the direction in the molecular frame is a function of time. Substituting $\omega_{xg} = \omega_0 - \delta/2$ and $\omega_{yg} = \omega_0 + \delta/2$ leads to

$$\langle \vec{\mu}(t) \rangle = (2\mu^2/\hbar) \sin(\omega_0 t) \cos(\delta t/2)[(\hat{\mathcal{E}} \cdot \hat{y})\hat{y} + (\hat{\mathcal{E}} \cdot \hat{x})\hat{x}]$$
$$+ (2\mu^2/\hbar) \cos(\omega_0 t) \sin(\delta t/2)[(\hat{\mathcal{E}} \cdot \hat{y})\hat{y} - (\hat{\mathcal{E}} \cdot \hat{x})\hat{x}]. \quad (1.6)$$

Ignoring the optical-frequency oscillations at ω_0, the dipole rotates with frequency $\delta/2$ around an ellipse in the molecular frame. This rotation, for a static energy gap δ, leads to[2]

$$r_{ESE}(t) = [4 + 3\cos(\delta t)]/10. \quad (1.7)$$

The factor of 2 frequency difference between Eqs. (1.6) and (1.7) arises because the anisotropy is sensitive only to alignment (up/down or left/right), and cannot distinguish between up and down at the half periods in Eq. (1.6).

If there is an inhomogeneous distribution of energy gaps, the anisotropy decays as the Fourier cosine transform of the energy gap distribution from $r_{ESE}(t=0) = 7/10$ to $r_{ESE}(t=\infty) = 4/10$ and does not fully equilibrate (in the absence of molecular rotation).[2] The anisotropy does not reach the equilibrium appropriate to delocalization in a plane ($r = 1/10$) because, for some molecular orientations (those with either $(\hat{\mathcal{E}} \cdot \hat{x}) = 0$ or $(\hat{\mathcal{E}} \cdot \hat{y}) = 0$) the dipole does not reorient at all through coherent beating without changes in the magnitude of the excited state coefficients. Similarly, off-diagonal couplings that cause change in the coefficients are not, by themselves, sufficient to equilibrate the anisotropy. Knox and Gülen[18] and Wynne and Hochstrasser[19] showed that "electronic dephasing" caused by stochastic fluctuations in the energy gap, in combination with population transfer caused by stochastic fluctuations in the off-diagonal coupling, would ultimately lead to electronic equilibration and $r_{ESE}(t=\infty) = 1/10$.

1.2. *Tuning and coupling coordinates*

All conical intersections and conical funnels involve two special vibrational coordinates[24–27]: the g coordinate "tunes" the energy gap between two electronic states and the h coordinate controls the coupling between two electronic states. Both energy gap and coupling are zero at a conical intersection. A "conical funnel" is a conical intersection or weakly avoided conical intersection characterized by passage between electronic states "so fast that there is no time for vibrational equilibration before the jump."[28] Just as a conical funnel can exist without a true conical intersection, a conical intersection may not act as an effective funnel if g or h lifts the degeneracy too weakly. Signatures of both g and h, or the electronic processes they drive, are required to prove that measured electronic dynamics arise from a conical funnel. Due to degeneracy at the intersection, the distinction between g and h is dependent on the electronic basis chosen: a $\pi/4$ rotation of the electronic basis vectors interchanges the role of these two vibrational coordinates.[26] Although g and h are individually dependent on the basis set, the "branching space" containing both coordinates that lift the degeneracy is not.[25] Other directions of motion leave the degeneracy at the conical intersection intact, forming the "seam space".[27]

In time-domain spectroscopy, rate theories of electronic relaxation (which we will see can be driven by conical funnels) may invoke three processes: population relaxation, coherence dephasing, and coherence transfer.[29] When a system is perturbed by an optical field, these processes restore thermal equilibrium. Population transfer restores the Boltzmann population distribution, while coherence dephasing destroys coherent phase relationships between states. For a two level system, population transfer and coherence dephasing are often characterized with the optical Bloch T_1 and T_2 time constants, respectively.[30] For larger systems, the density matrix element ρ_{mn} for a coherent superposition of states m and n may evolve into density matrix element ρ_{kl} for coherence between states k and l, a step usually described with a Redfield rate constant, R_{klmn}.[31,32] Redfield rate constants can also be used to describe population transfer (R_{kkmm}) and coherence dephasing (R_{klkl}). Although rate theories are not applicable to vibrationally correlated dynamics, as at conical intersections and funnels,[33] they connect terms in the Hamiltonian to relaxation processes and provide convenient labels for the nonexponential relaxation processes measured with femtosecond spectroscopy.

In quantum mechanics, the probability amplitudes for each basis state are constant unless there is an off-diagonal coupling in the Hamiltonian, therefore we identify population transfer between states in a conical intersection as the electronic process driven by the h coordinate.[34] Similarly, the relative phase between states in a coherent superposition evolves as $\delta t/\hbar$, where δ is the energy gap; the g coordinate will cause a randomization of this phase, known as dephasing, that damps oscillations of observables such as the dipole moment direction in Eq. (1.6).[34] Both dephasing and population transfer must occur at a conical funnel and will be defined more precisely in Sec. 2.

As g and h are basis-set dependent, so are dephasing and population transfer[20] (although the state of complete electronic relaxation is basis set independent). In the $\{|x\rangle, |y\rangle\}$ basis, a field projecting onto $(\hat{x}+\hat{y})/\sqrt{2}$ in the molecular frame excites $(|x\rangle+|y\rangle)/\sqrt{2}$ so that dephasing caused by the g-coordinate is needed for equilibration. If, however, the basis set had been chosen as $\{|+\rangle, |-\rangle = (|x\rangle \pm |y\rangle)/\sqrt{2}\}$, the excited state would be simply $|+\rangle$, and population transfer caused by the h-coordinate would be needed for equilibration. Of course, the g-coordinate for the $\{|x\rangle, |y\rangle\}$ basis is the h-coordinate for the $\{|+\rangle, |-\rangle\}$ basis. In an isotropic solution, molecules will be oriented randomly with respect to the light polarization, so that when a single basis set is used for all orientations both population transfer and dephasing are required for complete equilibration.

2. Theory

2.1. Model system and Hamilitonian

A diabatic electronic basis (which diagonalizes the nuclear kinetic energy but not coordinate dependent coupling[35]) is useful because the vibronic basis states and their linear combinations closely correspond to the states excited by a linearly polarized pulse. We use D_{4h} point group symmetry labels, but the discussion in this section applies more generally to four-fold symmetry and can be extended to symmetries lower than four-fold by inclusion of static vibrational displacements between potential surfaces (see Sec. 3.2). We choose $|x\rangle$ and $|y\rangle$ as doubly degenerate basis states with x and y polarized electronic transitions to the ground state in the molecular frame. In the $\{|x\rangle, |y\rangle\}$ basis, asymmetric vibrations of b_{1g} symmetry (rectangle deformations) tune the energy gap so that the g coordinate is a linear combination of Jahn–Teller active b_{1g} normal modes. Asymmetric vibrations of b_{2g} symmetry (diamond deformations) control the off-diagonal coupling so that the h coordinate is a linear combination of Jahn–Teller active b_{2g} normal modes. Franck–Condon active a_{1g} normal modes belong to the seam space. Because b_{1g} and b_{2g} vibrations are nondegenerate,[10,11,36] the electronic dephasing and population transfer can occur on different time scales, as in a reactive conical intersection.

For fourfold symmetry, the linear Jahn–Teller Hamiltonian in the diabatic electronic basis $\{|x\rangle, |y\rangle\}$, divided by $\hbar$, is,[36]

$$\hat{H} = \hat{H}_S + \hat{H}_B + \hat{H}_{SB},$$
$$\hat{H}_S = \omega_{eg}(|x\rangle\langle x| + |y\rangle\langle y|) = \omega_{eg}\hat{I},$$
$$\hat{H}_B = \left(\frac{1}{2}\omega_1(\hat{p}_1^2 + \hat{q}_1^2) + \frac{1}{2}\omega_2(\hat{p}_2^2 + \hat{q}_2^2) + \frac{1}{2}\omega_s(\hat{p}_s^2 + \hat{q}_s^2)\right)\hat{I}, \quad (2.1)$$
$$\hat{H}_{SB} = \begin{pmatrix} (\omega_s d_s \hat{q}_s + \omega_1 d_1 \hat{q}_1)|x\rangle\langle x| & (\omega_2 d_2 \hat{q}_2)|x\rangle\langle y| \\ (\omega_2 d_2 \hat{q}_2)|y\rangle\langle x| & (\omega_s d_s \hat{q}_s - \omega_1 d_1 \hat{q}_1)|y\rangle\langle y| \end{pmatrix}.$$

Here $\hat{H}_S$, $\hat{H}_B$, and $\hat{H}_{SB}$ are the Hamiltonians for the two-state electronic system, the harmonic vibrational bath, and the system-bath interaction, respectively, and ω_{eg} is the vertical electronic excitation energy (a constant). $\hat{p}_i$ and $\hat{q}_i$ are the momentum and position operators for dimensionless normal coordinates; the subscripts 1, 2, and s indicate b_{1g}, b_{2g} and a_{1g} symmetry vibrations. The coupling terms $\omega_i d_i$ in Eq. (2.1) are usually attributed to electrostatic stabilization.[36] For a harmonic oscillator with coupling ωd, the potential well is displaced by a distance $q = d$ and the bottom of the well is lowered by $(1/2)\omega d^2$. For a symmetric mode this stabilization energy of $(1/2)\omega d^2$ is commonly referred to as the Marcus[37] reorganization energy λ, while for an asymmetric mode it is commonly referred to as the Jahn–Teller[11] stabilization energy $(D\omega)$.

The Born–Oppenheimer electronic potential energy surfaces are generated by neglecting the nuclear kinetic energy terms $(\hat{p}_1, \hat{p}_2, \hat{p}_s)$ in Hamiltonian (2.1), treating the coordinates $(\hat{q}_1, \hat{q}_2, \hat{q}_s)$ as parameters, and diagonalizing to find the electronic eigenvalues as a function of the vibrational coordinates q_1 and q_2. Figure 2 shows a contour plot of the lower Born–Oppenheimer surface. The adiabatic character of the electronic eigenfunction for the lower surface is indicated by wavefunctions for a particle in a 2D box placed around the outside of the plot (the adiabatic wavefunction depends only on angle in the b_{1g}-b_{2g} coordinate system). One can see that starting at the top of the figure the adiabatic electronic wavefunction is y-polarized. Moving clockwise, x character builds until at the bottom of the

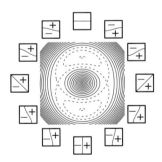

Fig. 2. Contour plot of the lower Born–Oppenheimer potential energy surface as a function of the two vibrational coordinates most active in driving electronic motion via the conical intersection in SiNc. The surface was generated from Hamiltonian (2.1) using $\omega_1/2\pi c = 140\,\mathrm{cm}^{-1}$, $\omega_1 d_1/2\pi c = 6\,\mathrm{cm}^{-1}$, $\omega_2/2\pi c = 138\,\mathrm{cm}^{-1}$, $\omega_2 d_2/2\pi c = 7\,\mathrm{cm}^{-1}$, which can reproduce the dynamics of Ref. 4 when other Jahn–Teller active vibrations are included. Contours are spaced at $0.5\,\mathrm{cm}^{-1}$, with dashed contours below the saddle points along b_{1g} to emphasize the minima along b_{2g}. Electronic eigenfunctions for a particle in a 2D box are drawn around the edges to illustrate how their character changes around the singularity at the conical intersection.

figure the wavefunction is x-polarized. Continuing the clockwise motion, the wavefunction mixes in more y character so that when returning to the top of the figure the wavefunction is y-polarized again, except that continuity requires a sign change of the electronic wavefunction. This is a manifestation of Berry's phase.[38] The requirement that the overall vibronic eigenfunction be single-valued imposes a compensating sign change on the vibrational wavefunction.

2.2. *Reduced density matrix and Redfield equations*

In condensed phase dynamics the molecular wavefunction is entangled with its environment and the molecule is only part of a system described by a density matrix[12] $\hat{\rho} = \sum_\psi p_\psi |\psi\rangle\langle\psi|$ with probabilities p_ψ for various wavefunctions $|\psi\rangle$. With $\hat{\rho}$, one can trace over the degrees of freedom that are not directly measured to obtain a reduced density matrix $\hat{\sigma}$ from which all measurements can be calculated using $\langle \hat{O} \rangle = Tr[\hat{O}\hat{\sigma}]$.[39,40] Impulsive pump-probe measurements in SiNc and H_2Nc can be calculated from the electronic reduced density matrix $\hat{\sigma}$ obtained from $\hat{\rho}$ by integration over all vibrational coordinates, $\hat{\sigma} = \int \langle \mathbf{R}_{\text{vib}}|\hat{\rho}|\mathbf{R}_{\text{vib}}\rangle d\mathbf{R}_{\text{vib}}$. Experimentally, this integration is complete when the ultrafast pulse spectrum uniformly covers the molecular absorption, emission, and excited state absorption and the pulse duration is transform limited in the time domain. Therefore, it is useful to view the adiabatic approximation from an electronic viewpoint (in which all vibrational coordinates are hidden but the electrons are fully described) rather than the more common vibrational viewpoint (in which a vibrational wavepacket on one electronic surface can be fully described, but the electronic state is hidden in a surface label). This full description of the electrons is necessary because electronic wavepacket motion, by itself, does not prove a conical intersection, or even vibronic coupling, is present. Indeed, electronic wavepackets can be excited in atoms.[8]

If two vibrational coordinate independent electronic states form an approximately complete basis for the doubly degenerate excited state over the range of vibrational coordinates accessed in the experiment, four elements of the electronic reduced density matrix are sufficient to describe the experiment: the two "populations" $\sigma_{xx} = \langle x|\hat{\sigma}|x\rangle$ and $\sigma_{yy} = \langle y|\hat{\sigma}|y\rangle$, plus the two "coherences" $\sigma_{xy} = \langle x|\hat{\sigma}|y\rangle$ and $\sigma_{yx} = \langle y|\hat{\sigma}|x\rangle$. After a linearly polarized pump pulse, the initial value of each of these elements is a function of the Euler angles that indicate molecular orientation, and their subsequent evolution is determined by the underlying vibronic dynamics of the density matrix $\hat{\rho}$.[40] The pump-probe experiments described in this chapter provide information on the orientation dependence of the relaxation dynamics of $\hat{\sigma}$.[4]

Electronic equilibration is defined by the approach of $\hat{\sigma}$ to equilibrium: it requires both dephasing (to send the off-diagonal elements to zero; this is the signature of the g-coordinate) and population transfer (to equilibrate the diagonal elements; this is the signature of the h-coordinate). In Redfield's theory[31] of density matrix relaxation, the Hamiltonian is separated into a reference Hamiltonian $\hat{H}_S$ (diagonal in the same basis as the density matrix), a bath Hamiltonian $\hat{H}_B$, and a perturbation $\hat{H}'$,

$$\hat{H}' = \begin{bmatrix} \hat{\Delta} & \hat{V} \\ \hat{V} & -\hat{\Delta} \end{bmatrix}, \tag{2.2}$$

where $\langle \hat{\Delta} \rangle = \langle \hat{V} \rangle = 0$. This perturbation describes the interaction of the system (the electrons) with the bath (the vibrations). It is this interaction that relaxes a nonequilibrium $\hat{\sigma}$ and leads to equations of motion for reduced density matrix elements.

The experiments are related to linear combinations of density matrix elements such as the population difference ($\sigma_{xx} - \sigma_{yy}$), the real coherence ($\sigma_{xy} + \sigma_{yx}$) and the imaginary coherence ($i\sigma_{xy} - i\sigma_{yx}$). Assuming $|x\rangle$ and $|y\rangle$ to be a degenerate basis set for which $\hat{H}_S$ is diagonal, taking $\hat{V}$ to be real, assuming $[\hat{\Delta}, \hat{V}] = 0$ and using Eq. (2.19) in Redfield[31] yields

$$\frac{d}{dt}(\sigma_{xx} + \sigma_{yy}) = 0,$$
$$\frac{d}{dt}(\sigma_{xx} - \sigma_{yy}) = -k_{VV}(\sigma_{xx} - \sigma_{yy}) + k_{DV}(\sigma_{xy} + \sigma_{yx}),$$
$$\frac{d}{dt}(\sigma_{xy} + \sigma_{yx}) = -k_{DD}(\sigma_{xy} + \sigma_{yx}) + k_{DV}(\sigma_{xx} - \sigma_{yy}), \tag{2.3}$$
$$\frac{d}{dt}(i\sigma_{xy} - i\sigma_{yx}) = -(k_{DD} + k_{VV})(i\sigma_{xy} - i\sigma_{yx}),$$

with rate constants

$$k_{VV} = 2\int_{-\infty}^{\infty} \langle \hat{V}(0)\hat{V}(\tau) \rangle d\tau,$$
$$k_{DD} = 2\int_{-\infty}^{\infty} \langle \hat{\Delta}(0)\hat{\Delta}(\tau) \rangle d\tau, \tag{2.4}$$
$$k_{DV} = 2\int_{-\infty}^{\infty} \langle \hat{\Delta}(0)\hat{V}(\tau) \rangle d\tau,$$

where $\hat{V}$ and $\hat{\Delta}$ have dimensions of radians per unit time and the angular brackets indicate ensemble averaging. If $\hat{\Delta} = 0$ or $k_{DV} = 0$, k_{VV} quantifies the decay of the population difference. Similarly, if $\hat{V} = 0$ or $k_{DV} = 0$, k_{DD} quantifies decay of the real coherence. k_{DV} couples real coherence with the population difference if $\hat{\Delta}$ and $\hat{V}$ are correlated.[41]

To apply Redfield theory to Hamiltonian (2.1), we move the symmetric modes from $\hat{H}_{SB}$ to $\hat{H}_S$ (since they do not contribute to the mixing between $|x\rangle$ and $|y\rangle$) and identify what remains in $\hat{H}_{SB}$ with $\hat{H}'$ in Eq. (2.2) yielding

$$k_{VV} = (2\omega_2 d_2)^2 \int_0^\infty \langle \hat{q}_2(0)\hat{q}_2(\tau)\rangle d\tau,$$
$$k_{DD} = (2\omega_1 d_1)^2 \int_0^\infty \langle \hat{q}_1(0)\hat{q}_1(\tau)\rangle d\tau, \qquad (2.5)$$
$$k_{DV} = (2\omega_1 d_1)(2\omega_2 d_2) \int_0^\infty \langle \hat{q}_1(0)\hat{q}_2(\tau)\rangle d\tau.$$

A key result in Eq. (2.5) is that vibrational motion, which can only decrease a coordinate's correlation function, *decreases* the rate of relaxation via the conical intersection. The rate constants in Eq. (2.5) are only applicable after the correlation function has decayed to zero.[31,33] The femtosecond spectroscopy described here probes the nonexponential dynamics at timescales faster than the vibrational motions which determine these correlation functions.[42]

2.3. *Effect of electronic relaxation on the anisotropy*

The experiments discussed in this chapter measure the polarization anisotropy of femtosecond pump-probe signals. The pump-probe signal S_{pp} is defined as the change in transmitted probe intensity caused by previous interaction with a pump and depends on the delay T between the pump and probe pulses.[7] This "transient transmission" pump-probe signal is the net sum of positive GSB, positive ESE, and negative ESA contributions (other methods of measuring pump-probe signals combine the three with different signs). As in vibrational spectroscopy, the term "doubly excited" is taken to indicate the level of excitation energy for the final states reached by ESA, and does not distinguish between electron configurations with one or two excited electrons, which are strongly mixed by configuration interaction in porphyrins and phthalocyanines.[43]

Since the first excited singlet state of SiNc has a symmetry required (Jahn–Teller[9–11,36]) conical intersection at the fourfold symmetric (D_{4h} point group) equilibrium geometry of the electronic ground state, an electronic wavepacket can be excited directly at the conical intersection.

This allows the dynamics occurring at the intersection to be measured without convolution with slower vibrational transport to and from the intersection. In each randomly oriented molecule, the pump laser coherently excites transitions to degenerate electronic states polarized along the two perpendicular axes.[2] Upon electronic excitation, asymmetric vibrations lower the total energy by elongating the molecule parallel to the electronic wavepacket momentum.[3,9–11,36,44] The asymmetric modes excited depend on the molecular orientation: vibrations that lower the symmetry to that of a rectangle (b_{1g}) are excited if the projection of the laser field onto the molecular frame lies along x or y; vibrations that lower the symmetry to that of a diamond (b_{2g}) are excited if that projection lies halfway between x and y. At a general orientation, both b_{1g} and b_{2g} symmetry vibrations are excited.[3]

Polarized pump-probe experiments depend on the elements of $\hat{\sigma}$ immediately after the pump *and* at the time of the probe pulse. To incorporate the former effect, we indicate the initially excited element of $\hat{\sigma}$ by a subscript in parentheses to the right of the current element: for example, $\sigma_{xx(yy)}$ indicates the current population of state x created by initial excitation of state y.[4] The evolution of the reduced density matrix for the doubly degenerate state is thus described by 16 = (4 current) × (4 initial) quantities for each fixed molecular orientation. Calculations for an isotropic sample require orientational averaging over all possible angles between molecular frame transition dipole vectors and laboratory frame electric field vectors of the polarized pulses.[2] Upon angular averaging, the excited state emission signal becomes

$$\begin{aligned} S_\parallel^{ESE} &= (1/5)\langle \sigma_{xx(xx)} + \sigma_{yy(yy)} \rangle + (1/15)\langle \sigma_{xy(xy)} + \sigma_{yx(yx)} \rangle \\ &\quad + (1/15)\langle \sigma_{yy(xx)} + \sigma_{xx(yy)} \rangle + (1/15)\langle \sigma_{yx(xy)} + \sigma_{xy(yx)} \rangle, \\ S_\perp^{ESE} &= (1/15)\langle \sigma_{xx(xx)} + \sigma_{yy(yy)} \rangle - (1/30)\langle \sigma_{xy(xy)} + \sigma_{yx(yx)} \rangle \\ &\quad + (2/15)\langle \sigma_{yy(xx)} + \sigma_{xx(yy)} \rangle - (1/30)\langle \sigma_{yx(xy)} + \sigma_{xy(yx)} \rangle. \end{aligned} \quad (2.6)$$

The angular brackets indicate ensemble averaging, while orientational averaging gives rise to the constant prefactors on each term. The information content of the experiment can be clarified by introducing four real quantities that survive angular averaging:[4] the excited state population,

$$p = \langle \sigma_{xx(xx)} + \sigma_{yy(yy)} \rangle + \langle \sigma_{yy(xx)} + \sigma_{xx(yy)} \rangle, \quad (2.7)$$

the normalized orientational population difference,

$$d_-(T) = [\langle \sigma_{xx(xx)} + \sigma_{yy(yy)} \rangle - \langle \sigma_{yy(xx)} + \sigma_{xx(yy)} \rangle]/p, \quad (2.8)$$

the normalized real coherence,

$$c_+(T) = [\langle\sigma_{xy(xy)} + \sigma_{yx(yx)}\rangle + \langle\sigma_{yx(xy)} + \sigma_{xy(yx)}\rangle]/p, \qquad (2.9)$$

and the normalized orientational coherence difference,

$$c_-(T) = [\langle\sigma_{xy(xy)} + \sigma_{yx(yx)}\rangle - \langle\sigma_{yx(xy)} + \sigma_{xy(yx)}\rangle]/p. \qquad (2.10)$$

For experiments with linearly polarized light, the signals depend on p, d_-, and c_+ (experiments with circularly polarized light access c_-). d_- and c_+ are interchanged by a $\pi/4$ rotation of the electronic basis, while c_- is unaffected. Both d_- (which decays through population transfer) and c_+ (which decays through dephasing) are initially one and decay to zero at equilibrium.[4] However, if $(D\omega)_2 = 0$ there is no population transfer and $d_- = 1$ is constant; similarly, if $(D\omega)_1 = 0$, there is no dephasing and $c_+ = 1$.

Decomposing each pair of parallel and perpendicular signals into a magic angle signal $S_{MA} = (S_\| + 2S_\perp)/3$ and an anisotropy leads to simple expressions in terms of these three new variables. For the ESE signal given in Eq. (2.6) we obtain

$$\begin{aligned} S^{ESE}_{MA} &= (1/9)p, \\ r_{ESE}(T) &= (1/10)[1 + 3d_-(T) + 3c_+(T)]. \end{aligned} \qquad (2.11)$$

The magic angle signal depends only on excited state population, and the anisotropy decays with population transfer and dephasing. The ground state bleach has $S^{GSB}_{MA} = (2/9)p$ and a constant anisotropy,

$$r_{GSB}(T) = (1/10). \qquad (2.12)$$

As discussed by Qian and Jonas,[2] the anisotropy for the ESA contribution depends on the symmetry of the final state through a cyclic set of 4 transition dipoles of the form $\langle a|\boldsymbol{\mu}|b\rangle\langle b|\boldsymbol{\mu}|c\rangle\langle c|\boldsymbol{\mu}|d\rangle\langle d|\boldsymbol{\mu}|a\rangle$ with an overall sign that is independent of the arbitrary wave function phases and physically significant.[2-4] The transition dipoles and ESA anisotropy for the four symmetry changes with transition dipoles in the xy plane are shown in Fig. 3. This anisotropy also applies to frequency resolved Raman scattering,[19] but not vibronic quantum beats. Vibronic quantum beats either involve a different time-ordering on the ground state or states not degenerate by symmetry on the excited state, aspects treated by Farrow et al.[5] The latter paths provide additional information on electronic

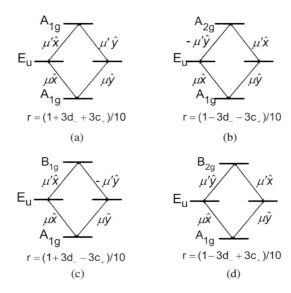

Fig. 3. Cyclic sets of transition dipoles in the x–y plane. The cyclic set of directions, relative magnitudes, and overall sign through a degenerate state are dictated by the overall change in symmetry from bottom to top of each diamond, but the individual transition dipoles are basis set dependent. In each diagram, the magnitudes of the two transition dipoles labeled $\mu(\mu')$ are equal by symmetry. The anisotropy $r(T)$ for a time-ordered process in which the pump pulse interacts with the lower two transition dipoles, followed by probe interaction with the upper two transition dipoles is given under each set in terms of the reduced density matrix observables $c_+(T)$ and $d_-(T)$. This anisotropy applies to excited state absorption and frequency resolved Raman scattering, but not vibronic quantum beats. (Vibronic quantum beats either involve states not degenerate by symmetry on the excited state or a different time-ordering on the ground state.)

relaxation processes.[5] The time-dependence of the anisotropy is entirely contained in signals which evolve on the excited state. The ESA contribution, depending on symmetry, can either reinforce or cancel out the contribution to the signal from either d_- or c_+. A consequence of this is that the possible symmetries of the doubly excited states *must* be considered when interpreting data. These contributions can be averaged once the relative strengths of GSB, ESE and ESA transitions are known (for example, from a spectrally resolved pump-probe experiment, which contains the linear absorption and emission spectra of the excited state).

To illustrate the critical importance of excited state absorption, and the symmetry of the doubly excited state, we consider three scenarios[4] For all three, relative signal strengths are approximated by those from a simple model: two non-interacting electrons in a 2D box. Just as in porphyrins and phthalocyanines, there are four doubly excited states in this model, one each with A_{1g}, B_{1g}, and B_{2g} symmetries plus one "extra" state of either B_{1g}, or B_{2g} symmetry. The ratio of signal strengths is $4:2:-1:-1:-1:-1$ for GSB:ESE:ESA:ESA:ESA:ESA.

If the "extra" state has B_{2g} symmetry,

$$r_{\text{tot}}(T) = (1/10)[1 + 3d_-]. \tag{2.13}$$

The anisotropy in this case decays from an initial value of 0.4 to a final value of 0.1. The anisotropy in ESA and ESE coming from coherences cancels out and only population transfer is observable in the electronic anisotropy. If the "extra" state instead had B_{1g} symmetry, we would have found

$$r_{\text{tot}}(T) = (1/10)[1 + 3c_+]. \tag{2.14}$$

In this case dephasing dynamics dominates the signal. For the third example, consider the case in which no excited state absorption occurs. In that case we find

$$r_{\text{tot}}(T) = (1/10)[1 + d_- + c_+]. \tag{2.15}$$

The initial anisotropy is only 0.3, the final value is still 0.1, and both relaxation processes are observable.

If the pulse is shorter than half a vibrational period, molecular vibrations can modulate the pump-probe signal by changing the frequency, strength, or direction of a transition.[7] In the impulsive limit, when the pulse spectrum covers the entire absorption, emission, and excited state absorption spectra, pump-probe should become insensitive to Frank–Condon active vibrations; however, quantum beats can arise from the breakdown of the adiabatic and Condon approximations.[45] In this way, Jahn–Teller active asymmetric vibrations may be observed indirectly via their effect on the electrons. These quantum beats can arise with different phases and/or different amplitudes in parallel and perpendicular signals, and have an anisotropy.[5] The anisotropy of each vibration is calculated with

$$r_{\text{vib}}(\omega) = \frac{A_\parallel^v - A_\perp^v \cos(\phi_\parallel^v - \phi_\perp^v)}{A_\parallel^v + 2A_\perp^v \cos(\phi_\parallel^v - \phi_\perp^v)}, \tag{2.16}$$

where $A_\perp^\nu (A_\parallel^\nu)$ indicates the amplitude of the oscillation at frequency ν in the perpendicular (parallel) signal and $\phi_\perp^\nu (\phi_\parallel^\nu)$ is its phase. Equation (2.16) has the same form as Eq. (1.1) except for the cosine, which accounts for the phase difference between the quantum beat in parallel and perpendicular signals. The vibrational quantum beat anisotropy allows determination of the vibrational symmetry (b_{1g} and b_{2g} vibrations have $r_{\text{vib}} = \infty$ vs. $r_{\text{vib}} = 1/10$ for a_{1g}). For vibrations on the excited state, r_{vib} probes vibronic relaxation in a different way than the "electronic" anisotropy decay. Because the amplitudes of the vibrational quantum beats depend on the couplings in the Hamiltonian, their measurement links the electronic and vibrational dynamics.

3. Experiment

3.1. *A D_{4h} silicon naphthalocyanine (SiNc)*

Farrow et al.[4] performed pump-probe measurements using linearly polarized pulses with 25 fs duration that spectrally covered the π to π^* $Q(0,0)$ transition of SiNc in benzonitrile solution. The $Q(0,0)$ transition connects the nondegenerate $^1A_{1g}$ ground state to the doubly degenerate ^1E_u symmetry first excited singlet state without any change in high frequency vibrational quantum numbers. Low pulse fluence, low absorption, and a complete sample change between laser shots were needed to prevent distortion of the fast anisotropy decay, while a high signal-to-noise ratio was required to recover low-amplitude quantum beat modulations. Naphthalocyanine's size makes diffusive rotation in the benzonitrile solvent slow (450 ps),[1] so that any faster changes in the polarization anisotropy must be attributed to vibronic dynamics without molecular rotation.

The pump-probe anisotropy in Fig. 4 decays from $r(0) \sim 2/5$ (slightly before time zero) to $r(\sim 2ps) = 0.1005 \pm 0.0008$, which is characteristic of randomized electronic alignment within the naphthalocyanine plane ($r = 1/10$).[4] The vibrational quantum beats can be fit at long times and then subtracted from the measured anisotropy to reveal that the fast dynamics are a complicated monotonic decay, which is mostly complete in ~ 100 fs. For convenience, this monotonically decaying component will be referred to as the electronic anisotropy and the quantum beats will be referred to as vibrational, without implying a separation of the vibronic dynamics. Subtracting a fit to the electronic anisotropy from the total anisotropy and taking a Fourier transform of the residuals shows the power spectrum of vibrational quantum beats (inset to Fig. 4).

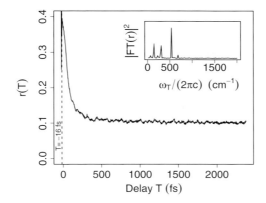

Fig. 4. Pump-probe polarization anisotropy in SiNc. The anisotropy is not shown before $T = -16$ fs, when low signal-to-noise in the denominator causes wild oscillations. The earliest reliable anisotropy is $r(-12\,\text{fs}) = 0.395$, and the anisotropy has decayed to $r(0) = 0.380$ at $T = 0$ (maximum pump-probe temporal overlap). Inset: power spectrum after subtraction of the monotonic anisotropy decay. The peaks at 147, 309, 538 and 684 cm^{-1} are reproducible.

Fitting[4] of the parallel, perpendicular, and magic-angle signals recovers two vibrations (686 cm^{-1} and 617 cm^{-1}) assigned to Frank–Condon active a_{1g} modes based on $r_{\text{vib}} = 1/10$.[4,5] The 686 cm^{-1} mode dominates the individual pump-probe traces, but is a minor peak in the total anisotropy power spectrum while the 617 cm^{-1} peak does not appear in the anisotropy power spectrum. Three Jahn–Teller active modes (535 cm^{-1}, 301 cm^{-1} and 140 cm^{-1}) are assigned based on anisotropies consistent with infinity, due to a vanishing denominator (within error) in Eq. (2.16).[4,5] These three modes are minor components of the individual pump-probe traces, but dominate the power spectrum of the anisotropy. The stabilization energies for each vibration were determined from the amplitude of the vibrational modulation of the signal, the pulse duration, the pulse spectrum, and the equilibrium absorption and emission spectra of the sample by extending established procedures[46] for determination of vibrational reorganization energies for Franck–Condon active modes.

If a single relaxation mechanism dominates the dynamics, the data enable direct calculation of the relevant asymmetric vibrational correlation function, $M(t)$. For harmonic potential surfaces with the same curvature, the Brownian oscillator model uses the vibrational correlation function and displacement between curves for Franck–Condon active vibrational modes to predict signals in linear and nonlinear electronic spectroscopy.[42] Smith and Jonas exploited symmetry to develop a Brownian oscillator model for Jahn–Teller active vibrational modes that applies whenever a single relaxation process dominates.[3] This Brownian oscillator model leads to the coherence decay[3] between components of the E_u state

$$c_+(t) = \exp(-4\text{Re}[g(t)]), \qquad (3.1)$$

where $g(t)$ is the absorption lineshape function for b_{1g} vibrations. [The factor of 4 arises because the E_u states are displaced twice as far from each other (at $q = \pm d$) as each is from the ground electronic state (at $q = 0$) and the lineshape function is proportional to the square of the displacement.] The lineshape function is zero at $t = 0$ and has zero derivative, so the fast dynamics are determined by its second derivative. In the high-temperature limit, this is

$$d^2\text{Re}[g(t)]/dt^2 = [2(D\omega)/(\hbar\beta)]M(t), \qquad (3.2)$$

where β is the inverse thermal energy and $(D\omega)$ is the Jahn–Teller stabilization energy (for several modes, the right hand side becomes a sum of such terms, one for each mode). $(D\omega)$ calculated from the initial anisotropy decay is $5\,\text{cm}^{-1}$, about that of the asymmetric $140\,\text{cm}^{-1}$ and $300\,\text{cm}^{-1}$ vibrations. $M(t)$ deduced from the anisotropy is shown in Fig. 5; it is initially underdamped, then suddenly settles to near zero at $\sim 150\,\text{fs}$, about when such a vibrational wavepacket first begins returning to the conical intersection. The oscillation and stabilization energy suggest coupling to these low frequency Jahn–Teller vibrations determines the fast electronic dynamics, while the sudden death of $M(T)$ suggests the single relaxation hypothesis is incorrect.

Model Hamiltonians with the $300\,\text{cm}^{-1}$ vibration assigned as b_{1g} symmetry and the $140\,\text{cm}^{-1}$ vibration assigned as b_{2g} symmetry (or vice versa) could not produce the fast anisotropy decay. Since the vibrations of porphyrins and phthalocyanines often occur in nearly degenerate pairs, a model in which the $140\,\text{cm}^{-1}$ vibrations are a near degenerate pair was tried and found sufficient to account for most of the anisotropy decay. This model was used to guide phenomenological inclusion of population relaxation into a Brownian oscillator type model that could be used to calculate finite pulse effects, such as the amplitude of the totally symmetric vibration at $686\,\text{cm}^{-1}$. This model fit, to within experimental error, the linear absorption spectrum, the excited state absorption and emission spectra, the electronic anisotropy decay, and the vibrational quantum beat amplitudes and

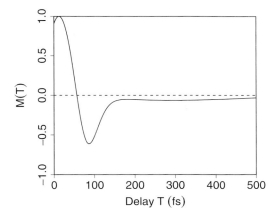

Fig. 5. Asymmetric vibrational correlation function $M(T)$ calculated from a smoothed fit to the anisotropy using Eqs. (2.14), (3.1) and (3.2). $M(T)$ has its maximum delayed to $T \sim 15\,\text{fs}$ by finite pulse duration and shows an initial underdamped oscillation (frequency $\sim 220\,\text{cm}^{-1}$) before suddenly dying away at $\sim 150\,\text{fs}$. [Reprinted with permission from Darcie A. Farrow, Wei Qian, Eric R. Smith, Allison A. Ferro, and David M. Jonas, *Journal of Chemical Physics* **128**, 144510 (2008). Copyright 2008, American Institute of Physics.]

anisotropies. Interestingly, the dominant totally symmetric quantum beat at $686\,\text{cm}^{-1}$ (for which the stabilization energy can be determined from the absorption spectrum) cannot be fit without invoking complete electronic relaxation. This suggests that the anisotropy of pump-probe quantum beats will be useful for characterizing general conical intersections where the g and h coordinates are not both asymmetric.[27]

3.2. *A D_{2h} free-base naphthalocyanine (H_2Nc)*

The SiNc experiments showed that low frequency vibrational modes with modest (less than $10\,\text{cm}^{-1}$) Jahn–Teller stabilization energies could drive fast ($\sim 100\,\text{fs}$) equilibration, and that studying the anisotropies of vibrational modes, even totally symmetric ones, could reveal signatures of the intersection. Smith *et al.*[6] studied a system where symmetry alone does not predict a conical intersection. Free-base naphthalocyanine (H_2Nc) has D_{2h} symmetry (ignoring the peripheral R groups which do not join the delocalized π system); the two central hydrogen atoms in free-base porphyrins and phthalocyanines are fixed in place by barriers of $\sim 10\,\text{kcal/mol}$.[47] Although D_{2h} molecules have no required degenerate electronic states (and therefore no required conical intersection), the LUMO and LUMO+1 in H_2Nc are close enough that that the first two excited states are expected to mix though the pseudo Jahn–Teller effect.[11,48]

The relevant electronic and vibrational symmetries can best be discussed by analogy to SiNc. Starting from D_{4h}, the hydrogens are a b_{1g} (rectangle) static perturbation, totally symmetric (a_g) in D_{2h}. The diamond deformation remains asymmetric (b_{2g} in D_{4h}, b_{1g} in D_{2h}); the electronic ground state remains totally symmetric (A_{1g} in D_{4h}, A_g in D_{2h}); the degenerate E_u state of D_{4h} splits into B_{2u} and B_{3u} states with y and x-polarized transitions from A_g in D_{2h}.[49] For many porphyrins and phthalocyanines this splits the Q band into Q_x and Q_y,[50] but as the size of the aromatic system increases, the Q_x–Q_y frequency gap decreases and the two transition strengths become equal.[50] For H_2Nc in liquid solutions, the splitting disappears, although splittings can be seen in solid naphthalene $(120\,\text{cm}^{-1})$[51] and low-density polyethylene $(243\,\text{cm}^{-1})$.[52]

The experiment was more difficult than for SiNc because the lack of axial ligands ("R" in Fig. 1) causes increased aggregation, so the concentration was lower and the pathlength longer. The transients were wavelength dependent, so Smith et al. measured once spectrally covering the $Q(0,0)$ absorption (blue tuned 29 fs pulses) and once on the low-energy edge of $Q(0,0)$ (red tuned 38 fs pulses). The red-tuned pulses excited vibronic wavepackets near the classical turning points of the ground state vibrations, allowing for larger quantum beat amplitudes.

Figure 6 shows the measured anisotropies. Both red and blue tuned pulses give an initial anisoptropy of ~0.3, which [from Eq. (2.15)] suggested that the experiment did not access any doubly-excited states. The lack of excited state absorption was confirmed by measuring a spectrally resolved pump-probe signal at 100 ps delay. The asymptotic blue-tuned anisotropy is 0.101 ± 0.003, which is 1/10 within error, suggesting the electronic wavepacket is fully delocalized in the plane of the molecule. The red-tuned anisotropy decays to 0.126 ± 0.003 during the experiment, which indicates incomplete equilibration. For a single relaxation mechanism, in the absence of excited state absorption, the anisotropy decays from 0.3 to 0.2 [Eq. (2.15) with either c_+ or d_- fixed to 1]. For both "red" and "blue" pulses, the anisotropy decays well below 0.2, which indicates that more than one relaxation mechanism is active, suggesting a conical funnel. For the "blue" pulses, both c_+ and d_- fully decay, and the anisotropy reaches 0.1 (within error) by about 300 fs. The anisotropy for the "red" pulses,

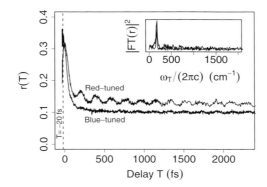

Fig. 6. Pump-probe polarization anisotropy in H_2Nc from $T = -20$ fs, with the earliest reliable anisotropy $r(-13\,\text{fs}) = 0.34$ for the blue-tuned pulses and $r(-13\,\text{fs}) = 0.31$ for red-tuned pulses [$r(0) = 0.32$ for blue and 0.31 for red-tuning]. The inset shows a power spectrum obtained by subtracting the monotonic anisotropy decay. Peaks at $163\,\text{cm}^{-1}$ (blue) and $174\,\text{cm}^{-1}$ (red) match within the discrete frequency resolution of the grid ($16\,\text{cm}^{-1}$ for blue and $22\,\text{cm}^{-1}$ for red).

however, behaves differently. It decays past 0.2, indicating that both relaxation mechanisms are active, but does not reach 0.1, indicating that one or both relaxations are incomplete.

The two excited states accessed by the experiment are nondegenerate and may have a well defined energy gap hidden under the $450\,\text{cm}^{-1}$ wide $Q(0,0)$ band and/or unequal transition strengths. As an energy gap is needed to explain the pulse dependent final anisotropy (and transition strengths are almost equalized in H_2Nc when splittings can be resolved), Smith et al. focused on characterizing the energy gap while assuming equal transition strengths. This energy gap arises from the b_{1g} (in D_{4h}) static perturbation of the hydrogens. The Hamiltonian also includes coupling through a b_{2g} (in D_{4h}) vibration analogous to the $140\,\text{cm}^{-1}$ vibration of SiNc; the pump-probe transients reveal a vibration at $176\,\text{cm}^{-1}$ with $r_{\text{vib}} \sim 8$, indicative of b_{1g} symmetry in D_{2h}. The simplest Hamiltonian that can account for the data is

$$\hat{H} = \hat{H}_S + \hat{H}_B + \hat{H}_{SB},$$
$$\hat{H}_S = \omega_{eg}(|x\rangle\langle x| + |y\rangle\langle y|),$$
$$\hat{H}_B = \left(\tfrac{1}{2}\omega(\hat{p}^2 + \hat{q}^2)\right)(|x\rangle\langle x| + |y\rangle\langle y|), \quad (3.3)$$
$$\hat{H}_{SB} = \begin{bmatrix} (-\delta/2)|x\rangle\langle x| & (\omega d\hat{q})|x\rangle\langle y| \\ (\omega d\hat{q})|y\rangle\langle x| & (\delta/2)|y\rangle\langle y| \end{bmatrix}.$$

The best fit to the data indicate an energy gap of $\delta/2\pi c \sim 100\,\text{cm}^{-1}$ and a stabilization energy of $(D\omega)/2\pi c = 9\,\text{cm}^{-1}$ for the $176\,\text{cm}^{-1}$ asymmetric vibration. The simplest model consistent with this data, therefore, is an avoided crossing that drives fast ($<300\,\text{fs}$) but incomplete equilibration. If the static energy gap arises from the zero-point vibration of the hydrogens, one is led to a picture of a conical funnel, possibly with an energetically inaccessible intersection displaced away from the Franck–Condon region.

Inspection of the vibronic energy levels for the above model in Fig. 7 shows that even weak nonadiabatic coupling (($D\omega)/\omega = 0.05$) can have dramatic effects, qualitatively inconsistent with the Born–Oppenheimer approximation, if Born–Oppenheimer vibronic levels associated with different electronic curves are nearly degenerate. The diabatic and Born–Oppenheimer potential curves are shown for the nonadiabatic Hamiltonian (3.3). The diabatic potential curves (dashed curves) and basis state energies (dashed lines) were calculated by neglecting the coordinate-dependent terms $\omega d\hat{q}$ of $\hat{H}_{SB}$. The Born–Oppenheimer potential curves (gray) were calculated by neglecting nuclear momentum, treating nuclear coordinate

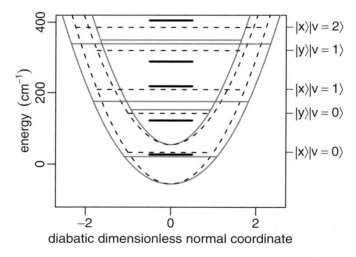

Fig. 7. Diabatic (dashed) and Born–Oppenheimer (gray) potential energy curves, associated basis state energies (lines attached to each curve), and exact nonadiabatic vibronic energy levels (thick solid black line segments, not attached to any curve) for the vibrational coordinate most active in electronic equilibration for H_2Nc. For each diabatic basis state energy, diabatic basis state quantum numbers are given on the right-hand side. Constructed from Hamiltonian (3.3) with parameters $\delta/2\pi c = 110\,\text{cm}^{-1}$, $(D\omega)/2\pi c = 9\,\text{cm}^{-1}$ and $\omega/2\pi c = 176\,\text{cm}^{-1}$.

operators as parameters, and diagonalizing to obtain the coordinate dependent electronic eigenvalues $U_\pm(q) = \frac{1}{2}\omega q^2 \pm \sqrt{(\delta/2)^2 + (\omega d q)^2}$ that make up the upper (+) and lower (−) Born–Oppenheimer potential curves. The adiabatic dimensionless normal coordinates[53] and harmonic vibrational frequencies are related to the diabatic ones by $q_\pm = \gamma_\pm q$ and $\omega_\pm = \gamma_\pm^2 \omega$, respectively, where $\gamma_\pm = [1 \pm (2\omega d^2/\delta)]^{1/4}$. The Born–Oppenheimer vibronic energies (gray lines) in Fig. 7 result from numerically diagonalizing to find the anharmonic vibrational energies on each potential energy surface.

The nonadiabatic vibronic eigenstate energy levels were found by numerical diagonalization in the both the diabatic direct product basis (with the off-diagonal vibronic coupling of Eq. (3.3) as the perturbation), and the Born–Oppenheimer vibronic basis (with the nonadiabatic coupling operator given by Eq. (VIII.7) of Ref. 54 as the perturbation). In diabatic dimensionless normal coordinates, the nonadiabatic coupling operator has matrix elements

$$\hat{\Lambda}_{kv,k'v'} = -\omega \langle \chi_{kv}| \langle \phi_k(q)| (\partial/\partial q) |\phi_{k'}(q)\rangle (\partial/\partial q) |\chi_{k'v'}\rangle \\ -\frac{1}{2}\omega \langle \chi_{kv}| \langle \phi_k(q)| (\partial^2/\partial q^2) |\phi_{k'}(q)\rangle |\chi_{k'v'}\rangle, \quad (3.4)$$

where $|\phi_k(q)\rangle$ is a Born–Oppenheimer electronic eigenfunction depending parametrically on the vibrational coordinate q and $|\chi_{kv}\rangle$ is a Born–Oppenheimer vibrational eigenfunction with v quanta of vibrational energy on electronic state k. The diagonal energy correction from the nonadiabatic coupling was included in this perturbation, not as part of the potential energy curves (see appendix VIII of Ref. 54); for this reason we refer to Born–Oppenheimer curves. This diagonal correction is relatively small and about the same size (ranging from $+5 \, \text{cm}^{-1}$ to $+15 \, \text{cm}^{-1}$) for the levels shown. The energies of the nonadiabatic vibronic eigenstates are indicated by solid black lines unassociated with any potential energy surface.

With respect to the diabatic curves, the lower Born–Oppenheimer curve is softened and the upper hardened. Therefore, all of the Born–Oppenheimer vibronic energies for the lower curve lie below the corresponding diabatic basis state energies, while all the Born–Oppenheimer vibronic energies for the upper curve lie above the corresponding diabatic basis state energies. However, on the lower diabatic curve, only the lowest nonadiabatic level (nominally $|x\rangle |v = 0\rangle$) is shifted down from the diabatic basis state energy, the two other levels shown have been shifted up. On the upper diabatic curve, the two nonadiabatic level shifts are both downward from the corresponding diabatic basis state energies, qualitatively opposite the Born–Oppenheimer prediction. Even without numerical diagonalization, such failure can be predicted from the interaction between diabatic states $|y\rangle|v=0\rangle$ and $|x\rangle|v=1\rangle$. The Born–Oppenheimer approximation qualitatively fails to describe the levels of this system.

It can be seen in Fig. 7 that the lowest energy level can be reasonably described by the $v=0$ vibration associated with either the lower diabatic or lower Born–Oppenheimer potential. However, higher energy basis functions for the Born–Oppenheimer basis occur in nearly-degenerate pairs that become strongly mixed by nonadiabatic interactions. For this system, adiabatic approximations qualitatively fail to describe the correct level structure, predicting nearly degenerate pairs of levels that do not occur; the diabatic basis offers a better starting point for nonadiabatic dynamics in H_2Nc. Modeling the anisotropy with the diabatic basis states and energy levels predicts some decay of the anisotropy, however the linear coupling term (which is the nonadiabatic term in the diabatic representation) is required to correctly reproduce the data.

4. Discussion

The semi-classical Landau–Zener theory of electronic curve crossing emphasizes the role of vibrational wavepacket velocity in making the transition from one adiabatic electronic potential energy curve to another at an avoided crossing. The Landau–Zener treatment predicts that the curve-crossing probability for a single passage through an avoided crossing is

$$P = \exp(-2\pi H^2/\hbar v |\Delta F|), \qquad (4.1)$$

where H is the perturbing coupling that produces the avoided crossing, v is the vibrational wavepacket velocity, and $|\Delta F|$ is the difference in force (potential slope) between the two unperturbed curves at the avoided crossing. According to Eq. (4.1), curve crossing becomes more probable for smaller couplings, higher wavepacket velocities, and larger differences in force. Although Eq. (4.1) suggests that the curve crossing probability vanishes at zero vibrational wavepacket velocity, it cannot be applied in this limit; a semi-classical theory inherently assumes that the vibrational wavepacket width is negligible, a condition that can only be met (according to the uncertainty principle) by wavepackets with high momentum components. The Landau–Zener theory is self-consistent in predicting curve crossing within its domain of validity, but says nothing about low velocity vibrational wavepackets, wavepacket width in curve crossing phenomena, or actual intersections between curves.

When a vibrational wavepacket's velocity is sufficiently low, its width will dominate the vibrational wavefunction derivative in the nonadiabatic coupling. Here, we expand on the discussion of electronic curve crossing dynamics at zero velocity in Refs. 3 and 4. The discussion will concentrate on short time results, but use the quantum mechanical form of the lineshape function $g(t)$ in Eq. (3.1) instead of the high temperature limit. In the high temperature limit, the distribution of vibrational coordinates can be regarded as an inhomogeneity, but the dynamics originating from excitation of the $v = 0$ wavefunction at zero temperature must be regarded as homogeneous. In the absence of damping, the second differential coefficient of the real part of the quantum lineshape function given by Eq. (5.23) of Ref. 55 is

$$d^2 \text{Re}[g(t)]/dt^2|_{t=0} = (D\omega)\omega\coth(\beta\hbar\omega/2)$$
$$= (\omega d)^2[(1/2)\coth(\beta\hbar\omega/2)],$$

where $(D\omega)$ is the stabilization energy (in frequency units), ω is the frequency, and β is the inverse temperature. The second line uses dimensionless normal coordinates, in which $q = \pm 1$ at the classical turning points of the $v = 0$ eigenstate. Here d is the displacement between harmonic curves and the width of the vibrational coordinate probability distribution is quantified by its variance $\sigma^2 = (1/2)\coth(\beta\hbar\omega/2)$. Recognizing that for $V(q) = (1/2)\hbar\omega(q\pm d)^2$, the difference in force between the two curves at the conical intersection is $|\Delta F| = 2\hbar\omega d$, the early time coherence decay can be written

$$c_+(t) = \exp[-|\Delta F_1/\hbar|^2 \sigma_1^2 (t^2/2)], \tag{4.2}$$

revealing a universal fast dephasing driven by the product of the variance and the square of the difference in force of the energy tuning coordinates (b_{1g} in D$_{4h}$, hence the subscript 1). If several normal modes are involved, each acts independently on this timescale so that $|\Delta F/\hbar|^2 \sigma^2$ should be replaced by $\sum |\Delta F_i/\hbar|^2 \sigma_i^2$ (a result independent of the vibrational basis). A rotation of the electronic basis, always possible around a conical intersection, allows the result for dephasing driven by b_{1g} modes [Eq. (3.1)] to be applied to population transfer driven by b_{2g} modes. Because the dynamics along g and h are also independent to order t^2,[4] one obtains

$$d_-(t) = \exp[-|\Delta F_2/\hbar|^2 \sigma_2^2 (t^2/2)], \tag{4.3}$$

a fast population transfer driven by the product of the variance and the square of the difference in force on the derivative coupling coordinates (b_{2g} in D$_{4h}$, hence the subscript 2).

The vibrational frequency does not appear in Eqs. (4.2) and (4.3) due to the neglect of the higher order terms in the Taylor expansion of the lineshape function. In fact, it is possible for the reduced density matrix to reach equilibrium before such terms, or the interaction between g-and h-coordinate driven electronic processes, become significant. Near a conical intersection, the difference in forces arises from the coordinate dependence of the vibronic coupling [see Eq. (2.1)]. Equations (4.2) and (4.3) show that the variation of vibronic coupling across the width of the vibrational wavepacket can drive electronic equilibration on a timescale faster than vibrational wavepacket motion. This role of wavepacket width is a homogeneous effect, and occurs even when a minimum uncertainty wavepacket is placed on the conical intersection by excitation from the $v = 0$ level at zero temperature [$\sigma^2 = (1/2)$]. Vibrational wavepacket width is important because the coupling and energy gap are zero at the intersection itself — the wings of the vibrational wavepacket feel the driving force. Equations (4.2) and (4.3) do not rely on the entire surface being harmonic, but rather on three local assumptions: 1) a harmonic approximation to the curves along g and h over the width of the wavepacket; 2) wavepacket centering on the conical intersection; and 3) the conical intersection reflection symmetries assumed in deriving Eq. (3.1).

The decay of population differences between diabatic basis states given by Eq. (4.3) is qualitatively different from the Landau–Zener probability for nonadiabatic curve crossing. However, neither formula is applicable in a parameter range where the other is valid. Comparisons involving both wavepacket width and wavepacket velocity effects may be possible by extending the harmonic approximation for the short-time dynamics to include off-center vibrational wavepackets using the multilevel nonlinear response functions derived by Sung and Silbey[56] or Mukamel and Abramavicius.[57]

5. Conclusion

Time-resolved measurements of electronic wavefunction alignment with polarized light are a promising way to characterize the coupled vibronic dynamics at conical intersections and funnels. Such measurements reveal vibrations coupled to the electronic motion through their effect on electronic alignment. The vibrational quantum beat amplitude reflects the vibronic coupling strength and the anisotropy determines the vibrational symmetry. This symmetry determination is based on the same principle as the Raman depolarization ratio, but provides new information about excited state processes that complements and can sometimes predict the initial electronic alignment dynamics. However, recent calculations indicate that asymmetric vibrational quantum beats on the excited electronic state (but not on the ground electronic state) can be rapidly suppressed by small environmental asymmetries.[58] Experiments on SiNc and H_2Nc have highlighted the role of vibrational wavepacket width in rapidly driving electronic motion via conical intersections and funnels with small couplings.

References

1. A. Albrecht Ferro and D.M. Jonas, *J. Chem. Phys.* **115**, 6281 (2001).
2. W. Qian and D.M. Jonas, *J. Chem. Phys.* **119**, 1611 (2003).
3. E.R. Smith, D.A. Farrow and D.M. Jonas, *J. Chem. Phys.* **123**, 044102 (2005); E.R. Smith, D.A. Farrow and D.M. Jonas, *J. Chem. Phys.* **123**, 179902 (2005); E.R. Smith, D.A. Farrow and D.M. Jonas, *J. Chem. Phys.* **128**, 109902 (2008).
4. D.A. Farrow, W. Qian, E.R. Smith, A.A. Ferro and D.M. Jonas, *J. Chem. Phys.* **128**, 144510 (2008).
5. D.A. Farrow, E.R. Smith, W. Qian and D.M. Jonas, *J. Chem. Phys.* **129**, 174509 (2008).
6. E.R. Smith, W.K. Peters and D.M. Jonas, in *Ultrafast Phenomena XVI*, edited by P. Corkum, S. de Silvestri, K.A. Nelson, E. Riedle and R.W. Schoenlein (Springer-Verlag, Berlin, 2009), p. 385.
7. D.M. Jonas, S.E. Bradforth, S.A. Passino and G.R. Fleming, *J. Phys. Chem.* **99**, 2594 (1995).
8. R.R. Jones, C.S. Raman, D.W. Schumacher and P.H. Bucksbaum, *Phys. Rev. Lett.* **71**, 2575 (1993).
9. H.A. Jahn and E. Teller, *Proc. Roy. Soc. (London) Series A* **161**, 220 (1937).
10. G.H. Herzberg, *Electronic Spectra of Polyatomic Molecules* (Krieger, Malabar, FL, 1991).
11. I.B. Bersuker, *The Jahn–Teller Effect* (Cambridge University Press, Cambridge, 2006).
12. L.D. Landau and E.M. Lifschitz, *Quantum Mechanics*, 3rd ed. (Pergamon Press, New York, 1977).
13. C. Zener, *Proc. Roy. Soc. (London) Series A* **137**, 696 (1932).
14. B.J. Berne and R. Pecora, *Dynamic Light Scattering* (Dover, Mineola, NY, 2000); P.M. Felker, *J. Phys. Chem.* **96**, 7844 (1992).
15. G.R. Fleming, *Chemical Applications of Ultrafast Spectroscopy* (Oxford University Press, New York, 1986).
16. R.N. Zare, *Angular Momentum: Understanding Spatial Aspects in Physics and Chemistry* (Wiley-Interscience, New York, 1988).

17. D.M. Jonas, M.J. Lang, Y. Nagasawa, T. Joo and G.R. Fleming, *J. Phys. Chem.* **100**, 12660 (1996).
18. R.S. Knox and D. Gülen, *Photochem. Photobiol.* **57**, 40 (1993).
19. K. Wynne and R.M. Hochstrasser, *Chem. Phys.* **171**, 179 (1993); K. Wynne and R.M. Hochstrasser, *Chem. Phys.* **173**, 539 (1993).
20. C. Galli, K. Wynne, S.M. LeCours, M.J. Therien and R.M. Hochstrasser, *Chem. Phys. Lett.* **206**, 493 (1993).
21. C.-K. Min, T. Joo, M.-C. Yoon, C.M. Kim, Y.N. Hwang, D. Kim, N. Aratani, N. Yoshida and A. Osuka, *J. Chem. Phys.* **114**, 6750 (2001).
22. D.C. Arnett, C.C. Moser, P.L. Dutton and N.F. Scherer, *J. Phys. Chem. B* **103**, 2014 (1999); M.H. Vos, J. Breton and J.-L. Martin, *J. Phys. Chem. B* **101**, 9820 (1997).
23. D.M. Jonas, *Annu. Rev. Phys. Chem.* **54**, 425 (2003).
24. J.v. Neumann and E. Wigner, *Phys. Z.* **30**, 467 (1929); E. Teller, *J. Phys. Chem.* **41**, 109 (1937); E. Teller, *Israel J. Chem.* **7**, 227 (1969); H. Köppel, W. Domcke and L.S. Cederbaum, *Adv. Chem. Phys.* **57**, 59 (1984); *Conical Intersections: Electronic Spectra, Dynamics and Spectroscopy*, edited by W. Domcke, D.R. Yarkony and H. Köppel (World Scientific, New Jersey, 2004).
25. G.J. Atchity, S.S. Xantheas and K. Ruedenberg, *J. Chem. Phys.* **95**, 1862 (1991); F. Bernardi, M. Olivucci and M.A. Robb, *Chem. Soc. Rev.* **25**, 321 (1996); G.A. Worth and L.S. Cederbaum, *Annu. Rev. Phys. Chem.* **55**, 127 (2004).
26. D.R. Yarkony, *J. Phys. Chem. A* **105**, 6277 (2001).
27. M.A. Robb and M. Olivucci, *J. Photochem. Photobiol. A* **144**, 237 (2001).
28. M. Klessinger and J. Michl, *Excited States and Photochemistry of Organic Molecules* (VCH Publishers, New York, 1995).
29. N. Bloembergen, *Nonlinear Optics* (Addison-Wesley, New York, 1992).
30. L. Allen and J.H. Eberly, *Optical Resonance and Two-Level Atoms* (Dover Publications, Inc., New York, 1987).
31. A.G. Redfield, *IBM J. Res. Dev.* **1**, 19 (1957).
32. P. de Bree and D.A. Wiersma, *J. Chem. Phys.* **70**, 790 (1979); A.G. Redfield, *Adv. Magn. Reson.* **1**, 1 (1965).
33. D. Chandler, *Introduction to Modern Statistical Mechanics* (Oxford University Press, New York, 1987).
34. B. Hoffman and M.A. Ratner, *Mol. Phys.* **35**, 901 (1978).
35. J.C. Tully, in *Dynamics of Molecular Collisions Part B*, edited by W.H. Miller (Plenum Press, New York, 1976), p. 217.
36. J.T. Hougen, *J. Mol. Spectrosc.* **13**, 149 (1964).
37. R.A. Marcus, *Angew. Chem. Int. Edit.* **32**, 1111 (1993).
38. M.V. Berry, *Proc. R. Soc. Lond. A* **392**, 45 (1984).
39. C. Cohen-Tannoudji, B. Diu and F. Laloë, *Quantum Mechanics* (Wiley-Interscience, Paris, 1977).
40. C. Cohen-Tannoudji, J. Dupont-Roc and G. Grynberg, *Atom-Photon Interactions: Basic Processes and Applications* (Wiley-Interscience, New York, 1992).

41. H.W.H. Lee and M.D. Fayer, *J. Chem. Phys.* **84**, 5463 (1986); R. Wertheimer and R. Silbey, *Chem. Phys. Lett.* **75**, 243 (1980).
42. S. Mukamel, *Principles of Nonlinear Optical Spectroscopy* (Oxford University Press, New York, 1995).
43. C. Weiss, H. Kobayashi and M. Gouterman, *J. Mol. Spectrosc.* **16**, 415 (1965).
44. W.L. Clinton and B. Rice, *J. Chem. Phys.* **30**, 542 (1959).
45. H. Kano, T. Saito and T. Kobayashi, *J. Phys. Chem. A* **106**, 3445 (2002).
46. W.T. Pollard and R.A. Mathies, *Annu. Rev. Phys. Chem.* **43**, 497 (1992).
47. D. Maity and T. Truong, *J. Porphyr. Phthalocya.* **5**, 289 (2001).
48. B.E. Applegate, T.A. Barckholtz and T.A. Miller, *Chem. Soc. Rev.* **32**, 38 (2003); T.S. Venkatesan, S. Mahapatra, H.D. Meyer, H. Köppel and L.S. Cederbaum, *J. Phys. Chem. A* **111**, 1746 (2007).
49. E.B. Wilson, J.C. Decius and P.C. Cross, *Molecular Vibrations: The Theory of Infrared and Raman Vibrational Spectra* (Dover Publications, New York, 1980).
50. N. Kobayashi, S. Nakajima, H. Ogata and T. Fukuda, *Chem.-Eur. J.* **10**, 6294 (2004).
51. S.M. Arabei, J.P. Galaup, K.N. Solovyov and V.F. Donyagina, *Chem. Phys.* **311**, 307 (2005).
52. I. Renge, H. Wolleb, H. Spahni and U.P. Wild, *J. Phys. Chem. A* **101**, 6202 (1997).
53. D. Papoušek and M.R. Aliev, *Molecular Vibrational-Rotational Spectra* (Elsevier, Amsterdam, 1982).
54. M. Born and K. Huang, *Dynamical Theory of Crystal Lattices* (Clarendon Press, Oxford, 1962).
55. Y. Tanimura and S. Mukamel, *Phys. Rev. E* **47**, 118 (1993).
56. J. Sung and R.J. Silbey, *J. Chem. Phys.* **118**, 2443 (2002).
57. S. Mukamel and D. Abramavicius, *Chem. Rev.* **104**, 2073 (2004).
58. E.R. Smith and D.M. Jonas, *J. Phys. Chem. A* **115**, 4101 (2011).

Index

ab initio molecular dynamics (AIMD), 348, 368, 369, 499
ab initio multiple spawning (AIMS), 111, 348, 362
adenine (Ade), 89, 96, 109, 110, 486–489, 604, 624–626, 628, 662, 663, 670, 678, 686
adenosine, 676
Aharonov–Bohm effect, 157
alignment, 718, 721, 732, 742, 743
aminobenzonitrile, 27
2-aminopurine, 678
2-aminopyridine, 682
analytic energy derivative, 504
analytic energy gradient, 434, 436, 439, 445, 453, 643
angular anisotropy parameter, 605
anisotropy, 716–719, 721, 727, 729–737, 740, 742
antisymmetrized square, 123
autocorrelation function, 262, 329
avoided crossing, 18–20, 29, 266, 267, 376, 402, 406, 409, 410
azobenzene, 672, 694, 696, 701

base pairing, 681, 701
base stacking, 681, 687, 701
benzene cation, 251, 263, 281, 285, 292–295
benzylideneaniline, 502, 508, 524, 527–531
Berry phase, 99, 156, 725

bimolecular reaction, 169
biradical state, 684
Bohr magneton, 124
Born–Huang expansion, 641
Born–Oppenheimer diagonal correction, 739
Born–Oppenheimer potential energy surface, 724
branching plane, 227, 235–237, 312, 314, 315, 317, 332, 377
branching space, 6, 12–15, 19, 20, 23, 25, 26, 29, 31–34, 36, 41, 42, 87–89, 95, 103, 158, 164, 635, 721
branching space coordinate, 14, 23, 25, 26, 29, 33, 42
Breit–Pauli approximation, 206
Breit–Pauli operator, 119, 120, 124
Brownian oscillator, 305, 319, 321, 327, 733, 734
butadiene, 16, 17

cancellation puzzle, 183, 185
charge transfer state, 67, 69, 70, 72, 73, 671
cis-butadiene, 16
cis-trans isomerization, 40–42, 44, 693
cis-trans photoisomerization, 585
classical path approximation, 383
classical phase space, 429, 430
closed-loop experiment, 571, 572
closed-loop learning control, 501

747

748

Index

coherence decay, 734, 741
coherence transfer, 722
coherent control, 570, 590, 591, 596
collective bath mode, 312
configuration interaction, 94, 104, 433
configuration state function (CSF), 92, 435, 643
conical funnel, 716, 721, 722, 736
conical intersection, 716, 719, 721–725, 727, 734–736, 741–743
continued fraction, 331, 333, 340, 342
control achievement, 575, 578, 579, 584, 585, 592
Coriolis coupling, 385
correlated bath, 309, 330
correlation function, 727, 733, 735
coupled coherent states (CCS), 352
coupled electron-proton transfer, 684, 685
coupled potential-energy surfaces, 376, 397
coupling coordinate, 721, 741
coupling mode, 55–57, 69, 308, 570, 587–591, 614, 635
cumulant, 315, 321, 322
cyclohexadiene, 18, 20
cyclopentadienyl, 222, 223
cytosine, 96–98, 101, 103, 109, 110

de Broglie wavelength, 191, 192
decatetraene, 650, 652, 657
decay
  nonradiative, 376, 378
  stochastic, 388, 390, 393
decay-of-mixing (DM) formalism, 397
decoherence, 303, 304, 334, 386, 392–395, 397–399, 401, 404–406, 410, 416, 425, 429
degeneracy space, 96
demixing, 397, 398
density matrix, 722, 725, 726, 728, 730, 742
derivative coupling, 85, 95, 100–105, 201, 214, 215, 228, 229, 254, 255, 424, 634, 642–644, 741

diabatic basis, 738–740, 742
diabatic electronic basis, 723
diagonal correction, 380
diarylethylenes, 18
dimethyladenine, 678
dimethylaminobenzonitrile (DMABN), 678
dinucleotide, 685–687, 701
Dirac equation, 119
Dirac–Frenkel variational principle, 261
DNA damage, 670
DNA repair, 671
double resonance spectroscopy, 682
double-cover space, 160
double-valued representation, 122
dual fluorescence, 678
Dyson orbital, 648, 649

E/Z isomerization, 671, 672, 694
effective core potential, 483
effective mode, 305, 311–319, 321, 322, 324–326, 328, 331–333, 341, 342
Ehrenfest method, 401, 422, 423, 425, 465
electron driven proton transfer (EDPT), 70
electron/proton transfer, 52, 75
electronic dephasing, 721, 723
electronic embedding, 484
electronic population dynamics, 659
electronic population transfer, 261, 272, 273, 280, 290–292, 721–723, 726, 729, 731, 741
electronic relaxation, 722, 727, 730, 735
electronic relaxation dynamics, 679
electronic wavepacket, 716, 725, 728, 736
electronic–vibrational coupling, 716
electrostatic JT forces, 122, 145, 151
emission anisotropy, 718, 719
encirclement angle, 160, 172, 175, 181, 188

Index

energy gap tuning, 720–723, 737, 738, 742
energy gradient, 214, 228, 230
energy-gap coordinate, 303
environmental mode, 304–306, 308, 325, 327, 333
evolutionary algorithm, 522
excited state intramolecular proton transfer (ESIPT), 59
excited-state quenching, 65, 76
excited-state absorption (ESA), 719, 725, 730, 731, 734, 736
excited-state H transfer, 629
excited-state intramolecular proton transfer (ESIPT), 660
excited-state proton transfer, 367
extended conical intersection seam, 5, 14, 23, 25, 31, 40

feedback control, 571
fewest switches algorithm (FSA), 416, 468, 469
fewest switching criterion, 500
Feynman path, 156, 157, 159, 164, 166–171, 174–176, 181, 188, 192
field-induced surface hopping (FISH) method, 502, 523, 561
first principles molecular dynamics, 347
first-order intersection space, 9
flavin mononucleotide, 503, 547, 551, 553
fluorescence anisotropy, 695
fluorescence depletion, 547, 552, 555, 556, 558
fluorescence lifetime, 618
fluorescence quantum yield, 291, 670, 675, 678
fluorescence quenching, 52, 64
fluorescence upconversion, 673, 674, 679, 690, 698
fluorobenzene cation, 260, 282, 284, 293
Fock matrix, 440, 448, 480–483

formaldehyde, 251, 258, 263, 264, 266, 270, 294
fragment ion signal, 646
frozen Gaussian, 350, 352, 378
frustrated hopping, 428
full multiple spawning (FMS), 349, 354, 359, 395
fulvene, 14–16, 18
funnel, 716, 721, 722, 736, 738, 742, 743
furylfulgide, 689, 690, 692–694, 701

g coordinate, 721–723
Gaussian wave packet, 304, 348, 349, 424
geometric phase, 85, 99, 100, 118, 133, 135, 136, 155, 465, 585, 590, 593, 595, 596
gradient difference vector, 315, 332, 472
green fluorescent protein (GFP), 303
guanine (Gua), 670

H_2Nc, 717, 725, 735–738, 740, 743
H-atom tunnelling, 618, 627
Ham reduction effect, 201
harmonic oscillator function, 656
h coordinate, 721–723, 735
Hellmann–Feynman theorem, 435, 446
hexafluorobenzene, 673
hierarchical electron-phonon (HEP) model, 319
homotope, 164
homotopy, 159, 164, 165
hop, frustrated, 390, 392–394, 401, 406
hopping path, 182
hopping probability, 427, 429, 506, 520
hydrogen abstraction reaction, 275
hydrogen bond, 52, 58, 59, 63–67, 70–75, 78
hydrogen detachment, 58
hydrogen-exchange reaction, 157, 183

4-hydroxyindole, 620, 621

imidazole, 604, 609–614, 616, 624, 627
imidazolyl, 223–225
impulsive excitation, 716
inhomogeneous distribution, 721
internal conversion, 58, 64, 70, 78
intersection adapted coordinates, 215, 236
intersection space, 4, 6, 8, 9, 12–14, 16, 18–20, 22, 42
intersection space coordinate, 6, 8, 9, 22, 42
intersection space Hessian, 10
interstate coupling gradient, 214
intramolecular charge transfer (ICT), 678
intramolecular vibrational distribution, 621
intramolecular vibrational redistribution (IVR), 657
intramolecular vibrational relaxation (IVR), 58
ion imaging method, 606, 616
isopropoxy, 223, 234, 236, 237, 242

Jacobi coordinate, 189
Jahn–Teller (JT) effect, 88, 90, 92, 118, 282, 286, 287, 292
Jahn–Teller E (b_{1g}, b_{2g}), 723
Jahn–Teller effect, 736
Jahn–Teller model, 187, 190, 191
Jahn–Teller selection rule, 120
Jahn–Teller stabilization energy, 724, 734
Jahn–Teller system, 183, 187, 189

Kekule structure, 29
Kernel, 163, 165, 166, 169, 171, 180, 181
Koopmans' correlation corresponding, 652
Koopmans' correlation, complementary, 649–653, 655, 657, 658, 663

Kramers degeneracy, 87, 128, 133, 141
Kramers degenerate pair, 204
Kramers doublet, 204

Lagrange multiplier, 436, 578, 580–583
Lagrangian, 94, 163, 165
Lanczos algorithm, 205, 218, 219
Lanczos diagonalization, 218, 241
Lanczos recursion relation, 210, 217, 218, 221
Lanczos vector, 219, 221
Landau–Zener theory, 740
Landau–Zener transition probability, 428
Langevin dynamics, 521
laser control, 569, 571, 572, 577, 579
learning algorithm, 571
linear vibronic coupling (LVC) model, 253, 286, 307, 308, 332
linearly interpolated transit path, 74
Liouville description, 465
local excited (LE) state, 678
locally excited state, 69

malonaldehyde, 95, 98, 111–113, 367–370
mapping procedure, 465
Marcus reorganization energy, 724
Markovian approximation, 304, 320, 325, 339
Markovian closure, 305, 320, 324–326, 330, 333, 340, 341
mean-field approximation, 396
methaniminium, 484, 486
methoxy, 222, 223, 234, 237, 239, 240
mixed quantum-classical approaches, 465, 466
mixed quantum-classical dynamics, 304, 500, 541
mixed quantum-classical method, 417, 465
mixed quantum-classical systems, 466
mixed quantum-classical treatments, 465

mixing angle, 255, 257, 258
MNDO method, 480, 481, 490
mode mixing, 619
molecular switch, 670, 671, 696, 701
Mori chain, 305, 319, 322, 325, 326
multiconfiguration time-dependent Hartree (MCTDH) method, 240, 261, 324, 353, 395, 465, 583
multiconfigurational method, 434
multilayer MCTDH, 465
multimode vibronic coupling (MMVC), 251
multimode vibronic coupling (MMVC) model, 304
multiphoton absorption, 629
multiphoton ionization, 646, 677
multiple spawning method, 199, 416, 466, 500
multireference method, 433
multistate dynamics, 290, 500, 507, 518, 552

N-methyl pyrrole, 615, 626
naphtalocyanine, silicon, 716
naphthalene, 653
naphthalocyanine, free base, 716, 735
naphthalocyanine, silicon, 717, 732
NDDO method, 480
non-Born–Oppenheimer (NBO) dynamics, 375, 376, 409
nonadiabatic chemistry, 5
nonadiabatic coupling, 118, 145, 249, 250, 281, 288, 466, 468, 469, 472–475, 477, 479, 480, 488, 500, 504, 505, 507, 508, 510, 512, 521–523, 552, 554, 635, 642, 644, 658, 720, 738, 739, 741
nonadiabatic coupling vector, 308, 315, 357, 380, 382, 389, 417, 450, 454
nonadiabatic dynamics, 363, 740
nonadiabatic dynamics simulation, 416, 417, 457, 465
nonadiabatic effect, 108

nonadiabatic interaction, 573, 574, 577, 579, 580
nonadiabatic phenomena, 84
nonadiabatic process, 416
nonadiabatic trajectory dynamics, 417, 425, 433, 434
nonadiabatic transition, 64, 74, 100, 257, 272, 273, 355, 358, 359, 361, 368, 483, 499, 513, 529, 531, 570, 574, 577–579, 585, 590, 595, 635, 640, 650–652, 670
nonadiabatic wave-packet dynamics, 641
noncrossing rule, 85, 86
nonradiative decay, 604, 626
nonradiative relaxation, 503, 541, 545, 546
nonradiative transition, 635
normal coordinate, 126, 129, 134
normal equations, 215, 228, 230, 241

o-hydroxybenzaldehyde, 660
o-hydroxyphenyl-(1,3,5)-triazine, 23
Ohmic bath, 321, 340
oligonucleotide, 676
optimal control simulation, 579, 581–583, 585, 586, 589, 596
optimal control theory, 517, 562, 572, 579
orthogonalization correction, 480–482, 490

parabolic approximation, 12
parallel transition, 605
parent ion signal, 646
path integral, 163, 178
Pauli approximation, 119
Pauli spin matrix, 124
perpendicular transition, 605
phase change theorem, 32
phenanthrene, 653–655, 658
phenol, 604, 608, 609, 615, 616, 618, 619, 621–624, 627–630
photoactive yellow protein (PYP), 44, 303

photochemical funnel, 18, 20, 670, 699
photochemistry, 377, 386
photochromic switch, 694
photochromism, 22
photodetachment spectroscopy, 240
photodissociation, 382, 387, 405–409, 570, 585, 591, 596
photoelectron angular distribution, 198, 513, 655, 663
photoelectron spectroscopy, 198, 200, 672, 673
photoelectron spectrum, 282, 289
photoelectron-photoion coincidence (PEPICO), 646
photofragment dynamics, 604, 606
photofragment imaging, 672
photofragment translational spectroscopy, 604
photofragment velocity, 608
photoinduced H-atom transfer, 672
photoionization cross section, 649
photoisomerization, 570, 585, 587, 596
photostability, 26, 58, 66, 70, 540, 661, 670
photoswitching, 502, 524, 527, 529, 530, 532, 547
Polanyi rules, 664
polarization, 716, 717, 719, 722, 727, 732, 733, 737
polarization anisotropy, 716, 717, 719, 727, 732, 733, 737
population relaxation, 722, 734
pre-target state, 578, 579, 581
prefulvene, 39, 42, 44
promoting mode, 604, 627
propagator, 178, 179, 181, 182
proton transfer, 18, 24–26, 51–53, 58, 59, 63–65, 67, 70, 72, 75, 76, 78, 671, 682
proton-transfer reaction, 634
protonated tryptophan, 628, 629
pseudo Jahn–Teller effect, 716
pseudo normal equations, 228
pseudopotential, 146
pulse design equations, 572, 580–584

pump-dump control, 503, 547
pump-probe anisotropy, 719, 732
pump-probe experiment, 636
pump-probe spectroscopy, 669, 675, 696, 701, 716
pyrazine, 503, 508, 532–535, 537–540
pyrazole, 609–611, 614, 623
pyrazolyl, 96, 101, 102, 106, 107
pyrrole, 251, 260, 263, 273–281, 294, 295, 604, 606–612, 614, 616, 624, 626–628
pyrrolyl, 223–227, 229, 231, 232, 235

quadratic model, 230
quadratic vibronic coupling (QVC) model, 253, 278
quantum beat, 716, 720, 729–732, 734–736, 743
  vibrational, 716, 720, 729–732, 734–736, 743
quantum dynamics, 155, 172, 178
quantum interferometer, 576
quantum Liouville equation, 583
quantum-classical Liouville description, 465
quasi-degenerate perturbation theory, 92

Rabi oscillation, 522, 549, 550
radiationless decay, 8, 26, 29, 77, 109, 110, 113
radiationless electronic deactivation, 671
radiationless process, 634
radiationless transfer, 604, 605, 612–614, 616
radiationless transition, 455, 579
Raman depolarization ratio, 743
rare-event sampling, 395, 410
reaction coordinate, 6, 14, 18, 20, 41, 586, 588–591
reaction path, 5, 6, 8, 16, 18–20, 22, 23, 26, 28–31, 40, 45, 52, 59, 61, 64, 67, 74, 75, 464
reactive scattering, 402

recoil anisotropy, 605, 609, 622
recoil velocity distribution, 605, 608, 617
Redfield approach, 465
Redfield theory, 727
reduced density matrix, 725, 726, 728, 730, 742
regularized diabatic state, 251, 254, 256, 260, 308
rejected hop, 469
relativistic JT forces, 120, 145, 151
relaxed scan, 54, 60, 61, 72, 76
residual bath mode, 314, 321, 326
riboflavin, 503, 547, 551, 553
Rumer function, 33, 34
Runge-Kutta procedure, 505, 521, 526
Rydberg character, 490, 610
Rydberg state, 263–267, 270, 271, 281, 294

saddle point approximation, 352, 419, 421, 425
scattering cross section, 186
scattering wave function, 156, 186
Schiff base, 18, 455, 484, 503, 527, 547
seam space, 86, 88, 91, 95, 105, 723
semiclassical approximation, 380, 384
semiempirical method, 463, 479, 489, 490
semiempirical molecular orbital method, 479
short time dynamics, 742
Silvester's formula, 476
SiNc, 717, 724, 725, 727, 732, 733, 735–737, 743
single-particle functions, 261
slow electron velocity-map imaging (SEVI), 198
spawning, 347–350, 354–362
spectral density, 305, 309, 310, 319, 321, 324–331, 333, 335
spectral intensity distribution function, 200, 208
spin-boson Hamiltonian, 304
spin-double group, 151

spin-orbit coupling, 87, 90, 210, 235, 239, 382
spin-orbit (SO) coupling, 118
spin-orbit (SO) splitting, 120, 131, 132, 140–145, 147, 150, 151
state-averaging procedure, 441, 445
statistical limit, 659
Stieltjes imaging, 503, 513, 514, 517, 533–535, 539, 540
stimulated emission, 717
Stokes shift, 678
strong orthogonality approximation, 648
surface hop, 388, 391–393, 404, 409
surface hopping, 423–426, 428, 429, 457
surface hopping method, 465, 467, 469, 473
switches
  coherent, 378, 396, 400
  fewest, 378, 400
symmetrized square, 121
system-bath coupling, 465
system-bath Hamiltonian, 304–313, 316, 318, 324–326, 331–334, 336, 341

target operator, 581–584
thiophenol, 621, 622, 628
three-state intersection, 99, 103, 105, 106, 108–111, 113, 317, 367
time reversal, 204, 210, 213, 241
time-dependent variational principle, 353
time-derivative nonadiabatic coupling, 450
time-resolved photoelectron spectroscopy (TRPES), 502, 503, 512, 531, 633, 635, 645, 663
time-resolved photoelectron spectrum, 626
time-reversal operator, 125, 127, 138, 139
time-reversal symmetry, 122, 136
time-uncertainty hopping, 394

topology, 157–159, 164
topology-adapted mode, 317
total kinetic energy release (TKER), 606
trajectory basis function, 349
trajectory calculations, 463
trajectory surface hopping (TSH), 388, 390, 466
trans-cis isomerization, 527, 529, 547–549
transformation angle, 254
transient absorption spectroscopy, 674
transient transmission, 727
transition-metal complex, 118, 119, 137
transition-metal trifluorides, 141
Tully's surface hopping (TSH) method, 500, 504, 505, 519
tuning coordinate, 741
tuning mode, 308, 326–328, 338
twisted intermolecular charge transfer, 27
two-state intersection, 96, 103, 110, 367–370

unimolecular decay, 606, 613
unimolecular reaction, 160, 167–169
uracil, 108–110
UV absorption cross section, 612
UV photolysis, 273

valence bond, 24, 26, 33, 37–40
valence excited state, 490
valence state, 263, 265, 266, 274
van Vleck–Gutzwiller formulation, 466
velocity adjustment, 471, 472, 474
velocity-Verlet algorithm, 526
velocity-Verlet integration, 474

vibrational coherence, 570, 589
vibrational quantum beat anisotropy, 716, 732, 734, 743
vibrational wavepacket, 716, 719, 734, 740–743
vibrationally mediated photodissociation, 618
vibronic coupling, 725, 739, 742
vibronic coupling constant, 251, 253, 254, 256, 258–260, 268, 272, 276–278, 284, 285, 287
vibronic coupling model, 200, 203, 220, 230
vibronic interaction, 611, 612
vibronic spectra, 141, 145
vibronic symmetry conservation, 614, 618, 627
vibronic wavepacket, 736
VUV absorption spectrum, 270, 294

Watson–Crick base pairing, 65, 67
Watson–Crick base pair, 682
wave packet, 167, 172–178, 180–182, 189, 191
wave packet cannon, 585, 589, 593, 596
wave packet shaping, 576, 596
wave packet
  electronic, 716, 725
  vibrational, 716, 719, 734, 740–743
Wigner distribution, 359, 362, 387, 430, 431, 477, 513, 514, 517, 520, 526, 548
Wigner rotation matrix, 185
winding number, 164–167, 178, 180, 181, 184

zero electron kinetic energy (ZEKE) spectrum, 198
zwitterionic structure, 29